selected conversion factors

$1 \text{ u} = 1.66 \times 10^{-27} \text{ kg}$

$1 \text{ eV} = 1.60 \times 10^{-19} \text{ J}$

$1 \text{ hp} = 746 \text{ W}$

$1 \text{ kcal} = 4.184 \text{ kJ}$

$1 \text{ atm} = 1.01 \times 10^{5} \text{ Pa}$

$1° = (\pi/180) \text{ rad}$

$1 \text{ Å} = 10^{-10} \text{ m}$

$1 \text{ ft} = 0.305 \text{ m}$

$1 \text{ mi} = 1.61 \text{ km}$

$1 \text{ mi/h} = 0.447 \text{ m/s}$

$1 \text{ km/h} = 0.278 \text{ m/s}$

$1 \text{ lb} = 4.45 \text{ N}$

greek letters used in the text

alpha	α			xi	ξ	Ξ
beta	β			pi	π	
gamma	γ			rho	ρ	
delta	δ ∂	Δ		sigma	σ	Σ
epsilon	ϵ			tau	τ	
theta	θ			phi	ϕ	Φ
kappa	κ			chi	χ	
lambda	λ	Λ		psi	ψ	Ψ
mu	μ			omega	ω	Ω
nu	ν					

university physics

for science and engineering

university physics

for science and engineering

Donald E. Tilley

Collège militaire royal de Saint-Jean
Saint-Jean, Quebec

Cummings Publishing Company, Inc.

Menlo Park, California · Reading, Massachusetts · London
Amsterdam · Don Mills, Ontario · Sydney

Other books by the same author
(with Walter Thumm)

Physics for College Students
Physics: A Modern Approach
College Physics

Copyright © 1976 by Cummings Publishing
Company, Inc. Philippines copyright 1976.

Printed in the United States of America.
Published simultaneously in Canada.
Library of Congress Catalog Card No. 75-14974

ISBN 0-8465-7536-1
ABCDEFGHIJKL- HA 79876

Cummings Publishing Company, Inc.
2727 Sand Hill Road
Menlo Park, California 94025

preface

This book has been written for students of science and engineering to provide the basis for a first university-level physics course using calculus. Every effort has been made to face the experimental fact that the majority find such courses to be extremely demanding. Priority has been given to helping students understand and use the basic principles of physics. The presentation and emphasis are designed to obtain the advantages of both a modern point of view and a traditional order of topics (Newtonian mechanics, fluids, heat, electricity and magnetism, waves, relativity, quantum physics). The only units the student is asked to master are those of the Système International (SI).

Newton's second law is discussed immediately after the introductory chapter on velocity and acceleration vectors. Demanding mathematical analyses and exercises are deliberately avoided in the first two chapters so that problems involving the application of Newton's second law to a variety of physical systems can be reached without delay; such problems are among the best for teaching the student how to use basic physical principles. A systematic approach to physics problems is stressed in Chapters 3 and 4.

As the book progresses, its mathematical level rises. Throughout Chapters 2 to 6—which discuss Newton's laws of motion, statics, and momentum—analysis using the components of vectors is the single mathematical technique required. Although definitions involving derivatives are given in the first chapter, calculus is not needed for derivations in the first six chapters. The arrangement of material makes it possible to delay integration without inconvenience until

Chapter 7, and, at the same time have the first encounter with integration occur while treating the topic that best brings out the significance of an integral. By deriving the usual equations for motion with constant acceleration in Chapter 7 instead of in Chapter 1, the student can understand how these equations (which he has usually seen emphasized and even overemphasized in high school) follow from the integration of the equation of motion.

Circular motion is treated in Chapter 8 in a way most appropriate for the discussion of simple harmonic motion given in Chapter 9. It is a decided advantage to have on hand the example of simple harmonic motion when studying potential energy in the following two chapters.

Phasor diagrams, first introduced in connection with simple harmonic motion, are later used to give a mathematically unified treatment of alternating current, interference phenomena with sound and light, and the quantum mechanical superposition principle. The study of wave motion has been organized into five consecutive chapters following the treatment of alternating current. (In the author's experience, students do not have difficulty in grasping the meaning of partial derivatives; therefore they have been used when required for clarity.) The important and popular subject of special relativity is developed to the extent that ample material is available for courses that emphasize the topics of modern physics. Quantum phenomena are introduced in a chapter on photons and energy levels, and basic features of quantum mechanics are outlined in the penultimate chapter, with emphasis placed on the quantum mechanical superposition principle. The book closes with an overview of quantum physics that provides the instructor with representative topics for lectures or assigned readings which can serve as points of departure for future studies.

The more than 500 illustrations should make the students' task easier, as should the summaries at the ends of chapters. The questions have been grouped into two sets: one which corresponds rather closely with the order in which the subject matter has been presented and which contains no exceptionally demanding problems, and one which includes more difficult problems, some general review questions, and questions that supplement topics merely introduced in the text.

Rearrangements and possible omissions are indicated in the chart. For example, Chapters 15 through 20 on fluids and heat can be skipped, in part or in their entirety, with minimal effect on most of the remaining material. Chapters 39 and 40 on special relativity can be introduced after the chapters on energy in Newtonian mechanics. A course concentrated on mechanics, electricity, and waves can be based on the chapters in the central stream of this chart.

I wish to acknowledge my indebtedness to Professor Walter Thumm of Queen's University who freely gave me permission to use whatever was applicable from *Physics for College Students* (D. Tilley and W. Thumm, Cummings, 1974). My thanks are extended to Professors G. D. Thaxton of Auburn University, J. Fetter of Stanford University, and W. H. Kelly of Michigan State University, for reading the manuscript and providing valuable comments and

SEQUENCE OF TOPICS

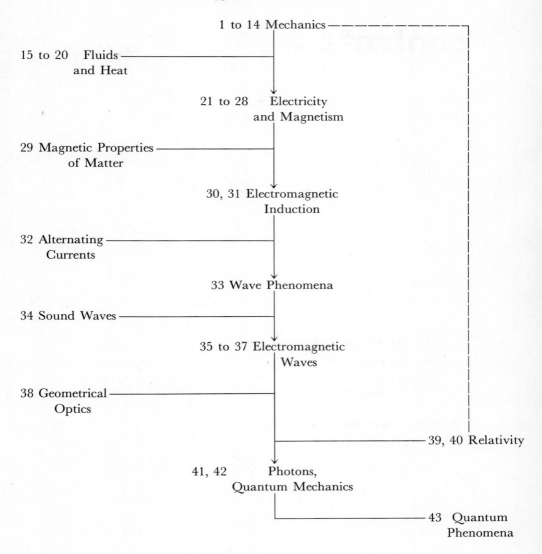

1 to 14 Mechanics

15 to 20 Fluids and Heat

21 to 28 Electricity and Magnetism

29 Magnetic Properties of Matter

30, 31 Electromagnetic Induction

32 Alternating Currents

33 Wave Phenomena

34 Sound Waves

35 to 37 Electromagnetic Waves

38 Geometrical Optics

39, 40 Relativity

41, 42 Photons, Quantum Mechanics

43 Quantum Phenomena

suggestions. The encouragement and dedication of the Cummings staff has been vital. Particular thanks are due to my wife, not only for typing the manuscript but also for assisting with many other aspects of manuscript preparation.

Donald E. Tilley
Saint-Jean, Quebec
January 1976

contents

I.

velocity and acceleration vectors

The imposing body of knowledge that constitutes present-day physics is to a large extent contained in five great theories:

1 *Newtonian mechanics*   The theory that describes the motion of tangible objects such as stones, spaceships, and planets with a more than adequate accuracy and with a simplicity that ensures its continued use.
2 *Thermodynamics*   The theory of heat and the behavior of systems comprising a large number of particles.
3 *Electromagnetism*   The theory of electricity, magnetism, and electromagnetic radiation.
4 *Special relativity*   A theory that includes modifications of Newtonian mechanics. These modifications become important at speeds so high that they are a significant fraction of the speed of light.
5 *Quantum mechanics*   The mechanics that has been developed to fit the facts of the atomic and subatomic world.

Each of these theories has arisen from the interplay between experiment and the mathematical expression of human imagination. Within the domains that are emphasized in this book, these theories have been so thoroughly tested and have proved so useful that it seems almost certain that, during our century, they will remain the theoretical foundation for much of science and technology.

We begin with Newtonian mechanics. This theory developed from a careful analytical study of relatively simple motions, those of a falling stone and of a planet in its orbit. In describing the motion of such objects we can sometimes ignore the dimensions of the object itself and represent it by what is called a particle—an object whose dimensions are negligible for the problem at hand and whose position is represented by a mathematical point. In this chapter we introduce what is called particle *kinematics,* which is limited to a description of particle motion without treating the factors that influence the motion, such as the particle's mass and the force acting on the particle. Force and mass are introduced in dynamics, a topic of succeeding chapters.

Details of particle motion are given in terms of the particle's position, velocity, and acceleration. The mathematical relationships between any two of these quantities involve derivatives and integrals, and therefore should be investigated in detail after the reader is familiar with the central ideas of calculus. For this reason we postpone much of the analytical treatment of kinematics until Chapter 7 and present here only an introduction intended primarily to give a pictorial appreciation of how velocity and acceleration determine the evolution of motion. To be able to describe motion with ease and clarity, we first discuss essential topics in vector algebra.

1.1 Vector Algebra

Physical quantities that have both *direction* and *magnitude* (and that sum according to the triangle rule indicated in Figure 1-1) are called *vector quantities* and are represented by mathematical entities called *vectors.* Vector quantities that we will soon study are: displacement, velocity, acceleration, force, and momentum.

A vector is represented on a diagram by an arrow drawn to scale so that the length of the arrow is proportional to the magnitude of the vector. The direction of the arrow represents the vector's direction in space. In this book we use a letter in boldface type, such as **a**, as a symbol representing a vector. When handwritten the symbol for a vector is most frequently a letter with an arrow placed above it, such as $\vec{a}$. The magnitude of the vector **a** is a positive number denoted by $|\mathbf{a}|$ or often simply by the italic letter symbol a. It is to be emphasized that statements involving the symbol **a** refer to **both** the **direction and** the **magnitude** of the vector, while statements concerning $|\mathbf{a}|$ or a refer *only* to the *magnitude* of the vector and not to its direction.

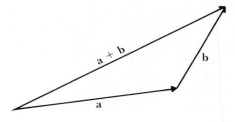

Figure 1-1   The triangle rule for vector addition. The vector **a** + **b** is defined to be the vector sum or resultant of **a** and **b**.

Equality of Vectors

Vector equations are statements about directions as well as magnitudes. The equation

$$\mathbf{a} = \mathbf{b}$$

means not only that these two vectors have the same magnitude ($a = b$ or $|\mathbf{a}| = |\mathbf{b}|$) but also that they point in the same direction.

All parallel vectors of the same magnitude, as in Figure 1-2, are regarded as equivalent in the following vector algebra. In other words, the vector represented by a certain arrow is unchanged if the arrow is moved to a new position, provided the direction and the length of the arrow are unchanged.

Vector Addition

The sum of two vectors, $\mathbf{a}$ and $\mathbf{b}$, is defined by the triangle rule as the vector denoted by $\mathbf{a} + \mathbf{b}$ in Figure 1-1. In words, this rule for defining the vector sum is as follows: Place the tail of the second vector at the tip of the first; the sum is the vector drawn from the tail of the first to the tip of the second. This rule can be extended to define the sum of any number of vectors, as is illustrated for three vectors in Figure 1-3. The vector sum is often called the *resultant* of the vectors that have been added.

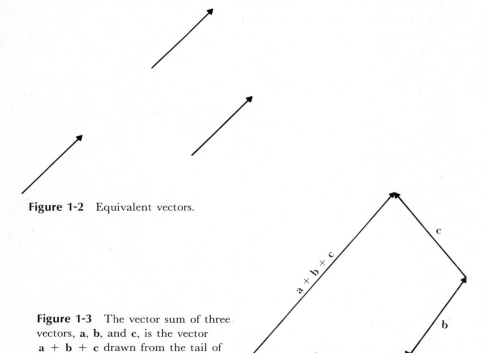

Figure 1-2 Equivalent vectors.

Figure 1-3 The vector sum of three vectors, $\mathbf{a}$, $\mathbf{b}$, and $\mathbf{c}$, is the vector $\mathbf{a} + \mathbf{b} + \mathbf{c}$ drawn from the tail of the first to the tip of the last.

From the definition of vector addition it is evident that the symbol + in an expression like **a** + **b** has a meaning different from that of the + in arithmetic, say 2 + 3. Consequently the rules for manipulation of vector sums must be worked out anew, using the definition just given for vector addition. From this definition it is not difficult to show that the order of the vectors in a sum can be interchanged,

$$\mathbf{a} + \mathbf{b} = \mathbf{b} + \mathbf{a} \qquad \text{(commutative law)}$$

and the grouping does not matter,

$$\mathbf{a} + (\mathbf{b} + \mathbf{c}) = (\mathbf{a} + \mathbf{b}) + \mathbf{c} \qquad \text{(associative law)}$$

so the parentheses can be omitted in a sum of several vectors (as has been done in Figure 1-3).

When we are concerned with the addition of only two vectors, it is sometimes convenient to place the two tails together. The resultant is then the diagonal of the parallelogram shown in Figure 1-4. This construction is known as the parallelogram rule for vector addition.

The difference, **a** − **b**, of the vectors **a** and **b** is defined as the vector which must be added to **b** to yield **a**:

$$\mathbf{b} + (\mathbf{a} - \mathbf{b}) = \mathbf{a}$$

as is illustrated in Figure 1-5. The negative of a vector is another vector of equal magnitude, but opposite direction (Figure 1-6).

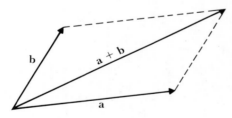

Figure 1-4   The parallelogram law for vector addition.

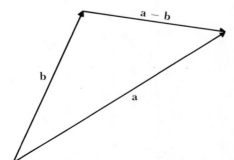

Figure 1-5   The difference (**a** − **b**) is the vector which must be added to **b** in order to obtain **a**. That is, **b** + (**a** − **b**) = **a**.

Multiplication of a Vector by a Scalar

In vector algebra, quantities that have a numerical value and possibly a unit, but *no direction*, are called *scalars*. Examples of scalars are real numbers such as 9.2, −74.8, and certain physical quantities such as mass, length, time, density, energy, temperature, and the *magnitude* of any vector.

We now extend the operations of vector algebra to include multiplication of a vector by a scalar. The significance of multiplication of a vector by an integer is revealed by considering the equation

$$2\mathbf{a} = \mathbf{a} + \mathbf{a}$$

Evidently 2**a** is a vector in the same direction as **a** but with a magnitude twice that of **a**. The product − 2**a** is a vector with a direction opposite to that of **a** and with a magnitude twice that of **a**. These facts lead us to the definition for the product of a vector and any scalar q: $q\mathbf{a}$ is a vector in the direction of **a** if q is positive, but in the direction opposite to that of **a** if q is negative (Figure 1-7). The magnitude of this product is the product of the magnitudes of q and of **a**.

The following distributive laws can be derived from the definition of a vector sum and the definition of multiplication of a vector by a scalar:

$$q(\mathbf{a} + \mathbf{b}) = q\mathbf{a} + q\mathbf{b}$$

$$(q_1 + q_2)\mathbf{a} = q_1\mathbf{a} + q_2\mathbf{a}$$

The vector equations in this section show that the sums and products we have defined have most of the algebraic properties of the sums and products of

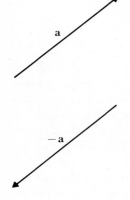

Figure 1-6   The vectors **a** and −**a** have equal magnitudes and opposite directions.

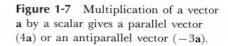

Figure 1-7   Multiplication of a vector **a** by a scalar gives a parallel vector (4**a**) or an antiparallel vector (−3**a**).

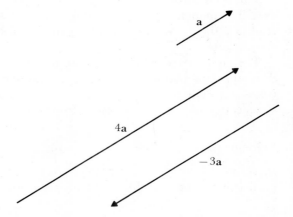

ordinary numbers. Consequently, we can manipulate these sums and products just as we manipulate numbers in ordinary algebra, with the exception that the product we have defined can be formed only between a scalar and a vector, and this product is itself a vector.

All expressions involving vectors should be checked to be sure they are meaningful. For instance, it is incorrect to equate a vector to a scalar, as in

$$\mathbf{a} = q \qquad \text{(wrong)}$$

Since the vector on the left-hand side has a direction, the quantity on the right-hand side must also have this same direction—it cannot be a scalar. We have assigned no meaning to the process of adding a scalar to a vector; consequently, an expression such as $q + \mathbf{a}$ is meaningless in vector algebra.

1.2 Particle Kinematics

Details of particle motion are described using terms that are very familiar in daily life such as speed, velocity, and acceleration. The speedometer of a car measures speed. *Velocity is a vector with a magnitude that is the speed and with a direction that is the direction of motion.* The position of a moving car changes as time elapses and the car's velocity is the time rate of change of its position. The car's velocity itself may be changing in magnitude because the car is speeding up or slowing down, or the velocity may be changing direction because the car is going around a curve. *The time rate of change of velocity is called the acceleration* and the change in velocity as time elapses is determined by the acceleration. A main endeavor in this section is to give a useful precision to these preliminary notions of velocity and acceleration.

Position Vector

Position is specified relative to some definite *frame of reference*—a set of coordinate axes with the position of the origin and the orientation of the axes determined relative to identified physical objects. Figure 1-8 shows the path of a particle as it moves relative to the frame of reference provided by the *x*- and *y*-coordinate axes. The motion represented could be that of a car traveling along a curved road and the frame of reference a pair of perpendicular lines marked on the earth's surface.

The position vector **r**, with its tail at the origin and its tip at the particle, determines the position at any instant. As time goes on, the tip of the vector traces out the path of the particle. A description of the motion consists of a specification of the position vector **r** for each instant of time.

Velocity

In Figure 1-8 we show the position vector **r** at a certain instant *t* when the particle is at a point *A*. In a time interval Δt the particle moves to point *B* and

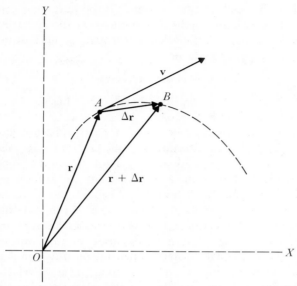

Figure 1-8   Particle moves from A to B in time interval Δt, undergoing a displacement $\Delta \mathbf{r}$. At the instant t when the particle is at A its velocity is $\mathbf{v}$.

the particle's position vector changes to a new value, $\mathbf{r} + \Delta \mathbf{r}$. The change in the position vector, the difference between the new value and the old value, is the *displacement* $\Delta \mathbf{r}$. The *average velocity* of the particle for this time interval is defined by

$$\overline{\mathbf{v}} = \frac{\Delta \mathbf{r}}{\Delta t} \tag{1.1}$$

where $\Delta \mathbf{r}$ is the particle's displacement during the time interval Δt. (A bar over a symbol is used to indicate that the quantity is an average value.)

For example, if, during a 4.0-s interval,* the particle's displacement is 80 m southeast, then the average velocity for this time interval is 20 m/s southeast. Notice that since $\overline{\mathbf{v}}$ is the product of a positive scalar $1/\Delta t$ and a vector $\Delta \mathbf{r}$, $\overline{\mathbf{v}}$ is a vector in the direction of the displacement $\Delta \mathbf{r}$. The magnitude $|\overline{\mathbf{v}}|$ of the average velocity has a unit which is a distance unit divided by a time unit such as mi/h, ft/s, km/h, or m/s. Conversion from one system of units to the other can be effected from the approximate relations:

$$60 \text{ mi/h} = 88 \text{ ft/s} = 96.5 \text{ km/h} = 26.8 \text{ m/s}$$

We are usually interested in a more detailed description of motion than is provided by the average velocity over a sizable time interval. We would like to

*In this book we follow the conventions of the Système International d'Unités (SI) described in Section 2.3 and in Appendix A. The SI unit of length, the metre, is denoted by the letter m and the SI unit of time, the second, is denoted by the letter s.

associate a velocity with each instant of time. In the particular case when a particle moves along a straight line in such a way that the average velocity $\overline{\mathbf{v}}$ is the same for any time interval, the velocity $\mathbf{v}$ at any instant naturally should be defined to be this constant average velocity:

$$\mathbf{v} = \overline{\mathbf{v}} = \textbf{constant vector} \tag{1.2}$$

In such a motion with *constant velocity,* the velocity vector has a constant magnitude and a constant direction.

In many motions, the average velocity of a particle is not constant but changes in both magnitude and direction for different time intervals. In order to give meaning to the idea of velocity at the instant t when the particle is at a point A, we consider the values of the average velocity $\Delta\mathbf{r}/\Delta t$ for a sequence of ever decreasing time intervals Δt, each interval beginning at the same instant t when the particle is at A. The average velocity $\Delta\mathbf{r}/\Delta t$ is generally different for different values of the time interval Δt, but, as Δt approaches zero, the average velocity approaches a limiting value independent of the time interval Δt. It is this limiting value of the average velocity, independent of Δt but characteristic of the instant t, that is the appropriate quantity to define as *the* velocity $\mathbf{v}$ at the instant t. This definition of the velocity $\mathbf{v}$ at the instant t (often termed the instantaneous velocity) is symbolized by

$$\mathbf{v} = \lim_{\Delta t \to 0} \overline{\mathbf{v}} = \lim_{\Delta t \to 0} \frac{\Delta \mathbf{r}}{\Delta t} = \frac{d\mathbf{r}}{dt} \tag{1.3}$$

where $d\mathbf{r}/dt$ is defined as the limit in question and is called the derivative of $\mathbf{r}$ with respect to t. When the word "velocity" is used without qualification, it is understood to mean the velocity at the instant t, rather than the average velocity for some time interval.

Since $d\mathbf{r}/dt$ is the limiting value of a *change* in $\mathbf{r}$ divided by the time required for the change, this derivative is called the time rate of change of $\mathbf{r}$. This leads to the verbal definition of $\mathbf{v}$: *The velocity is the time rate of change of the position vector.*

The definition of $\mathbf{v}$ implies that the direction of $\mathbf{v}$ is the limiting direction of $\Delta\mathbf{r}$, as Δt approaches zero, or as point B approaches point A. This limiting direction of $\Delta\mathbf{r}$ is the direction of the tangent to the path at A. Consequently, the direction of the velocity vector $\mathbf{v}$ at the instant t is that of the tangent to the path at point A.

The magnitude of the velocity vector $\mathbf{v}$ is called the *speed* and is denoted by $|\mathbf{v}|$:

$$speed = |\mathbf{v}| = \left|\frac{d\mathbf{r}}{dt}\right| = \lim_{\Delta t \to 0}\left|\frac{\Delta\mathbf{r}}{\Delta t}\right| \tag{1.4}$$

Determination of Δr from v

We can often measure or calculate values of a particle's velocity at different instants of time, and from these data we can determine the displacement at various instants and thereby describe the particle's motion. The displacement

$\Delta\mathbf{r}$ which occurs in any time interval Δt is (from Eq. 1.1)

$$\Delta\mathbf{r} = \bar{\mathbf{v}}\Delta t$$

where $\bar{\mathbf{v}}$ is the average velocity during the time interval Δt. This equation does not solve our problem, however, because if the velocity varies during the time interval Δt, it is not obvious what value we should use for the average velocity $\bar{\mathbf{v}}$. If we consider a time interval small enough so that the velocity does not vary significantly within the interval, the displacement will be given approximately by

$$\Delta\mathbf{r} = \mathbf{v}\Delta t \qquad\qquad \text{(approximately)} \quad (1.5)$$

where $\mathbf{v}$ is any velocity value occurring within the short time interval Δt.

A general procedure, based on repeated application of Eq. 1.5 to a succession of small time intervals of length Δt, is displayed in Figure 1-9. To estimate the displacement occurring in any one of these small time intervals we select, from the known velocity values, a value appropriate to the interval in question. A simple choice is the velocity at the beginning of the interval. Then, at the instant $t_n = n\Delta t$, after n time intervals of length Δt, the estimated displacement is

$$\mathbf{r}_n - \mathbf{r}_0 = \mathbf{v}_0\Delta t + \mathbf{v}_1\Delta t + \cdots + \mathbf{v}_{n-1}\Delta t \qquad (1.6)$$

where $\mathbf{v}_{n-1}$ is the velocity at the beginning of the nth interval.

When a computer is used to perform the large number of routine calculations necessary, this method becomes an accurate and practical way of determining motion. Still more important from a theoretical point of view is the fact that the extension of this process leads to the concept of an integral. Integrals and integration will be discussed in Chapter 7.

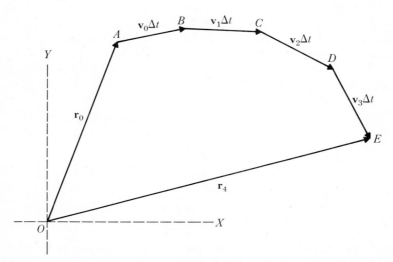

Figure 1-9 The approximate path $ABCDE$ is found by computing the displacements for a succession of small time intervals using $\Delta\mathbf{r} = \mathbf{v}\Delta t$.

Acceleration

Acceleration is the *time rate of change of velocity.* When a particle is speeding up or slowing down, or simply changing the direction of its motion as it travels along a curved path, it has an acceleration, according to the usage of the word in physics. In each case the velocity is changing.

Because acceleration is the time rate of change of velocity and velocity is the time rate of change of the position vector, the relationship between acceleration and velocity is mathematically the same as the relationship between velocity and the position vector. We present the formal definition of acceleration in a fashion precisely analogous to that followed in defining velocity. Figure 1-10 shows the velocity vector **v** of the particle at the instant t when the particle is at point A. As the particle moves along a path, its velocity can change in both magnitude and direction. In a time interval Δt, the particle moves to point B and its velocity vector changes to a new value **v** + Δ**v**. The average acceleration $\overline{\mathbf{a}}$ for the time interval Δt is defined by

$$\overline{\mathbf{a}} = \frac{\Delta \mathbf{v}}{\Delta t} \tag{1.7}$$

The acceleration **a** *at the instant t is defined as the limiting value of the average acceleration as the time interval* Δt *approaches zero:*

$$\mathbf{a} = \lim_{\Delta t \to 0} \frac{\Delta \mathbf{v}}{\Delta t} = \frac{d\mathbf{v}}{dt} \tag{1.8}$$

This equation is the mathematical statement of the verbal definition of acceleration as the time rate of change of velocity. Since **v** $= d\mathbf{r}/dt$, the acceleration

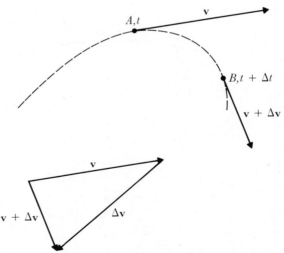

Figure 1-10 Δ**v** is the change in the velocity vector that occurs in the time interval Δt as the particle moves from A to B.

can be expressed directly in terms of the position vector **r** by

$$\mathbf{a} = \frac{d}{dt}\left(\frac{d\mathbf{r}}{dt}\right) = \frac{d^2\mathbf{r}}{dt^2} \tag{1.9}$$

where the symbol $d^2\mathbf{r}/dt^2$ is the common calculus notation for the derivative of a derivative (called a second derivative).

The unit for measurement of acceleration is a velocity unit divided by a time unit. The SI unit is the (m/s)/s which is written m/s^2.

The *direction* of the acceleration vector at instant t, according to the definition given by Eq. 1.8, is the direction approached by the *change in velocity* $\Delta\mathbf{v}$ as Δt approaches zero. From this observation we can deduce the important results that are displayed in Figure 1-11. The acceleration in motion along a curved path will be examined in detail in Section 8.3.

In the particular case when the average acceleration $\bar{\mathbf{a}}$ has the same value for all time intervals, the definition of **a** implies

$$\mathbf{a} = \bar{\mathbf{a}} = \text{constant vector} \tag{1.10}$$

In such a motion with *constant acceleration,* the acceleration vector has not only a constant magnitude but also a constant direction which implies that the direction of the velocity *change* $\Delta\mathbf{v}$ is the same no matter what time interval is considered.

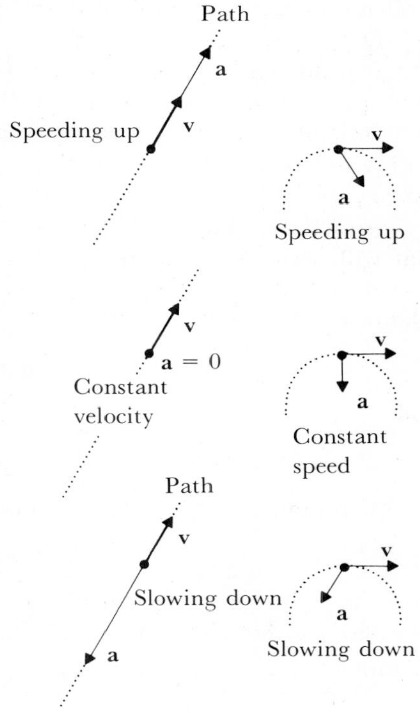

Figure 1-11 The direction of the acceleration vector for motion along a straight line and along a curved path.

Determination of Δv from a

The change in the velocity which takes place during a time interval Δt can be calculated, according to Eq. 1.7, from

$$\Delta v = \bar{a}\Delta t$$

where $\bar{a}$ is the average value of a during the interval Δt. This equation is directly useful only when the average acceleration is known. In general, to determine a velocity change, we first consider a time interval Δt so small that the acceleration does not alter significantly during this interval. Then the velocity change Δv is given approximately by

$$\Delta v = a\Delta t \qquad \text{(approximately)} \quad (1.11)$$

where a is any value of the acceleration occurring within the time interval Δt. A large time interval can be regarded as a succession of many short time intervals, and Eq. 1.11 can be applied to each short interval, as will be discussed in the following paragraphs.

Determination of Motion

In physics, motion is discussed using the position vector r, the velocity v, and the acceleration a, where $v = dr/dt$ and $a = dv/dt$. It is not necessary to continue in this manner and consider next the time rate of change of acceleration. The acceleration itself is the quantity that is important in dynamics.

When we take up the study of Newton's second law of motion, it will become apparent that our laws of nature specify a particle's acceleration as a function of time, position, and velocity. However, we must know something more than the acceleration before we can say what motion will occur, for if the particle is started differently, different motions will result. If we know the *initial position vector* r_0 and the *initial velocity* v_0, and if we also have complete information about the *acceleration* a, *then the entire motion is determined.* We can compute the velocities that will occur and find the positions reached at the succession of instants $t_0 = 0, t_1 = \Delta t, t_2 = 2\Delta t, \ldots, t_n = n\Delta t$, by repeated use of the approximate relations

$$\Delta r = v\Delta t \qquad (1.5)$$

and

$$\Delta v = a\Delta t \qquad (1.11)$$

At the instant $t_n = n\Delta t$, the estimated velocity is

$$v_n = v_0 + (a_0\Delta t + a_1\Delta t + \cdots + a_{n-1}\Delta t) \qquad (1.12)$$

where a_{n-1} is the acceleration at the beginning of the nth interval. Velocity estimates can be used in Eq. 1.6 to find the estimated position:

$$r_n = r_0 + (v_0\Delta t + v_1\Delta t + \cdots + v_{n-1}\Delta t) \qquad (1.13)$$

The calculation must proceed step by step because, in order to know which of the given values of acceleration to use at a given instant t_n, we must generally know the particle's location $\mathbf{r}_n$ as well as its velocity $\mathbf{v}_n$. It is worthwhile to trace a few steps to see how the acceleration and the initial conditions ($\mathbf{r}_0$ and $\mathbf{v}_0$) determine the evolution of the motion.

The particle starts with position vector $\mathbf{r}_0$, velocity $\mathbf{v}_0$, and acceleration $\mathbf{a}_0$. After the first time interval, at the instant $t_1 = \Delta t$, the particle's position vector is estimated to be

$$\mathbf{r}_1 = \mathbf{r}_0 + \Delta\mathbf{r} = \mathbf{r}_0 + \mathbf{v}_0\Delta t$$

and its velocity is estimated to be

$$\mathbf{v}_1 = \mathbf{v}_0 + \Delta\mathbf{v} = \mathbf{v}_0 + \mathbf{a}_0\Delta t$$

This velocity $\mathbf{v}_1$ is used to determine the displacement for the next time interval. Next we select, from the known values of the acceleration, the value $\mathbf{a}_1$ corresponding to the instant t_1, position $\mathbf{r}_1$, and velocity $\mathbf{v}_1$. This acceleration $\mathbf{a}_1$ is used to estimate the velocity change that occurs during the second interval. Then, at the instant $t_2 = 2\Delta t$, the estimates are

$$\mathbf{r}_2 = \mathbf{r}_1 + \Delta\mathbf{r} = \mathbf{r}_1 + \mathbf{v}_1\Delta t = \mathbf{r}_0 + (\mathbf{v}_0\Delta t + \mathbf{v}_1\Delta t)$$

and

$$\mathbf{v}_2 = \mathbf{v}_1 + \Delta\mathbf{v} = \mathbf{v}_1 + \mathbf{a}_1\Delta t = \mathbf{v}_0 + (\mathbf{a}_0\Delta t + \mathbf{a}_1\Delta t)$$

We can continue this process indefinitely, at each step using the latest values of velocity to estimate the subsequent displacement and the relevant value of the acceleration to estimate the velocity change in the forthcoming time interval.

Although the procedure just outlined is obviously cumbersome for hand calculation, it is the basis of programs for digital computers which are employed to get numerical answers to practical problems in mechanics such as rocket trajectories and satellite orbits.

In certain motions the acceleration is a simple function of $\mathbf{r}$, $\mathbf{v}$, and t. It then may be possible to describe the motion by finding a formula which expresses $\mathbf{r}$ as a function of t. Two examples are given in Table 1-1. The first example, motion with constant acceleration, which the reader may have studied in detail in

Table 1-1

Name	Acceleration function	Initial conditions	Mathematical description of the motion: position as a function of time
Motion with constant acceleration	$\mathbf{a} = $ constant vector	at $t = 0$ $\mathbf{r} = \mathbf{r}_0$ $\mathbf{v} = \mathbf{v}_0$	$\mathbf{r} = \mathbf{r}_0 + \mathbf{v}_0 t + \frac{1}{2}\mathbf{a}t^2$
Simple harmonic motion	$\mathbf{a} = -\omega^2\mathbf{r}$ (ω is a constant)	at $t = 0$ $\mathbf{r} = \mathbf{A}$ ($\mathbf{A}$ is a constant vector) $\mathbf{v} = 0$	$\mathbf{r} = \mathbf{A}\cos\omega t$

earlier courses, will be reviewed in Chapter 7 using the language and techniques of calculus. The second example is one particular case of simple harmonic motion, the topic of Chapter 9.

Summary

☐ Vectors have both magnitude and direction and sum according to the triangle rule.

☐ Scalars have a numerical value and possibly a unit but no direction.

☐ The motion of a particle is described by giving its position vector $\mathbf{r}$ as a function of time. The particle's velocity $\mathbf{v}$ is defined as the time rate of change of its position vector by the equation

$$\mathbf{v} = \frac{d\mathbf{r}}{dt}$$

and the particle's acceleration $\mathbf{a}$ is defined as the time rate of change of its velocity by

$$\mathbf{a} = \frac{d\mathbf{v}}{dt} = \frac{d^2\mathbf{r}}{dt^2}$$

Questions*

1 **(a)** The vector $\mathbf{a}$ is defined as 40 units west. What is its magnitude?
   (b) Another vector $\mathbf{b}$ is 40 units north. Can you say $\mathbf{a} = \mathbf{b}$?
   (c) Does $a = b$?

2 What is the sum of a displacement (a change of position) of 30 m east and a displacement of 40 m south?

3 Find the resultant of the following displacement vectors: 10.0 m west; 20.0 m north; 30.0 m south; 20.0 m east.

4 Sketch a diagram showing three vectors $\mathbf{a}$, $\mathbf{b}$, and $\mathbf{c}$ for which

$$\mathbf{a} + \mathbf{b} + \mathbf{c} = 0$$

   in the case when:
   (a) $\mathbf{a}$ and $\mathbf{c}$ have the same direction.
   (b) No two vectors have the same direction.

5 A car experiences the following displacements:
   $\mathbf{a} = 4.00$ km, $10.0°$ south of east.
   $\mathbf{b} = 10.00$ km, $40.0°$ east of north.
   $\mathbf{c} = 3.00$ km, $10.0°$ north of east.
   Using a vector diagram drawn to scale, determine the magnitude and direction of the vector sum of these displacements.

*In this text, in the questions and examples, we usually present data with two or three significant figures (Appendix B). After calculating with these data, the student should retain in his results only the number of significant figures consistent with the uncertainty of the original data.

6 Two displacements are **a** = 30.0 m east; **b** = 30.0 m north:
   (a) Use the parallelogram law to find the magnitude and direction of **a** + **b**.
   (b) Find the magnitude and direction of **a** − **b**.

7 The vector **b** is a displacement of 10 m northeast. Find the magnitude and direction of the vector − 5**b**.

8 Explain the defect in each of the following equations:
   (a) **a** = 10 m (wrong).
   (b) **a** + q = **b** (wrong).

9 Point P_1 has position vector r_1 and point P_2 has position vector r_2. Verify that the vector directed from P_1 to P_2 is $r_2 - r_1$.

10 Using the result of the preceding question show that the midpoint M of the line P_1P_2 has the position vector given by $r_M = \frac{1}{2}(r_1 + r_2)$.

11 A car's position vector **r** is 30 m east at t = 10 s. It is traveling with a constant velocity vector **v** which is 10 m/s east. Find the change in the position vector (**Δr**) which occurs in the next 2.0-s interval. What is the position vector of the car at t = 12 s?

12 Define the position vector, velocity vector, and acceleration vector of a particle.

13 A record of a car's position vector at certain instants is shown in the following table:

Reading t of watch s	Position vector m
0.00	0
6.00	100 east
8.00	100 east + 100 north
32.00	0

Calculate the displacements and average velocities for the time intervals indicated in the following table:

Time interval s	Displacement m	Average velocity m/s
0.00 → 6.00		
6.00 → 8.00		
0.00 → 8.00		
0.00 → 32.00		

14 A record of speedometer readings for a car traveling east is given in the following table:

Time (s)	8.0	9.0	10.0	11.0	12.0	13.0	14.0
Speed (m/s)	10.0	11.0	13.0	9.0	4.0	2.0	1.0

   (a) Estimate the car's displacement for the time interval extending from t = 8.0 s to t = 14.0 s.
   (b) Estimate the average velocity for the time interval of part (a).

15 A record of an automobile's velocity at certain instants is shown in the following table:

Reading t of watch s	Velocity m/s
0.00	0
6.00	20 east
8.00	20 east + 20 north
32.00	40 north

Calculate the velocity changes and average accelerations required to complete the following table:

Time interval s	Change in velocity m/s	Average acceleration m/s^2
0.00 → 6.00		
6.00 → 8.00		
0.00 → 32.00		

16 **(a)** A car's speed increases as it is driven straight north. Draw a diagram showing the directions of the velocity and acceleration vectors.

(b) The driver of the car steps on the brakes. Draw the velocity and acceleration vectors.

(c) Justify the relative directions of the acceleration and velocity vectors given in parts (a) and (b) using $\Delta\mathbf{v} = \mathbf{a}\Delta t$ and assuming for simplicity that the acceleration is constant during the time interval Δt.

Supplementary Questions

S-1 Consider a triangle OP_1P_2 with position vectors of P_1 and P_2 (relative to the origin O) given by $\mathbf{r}_1$ and $\mathbf{r}_2$, respectively. M_1 and M_2 are the midpoints of the lines OP_1 and OP_2, respectively. Show that the displacement from M_1 to M_2 is given by the vector $\frac{1}{2}(\mathbf{r}_2 - \mathbf{r}_1)$ and thereby prove that the line M_1M_2 joining the midpoints of the sides of a triangle is parallel to the remaining side P_1P_2 of the triangle.

S-2 Consider a parallelogram $OP_1P_3P_2$ where the position vectors of points P_1 and P_2 (relative to the origin O) are $\mathbf{r}_1$ and $\mathbf{r}_2$, respectively:

(a) Show that the diagonals of this parallelogram are represented by $\mathbf{r}_1 + \mathbf{r}_2$ and $\mathbf{r}_1 - \mathbf{r}_2$.

(b) Show that the midpoints of the diagonals coincide, that is, that the diagonals bisect each other.

S-3 A motion is described by

$$\mathbf{r} = \mathbf{r}_0 + \mathbf{v}_0 t + \tfrac{1}{2}\mathbf{a}t^2$$

where $\mathbf{r}_0$, $\mathbf{v}_0$, and $\mathbf{a}$ are constant vectors. By differentiation, find the velocity and the acceleration functions for this motion.

S-4 Consider the motion described by the polynomial in t

$$\mathbf{r} = \mathbf{b}_0 + \mathbf{b}_1 t + \mathbf{b}_2 t^2 + \mathbf{b}_3 t^3$$

where $\mathbf{b}_0$, $\mathbf{b}_1$, $\mathbf{b}_2$, and $\mathbf{b}_3$ are constant vectors. Deduce expressions for the velocity and acceleration as functions of time. Find the physical significance of each of the constant vectors, $\mathbf{b}_0$, $\mathbf{b}_1$, $\mathbf{b}_2$, and $\mathbf{b}_3$.

S-5 At a certain point on a particle's trajectory, the particle has zero velocity. Explain how the particle can move away from this point.

2.

force, mass,
and acceleration

Up to this point we have been concerned only with kinematics, a geometrical description of motion given in terms of kinematical quantities—position vector, velocity, and acceleration. We have seen that the motion of a particle relative to a specified frame of reference is completely determined by the acceleration and the initial conditions. Therefore, to predict a particle's motion, we want to know the factors that determine the particle's acceleration. This brings us to the branch of physics called *dynamics,* the study of the forces acting on a body and the associated motion of the body.

The entire influence of the environment on the motion of a particle is assumed to be represented by forces exerted on the particle by the objects of the environment. In physics a *force is a push or a pull;* any other meanings that the word force has in daily life are excluded.

2.1 Law of Inertia

Motion with No Force

We begin the study of the relationship between force and motion by considering the motion of a particle when no force is exerted on it. The type of motion that occurs when there is no force cannot be ascertained by direct experiment because we have no means of ensuring the absence of external forces. We can, however, do experiments in which the forces that obviously affect an object's motion are progressively reduced and in this way infer what the motion would be with no force.

When a smooth block is projected along a smooth horizontal surface we observe that the block slides along in a straight line, but it gradually slows down and finally stops. If the experiment is repeated with smoother surfaces, and with a lubricant provided, we find that the block's speed decreases more slowly with the result that it keeps moving for a longer time and travels further before coming to rest. Such experiments suggest that if a moving particle were truly left to itself its velocity would not change at all. This was the conclusion reached by Galileo Galilei (1564–1642) after a similar series of experiments.

Newton's First Law

Profiting from Galileo's experiments and reflections, Sir Isaac Newton (1642–1727) took as his starting point what we call *Newton's first law of motion* or the *law of inertia: Every body continues in its state of rest or of uniform motion in a straight line, except insofar as it is compelled by external forces to change that state.*

Notice that this law refutes the commonly held Aristotelian notion that rest is the "natural state" and that force is required to maintain motion. Galileo and Newton asserted that *constant velocity* is the natural state of affairs and that only *changes in velocity* have to be explained by the presence of forces. "Rest" is just one special kind of constant velocity. A body has an inherent tendency to keep its velocity vector constant: we say the body has inertia.

Inertial Frames

The law of inertia is certainly not true in all frames of reference. To see this one has only to think of the motion of loose objects inside a car while the brakes are being applied. With the car as the frame of reference, any free object that accelerates forward before it slams into the dashboard is changing its velocity in gross violation of the law of inertia. Evidently the laws of mechanics are not the same relative to all frames of reference.

A reference frame in which the law of inertia holds good is called an *inertial frame.* In all our work, unless we mention to the contrary, we assume that measurements of position, velocity, and acceleration are made relative to an inertial frame. Experiments are necessary to decide whether or not a given reference frame is inertial, and, as is suggested by the motion of a puck on a sheet of ice, any frame fixed in our planet is inertial, at least to an approximation adequate for the description of many mundane motions. A better approximation to an inertial frame is the astronomical frame, a frame in which motion of the distant stars is not apparent.

To emphasize that Newton's first law is really a statement about reference frames, we rephrase it to give a modern statement of the *law of inertia: There exist certain reference frames, called inertial frames, relative to which any particle has a constant velocity vector (perhaps zero) when free of all external forces.*

2.2 Force

An Operational Definition of Force

We have stated that a force is a push or a pull. Although such a statement helps relate force to familiar concepts, it is hardly a definition adequate for scientific work.

A physically significant definition, called an operational definition, prescribes a sequence of operations to *measure* the physical quantity in question. Such a definition need not contain the details of a practical procedure but merely show how, at least in principle, the quantity can be measured.

An operational definition of force can be given in terms of spring balances, with the force under investigation applied to a hook as shown in Figure 2-1. We shall call the force which produces a certain definite extension a force of unit magnitude. Two such forces acting in the same direction define a force with a magnitude of two units; two equal forces in the same direction which combine to give a force of unit magnitude are defined to be each of $\frac{1}{2}$ unit magnitude, and so on. Although instruments more sensitive and accurate than spring balances are required for scientific work, this method is adequate in principle to define the magnitude and direction of a force.

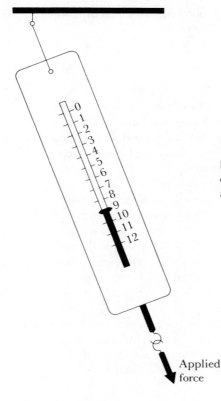

Figure 2-1   A spring balance which can be used to define the magnitude and direction of force.

Applied
force

Superposition Principle for Forces

A force has both magnitude and direction. A most important experimental result is that forces sum according to the triangle rule for addition of vectors. This is verified by applying simultaneously several forces to the same spring balance and discovering the single force to which they are equivalent. Forces are thus *vector quantities*. We summarize these experimental results in a statement known as the *superposition principle for forces* (Figure 2-2): *When several forces act simultaneously on a particle, they are equivalent to a single force which is their vector sum or resultant.*

To understand motion we must think about the vector sum of the forces acting on a particle. In particular, we must realize that when several forces act on a particle their vector sum may be zero, and in this case the motion proceeds as if there were no forces acting on the particle. The particle's velocity is then constant; in other words, its acceleration is zero.

Force and Acceleration

The relationship between the motion of a particle and the forces acting on the particle can be investigated experimentally. Known forces, perhaps exerted by pulling with spring balances (Figure 2-3), are applied and the object's motion is observed.

We focus attention on the acceleration vector, the time rate of change of velocity, rather than on velocity itself. Experiments show that for a given particle the resultant force vector and the acceleration vector are proportional. Doubling the resultant force doubles the acceleration, and tripling the resultant force triples the acceleration. This relationship between force and acceleration is independent of the velocity of the particle under investigation. Moreover, the accel-

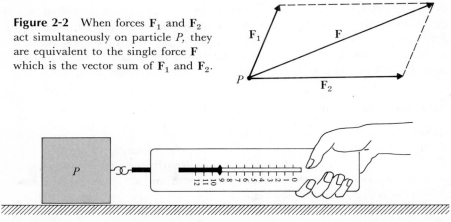

Figure 2-2   When forces $\mathbf{F}_1$ and $\mathbf{F}_2$ act simultaneously on particle P, they are equivalent to the single force $\mathbf{F}$ which is the vector sum of $\mathbf{F}_1$ and $\mathbf{F}_2$.

Figure 2-3   "Particle" P is accelerated by the force exerted on it by the spring balance.

eration of the particle is always in the direction of the resultant force, regardless of the direction of the velocity vector. The results of all experiments that it is feasible to perform with tangible particles are in accord with the vector proportionality that is expressed by

$$\mathbf{F} \propto \mathbf{a}$$

where $\mathbf{F}$ is the resultant force acting on the particle and $\mathbf{a}$ is the particle's acceleration. (The symbol $\propto$ means "is proportional to.")

2.3 Mass

The proportionality $\mathbf{F} \propto \mathbf{a}$ implies that we can write an equation

$$\mathbf{F} = m\mathbf{a} \tag{2.1}$$

where m, the constant of proportionality, is characteristic of the particle being accelerated. Since $\mathbf{F}$ and $\mathbf{a}$ are in the same direction, m is positive. This positive number m, given by

$$m = \frac{F}{a}$$

is defined to be the mass of the particle.

The mass of a particle is the intrinsic characteristic of the particle that determines its response to the resultant force. If we apply forces of the same magnitude F to different masses, the larger mass will have the smaller acceleration. We say that the larger mass has greater *inertia*, and that mass is a *quantitative measure of inertia*. The rather vague term inertia refers to the tendency of a particle to continue to move at the same speed in the same direction, that is, to maintain a constant velocity vector.

By experimenting with different tangible particles of masses m_1 and m_2, we find that, when they are fastened together and accelerated, they behave as a single particle of mass $m_1 + m_2$. This additive property of mass is in accord with extremely accurate experimental observations dealing with combinations of atoms, molecules, and larger bodies. The mass of an object is the most appropriate measure of the "amount of matter" in it, because the mass of an object made up of many parts is the sum of the masses of its parts,* and the object's mass will not change no matter how its shape and volume are altered.

Standard of Mass, Units

Measurements of physical quantities are expressed in terms of a system of units which is established by choosing arbitrary standard units for a set of so-called fundamental physical quantities such as mass, length, and time. Further derived units are expressed in terms of these fundamental units. For example, the unit of velocity is a length unit per unit of time.

---

*This statement is re-examined and qualified in Section 11.6.

In this book we shall use the system of units described in Appendix A and universally designated by the abbreviation SI (Système International). The fundamental physical quantities of mechanics are chosen to be mass, length, and time. The units of length and time, named the metre and the second, are defined in terms of standards described in Appendix A. To establish a scale of mass, a definite object, a cylinder of platinum-iridium alloy stored at the International Bureau of Weights and Measures at Sèvres, France, is selected as a standard body and its mass is defined to be exactly 1 kilogram (1 kg).

In the SI system, the unit of force is a derived unit. In accordance with the equation $F = ma$, the force which accelerates a 1-kg mass at the rate of 1 m/s^2 is the unit of force. This force unit is called a *newton* (N) and

$$1 \text{ N} = 1 \text{ kg} \cdot \text{m/s}^2$$

Table 2-1 gives several SI units together with the corresponding British engineering units.

Table 2-1

Physical quantity	SI unit	British engineering unit	Conversion	
Length	metre (m)	foot (ft)	1 ft = 0.3048 m	1 m = 3.281 ft
Time	second (s)	second (s)		
Mass	kilogram (kg)	slug (sl)	1 sl = 14.59 kg	1 kg = 0.06852 sl
Force	newton (N)	pound (lb)	1 lb = 4.448 N	1 N = 0.2248 lb

2.4 Newton's Second Law, $\mathbf{F} = m\mathbf{a}$

The definitions and experimental results that have been given in this chapter are summarized in what is called *Newton's second law*

$$\mathbf{F} = m\mathbf{a} \tag{2.1}$$

where **F** *is the resultant of all the forces acting on a particle of mass m, and* **a** *is the particle's acceleration relative to an inertial frame of reference.* This is the central law of Newtonian mechanics, the key to Newton's synthesis of much of the natural philosophy of his time, and the law that is the starting point for the comprehension of a large realm of modern engineering and science.

Newtonian mechanics gives a simple and accurate description of a tremendous range of phenomena. Of course in this century experiments and new theories have led to deep-seated changes in our ideas about the most fundamental things, and the domain of applicability of Newtonian mechanics has been staked out. Modifications required by Einstein's special theory of relativity are significant when particle speeds are so large that they are an appreciable fraction of the speed of light (3×10^8 m/s). Quantum mechanics requires radical departures from the viewpoint of Newtonian mechanics, but these considerations become important only when very small magnitudes are to be measured. For the motion of tangible objects at speeds small compared to the speed of light, the predictions of these modern theories are imperceptibly different from those based on Newton's second law. Newtonian mechanics remains the basic subject for much of modern science and engineering, and the necessary path for the understanding of more recent theories.

Force Laws

To evaluate the acceleration of a particle using

$$\mathbf{a} = \frac{\mathbf{F}}{m}$$

we must know the resultant of the forces exerted on the particle by other objects. This requires an experimental investigation of the different types of forces that can be exerted by one object on another. The success of Newtonian mechanics is associated with the discovery of *simple* laws of force. The theory then becomes both a convenient vehicle for thought and a method of precise calculation.

2.5 Fundamental Forces

The interactions of different chunks of matter arise from the interaction of the particles comprising that matter. Investigations of the interactions between particles such as electrons, protons, and neutrons and many other "elementary particles" that will be described later suggest that there are just the four fundamental types of forces or interactions that are listed in Table 2-2 in order of increasing strength.

The very strong nuclear forces that are involved in holding together the nucleus of an atom, and the weak interaction that can lead to the sudden spontaneous change of some nuclei (β radioactivity), will be discussed only near the end of the book. In our daily life we are completely unaware of the existence of such forces.

Table 2-2 Fundamental Forces

Name	State of knowledge
Gravitational force	Law known
Weak interaction	Knowledge incomplete
Electromagnetic interaction	Law known
Nuclear forces	Knowledge incomplete

An electromagnetic force is exerted by one electrically charged particle on another. This electromagnetic interaction will be studied throughout several later chapters. We find that electromagnetic forces determine the structure of an atom outside of its nucleus and thereby govern the interactions of one atom with other atoms. Forces exerted when one laboratory-size object is in contact with another, such as the forces described by the empirical laws of friction (Chapter 3) or by Hooke's law (Chapter 9), arise from the interaction of billions of atoms, and they are therefore ultimately determined by electromagnetic forces between the electrons and protons.

2.6 Gravitational Field g and Weight *m*g

In the presence of any massive object like our planet earth, any other mass m, say a marble, experiences a measurable attractive force called a gravitational force. Measurement of the gravitational force exerted on different masses m at a given location shows that this gravitational force is strictly proportional to the mass m; that is, we can write for the gravitational force **F**

$$\mathbf{F} = m\mathbf{g} \tag{2.2}$$

where the proportionality factor is denoted by a vector **g** in the direction of **F**. *This vector* **g** *is called the gravitational field at the point in question and we say that the gravitational field* **g** *exerts a force* **F** $= m\mathbf{g}$ *on the mass m.* This gravitational force $m\mathbf{g}$ has a name that is very familiar: it is called the *weight* of the object.

At different points on the earth or elsewhere in space, **g** generally has a different magnitude and a different direction, as shown in Figure 2-4. This figure gives us an idea of the gravitational field that will be encountered at different points in the vicinity of our planet. Notice that **g** always points to the center of the earth and that the magnitude of **g** decreases as one moves away from the earth. Near the earth's surface the magnitude of **g** is approximately 9.8 N/kg, varying slightly from place to place. At an altitude equal to the earth's

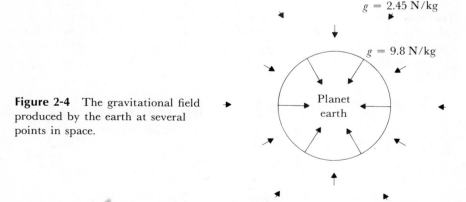

Figure 2-4 The gravitational field produced by the earth at several points in space.

$g = 2.45 \text{ N/kg}$

$g = 9.8 \text{ N/kg}$

Planet earth

radius (approximately 6400 km), g is 2.45 N/kg, merely one-fourth of its value at the earth's surface. Thus an astronaut whose mass is 68 kg will weigh (68 kg) (9.8 N/kg) $= 6.7 \times 10^2$ N (about 150 lb) at the earth's surface, and at an altitude of 6400 km his weight will have decreased to (68 kg) (2.45 N/kg) $= 1.67 \times 10^2$ N (37.5 lb).

The important point is that the mass m of an object does not depend on where the object is located, but its weight $m\mathbf{g}$, being just the gravitational force exerted by the gravitational field $\mathbf{g}$ on the object, decreases if the object is moved to a position where the gravitational field $\mathbf{g}$ is smaller. By moving to a high altitude someone who is overweight can indeed decrease his weight, but this probably is not really what he is interested in; his mass will not be changed by the trip, and there will be just as much of him as before.

Convenient practical methods of measuring mass are based on the fact that the weights of different objects are proportional to their masses, provided the weighings are made at the same location in space. When g is known we can calculate the mass m of an object by measuring the gravitational force $m\mathbf{g}$ on it, perhaps using a spring balance. Alternatively the object's mass can be compared to the mass of a standard object by using a beam or platform balance to compare gravitational forces.

Acceleration Due to Gravity

Let us consider the motion of a particle of mass m in a region where there is a gravitational field $\mathbf{g}$. To determine the acceleration we use Newton's second law, $\mathbf{F} = m\mathbf{a}$. After the particle has been projected, if air resistance is negligible, the only force acting on the particle is the gravitational force $m\mathbf{g}$. Hence this force is the resultant force acting on the particle, and application of Newton's second law yields

$$m\mathbf{g} = m\mathbf{a}$$

The mass cancels out and we find that

$$\mathbf{a} = \mathbf{g} \tag{2.3}$$

regardless of the mass of the particle. This tells us that at any point along its path, the particle's acceleration is given by the value of $\mathbf{g}$ at that point (Figure 2-5). We see that $\mathbf{g}$ can be called the *acceleration due to gravity* as well as the gravitational field.

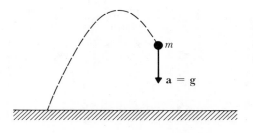

Figure 2-5 A particle has an acceleration $\mathbf{g}$ when the only force acting on it is its weight $m\mathbf{g}$.

Summary

☐ There exist certain reference frames, called inertial frames, relative to which any particle has a constant velocity vector when free of all external forces.

☐ Newton's second law states that

$$\mathbf{F} = m\mathbf{a}$$

where $\mathbf{F}$ is the resultant force acting on a particle of mass m, and $\mathbf{a}$ is the particle's acceleration relative to an inertial frame of reference. In SI units, we measure the resultant force $\mathbf{F}$ in newtons (N), the mass m in kilograms (kg), and the acceleration $\mathbf{a}$ in m/s². For the special case when the resultant force is zero, Newton's second law gives $\mathbf{a} = 0$ which implies that the particle's velocity is constant.

☐ The weight of a body is the gravitational force exerted on the body. A particle of mass m has a weight $m\mathbf{g}$ at a location where the gravitational field is $\mathbf{g}$. If the only force acting on the particle is the gravitational force $m\mathbf{g}$, the particle's acceleration will be $\mathbf{g}$.

Questions

1 Compute the vector sum of the following forces:
   (a) 3 N east; 2 N east.
   (b) 3 N east; 2 N west.
   (c) 3 N east; 3 N west.
   (d) 3 N east; 4 N north.
   (e) 5 N north; 6 N east; 5 N south; 2 N west.

2 **(a)** While a car is traveling east, the horizontal force exerted on the car by the road is 300 N east. The air exerts a force of 200 N west on the car. Find the vector sum of these forces acting on the car.
   (b) The brakes of the car are applied. Then the horizontal force exerted on the car by the road is 1200 N west. The air still exerts a force of 200 N west on the car. What is the vector sum of these forces acting on the car?

3 Shortly after jumping from an aircraft a man experiences an upward force of 200 N exerted by the air. The downward gravitational force acting on the man is 800 N. What is the resultant force acting on the man?

4 You are pushing a wagon with a force of 200 N and it is traveling with a constant velocity of 3.0 m/s. What is the total retarding force acting on the wagon?

5 The force acting down (weight, the gravitational force exerted by the earth on the object) on a paratrooper and his parachute is 1000 N. A short time after jumping, his velocity is constant. What can be said about the force exerted by the air on the parachute and on the man?

6 To measure the mass of a trunk, we push it along a sheet of ice exerting a measured horizontal force of 100 N. The acceleration is observed to be 2.0 m/s². What is the mass of the trunk?

7 Find the acceleration vector of a 15-kg particle acted on by the forces of Question 1(d).

8 The car in Question 2 has a mass of 1.40×10^3 kg. Find the magnitude and direction of its acceleration vector in Question 2(a) and also in Question 2(b).

9 A 20-kg box sliding across the floor slows down from 2.5 m/s to 1.0 m/s in 3.0 s. Assuming that the force (and therefore the acceleration) is constant, find the resultant force acting on the box. Give the magnitude and the direction of the force relative to the velocity vector of the box.

10 A 2.0-kg stone falls with an acceleration of 9.8 m/s^2. What is the resultant force acting on the stone?

11 At a certain instant, a falling 0.50-kg cardboard box has an acceleration of 8.0 m/s^2. What is the resultant force acting on the box at this instant?

12 **(a)** Define weight.
(b) What is the definition of the gravitational field **g** at a point in space?

13 At the surface of a certain planet the gravitational field **g** has a magnitude of 2.0 N/kg. A 4.0-kg brass ball is transported to this planet. Give:
(a) The mass of the brass ball on the earth and on the planet.
(b) The weight of the brass ball on the earth and on the planet.

14 **(a)** What is the gravitational field at a point where a 2.00-kg mass experiences a gravitational force of 18.0 N?
(b) What gravitational force would be exerted on a 6.00-kg mass placed at this point?

15 Does the acceleration of a baseball, after it has been struck and is traveling toward the outfield, depend on who hit it? Does this acceleration depend on whether the baseball is going up or down? (Neglect the effect of the force exerted on the baseball by the air.)

16 Show that $\mathbf{F} = m\mathbf{g}$ and Newton's second law imply that particles of different masses have the same acceleration in a gravitational field.

17 What happens to refuse that is released from an earth satellite? Does it fall straight down to earth? Explain.

Supplementary Questions

S-1 A resultant force of 10 N gives a mass m_1 an acceleration of 4.0 m/s^2, and a mass m_2 an acceleration of 12 m/s^2. What acceleration will this same force impart to the two masses when they are fastened together?

S-2 Do particles always move in the direction of the resultant force acting upon them? That is, in a short time interval Δt, is $\Delta \mathbf{r}$ always in the direction of **F**? Is $\Delta \mathbf{v}$ in the direction of **F**? Explain.

S-3 An empty box acquires an acceleration of 1.50 m/s^2 when acted upon by a certain resultant force **F**. When a brick is placed in the box, the acceleration, with the same resultant force acting on the box, is observed to be 0.50 m/s^2. If four such bricks are placed in the box, what acceleration will be imparted to the box by the same resultant force **F**?

3.

forces acting
on a particle

In this chapter we develop systematic procedures for the application of Newton's second law in cases when several forces act on a single particle. Analysis using *components* of vectors is stressed. Contact forces are discussed and the laws of friction are presented.

3.1 Components of a Vector

It is often convenient to express a given vector in a plane as the sum of two perpendicular vectors in the plane which are then called *vector components* of the given vector. The vector is said to be *resolved into components* along the two perpendicular directions of these vector components.

If we introduce in the plane two perpendicular coordinate axes, OX and OY, any vector $\mathbf{a}$ in the plane can be resolved into vector components $\mathbf{a}_x$ and $\mathbf{a}_y$ parallel to OX and OY, respectively, such that

$$\mathbf{a} = \mathbf{a}_x + \mathbf{a}_y$$

as illustrated in Figure 3-1.

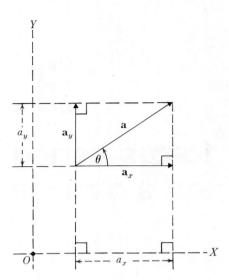

Figure 3-1 The vector components of **a** resolved along OX and OY are the vectors $\mathbf{a}_x$ and $\mathbf{a}_y$. The x component of **a** is $a_x = a \cos \theta$ and the y component of **a** is $a_y = a \sin \theta$. When the components are known, the magnitude of the vector can be determined from $a = \sqrt{a_x^2 + a_y^2}$ and the angle from $\tan \theta = a_y/a_x$.

Figure 3-2 The vectors of Example 1.

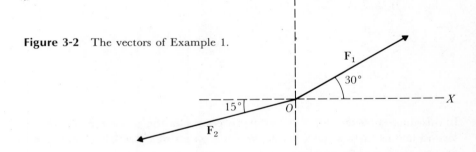

The *positive or negative number a_x, called the x component* of **a**, has a magnitude which is just the projection of **a** along OX. From the triangle in Figure 3-1, we see that

$$a_x = a \cos \theta \qquad (3.1)$$

where θ is the angle measured counterclockwise from the direction of OX to the vector **a**. The number a_x is positive if the vector component $\mathbf{a}_x$ points in the positive x direction. Negative a_x corresponds to the vector component $\mathbf{a}_x$ in the negative x direction.

The y component of **a**, the number a_y, is related in a similar way to the direction of the Y axis:

$$a_y = a \cos (90° - \theta) = a \sin \theta \qquad (3.2)$$

It is worth noting that if a vector is *perpendicular* to an axis, its component along that axis is *zero*. If the direction of the vector **a** is parallel to the positive direction of an axis, the component along this axis is simply the positive number a; this component is $-a$ if the vector points in the direction opposite to the positive direction of the axis. The ability to recognize immediately the value of a component in these particular cases is necessary for facility in analyzing many physical situations.

Example 1   A coordinate system is selected with east as the positive x direction and north as the positive y direction. Find the components of the following forces (Figure 3-2):

$$\mathbf{F}_1 = 8.00 \text{ N}, 30.0° \text{ north of east}$$

$$\mathbf{F}_2 = 10.0 \text{ N}, 15.0° \text{ south of west}$$

Solution   Particular care must be taken to specify the correct sign for each component. A simple procedure is first to determine the magnitude of the component and then to affix a positive or negative sign according to whether the *vector component* points in the positive or negative direction of the axis in question. To find the magnitude of a component along a given axis, we need to compute only the cosine of the *acute* angle that the vector makes with the axis and then to multiply this positive number by the magnitude of the vector. Proceeding in this way we obtain

$$F_{1x} = (8.00 \text{ N}) \cos 30.0° = 6.93 \text{ N}$$

$$F_{1y} = (8.00 \text{ N}) \cos 60.0° = 4.00 \text{ N}$$

$$F_{2x} = -(10.0 \text{ N}) \cos 15.0° = -9.66 \text{ N}$$

$$F_{2y} = -(10.0 \text{ N}) \cos 75.0° = -2.59 \text{ N}$$

Notice that in this method we always work with cosines and acute angles. This is the simplest procedure for the *numerical* evaluation of components.

Often we know the components (a_x, a_y) of a vector and are confronted with the problem of using this information to find the vector's magnitude and direction. The direction of the vector is usually specified by giving the angle θ that the vector makes with the X axis. In Figure 3-1, using the definition of $\tan \theta$, we see that

$$\tan \theta = \frac{a_y}{a_x} \tag{3.3}$$

Knowledge of $\tan \theta$ and of the signs of a_x and a_y is sufficient to determine θ.

The Pythagorean theorem applied to the right triangle in Figure 3-1 with sides of lengths a_x and a_y and hypotenuse of length a gives

$$a = \sqrt{a_x^2 + a_y^2} \tag{3.4}$$

Evidently the vector **a** can be specified just as well by giving its components (a_x, a_y) as by giving its magnitude a and the angle θ, and we can pass from one description of the vector to the other by routine calculations.

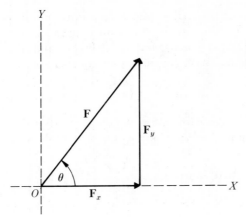

Figure 3-3   The vector components F_x and F_y of the vector **F** of Example 2.

Example 2   Find the magnitude and direction of the force **F** which has the components

$$F_x = 3.00 \text{ N} \quad \text{and} \quad F_y = 4.00 \text{ N}$$

Solution   A rough sketch (Figure 3-3) showing the geometrical addition of the vector components F_x and F_y makes evident the approximate direction of $F = F_x + F_y$. From this figure

$$\tan \theta = \frac{F_y}{F_x} = \frac{4.00}{3.00} = 1.33$$

Therefore $\theta = 53.1°$. The vector **F** makes an angle of $53.1°$ with the X axis.
The magnitude of the vector can be calculated from Pythagoras' theorem applied to Figure 3-3:

$$F = \sqrt{F_x^2 + F_y^2} = \sqrt{(3.00 \text{ N})^2 + (4.00 \text{ N})^2} = 5.00 \text{ N}$$

(An alternative procedure, after the angle θ has been determined, is to evaluate $\cos \theta$ and then to compute the magnitude of the vector from

$$F = \frac{F_x}{\cos \theta}$$

In this way we can find the magnitude without having to extract a square root.)

Vector Equations

Two vectors **a** and **b** are equal *if and only if* their corresponding components are equal. Consequently, for vectors in the XY plane, the one vector equation

$$\mathbf{a} = \mathbf{b}$$

is equivalent to two equations relating components:

$$a_x = b_x \quad \text{and} \quad a_y = b_y$$

3.2 Addition of Vectors by Adding Components

One major reason that components are so useful is that the component along a given direction of the vector sum of any number of vectors is just the algebraic sum of the components of these vectors in the specified direction. This is illustrated in Figure 3-4 for the vector addition of two vectors **a** and **b** to form a sum $\mathbf{s} = \mathbf{a} + \mathbf{b}$. Along the X axis, the x component of the vector sum s_x is the algebraic sum $a_x + b_x$ of the x components of the two vectors that have been added; that is

$$s_x = a_x + b_x \tag{3.5}$$

Similarly, along the Y axis we have

$$s_y = a_y + b_y \tag{3.6}$$

Knowledge of the components (s_x, s_y) completely determines the vector **s**. If we wish to find the numerical value of s and the angle θ that **s** makes with the X axis, we can use $s = \sqrt{s_x^2 + s_y^2}$, $\tan \theta = s_y/s_x$.

It is evident that, by working with components, the addition of any number of vectors can be performed analytically simply by computing algebraic sums. Because the problem of adding vectors arises in nearly every branch of physics, the topic merits special attention. The following is a systematic general procedure for adding vectors analytically:

1 Introduce rectangular coordinate axes. Try to select an orientation of the axes that will minimize the labor necessary to compute the components along the axes.
2 Resolve each vector into its x and y components.

Figure 3-4 Geometric proof that $s_x = a_x + b_x$ and $s_y = a_y + b_y$.

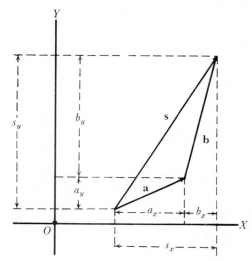

3   Calculate the algebraic sum of the x components of the different vectors to determine the x component of their vector sum. The y components are treated similarly.

4   When necessary, the magnitude s and the direction of the vector sum $\mathbf{s}$ can be obtained from (s_x, s_y).

5   A check that the evaluation of $\mathbf{s}$ is at least approximately correct should be made by sketching a rough geometric head-to-tail addition of the vectors according to the triangle rule or its extension (Figures 1-1 and 1-3).

Example 3   Figure 3-5 shows two forces acting on a particle. The force $\mathbf{F}_1$ is directed vertically downward and has a magnitude of 20.0 N. The force $\mathbf{F}_2$, of magnitude 50.0 N, makes an angle of 36.9° above a horizontal direction. Find the single force $\mathbf{F}$ equivalent to this combination of forces.

Solution   From the superposition principle for forces, the forces $\mathbf{F}_1$ and $\mathbf{F}_2$ are together equivalent to the single force $\mathbf{F}$, the *resultant force*, which is given by

$$\mathbf{F} = \mathbf{F}_1 + \mathbf{F}_2$$

We can evaluate this vector sum by adding corresponding components. With axis OX horizontal and OY vertically upward we find

$$F_x = F_{1x} + F_{2x}$$
$$= 0 + (50.0 \text{ N}) \cos (36.9°)$$
$$= 40.0 \text{ N}$$

$$F_y = F_{1y} + F_{2y}$$
$$= -20.0 \text{ N} + (50.0 \text{ N}) \cos (53.1°)$$
$$= 10.0 \text{ N}$$

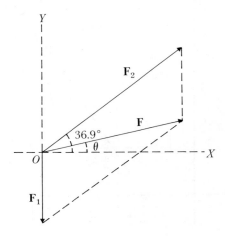

Figure 3-5   Vector addition of the two forces of Example 3. $\mathbf{F}$ is the resultant of $\mathbf{F}_1$ and $\mathbf{F}_2$.

Therefore

$$F = \sqrt{F_x^2 + F_y^2} = \sqrt{(40.0 \text{ N})^2 + (10.0 \text{ N})^2} = 41.2 \text{ N}$$

and

$$\tan \theta = \frac{F_y}{F_x} = \frac{10.0 \text{ N}}{40.0 \text{ N}} = 0.250$$

$$\theta = 14.0°$$

The resultant force, the single force equivalent to the combination of F_1 and F_2, is a force of magnitude 41.2 N at an angle 14.0° above the horizontal X axis.

3.3 Contact Forces, Friction

We are often concerned in daily experience not only with gravitational forces but also with the force exerted on one object by another when two solid objects are in contact, such as the block and the inclined plane in Figure 3-6. The force exerted on the block by the plane can be resolved into two vector components: a vector component **N** perpendicular to the surfaces in contact, called the *normal component,* and a vector component parallel to the surfaces in contact, called the *force of friction* **f**.

The direction of the friction force **f** is such that it always opposes the slipping or the tendency to slip of the surfaces in contact. If the block were sliding down the plane, the frictional force **f** would have the direction shown in Figure 3-6.

It is found that the magnitude f of the frictional force is proportional to the magnitude N of the normal component. Experimental results are most conveniently summarized in the following empirical relations called the laws of friction:

1 While sliding occurs

$$f = \mu_k N \tag{3.7}$$

where f is called the force of *sliding friction,* and the constant of proportionality μ_k is called the *coefficient of kinetic (or sliding) friction.* This coefficient has a value

Figure 3-6 Block sliding or tending to slide down the plane. The force exerted on the block by the plane is resolved into a component **N** normal to the surfaces in contact and a component parallel to these surfaces called the force of friction **f**.

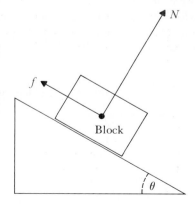

which depends on the nature of the surfaces in contact, being small if the surfaces are smooth and large if the surfaces are rough. Although the coefficient of kinetic friction for two given surfaces has slightly different values at different relative speeds, in our work we shall assume for simplicity that μ_k is independent of speed.

2   If the surfaces in contact are at rest relative to each other, the frictional force **f,** now named a force of *static friction,* has a magnitude which is not greater than

$$f_{\max} = \mu_s N \qquad (3.8)$$

where the constant of proportionality μ_s, called the *coefficient of static friction,* is characteristic of the surfaces in contact and has a value greater than μ_k. When the surfaces in contact are at the point of slipping, $f = f_{\max} = \mu_s N$. Otherwise, the force of static friction is less than this value:

$$f < \mu_s N \qquad (3.9)$$

Experiments show that μ_k and μ_s are nearly independent of the apparent area of contact between the two surfaces.

Microscopic examination reveals that even apparently smooth surfaces actually appear as shown in Figure 3-7. Since the surfaces touch only at isolated high spots, the actual contact area is very much smaller than the apparent contact area. If the apparent contact area is decreased while N is kept constant, plastic deformation of high spots occurs in such a way that the actual contact area remains the same. At the contact points, molecules of the different bodies are close enough together to exert strong attractive intermolecular forces on one another, leading to the formation of "cold welds." The force of kinetic friction is associated with the continual rupture and formation of thousands of such welds as one surface slides over the other.

Typical numerical values of coefficients of friction are given in Table 3-1. These tabulated values are obtained for surfaces exposed to air and therefore coated with a film of oxide. When metal surfaces are carefully cleaned in a vacuum and then placed in contact, the surfaces become welded together and the corresponding value of the coefficient of friction is enormous.

Occasionally we may wish to ignore the effects of friction. We shall then describe the surfaces in contact as *smooth* and assume that frictional force is absent. For example, with a block on a *smooth* inclined plane (Figure 3-8) the force exerted on the block by the plane has only a normal component N.

Table 3-1 Coefficients of Friction

Surfaces in contact	Static, μ_s	Kinetic, μ_k
Steel on steel	0.74	0.57
Copper on steel	0.53	0.36
Glass on glass	0.94	0.4
Teflon on steel	0.04	0.04

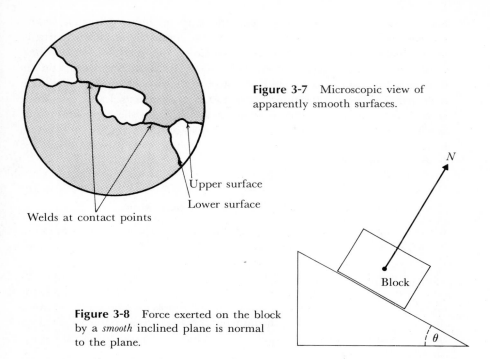

Figure 3-7 Microscopic view of apparently smooth surfaces.

Welds at contact points
Upper surface
Lower surface

N

Block

Figure 3-8 Force exerted on the block by a *smooth* inclined plane is normal to the plane.

θ

It is instructive to use Newton's second law to relate the forces acting on the block in Figures 3-6 and 3-8 to the acceleration of the block. This is done in the following section as an illustration of a very general procedure for the solution of problems in mechanics.

3.4 Using Newton's Second Law

It is simple enough to remember the formula $\mathbf{F} = m\mathbf{a}$, but a student must also learn and practice a systematic procedure for applying this law in order to analyze with confidence a great variety of problems. The following procedure will, if mastered, assist the student in solving many different types of problems not only in mechanics but also in all the various branches of physics:

1 *Preliminary sketch*   First draw a rough schematic diagram of the objects of interest. Visualize what will happen.
2 *System selection*   Select the object which will comprise the system or "particle" whose motion you are going to study. (Problems treated in the next chapter illustrate that this can often be done in different ways, so for clarity of thought it is vital that you specify the system.)
3 *Coordinate axes*   Choose perpendicular coordinate axes fixed in an inertial frame of reference. Orient the axes in such a way as to minimize future labor. When the acceleration vector has a known constant direction, it is usually convenient to have a coordinate axis in this direction.

4   *Free-body diagram*   Draw a new schematic diagram showing *only the particle* you have selected. Draw in vectors for *all* the forces acting *on* this particle. Do *not* include forces exerted *by* this particle on other objects.

5   *Finding an unknown acceleration or an unknown force*   Each force appearing on the free-body diagram is to be resolved into components along the coordinate axes and then deleted because it is *replaced by its components*. Label each force component, whether or not its value is known. Use the convention that symbols like f and N represent the magnitudes of the corresponding vectors and so are always positive. Then use these components to write down the following equations:

$$\text{algebraic sum of } x \text{ components of forces} = ma_x$$

$$\text{algebraic sum of } y \text{ components of forces} = ma_y$$

(3.10)

where a_x and a_y are the components of the particle's acceleration. For motion in the XY plane, these equations are equivalent to Newton's second law, $\mathbf{F} = m\mathbf{a}$. After equations expressing other information (such as the law of sliding friction, $f = \mu_k N$) have been used, Eq. 3.10 can be solved to find the unknown force or the acceleration. In many physics problems, it is advantageous to carry out the analysis using literal symbols for all quantities rather than numerical values (unless the numerical value is zero). Numerical values can be substituted after a general expression has been found for the unknown quantity of interest.

6   *Check*   Always, in any physics problem, think about the answer. If it is a numerical answer, is it a reasonable size? Is the correct number of significant figures given?* Are the units correct? Procedures for checking a formula will be illustrated after Example 4.

7   *Shortcuts*   Experienced and confident students may skip through steps 1 and 2, with the picture in their mind's eye, but should always draw the *free-body diagram* and ensure that it is correct.

In the following examples, Newton's second law is applied using the procedure just presented, even though each step may not be made explicit in every example.

Example 4   Find the acceleration of a block which is sliding down a smooth plane inclined at an angle θ with the horizontal.

Solution

(*Step 1*) Figure 3-9 is the schematic diagram. The block is moving in a straight line with its acceleration vector **a** directed down the plane.

(*Step 2*) The block is selected as the system and will be represented by a particle of mass m on the free-body diagram.

*Significant figures are discussed in Appendix B.

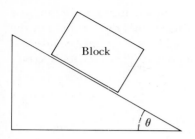

Block

Figure 3-9 Schematic diagram for Example 4.

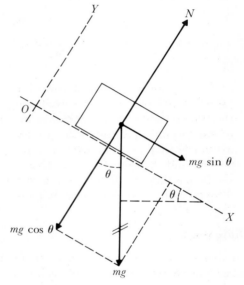

Figure 3-10 Free-body diagram for Example 4.

(*Step 3*) As an inertial frame of reference, we choose coordinate axes fixed in the plane with the X axis directed down the plane. Then $a_y = 0$ and **a** is determined by the value of just one component, a_x.

(*Step 4*) The free-body diagram is shown in Figure 3-10.* There are just two forces acting on the block:

(i) The force **N** exerted on the block by the smooth plane. The direction of **N** is normal (perpendicular) to the plane.

(ii) The gravitational force exerted on the block by the planet earth. This is the force *m***g** that is called the weight of the block.

---

*Notice that, in Figure 3-10, the symbol or mathematical expression associated with an arrow indicates only the *magnitude* (and not the direction) of the vector represented by the arrow. We follow this practice in figures wherever it is convenient.

(*Step* 5) The weight $m\mathbf{g}$ is resolved into an x component, $mg \sin \theta$, and a y component of magnitude $mg \cos \theta$. The two short lines drawn through the vector $m\mathbf{g}$ indicate that this force is to be deleted because it has been *replaced by its components*. The vector $\mathbf{N}$ has only a y component. Then Newton's second law for x components (Eq. 3.10) gives

$$mg \sin \theta = ma_x$$

For the y components, Newton's second law (Eq. 3.10) gives

$$N - mg \cos \theta = ma_y = 0$$

This equation merely determines the normal force which must be exerted by the plane for this motion parallel to the plane to occur:

$$N = mg \cos \theta$$

The block's acceleration, determined by the equation for x components, is

$$a_x = g \sin \theta$$

and the vector $\mathbf{a}$ is directed down the plane in the direction OX.

(*Step* 6) When problems are solved using symbols instead of numerical values, there are several checks that can be made. We can verify that the predictions of our formulas are correct for values of the variables which correspond to extremely simple situations. For instance, a horizontal plane corresponds to $\theta = 0$, and then our formulas give $a = 0$ and $N = mg$. We can mentally verify that these predictions are correct. For a vertical plane, corresponding to $\theta = 90°$, our formulas correctly predict that $a = g$ and $N = 0$.

Dimensions

Another way to spot some obvious errors in an equation is to check what are called the physical *dimensions* of each term in the equation. All the quantities that are introduced in mechanics can be expressed in terms of products or quotients of certain fundamental physical quantities such as length (L), mass (M), and time (T). For instance, a *velocity* is a *length* divided by a *time*. We say velocity has the physical dimensions $[L/T]$. A dimensional formula for a physical quantity is an expression, such as $[L/T]$ for velocity, which shows how length, mass, and time enter into any formula for the physical quantity. The brackets indicate that *only* the dimensions are represented, and that numerical values have been disregarded. Thus both 60 m/s and 2 cm/s have the dimensions $[L/T]$ since each is a length divided by a time.

The dimensions of any area are $[L^2]$ and of any volume $[L^3]$. Since an acceleration is a length divided by a time squared, the dimensional formula for an acceleration is $[L/T^2]$ which can be written $[LT^{-2}]$. The dimensions of force are the dimensions of mass times acceleration, or $[MLT^{-2}]$.

In science we start mathematical work with equations in which all terms have the same dimensions. After we perform legitimate manipulations of these equations to deduce new equations, every term in any new equation must have the same dimensions. A mathematical error can lead to a proposed equation with terms of different dimensions. Such errors can easily be detected by checking the dimensions of all terms in an equation.

For example, in the equation

$$x = x_0 + v_{0x}t + \tfrac{1}{2}a_x t^2$$

(which gives the position coordinate x for a motion with constant acceleration a_x, initial position x_0, and initial velocity v_{0x}) the dimensions are

$$[L] = [L] + [LT^{-1}T] + [LT^{-2}T^2]$$

Each term has the dimensions of a length, so the equation is dimensionally correct. A number like the quantity $\tfrac{1}{2}$ in this equation has no physical dimensions —it is called a dimensionless number. Notice that the equation would still have been dimensionally correct even if the number $\tfrac{1}{2}$ had been erroneously omitted. Obviously dimensional checks will not reveal all types of errors, but certain errors can easily be spotted. Suppose that after some manipulation we found the acceleration of Example 4 to be

$$a = \frac{m}{g}$$

Here the right-hand side has the dimensions

$$\left[\frac{M}{LT^{-2}}\right] = [ML^{-1}T^2]$$

which are not the dimensions $[LT^{-2}]$ of acceleration. Because the proposed equation is not dimensionally correct, we can conclude that we made an error. To check the dimensions for the expression for the acceleration actually found in Example 4, we first note that $\sin\theta$ is a dimensionless quantity since it is the ratio of two lengths. Then we obtain

$$[g\sin\theta] = [LT^{-2}]$$

which are the dimensions of acceleration. Therefore our expression for acceleration is at least dimensionally correct.

Example 5   A block of mass m slides down a plane inclined at an angle θ with the horizontal. The coefficient of kinetic friction between the block and the plane is μ_k. Find the acceleration of the block.

Solution   In Steps 1 through 3, we follow the same procedure as for Example 4. (*Step 4*) In the free-body diagram of Figure 3-11, the frictional force **f** exerted on the block by the plane is parallel to the plane and in the direction opposite to the direction of motion of the block. The force **N** is the normal component of the force exerted on the block by the plane. The only other force acting on the block is its weight $m\mathbf{g}$.

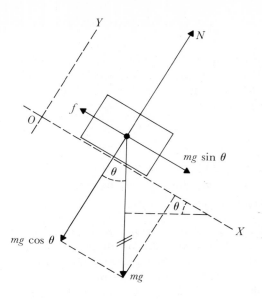

Figure 3-11   Free-body diagram for Example 5.

(*Step* 5) The weight *mg* is resolved into an *x* component, *mg* sin θ, and a *y* component of magnitude *mg* cos θ. The vector *mg* is replaced by its components. Then Newton's second law for the *x* components (Eq. 3.10) gives

$$mg \sin \theta - f = ma_x$$

For the *y* components, Eq. 3.10 gives

$$N - mg \cos \theta = ma_y = 0$$

where $a_y = 0$ because the acceleration **a** is parallel to the plane. Consequently,

$$N = mg \cos \theta$$

The law of sliding friction relates *f* and *N*.

$$f = \mu_k N$$

Substituting $\mu_k N$ for *f* in the equation for a_x and then using $N = mg \cos \theta$ we obtain

$$a_x = g \sin \theta - \mu_k g \cos \theta$$

(*Step* 6) Since sin θ, cos θ, and μ_k are dimensionless ratios, we can see that each term in this equation has the dimensions of acceleration. The equation is therefore dimensionally correct.

   If the plane is vertical, the block falls with an acceleration **g**, and our equation for the acceleration does correctly predict that $a = g$ when $\theta = 90°$ (then sin $\theta = 1$ and cos $\theta = 0$). If the plane is smooth, $\mu_k = 0$ and the equation reduces to $a_x = g \sin \theta$ in agreement with the result of Example 4.

   (A word of caution is necessary. Our result was derived under the assumption that the frictional force has a negative *x* component of magnitude $f = \mu_k N$. Since this is only true when the block is sliding in the positive *x* direction, our result can be used only for this case.)

Example 6   The coefficient of static friction between the block and the inclined plane in Figure 3-9 is given by $\mu_s = 0.75$. With the block at rest on the plane and θ initially zero, the left-hand end of the inclined plane is raised very slowly and the angle θ increases. At what value of θ will the block start to slip?

Solution

(Step 1) Figure 3-9 is the schematic diagram. We are concerned here with a body at rest. Such a problem in *statics* can be treated by the methods used for dynamics, using zero for the value of the acceleration. (Problems in which we must also consider the conditions for rotational equilibrium will be treated in Chapter 5.) While the block is at rest on the inclined plane, the force of friction acting on the block is a force of *static friction.*

(Step 2) We select the block as the system.

(Step 3) The coordinate axis OX is fixed on the inclined plane and points down the plane.

(Step 4) Figure 3-12 shows the free-body diagram. The force of static friction exerted on the block by the plane is directed up the plane, opposing the tendency of the block to slip.

(Step 5) We resolve the weight $m\mathbf{g}$ into x and y components and replace the vector $m\mathbf{g}$ by these components. Then Newton's second law (for the special case when the acceleration is zero) gives for the x components

$$mg \sin \theta - f = ma_x = 0$$

and for the y components

$$N - mg \cos \theta = ma_y = 0$$

Consequently,

$$f = mg \sin \theta$$

$$N = mg \cos \theta$$

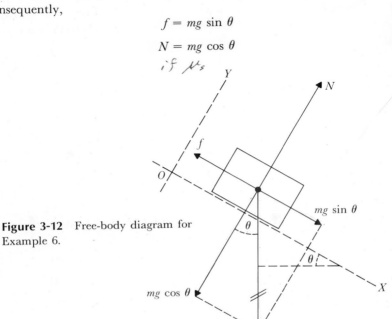

Figure 3-12   Free-body diagram for Example 6.

Dividing the first of these equations by the second gives

$$\frac{f}{N} = \frac{\sin \theta}{\cos \theta} = \tan \theta$$

Suppose the block slips if θ is increased above the value θ_{max} called the angle of repose. Then when the angle of inclination is θ_{max} the block is on the point of slipping, and the force of friction attains its maximum value given by

Then
$$f_{max} = \mu_s N$$

$$\tan \theta_{max} = \frac{f_{max}}{N} = \mu_s = 0.75$$
and
$$\theta_{max} = 36.9°$$

(At angles of inclination less than θ_{max}, the force of static friction will be less than $\mu_s N$. From the preceding analysis we see that f then will simply assume a value that cancels the component of the weight along the plane: that is, $f = mg \sin \theta$.)

Example 7 Find the magnitude N of the force exerted by the floor of an elevator on a man of mass m when the elevator has a vertical acceleration **a**.

Solution In the free-body diagram (Figure 3-13) the man is represented by a particle of mass m. There are two forces acting on the man: the force N exerted by the floor and the man's weight mg. We select an *inertial* frame of reference *fixed in the ground* with axis OY vertical. (Notice that since the elevator is accelerated relative to an inertial frame, the elevator itself does *not* constitute an inertial frame of reference.) Newton's second law (Eq. 3.10) for y components gives

$$N - mg = ma_y$$
Therefore
$$N = m(g + a_y)$$

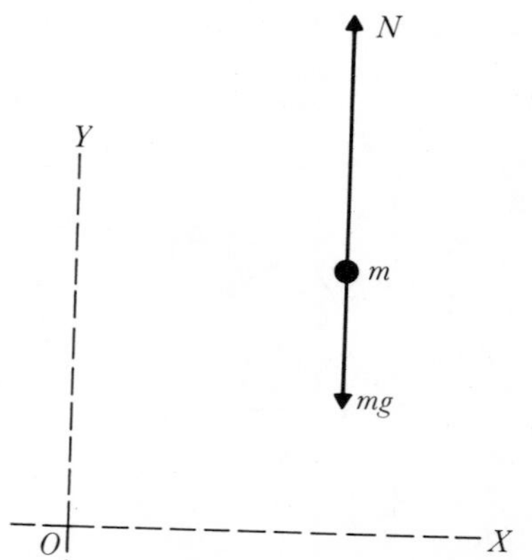

Figure 3-13 Free-body diagram for Example 7.

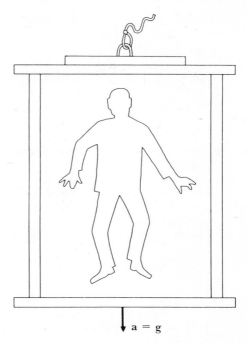

Figure 3-14 Elevator falling with acceleration **g**.

$$a = g$$

When the elevator has no acceleration, N is equal to the magnitude mg of the man's weight, but in order to give the man an upward acceleration ($a_y > 0$) a larger value of N is required. A downward acceleration ($a_y < 0$) is associated with a reduced value of N. In fact, when the elevator is falling freely with an acceleration $a_y = -g$, we find $N = 0$ and no contact with the floor is required for the man to maintain a fixed position relative to the elevator (Figure 3-14). The magnitude N of the upward force exerted on an object is often called its *apparent weight* and the condition corresponding to zero apparent weight is then referred to as *weightlessness,* even though the object actually continues to have a weight mg.

Summary

☐   The x and y components of a vector **a** are the positive *or negative* numbers defined by

$$a_x = a \cos \theta$$

$$a_y = a \cos (90° - \theta) = a \sin \theta$$

where θ is the angle measured counterclockwise from the X axis to the vector **a** and $\mathbf{a} = \mathbf{a}_x + \mathbf{a}_y$.

☐   The x component of the resultant of several vectors is the sum of their x components. The y components are similarly related.

☐   When two objects are in contact, the force exerted on one object by the other has a component **N** normal to the surfaces in contact and a component **f**, parallel to the surfaces in contact, called the force of friction which is directed so as to oppose the tendency to slip or the slipping of one surface relative to the other. The following approximate empirical laws relate f to N:

$$f = \mu_k N \qquad \qquad \text{(sliding)}$$

$$f \leq \mu_s N \qquad \qquad \text{(no sliding)}$$

where μ_k and μ_s are the coefficients of kinetic and static friction, respectively.

☐   In applying Newton's second law to analyze the motion of a particle, draw a free-body diagram showing the particle and all the forces acting *on* this particle. Replace each force by its components along two conveniently chosen perpendicular directions and then use these components in the equations

algebraic sum of x components of forces $= ma_x$

algebraic sum of y components of forces $= ma_y$

Employ these equations and any other information available to solve for the unknowns, preferably using literal symbols for all the quantities. Check results by ensuring that all equations are dimensionally correct and by inserting values of variables that correspond to particularly simple situations.

Questions

1   Find the x and y components of the forces in Example 1 in a coordinate system oriented so that the positive x direction is 30° north of east and the positive y direction is 30° west of north.

2   Find the x and y components of the forces in Example 1 in a coordinate system oriented so that the positive y direction is the direction of $\mathbf{F}_2$ and the positive x direction is the direction 15° west of north.

3   A force has the following components: $F_x = 5.00$ N and $F_y = 12.00$ N. Find the magnitude and direction of this force vector.

4   Find the magnitude and direction of the force vectors $\mathbf{F}_1$ and $\mathbf{F}_2$ if the components in a given coordinate system are:

$$F_{1x} = 4.00 \text{ N} \qquad \text{and} \qquad F_{1y} = -3.00 \text{ N}$$

$$F_{2x} = -5.00 \text{ N} \qquad \text{and} \qquad F_{2y} = 12.00 \text{ N}$$

5   A point Q is 4.00 m from the origin O and the angle measured counterclockwise from the X axis to OQ is 30°. Find the x and y components of the position vector of Q.

6   Find the resultant of the forces given in Example 1.

7   By adding components, find the vector sum of the following forces: 6.00 N north; 4.00 N east; 5.00 N, 30.0° south of east.

8   Find the resultant of the following forces: 20.0 N, 10° east of north; 40.0 N, 45° east of north; 30.0 N, 45° west of north.

9   A trunk, pulled by two different ropes, is acted upon by a force of 200 N east and a force of 400 N, 53.1° north of east. What is the vector sum of these two forces?

10 A stone is sliding down a board which is inclined at an angle of 30° with the horizontal. The force exerted on the stone by the board acts vertically upward and has a magnitude of 60 N. Find the force of friction acting on the stone. What is the value of the component normal to the board of the force exerted on the stone by the board?

11 A 10.0-kg block moves on a horizontal table. The coefficient of kinetic friction between the block and the table is 0.30. A horizontal force of 49.4 N pushes the block:
(a) Find the normal component of the force exerted on the block by the table.
(b) Find the frictional force exerted on the block by the table.
(c) Compute the acceleration of the block.
(d) While the block is moving the 49.4-N force is removed. Now what is the acceleration of the block?

12 A particle of mass m slides along a horizontal table top as it is pushed by a force F which makes an angle ϕ with the horizontal (Figure 3-15). The coefficient of kinetic friction between the table and the particle is μ_k. Show that the normal force exerted on the particle by the table is given by

$$N = mg + F \sin \phi$$

Find the particle's acceleration in terms of F, ϕ, μ_k, m, and g.

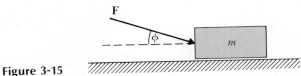

Figure 3-15

13 A 20.0-kg trunk is pushed by a *horizontal* force of 200 N as it slides up a smooth plane inclined at an angle of 36.9° with the horizontal. Find the acceleration of the trunk.

14 A 30-kg box is pushed by a horizontal force of 200 N as it slides up a plane inclined at an angle of 30° with the horizontal. The coefficient of kinetic friction between the box and the plane is 0.20. What is the acceleration of the box?

15 The coefficient of static friction between a 60-kg box and the floor is 0.50. A horizontal push is applied:
(a) What magnitude of the applied force is required to start the box moving?
(b) What is the initial acceleration of the box if the coefficient of kinetic friction between the box and the floor is 0.30?

16 What is the minimum magnitude of the horizontal force that will hold a 60-kg crate at rest on a plane inclined at an angle of 36.9° with the horizontal if the coefficient of static friction between the crate and the plane is 0.40?

17   A 10.0-kg particle is suspended at rest by two ropes. One rope makes an angle of 53.1° with the vertical and the other rope makes an angle of 45° with the vertical. Find the force exerted on the 10.0-kg particle by each rope.

18   A 5.0-kg particle is suspended from the ceiling by a rope. A horizontal string holds the particle in a position such that the rope makes an angle of 30° with the vertical. Find the forces exerted on the particle by the rope and by the string.

Supplementary Questions

S-1   The coefficient of static friction between a block and a vertical wall is 0.30. What horizontal push is required to hold the 10.0-kg block against the wall and prevent it from sliding down?

S-2   A block of mass 10.0 kg is projected up a plane inclined at an angle of 36.9° with the horizontal. The coefficients of kinetic and static friction between the block and the plane are 0.20 and 0.30, respectively:
   (a) What is the acceleration of the block as it moves up?
   (b) Show that the block will start down again after coming to rest.
   (c) What is its acceleration down the plane?

S-3   Show that the horizontal force F in Figure 3-16 required to keep the block of mass m from slipping down the plane is given by

$$F = mg \, \frac{\sin \theta - \mu_s \cos \theta}{\cos \theta + \mu_s \sin \theta}$$

where μ_s is the coefficient of static friction between the block and the plane. Verify that when θ is the angle of repose θ_{max} calculated in Example 6, F is zero.

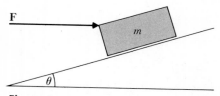

Figure 3-16

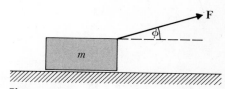

Figure 3-17

S-4   A force F, directed at an angle ϕ above the horizontal, is applied to a block of mass m which is resting on a rough horizontal table (Figure 3-17). The magnitude of this applied force is slowly increased until the block is about to slip. Show that the maximum value of F is given by

$$F = \frac{\mu_s mg}{\cos \phi + \mu_s \sin \phi}$$

S-5   A particle of mass m moves up a rough plane inclined at an angle θ with the horizontal. The particle is pulled by a force F which is parallel to the plane (Figure 3-18). Show that the particle's acceleration up the plane is given by

$$a = \frac{F}{m} - g(\sin \theta + \mu_k \cos \theta)$$

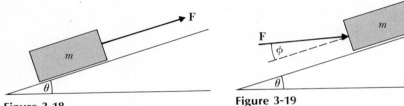

Figure 3-18

Figure 3-19

S-6 A particle, which slides up a *smooth* plane inclined at an angle θ with the horizontal, is pushed by a force F which makes an angle ϕ with the inclined plane (Figure 3-19). Express both the normal force N exerted by the plane and the particle's acceleration a as functions of F, m, θ, ϕ, and g.

S-7 A dog that weighs 100 N has a force of 125 N exerted upward on his feet by the floor of an elevator. Find:

(a) His mass.

(b) His acceleration.

S-8 A *unit vector* is a vector with magnitude equal to 1 (for example, if $\mathbf{i}$ is a unit vector, $|\mathbf{i}| = 1$). In a rectangular coordinate system the symbols $\mathbf{i}$, $\mathbf{j}$, and $\mathbf{k}$ are used to denote unit vectors in the positive x, y, and z directions, respectively:

(a) Verify that $\mathbf{A}_x = A_x\mathbf{i}$ and that $\mathbf{A}_y = A_y\mathbf{j}$.

(b) Verify that the vector $\mathbf{A}$ in Figure 3-1 can be expressed in terms of its components A_x and A_y by the vector equation

$$\mathbf{A} = A_x\mathbf{i} + A_y\mathbf{j}$$

(c) Draw a diagram to verify that a general vector $\mathbf{A}$ with components A_x, A_y, and A_z along rectangular coordinate axes OX, OY, and OZ can be expressed as the following sum of products of its components with $\mathbf{i}$, $\mathbf{j}$, and $\mathbf{k}$:

$$\mathbf{A} = A_x\mathbf{i} + A_y\mathbf{j} + A_z\mathbf{k}$$

(d) Vectors $\mathbf{B}$ and $\mathbf{C}$ are defined by

$$\mathbf{B} = 2\mathbf{i} - 3\mathbf{j}$$

$$\mathbf{C} = -5\mathbf{i} + 4\mathbf{j} + 2\mathbf{k}$$

Identify the x, y, and z components of $\mathbf{B}$ and of $\mathbf{C}$ and find the vector $\mathbf{A}$ defined by

$$\mathbf{A} = \mathbf{B} + \mathbf{C}$$

expressing $\mathbf{A}$ in the form displayed in part (c).

4.

forces and
systems of particles

In the preceding chapter we applied Newton's second law, $\mathbf{F} = m\mathbf{a}$, to a single particle of mass m. We now begin the study of the mechanics of a system which may consist of many particles. The system need not be a rigid body or even only one large body. In the analysis of a certain collision we might find it convenient to define the system as the combination of a train traveling north and a car traveling east. One of the surprising discoveries by physicists is that even such complex systems obey simple mechanical laws. Our analysis of the mechanics of systems is based on the important relationship between the forces that two objects exert on each other—the topic of the first section.

4.1 Newton's Third Law

A very general statement about the nature of forces is known as Newton's third law: *The force exerted on a particle A by another particle B is equal in magnitude and opposite in direction to the force exerted on particle B by particle A. These forces lie along the line joining the particles.* The situation is illustrated in Figure 4-1(a) and can be written as the vector equation

$$\mathbf{F}_{\text{on } A \text{ by } B} = -\mathbf{F}_{\text{on } B \text{ by } A} \tag{4.1}$$

(The significance of the minus sign is, as the student will recall from the discussion of vectors, that the two forces are parallel but act in opposite directions.) The same result also applies to extended objects such as those in Figure 4-1(b). This can be shown by considering the vector sum of the forces exerted on the particles of one object by the particles of the other.

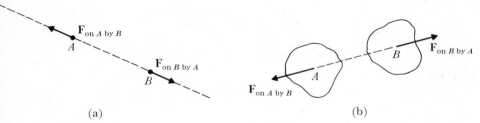

Figure 4-1   Newton's third law. (a) Particles. (b) Extended objects.

It is essential for us to understand that the two forces referred to in the third law always act on *different* objects. This is of the utmost importance. Generally when difficulties arise where Newton's third law is involved, they can almost invariably be traced to the failure to realize that each of the pair of forces involved acts on a different body.

Two forces related by Newton's third law are often called an *action-reaction pair*. Either of the forces can be called the action and the other the reaction. We should not think that the action is the cause and the reaction is the effect— the interactions that are in accord with Newton's third law are such that, at a given instant, the action and the reaction are precisely equal in magnitude.

At the surface of contact between two objects we meet such pairs of forces. Consider the brick which is sliding down the rough inclined plane of Figure 4-2. The brick exerts a force **P** on the plane and the plane exerts a force **Q** on the brick. Newton's third law gives **P** = −**Q**. Suppose we wish to determine the acceleration of the brick. According to Newton's second law, **F** = *m***a**, we must determine the resultant of all the forces acting *on* the brick; the free-body diagram, Figure 4-3, must therefore include the force **Q** that acts on the brick, but (and this is the point where it is easy to err) it must not include the force **P** because this force does *not* act on the brick; instead it acts on the plane.

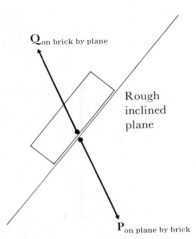

Figure 4-2   Two forces which are related by Newton's third law.

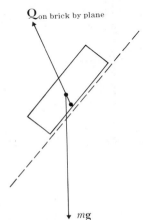

Figure 4-3   The forces which determine the acceleration of the brick. Resultant force = **Q** + *m***g**.

Example 1   A man (Figure 4-4), in throwing a 0.50-kg stone vertically upward, exerts an upward force of 20.0 N on the stone:

(*a*) How large is the downward force exerted on the man's hand by the stone?

(*b*) What, if any, is the acceleration of the stone?

Solution

(*a*) According to Newton's third law, the force exerted on the man's hand by the stone is 20.0 N downward.

(*b*) The acceleration of the stone is determined by the forces acting *on* it (shown in the free-body diagram, Figure 4-5), which are its weight, (0.50 kg) (9.8 N/kg) = 4.9 N, acting downward and the upward force of 20.0 N exerted *on* the stone by the man. Since these forces are not equal there is an acceleration. The acceleration is determined from Newton's second law for *y* components:

$$20.0 \text{ N} - 4.9 \text{ N} = (0.50 \text{ kg})a_y$$

This yields

$$a_y = 30.2 \text{ m/s}^2$$

Two points are to be noted. First, the force exerted on the man by the stone is *not* determined by the *weight* of the stone. *It is always exactly equal in magnitude to the force exerted on the stone by the man,* whatever this force happens to be. Second, the force exerted *by* the stone is *not* one of the forces to be considered in determining the acceleration of the stone.

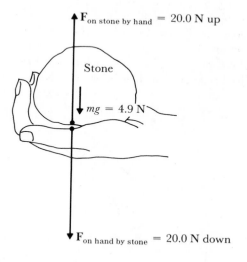

$\mathbf{F}_{\text{on stone by hand}} = 20.0 \text{ N up}$

Stone

$mg = 4.9 \text{ N}$

$\mathbf{F}_{\text{on hand by stone}} = 20.0 \text{ N down}$

Figure 4-4   The force exerted on the stone by the hand and the force exerted on the hand by the stone constitute an action-reaction pair.

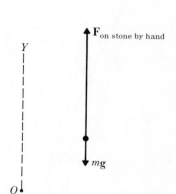

$\mathbf{F}_{\text{on stone by hand}}$

Y

mg

O

Figure 4-5   The free-body diagram for the stone of Example 1.

4.2 Center of Mass of a System

Observation of the motion of any system of particles shows that, even though the system undergoes wild gyrations and perhaps explosions, there is a representative point whose motion gives the average motion for the entire system (Figure 4-6). This representative point, called the *center of mass* of the system, is defined to be the point C with position vector $\mathbf{r}_c$ relative to a definite frame of reference given by

$$M\mathbf{r}_c = m_1\mathbf{r}_1 + m_2\mathbf{r}_2 + \cdot \cdot \cdot + m_n\mathbf{r}_n \tag{4.2}$$

where the particles of the system have masses $m_1, m_2, \ldots, m_n$ and position vectors $\mathbf{r}_1, \mathbf{r}_2, \ldots, \mathbf{r}_n$, respectively. The system's total mass is

$$M = m_1 + m_2 + \cdot \cdot \cdot + m_n$$

For particles distributed in the XY plane, this definition implies that the center of mass has coordinates x_c and y_c given by

$$Mx_c = m_1x_1 + m_2x_2 + \cdot \cdot \cdot + m_nx_n \tag{4.3}$$

$$My_c = m_1y_1 + m_2y_2 + \cdot \cdot \cdot + m_ny_n$$

where x_1, y_1 are the components of the particle m_1 with position vector $\mathbf{r}_1$, and so on.

An investigation of the consequences of Eq. 4.2, which defines the location of the center of mass, shows that a uniform regular solid has its center of mass at its geometrical center. (We shall show in Section 5.2 that the center of mass of a system coincides with what is called the center of gravity. In a uniform gravitational field, the weight of the entire system can be considered as acting at the center of gravity.)

Figure 4-6 The point C is the center of mass of the system comprising the masses m_1, m_2, and m_3.

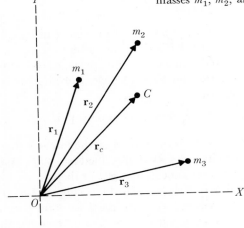

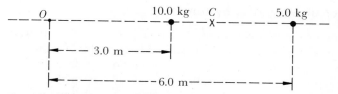

Figure 4-7   The center of mass C of a system of two particles.

Example 2   Locate the center of mass of the system of the two particles shown in Figure 4-7.

Solution   Equation 4.3 gives

$$(15.0 \text{ kg})x_c = (10.0 \text{ kg}) (3.0 \text{ m}) + (5.0 \text{ kg}) (6.0 \text{ m})$$

Hence

$$x_c = 4.0 \text{ m}$$

The center of mass lies on the line joining the particles and is closer to the more massive particle. Notice that there does not have to be any actual matter at the location of the center of mass of the system.

4.3 Law of Motion of the Center of Mass

The center of mass moves with a velocity $V_c = dr_c/dt$ which is given by

$$MV_c = m_1v_1 + m_2v_2 + \cdot \cdot \cdot + m_nv_n \tag{4.4}$$

This result is obtained by differentiating both sides of Eq. 4.2 with respect to t, noting that the masses are constant, and then recognizing that each time derivative of a position vector is a velocity of the corresponding particle. Similarly, differentiation of Eq. 4.4 yields

$$Ma_c = m_1a_1 + m_2a_2 + \cdot \cdot \cdot + m_na_n \tag{4.5}$$

which relates the acceleration $a_c = dV_c/dt$ of the center of mass and the accelerations $a_1, a_2, \ldots , a_n$ of the particles of the system. According to Newton's second law, the resultant force acting on the first particle is given by $F_1 = m_1a_1$, and there is a similar equation for each particle. Substituting these resultant forces into Eq. 4.5, we obtain

$$Ma_c = F_1 + F_2 + \cdot \cdot \cdot + F_n \tag{4.6}$$

The sum $F_1 + F_2 + \cdot \cdot \cdot + F_n$ is the sum of *all* the forces acting on the particles of the system. This sum includes what are called *internal forces,* forces exerted by one particle of the system on another particle of the system, but such forces cancel out in pairs according to Newton's third law (Figure 4-8). There remains only the vector sum of the forces exerted *on* the particles of the system by particles *outside* the system: such forces are called *external forces.* Denoting the vector sum of these external forces by $\mathbf{F}_{\text{ext}}$ we obtain from Eq. 4.6 the remarkable result known as the *law of motion of the center of mass,*

$$\mathbf{F}_{\text{ext}} = M\mathbf{a}_c \qquad (4.7)$$

which shows that *the center of mass of a system moves like a particle of mass equal to the total mass of the system, acted upon by the vector sum of the external forces.* We see that the problem of describing the average motion of an entire complex system is simply the problem of determining the motion of a single particle of mass M at the center of mass. In this sense we can say that Newton's second law applies to a system of particles as well as to a single particle.

Example 3   A shell explodes in midair (Figure 4-9). Describe the motion of the center of mass of the system of particles which comprise the shell.

Solution   Disregarding air resistance, the vector sum of the external forces on the particles of the shell is just $M\mathbf{g}$. (The explosion involves only *internal forces.*) Thus the law of motion of the center of mass gives

$$M\mathbf{g} = M\mathbf{a}_c$$

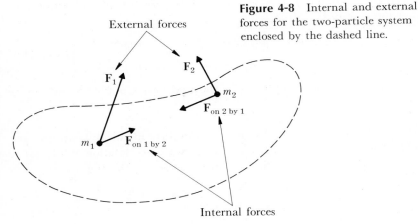

External forces

Figure 4-8   Internal and external forces for the two-particle system enclosed by the dashed line.

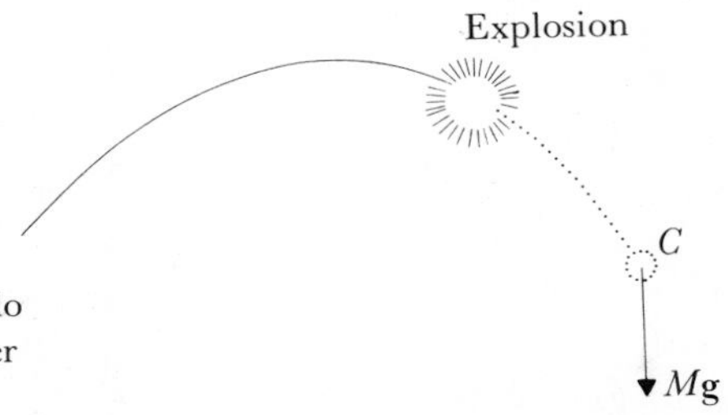

$\mathbf{F}_1$

$\mathbf{F}_2$

m_2

$\mathbf{F}_{\text{on 2 by 1}}$

m_1

$\mathbf{F}_{\text{on 1 by 2}}$

Internal forces

$\bullet\, m_3$

Explosion

C

$M\mathbf{g}$

Figure 4-9   The internal forces involved in an explosion of a shell do not influence the motion of its center of mass C.

Hence, $\mathbf{a}_c = \mathbf{g}$ before and after the explosion. The center of mass moves on its trajectory as if nothing had happened. Note, however, that we have not said anything about the individual motions of the various shell fragments.

4.4 Using Newton's Laws

The law of motion of the center of mass, $\mathbf{F}_{ext} = M\mathbf{a}_c$, allows us to treat the motion of the center of mass of any sizable object or collection of objects by the same procedure as that emphasized in Section 3.4 for the application of Newton's second law to particle motion. However, for systems more complex than a single particle, there are two steps that require particular care when using $\mathbf{F}_{ext} = M\mathbf{a}_c$:

1 *System selection* It is essential that there be no ambiguity about the identity of the system to which $\mathbf{F}_{ext} = M\mathbf{a}_c$ is to be applied. Therefore, it is common practice to draw a dotted line about the system that has been selected (Figure 4-10).
2 *Free-body diagram* This diagram should show only the system that has been selected and the *external* forces acting *on* all the objects of the system. Do *not* include forces exerted *by* the objects of the system. After each force has been replaced by its components along convenient perpendicular directions, we obtain the equations:

$$\text{algebraic sum of } x \text{ components of external forces} = Ma_x$$
$$\text{algebraic sum of } y \text{ components of external forces} = Ma_y$$

(4.8)

where a_x and a_y are the components of the acceleration of the system's center of mass.

Example 4 A train consists of a locomotive and three identical passenger cars, each of mass 10×10^3 kg. The train has a forward acceleration of 0.15 m/s². Find the force $\mathbf{F}_1$ exerted on the second passenger car by the first.

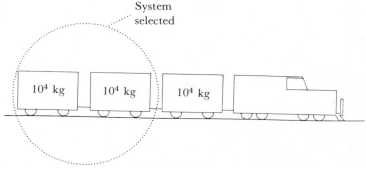

Figure 4-10 The system that has been selected is enclosed by a dotted line.

Solution

(*Step 1*) Figure 4-10 shows the train and the masses of the cars. Each car has the same acceleration. The second car is pulled forward by the first car and backward by the third car.

(*Step 2*) The system selected has been circled in Figure 4-10. It comprises the second and the third cars. For this system $M = (10 \times 10^3 \text{ kg}) + (10 \times 10^3 \text{ kg}) = 20 \times 10^3 \text{ kg}$.

(*Step 3*) Any frame of reference fixed in the earth is an appropriate inertial frame. Since the acceleration is obviously horizontal and to the right in Figure 4-10, we pick an axis OX in this direction, with OY vertical.

(*Step 4*) The free-body diagram for the system we have selected is shown in Figure 4-11. The external forces acting on this system are: the weight Mg of the two cars, the force N exerted on the cars by the rails, and the force F_1 exerted on the second car by the first. (For the purpose of determining the motion of the center of mass C of this system, all these forces can be represented as if they were applied at C.) Notice that the forces acting between the second and third cars do not appear on this free-body diagram because they are *internal* forces for the system we have selected.

(*Step 5*) The acceleration of the center of mass C has the components $a_x = 0.15 \text{ m/s}^2$ and $a_y = 0$. The law of motion of the center of mass for x components (Eq. 4.8) yields

$$F_1 = Ma_x = (20 \times 10^3 \text{ kg}) (0.15 \text{ m/s}^2) = 3.0 \times 10^3 \text{ N}$$

so the force $\mathbf{F}_1$ exerted on the second car by the first is in the positive x direction and has a magnitude of 3.0×10^3 N. (The equation for y components gives only $N - Mg = Ma_y = 0$ which implies $N = Mg$.)

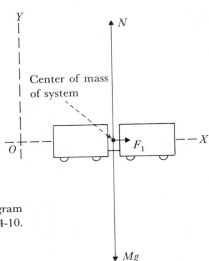

Figure 4-11 The free-body diagram for the system circled in Figure 4-10.

Example 5   A mass m_1, on a smooth plane inclined at an angle θ with the horizontal, is attached to a light rope which passes around a light frictionless pulley and suspends a second mass m_2. Find the tension in the rope and the acceleration of the two masses.

Solution

(*Step 1*) The schematic diagram is shown in Figure 4-12. We assume that the rope does not stretch. Then the two masses move with the same speed at every instant and consequently they have accelerations of the same magnitude. Whether m_2 is accelerated up or down will depend upon the values m_1, m_2, and the angle θ.

To understand the forces involved in this problem, we must know something about ropes and pulleys. The tension in the rope at any cross section is the force T exerted by the portion of the rope on one side of that cross section upon the portion of the rope on the other side (Figure 4-13). The key fact about a "light" rope and a "light frictionless" pulley is that the tension has the same value T throughout all segments of the rope; that is, the tension does not change when the rope passes around such an idealized pulley. (The reasons for this are examined in Question 23 and in Chapter 12, Question S-4.)

(*Step 2*) We select the mass m_1 as the system and study its motion.

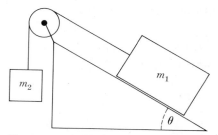

Figure 4-12   Schematic diagram for Example 5.

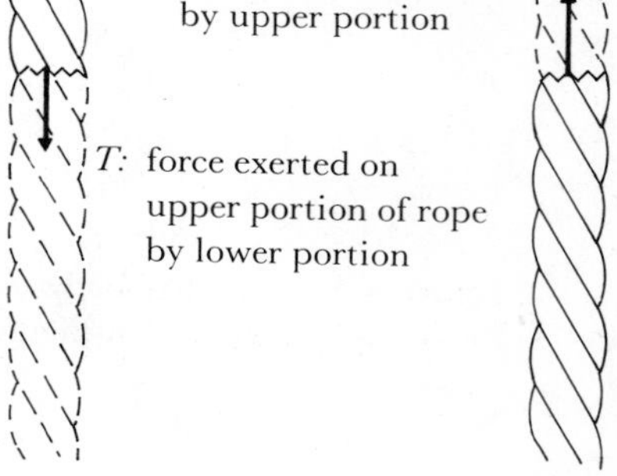

T: force exerted on lower portion of rope by upper portion

T: force exerted on upper portion of rope by lower portion

Figure 4-13   Tension in the rope is a force of magnitude *T*. Newton's third law requires that the two forces shown above have equal magnitudes and opposite directions.

(*Step 3*) We choose an axis OX fixed on the inclined plane and pointing down the plane.

(*Step 4*) Figure 4-14 is the free-body diagram showing the external forces acting on m_1. The rope exerts a force on m_1 that has a magnitude which we have called the tension T in the rope. This force is a pull (you cannot push anything with a rope), so it acts up the plane along the line of the rope.

(*Step 5*) The law of motion of the center of mass for x components (Eq. 4.8) yields

$$m_1 g \sin \theta - T = m_1 a \tag{4.9}$$

where a denotes the x component of the acceleration of m_1. Then the acceleration of m_1 is down or up the plane according to whether a is positive or negative.

We must now repeat the entire procedure for the system consisting of the mass m_2. For a coordinate axis directed vertically upward and the free-body diagram as in Figure 4-15, the law of motion of the center of mass for vertical components yields

$$T - m_2 g = m_2 a \tag{4.10}$$

Notice that the tension is *not* equal to $m_2 g$ unless the acceleration of the suspended mass is zero.

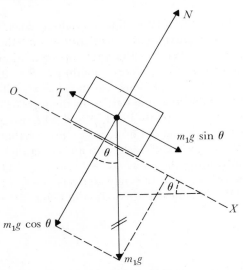

Figure 4-14 Free-body diagram for the block of mass m_1 of Example 5.

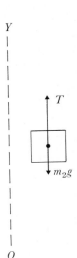

Figure 4-15 Free-body diagram for the block of mass m_2 of Example 5.

Equations 4.9 and 4.10 are a pair of simultaneous equations in two unknowns, T and a. We can eliminate T by adding these equations. This yields

$$m_1 g \sin \theta - m_2 g = m_1 a + m_2 a$$

which can be solved for a, giving

$$a = \frac{m_1 g \sin \theta - m_2 g}{m_1 + m_2} \tag{4.11}$$

Substituting this value for a in Eq. 4.10 and solving for T we obtain

$$T = \frac{m_1 m_2 g (1 + \sin \theta)}{m_1 + m_2} \tag{4.12}$$

(*Step 6*) To check Eq. 4.11 we note that the right-hand side has the dimensions

$$\left[\frac{(m_1 \sin \theta - m_2) g}{m_1 + m_2} \right] = \frac{[M]\,[LT^{-2}]}{[M]} = [LT^{-2}]$$

which are the dimensions of acceleration. Therefore this expression for the acceleration is at least dimensionally correct.

The expression for the tension (Eq. 4.12) has dimensions

$$\left[\frac{m_1 m_2 g (1 + \sin \theta)}{m_1 + m_2} \right] = \frac{[M^2]\,[LT^{-2}]}{[M]} = [MLT^{-2}]$$

Since these are the dimensions of force, Eq. 4.12 is dimensionally correct.

A detailed check can be made by evaluating the acceleration and the tension for particular values of the masses and of the angle of inclination of the plane. For instance, if the plane is horizontal corresponding to $\theta = 0$, and if the suspended mass m_2 is zero, then it is obvious that the mass m_1 will remain at rest on the horizontal board and that the tension in the rope will be zero. And when we substitute $\theta = 0$, $m_2 = 0$ into Eq. 4.11 and 4.12, using the fact that $\sin \theta = 0$, we find that these equations do correctly predict that $a = 0$ and $T = 0$. And if $m_1 = 0$, then m_2 will fall freely with an acceleration g and the tension will be zero. Substituting the value $m_1 = 0$ into Eq. 4.11 and 4.12, we find that these equations do give the right answers, $a = -g$ and $T = 0$. If we give θ the value $90°$, we achieve the situation in Figure 4-16

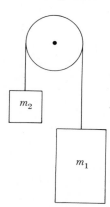

Figure 4-16   Atwood's machine. Compare with Figure 4-12 for the case when $\theta = 90°$.

in which the masses are suspended at the ends of a rope which passes over a pulley. In this device, called Atwood's machine, the acceleration of the masses, from Eq. 4.11 with $\sin \theta = \sin 90° = 1$, is given by

$$a = \left(\frac{m_1 - m_2}{m_1 + m_2}\right)g \tag{4.13}$$

Now if these suspended masses are equal we know that their acceleration must be zero, and this is easily seen to be the prediction of Eq. 4.13 when $m_1 = m_2$.

Summary

☐ Newton's third law: The force exerted on a particle A by another particle B is equal in magnitude and opposite in direction to the force exerted on particle B by particle A:

$$\mathbf{F}_{\text{on } A \text{ by } B} = -\mathbf{F}_{\text{on } B \text{ by } A}$$

and these forces lie along the line joining the particles.

☐ The center of mass C of a system of particles of masses $m_1, m_2, \ldots, m_n$ has a position vector $\mathbf{r}_c$ given by

$$M\mathbf{r}_c = m_1\mathbf{r}_1 + m_2\mathbf{r}_2 + \cdots + m_n\mathbf{r}_n$$

a velocity vector $\mathbf{V}_c$ such that

$$M\mathbf{V}_c = m_1\mathbf{v}_1 + m_2\mathbf{v}_2 + \cdots + m_n\mathbf{v}_n$$

and an acceleration $\mathbf{a}_c$ which is related to the resultant $\mathbf{F}_{\text{ext}}$ of the external forces acting on the particles of the system by the law of motion of the center of mass:

$$\mathbf{F}_{\text{ext}} = M\mathbf{a}_c$$

where $M = m_1 + m_2 + \cdots + m_n$.

☐ In applying $\mathbf{F}_{\text{ext}} = M\mathbf{a}_c$ to a complex system, we must make clear exactly which particles are to be included in the system and which are to be excluded. The free-body diagram shows the system selected and all the *external* forces acting *on* the particles of the system.

Questions

1 A painter weighs 1000 N. Find the magnitude and direction of the force exerted on the planet earth by the painter.

2 A car towing a glider accelerates down a runway. The force exerted on the glider by the towline is 800 N. Find the magnitude and direction of the force exerted on the towline by the glider.

3 The air exerts a horizontal retarding force of 200 N on the glider of Question 2. Draw a diagram showing the forces that determine the glider's acceleration. Find the magnitude and direction of this acceleration if the glider and its contents have a mass of 150 kg.

4 If, in Example 1, the man changes the force he exerts on the stone so that the *velocity of the stone is constant*, then find:

    (a) The acceleration of the stone.

    (b) The resultant force acting on the stone.

    (c) The force exerted on the stone by the man's hand.

    (d) The force exerted on the man's hand by the stone.

5 A stone weighing 40 N is shot vertically upward by a catapult. At an instant when the stone is still in contact with the catapult, the stone's acceleration is 80 m/s^2 upward. At this instant find:

    (a) The resultant force acting on the stone.

    (b) The force exerted on the stone by the catapult.

    (c) The force exerted on the catapult by the stone.

6 In Figure 4-7, move the origin of the coordinate system 2.0 m to the left. Determine the new coordinates of the 10.0-kg particle and the 5.0-kg particle and use Eq. 4.3 to locate the center of mass of the system. Verify that the *same* point C is obtained. (This example illustrates that the location of the center of mass of a system is independent of the choice of the coordinate system.)

7 Verify that, if each particle in Figure 4-7 is moved 2.0 m to the right, the center of mass of the system moves 2.0 m to the right. (As illustrated in this question, when all the particles of a system are given the same displacement, the center of mass also experiences this displacement; in other words, the center of mass moves as if rigidly attached to the system.)

8 Show that if you select the origin at the center of mass of a system, so that the position vector of the center of mass is zero, then $m_1r_1 = m_2r_2$ for a two-particle system.

9 Prove that the center of mass of two particles of equal mass is midway between them.

10 The mass of the earth is 6.0×10^{24} kg and the mass of the moon is 7.3×10^{22} kg. The separation of the earth and the moon is approximately 3.8×10^5 km. Show that the center of mass of the earth-moon system is 4.6×10^3 km from the earth's center.

11 Locate the center of mass of the system composed of three particles placed at the vertices of the right triangle of Figure 4-17.

12 In calculating the location of the center of mass of a composite system, we can replace any portion with mass M and center of mass C by a single particle of mass M located at C. (This can be proved from the defining equation of the center of mass,

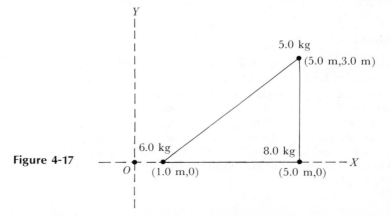

Figure 4-17

Eq. 4.2.) Use this fact to find the location of the center of mass of a letter T made from two thin uniform bars. The vertical bar has length 0.60 m and mass 3.0 kg. The horizontal bar has length 0.40 m and mass 2.0 kg.

13 A dumbbell is constructed from a 5.0-kg sphere and a 3.0-kg sphere joined together by a light rod. The distance between the centers of the spheres is 0.25 m:

(a) Find the center of mass of this assembly.

(b) This rigid dumbbell is placed on a sheet of ice and is pulled by a 40-N horizontal force applied to the smaller sphere. Find the acceleration of the center of mass of the dumbbell.

14 (a) What is the law of motion of the center of mass?

(b) For a system consisting of a single particle of mass m, show that the law of motion of the center of mass is the same as Newton's second law.

15 A 10-kg particle is fastened to a 30-kg particle by a light spring. This system is hurled through the air. Find the resultant external force acting on this system and the acceleration of its center of mass.

16 A 5.0-kg block and a 10-kg block are placed on a smooth table and connected by a spring. The 10-kg block is then pushed east by a horizontal force of 60 N:

(a) Find the acceleration of the center of mass of the two blocks.

(b) The velocity of the center of mass after 2.0 s have elapsed is 8.0 m/s east. At this time the 10-kg block has a velocity of 6.0 m/s east. What is the velocity of the 5.0-kg block?

17 (a) In Figure 4-10, find the force exerted on the first car of the train by the locomotive.

(b) Find the force exerted on the third car by the second.

18 A 3.0-kg wooden block is connected by a string to a 5.0-kg iron block. This assembly is pulled across a smooth table top by a wire fastened to the iron block which exerts a constant horizontal force of 24 N. Find:

(a) The acceleration of the blocks.

(b) The force exerted on the wooden block by the string.

19 A locomotive pulls two cars, each of mass 1.50×10^4 kg. The tension in the connecting link between the locomotive and the first car is 3.00×10^3 N. What is the tension in the connecting link between the cars?

20 A 10.0-kg block is placed on a frictionless horizontal surface. It is connected by a cord passing over a light frictionless pulley to a suspended block with a mass of 5.0 kg (Figure 4-18):

(a) Draw the free-body diagram for the 10.0-kg block and then for the 5.0-kg block.

(b) Find the magnitude of the acceleration of the blocks.

(c) What is the tension in the cord?

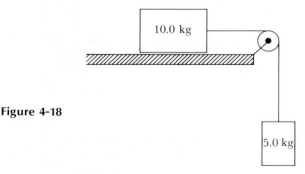

Figure 4-18

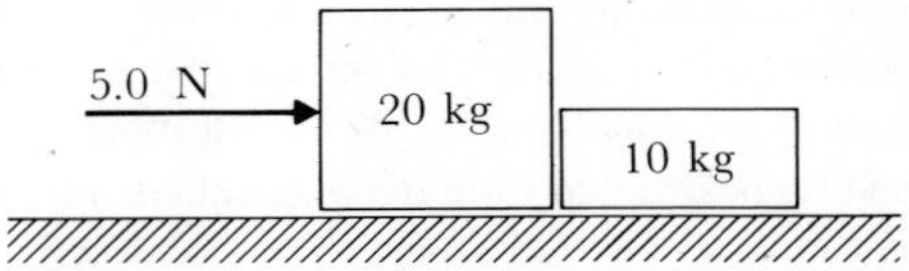

$\cdot\, 1\, 6\, 7\, \gamma_S\, 2$

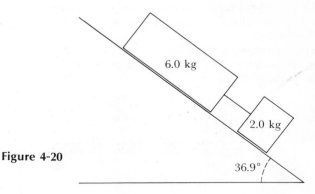

5.0 N   20 kg

10 kg

Figure 4-19

21 A 20-kg block and a 10-kg block are placed in contact on a smooth horizontal table as shown in Figure 4-19:

   (a) A horizontal force of 5.0 N is applied to the 20-kg block. Find the force exerted on the 10-kg block by the 20-kg block.

   (b) The 5.0-N horizontal force is now applied (acting to the left) to the 10-kg block rather than to the 20-kg block. Once again, find the force exerted on the 10-kg block by the 20-kg block.

   (c) Explain why the answers to parts (a) and (b) are different.

22 Check the dimensions of the following proposed formula for the tension T of the rope in Atwood's machine:

$$T = \frac{2m_1 m_2 g}{m_1 + m_2}$$

23 A heavy rope is given an acceleration a directed vertically upward. Consider a segment of the rope of mass m:

   (a) Show that the tension T_1 at the uppermost cross section of this segment and the tension T_2 at the lowest portion of this segment are related by

$$T_1 - T_2 = mg + ma$$

   (b) If this rope were being accelerated horizontally, what would be the difference in the tensions at the two extremities of the segment of mass m?

   (c) What must be assumed about the rope (as in the solution to Example 5) for the tension to have the same value throughout all segments of the rope?

24 In Figure 4-20, the coefficient of kinetic friction between the 6.0-kg block and the inclined plane is 0.40. The plane forms an angle of 36.9° with the horizontal. Find the tension in the rope connecting the 6.0-kg block to the smooth 2.0-kg block.

6.0 kg

2.0 kg

36.9°

Figure 4-20

25 In Figure 4-20, assume that the coefficient of static friction μ_s between the 6.0-kg block and the plane is such that the block is at rest but is on the point of slipping. The 2.0-kg block is smooth. Find the tension in the rope and the value of μ_s.

26 What is the minimum acceleration with which a 50-kg child can safely slide down a rope that will break at a tension of 400 N?

27 Find the tension in each string in Figure 4-21 when each block has an acceleration of magnitude a directed vertically upward.

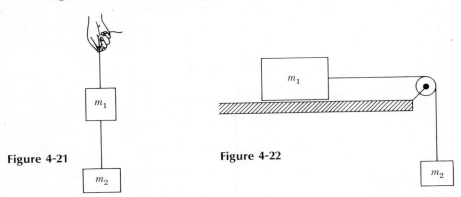

Figure 4-21   **Figure 4-22**

28 Assume that the block of mass m_1 in Figure 4-22 is retarded by friction as it slides over the table top. Find the acceleration of the block in terms of m_1, m_2, g, and the coefficient μ_k of kinetic friction between the sliding block and the table.

Supplementary Questions

S-1 In Figure 4-23 find the acceleration of the cart that is required to prevent block B from falling. The coefficient of static friction between the block and the cart is μ_s.

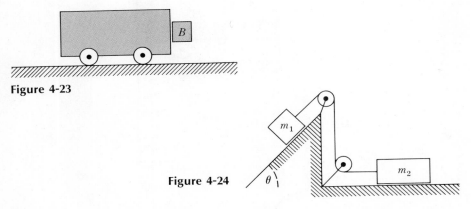

Figure 4-23

Figure 4-24

S-2 The blocks in Figure 4-24 slide over smooth surfaces and the string passes over light frictionless pulleys. Find the acceleration of the blocks in terms of m_1, m_2, g, and θ.

S-3 A mass is suspended from the roof of a car by a string. While the car has a constant acceleration, the string makes a constant angle θ with the vertical. Show how the car's acceleration can be determined from measurement of this angle.

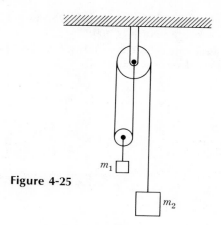

Figure 4-25

S-4 Assume that the pulleys in the block and tackle in Figure 4-25 are light and frictionless. Find the acceleration of the greater mass m_2 in terms of m_1, m_2, and g.

S-5 In Figure 4-25, what is the force exerted on the ceiling by the entire block and tackle system?

S-6 Find the tension in the cord of Figure 4-26, as mass m_1 slides over the smooth table and mass m_2 descends.

S-7 A uniform heavy rope is 10 m long and has a mass of 20 kg. When this rope is hanging vertically and is pulled upward with an acceleration of 5.0 m/s², what is the tension in the rope at a point 3.0 m from the free lower end of the rope?

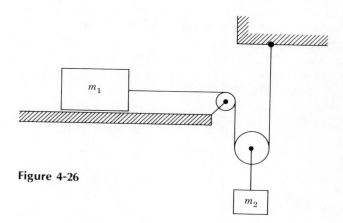

Figure 4-26

5.

statics of rigid bodies

In the construction of bridges, buildings, and most experimental apparatus, we are concerned with objects at rest. If these objects do not bend or vibrate appreciably, it is a useful approximation to consider that they are rigid bodies; that is, their parts all maintain a fixed location with respect to each other when external forces are applied.

In this chapter we define *torque* and then examine the forces and torques that act on a rigid body when it is at rest. The discussion will be limited to situations where the forces are all parallel to the XY plane.

5.1 Torque About an Axis

The effect of a force on the rotational motion of an extended object depends not only upon the magnitude and direction of the force but also upon its line of action (Figure 5-1). The quantity that is the significant measure of the turning effect of a force about an axis is the *torque,* denoted by τ (Greek letter tau), which we now define.

Consider a force $\mathbf{F}$ lying in a plane and an axis perpendicular to the plane through the point A in that plane (Figure 5-1). The perpendicular distance from the line of action of $\mathbf{F}$ to the axis is called the *moment arm D*. The magnitude

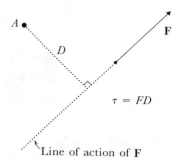

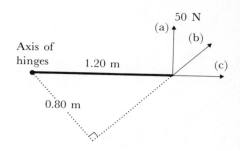

Figure 5-1 The torque τ of a force **F** about an axis through A and perpendicular to the plane of the paper is equal to FD. The moment arm is D.

Figure 5-2 The torque changes when the direction of the force changes because then the moment arm is different. The view is that seen looking down at the top of the door. The plane of the paper is horizontal.

of the torque τ of **F** about the axis is defined to be the *product of the magnitude of the force and the moment arm:* that is

$$\tau = FD \tag{5.1}$$

Torque can have either a clockwise or a counterclockwise sense. To distinguish between the two possible senses of rotation we adopt the convention that a *counterclockwise torque is positive* and that a *clockwise torque is negative.*

Example 1 A 50-N horizontal force is applied to the outer edge of a door (Figure 5-2) which is 1.20 m wide. Find the torque of this force about the vertical axis of the hinges if:

(*a*) The force is directed perpendicular to the door.

(*b*) The force is directed in such a way that the moment arm is 0.80 m.

(*c*) The force is directed so that its line of action passes through the axis of the hinges.

Solution

(*a*) Here the moment arm D is simply the width of the door, so

$$\tau = (50 \text{ N})(1.20 \text{ m}) = 60 \text{ N·m}$$

(*b*) The torque is now given by

$$\tau = (50 \text{ N})(0.80 \text{ m}) = 40 \text{ N·m}$$

(*c*) In this circumstance the moment arm D is 0. Therefore

$$\tau = (50 \text{ N}) (0) = 0$$

We can see from this example that the torque of the 50-N force applied at the edge of the door can be varied from a maximum value when the moment arm is equal to the door width to a minimum value of zero when the moment arm is zero, simply by varying the direction of the force.

5.2 Center of Gravity

Consider a system (Figure 5-3) of two particles m_1 and m_2 in a uniform gravitational field **g**. The definition of the center of mass (Eq. 4.3) of this system gives

$$Mx_c = m_1x_1 + m_2x_2$$

Multiplying each side of this equation by $-g$ we obtain

$$-Mgx_c = (-m_1gx_1) + (-m_2gx_2) \tag{5.2}$$

This equation has an interesting interpretation. Each term is a torque of a gravitational force mg with a moment arm x about an axis through the origin and perpendicular to the XY plane. Therefore Eq. 5.2 states that the sum of the torques of the gravitational forces acting on the system is the same as if the entire weight Mg were applied at the center of mass C. The point of application C of this equivalent resultant force Mg is called the *center of gravity*. Our result is that in a uniform gravitational field, the *center of mass is the center of gravity*.

This proof for a two-particle system can be extended without difficulty to systems comprising any number of particles. Therefore, in any problem involving a system of particles in the uniform gravitational field near the earth's surface, we can place the entire weight Mg at the center of mass and know that this will give the correct answer as far as gravitational torques are concerned.

The fact that the center of mass coincides with the center of gravity can be used to locate experimentally the center of mass of irregularly shaped objects. It is only necessary to suspend the object from different points and note that in each case the center of gravity must be vertically below the point of suspension as illustrated in Figure 5-4 (see Question 5).

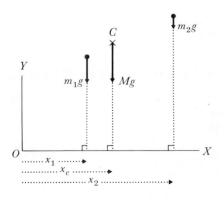

Figure 5-3 The sum of the torques $-m_1gx_1$ and $-m_2gx_2$ is equal to the torque $-Mgx_c$. The axis about which these torques are computed is an axis OZ through the origin and perpendicular to the XY plane.

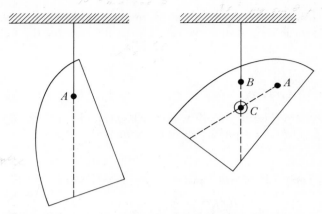

Figure 5-4   Locating the center of gravity of a flat object.

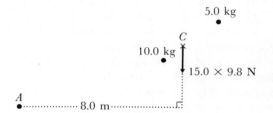

Figure 5-5   The gravitational torque about A can be computed by treating the system as if the entire weight of the system were applied at the center of gravity C.

Example 2   Consider the system (Example 2 in Chapter 4) which comprises a 10.0-kg particle and a 5.0-kg particle separated by 3.00 m with a center of mass 1.00 m from the 10.0-kg particle (Figure 5-5). Find the sum of the gravitational torques of this system about a perpendicular axis through the point A which is a horizontal distance of 8.0 m from the center of mass.

Solution   With the entire weight considered as concentrated at the center of mass, we obtain for this clockwise torque

$$-(15.0 \text{ kg})(9.8 \text{ m/s}^2)(8.0 \text{ m}) = -1.18 \times 10^3 \text{ N·m}$$

5.3 Conditions of Equilibrium for a Rigid Body

The forces acting on a rigid body which remains at rest (or, more generally, has no translational or rotational acceleration) must satisfy two conditions of equilibrium. First, from the law of motion of the center of mass,

$$\mathbf{F}_{\text{ext}} = M\mathbf{a}_c \tag{4.7}$$

we conclude, since $\mathbf{a}_c$ is zero, that

$$\mathbf{F}_{\text{ext}} = 0 \tag{5.3}$$

In words, the first condition of equilibrium is this: *The vector sum of all the external forces acting on a body in equilibrium is zero.* When all the forces acting on the body are parallel to the *XY* plane, Eq. 5.3 is equivalent to two equations:

$$\text{algebraic sum of x components of external forces} = 0 \qquad (5.4)$$

$$\text{algebraic sum of y components of external forces} = 0$$

Even when the resultant of the external forces is zero, these external forces may cause a rigid body to spin more rapidly or more slowly. It is the second condition of equilibrium that guarantees rotational equilibrium and, although we have not yet studied the dynamics of rotation, intuition suggests that torques are involved. The analysis to be given in Chapters 12 and 13 shows that the second condition of equilibrium may be stated as follows: For *any* axis

$$\text{algebraic sum of external torques} = 0 \qquad (5.5)$$

When the forces are all parallel to the *XY* plane, it is most convenient to consider torques about an axis perpendicular to the *XY* plane.

To apply the conditions of equilibrium in problems in the statics of rigid bodies, we follow the general procedure stressed in Sections 3.4 and 4.4. However, since torques are now important, we must ensure that the *line of action* of each force is correct on the free-body diagram. (In the preceding chapters, where we were interested only in the motion of the center of mass, the forces acting on the system could be represented as applied at the center of mass.)

Example 3 A 9.00-m uniform log weighing 800 N is supported at a point *A* 2.00 m from its left end and at a point *B* 6.00 m from the same end. Find the forces F_A and F_B exerted on the log by the supports.

Solution
(*Step 1*) Figure 5-6 is the schematic diagram.
(*Step 2*) The log is selected as the system in equilibrium.
(*Step 3*) As an inertial frame of reference, we choose coordinate axes fixed on the earth with the *Y* axis directed vertically upward.

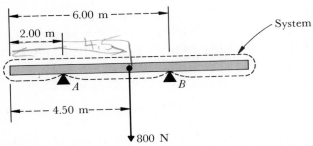

Figure 5-6 Schematic diagram for Example 3. The system which has been selected is enclosed by the dashed line.

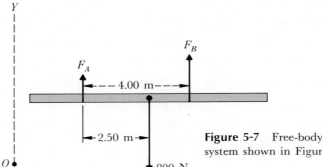

Figure 5-7 Free-body diagram for system shown in Figure 5-6.

(*Step 4*) The free-body diagram (Figure 5-7) shows the log and all the external forces acting on the log. These include the unknown forces F_A and F_B pushing up on the log. The log's weight is represented as acting at its center of mass, which, for a uniform log, is at its geometrical center.

(*Step 5*) The conditions of equilibrium are now applied. Algebraic manipulations are often simplified by using the torque equation first. *If one selects the axis about which torques are to be taken so that this axis passes through the line of action of an unknown force, this force will have zero moment arm and therefore zero torque, and so will not appear in the torque equation.* Thus, if we take torques about a horizontal axis perpendicular to the log and passing through the point A, the torque of F_A about this axis is zero, and the torque equation (the second condition of equilibrium, Eq. 5.5) yields

$$F_B(4.00 \text{ m}) - (800 \text{ N}) (2.50 \text{ m}) = 0$$

This gives

$$F_B = 500 \text{ N}$$

The first condition of equilibrium (Eq. 5.4) requires that the y components of the forces acting on the log satisfy

$$F_A + F_B - 800 \text{ N} = 0$$

Therefore

$$F_A = 300 \text{ N}$$

(*Step 6*) The answers make sense: a total force of 500 N + 300 N = 800 N is exerted in an upward direction. The answers can be checked by examining the torques about a different axis, say an axis through the center of gravity of the log. Then the algebraic sum of the torques is

$$(500 \text{ N}) (1.50 \text{ m}) - (300 \text{ N}) (2.50 \text{ m}) = 0$$

which verifies that the second condition of equilibrium is satisfied for this axis.

Example 4 A uniform beam, which weighs 400 N and is 5.00 m long, is hinged to a wall at its lower end by a frictionless hinge. A horizontal rope 3.00 m long is fastened between the upper end of the beam and the wall. Find the force T exerted on the beam by the rope and also find the horizontal and vertical components of the force exerted on the beam by the hinge.

Solution

(*Step 1*) The schematic diagram is shown in Figure 5-8.

(*Step 2*) The system in equilibrium that has been selected is the beam.

(*Step 3*) We choose an inertial frame of reference with the X axis directed horizontally to the right and the Y axis directed vertically upward.

(*Step 4*) The free-body diagram for the beam is given in Figure 5-9. The force exerted on the beam by the hinge has been replaced by its horizontal component F_x and its vertical component F_y. Again note that the weight (400 N) is represented as acting at the center of mass of the beam. The rope pulls on the beam and therefore exerts a force on the beam which acts to the left.

(*Step 5*) We consider torques about a horizontal axis perpendicular to the beam and passing through the hinge. Then, since the moment arms of F_x and F_y are zero, these forces have zero torque about this axis. The 400-N

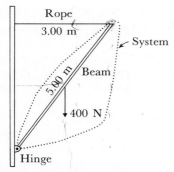

Figure 5-8   Schematic diagram for Example 4. The system selected has been enclosed by the dotted line.

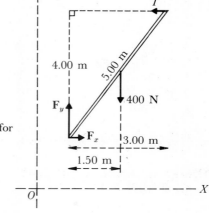

Figure 5-9   Free-body diagram for system of Figure 5-8.

weight has a moment arm of 1.50 m. The moment arm for the force T is evidently the distance along the wall from the hinge to the rope which is

$$\sqrt{(5.00 \text{ m})^2 - (3.00 \text{ m})^2} = 4.00 \text{ m}$$

Therefore the second condition of equilibrium gives

$$T(4.00 \text{ m}) - (400 \text{ N})(1.50 \text{ m}) = 0$$

Consequently

$$T = 150 \text{ N}$$

For the x components of the forces acting on the beam, the first condition of equilibrium yields

$$F_x - T = 0$$

Therefore,

$$F_x = 150 \text{ N}$$

For the y components, the first condition of equilibrium implies that

$$F_y = 400 \text{ N}$$

(*Step 6*) These answers can be checked by computing torques about some other axis, say an axis through the center of mass, and verifying that the second condition of equilibrium is satisfied for this axis.

Example 5   A uniform ladder of mass m and length L leans against a smooth wall and makes an angle θ with the horizontal (Figure 5-10). How large a frictional force f must be provided by the floor to prevent the ladder from slipping?

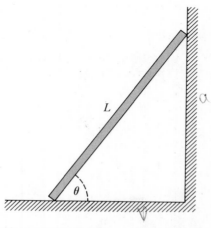

Figure 5-10   Schematic diagram of ladder of Example 5.

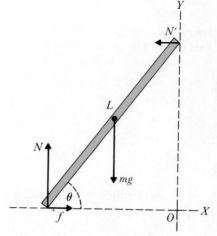

Figure 5-11   Free-body diagram of ladder of Example 5.

Solution   Figure 5-11 is the free-body diagram for the ladder. For the torques about the foot of the ladder, the second condition of equilibrium gives

$$N'L \sin \theta - mg\left(\frac{L}{2}\right) \cos \theta = 0$$

The first condition of equilibrium requires

$$f - N' = 0$$

Therefore

$$f = N' = \tfrac{1}{2}mg \cot \theta$$

Summary

☐   The magnitude of the *torque* τ of a force **F** about an axis is given by

$$\tau = FD$$

where the moment arm D is the perpendicular distance from the line of action of **F** to the axis.

☐   In a uniform gravitational field, for the purpose of computing gravitational torques, the entire weight $M\mathbf{g}$ of the system can be considered to act at the center of mass.

☐   If a rigid body is in equilibrium, then both the first and second conditions of equilibrium are satisfied. The first condition of equilibrium is

algebraic sum of x components of external forces $= 0$

algebraic sum of y components of external forces $= 0$

The second condition of equilibrium is

algebraic sum of external torques $= 0$

Solutions to problems are simplified by using the torque equation first and considering torques about an axis which passes through the line of action of an unknown force.

Questions

1   A 1000-N weight is placed on a horizontal seesaw at a point which is 2.0 m from its supporting axis. What is the magnitude of the torque of the weight about this axis?

2   A uniform door weighs 60 N and is 0.80 m wide and 2.4 m high. Find the torque of its weight about a horizontal axis perpendicular to the door and passing through a lower corner.

3 A 0.20-kg metre-stick lies along the positive horizontal X axis with the zero mark at the origin:

   (a) Find the torque of its weight about a horizontal axis OZ which is perpendicular to OX.

   (b) Find the torque of its weight about an axis parallel to OZ and passing through the 70-cm mark on the stick.

   (c) The end marked 100 cm is raised until the metre-stick makes an angle of 60° with the horizontal. Find the torque of its weight about the axis OZ.

4 A 2.0×10^3-kg uniform sphere with a radius of 0.50 m touches a vertical wall. What is the magnitude of the torque of the sphere's weight about a horizontal axis along the base of the wall?

5 In the method illustrated in Figure 5-4 for locating the center of gravity of an object, we assume that, when the object is in equilibrium, its center of gravity is vertically below the point of suspension. Use the second condition of equilibrium to prove that this assumption is correct.

6 A load weighing 160 N is hung 3.0 m from one end of an 8.0-m horizontal pole which weighs 80 N. Two men carry the pole, one at each end. Find the forces exerted on the pole by the men.

7 A metre-stick which weighs 20 N is suspended in a horizontal position from a vertical string attached at the 100-cm mark and from another vertical string attached at the 40-cm mark. Find the tension in each string.

8 A horizontal plank, 6.00 m long and weighing 200 N, rests on two supports placed 1.2 m from each end. A carpenter who weighs 800 N walks along this plank. How close can he go to the end of the plank without upsetting it?

9 A uniform beam, which weighs 600 N and is 5.00 m long, is hinged to a wall at its lower end by a frictionless hinge (as in Figure 5-8). A horizontal rope, 4.00 m long, is fastened between the upper end of the beam and the wall. Find the force T exerted on the beam by the rope and also find the horizontal and vertical components of the force exerted on the beam by the hinge.

10 A uniform trapdoor which is 2.00 m long and 2.00 m wide is hinged at one edge and held open at an angle of 60° with the horizontal by a rope attached at the opposite edge and pulling perpendicular to the door. The door weighs 180 N. Find the tension in the rope as well as the horizontal and vertical components of the force exerted on the door at the hinge.

11 The door of Question 2 is supported at one lower corner by a force which has a horizontal component F_x and a vertical component F_y:

   (a) What horizontal force H applied at an upper corner will hold the door in equilibrium?

   (b) Find F_x and F_y.

12 One end of a 600-N beam is hinged at a vertical wall and held horizontal by a rope 5.00 m long fastened to the wall and to the beam at a point 3.00 m from the wall (Figure 5-12). A 200-N weight is suspended at the free end of the beam, 4.00 m from the wall. Find the tension in the 5.00-m rope and the horizontal and vertical components of the force exerted on the beam by the hinge.

13 A 13.0-m uniform ladder weighing 300 N rests against a smooth wall at a point 12.0 m above the floor. The lower end of the ladder rests on a rough floor. Find the (horizontal) frictional force f and the (vertical) normal force N exerted on the ladder by the floor, as well as the (horizontal) normal force N' exerted on the ladder by the wall.

14 A light uniform ladder is supported on a rough floor and leans against a smooth wall, touching the wall at a point which is at height b above the floor. A woman climbs up the ladder until the base of the ladder is on the point of slipping. The coefficient of static friction between the foot of the ladder and the floor is μ. Show that the *horizontal* distance moved by the woman is given by $x = \mu b$.

Supplementary Questions

S-1 A uniform rod 2.00 m long has one end on a rough floor. The rod, resting against the smooth edge of a table which is 1.50 m high, is inclined at a 60° angle with the horizontal. If the rod is on the point of slipping, what is the coefficient of static friction between the floor and the rod? (The force exerted on the rod by the smooth edge of the table is perpendicular to the rod.)

S-2 A 13.0-m uniform ladder weighing 300 N rests against a smooth wall at a point 12.0 m above the floor. The ladder is about to slip after a painter weighing 900 N has climbed a distance of 3.90 m measured along the ladder. Find the coefficient of static friction (μ_s) between the foot of the ladder and the floor.

S-3 A light stepladder (Figure 5-13) is constructed from two ladders, each 3.00 m long, which are hinged at the top and joined by a rope 1.00 m long fastened to each ladder at points 1.00 m from the hinge. A carpenter weighing 800 N stands at the midpoint of the ladder on the right. Assuming that the floor is frictionless, find the tension in the rope.

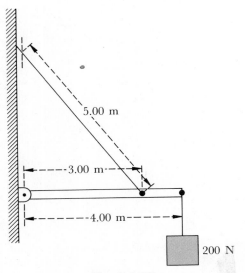

5.00 m

-----3.00 m-----

-----4.00 m-----

200 N

Figure 5-12

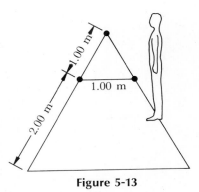

1.00 m

1.00 m

2.00 m

Figure 5-13

S-4 Show that if a body is in equilibrium when acted upon by three nonparallel forces, the three forces must all pass through a common point. Locate this point for the ladder of Figure 5-10.

S-5 Two forces have equal magnitudes F and opposite directions with a distance D between the lines of action of the forces. Such a pair of forces constitutes a *couple*. Show that the torque exerted by this couple about any axis perpendicular to the plane of the forces has a magnitude FD, and hence is independent of the position of the axis.

6.

momentum

6.1 Momentum and Newton's Second Law

The product $m\mathbf{v}$ of the mass of a particle and its velocity vector is defined to be the *momentum* $\mathbf{p}$ of the particle:

$$\mathbf{p} = m\mathbf{v} \tag{6.1}$$

Since momentum is a product of a positive scalar m and a vector $\mathbf{v}$, $\mathbf{p}$ is itself a vector in the direction of $\mathbf{v}$ with a magnitude which is the product mv of the mass and the speed. A 15×10^3-kg freight car moving south at 3.0 m/s has a momentum of 45×10^3 kg·m/s south; an automobile with one-tenth the mass but moving in the same direction with ten times the speed will have the same momentum as the freight car.

Newton's statement of what we have called Newton's second law, $\mathbf{F} = m\mathbf{a}$, was actually expressed in terms of momentum. In modern terminology this formulation of Newton's second law is: *The time rate of change of momentum of a particle is equal to the resultant force acting on the particle,* that is

$$\mathbf{F} = \frac{d\mathbf{p}}{dt} \tag{6.2}$$

In Newtonian mechanics the equation $\mathbf{F} = d\mathbf{p}/dt$ implies (since the mass m of a particle is constant)

$$\mathbf{F} = \frac{d}{dt}(m\mathbf{v}) = m\frac{d\mathbf{v}}{dt} = m\mathbf{a}$$

so the two expressions of Newton's second law, $\mathbf{F} = d\mathbf{p}/dt$ and $\mathbf{F} = m\mathbf{a}$, are equivalent.

However, as we shall discuss in Chapter 40, in the special theory of relativity the equation $\mathbf{F} = m\mathbf{a}$ is no longer valid, but it is still true that $\mathbf{F} = d\mathbf{p}/dt$ provided we change the definition of momentum to be $\mathbf{p} = m\mathbf{v}/\sqrt{1 - v^2/c^2}$. The factor $\sqrt{1 - v^2/c^2}$ is imperceptibly different from the number 1 unless the speed v is an appreciable fraction of the speed c of light; thus the particle motion predicted by the special theory of relativity is essentially the same as that predicted by Newtonian mechanics for the speeds encountered in everyday life and in most engineering practice.

6.2 Momentum of a System of Particles

The total momentum $\mathbf{P}$ of a system of n particles is defined to be the vector sum of the momentum vectors $\mathbf{p}_1, \mathbf{p}_2, \ldots, \mathbf{p}_n$ of the individual particles; that is

$$\mathbf{P} = \mathbf{p}_1 + \mathbf{p}_2 + \cdots + \mathbf{p}_n \tag{6.3}$$

As forces that are internal and external to the system act on the individual particles of the system, the momenta of these particles change. Thus, $d\mathbf{p}_1/dt = \mathbf{F}_1$ where $\mathbf{F}_1$ is the resultant force acting on the first particle, and there is a similar equation for each particle of the system. Therefore, by differentiating each side of Eq. 6.3, we find that the rate of change of the system's total momentum is given by

$$\frac{d\mathbf{P}}{dt} = \frac{d\mathbf{p}_1}{dt} + \frac{d\mathbf{p}_2}{dt} + \cdots + \frac{d\mathbf{p}_n}{dt}$$

$$= \mathbf{F}_1 + \mathbf{F}_2 + \cdots + \mathbf{F}_n$$

The internal forces included in this sum cancel out in pairs according to Newton's third law, and the sum reduces to $\mathbf{F}_{\text{ext}}$, the vector sum of *external* forces acting on the particles of the system. The result, known as the *principle of momentum,* is that

$$\mathbf{F}_{\text{ext}} = \frac{d\mathbf{P}}{dt} \tag{6.4}$$

exactly as if the system were a single particle with momentum $\mathbf{P}$.

Notice that the *internal forces,* the forces exerted by one particle of the system on other particles of the system, do not change the total momentum of the system. The momentum changes of a pair of particles produced by their interaction are always vectors of equal magnitudes and opposite directions. When summed, such momentum changes cancel out and therefore cannot change the value of the system's total momentum $\mathbf{P}$.

The total momentum of a system can be evaluated if the velocity $\mathbf{V}_c$ of the system's center of mass is known. Equation 4.3 is

$$M\mathbf{V}_c = m_1\mathbf{v}_1 + m_2\mathbf{v}_2 + \cdots + m_n\mathbf{v}_n$$

where M is the total mass of the system. The vector sum on the right-hand side of this equation is the system's total momentum. Therefore

$$\mathbf{P} = M\mathbf{V}_c \tag{6.5}$$

The system has a total momentum equal to that of a particle of mass M moving with the center of mass.

6.3 Conservation of Momentum

A most significant consequence of the equation

$$\mathbf{F}_{\text{ext}} = \frac{d\mathbf{P}}{dt}$$

is that if $\mathbf{F}_{\text{ext}}$ is zero, the total momentum $\mathbf{P}$ of a system is a constant vector. We have been led from Newton's second and third laws to assert the *law of conservation of momentum: If the vector sum of the external forces acting on a system is zero, the total momentum $\mathbf{P}$ of the system remains constant;* that is

$$\mathbf{F}_{\text{ext}} = 0 \text{ implies } \mathbf{P} = \text{constant vector} \tag{6.6}$$

(The converse is also true.)

In the terminology of physics a quantity whose total amount does not change while individual amounts do alter is said to be conserved, and we speak of a law of conservation of this quantity. Conservation laws are regarded by physicists as particularly important for several reasons. For one thing, it is natural to suspect that a quantity which remains the same throughout complex and perhaps violent changes is somehow of fundamental significance. Certainly a conservation law has the great merit of being easy to apply; in any process in which a certain quantity is conserved one has merely to equate the total of the conserved quantity at the beginning of the process to its total at the end. It is, moreover, interesting to note that in the history of physics, conservation laws, which arose within the context of a particular theory, have in several instances turned out to be valid even in domains where the original theory was no longer successful. [For example, in relativistic and quantum physics we find that the law of conservation of momentum survives every experimental test, although in the special theory of relativity (Chapters 39 and 40) forces no longer obey Newton's third law, and in quantum mechanics (Chapter 42) Newton's second and third laws can no longer be used. In the derivation of the law of the conservation of momentum from Newton's third law we built better than we knew. The law of conservation of momentum seems to be one of nature's most fundamental laws.]

When $\mathbf{F}_{\text{ext}}$ is zero, and consequently $\mathbf{P}$ remains constant, the motion of the center of mass is extremely simple. No matter how the parts of the system fly about, the *center of mass moves with a constant velocity* given from Eq. 6.5 by $\mathbf{V}_c = \mathbf{P}/M$.

For any process occurring within a system we can conveniently express the law of conservation of momentum as follows: if $\mathbf{F}_{\text{ext}} = 0$

$$\mathbf{P}_{\text{initial}} = \mathbf{P}_{\text{final}} \tag{6.7}$$

or equivalently, for motion parallel to the XY plane,

$$P_{x\,\text{initial}} = P_{x\,\text{final}}$$
$$P_{y\,\text{initial}} = P_{y\,\text{final}} \tag{6.8}$$

where $P_{x\,\text{initial}}$ is the algebraic sum of the x components of the momenta of the particles of the system *before* the process, and $P_{x\,\text{final}}$ is the algebraic sum of the system's x components of momenta *after* the process.

Even when $\mathbf{F}_{\text{ext}}$ is not zero, these equations expressing conservation of momentum still can be applied as an excellent approximation provided two conditions are satisfied:

1   The process is of short duration as is the case for a collision or explosion. We then evaluate $\mathbf{P}_{\text{initial}}$ just before the process and $\mathbf{P}_{\text{final}}$ just after the process. In a very short time interval Δt an external force such as the force of gravity $M\mathbf{g}$ will change the system's momentum by only the small amount $M\mathbf{g}\Delta t$.

2   There are no *external impulsive forces*. An impulsive force is a force that has an extremely large value during a short time interval and a negligible value at other times. For example, when a baseball is hit, the bat exerts an impulsive force on the baseball. Even for a time interval as short as the duration of a collision or an explosion, an impulsive force acting on an object can cause an appreciable change in the object's momentum. To apply the law of conservation of momentum to a given process, we must consider a system large enough to include all the objects which exert impulsive forces.

Example 1   A cannon (Figure 6-1) of mass m_1 fires a cannonball of mass m_2 which emerges from the cannon traveling horizontally with a velocity $\mathbf{v}_2$. Find the recoil velocity $\mathbf{v}_1$ of the cannon.

Cannon's   Ball's
momentum   momentum

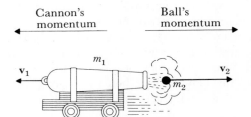

Figure 6-1   The cannon recoils with a momentum equal and opposite to that of the cannon ball.

Solution During the firing, the cannon and the ball exert on each other horizontal impulsive forces. For the system comprising both the ball and the cannon, there are no external impulsive forces. Consequently, conservation of momentum as expressed by Eq. 6.7 can be applied to this system to give

$$0 + 0 = m_1\mathbf{v}_1 + m_2\mathbf{v}_2$$

Therefore

$$\mathbf{v}_1 = -\frac{m_2}{m_1}\,\mathbf{v}_2$$

The negative sign means that $\mathbf{v}_1$ and $\mathbf{v}_2$ have opposite directions. It is worth noting that the center of mass of the cannon and the ball is at rest before the cannon is fired and therefore remains at rest, since $\mathbf{P}$ does not change and $\mathbf{V}_c = \mathbf{P}/(m_1 + m_2)$.

Example 2 A 1.20×10^3-kg car is traveling east at 30.0 m/s and collides with a 3.60×10^3-kg truck traveling at 20.0 m/s in a direction 60° north of east. The vehicles interlock and move off together. Find their common velocity $\mathbf{v}$.

Solution During the collision the car and the truck exert impulsive forces on each other. For the system comprising both the car and the truck, there are no external horizontal impulsive forces. Therefore the total horizontal momentum of this system is conserved in the collision. We measure each momentum relative to an inertial frame fixed on the earth with axis OX pointing east and OY pointing north (Figure 6-2). After collision the interlocked vehicles have

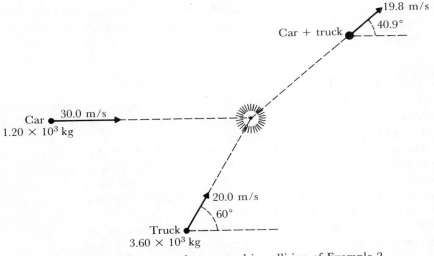

Figure 6-2 Momentum is conserved in collision of Example 2.

a velocity **v** with components v_x and v_y. From Eq. 6.8 the sum of the x components of the two vehicles' momenta before collision is equal to the x component of the momentum of the interlocked vehicles after collision:

$$(1.20 \times 10^3 \text{ kg}) (30.0 \text{ m/s}) + (3.60 \times 10^3 \text{ kg}) (20.0 \text{ m/s}) \cos 60°$$

$$= (4.80 \times 10^3 \text{ kg})v_x$$

Similarly, the y components of momentum before and after collision are related by

$$0 + (3.60 \times 10^3 \text{ kg})(20.0 \text{ m/s}) \cos 30° = (4.80 \times 10^3 \text{ kg})v_y$$

Solving these equations for v_x and v_y we obtain

$$v_x = 15.0 \text{ m/s} \qquad \text{and} \qquad v_y = 13.0 \text{ m/s}$$

This velocity vector makes an angle θ with the X axis such that

$$\tan \theta = \frac{v_y}{v_x} = \frac{13.0}{15.0} = 0.866$$

and

$$\theta = 40.9°$$

The speed is

$$v = \sqrt{v_x^2 + v_y^2} = \sqrt{(15.0)^2 + (13.0)^2} \text{ m/s} = 19.8 \text{ m/s}$$

The interlocked vehicles move off at a speed of 19.8 m/s in a direction 40.9° north of east.

One should not get the impression that the law of conservation of momentum alone is sufficient to determine what happens after a collision. Detailed prediction requires a knowledge of the forces acting during the collision. However, whatever does happen must conserve momentum. Thus the car and the truck might very well bounce off one another and go off in different directions. Conservation of momentum merely tells us that the vector sum of the momentum of the truck and the momentum of the car is not changed by the collision.

Rocket Propulsion

The law of conservation of momentum enables us to understand rocket propulsion. A rocket increases its forward momentum simply by ejecting fuel backwards with as large a momentum as possible.

For simplicity, we analyze the motion of a rocket in remote space where gravitational forces can be ignored. It is convenient to view the rocket (Figure 6-3) from an inertial frame of reference in which the rocket is at rest at a certain instant t. Consider the system of mass m comprising the rocket and its remaining fuel at this instant t. The total momentum **P** of this system is zero and, because there are no external forces acting on this system, its total momentum **P** will remain zero. The rocket ejects fuel backward at a rate R (mass per unit time) with a velocity **v**' relative to the rocket. During the time interval from t to

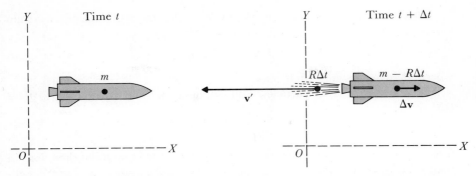

Figure 6-3 A rocket increases its forward momentum by ejecting fuel backward with as large a momentum as possible.

$t + \Delta t$, a mass $R\Delta t$ of fuel has acquired a backward momentum $(R\Delta t)\mathbf{v}'$ and, since momentum is conserved, we know that during the fuel ejection the fuel must have exerted a force forward on the rocket (and its remaining fuel) sufficient to impart forward momentum of equal magnitude. If the change in the rocket's velocity is $\Delta\mathbf{v}$, the forward momentum increase is $(m - R\Delta t)\Delta\mathbf{v}$ which we can approximate by simply $m\Delta\mathbf{v}$ provided that Δt is small. The law of conservation of momentum for the system therefore requires

$$m\Delta\mathbf{v} + R\Delta t\mathbf{v}' = 0$$

which yields

$$-R\mathbf{v}' = m\frac{\Delta\mathbf{v}}{\Delta t}$$

Taking the limit as Δt approaches zero, we find

$$-R\mathbf{v}' = m\mathbf{a}$$

This result gives the rocket's acceleration $\mathbf{a}$ ($= d\mathbf{v}/dt$) and shows that $R\mathbf{v}'$ plays the role of the force in $F = ma$; this force is called the rocket *thrust:*

$$thrust = Rv'$$

In a rocket engine with a large thrust, mass is ejected rapidly (R large) with a high ejection speed v'.

6.4 Center-of-Momentum Frame

A simple description of the motion of a system is obtained by choosing a frame of reference in which the total momentum $\mathbf{P}'$ of the system is zero. Such a frame is called the *center-of-momentum frame* (CM-frame).*

*In Newtonian mechanics, the CM-frame is often called the center-of-mass frame.

Relative to the CM-frame, the velocity $\mathbf{V}'_c$ of the center of mass of the system is zero, since

$$MV'_c = \mathbf{P}' = 0 \tag{6.9}$$

where M is the system's total mass.

For a system of two particles with momenta $\mathbf{p}'_1$ and $\mathbf{p}'_2$ relative to their CM-frame,

$$\mathbf{p}'_1 + \mathbf{p}'_2 = 0$$

Since $\mathbf{p}'_1 = -\mathbf{p}'_2$ at any instant, the two particles move in opposite directions in their CM-frame and always have momenta of equal magnitudes. For the collisions shown in Figure 6-4, this implies that the initial momenta are related by

$$\mathbf{p}'_{1i} = -\mathbf{p}'_{2i}$$

and the final momenta by

$$\mathbf{p}'_{1f} = -\mathbf{p}'_{2f}$$

The usual laws of Newtonian mechanics can be used for the analysis of motion relative to the CM-frame only if this frame is an *inertial* frame of reference. In the last section of this chapter it will be shown that the CM-frame is inertial whenever it has a *constant* velocity $\mathbf{V}_c$ relative to an inertial frame (that is, whenever the resultant external force acting on the system is zero).

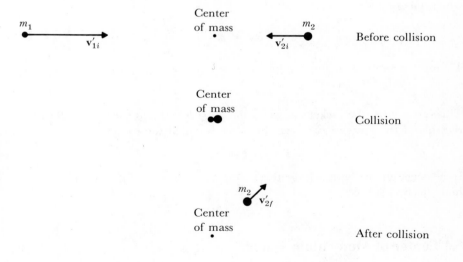

Figure 6-4 In the CM-frame the particles move in opposite directions and have momenta of equal magnitudes.

6.5 Frames in Relative Motion

Navigators, mechanical engineers, and physicists are often confronted with the problem of relating the velocity of an object relative to a frame S' to the velocity of the same object relative to another frame S, when the frame S' is moving relative to S with a velocity $\mathbf{V}_{S'}$. Figure 6-5 shows a particle's position vector $\mathbf{r}'$ relative to S' and $\mathbf{r}$ relative to S, with the position vector of S' relative to S denoted by $\mathbf{r}_{S'}$. From this figure it is seen that $\mathbf{r}$ is the *vector* sum of $\mathbf{r}_{S'}$ and $\mathbf{r}'$:

$$\mathbf{r} = \mathbf{r}_{S'} + \mathbf{r}'$$

Differentiating both sides of this equation, we have

$$\frac{d\mathbf{r}}{dt} = \frac{d\mathbf{r}_{S'}}{dt} + \frac{d\mathbf{r}'}{dt}$$

or

$$\mathbf{v} = \mathbf{V}_{S'} + \mathbf{v}' \qquad (6.10)$$

where $\mathbf{v}$ is the particle's velocity relative to S and $\mathbf{v}'$ is the particle's velocity relative to S'. This relationship is shown in the "triangle of velocities" used for the graphical solution of problems in navigation (Figure 6-6). In analytical

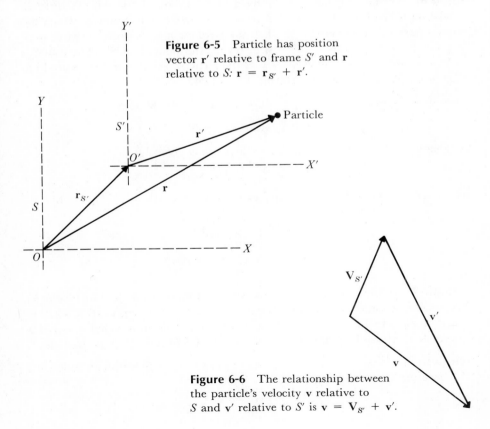

Figure 6-5 Particle has position vector $\mathbf{r}'$ relative to frame S' and $\mathbf{r}$ relative to S: $\mathbf{r} = \mathbf{r}_{S'} + \mathbf{r}'$.

Figure 6-6 The relationship between the particle's velocity $\mathbf{v}$ relative to S and $\mathbf{v}'$ relative to S' is $\mathbf{v} = \mathbf{V}_{S'} + \mathbf{v}'$.

work, for motion parallel to the XY plane, we often find it convenient to use equations for the components:

$$v_x = V_{S'x} + v'_x \tag{6.11}$$
$$v_y = V_{S'y} + v'_y$$

Differentiation of Eq. 6.10 yields the relationship between the particle's acceleration $\mathbf{a}'$ relative to S' and $\mathbf{a}$ relative to S:

$$\mathbf{a} = \mathbf{a}_{S'} + \mathbf{a}' \tag{6.12}$$

where $\mathbf{a}_{S'}$ is the acceleration of frame S' relative to S.

The equations of this section are called *transformation* equations because they show how a description of motion in one frame of reference can be transformed into a description of motion relative to another frame.

Example 3 An airplane's airspeed indicator reads 100 m/s and its compass gives its heading as 37° east of north. The meteorological information provided to the navigator is that the wind velocity is 20 m/s due east. Find the velocity of the airplane relative to the ground.

Solution Frame S is chosen to be fixed in the ground with axis OX directed east and OY directed north. The air is at rest relative to frame S' which has axes parallel to the axes of S. Then the airplane has a velocity relative to S' given by $\mathbf{v}' = 100$ m/s, 37° east of north; the velocity of S' relative to S is $\mathbf{V}_{S'} = 20$ m/s east; the velocity $\mathbf{v}$ of the airplane relative to the ground has components v_x and v_y, given from Eq. 6.11 by

$$v_x = (20 \text{ m/s}) + (100 \text{ m/s}) \cos 53° = 80 \text{ m/s}$$
$$v_y = 0 + (100 \text{ m/s}) \cos 37° = 80 \text{ m/s}$$

Therefore the ground speed of the airplane is given by

$$v = \sqrt{(80 \text{ m/s})^2 + (80 \text{ m/s})^2} = 113 \text{ m/s}$$

The airplane's velocity vector $\mathbf{v}$ makes an angle θ north of east given by

$$\tan \theta = \frac{v_y}{v_x} = \frac{80}{80} = 1$$
$$\theta = 45°$$

Example 4 In the laboratory, we observe that a mass m_1 moving with velocity $\mathbf{v}_{1i}$ strikes a stationary mass m_2 (Figure 6-7). The two particles stick together and move off with a common velocity. (Such a collision is called a *completely inelastic collision.*) Describe the motion in the CM-frame of these particles.

Solution The velocity $\mathbf{V}_c$ of the center of mass relative to a frame fixed in the laboratory (L-frame) is given by

$$\mathbf{V}_c = \frac{\mathbf{P}}{m_1 + m_2} = \frac{m_1\mathbf{v}_{1i}}{m_1 + m_2}$$

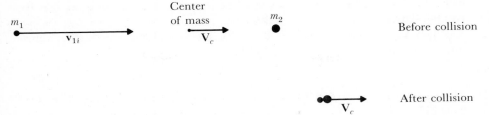

Figure 6-7 The completely inelastic collision of Example 4, viewed from the laboratory frame.

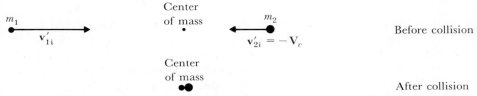

Figure 6-8 The completely inelastic collision of Example 4, viewed from the CM-frame.

The law of conservation of momentum implies that $\mathbf{P}$ and consequently $\mathbf{V}_c$ will not be changed by the collision. After collision the common velocity of the particles is $\mathbf{V}_c$ relative to the L-frame. Relative to the CM-frame the particles are at rest after collision (Figure 6-8). Before collision m_1 has a velocity relative to the CM-frame which, from Eq. 6.10, is given by

$$\mathbf{v}'_{1i} = \mathbf{v}_{1i} - \mathbf{V}_c = \frac{m_2}{m_1 + m_2}\mathbf{v}_{1i}$$

Similarly the velocity relative to the CM-frame of m_2 before collision is

$$\mathbf{v}'_{2i} = 0 - \mathbf{V}_c = \frac{-m_1}{m_1 + m_2}\mathbf{v}_{1i}$$

In the CM-frame, the particles first approach each other with momentum vectors of equal magnitudes and then meet and remain at the origin.

6.6 Principle of Relativity for Newtonian Mechanics

Suppose that S is a frame fixed in a laboratory and is therefore an *inertial* frame (at least to an approximation sufficient for most analyses). Relative to an *inertial frame of reference,* a force-free object moves with a constant velocity. The velocity transformation equation, $\mathbf{v} = \mathbf{V}_{S'} + \mathbf{v}'$, shows that if an object has a constant velocity $\mathbf{v}$ relative to S, then the velocity $\mathbf{v}'$ of this object relative to S' will also be constant, provided $\mathbf{V}_{S'}$ is constant. The conclusion is that S' constitutes another inertial frame of reference as long as its velocity $\mathbf{V}_{S'}$ remains constant. In fact we now see that, given one inertial frame, we can find an infinity of other inertial frames, namely, all frames of reference which have a constant velocity vector relative to the given inertial frame.

When $\mathbf{V}_{S'}$ is constant, $\mathbf{a}_{S'}$ ($= d\mathbf{V}_{S'}/dt$) is zero and the acceleration transformation equation, $\mathbf{a} = \mathbf{a}_{S'} + \mathbf{a}'$, reduces to

$$\mathbf{a} = \mathbf{a}' \tag{6.13}$$

The *acceleration* of a particle is the same relative to all inertial frames. Therefore Newton's second law for motion relative to frame S, $\mathbf{F} = m\mathbf{a}$, may be transformed to $\mathbf{F} = m\mathbf{a}'$. This shows that Newton's second law and all its consequences will hold true when S' is used as the frame of reference. This discovery is emphasized by formally stating the *principle of relativity for Newtonian mechanics: The laws of mechanics are the same in all inertial frames of reference.*

Common experience accords with the relativity principle. Any mechanics experiment has the same outcome when it is performed in a bus traveling smoothly at 80 km/h in a straight line as it has when the experiment is repeated with the bus parked on the highway. We are not conscious of motion with constant speed in a straight line and become aware that there is something different about a bus as a frame of reference only when it acquires an acceleration by going over a bump, speeding up, slowing down, or rounding a turn.

This relativity principle implies that all inertial frames are on an equal footing, at least as far as mechanics is concerned. *There is no single preferred frame of reference.* Nature does *not* seem to single out one particular "absolute frame" that is "really at rest" and distinguish this frame from other inertial frames because they are moving relative to this "absolute frame."

According to the relativity principle, mechanical laws such as the law of conservation of momentum will be valid for motion relative to the CM-frame, provided this is an inertial frame, that is, provided $\mathbf{V}_c$ is constant. This will be the case whenever the vector sum of the external forces acting on the system is zero.

Summary

☐   The momentum of a particle of mass m moving with velocity $\mathbf{v}$ is

$$\mathbf{p} = m\mathbf{v}$$

☐   Newton's second law is: *The time rate of change of momentum of a particle is equal to the resultant force acting on the particle,* that is

$$\mathbf{F} = \frac{d\mathbf{p}}{dt}$$

☐   The total momentum $\mathbf{P}$ of a system of particles is the vector sum of the momentum vectors of the individual particles: $\mathbf{P} = \mathbf{p}_1 + \mathbf{p}_2 + \cdots + \mathbf{p}_n = M\mathbf{V}_c$.

$$\mathbf{F}_{ext} = \frac{d\mathbf{P}}{dt}$$

where $\mathbf{F}_{ext}$ is the vector sum of the external forces acting on the particles of the system.

☐ The law of conservation of momentum states:

if $\mathbf{F}_{ext} = 0$, then $\mathbf{P} = $ **constant vector**

This implies that, for any process occurring within the system,

$$\mathbf{P}_{initial} = \mathbf{P}_{final}$$

☐ A frame of reference in which a system's total momentum $\mathbf{P}'$ is zero is called the center-of-momentum frame (CM-frame) for the system. Relative to this frame,

$$\mathbf{V}'_c = 0 \qquad \text{and} \qquad \mathbf{P}' = 0$$

☐ If a particle has a velocity $\mathbf{v}$ relative to frame S and a velocity $\mathbf{v}'$ relative to frame S', then

$$\mathbf{v} = \mathbf{V}_{S'} + \mathbf{v}'$$

where $\mathbf{V}_{S'}$ is the velocity of frame S' relative to frame S.

☐ If S is an inertial frame and if $\mathbf{V}_{S'}$ is constant, then S' is also an inertial frame and a particle has the same acceleration relative to both frames. The principle of relativity for Newtonian mechanics states that the laws of mechanics are the same in all inertial frames of reference.

Questions

1   A system consists of three particles: a 5.0-kg mass moving east at 2.0 m/s, a 3.0-kg mass moving east at 10 m/s, and a 2.0-kg mass moving west at 8.0 m/s:
   (a) Find the magnitude and direction of the total momentum vector $\mathbf{P}$ of the system.
   (b) Find the magnitude and direction of the velocity of the center of mass of this system.

2   **(a)** From $\mathbf{P} = M\mathbf{V}_c$ and $\mathbf{F}_{ext} = d\mathbf{P}/dt$, deduce the law of motion of the center of mass, $\mathbf{F}_{ext} = M\mathbf{a}_c$.
   (b) From $\mathbf{P} = M\mathbf{V}_c$ and $\mathbf{F}_{ext} = M\mathbf{a}_c$, deduce that $\mathbf{F}_{ext} = d\mathbf{P}/dt$.

3   A 2.0-kg piece of putty travels east at 4.0 m/s and collides with a 3.0-kg piece of putty traveling west at 5.0 m/s. After collision they stick together. Find the magnitude and direction of their velocity vector after collision.

4   A 0.010-kg bullet traveling 1000 m/s east hits a stationary 100-kg block, passes through undeviated, but emerges with a velocity reduced to 200 m/s. Find the block's momentum and velocity as the bullet emerges.

5   If, in the preceding question, the bullet bounces off the block and retraces its original path with a speed of 200 m/s, what is the block's velocity?

6   You are standing at rest at point A on a perfectly smooth sheet of ice armed with a revolver containing two bullets. How can you get from A to a distant point B and stop at B? Explain.

7   A white cue ball strikes a red ball head on and stops dead. Prove that if the two balls have the same mass, the red ball acquires the velocity that the cue ball had just before impact.

8   An 80-kg fullback running north at 10 m/s is tackled by a 100-kg lineman moving south at 4.0 m/s. Find the speed and direction of these players after the tackle.

9   A 30-kg child dives from the stern of a 45-kg boat with a horizontal component of velocity of 3.0 m/s north. Initially the boat was at rest. Find the magnitude and direction of the velocity acquired by the boat.

10   A 2.70×10^3-kg Cadillac, traveling east at 3.00 m/s, collides with a 900-kg Volkswagen traveling north at 30.0 m/s. They interlock and move off together. Find the magnitude and direction of their common velocity.

11   A 100-kg teacher and an 80-kg student are at rest at opposite ends of a 30-m rope on a sheet of ice. They each pull on the rope. How far from the original position of the teacher will they be when they meet?

12   During a burn, a rocket ejects gases at the rate of 150 kg/s and at a speed of 2.0×10^3 m/s relative to the rocket:
  (a) What is the thrust of this rocket?
  (b) In a region where the gravitational field is negligible, what is the acceleration of this rocket at the instant when the mass of the rocket and its remaining fuel is 60×10^3 kg?

13   A uniform canoe of mass M has seats that are a distance s apart. The canoe is at rest with a woman of mass m seated in the bow. She then moves to the stern and sits down. Assuming that the horizontal component of the force exerted on the canoe by the water is negligible, find the displacement of the canoe.

14   While at rest, a 6.0-kg package explodes into three fragments. A 1.0-kg fragment moves west at 40 m/s and a 2.0-kg fragment moves south at 15 m/s. Find the velocity of the remaining fragment.

15   A neutron of mass M traveling with a speed of $\frac{1}{30}c$ strikes a stationary uranium nucleus of mass $235M$. The neutron is captured by the uranium nucleus and they move off together as a compound nucleus:
  (a) What is the velocity of this compound nucleus?
  (b) While moving at this velocity, the compound nucleus decays into two fragments of masses $128M$ and $108M$. The heavier fragment travels with a velocity of $\frac{1}{40}c$ in a direction perpendicular to that of the incident neutron. Find the magnitude and direction of the velocity of the lighter fission fragment. (c denotes the speed of light in a vacuum.)

16   For which of the preceding questions is $V_c = 0$ so that the CM-frame and the L-frame coincide?

17   **(a)** A river flows east at a speed of 2.0 m/s. A man rows a boat east at a speed of 3.0 m/s relative to the water. Find the velocity vector of the boat relative to the land. (Consider a frame of reference S' moving with the water in the river at a speed $V = 2.0$ m/s relative to a frame S fixed on the land.)
  (b) In the preceding situation, the man turns the boat around and rows west at the same speed relative to the water. Find the velocity vector of the boat relative to the land.

18   The stairs of an escalator extend 15.0 m up an incline. The escalator moves upward at a speed of 1.5 m/s. A boy walks with a speed relative to the escalator of 2.0 m/s.

Find how long the boy will be on the escalator when:

(a) He walks up the escalator.

(b) He walks down the escalator.

19 An airplane flies east with an airspeed of 100 m/s (airspeed is speed relative to a frame S' floating along with the air), and the wind velocity is 20 m/s east (the wind velocity is the velocity of the frame S' relative to earth). Find the ground speed of the aircraft (ground speed is the speed relative to earth).

20 Where is the "point of no return" for an aircraft with a flying time of 3.0 h, when the aircraft is heading east with an airspeed of 100 m/s and the wind velocity is 20 m/s east?

21 A boat moves at a constant speed relative to the water in a river which flows at a constant velocity relative to the land (a constant current). On a trip between two towns which are 60 km apart, the boat takes 4.0 h to travel upstream and 3.0 h to travel downstream. Find the current and the speed of the boat relative to the water.

22 A bus travels north at a speed of 20 m/s. Inside the bus a ball is rolled straight across the bus from the west side to the east side, traveling 3.0 m in 0.30 s. Find the magnitude and direction of the ball's velocity vector relative to the bus and also relative to the ground.

23 An aircraft heading 30° north of east has an airspeed of 90 m/s. The wind velocity is 30 m/s east. Find the magnitude and direction of the aircraft velocity relative to the ground.

24 A river which is 60 m wide flows east at 2.0 m/s. A motorboat starts from the south bank traveling at a speed of 4.0 m/s relative to the river:

(a) If the boat is heading 30° north of east, find the magnitude and direction of its velocity relative to the land. Where and when will the boat reach the opposite bank?

(b) In what direction should the boat be headed in order to cross in the shortest possible time? What is this time?

(c) If the boat reaches a point on the opposite bank due north of its starting point, in what direction was the boat heading as it crossed the river?

25 A car travels at 30 m/s in a direction 36.9° north of east. Find the magnitude and direction of the car's velocity vector relative to a bus traveling north at 20 m/s.

26 A bus, heading east, speeds up with an acceleration of 2.0 m/s². A baseball is pushed forward in the bus with an acceleration of 40 m/s² relative to the bus. What is the acceleration of the baseball relative to the ground?

27 The bus of Question 26 stops accelerating and travels with a constant velocity of 30 m/s. The baseball is again given an acceleration relative to the bus of 40 m/s². What is the acceleration of the baseball relative to the ground?

28 While a bus is traveling at a constant velocity of 30 m/s, an ice cube slides along the bus floor in a straight line with constant speed relative to the bus. Will the motion of the ice cube, as viewed from the highway, be motion in a straight line with constant speed? Will this speed be different from the speed relative to the bus? What is the acceleration of the ice cube relative to the bus? To the highway?

29 On a certain trip, a car:

(a) Speeds up.

(b) Travels with constant speed in a straight line.

(c) Goes around a turn.

(d) Jams on the brakes.

For each portion of the trip, decide whether or not the car is an inertial frame of reference, assuming that the highway is an inertial frame.

Supplementary Questions

S-1 **(a)** A train 200 m long takes 100 s to pass another train 800 m long which is traveling in the same direction. What is the velocity of the shorter train relative to the longer train?

(b) When these two trains are traveling with the same speeds as in part (a) but in opposite directions, it takes 10.0 s for them to pass. Find the speed of each train relative to the ground.

S-2 A fisherman rowing upstream passes a floating bottle and keeps on rowing upstream for 15 min. He then turns around and rows downstream, always maintaining the same speed relative to the water. When he overtakes the bottle he finds it has drifted 2 km downstream. What is the current in the river?

S-3 Rain is falling vertically. When a car is traveling at 10 m/s, the tracks made by the raindrops on the vertical side windows are inclined at an angle of 60° with the vertical. Find the speed of a raindrop relative to the car and also relative to the ground.

S-4 Two ships move toward each other on intersecting lines at speeds which will cause them to collide. Examine this situation from a frame of reference fixed on one of these ships. Explain how observers on this ship can tell that they are on a collision course by successive observations of the bearing of the other ship.

S-5 A ship travels north at 4.0 m/s. The wind velocity is 3.0 m/s east. A pennant flies from the masthead. In what direction does this pennant point?

S-6 In Question 7, the white cue ball is incident with a velocity $\mathbf{v}_{1i}$ relative to the L-frame. Find the velocity of the center of mass of the system of the two balls and describe the collision in the CM-frame, giving the velocities relative to this frame of each ball before and after collision.

S-7 A girl of mass m is standing on a toboggan of mass M which is moving with constant velocity $\mathbf{V}_i$ as it slides along the ice on the surface of a lake. The girl then runs along the toboggan in the direction opposite to $\mathbf{V}_i$ and acquires a speed v *relative to the toboggan* as she jumps off the rear of the toboggan. Show that the final speed of the toboggan relative to the ice is

$$V_f = V_i + \frac{mv}{M + m}$$

S-8 Assume that the toboggan of the preceding question is initially at rest. Show that if the toboggan carries two girls, each of mass m, and if each girl runs and jumps off in succession with a speed v relative to the toboggan, the final speed of the toboggan is

$$V_f = \left(\frac{m}{M + 2m} + \frac{m}{M + m} \right) v$$

Show that this speed is greater than the speed the toboggan would acquire if the two girls ran and jumped off simultaneously (again with a speed v relative to the toboggan).

7.

integration of the equations of motion

The mathematical process of integration is often required for the formulation of physical theories and the solution of practical problems. Probably the best topic in physics for a first encounter with integrals is the problem of determining the position coordinate x as a function of time t when the velocity component v_x is a known function of time. A mathematically identical problem is the determination of v_x when the acceleration a_x is a known function. Both are fundamental in mechanics because, when the resultant force component F_x is a known function, Newton's second law gives what is called the equation of motion, $a_x = F_x/m$. To describe the motion, v_x and x must be determined from the acceleration function specified in the equation of motion.

In this chapter we show how integrals arise in these problems. Solutions are found for the simple case when the acceleration is constant, and these solutions are used to discuss the motion of projectiles.

7.1 One-Dimensional Motion, Definitions

Velocity

Many interesting motions take place along a straight line, which then can be chosen as the direction of a coordinate axis, say the X axis, and the entire discussion of the motion can be given in terms of the x components, x, Δx, $\bar{v}_x$, and v_x of the vectors $\mathbf{r}$, $\Delta\mathbf{r}$, $\bar{\mathbf{v}}$, and $\mathbf{v}$, respectively.

The direction of each of these vectors is determined by the sign of its x component in the usual way. A vector points in the positive or negative x direction according as its x component is a positive or a negative number. In the usual terminology for a one-dimensional motion, the words "x component of" are omitted and x is called the position coordinate, v_x is referred to briefly as the velocity, $\bar{v}_x$ as the average velocity, and Δx as the displacement.

From the definitions given in Chapter 1, $\bar{\mathbf{v}} = \Delta\mathbf{r}/\Delta t$ and $\mathbf{v} = d\mathbf{r}/dt$, it follows that

$$\bar{v}_x = \frac{\Delta x}{\Delta t} \tag{7.1}$$

and

$$v_x = \lim_{\Delta t \to 0} \frac{\Delta x}{\Delta t} = \frac{dx}{dt} \tag{7.2}$$

The motion of a particle along the X axis is described by giving x as a function of time. Then the derivative of this function gives the velocity as a function of time. A graphical determination of the velocity at any instant can be made by plotting a graph of x versus t and measuring the slope dx/dt at the point on this graph corresponding to the instant in question (Figure 7-1).

Acceleration

When the motion takes place along the X axis, the acceleration vector is completely determined by its component a_x, which then often is called simply "the acceleration."

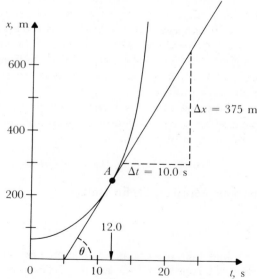

Figure 7-1   At 12.0 s, *velocity = slope at point A =* 375 m/10.0 s = 37.5 m/s. (Notice that the slope is *not* equal to tan θ. The value of θ is not significant because it depends on the scale used for the different physical quantities plotted as abscissa and ordinate.)

The direction of **a** is in the positive or negative x direction according as a_x is a positive or a negative number. The definition of **a** implies that

$$a_x = \lim_{\Delta t \to 0} \frac{\Delta v_x}{\Delta t} = \frac{dv_x}{dt} \tag{7.3}$$

and, since $v_x = dx/dt$, we have

$$a_x = \frac{d^2x}{dt^2} \tag{7.4}$$

Another useful expression for the acceleration can be obtained by using the chain rule:

$$a_x = \frac{dv_x}{dt} = \frac{dv_x}{dx}\frac{dx}{dt}$$

which gives

$$a_x = v_x \frac{dv_x}{dx} \tag{7.5}$$

Notice that the variable t does not appear in this formula.

When the velocity is a known function of time, the acceleration can be found by taking the derivative dv_x/dt of this function. A graphical determination of the acceleration at any instant can be effected by plotting a graph of v_x versus t and finding the slope dv_x/dt of this graph at the point corresponding to the instant in question.

Example 1   A motion is described in SI units by

$$x = 10 + 20t + 15t^3$$

Find the velocity and the acceleration as functions of time.

Solution

$$v_x = \frac{dx}{dt} = 20 + 45t^2$$

$$a_x = \frac{dv_x}{dt} = 90t$$

(The numbers in each equation have the physical dimensions required to make the expressions dimensionally correct. For example

$$x = (10 \text{ m}) + (20 \text{ m/s})t + (15 \text{ m/s}^3)t^3$$

In such expressions involving literal symbols and numerical values, we omit the units when no misinterpretation can arise.)

7.2 Finding the Coordinate by Integration

The value of the velocity at the instant t is denoted by $v_x(t)$ using the conventional notation for functions. Suppose that $v_x(t)$ is a known function of the time t. Then the function $x(t)$ which gives the position coordinate at any instant t can be found by an extension of the process discussed in Chapter 1. For any short time interval Δt, the displacement is given approximately by

$$\Delta x = v_x \Delta t \qquad \text{(approximately)} \qquad (7.6)$$

If a sizable time interval $t - t_0$ is divided into n short time intervals, each of duration $\Delta t'$, then the displacement is *approximately*

$$x(t) - x(t_0) = v_x(t_0')\Delta t' + v_x(t_1')\Delta t' + \cdots + v_x(t_{n-1}')\Delta t'$$

$$= \sum_{i=0}^{n-1} v_x(t_i')\Delta t' \qquad (7.7)$$

where t_0' is the instant at the beginning of the first time interval, t_1' is the instant at the beginning of the second time interval, and so on. From this approximation, an exact expression will be obtained by evaluating the limit approached by the approximating sum as the number n of intervals between t_0 and t increases indefinitely while the lengths of the individual time intervals approach zero. The limit of this approximating sum is denoted by $\int_{t_0}^{t} v_x(t')\,dt'$ and is called the definite integral of the function v_x between t_0 and t. We have

$$x(t) - x(t_0) = \int_{t_0}^{t} v_x(t')\,dt' \qquad (7.8)$$

(The integration sign $\int$ is a distorted S suggesting a *sum.*)*

Computation of the approximating sum displayed in Eq. 7.7 gives an approximate evaluation of the integral $\int_{t_0}^{t} v_x(t')\,dt'$. By subdividing the interval $t - t_0$ into many small time intervals, we can make a very good approximation indeed. This method involves a large number of routine calculations, but these can be handled easily by modern computers.

Another method of evaluating the integral $\int_{t_0}^{t} v_x(t')\,dt'$ consists of measuring an area on a velocity-time graph. This method is based on the geometric interpretation of a definite integral. On a graph showing the velocity as a function of time, this integral is equal to the *area* shown in Figure 7-2 under the curve between t_0 and t.† We can understand this by first considering the geometric interpretation of the individual terms in the sum in Eq. 7.7. A term such as $v_x(t_1')\Delta t'$ is equal to the area of a rectangle of height $v_x(t_1')$ and width $\Delta t'$. The sum of all such thin rectangles approximates the area under the curve in the velocity-time graph. The approximation generally improves as the number of subdivisions increases and the width of each narrows. And the limit of the sum of all these rectangular areas is the area under the smooth curve.

---

*We have denoted the "dummy variable" of integration by t' to distinguish it from the particular value t that is assumed by the integration variable at the upper limit.
†On the velocity-time graph the "area" must be measured in the unit of the product, velocity × time, which is a length unit.

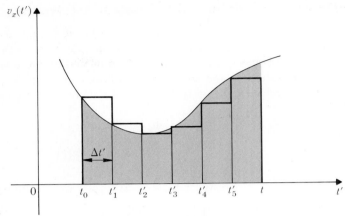

Figure 7-2 The integral $\int_{t_0}^{t} v_x(t')\,dt'$ is equal to the shaded area under the graph showing $v_x(t')$ as a function of t'. The approximating sum, obtained when $t - t_0$ is divided into six equal time intervals, is equal to the sum of the areas of the six rectangles.

The mathematical relationship between acceleration and velocity is the same as the relationship between velocity and coordinate. Consequently the change in velocity in the time interval $t - t_0$ is given by

$$v_x(t) - v_x(t_0) = \int_{t_0}^{t} a_x(t')\,dt' \qquad (7.9)$$

and this definite integral is equal to the area between an acceleration-time graph and the time axis, bounded by vertical lines at the beginning, $t' = t_0$, and at the end, $t' = t$, of the time interval $t - t_0$.

Average Velocity

In the time interval $t - t_0$ the displacement is $x(t) - x(t_0)$. The average velocity for this time interval (from Eq. 7.1) is given by

$$\bar{v}_x = \frac{x(t) - x(t_0)}{t - t_0}$$

Expressing the displacement as the integral of the velocity (Eq. 7.8), we have

$$\bar{v}_x = \frac{\int_{t_0}^{t} v_x(t')\,dt'}{t - t_0} \qquad (7.10)$$

The average value $\bar{f}$ of any function $f(u)$ in an interval $\Delta u = u - u_0$ is defined by an expression analogous to Eq. 7.10:

$$\bar{f}\Delta u = \int_{u_0}^{u} f(u')\,du' \qquad (7.11)$$

7.3 One-Dimensional Motion with Constant Acceleration

When the resultant force F_x acting on a particle is *constant*, the particle experiences a *constant acceleration*, $a_x = F_x/m$. In this special case it is a rather simple matter to determine the velocity $v_x(t)$ and the position coordinate $x(t)$ in terms of the *constant* acceleration a_x and the initial conditions, the values x_0 and v_{0x} of the position and velocity at $t = 0$. Equation 7.9 with $t_0 = 0$ gives

$$v_x(t) - v_{0x} = \int_0^t a_x\, dt' = a_x t$$

Therefore

$$v_x(t) = v_{0x} + a_x t \qquad (a_x \text{ constant}) \quad (7.12)$$

Using this velocity function in Eq. 7.8, we find

$$x(t) - x_0 = \int_0^t (v_{0x} + a_x t')\, dt' = v_{0x} t + \tfrac{1}{2} a_x t^2$$

Therefore

$$x(t) = x_0 + v_{0x} t + \tfrac{1}{2} a_x t^2 \qquad (a_x \text{ constant}) \quad (7.13)$$

The motion is described by this function. Notice that, for the particular case of constant acceleration, we have verified the assertion made in Chapter 1 that the motion is completely determined by a specification of the acceleration a_x and the initial conditions x_0 and v_{0x}. Given these data, we can now find x and v_x for any value of t.

It is sometimes useful to be able to relate the velocity to position without explicit reference to the time. The time does not appear in Eq. 7.5:

$$v_x \frac{dv_x}{dx} = a_x$$

This equation can be expressed in differential form as

$$v_x\, dv_x = a_x\, dx$$

Integrating both sides and using corresponding quantities for the limits of the definite integrals, we obtain

$$\int_{v_{0x}}^{v_x} v_x'\, dv_x' = \int_{x_0}^{x} a_x\, dx'$$

This gives

$$\tfrac{1}{2} v_x^2 - \tfrac{1}{2} v_{0x}^2 = a_x (x - x_0)$$

a result which is usually written as

$$v_x^2 = v_{0x}^2 + 2 a_x (x - x_0) \qquad (a_x \text{ constant}) \quad (7.14)$$

The time has been eliminated in this relationship involving position, velocity, and acceleration.

The average velocity for this motion can be evaluated from Eq. 7.10:

$$\bar{v}_x = \frac{\int_0^t (v_{0x} + a_x t') \, dt'}{t}$$

$$= v_{0x} + \tfrac{1}{2} a_x t$$

$$= \tfrac{1}{2}(v_{0x} + v_{0x} + a_x t)$$

$$= \tfrac{1}{2}(v_{0x} + v_x) \qquad\qquad (a_x \text{ constant}) \quad (7.15)$$

which states that the average velocity is one-half the sum of the initial and final velocities for motion with constant acceleration. When this result is used in conjunction with the definition of the average velocity, $\bar{v}_x = \Delta x/\Delta t = (x - x_0)/t$, we obtain

$$x - x_0 = \tfrac{1}{2}(v_{0x} + v_x)t \qquad\qquad (a_x \text{ constant}) \quad (7.16)$$

a relationship in which the acceleration does not appear.

Example 2 While traveling at a speed of 30.0 m/s we apply the brakes of a car in such a way that the car has a constant deceleration of 3.00 m/s²:
(a) Find the time required for the car to be brought to rest after the application of the brakes.
(b) How far does the car travel after the brakes are applied?

Solution We select the axis OX fixed on the road with the origin at the point where the brakes are applied and the positive direction in the direction of the car's velocity. Let $t = 0$ at the instant the brakes are applied. Then the initial conditions are $x_0 = 0$ and $v_{0x} = 30.0$ m/s. While the car is slowing down, the acceleration is in the direction *opposite* to that of the velocity. The velocity is in the positive x direction. Therefore the acceleration is in the negative x direction and $a_x = -3.00$ m/s²:
(a) At the instant t when the car is brought to rest, its velocity v_x is zero. Therefore the equation

$$v_x = v_{0x} + a_x t$$

gives

$$0 = v_{0x} + a_x t$$

This yields

$$t = \frac{-v_{0x}}{a_x} = \frac{-30.0 \text{ m/s}}{-3.00 \text{ m/s}^2} = 10.0 \text{ s}$$

(b) The equation

$$x = x_0 + v_{0x}t + \tfrac{1}{2}a_x t^2$$

gives

$$x = 0 + (30.0 \text{ m/s}) (10.0 \text{ s}) + \tfrac{1}{2}(-3.00 \text{ m/s}^2) (10.0 \text{ s})^2 = 150 \text{ m}$$

This position coordinate can be found directly without first finding the elapsed time by using

$$v_x^2 = v_{0x}^2 + 2a_x(x - x_0)$$

which gives

$$0 = v_{0x}^2 + 2a_x x$$

Therefore

$$x = \frac{-v_{0x}^2}{2a_x} = \frac{-(30.0 \text{ m/s})^2}{2(-3.00 \text{ m/s}^2)} = 150 \text{ m}$$

7.4 Projectiles near the Earth

We consider the motion of a baseball or of any projectile whose trajectory is not more than a few kilometres long. This situation is relatively simple because the gravitational field **g** does not change appreciably in magnitude over the entire region of interest (Section 2.6). This is so because the percentage change in distance from the center of the earth is small (0.1% for a 6 km change in altitude). Over such a limited region we can consider the earth's surface as flat, with **g** a constant vector pointing vertically downward.

Suppose that a particle of mass m is projected with an initial velocity $\mathbf{v_0}$. We choose a coordinate system fixed in the earth with the origin at the initial position and with the axis OX horizontal and OY directed vertically upward (Figure 7-3). The initial position is then given by $x_0 = 0, y_0 = 0$, and the initial velocity $\mathbf{v_0}$ has components v_{0x}, v_{0y}.

Newton's second law, $\mathbf{F} = m\mathbf{a}$, is equivalent to the two equations

$$F_x = ma_x \quad \text{and} \quad F_y = ma_y$$

The gravitational field **g** at the location of the particle exerts a force $m\mathbf{g}$ on the particle (Section 2.6). We assume that the particle is falling freely, that is, the only force acting on the particle is the gravitational force. With $\mathbf{F} = m\mathbf{g}$ we have

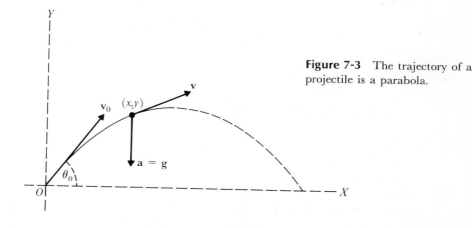

Figure 7-3 The trajectory of a projectile is a parabola.

$F_x = 0$ and $F_y = -mg$ where g is the positive number 9.8 m/s². Therefore Newton's second law yields

$$a_x = 0 \quad \text{and} \quad a_y = -g$$

The horizontal motion is evidently extremely simple. Since $a_x = 0$, v_x has the constant value v_{0x} and, at the instant t, the x coordinate of the particle's position is given by

$$x = v_{0x}t$$

As the horizontal motion continues, there is also vertical motion with a changing vertical component of velocity v_y, but with a constant vertical component of acceleration, $a_y = -g$. Because this acceleration is constant, we can use equations for motion parallel to the Y axis analogous to those that have been derived for motion along the X axis in Section 7.3. At the instant t,

$$v_y = v_{0y} - gt$$
$$y = v_{0y}t - \tfrac{1}{2}gt^2$$

From these results we can find the velocity and position at any instant and, by plotting the positions attained at various instants, we can trace out the trajectory. It turns out to be a parabola. (This can be seen from the equations for x and y, either by recognizing these equations as parametric equations of a parabola in terms of the parameter t, or by eliminating t between these equations and showing that y is a quadratic function of x.)

Example 3   A man leans out a window and throws a ball vertically upward with a speed of 29.4 m/s:
(*a*) What is the maximum height attained by the ball?
(*b*) How long does the ball take to reach maximum height?
(*c*) Where is the ball after 7.0 s have elapsed?
(*d*) When is the ball 24.5 m above the man?

Solution   We use the axes in Figure 7-3. With the origin at the initial position, the initial conditions are $y_0 = 0$ and $v_{0y} = 29.4$ m/s. Throughout the motion, the acceleration is directed downward and has a magnitude of 9.8 m/s². Therefore $a_y = -9.8$ m/s²:
(*a*) At maximum height, where the ball reverses the direction of its motion from straight up to straight down, the velocity component v_y must be zero. Therefore

$$0 = v_{0y}^2 + 2a_y y$$

$$y = \frac{-v_{0y}^2}{2a_y} = \frac{-(29.4 \text{ m/s})^2}{2(-9.8 \text{ m/s}^2)} = 44 \text{ m}$$

(*b*) From the equation

$$v_y = v_{0y} + a_y t$$

we obtain

$$t = \frac{v_y - v_{0y}}{a_y}$$

At maximum height, $v_y = 0$. Therefore

$$t = \frac{0 - 29.4 \text{ m/s}}{-9.8 \text{ m/s}^2} = 3.0 \text{ s}$$

(c) At $t = 7.0$ s, we have

$$y = 0 + (29.4 \text{ m/s})(7.0 \text{ s}) + \tfrac{1}{2}(-9.8 \text{ m/s}^2)(7.0 \text{ s})^2 = -34 \text{ m}$$

The ball is 34 m *below* its initial position.
(d) When $y = 24.5$ m, the time t satisfies

$$24.5 \text{ m} = (29.4 \text{ m/s})t - (4.9 \text{ m/s}^2)t^2$$

or

$$(4.9 \text{ m/s}^2)t^2 - (29.4 \text{ m/s})t + (24.5 \text{ m}) = 0$$

There are two solutions to this quadratic equation: $t = 1.0$ s and $t = 5.0$ s. The ball passes through the point $y = 24.5$ m on its way up at $t = 1.0$ s and on its way down at $t = 5.0$ s (Figure 7-4).

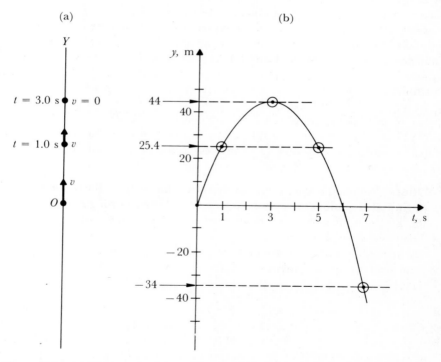

(a)   (b)

Figure 7-4   (a) Motion along Y axis with constant acceleration $a_y = -9.8 \text{ m/s}^2$. Position and the corresponding velocity are shown at 0.0 s, 1.0 s, and 3.0 s. (b) Graph showing y as a function of t.

Example 4   A stone is projected from the surface of a flat field with a speed of 20.0 m/s at an angle 53.1° above the horizontal (in Figure 7-3, $\theta_0 = 53.1°$). When and where will the stone hit the earth?

Solution   With the coordinate system of Figure 7-3, components of the initial velocity are

$$v_{0x} = v_0 \cos \theta_0 = (20.0 \text{ m/s})(0.600) = 12.0 \text{ m/s}$$

$$v_{0y} = v_0 \sin \theta_0 = (20.0 \text{ m/s})(0.800) = 16.0 \text{ m/s}$$

At the instant t when the stone hits the earth, $y = 0$. This occurs for a time t such that

$$0 = v_{0y}t - \tfrac{1}{2}gt^2$$

Therefore

$$t = \frac{2v_{0y}}{g} = \frac{2(16.0 \text{ m/s})}{9.8 \text{ m/s}^2} = 3.27 \text{ s}$$

In this time the stone travels a horizontal distance

$$x = v_{0x}t = (12.0 \text{ m/s})(3.27 \text{ s}) = 39.2 \text{ m}$$

Therefore the time of flight is 3.27 s and the range is 39.2 m.

Example 5   At what angle θ_0 should a stone be projected in order to achieve maximum range for a given projection speed v_0? What is the maximum range?

Solution   We can use the notation and the results of the preceding example. The time of flight is $t = 2v_{0y}/g$, so the range is

$$x = v_{0x}t = \frac{2v_{0x}v_{0y}}{g} = \frac{v_0^2(2 \sin \theta_0 \cos \theta_0)}{g} = \frac{v_0^2(\sin 2\theta_0)}{g}$$

The sine function attains the value 1, its maximum value, for an angle of 90°. Therefore, for maximum range $2\theta_0 = 90°$ or $\theta_0 = 45°$ and

$$x_{\text{max}} = \frac{v_0^2}{g}$$

7.5 Galileo's Experiments

Efforts to understand the motion of falling bodies loom large in the history of physics. Major contributions to this study were made by Galileo, who is widely regarded as the father of physics. Galileo discovered that when a ball started from rest and rolled down an inclined plane, the distance x traversed was proportional to the square of the elapsed time t, in agreement with Eq. 7.13, which for this case is simply

$$x = \tfrac{1}{2}a_x t^2$$

Galileo concluded that the ball moved with constant acceleration down the plane, and he inferred correctly that a falling body moves with constant acceleration. Comparing the motion of falling bodies of different weights, he showed that in free fall all bodies in the same region of space experience the same downward acceleration.

Two relevant and very significant attitudes adopted by Galileo deserve emphasis. First, although the acceleration of an object falling through the air depends upon the size and the shape of the object as well as upon the atmospheric conditions and even upon the precise location on the earth, Galileo realized that the *essential* point was a constant acceleration downward with deviations caused by perhaps quite complicated factors. His very instructive simple description of the motion was obtained only by ignoring these complications. Galileo thus demonstrated the great value of *idealizations* of reality. Second, we note that Galileo was not daunted by the fact that his experimental results were in conflict with Aristotle's widely accepted teaching that heavier bodies should fall faster. Such *acceptance of experimental results,* even though they disagree with the perhaps plausible ideas of the most renowned men, has been the crucial factor in the development of all areas of science since Galileo's time.

7.6 Impulse

Newton's second law for x components can be expressed in the form

$$F_x = \frac{dp_x}{dt}$$

where F_x is the x component of the resultant force acting on a particle with x component of momentum p_x. Integration of this equation leads to a new and useful relationship.

The equation in differential form is

$$F_x \, dt = dp_x$$

Integrating over a time interval from t_0 to t, we obtain

$$\int_{t_0}^{t} F_x(t') \, dt' = p_x(t) - p_x(t_0) \tag{7.17}$$

The integral is called the x component of the *impulse* of the force **F** in the time interval from t_0 to t. This integral is represented by the area under the force-time graph between t_0 and t (Figure 7-5). Equation 7.17 can be cast into the simple form

$$\overline{F}_x \Delta t = \Delta p_x \tag{7.18}$$

where $\overline{F}_x$ is the average value of $F_x(t)$ in the time interval $\Delta t = t - t_0$ and $\Delta p_x = p_x(t) - p_x(t_0)$.

The **impulse** of a force $\mathbf{F}$ in a time interval $t - t_0$ is the **vector** defined by

$$\mathbf{impulse} = \int_{t_0}^{t} \mathbf{F}(t') \, dt' \tag{7.19}$$

In a three-dimensional motion there are, for each component, equations (similar to Eq. 7.18) which together are equivalent to the vector equation

$$\overline{\mathbf{F}} \Delta t = \Delta \mathbf{p} \tag{7.20}$$

or

$$\int_{t_0}^{t} \mathbf{F}(t') \, dt' = \mathbf{p}(t) - \mathbf{p}(t_0) \tag{7.21}$$

which state that, for a specified time interval, the *impulse of the resultant force acting on a particle is equal to the change in the particle's momentum.* Impulse is a vector. The SI unit is the newton-second ($\mathbf{N \cdot s}$).

Figure 7-6 shows the force-time graph for an impulsive force, the type of force mentioned in Section 6.3 in connection with collisions or explosions. Because an impulsive force attains such a large magnitude, the impulse $\overline{F} \Delta t$ of

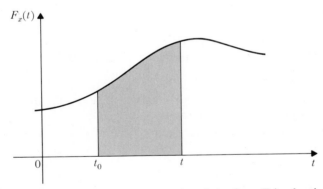

Figure 7-5 The x component of the impulse of the force $\mathbf{F}$ in the time interval from t_0 to t is represented by the shaded area under this force-time graph.

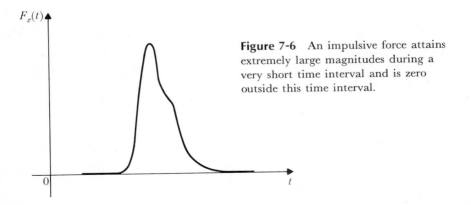

Figure 7-6 An impulsive force attains extremely large magnitudes during a very short time interval and is zero outside this time interval.

such a force can be appreciable even though the time interval Δt is extremely short. The average value of the resultant force acting on a particle during a collision of measured duration Δt can be calculated from Eq. 7.20 if the change in the particle's momentum is measured.

When the interaction of two particles is re-examined in terms of the impulses involved, it is easy to understand why their total momentum is not changed by their interaction. According to Newton's third law, the forces that two particles exert on each other are equal in magnitude and opposite in direction at every instant. The impulses of these forces in any time interval and the momentum changes associated with these impulses are therefore of equal magnitude but opposite direction. Consequently, the vector sum of these changes of momentum is zero.

Summary

☐   The x components of $\mathbf{r}$, $\mathbf{v}$, and $\mathbf{a}$ are x, $v_x = dx/dt$, $a_x = dv_x/dt = d^2x/dt^2$ $= v_x(dv_x/dx)$, respectively. When $x(t)$ is a known function of time, the functions $v_x(t)$ and $a_x(t)$ can be found by differentiation.

☐   A definite integral is a limit of a *sum*.

$$x(t) - x(t_0) = \int_{t_0}^{t} v_x(t') \, dt'$$

$$v_x(t) - v_x(t_0) = \int_{t_0}^{t} a_x(t') \, dt'$$

If $a_x(t)$ is a known function of time and $v_x(t_0)$ and $x(t_0)$ are known, then v_x and x can be found by integration.

☐   The average value $\bar{f}$ of $f(u)$ in the interval $\Delta u = u - u_0$ satisfies

$$\bar{f}\Delta u = \int_{u_0}^{u} f(u') \, du'$$

For example, the average velocity in a time interval $t - t_0$ is

$$\bar{v}_x = \frac{\int_{t_0}^{t} v_x(t') \, dt'}{t - t_0}$$

☐   The equations for motion with constant acceleration are displayed in Table 7-1.

Table 7-1

Equation	Quantity eliminated
$x = x_0 + v_{0x}t + \frac{1}{2}a_x t^2$	v_x
$v_x = v_{0x} + a_x t$	$x - x_0$
$v_x^2 = v_{0x}^2 + 2a_x(x - x_0)$	t
$x - x_0 = \frac{1}{2}(v_{0x} + v_x)t$	a_x

The significance of the *signs* of x, v_x, and a_x follows from their definitions as x components of the vectors $\mathbf{r}$, $\mathbf{v}$, and $\mathbf{a}$, respectively. A vector's x component is a positive number when the vector points in the positive x direction and a negative number when the vector points in the negative x direction.

☐ The **impulse** of a force $\mathbf{F}$ in a time interval $t - t_0$ is the **vector** defined by

$$\text{impulse} = \int_{t_0}^{t} \mathbf{F}(t') \, dt' = \overline{\mathbf{F}} \Delta t$$

where $\overline{\mathbf{F}}$ is the average value of $\mathbf{F}$ during the time interval $t - t_0$. The impulse of the resultant force acting on a particle is equal to the change in the particle's momentum.

Questions

1 The reading of a stopwatch is 0.00 s when a particle reaches point A in Figure 7-7, 0.50 s when it reaches point B, and 2.50 s when it reaches point C. Calculate the displacement Δx and the average velocity $\Delta x / \Delta t$ for each part of the trip.

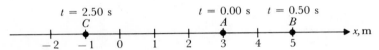

Figure 7-7

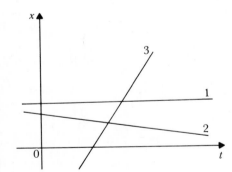

Figure 7-8

2 Figure 7-8 shows a graph of x as a function of time for three different motions. Identify the graph corresponding to:
(a) The greatest velocity.
(b) Zero velocity (particle stationary).
(c) Particle moving in the positive x direction.
(d) Particle moving in the negative x direction.

3 A motion along the X axis is given in SI units by

$$x = 3.00t^2$$

(a) Determine the average velocity for the time intervals: 2.00 s to 4.00 s; 2.00 s to 3.00 s; 2.00 s to 2.50 s; 2.00 s to 2.10 s; 2.00 s to 2.01 s.

(b) Determine the velocity at the instant $t = 2.00$ s, first by evaluation of the derivative dx/dt at this instant, and then by graphical measurement of the slope of an x versus t graph at the point corresponding to $t = 2.00$ s.

(c) What is the relationship between the results of part (a) and that of part (b)?

4 A motion is described by

$$x = bt^3$$

(a) Deduce an expression for the average velocity for the time interval from t to $t + \Delta t$. [Use the binomial expansion, $(t + \Delta t)^3 = t^3 + 3t^2\Delta t + 3t(\Delta t)^2 + (\Delta t)^3$.]

(b) What is the limiting value, as Δt approaches zero, of the average velocity deduced in part (a)? What is the significance of this limiting value?

5 State whether the acceleration a_x is positive, negative, or zero and give the corresponding direction of the acceleration vector $\mathbf{a}$ at each of the five instants indicated in Figure 7-9.

Figure 7-9

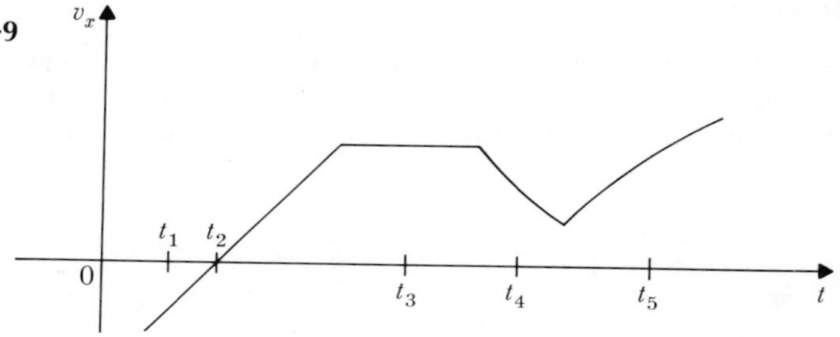

6 The velocity of a particle is given in SI units by

$$v_x = 3.00t^2$$

(a) What is the average velocity for the interval between $t = 4.00$ s and $t = 6.00$ s?

(b) What is the acceleration at $t = 4.00$ s?

7 Speedometer readings for a car traveling along a straight road are tabulated as follows:

Speed m/s	40.0	25.6	17.8	13.1	10.0	7.9
Time s	4.00	5.00	6.00	7.00	8.00	9.00

Plot a graph of v_x versus t and, from this graph, determine the acceleration at $t = 7.00$ s.

8 In a certain motion the velocity of a particle is given in SI units by

$$v_x = \frac{10}{t^2 + 1.00}$$

Find the particle's acceleration at $t = 2.00$ s.

9 A particle's position coordinate at the instant t is given in SI units by

$$x = 27t - t^3$$

Assume that t is positive. Where and when does the particle come to rest? What is its acceleration at this point?

10 A record of speedometer readings for a car traveling east is given in the following table:

Speed m/s	10.0	11.0	13.0	9.0	4.0	2.0	1.0
Time s	8.0	9.0	10.0	11.0	12.0	13.0	14.0

(a) Use approximating sums (Eq. 7.7) to estimate the car's displacement for the time interval extending from $t = 8.0$ s to $t = 14.0$ s.

(b) Estimate the average velocity for the time interval of part (a).

11 Use approximating sums (Eq. 7.7) to estimate the displacement of the car of Question 7 for the time interval from $t = 4.00$ s to $t = 9.00$ s. Check the answer by using the graph drawn for Question 7.

12 A particle, initially at rest at the origin, moves along the X axis with an acceleration given in SI units by

$$a_x = 6.0t^2$$

Where is the particle at $t = 2.0$ s and what is its velocity at this instant?

13 At $t = 0$, a particle passes through the origin, moving in the positive x direction with a speed of 16 m/s. The particle's acceleration is given in SI units by

$$a_x = -8.0t$$

Where and when does the particle's x coordinate reach its maximum value? Sketch a graph showing x as a function of time.

14 At the instant that a policeman's stopwatch reads 2.00 s, a car is moving in the positive x direction with a velocity of 22 m/s. The car is slowing down (decelerating), the magnitude of its acceleration having the constant value of 2.0 m/s²:

(a) What change in velocity has occurred when the stopwatch reads 5.00 s?

(b) What is the velocity when the stopwatch reads 5.00 s?

(c) What will the stopwatch read when the car is at rest?

15 A police officer starts a stopwatch when a car is 40 m beyond him, traveling at a speed of 30 m/s. The car begins a constant deceleration of 3.0 m/s²:

(a) How far is the car from the police officer at the instants $t = 2.00$ s, 4.00 s, 6.00 s, 8.00 s, 10.00 s? Plot a graph showing the distance of the car from the police officer as a function of time for stopwatch readings from 0.00 to 10.00 s.

(b) Plot a graph showing speed versus time for the stopwatch readings considered in part (a).

16 A train starts at a station and maintains a constant acceleration for 12.0 km to the next station where its speed is 36 m/s. Find the acceleration of the train and the elapsed time.

17 A train starts from rest and, after a constant acceleration for 5.00 min, reaches a speed of 20 m/s. How far did the train go in the fifth minute?

18 In the electron gun of a television tube, an electron is accelerated through a distance of 1.20×10^{-2} m and emerges with a speed of 8.0×10^6 m/s. The initial velocity is negligible compared to the final velocity. Find the electron's acceleration within the gun, assuming this acceleration to have been constant. For what time interval was the electron accelerated?

19 In an investigation of an accident that occurred in a 40 km/h (about 11 m/s) zone, skid marks made by one automobile were 29 m long. If we assume a maximum deceleration of 5.0 m/s², can we conclude that the automobile was speeding?

$v_f^2 = v_i^2 + 2a \cdot d$

20 A rocket-driven sled, used in experiments on the physiological effects of large accelerations, runs on a straight horizontal track. Starting from rest, it attains a speed of 450 m/s in 1.8 s:

(a) Calculate the acceleration, assuming it to be constant. Compare this value of acceleration with that of a freely falling body.

(b) How far does this sled travel during the 1.8-s interval?

21 A subway train travels a distance of 1000 m between stations, accelerating at a constant rate of 1.2 m/s² for the first half of the trip and decelerating at this same rate for the last half. Find the maximum speed attained and the time required for the trip.

22 A 1.0-kg gold ball and a 10-kg lead ball are dropped at the same instant from the top of a tower:

(a) Which hits the ground first? Why?

(b) What is the acceleration of the lead ball as it starts? What is the acceleration of this ball halfway down?

(c) What is the velocity of the lead ball 2.0 s after it is released?

(d) How far does the lead ball fall in the first 2.0-s interval?

(e) What is the velocity of the lead ball after it has traveled 4.0 m downward?

23 A ball is dropped from the top of a building 39.2 m above the sidewalk:

(a) How long will it take the ball to reach the sidewalk?

(b) What is the ball's speed just before it hits the sidewalk?

(c) What is the average velocity of the ball during the interval from $t = 1.00$ s to $t = 2.00$ s, assuming that the ball was dropped at $t = 0$? How far did the ball travel in this time interval?

24 If the ball in the preceding question, instead of being released at rest, were thrown vertically downward with a speed of 19.6 m/s, how long would it take to reach the sidewalk and what would be its speed just before impact?

25 A portion of a rocket is detached at a height of 4.9×10^2 m while the rocket is climbing vertically with a speed of 98 m/s. Neglecting air resistance, estimate the time required for this portion to hit the ground.

26 A ball is thrown vertically upward with an initial speed of 19.6 m/s:

(a) How long will the ball take to reach maximum height?

(b) What is the maximum height it will attain?

(c) At what times will the speed be 4.9 m/s?

(d) At what times will the ball be 14.7 m above the point of projection?

27 Show that when the brakes are applied in a car moving with a speed v_0, the shortest distance in which the car can be stopped is

$$x = \frac{v_0^2}{2g\mu_s}$$

(Notice that if there is no sliding between the tires and the road, the frictional force is a force of *static* friction.)

28 A 30-kg box is pushed by a horizontal force of 900 N as it slides up a plane inclined at an angle of 30° with the horizontal. The coefficient of kinetic friction between the box and the plane is 0.20. How far will the box travel along the plane in 2.0 s if it starts from rest?

29 A girl on a toboggan slides down a hill and attains a speed of 15 m/s as she reaches the ice at the edge of a lake. The toboggan comes to rest after sliding a distance of 60 m. What is the coefficient of friction between the ice and the toboggan?

30 A bullet leaves a rifle traveling horizontally with a speed of 1000 m/s and hits a wall which is 50 m away:
   (a) How long does the bullet's flight last?
   (b) How far will it have fallen by the time it hits the wall?

31 A baseball is projected horizontally with a speed of 30.0 m/s from the top of a building which is 40.0 m high. What is the speed of the baseball just before it hits the ground?

32 A car, traveling horizontally, drives over the edge of a cliff which is 19.6 m high and lands at a horizontal distance of 70 m from the base of the cliff. What was the speed of the car as it went over the edge of the cliff?

33 A ball is thrown at a speed of 25.0 m/s at an angle of 53.1° above the horizontal. At what height will it strike a vertical wall which is 30.0 m away?

34 A stone is projected from the top of a building with a speed of 20.0 m/s at an angle of 36.9° above the horizontal. The building is 50.0 m high. What horizontal distance will the stone travel before striking the ground?

35 A boy is in an elevator going down with a constant speed. He throws a ball straight up with a speed of 5.0 m/s relative to the elevator and catches it when it comes down. What is the time of flight of the ball?

36 **(a)** During a 10-s time interval, what is the impulse of a constant force of 3.0 N directed east?
   (b) The force of part (a) is the resultant force acting on a 5.0-kg particle. If the particle has an initial velocity of 8.0 m/s north, what is its velocity at the end of the 10-s interval?

37 A 0.20-kg billiard ball is struck by a cue and acquires a speed of 3.0 m/s. The cue was in contact with the ball for 10×10^{-3} s. Find the average force exerted on the ball by the cue.

38 A ball of mass m and speed v moves perpendicular to a wall. After collision with the wall, the ball moves with speed v in the opposite direction. Find the average force exerted on the wall by the ball in terms of m, v, and the duration Δt of the collision.

Supplementary Questions

S-1 The motion of a particle is given in SI units by

$$x = 20t - 2.0t^3$$

   (a) What is the velocity at $t = 3.0$ s?
   (b) When is the speed zero?
   (c) When does x reach its maximum value?
   (d) What is the maximum value of x?
   (e) When does the particle return to the origin and what is its velocity at this instant?
   (f) For the time interval from $t = 2.0$ s to $t = 3.0$ s, what are the displacement of the particle and its average velocity?

S-2 The motion of a particle is given in SI units by

$$x = \frac{1}{1 + t^2}$$

   (a) Find the functions v_x and a_x.
   (b) Sketch rough graphs of x, v_x, and a_x versus t. Show the limiting value as t approaches infinity for each of these three functions.
   (c) How far does the particle move in the third second?

S-3 *Amber light problem*   A motorist is approaching a traffic light with a speed v_0 when the light turns from green to amber. Suppose that it takes a time t_0 for the motorist to make a decision and then to apply the brakes, and that the maximum braking deceleration has a magnitude a:
   (a) Show that the motorist cannot stop without passing the light if his distance from the light is less than $v_0 t_0 + (v_0^2/2a)$.
   (b) Suppose the amber light remains on for a time t_1. If the motorist wants to continue at speed v_0 and still pass the light before it turns red, what is the maximum distance he can be from the light as it turns from green to amber?
   (c) Show that there is a zone in which the motorist will pass through a red light, whether he continues at constant speed or attempts to stop, if his speed v_0 is greater than $2a(t_1 - t_0)$.
   (d) Evaluate this critical speed when $a = 5.0$ m/s², $t_1 = 4.0$ s, and $t_0 = 1.0$ s.

S-4 A rocket ascends vertically with an acceleration of 19.6 m/s² for 80 s. Then, with its fuel supply exhausted, it continues upward. What is the maximum height attained and the time required to reach this height?

S-5 A stream of water has its velocity reversed but its speed v unchanged by a curved turbine blade. Find the force exerted on the blade in terms of v and R, where R is the mass per unit time of water striking the blade.

S-6 Water is poured into a barrel at the rate of 0.80 kg/s from a height of 5.0 m. The 10.0-kg barrel rests on a scale. Find the reading of the scale 5.0 s after the barrel begins to fill with water.

S-7 Show that projection at the angle $(45° - \phi)$ gives the same horizontal range as projection at the angle $(45° + \phi)$.

8.

motion along a curved path

In this chapter we study motion along a curved path, starting with the relatively simple case of circular motion at constant speed, next treating circular motion with varying speed, and finally discussing a general motion along any path. These topics introduce terms and ideas that recur in many other branches of mechanics.

The most significant point of immediate interest is a fact illustrated in Figure 8-1 which shows the acceleration vector of a particle that is moving along a curved path. Notice that **a** always points inward and that if the speed is constant, **a** points toward the center of the circular arc. The fact that there is an *acceleration* for circular motion with constant *speed* often seems surprising to students. The point is that the velocity vector changes *direction* even though its magnitude, the speed, is unchanged. Figure 8-2 indicates that the *direction* of the velocity change $\Delta \mathbf{v}$, in a small time interval Δt, is toward the center of the circular arc; therefore this is the direction of the acceleration (**a** is the limiting value of $\Delta \mathbf{v}/\Delta t$). This acceleration, directed toward the center of the arc, is called the *centripetal acceleration* ("center-seeking" acceleration). The magnitude of the centripetal acceleration is v^2/r, a fact that follows from the geometry of the situation. In Section 8.1 this result is obtained by an analytical method based on relationships that will

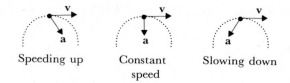

Speeding up   Constant   Slowing down
                   speed

Figure 8-1   The direction of the acceleration vector in motion along a curved path.

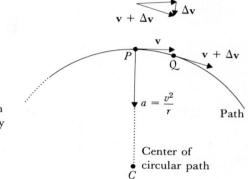

Figure 8-2   Circular motion with constant speed v. For point Q very close to P, the vector $\Delta \mathbf{v}$ is in the direction PC.

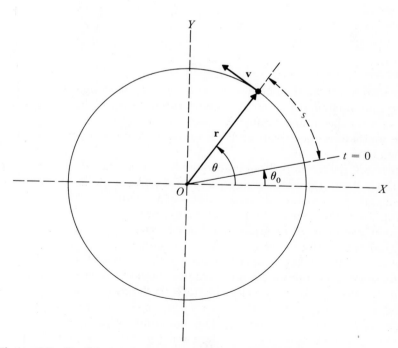

Figure 8-3   Particle moves with a constant speed v in a circle of radius r.

be found useful in later work, particularly in the study of simple harmonic motion.

8.1 Circular Motion with Constant Speed

The particle in Figure 8-3 moves with a constant speed v in a circle of radius r. The angle θ between the particle's position vector and the X axis is a function of time with initial value θ_0. It is convenient to measure θ in *radians* (rad).* Then the angular displacement $\theta - \theta_0$ is related to the distance s (measured along the path from the initial position to the particle) by

$$\theta - \theta_0 = \frac{s}{r} \tag{8.1}$$

For an angular displacement of 1 revolution or $360°$, $s = 2\pi r$ and $\theta - \theta_0 = 2\pi r/r = 2\pi$ rad. Consequently,

$$2\pi \text{ rad} = 360° = 1 \text{ rev}$$

The rate of change of θ is defined as the angular speed ω:

$$\omega = \frac{d\theta}{dt}$$

The SI unit of ω is the radian per second (rad/s). Often the number of revolutions per unit time, called the *frequency f* of revolution, is used to specify angular speed. Conversion is effected using

$$\omega = 2\pi f \tag{8.2}$$

Differentiation of Eq. 8.1 gives the relation between angular speed ω and linear speed v for circular motion:

$$\omega = \frac{v}{r} \qquad (r \text{ constant}) \tag{8.3}$$

where we have used $v = ds/dt$.† Equation 8.1, with $s/r = vt/r = \omega t$, gives

$$\theta - \theta_0 = \omega t \qquad (\omega \text{ constant}) \tag{8.4}$$

The time T required for one complete revolution is called the *period* of the motion. Since there are f revolutions per unit time, we have

$$T = \frac{1}{f} = \frac{2\pi}{\omega} \qquad (\omega \text{ constant}) \tag{8.5}$$

---

*We adopt the convention followed in trigonometry: if no unit is specified it is understood that an angle is measured in radians.
†For a small time interval from t to $t + \Delta t$, the magnitude $|\Delta \mathbf{r}|$ and the distance Δs are approximately equal and they approach equality as Δt approaches zero. This observation, together with the definition (Eq. 1.4) of the speed v (the *positive* quantity $|\mathbf{v}|$), gives

$$\frac{ds}{dt} = v = speed$$

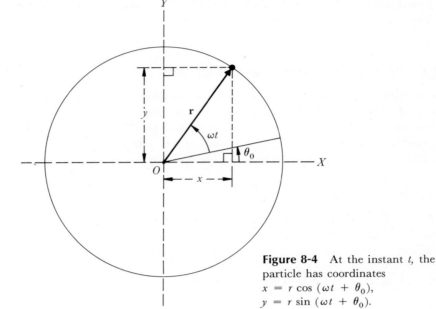

Figure 8-4 At the instant t, the particle has coordinates
$x = r \cos (\omega t + \theta_0)$,
$y = r \sin (\omega t + \theta_0)$.

The coordinates x and y (Figure 8-4) of the particle are related to r and θ (which is $\omega t + \theta_0$) by

$$x = r \cos (\omega t + \theta_0)$$
$$y = r \sin (\omega t + \theta_0)$$

(8.6)

Centripetal Acceleration

The components a_x, a_y of the acceleration **a**, can be found by differentiating the expressions for the coordinates. Using the chain rule we obtain

$$v_x = \frac{dx}{dt} = -r \sin (\omega t + \theta_0) \frac{d}{dt} (\omega t + \theta_0)$$
$$= -\omega r \sin (\omega t + \theta_0)$$

(8.7)

$$a_x = \frac{dv_x}{dt} = -\omega r \cos (\omega t + \theta_0) \frac{d}{dt} (\omega t + \theta_0)$$
$$= -\omega^2 r \cos (\omega t + \theta_0)$$

(8.8)

This shows that

$$a_x = -\omega^2 x \qquad (r \text{ constant}) \quad (8.9)$$

and by analogy

$$a_y = -\omega^2 y \qquad (r \text{ constant}) \quad (8.10)$$

Since x and y are the x and y components of the position vector $\mathbf{r}$, Eq. 8.9 and 8.10 are together equivalent to the *vector* equation

$$\mathbf{a} = -\omega^2 \mathbf{r} \qquad (r \text{ constant}) \quad (8.11)$$

The minus sign shows that the acceleration has a direction opposite to that of $\mathbf{r}$, that is, a direction toward the center of the circle. The magnitude of the acceleration is $a = \omega^2 r = (v/r)^2 r = v^2/r$. In summary: *A particle moving with speed v in a circular path of radius r has a centripetal acceleration (directed toward the center) of magnitude*

$$a = \frac{v^2}{r} \qquad (8.12)$$

Example 1   A propeller, 2.20 m long, rotates about a fixed axis through its center. Find the speed and the acceleration of a particle at the tip of the propeller when the angular speed of the propeller is 18.0×10^3 rev/min.

Solution   The frequency of revolution of the particle is $f = 300$ rev/s. The particle's speed is given by

$$v = \omega r = 2\pi f r = (600\pi \text{ rad/s})(1.10 \text{ m}) = 2.07 \times 10^3 \text{ m/s}$$

The particle's acceleration vector is directed toward the propeller shaft and has a magnitude given by

$$a = \frac{v^2}{r} = \frac{(2.07 \times 10^3 \text{ m/s})^2}{1.10 \text{ m}} = 3.90 \times 10^6 \text{ m/s}^2$$

8.2 Circular Motion with Varying Speed

In a general circular motion, a particle's speed is a varying function of time. The particle's acceleration vector then has a component along the path, called the *tangential component,* defined by

$$a_s = \frac{dv}{dt} = \frac{d^2s}{dt^2} \qquad (8.13)$$

This definition implies that a_s is positive when the speed is increasing and negative when the speed is decreasing. Evidently the tangential component of acceleration is the component of $\mathbf{a}$ in the direction of $\mathbf{v}$ (the direction of motion).

When the instantaneous speed is v, the particle has a centripetal acceleration v^2/r whether or not the speed is changing. This centripetal acceleration constitutes the normal component a_N of the acceleration vector

$$a_N = \frac{v^2}{r} \qquad (8.14)$$

The acceleration vector is the vector sum of its normal and tangential components.

If the particle's linear speed v is a varying function of time, the same will be true of its angular speed ω. The time rate of change of the angular speed is called the angular acceleration α:

$$\alpha = \frac{d\omega}{dt} = \frac{d^2\theta}{dt^2} \qquad (8.15)$$

By differentiating the equation $\omega = v/r$, we obtain the relationship between the angular acceleration and the linear acceleration:

$$\alpha = \frac{a_s}{r} \qquad (r \text{ constant}) \quad (8.16)$$

When a_s and α are constant, Eq. 8.13 and 8.15 can be integrated. Following the procedures of Section 7.3 we are led to the analogous results displayed in Table 8-1.

Table 8-1 Motion with Constant Acceleration

a_s constant	α constant
$s = v_0 t + \frac{1}{2} a_s t^2$	$\theta = \theta_0 + \omega_0 t + \frac{1}{2}\alpha t^2$
$v = v_0 + a_s t$	$\omega = \omega_0 + \alpha t$
$v^2 = v_0^2 + 2a_s s$	$\omega^2 = \omega_0^2 + 2\alpha(\theta - \theta_0)$
$s = \frac{1}{2}(v_0 + v)t$	$\theta - \theta_0 = \frac{1}{2}(\omega_0 + \omega)t$

Kinematics of Rigid Body Rotation

The angular quantities θ, ω, and α are particularly useful in describing the rotational motion of a rigid body. Figure 8-5 shows a rigid body rotating about a fixed axis passing through O and perpendicular to the XY plane. The orientation of the rigid body at any instant is given by specifying the angle θ between a fixed direction, say the X axis, and a line OP painted on the body. Since the body is rigid, all lines painted on it receive the *same angular displacement* in any rotation. Therefore $\theta - \theta_0$, ω, and α characterize the rotation of the entire body. At a given instant, every particle of the rigid body has the same value of $\theta - \theta_0$, of ω, and of α. By way of contrast, the distance s, the linear speed v, and the acceleration a_s of a particle P of the rotating body are each proportional to the particle's distance r from the axis of rotation.

Example 2   A flywheel with a diameter of 0.60 m is mounted on a horizontal axis and given a constant angular acceleration of 0.50 rad/s². The flywheel is initially at rest. After it has turned through π rad, find the velocity and the acceleration of the particle on the rim of the flywheel which, at that instant, is vertically below the axis (Figure 8-6).

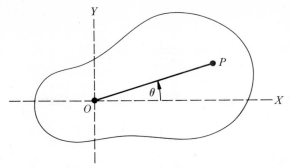

Figure 8-5 The orientation of the rigid body is determined by the angle θ.

Solution After turning through π rad from rest, the flywheel has an angular velocity ω given by

$$\omega^2 = 2\alpha(\theta - \theta_0) = 2(0.50 \text{ rad/s}^2)(\pi \text{ rad})$$

$$\omega = 1.77 \text{ rad/s}$$

The particle's speed is

$$v = \omega r = (1.77 \text{ rad/s})(0.30 \text{ m}) = 0.53 \text{ m/s}$$

The direction of the particle's velocity vector $\mathbf{v}$ is tangent to its path and therefore, at the instant in question, is horizontal (and perpendicular to the flywheel axis). The tangential component of this particle's acceleration is

$$a_s = \alpha r = (0.50 \text{ rad/s}^2)(0.30 \text{ m}) = 0.15 \text{ m/s}^2$$

This component of $\mathbf{a}$ points in the direction of $\mathbf{v}$. The particle has a normal component of acceleration, directed upward at the instant in question, and of magnitude

$$a_N = \frac{v^2}{r} = \omega^2 r = 0.94 \text{ m/s}^2$$

Figure 8-6 Flywheel of Example 2.

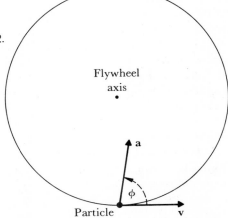

The acceleration vector has a magnitude

$$a = \sqrt{a_s^2 + a_N^2} = 0.95 \text{ m/s}^2$$

and is inclined at an angle ϕ above the horizontal given by

$$\phi = \tan^{-1}\left(\frac{a_N}{a_s}\right) = \tan^{-1}(6.27) = 81°$$

8.3 Kinematics of Motion Along a Curved Path

The general motion of a particle along any path (Figure 8-7) can be conveniently described by the positive distance s that the particle has traveled, measured along the path of the particle from its position at $t = 0$. Then

$$\frac{ds}{dt} = v = speed \qquad (8.17)$$

$$\frac{d^2s}{dt^2} = \frac{dv}{dt} = a_s = tangential\ component\ of\ acceleration \qquad (8.18)$$

These equations for the linear quantities in a general motion are exactly the same as those for a circular motion. (However, the equations for circular motion, $\omega = v/r$ and $\alpha = a_s/r$, must be replaced by more complicated expressions when the distance r from the origin to the particle is not constant.)

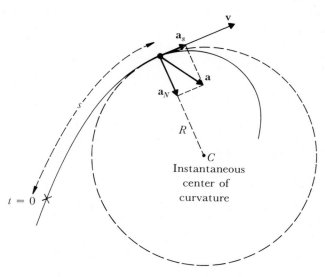

Figure 8-7   Particle moving along a curved path has a normal component of acceleration $a_N = v^2/R$.

Where the path of the particle is curved, the particle has a centripetal or normal component of acceleration. In Figure 8-7, the segment of the curved path at the location of the particle is approximated by the arc of a circle with center C, called the instantaneous center of curvature, and a radius R, called the radius of curvature of the particle's path. At the instant depicted in Figure 8-7, the particle has precisely the centripetal acceleration it would have if it were always traveling with speed v on a circular path with center C and radius R; that is, the *normal component a_N of the particle's acceleration is directed toward the instantaneous center of curvature and has a magnitude given by*

$$a_N = \frac{v^2}{R} \qquad (8.19)$$

where v is the particle's instantaneous speed and R is the radius of curvature of the path at the location of the particle.* At a given speed, the largest value of a_N occurs where the radius of curvature of the path is the smallest, that is, where the turn is sharpest. Where the path is straight, $a_N = 0$.

8.4 Newton's Second Law for Normal and Tangential Components

Newton's second law, $\mathbf{F} = m\mathbf{a}$, is equivalent to two equations, one for normal components and another for tangential components.

The equation of motion for normal components is

$$F_N = ma_N = \frac{mv^2}{R} \qquad . \quad (8.20)$$

where F_N is the normal (or centripetal) component of the resultant force acting *on* the particle of mass m.

If the resultant external force acting on the particle has a tangential component F_s, then there will be a tangential component of acceleration a_s such that

$$F_s = ma_s \qquad (8.21)$$

This equation of motion determines the acceleration function a_s when the force is known. Then the speed v and the distance s can be evaluated. If F_s is constant, then a_s is constant and the integrations can be performed yielding the results given on the left in Table 8-1.

Example 3   A particle of mass m is attached to a string of length L and whirled in a vertical circle about a fixed point O to which the other end of the string is attached. Find an expression for the tension T in the string at the instant that the particle's speed is v and the string makes an angle θ with the vertical.

---

*Equations 8.18 and 8.19 for a general motion have been made plausible by the preliminary study of circular motion. The outlines of a general proof are given in Question S-4.

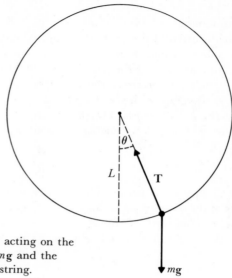

Figure 8-8 The forces acting on the particle are its weight $m\mathbf{g}$ and the force $\mathbf{T}$ exerted by the string.

Solution   Figure 8-8 is the free-body diagram for the mass m. We resolve the forces into normal and tangential components. Newton's second law for normal components with $a_N = v^2/L$, gives

$$T - mg \cos \theta = \frac{mv^2}{L}$$

Therefore

$$T = \frac{mv^2}{L} + mg \cos \theta$$

Summary

☐   A particle moving with *constant* angular speed ω in a circle of radius r has

$$period \ T = \frac{1}{f} = \frac{2\pi}{\omega}$$

$$coordinate \ x = r \cos (\omega t + \theta_0)$$

$$acceleration \ component \ a_x = -\omega^2 x$$

☐   Angular motion is described by

$$angular \ displacement = \theta - \theta_0$$

$$angular \ speed = \omega = \frac{d\theta}{dt}$$

$$angular \ acceleration = \alpha = \frac{d\omega}{dt} = \frac{d^2\theta}{dt^2}$$

In a motion parallel to the XY plane, the rotation of an entire rigid body is characterized by these angular quantities.

☐ A particle a constant distance r from a fixed axis of rotation has

$$s = (\theta - \theta_0)r$$

$$v = \omega r$$

$$a_s = \alpha r$$

☐ In a general motion described by giving a particle's distance s traveled along a curved path, the particle has

$$speed = v = \frac{ds}{dt}$$

$$tangential\ component\ of\ acceleration = a_s = \frac{d^2s}{dt^2}$$

$$normal\ component\ of\ acceleration = a_N = \frac{v^2}{R}$$

where R is the radius of curvature of the path. Newton's second law gives

$$F_N = ma_N \qquad and \qquad F_s = ma_s$$

where the resultant force acting on the particle has a normal component F_N and a tangential component F_s.

Questions

1   A car, traveling at a constant speed of 30 m/s, goes clockwise about a circular race-track of radius 150 m. Find the angular speed of the car about the center of the track. Calculate the period of the car's motion. Find the magnitude and direction of the car's acceleration vector at the instant that the car is heading due south.

2   A watch has a second hand which is 1.0×10^{-2} m long. What is the speed of the tip of the second hand relative to the watch? What is the frequency of revolution?

3   To describe the motion of the car in Question 1, use a coordinate system with origin at the center of the track, axis OX pointing east and OY pointing north:
   (a) Evaluate the component v_x of the car's velocity when the car's direction from the center is east, north, west, and south.
   (b) At these four locations, find the component a_x of the car's acceleration.

4   A propeller has a *constant* angular speed of 1.80×10^3 rad/s:
   (a) What is the propeller's angular displacement in 0.50×10^{-3} s?
   (b) How long will it take the propeller to make 10 revolutions?

5   The orbit of the earth about the sun is approximately circular, with a radius of 1.5×10^{11} m. Consider the earth as a particle and calculate both its angular velocity about the sun and its speed as it moves in its orbit around the sun.

6   A circular saw with a 0.30-m radius is turning at 1600 rev/min. What is the speed of a tooth of the saw?

7 A propeller of 0.80-m radius is rotating counterclockwise with a constant angular speed of 2.0×10^3 rad/s. At the instant when the propeller is vertical, find:
(a) The magnitude and direction of the velocity vector of the propeller tip.
(b) The magnitude and direction of the acceleration vector of the propeller tip.

8 A particle on the rim of a grindstone is traveling at a speed of 10 m/s. The diameter of the grindstone is 0.40 m. What is the angular speed of the grindstone?

9 (a) A car is going around a circular track of 100-m radius with a constant speed of 20 m/s. What is the normal component of acceleration at the instant when the car is due north from the center of the circle and traveling east? Draw a diagram showing the path, the velocity vector, and the acceleration vector.
(b) Just after the instant mentioned in part (a), the driver steps on the gas and acquires a tangential component of acceleration equal to 3.0 m/s². Find the acceleration vector.

10 A car passes the starting line with a speed of 10.0 m/s and travels around a circular racetrack of 200-m radius with a constant tangential component of acceleration of 2.00 m/s². After 20 s have elapsed, find:
(a) The distance traveled.
(b) The speed.
(c) The tangential and normal components of the acceleration.

11 Deduce an expression for the time it takes a particle to travel once around a circular path of radius R if:
(a) The speed has the constant value v_0.
(b) The tangential component a_s of the acceleration is constant and the particle starts from rest.

12 A particle starts from rest and moves with a constant tangential component a_s of acceleration in a circular path. Show that after going once around the circle, the particle has a normal component of acceleration given by $a_N = 4\pi a_s$.

13 The angular speed of a flywheel is increased from 5.0 rad/s to 15.0 rad/s in a 3.0-s interval. Find the angular acceleration (assuming it to be constant) for this time interval.

14 Find the magnitude and direction of the acceleration of a propeller tip at an instant when the propeller's angular speed is 2.0×10^3 rad/s and its angular acceleration is 0.50 rad/s². The propeller radius is 0.80 m.

15 The particle in Question 8 has a constant tangential component of acceleration of 20 m/s². What will be the increase in the grindstone's angular speed in a 2.5-s interval?

16 A merry-go-round, initially at rest, is given a constant angular acceleration of 0.010 rad/s². Find the angular speed attained and the number of revolutions that have been made after 30 s have elapsed.

17 The turntable of a record player, initially revolving at 33 rev/min, slows down and stops in 15 s. Find the angular acceleration (assuming it to be constant) and the number of revolutions of the turntable.

18 A brake is applied to a wheel which has an angular speed of 600 rev/min. The wheel is brought to rest after making 100 revolutions. Find the angular acceleration, assuming it to be constant.

19 The shaft of an automobile engine, which is rotating at 1200 rev/min, is given an angular acceleration of 2.0 rad/s² for a 10-s interval:

(a) What will be the angular displacement of the shaft during this interval?

(b) What will be the final angular speed?

20 A wheel rotates through 160 rad in 4.0 s. At the end of this time interval it has an angular speed of 50 rad/s. Find its constant angular acceleration.

21 A flywheel rotates with a constant angular acceleration of 4.0 rad/s^2 and has an angular displacement of 70 rad in a certain 5.0-s time interval. If the flywheel starts from rest, how long does it rotate before this 5.0-s interval begins?

22 Assume the earth to be a sphere of radius 6.4 × 10^6 m. Relative to a frame which does not rotate relative to the distant stars, the earth itself rotates about a polar axis with an angular speed of 1 revolution per sidereal day (1 year = 366.2 sidereal days):

(a) Evaluate this angular speed in radians per second.

(b) Calculate the velocity and the acceleration of a point on the equator.

(c) Calculate the velocity and the acceleration of a point at latitude 45°.

23 **(a)** Traveling at a constant speed, a car first goes down a straight road, next around a gentle curve, and finally around a sharp turn. Compare the normal components of acceleration for these three cases and state which is greatest and which is least.

(b) A car speeds up as it goes around a curved portion of a racetrack. Draw the path, and then draw the velocity and acceleration vectors at one instant.

24 A roller coaster starts from rest and goes down a straight incline, 70 m long, with a constant acceleration of 1.5 m/s^2. The roller coaster then moves with constant speed along a circular section of horizontal track which has a radius of curvature of 30 m. Find the tangential and normal components of the roller coaster's acceleration when it is traveling along the curved horizontal track.

25 A car at point A (Figure 8-9) has a speed of 30.0 m/s. It slows down at the constant rate of 2.00 m/s^2. What are the tangential and normal components of the car's acceleration at:

(a) Point B, a distance of 90 m from A?

(b) Point D, a distance of 150 m from A, measured along the road? At point D the radius of curvature of the road is 100 m.

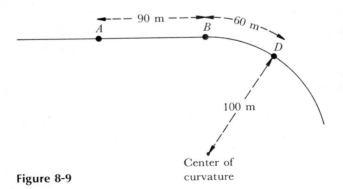

Figure 8-9

26 Find the magnitude and direction of the horizontal force acting on a 1500-kg car which is traveling at a constant speed of 20 m/s around a curve of radius 80 m.

Supplementary Questions

S-1 Consider a car moving at speed v along a road which is curved into a circular arc of radius r. The road is banked at an angle θ to the horizontal. Show that if the car is traveling at a speed such that there is no frictional force exerted on the car by the road, then

$$v^2 = rg \tan \theta$$

S-2 Figure 8-10 shows a conical pendulum—a particle of mass m revolving in a horizontal circle with a constant speed. The particle is attached to the end of a string of length L which makes an angle θ with the vertical. Show that the time required for one complete revolution of the mass is $2\pi\sqrt{L \cos \theta / g}$.

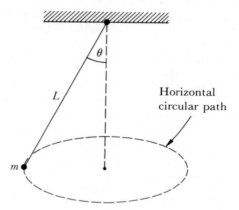

Figure 8-10

S-3 A particle of mass m, initially at rest, moves in a circular path of radius r. The resultant force acting on the particle has a tangential component given by

$$F_s = Kt$$

Express the time required for the particle to return to its starting point in terms of r, K, and m.

S-4 For a general motion of a particle along a curved path, show that the acceleration has a tangential component $a_s = dv/dt$ and a normal component $a_N = v^2/R$, where R is the radius of curvature of the path. This can be done in the following steps:

(a) Introduce vectors $\mathbf{u}_s$ and $\mathbf{u}_N$ which, at any point on the particle's path, are defined to have unit magnitude ($|\mathbf{u}_s| = 1$, $|\mathbf{u}_N| = 1$) with $\mathbf{u}_s$ in the direction of $\mathbf{v}$ and $\mathbf{u}_N$ pointing toward the center of curvature of the path (Figure 8-11). If the angle from the axis OX to the direction of $\mathbf{u}_s$ is θ, $\mathbf{u}_s$ has the components

$$u_{sx} = \cos \theta \quad \text{and} \quad u_{sy} = \sin \theta$$

Find the components of $\mathbf{u}_N$ and show that

$$\frac{d\mathbf{u}_s}{d\theta} = \mathbf{u}_N$$

(b) The velocity $\mathbf{v}$ is correctly given in magnitude and direction by

$$\mathbf{v} = v\mathbf{u}_s$$

Using the usual rule for the differentiation of a product, and then using the chain rule, show that

$$\mathbf{a} = \frac{dv}{dt}\mathbf{u}_s + v\frac{d\mathbf{u}_s}{d\theta}\frac{d\theta}{ds}\frac{ds}{dt}$$

$$= \frac{dv}{dt}\mathbf{u}_s + v^2\frac{d\theta}{ds}\mathbf{u}_N$$

(c) Show that $d\theta/ds = 1/R$. Then the result of part (b) is

$$\mathbf{a} = \frac{dv}{dt}\mathbf{u}_s + \frac{v^2}{R}\mathbf{u}_N$$

(d) The acceleration vector can be expressed in terms of its tangential and normal components by

$$\mathbf{a} = a_s\mathbf{u}_s + a_N\mathbf{u}_N$$

Comparison with the result of part (c) gives a_s and a_N.

Figure 8-11 The unit tangential and normal vectors for a particle's path which has center of curvature C and radius of curvature R. The velocity vector is $v\mathbf{u}_s$.

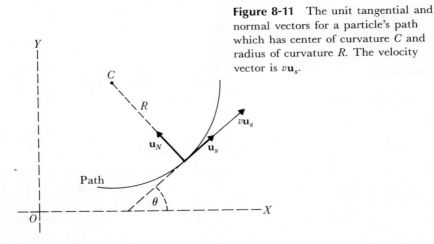

9.

simple harmonic motion

The bobbing motion of a mass suspended from a spring, and the small oscillations of a pendulum, are examples of what is called *simple harmonic motion*. An understanding of simple harmonic motion is particularly important, not only because this motion occurs in a variety of circumstances, but also because the study of this type of motion involves mathematical terms and relationships necessary for the analysis of sound, light, alternating electric current, and phenomena of quantum physics.

Any motion that is repeated in equal time intervals of duration T is called periodic: T is called the *period of the motion*. The motion is oscillatory if it takes place back and forth over the same path. One complete execution of the motion is called a *cycle*. For example, the mass m in Figure 9-1 completes one cycle in any round trip, say from point O to $x = A$, back through O to $x = -A$, and then back again to O. Since one cycle is executed in a time interval of one period T, the number of cycles per unit time is $1/T$ which is called the *frequency f*:

$$f = \frac{1}{T} \tag{9.1}$$

The SI unit of frequency is the cycle per second, named the *hertz* (Hz).

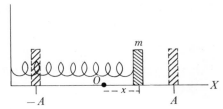

Figure 9-1 Simple harmonic motion. The mass m oscillates back and forth on a smooth table along the axis OX between the points $x = A$ and $x = -A$. The amplitude of this simple harmonic motion is A.

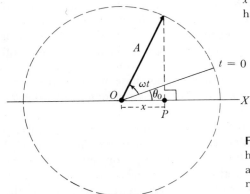

Figure 9-2 Point P executes simple harmonic motion along OX as the arrow tip moves in a circle of radius A with a constant angular velocity ω. The arrow A represents a *phasor* and this diagram is called a *phasor diagram*.

9.1 Position Coordinate in Simple Harmonic Motion

Simple harmonic motion is the particular type of oscillatory motion executed by the point P in Figure 9-2 as the arrow of constant length A rotates about the origin with a constant angular speed ω. The point P is always at the foot of the perpendicular from the tip of the arrow to the X axis. At the instant t, the arrow makes an angle

$$\theta = \omega t + \theta_0 \tag{9.2}$$

with the X axis.

The arrow in Figure 9-2 does *not* have a direction in space. It is an arrow drawn on a diagram to represent the length A and the angle θ. Such a quantity is called a *phasor* to avoid confusion with vectors that do have direction in space. Formally, *we define a phasor to be a mathematical quantity characterized by a magnitude (A) and an angle (θ)* with algebraic properties that will be presented as required. (In mathematics, phasors are called complex numbers.)

As time elapses, θ changes and the phasor rotates about a perpendicular axis through its tail, which is fixed at the point O. The point P moves along the

X axis, and its position coordinate x has the time dependence that is characteristic of simple harmonic motion:

$$x = projection\ of\ A\ on\ X\ axis$$

$$= A\ \cos \theta \tag{9.3}$$

$$= A\ \cos (\omega t + \theta_0)$$

where A, θ_0, and ω are constants. Figure 9-3 shows the corresponding graph of x as a function of time.

The positive constant A is named the *amplitude* of the simple harmonic motion executed by the point P. As the phasor of length A rotates, point P moves back and forth along the X axis between the points $x = A$ and $x = -A$. We see that the amplitude is the maximum value attained by the position coordinate x. The total range of the motion is the distance $2A$. The point O, at the center of the range of motion, is called the equilibrium position (for a reason that will become clear as we investigate the dynamics of this motion).

The angle θ is called the *phase angle* or, briefly, the phase of the coordinate x. The phase angle changes at the constant rate ω. The value of θ at $t = 0$ is θ_0, which is called the *initial phase*.

The point P completes one cycle of its simple harmonic motion as the rotating phasor makes exactly one complete revolution. Therefore the period T of the simple harmonic motion, being the time required for one cycle, is also the time required for one revolution (2π rad) of the phasor, which is rotating with constant angular speed ω. The frequency f of the simple harmonic motion of the point P is the same as the number of revolutions per unit time of the rotating phasor. We have

$$f = \frac{1}{T} \tag{9.1}$$

and, from Eq. 8.2

$$\omega = 2\pi f \tag{9.4}$$

which gives ω in *radians per unit time*. This constant ω, the angular speed of the phasor, is called the *angular frequency* of the simple harmonic motion executed by P.

In summary, *a simple harmonic motion of angular frequency ω, amplitude A, and initial phase θ_0 is a motion that can be considered as the projection on a diameter (say the X axis) of the tip of a phasor. The phasor has a length A, makes an angle (the phase angle) $\theta = \omega t + \theta_0$ with the X axis, and rotates about its fixed tail with a constant angular speed ω.*

Example 1   A phasor of length 2.00 m makes 4.00 rev/s. The initial phase angle is $-\pi/2$ rad:

(*a*) Describe the simple harmonic motion of the point P which is the projection of the tip of the phasor on the X axis (Figure 9-4).

(*b*) Locate P, first at the initial instant ($t = 0$) and next at an instant one-quarter of a period later.

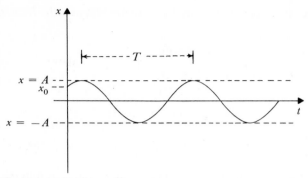

Figure 9-3 The graph shows the coordinate x as a function of time for a particle oscillating with simple harmonic motion of amplitude A, with the initial position given by $x_0 = A \cos \theta_0$ and the period by $T = 2\pi/\omega$. Such a graph of a cosine or sine function is called a *sine curve*, and we say that x is a *sinusoidal function* of time.

Figure 9-4 The initial phase angle is $-\pi/2$ rad, and the amplitude is 2.00 m.

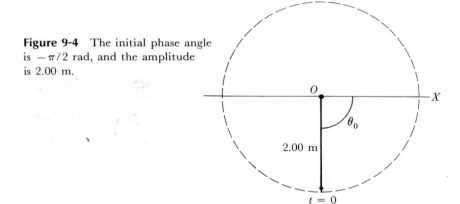

Solution

(a) The initial position of the phasor is shown in Figure 9-4. The number of revolutions per second of the phasor is the number of cycles per second (f) of the associated harmonic oscillation. The simple harmonic motion executed by P therefore has a frequency $f = 4.00$ Hz, a period $T = 1/f = 0.250$ s, and an angular frequency $\omega = (2\pi \text{ rad})(4.00 \text{ Hz}) = 25.1$ rad/s. The amplitude of this simple harmonic motion is the length of the phasor: $A = 2.00$ m. In summary, P executes simple harmonic motion with period 0.250 s, amplitude 2.00 m, and initial phase $-\pi/2$ rad.

(b) From the initial position of the phasor in Figure 9-4, we see that its projection on the X axis is zero, so initially the point P is at the origin O. One-quarter of a period later the phasor will have completed one-quarter of a revolution and rotated to the position of the axis OX. Projection of the phasor on the X axis gives the position of P as $x = 2.00$ m. At this instant P reaches maximum displacement from the equilibrium position O.

9.2 Velocity and Acceleration in Simple Harmonic Motion

The velocity in the simple harmonic motion of the point P in Figure 9-2 is found by differentiation of the expression for x:

$$v_x = \frac{dx}{dt} = -A \sin(\omega t + \theta_0)\frac{d}{dt}(\omega t + \theta_0) = -\omega A \sin(\omega t + \theta_0) \qquad (9.5)$$

The acceleration in simple harmonic motion is

$$a_x = \frac{dv_x}{dt} = -\omega A \cos(\omega t + \theta_0)\frac{d}{dt}(\omega t + \theta_0) = -\omega^2 A \cos(\omega t + \theta_0) \qquad (9.6)$$

An instructive alternative method of finding v_x and a_x is to refer to the rotating phasor (Figure 9-5). Since the tip of the phasor moves in a circle of radius A with an angular speed ω, its velocity has a magnitude, from Eq. 8.3, of

$$v = \omega A$$

and a direction (on the phasor diagram) tangent to its circular path. The angle between this velocity and the vertical is the phase angle θ. The point P moves along the X axis with a velocity v_x which is just the x component of the velocity of the phasor tip:

$$v_x = -v \sin \theta = -\omega A \sin(\omega t + \theta_0) \qquad (9.5)$$

The negative sign in this expression corresponds to the fact that for θ between $0°$ and $180°$, where $\sin \theta$ is positive, the point P moves in the negative x direction; for θ between $180°$ and $360°$, the point P moves in the positive x direction.

Our result shows that the point P moves back and forth with varying velocity, reaching a maximum speed ωA as it passes through the equilibrium position ($\theta = \pi/2$ rad or $-\pi/2$ rad), and coming to rest at $x = A$ (where $\theta = 0$) and at $x = -A$ (where $\theta = \pi$ rad).

The acceleration of the phasor tip (Figure 9-6) is just a centripetal acceleration directed (on the phasor diagram) toward O with a magnitude, from Eq. 8.11, given by

$$a = \omega^2 A$$

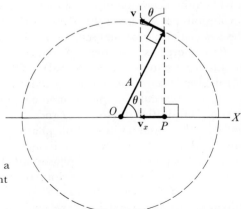

Figure 9-5   Point P moves with a velocity which is the x component of the velocity of the arrow tip.

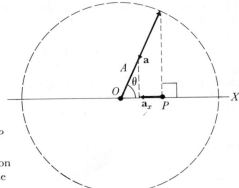

Figure 9-6 The arrow tip has centripetal acceleration **a**. Point P has an acceleration which is the x component of **a**. The acceleration of P is always directed toward the equilibrium position O.

The x component of the acceleration of the phasor tip is the acceleration a_x of the point P as it moves along the X axis:

$$a_x = -\omega^2 A \cos \theta = -\omega^2 A \cos(\omega t + \theta_0) \qquad (9.6)$$

Now since $x = A \cos \theta$, we reach the important result

$$a_x = -\omega^2 x \qquad (9.7)$$

The minus sign corresponds to the fact that the acceleration of P is always directed toward O, so that a_x is negative when x is positive, and vice versa.

The equation $a_x = -\omega^2 x$ states that *in simple harmonic motion, a point P moves back and forth over the same path in such a way that its acceleration is proportional to its distance from a fixed point O and is always directed toward O*. This statement is an alternative definition of simple harmonic motion. The magnitude of the acceleration of the point P is greatest when x reaches its maximum magnitude, that is, $x = A$ or $x = -A$. As the point P sweeps through the equilibrium position, its acceleration is zero.

The relationship between a_x and x is to be contrasted with the relationship between a_x and v_x. At the equilibrium position where x and a_x are zero, the speed is a maximum. Where x and a_x reach their maximum magnitudes, the speed is zero.

9.3 Initial Conditions

The amplitude A and the initial phase θ_0 determine the initial position x_0 and the initial velocity v_{0x} of the simple harmonic motion executed by the particle P. Thus, putting $t = 0$ in Eq. 9.3 and 9.5, we obtain

$$x_0 = A \cos \theta_0 \qquad (9.8)$$

$$v_{0x} = -\omega A \sin \theta_0$$

Conversely, the particle's initial position and initial velocity determine the amplitude and the initial phase of the simple harmonic motion. Squaring and then adding Eq. 9.8, we find

$$A^2 = x_0^2 + \frac{v_{0x}^2}{\omega^2} \tag{9.9}$$

Division of the second of Eq. 9.8 by the first yields

$$\tan \theta_0 = -\frac{v_{0x}}{\omega x_0} \tag{9.10}$$

These relations can be obtained by inspection from the triangle shown in Figure 9-7.

When finding the initial phase θ_0 corresponding to known values of x_0 and v_{0x}, we can determine the correct quadrant for θ_0 most easily by considering the orientation of the phasor that is required to give the correct signs for x_0 and v_{0x}. For example, if x_0 is a negative number and v_{0x} is positive, then the rotating phasor must be in the third quadrant and θ_0 is between 180° and 270°.

Example 2   A particle executes simple harmonic motion with an angular frequency of 2.0 rad/s. Initially the particle is 3.0 m to the right of its equilibrium position and is traveling to the right with a speed of 8.0 m/s. Where will the particle be after 0.80 s have elapsed?

Solution   With the origin at the equilibrium position and the axis OX directed to the right, the initial conditions are $x_0 = 3.0$ m and $v_{0x} = 8.0$ m/s. From Eq. 9.9 and 9.10 we obtain

$$A = \sqrt{(3.0 \text{ m})^2 + \left(\frac{8.0 \text{ m}}{2.0}\right)^2} = 5.0 \text{ m}$$

$$\theta_0 = \tan^{-1}\left(-\frac{8.0}{6.0}\right) = -53° \text{ or } (180° - 53°)$$

The value $\theta_0 = -53°$ gives the phasor orientation corresponding to positive x_0 and positive v_{0x}. At any instant t, the position coordinate of the oscillating particle P is given in SI units by

$$x = 5.0 \cos(2.0t + \theta_0)$$

At $t = 0.80$ s, $\omega t = 1.60$ rad $= 92°$.

$$x = (5.0 \text{ m}) \cos(92° - 53°) = 3.9 \text{ m}$$

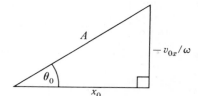

Figure 9-7   Relationships between (A, θ_0) and (x_0, v_{0x}) can be determined from this triangle.

9.4 Hooke's Law

Up to this point we have been concerned only with the description of simple harmonic motion. We now ask: What are the circumstances under which a particle will execute such a motion? What are the characteristics of the resultant force acting on a particle that cause it to oscillate in this particular way? In a sense we have a solution in search of a problem.

When a particle P of mass m executes simple harmonic motion along the X axis, its acceleration is

$$a_x = -\omega^2 x \tag{9.7}$$

Then Newton's second law, $F_x = ma_x$, implies that the resultant force F_x acting on this particle must be

$$F_x = -m\omega^2 x \tag{9.11}$$

This answers our question. The resultant force that causes simple harmonic motion is a restoring force always directed toward O (as is the acceleration), and this force is proportional to the particle's distance from O. The point O where this resultant force is zero is therefore a position of *stable equilibrium.*

The magnitude of the constant of proportionality in Eq. 9.11 is called the force constant k; then

$$F_x = -kx \tag{9.12}$$

and

$$\omega^2 = \frac{k}{m} \tag{9.13}$$

The period T of the motion is, from Eq. 9.1 and 9.4

$$T = \frac{2\pi}{\omega} = 2\pi\sqrt{\frac{m}{k}} \tag{9.14}$$

This shows that the period is determined entirely by the force constant and the particle's mass. In any simple harmonic motion the period is independent of the amplitude A and of the initial phase θ_0.

The force law

$$F_x = -kx \tag{9.12}$$

is known as Hooke's law. Whenever the forces acting on a particle are such that the particle has a position of stable equilibrium, then, at least for small displacements x from this equilibrium position, the resultant force F_x acting on the particle must obey Hooke's law. To understand the reasons for this, we consider the graph (Figure 9-8) of a possible resultant force F_x plotted as a function of x when the point O is assumed to be a position of stable equilibrium. This assumption implies that, at least for small displacements, F_x must be a *restoring* force always directed toward O. Then, when x is positive, $\mathbf{F}$ must be in the negative x direction, which implies that F_x is negative. Also, when x is nega-

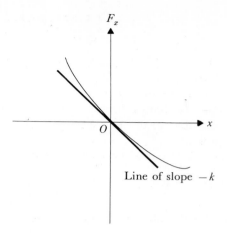

F_x

O

x

Line of slope $-k$

Figure 9-8 Graph of F_x versus x when O is a position of stable equilibrium. Short segment of any smooth curve can be approximated by a straight line.

tive, F_x must be positive. Therefore the graph of any possible F_x must pass through the origin in the manner indicated in Figure 9-8. If we now assume that the graph is smooth at the origin, the curve can be approximated for a short section by a straight line. The equation for this straight line graph is

$$F_x = -kx$$

where $-k$ is the slope of the straight line. Therefore the resultant force acting on a particle close to a position of stable equilibrium always obeys Hooke's law. And a small oscillation about an equilibrium position will always be a simple harmonic motion because this is the motion that occurs whenever the resultant force obeys Hooke's law. Since so many objects have an equilibrium position and are usually in the vicinity of this equilibrium position, simple harmonic motion occurs frequently in nature.

Figure 9-1 shows a mass m fastened to one end of a spring. The force exerted on this mass by the spring obeys Hooke's law even when the stretch or compression x is quite large. The *force constant* k gives the force per unit elongation, so the greater the value of k, the stiffer the spring.

Even for springs there are departures from Hooke's law if the stretch is too large. In fact, with any solid object, if the deformation is so large that what is known as the *elastic limit* is exceeded, the deformation ceases to be proportional to the deforming force. We then observe that, when the deforming force is removed, the object has acquired a permanent deformation.

Example 3 The force constant k of a spring is measured by observing that the spring is stretched 0.100 m by a 10.0-N force applied at one end. A mass m equal to 0.250 kg is then fastened to one end of the spring, while the other end is fixed, with the assembly placed on a smooth table as in Figure 9-1. The spring is compressed by moving the mass m a distance 0.200 m to the left of its equilibrium position. The mass is then released. Describe the subsequent motion of m.

Solution The force constant k is given by

$$k = \frac{F}{x}$$

where F is the *magnitude* of the force associated with a stretch x. (Note that k is a positive number by definition.) For this spring we have

$$k = \frac{10.0 \text{ N}}{0.100 \text{ m}} = 100 \text{ N/m}$$

The resultant force acting on the mass m is the force exerted by the spring, $F_x = -kx$. Therefore, the mass m executes simple harmonic motion with an angular frequency determined from Eq. 9.13 by m and the spring's force constant k:

$$\omega^2 = \frac{k}{m} = \frac{100 \text{ N/m}}{0.250 \text{ kg}} = 400 \text{ s}^{-2}$$

Therefore, the angular frequency is $\omega = 20.0$ rad/s, the period is $T = 2\pi/\omega = 0.314$ s, and the frequency is $f = 1/T = 3.18$ Hz. The amplitude A and the initial phase θ_0 are determined by the initial conditions which, for this motion, are $v_{0x} = 0$ and $x_0 = -0.200$ m. By considering the rotating phasor tip (Figure 9-2), whose projection on the X axis would coincide with the motion of m, we can see that, in its initial position, the phasor must point horizontally to the left. The initial phase is therefore π rad, and the amplitude is 0.200 m. Since

$$x = A \cos(\omega t + \theta_0) \quad 2\pi\sqrt{\tfrac{m}{k}} \tag{9.3}$$

for this motion

$$x = (0.200 \text{ m}) \cos[(20.0 \text{ rad/s})t + (\pi \text{ rad})]$$

9.5 Dynamics of Simple Harmonic Motion

The particle motion described by $x = A \cos(\omega t + \theta_0)$ has been defined as simple harmonic motion and it has been found that the resultant force required to produce such a motion obeys Hooke's law, $F_x = -kx$. We now consider the inverse problem: When the resultant force acting on a particle obeys Hooke's law, what is the motion of the particle? Newton's second law, $ma_x = F_x$, with $a_x = d^2x/dt^2$ and $F_x = -kx$, gives

$$\frac{d^2x}{dt^2} = -\frac{k}{m}x \tag{9.15}$$

Since k/m is positive, it can always be written as the square of some real number ω; that is, we simply define a quantity ω by

$$\omega^2 = \frac{k}{m}$$

Then the equation of motion becomes

$$\frac{d^2x}{dt^2} = -\omega^2 x \tag{9.16}$$

The mathematical problem is to find a function of time $x(t)$ that is the general solution to this differential equation. The highest-order derivative appearing is a second derivative, so the differential equation is what is called a second-order differential equation and it can be proved that the general solution to such a differential equation contains two arbitrary constants. (Values of these two arbitrary constants can then be selected so that the solution fits two initial conditions.) There is no standard method that can be applied with success to solve all types of differential equations. One method that is not very subtle, but is quite acceptable, is to proceed by trial and error until we find a function which satisfies the equation and contains the correct number of arbitrary constants. There is no danger of stumbling upon the wrong solution because there is a "uniqueness theorem" which guarantees that there is only one solution that satisfies a given second-order differential equation and also fits two given initial conditions. From the previous work in this chapter we have no difficulty guessing that the general solution to the differential equation, Eq. 9.16, is the function

$$x = A \cos(\omega t + \theta_0)$$

where A and θ_0 are arbitrary constants. We find

$$\frac{d^2}{dt^2}[A \cos(\omega t + \theta_0)] = -\omega^2 A \cos(\omega t + \theta_0)$$

which verifies that $A \cos(\omega t + \theta_0)$ is a solution to Eq. 9.16. Since this solution contains two arbitrary constants, A and θ_0, it is the general solution to the second-order differential equation. Therefore, the answer to the problem of determining the motion of a particle of mass m acted on by a resultant force, $F_x = -kx$, is that the particle executes simple harmonic motion

$$x = A \cos(\omega t + \theta_0)$$

with an angular freqency $\omega = \sqrt{k/m}$.

In a multitude of different physical situations, we encounter a differential equation of the form

$$\frac{d^2}{dt^2}(variable) = -(positive\ constant)(variable) \tag{9.17}$$

with the same dependent variable on each side of the equation. Such an equation is mathematically identical to Eq. 9.16, and we can immediately state that the variable is a sinusoidal function of time, $A \cos(\omega t + \theta_0)$, with an angular frequency ω which is the square root of the positive constant in Eq. 9.17.

9.6 Simple Pendulum

Figure 9-9 shows a particle of mass m suspended from a fixed point by a light inextensible string of length L. Such an idealized system is called a *simple pendulum*. The lowest point accessible to the particle, point O, is obviously the position of stable equilibrium.

To analyze the motion of the mass m we consider the free-body diagram for the particle when it is displaced a horizontal distance x from the equilibrium position (Figure 9-10). Denoting the tension in the string by P we find that the x component of the resultant force acting on m is given by

$$F_x = -P \sin \beta = -\frac{Px}{L}$$

As the particle swings back and forth, the tension P will vary, but, provided x/L remains small, the value of P will not depart appreciably from the value mg that it would have if the particle were at rest at O. Using this approximate value, $P = mg$, we obtain

$$F_x = -\frac{mg}{L} x$$

Newton's second law therefore gives

$$-\frac{mg}{L}x = m\frac{d^2x}{dt^2}$$

or

$$\frac{d^2x}{dt^2} = -\frac{g}{L}x$$

This differential equation has the form of Eq. 9.17. Therefore (provided x/L remains small) the particle will execute simple harmonic motion with an angular frequency ω determined from

$$\omega^2 = \frac{g}{L}$$

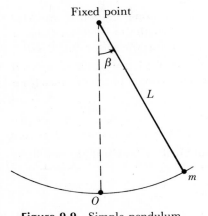

Fixed point

Figure 9-9   Simple pendulum.

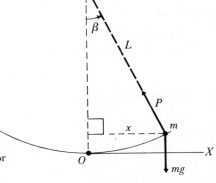

Figure 9-10   Free-body diagram for mass m.

Consequently, the period T of the motion is

$$T = \frac{2\pi}{\omega} = 2\pi\sqrt{\frac{L}{g}} \tag{9.18}$$

Because of its periodic motion, the pendulum is used as the basic timekeeper in pendulum clocks. An exact analysis of pendulum motion shows that the motion is not precisely simple harmonic motion and that the period increases slightly if the amplitude of oscillation is increased. But even for amplitudes so large that the angle β reaches $15°$, the error in the expression $T = 2\pi\sqrt{L/g}$ is less than 0.5%.

The period of a pendulum can be measured with great accuracy and we can then determine an accurate value of the gravitational field g acting on the pendulum. This is the principle underlying the precise "g-meters" that are so valuable in geophysical detection of possible deposits of ore or oil. Because their density is appreciably different from that of their surroundings, such deposits give rise to local variations in the value of the gravitational field.

Summary

☐ A particle, with x coordinate given as a function of time by

$$x = A\,\cos(\omega t + \theta_0)$$

where ω, A, and θ_0 are constants, executes simple harmonic motion of angular frequency ω, amplitude A, and initial phase θ_0. The period T and frequency f are given by $T = 1/f = 2\pi/\omega$.

☐ A convenient representation of this simple harmonic motion is provided by a phasor of magnitude A at an angle $\omega t + \theta_0$ with the X axis on a phasor diagram. The coordinate x is the projection of this phasor on the X axis.

☐ In simple harmonic motion, the particle's acceleration vector is directed toward the equilibrium position O and is proportional to the distance from O:

$$a_x = -\omega^2 x$$

☐ A particle executes simple harmonic motion if and only if the resultant force acting on the particle obeys Hooke's law:

$$F_x = -kx$$

☐ The period of a simple pendulum of length L is

$$T = 2\pi\sqrt{\frac{L}{g}}$$

Questions

1 (a) In the motion of the mass m in Figure 9-1, identify the trip that constitutes a cycle starting at the point $x = -A$.
   (b) If the frequency is 4.0 Hz, how long will it take to complete the cycle specified in part (a)?

2 In one step of an experiment designed to measure the gravitational field, a precise determination of the period of a certain pendulum is to be made from the observation that the pendulum completes 200.0 cycles in 440.0 s. Calculate the frequency and the period.

3 A particle oscillates with a frequency of 0.400 Hz. How long will it take to execute 50 cycles?

4 Show on a phasor diagram representing the simple harmonic motion of a point P the significance of the following terms: phase angle, initial phase angle, amplitude.

5 A phasor 0.30 m long rotates with an angular speed of 0.40 rad/s:
   (a) Find the period and the amplitude of the simple harmonic motion executed by the projection P of the tip of the phasor on a diameter of its path.
   (b) The initial phase angle is π rad. How long will it take point P to move from its initial position to the equilibrium position?

6 The motion of a particle P is described in SI units by

$$x = 2.00 \cos\left(5.0t + \frac{\pi}{2}\right)$$

   For this simple harmonic motion find:
   (a) The angular frequency.
   (b) The frequency.
   (c) The period.
   (d) The initial phase angle.
   (e) The amplitude.

7 Draw a diagram showing the length and the initial phase angle of the phasor whose projection on the X axis will execute the simple harmonic motion described in the preceding question.

8 Give the displacement x as a function of time t for a particle which executes a simple harmonic motion with a period of 0.50 s. The total range of the particle's motion is 0.16 m. The initial phase angle is zero.

9 (a) What is the speed of the tip of the rotating phasor in Question 5?
   (b) What is the maximum speed attained by the projection P of this phasor tip on a diameter of its circular path?
   (c) Where in its motion does P reach maximum speed?
   (d) What is the minimum speed of P, and where is this attained?

10 A particle executes simple harmonic motion with an *angular* frequency of 5.0 rad/s. As it passes through the equilibrium position its speed is 2.5 m/s. What is the amplitude of this simple harmonic motion?

11 (a) Find the magnitude and the direction of the acceleration of the tip of the phasor of Question 5.
   (b) What is the maximum acceleration of the projection P of this phasor tip on a diameter of its circular path?
   (c) Where in its motion does P have maximum acceleration?

(d) What is the minimum magnitude of the acceleration of P, and where will this occur?

12 The position coordinate of an oscillating particle is given in SI units by

$$x = 5.0 \cos\left(0.50\pi t + \frac{\pi}{3}\right)$$

(a) Find the angular frequency, the frequency, and the period.
(b) Find the initial phase angle and the amplitude.
(c) At the instant $t = 2.00$ s, find the position, the velocity, and the acceleration.

13 The end of a prong of a tuning fork executes simple harmonic motion with a frequency of 256 Hz and an amplitude of 0.40×10^{-3} m. Find the maximum acceleration and the maximum speed of the end of the prong.

14 **(a)** What is the force constant of a spring that is stretched 5.0×10^{-2} m by a force of 10.0 N?
(b) What is the period of oscillation of a 0.80-kg mass which is fastened to this spring as in Figure 9-1?

15 **(a)** When the oscillating mass in Figure 9-1 is 2.0 kg, the period of oscillation is 1.5 s. If a 6.0-kg mass is fastened to the 2.0-kg mass and the same spring is used, what will be the period of oscillation?
(b) What is the force constant of the spring?
(c) What force is exerted by this spring on the mass fastened to one end when the spring is stretched a distance of 0.10 m?

16 A block rests on a rough board which is moved horizontally back and forth in a simple harmonic motion of amplitude 0.15 m. The frequency is slowly increased until, at a frequency of 0.25 Hz, the block just begins to slide. Compute the coefficient of friction between the block and the board.

17 What is the period of small oscillations of a simple pendulum which is 1.00 m long?

18 A clock timekeeper is desired with a period of 2.00 s. If a simple pendulum is to be used, what length should it have?

19 What is the value of the gravitational field at the location of the pendulum of Question 2? The length of the pendulum is 1.20 m.

Supplementary Questions

S-1 Suppose a mass m is suspended from a vertical spring with a force constant k (Figure 9-11). When the spring stretches a distance s, the suspended mass m is in equilibrium. We select an axis OX directed downward and the origin O at the equilibrium position. When the mass m has a position coordinate x, the total stretch of the spring is $s + x$. Show that the resultant force acting on the suspended mass is given by

$$F_x = -k(s + x) + mg$$

and that this reduces to

$$F_x = -kx$$

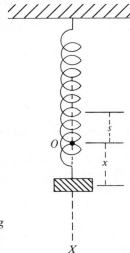

Figure 9-11 *Resultant* force acting on a suspended mass is given by $F_x = -kx$.

S-2 When a 2.0-kg mass is suspended in *equilibrium* from a vertical spring, we observe that the spring has stretched by 4.0×10^{-2} m. The suspended mass is then pulled down 1.5×10^{-2} m further and released at rest. What is:
  (a) The force constant of the spring?
  (b) The period of oscillation of the mass?
  (c) The amplitude of the simple harmonic motion executed by the mass?

S-3 A spring with a force constant of 39.2 N/m is suspended from the ceiling. A 0.40-kg mass is attached to the lower end of the spring and then released at rest:
  (a) How far will the mass descend before reaching the equilibrium position?
  (b) How far below the equilibrium position will the mass descend?
  (c) Find the frequency and the amplitude of the simple harmonic motion executed by the oscillating mass.

S-4 The 1400-kg body of a car, suspended on springs, executes vertical oscillations with a frequency of 4.0 Hz:
  (a) What is the force constant of this system of springs?
  (b) What will be the frequency of oscillation when there are four passengers in the car, each with a mass of 100 kg?

S-5 A block is placed on top of a piston which executes simple harmonic motion along a vertical line with a frequency of 2.0 Hz. The amplitude of oscillation gradually increases. At what amplitude of motion will the block and piston separate?

S-6 A mass is supported by two springs with force constants k_1 and k_2:
  (a) Show that when one end of each spring is connected to the ceiling and the other end is connected to a suspended mass, the pair of springs exerts the same force that would be exerted by a single spring with a force constant k given by

$$k = k_1 + k_2$$

(b) Show that when an end of one spring is joined to an end of the other to form a single long spring, this combination has a force constant k determined by

$$\frac{1}{k} = \frac{1}{k_1} + \frac{1}{k_2}$$

and that this implies that

$$k = \frac{k_1 k_2}{k_1 + k_2}$$

S-7 **(a)** At what position will the tension in an oscillating simple pendulum reach its maximum value?

(b) Show that this maximum tension in a pendulum of length L, for small oscillations of amplitude A, has the value $mg[1 + (A/L)^2]$.

S-8 A small block, placed inside a smooth hemispherical bowl of 0.40-m diameter, slides back and forth executing oscillations about its lowest point. What is the period of these oscillations?

S-9 When a particle executes simple harmonic motion, its velocity and acceleration are each sinusoidal functions of time and therefore can be represented by phasors. Draw the phasors representing x, v_x, and a_x on one diagram. The angle between two phasors gives the phase difference between the quantities represented by the phasors. From the phasor diagram, determine the phase difference between x and v_x, x and a_x, v_x and a_x.

S-10 A particle of mass M is attached to the midpoint of a light wire of length L which is stretched tightly between two fixed points with a tension P. The mass M is given a small displacement perpendicular to the wire and released. Show that the period of the oscillations executed by this mass is $\pi\sqrt{ML/P}$. (For simplicity, assume that the oscillations take place along a horizontal line OX.)

IO.

energy of
a particle

The concept of energy as a conserved quantity that can be transformed but never created or destroyed has become perhaps the most useful idea in all science. In this first chapter on energy, the system under consideration is a single particle acted upon by external forces. After defining work, we introduce energy of motion (kinetic energy) and energy associated with position or configuration (potential energy); we then examine the conservation of the particle's total energy, the sum of its kinetic and potential energies.

10.1 Work

Work Done by a Constant Force

The term *work* is used in physics as a technical term with a precise meaning that must be distinguished from its usage in daily life. The work W_{AB} done by a constant force $\mathbf{F}$, when the particle on which it acts moves a distance Δs along a straight path from A to B (Figure 10-1), is defined by

$$W_{AB} = F_s \Delta s \qquad \text{($\mathbf{F}$ constant)} \quad (10.1)$$

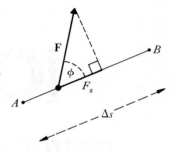

Figure 10-1 Force **F** does work $W_{AB} = F_s \Delta s$ on particle which goes from A to B.

F_s denotes the component of **F** in the direction of the particle's motion and can be positive, zero, or negative depending on the direction of **F**. From Figure 10-1 we see that $F_s = F \cos \phi$. Consequently, Eq. 10.1 can be written

$$W_{AB} = F \Delta s \cos \phi \qquad \text{(\textbf{F} constant)} \quad (10.2)$$

Thus:

1   If **F** is in the direction of motion,

$$F_s = F \quad \text{and} \quad W_{AB} = F \Delta s$$

2   If **F** is perpendicular to the direction of motion,

$$F_s = 0 \quad \text{and} \quad W_{AB} = 0$$

3   If **F** is opposite to the direction of motion,

$$F_s = -F \quad \text{and} \quad W_{AB} = -F \Delta s$$

Work is a *scalar* quantity that can be positive, zero, or negative. It must be emphasized that if there is no motion, $\Delta s = 0$ and the force does no work in the sense that physicists use the term. No work is done on a suitcase by a person who merely holds it without moving it up or down.

In SI units, the unit of work is named the *joule* (J) and

$$1 \text{ J} = 1 \text{ N} \cdot \text{m}$$

Thus, a force of 6.0 N acting on a particle which moves 4.0 m in the direction of the force does a work of 24 J on the particle.

Example 1   In Figure 10-2, evaluate the work done by the gravitational force (the weight) $m\mathbf{g}$ acting on a particle of mass m as the particle is moved (by the application of other forces) from:

(*a*) A to B (*d*) A to C directly
(*b*) B to A (*e*) A to B to C to A
(*c*) A to B to C

Solution

(*a*) For the path AB, the force $m\mathbf{g}$ is in a direction opposite to the direction of motion. Therefore, denoting the tangential component of $m\mathbf{g}$ by $(mg)_s$, we have $W_{AB} = (mg)_s \Delta s = (-mg)y$.

(b) $W_{BA} = (m\mathbf{g})_s \Delta s = mgy.$

(c) $W_{ABC} = W_{AB} + W_{BC} = -mgy + (m\mathbf{g})_s x = -mgy + 0$, since $m\mathbf{g}$ is perpendicular to the path BC.

(d) For the straight path AC: $(m\mathbf{g})_s = -mg \cos \phi$, length $AC = \Delta s = y/\cos \phi$. Therefore $W_{AC} = (m\mathbf{g})_s \Delta s = (-mg \cos \phi)\Delta s = -mgy$. Notice that the work done by $m\mathbf{g}$ in going from A to C has the same value for different paths connecting these points.

(e) $W_{CA} = (m\mathbf{g})_s \Delta s = (mg \cos \phi)\Delta s = mgy$. Therefore $W_{ABCA} = W_{AB} + W_{BC} + W_{CA} = -mgy + 0 + mgy = 0$. The work done by the force $m\mathbf{g}$ is zero for the closed path.

Example 2 A block slides on a rough plane along a path of length Δs:

(a) Find the work done by the normal component $\mathbf{N}$ of the force exerted on the block by the plane.

(b) Find the work done by the force of kinetic friction $\mathbf{f}$ exerted on the block by the plane.

Solution

(a) Because $\mathbf{N}$ is normal to the path, $N_s = 0$ and $W = N_s\Delta s = 0$.

(b) Because $\mathbf{f}$ is always in the direction opposite to the direction of motion of the block, $f_s = -f$ and $W = f_s\Delta s = -f\Delta s$.

Work When F_s Varies Along the Path

In the general case when F_s has different values at different points on the path from A to B (Figure 10-3), we divide the path into n segments, each of a length Δs that is so short that within each segment F_s does not alter appreciably. For each segment we compute a product $F_s\Delta s$ (using any value of F_s occurring within the segment). The sum of the values of $F_s\Delta s$ for each segment of the path AB is called an approximating sum. The work W_{AB} is defined as the limit of such approximating sums, as n increases indefinitely. In mathematics, such a

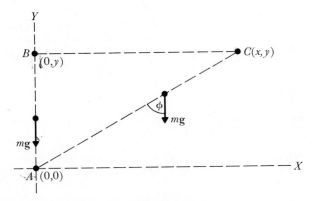

Figure 10-2 Particle of mass m is moved along the path $ABCA$.

limit of sums is called the line integral (or the integral along the path) of **F** and is denoted by $\int_A^B F_s\,ds$. We have

$$W_{AB} = \int_A^B F_s\,ds \tag{10.3}$$

as the general definition of the work done by a force **F** on a particle which moves from A to B along a specified path.

As an example, consider the work done by the Hooke's law force acting on the particle of Figure 10-4 as it moves from the origin to a coordinate x. For this path in the positive x direction, $ds = dx'$ and $F_s = -kx'$. Therefore

$$W = \int_0^x -kx'\,dx' = -\tfrac{1}{2}kx^2 \tag{10.4}$$

The average value $\overline{F}_s$ of the tangential component along the path from A to B is defined by

$$\overline{F}_s = \frac{1}{\Delta s}\int_A^B F_s\,ds \tag{10.5}$$

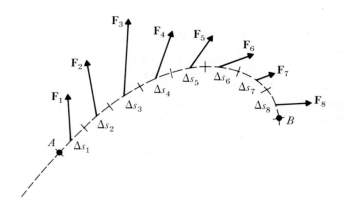

Figure 10-3 The sum of the values of $F_s\Delta s$ for each segment of the path AB is called an approximating sum. The work W_{AB} is defined as the limit of the approximating sums as the number of segments increases indefinitely.

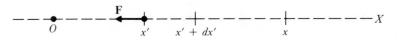

Figure 10-4 The Hooke's law force, $F_x = -kx'$, varies as the particle moves from the origin to x. The work done by this force is

$$\int_0^x -kx'\,dx'$$

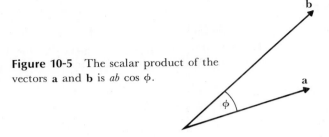

Figure 10-5 The scalar product of the vectors **a** and **b** is $ab \cos \phi$.

where Δs is the distance measured along the path from A to B. Using this definition of $\overline{F}_s$, Eq. 10.3 can be written

$$W_{AB} = \overline{F}_s \Delta s \tag{10.6}$$

This expression shows that we can always think of work as

force × distance

but when the tangential component of the force is not constant along the path, we must use its average value.

Work Expressed as a Scalar Product*

In computing the work done by a constant force, we multiply the magnitude of the vector Δ**s** by the component of another vector **F** in the direction of Δ**s**. This product is called the *scalar* or the *dot product* of the vectors **F** and Δ**s**. For any two vectors, **a** and **b**, their scalar or dot product, written as **a·b** and read as "**a** dot **b**," is defined by

$$\mathbf{a \cdot b} = ab \cos \phi \tag{10.7}$$

where a is the magnitude of the vector **a**, b is the magnitude of the vector **b**, and ϕ is the (smaller) angle between **a** and **b** (Figure 10-5). According to this definition, **a·b** is a scalar quantity which is positive if ϕ is an acute angle and negative if ϕ is an obtuse angle. When **a** and **b** are orthogonal (perpendicular), ϕ is 90°, $\cos \phi = 0$, and

$$\mathbf{a \cdot b} = 0 \qquad \text{(orthogonal vectors)} \tag{10.8}$$

Using this notation, Eq. 10.2 for the work done by a constant force **F** in a displacement Δ**s** from A to B can be written

$$W_{AB} = \mathbf{F \cdot \Delta s} \tag{10.9}$$

and in the general case (Eq. 10.3) where **F** may vary along the path from A to B,

$$W_{AB} = \int_A^B \mathbf{F \cdot} d\mathbf{s} \tag{10.10}$$

*This material can be omitted without loss of continuity.

10.2 Power

In applications it is often more important to know the *rate* at which a machine can do work rather than the total amount of work that the machine can do. The *rate* at which work is done is called *power*. The instantaneous power P is defined by

$$P = \frac{dW}{dt} \tag{10.11}$$

The average power $\overline{P}$ for a time interval Δt during which the work ΔW is done is given by

$$\overline{P} = \frac{\Delta W}{\Delta t} \tag{10.12}$$

The SI unit of power, the joule per second, is named the *watt* (W):

$$1 \text{ W} = 1 \text{ J/s}$$

A commonly used multiple of this unit is the *kilowatt* (kW): 1 kW = 1000 W. In engineering, power is often given in *horsepower* (hp): 1 hp = 746 W.

Any power unit multiplied by a time unit is a unit of work. Thus, *one kilowatt hour* (kW·h) is the *work done in one hour at a constant rate of one kilowatt:*

$$1 \text{ kW·h} = (1000 \text{ W})(3600 \text{ s}) = 3.6 \times 10^6 \text{ J}$$

The instantaneous power supplied by a force **F** to a particle moving with a speed $v = ds/dt$ is given by

$$P = \frac{dW}{dt} = F_s \frac{ds}{dt} = F_s v \tag{10.13}$$

When F_s is positive, power is supplied to the particle, but if F_s is negative, the particle delivers power to the agent exerting the force **F**.

10.3 Kinetic Energy and Potential Energy of a Particle

The Work-Energy Principle

We begin the study of energy by considering a system which is a single particle of mass m acted on by external forces, such as a gravitational force and a Hooke's law force. Newton's second law for tangential components states that

$$F_s = ma_s \tag{8.21}$$

where F_s is the tangential component of the resultant force acting on the particle. The tangential component a_s of the particle's acceleration can be expressed in terms of the particle's speed $v \, (= ds/dt)$ and the distance s:

$$a_s = \frac{dv}{dt} = \frac{dv}{ds}\frac{ds}{dt} = v\frac{dv}{ds}$$

Then Newton's second law for tangential components becomes

$$F_s = mv \frac{dv}{ds}$$

or, written in terms of differentials,

$$F_s \, ds = mv \, dv$$

This can be integrated along the particle's path from any point A to any other point B:

$$\int_A^B F_s \, ds = \int_A^B mv \, dv$$

When the integral on the right is evaluated (using the fact that m is constant) this equation becomes

$$\int_A^B F_s \, ds = \tfrac{1}{2}mv_B^2 - \tfrac{1}{2}mv_A^2 \tag{10.14}$$

where v_A and v_B are the speeds of the particle at the points A and B, respectively. The quantity $\tfrac{1}{2}mv^2$ is called the kinetic energy K of the particle:

$$K = \tfrac{1}{2}mv^2 \qquad \text{(particle)} \quad (10.15)$$

Equation 10.14 therefore gives

$$W_{AB} = K_B - K_A \tag{10.16}$$

a result known as the *work-energy principle: The work done by the resultant force acting on a particle is equal to the change of the particle's kinetic energy.*

Kinetic Energy

The kinetic energy of a particle is a scalar quantity that depends only on the particle's speed and not on the direction of motion. The physical dimensions of kinetic energy are the same as those of work $[ML^2T^{-2}]$ and the same unit (the joule in the SI system) is used for both quantities.

The kinetic energy $\tfrac{1}{2}mv^2$ of a moving particle is equal to the work that must be done on the particle in order to accelerate it from rest to the speed v. (Equation 10.16, $W_{AB} = K_B - K_A$, becomes $W = K - 0$ for a particle accelerated from rest to kinetic energy K.) And if a particle has a kinetic energy $\tfrac{1}{2}mv^2$, this is the amount of work the particle can do on other objects in the process of being brought to rest by a resultant force $\mathbf{F}$. (Equation 10.16, $W_{AB} = K_B - K_A$, becomes $W = 0 - K$ for a particle which starts with kinetic energy K and is brought to rest. Therefore $K = -W$, which is the work done *by* the particle which exerts a force $-\mathbf{F}$, equal in magnitude and opposite in direction to the force $\mathbf{F}$ exerted *on* the particle.)

The following is a general definition of energy that is often adopted: *The energy of an object is a measure of its ability to do work.* We have found that work can be done by a moving particle. *Kinetic energy is defined as energy associated with motion.* When we take this definition as the starting point, the preceding analysis justifies the identification of the term $\frac{1}{2}mv^2$ as the kinetic energy of a particle.

Potential Energy

For a force which is a function of position, it may be possible to find a related function U such that

$$F_s = -\frac{dU}{ds} \qquad (10.17)$$

Then the work done by such a force is

$$W_{AB} = \int_A^B F_s \, ds = \int_A^B -\frac{dU}{ds} \, ds = -(U_B - U_A) \qquad (10.18)$$

If the work done by the *resultant* force acting on a particle is $-(U_B - U_A)$, the work-energy principle (Eq. 10.16) gives

$$-(U_B - U_A) = K_B - K_A$$

or

$$K_A + U_A = K_B + U_B \qquad (10.19)$$

which is a statement in symbols of the *law of conservation of energy for a particle: The particle's total energy, K + U, is conserved.* It must be noted that this law holds only when the work done by the *resultant* force is given by Eq. 10.18, $W_{AB} = -(U_B - U_A)$.

The quantity U, a function of position only, is called the particle's *potential energy.* In other words, the *potential energy of a particle is defined as the energy the particle possesses because of its position.* Potential energy is a scalar quantity with the same physical dimensions and the same unit (the joule in the SI system) as kinetic energy and work. During the motion of the particle, while the particle is slowing down, its kinetic energy is transformed into potential energy, and while the particle is speeding up, its potential energy is transformed into kinetic energy, always in such a way that the sum $K + U$ remains constant if the forces are such that energy is *conserved.*

A force **F** that can be derived from a potential energy function U (by $F_s = -dU/ds$) is called a *conservative* force, because if this is the only force doing work on a particle, the total energy $K + U$ will be conserved. The function U is the potential energy associated with the force **F**. If there are several different conservative forces acting on a particle, its potential energy will be the sum of the potential energy functions associated with these different conservative forces.

10.4 Potential Energy Functions

In order to make use of the law of conservation of energy in mechanics, we must find the potential energy functions associated with the conservative forces acting on a particle. A potential energy function U associated with a force $\mathbf{F}$ is a function of position such that, at any point

$$F_s = -\frac{dU}{ds} \qquad (10.20)$$

where F_s is the component of the force in the direction of a displacement of magnitude ds away from the point.

One-Dimensional Motion

For a one-dimensional motion along the X axis, $ds = dx$ and Eq. 10.20 simplifies to

$$F_x = -\frac{dU}{dx} \qquad (10.21)$$

As an example consider the function

$$U = \tfrac{1}{2}kx^2 \qquad (10.22)$$

The force derived from this potential energy function is

$$F_x = -\frac{dU}{dx} = -\frac{d}{dx}\left(\tfrac{1}{2}kx^2\right) = -kx$$

which is the Hooke's law force.

Any constant can be added to the potential energy without changing the associated force, because the derivative of a constant is zero. For example, the force derived from

$$U = \tfrac{1}{2}kx^2 - \tfrac{1}{2}kx_0^2$$

where x_0 is a constant, is again $F_x = -kx$.

Given a potential energy function, we can find the associated force by differentiation. The inverse problem, that of finding the potential energy function for a given force, is solved by integration. From Eq. 10.21

$$dU = -F_x\,dx$$

and

$$\int_{x_0}^{x} dU = -\int_{x_0}^{x} F_x\,dx'$$

where x_0 is any arbitrary reference point. Then

$$U(x) = -\int_{x_0}^{x} F_x(x')\,dx' + U(x_0) \qquad (10.23)$$

which shows that the potential energy associated with a force F is the negative of the work done by this force when the particle goes from x_0 to x. The additive constant $U(x_0)$ can be assigned any convenient value, usually zero.

For the Hooke's law force, this procedure gives

$$U(x) = -\int_{x_0}^{x} (-kx')\,dx' + U(x_0)$$

$$= \tfrac{1}{2}kx^2 - \tfrac{1}{2}kx_0^2 + U(x_0)$$

which assumes its simplest form by taking the arbitrary reference point to be the origin ($x_0 = 0$) and the arbitrary value of the potential energy at this reference point to be zero [$U(x_0) = 0$].

Three-Dimensional Motion

When a particle moves in three dimensions, a potential energy function is a function of the particle's three position coordinates (x,y,z):

$$U = U(x,y,z)$$

At a point (x,y,z) the x component of the force $\mathbf{F}$ derived from this potential energy function is found by applying $F_s = -dU/ds$ to a displacement of magnitude ds in the positive x direction. For such a displacement, $dx = ds$, $dy = 0$, and $dz = 0$. We write

$$F_x = -\frac{\partial U}{\partial x} \tag{10.24}$$

using curved symbols for the derivative to indicate that the variables y and z are to be held constant in evaluating this derivative. $\partial U/\partial x$ is called the partial derivative of $U(x,y,z)$ with respect to x. Thus if $U(x,y,z) = -xy^2 + x^3yz^2$, we have

$$\frac{\partial U}{\partial x} = -y^2 + 3x^2yz^2$$

The y and z components of the force $\mathbf{F}$ derived from $U(x,y,z)$ are found in a similar way:

$$F_y = -\frac{\partial U}{\partial y} \quad \text{and} \quad F_z = -\frac{\partial U}{\partial z} \tag{10.25}$$

Example 3   Find the components of the force $\mathbf{F}$ derived from the potential energy function

$$U(x,y,z) = -xy^2 + x^3yz^2$$

Solution   The components of $\mathbf{F}$ are

$$F_x = -\frac{\partial U}{\partial x} = y^2 - 3x^2yz^2$$

$$F_y = -\frac{\partial U}{\partial y} = 2xy - x^3z^2$$

$$F_z = -\frac{\partial U}{\partial z} = -2x^3yz$$

The relationship, $F_s = -dU/ds$, between the force component and the potential energy determines not only the magnitude but also the *direction* of the force. At a given point, the direction for which F_s is a maximum is the direction of **F**. It follows that the *direction of the force at a given point in space is the direction in which the potential energy decreases most rapidly.*

A simple but important potential energy function is $U = mgy$, where the positive y direction is vertically upward. The force **F** derived from this potential energy function has components

$$F_x = -\frac{\partial}{\partial x}(mgy) = 0 \qquad F_y = -\frac{\partial}{\partial y}(mgy) = -mg \qquad F_z = -\frac{\partial}{\partial z}(mgy) = 0$$

which shows that $\mathbf{F} = m\mathbf{g}$. Therefore the function

$$U = mgy$$

is the potential energy of a particle of mass m in a uniform gravitational field $\mathbf{g}$ directed vertically downward. This potential energy is called *gravitational potential energy.*

For a given conservative force **F**, the potential energy function $U(x,y,z)$ can be determined by integration:

$$U(x,y,z) = -\int_A^{x,y,z} F_s\, ds + U_A \tag{10.26}$$

The integral is the work done by **F** when the particle moves from the arbitrary reference point A to the point with coordinates x,y,z. For conservative forces, since the function $U(x,y,z)$ exists, this work must be independent of the path of integration selected to go from A to the point x,y,z (Figure 10-6). The fact that the work done by a conservative force has the same value for different paths joining the same end points was illustrated in Example 1 for the conservative gravitational force $m\mathbf{g}$. The student should verify that the calculations of that example confirm that the gravitational potential energy function is given by $U = mgy$, if the reference point A is selected in the plane $y = 0$ and $U_A = 0$.

Figure 10-6 The work done by a conservative force on a particle that moves from A to the point (x,y,z) has the same value for different paths.

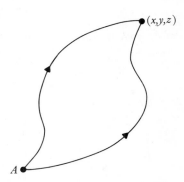

Let us now consider the work done by a conservative force **F** on a particle which moves in a *closed path* so as to return to its original position (Figure 10-7). Then in Eq. 10.26, the final point (x,y,z) is the same as the initial point A, and this equation yields

$$\oint F_s \, ds = 0 \qquad (10.27)$$

where the circle around the integration sign is used to indicate that the path of integration is a closed curve and that the initial point coincides with the final point. We have found that a conservative force has the following remarkable property: *If a particle moves on any closed path and returns to its starting point, the work done on the particle by a conservative force is zero.* This has been illustrated by the calculation given in Example 1(e) for the case of the conservative force $m\mathbf{g}$ and the closed path $ABCA$ of Figure 10-2.

Nonconservative Forces

Not all forces are conservative. Because the force of kinetic friction acting on a particle depends on the direction of the particle's motion, this force is not simply a function of the particle's position. Consequently, the force cannot be derived from a potential energy function. Such a force is a *nonconservative force.* The work done by a frictional force acting on a particle that moves on a path of length Δs is given by $W = -f\Delta s$ and therefore depends on the path joining the end points. This feature of a path-dependent work is characteristic of nonconservative forces and the work done by a nonconservative force when the path is closed is generally *not* equal to zero.

For a one-dimensional motion, any force which is a definite function of position is conservative. But in two or three dimensions, not all functions of position correspond to conservative forces.

10.5 Conservation of Energy of a Particle

The *law of conservation of energy for a particle* is: *For any two points A and B on the particle's path*

$$K_A + U_A = K_B + U_B \qquad (10.19)$$

provided no work is done by nonconservative forces acting on the particle. U is the sum

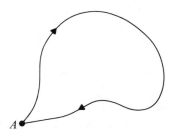

Figure 10-7 If a particle moves in a closed path so as to return to its original position, the work done on the particle by a conservative force is zero.

A

of the potential energies associated with the conservative forces acting on the particle. This law relates the particle speed to its position but makes no reference to the time elapsed in going from A to B. Questions involving speed and position that might appear formidable if approached by direct use of Newton's second law can be answered with relative ease by applying the law of conservation of energy.

Example 4   A stone is projected with a speed v_0. Find its speed at altitude y (Figure 10-8).

Solution   If we neglect air resistance, the only force acting on the stone is its weight mg, and the associated potential energy function is the gravitational potential energy, $U = mgy$. Equating the stone's total energy $K + U$ at altitude y to the initial value of $K + U$, we obtain

$$\tfrac{1}{2}mv^2 + mgy = \tfrac{1}{2}mv_0^2 + 0$$

$$v = \sqrt{v_0^2 - 2gy}$$

For a given projection speed v_0, the speed for different trajectories is the same at points at the same altitude, as indicated in Figure 10-8.

Example 5   A stone is projected vertically upward with a speed v_0. What is the maximum altitude y_{max} reached by the stone?

Solution   At maximum altitude the speed of the stone is zero. To find this maximum altitude we equate the stone's total energy $K + U$ at this point to its initial value:

$$0 + mgy_{max} = \tfrac{1}{2}mv_0^2 + 0$$

which gives

$$y_{max} = \frac{v_0^2}{2g}$$

Energy transformations occur in all physical processes. The motion considered in Example 5 provides an opportunity to learn to describe energy transformations in a very simple situation. Initially the total energy $K + U$ of the stone is entirely kinetic energy, but as the stone rises its kinetic energy decreases and its gravitational potential energy increases. At maximum altitude, for a vertical trajectory, the stone has no kinetic energy; its gravitational potential energy has reached a maximum equal to the initial value of its kinetic energy.

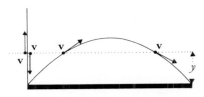

Figure 10-8   For a given projection speed v_0, the speed v on different trajectories is the same at points at the same height.

As the stone falls, kinetic energy is acquired at the expense of gravitational potential energy. Throughout the entire flight, assuming that negligible work is done by the frictional forces exerted on the stone by the air, the total energy $K + U$ of the stone is constant.

Example 6   A stone of mass m is placed on the upper free end of a vertical spring that has been compressed a distance A (Figure 10-9). When the spring is released it projects the stone vertically. Find an expression for the vertical distance h traveled during the ascent of the stone in terms of m, g, A, and the force constant k of the spring.

Solution   Using the coordinate system in Figure 10-9, and equating the value of the stone's total energy $K + U$ at maximum altitude, $y = h - A$, to the initial value of $K + U$, we obtain

$$0 + mg(h - A) = 0 + \tfrac{1}{2}kA^2 + mg(-A)$$

$$h = \frac{kA^2}{2mg}$$

Energy in Simple Harmonic Motion

The resultant force acting on a particle of mass m that executes simple harmonic motion is a Hooke's law force, $F_x = -kx$, and the particle has a potential energy

$$U = \tfrac{1}{2}kx^2$$

The particle has a total energy

$$K + U = \tfrac{1}{2}mv_x^2 + \tfrac{1}{2}kx^2$$

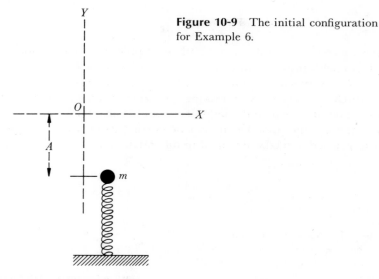

Figure 10-9   The initial configuration for Example 6.

This sum of the particle's kinetic and potential energies is conserved throughout the motion. Where x attains its maximum magnitude A, the speed is zero. Evaluating $K + U$ at this point, we find

$$K + U = 0 + \tfrac{1}{2}kA^2$$

Using Eq. 9.13, $\omega^2 = k/m$, we obtain

$$K + U = \tfrac{1}{2}m\omega^2 A^2 \tag{10.28}$$

which shows that the total energy $K + U$ of a particle executing simple harmonic motion is proportional to the square of the amplitude and to the square of the angular frequency.

Example 7   A particle executes simple harmonic motion with an angular frequency ω. The initial position is x_0 and the initial velocity is v_{0x}. Find the amplitude of the motion.

Solution   Equating the value of the total energy $K + U$ at $x = A$ to its initial value, we obtain

$$0 + \tfrac{1}{2}kA^2 = \tfrac{1}{2}mv_{0x}^2 + \tfrac{1}{2}kx_0^2$$

Using $k = m\omega^2$, we find

$$A = \sqrt{\frac{v_{0x}^2}{\omega^2} + x_0^2}$$

Energy Graphs

Some of the principal features of the motion that takes place when a particle is acted upon by a complicated conservative force can be most easily understood by considering a graph of the potential energy function. Let us consider the one-dimensional motion of a particle whose potential energy function is plotted in Figure 10-10. The total energy $K + U$ of the particle can be determined from the initial position x_0 and the initial speed v_{0x}:

$$K + U = \tfrac{1}{2}mv_{0x}^2 + U(x_0)$$

Since the total energy $K + U$ is constant throughout the motion, its graph is a horizontal line. At any x, the kinetic energy is represented by the vertical distance measured upward from the potential energy curve to the horizontal line representing the value of the total energy $K + U$. The intersections of this line with the potential energy curve occur at the boundaries x_1 and x_2 of the particle's motion. Here the kinetic energy is zero. Regions to the left of x_1 and between x_2 and x_3 are not accessible to the particle because they correspond to a potential energy greater than the total energy $K + U$.

The force **F** acting on the particle is determined by

$$F_x = -\frac{dU}{dx}$$

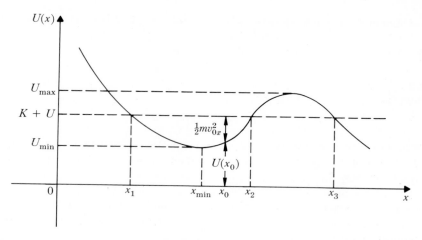

Figure 10-10   Graph of a potential energy function $U(x)$. Particle which starts at x_0 with $K + U = \frac{1}{2}mv_{0x}^2 + U(x_0)$ will oscillate back and forth between x_1 and x_2.

Examination of the slope dU/dx of the potential energy curve reveals that F_x is positive between x_1 and x_{min}, zero at x_{min}, and negative between x_{min} and x_2. This implies that, at all points between x_1 and x_2, the force vector **F** is directed toward x_{min}.

The point x_{min}, where the potential energy reaches a minimum value, is a position of equilibrium because $F_x = -dU/dx = 0$. This position is evidently one of *stable equilibrium*, since, if the particle is displaced from this equilibrium position, it will experience a force directed so as to tend to *restore* the particle to the equilibrium position. At a maximum of the potential energy function, we again have $F_x = -dU/dx = 0$, so such a position is also an equilibrium position. But at a maximum of potential energy, we have a position of *unstable equilibrium* because, if a particle is displaced from such a position, it experiences a force directed so as to increase the magnitude of the displacement.

In the particle motion with total energy $K + U$, the point x_1 is a position where the kinetic energy is zero and the particle is at rest. The force acting on the particle is directed toward the right and the particle is accelerated in this direction reaching a maximum speed at the equilibrium position x_{min}. As the particle continues to the right of x_{min}, it experiences a force directed to the left. The particle slows down until it comes to rest at x_2. It then moves from x_2 toward x_{min}. We see that the particle oscillates back and forth between x_1 and x_2 with a period that is twice the time that it takes to go from x_1 to x_2. If the potential energy curve is parabolic between x_1 and x_2 (Figure 10-11), the motion will be simple harmonic.

If a particle with the potential energy curve of Figure 10-10 starts at infinity with a speed v_∞ and then moves to the left, its speed will decrease and reach zero at x_3 where the particle will turn around and go back to infinity. The point x_3 is the point where the horizontal line representing the value of the particle's total energy $K + U$ intersects the potential energy curve; that is, x_3 is such that

$$0 + U(x_3) = \tfrac{1}{2}mv_\infty^2 + 0$$

The region between x_2 and x_3, which is inaccessible for the given value of total energy $K + U$, is called a *potential barrier* and the region between x_1 and x_2, where the motion is oscillatory, is called a *potential well*. The higher the value of $K + U$, the smaller the width $x_3 - x_2$ of the potential barrier. For $K + U \geq U_{\text{max}}$, there is no barrier and no potential well. In this case oscillatory motion cannot occur and the particle must eventually go off to infinity.

Particle interactions in atoms or in molecules are often described using a potential energy function of the general form illustrated in Figure 10-12. Here the potential energy curve approaches zero as x approaches infinity. In this case, motion with negative total energy $K + U$ is confined to finite values of x and we say that the particle is bound. If $K + U$ is positive, the particle will move off to infinity.

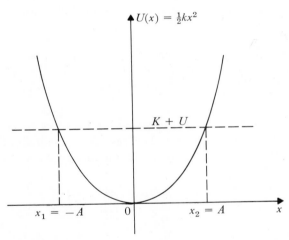

Figure 10-11   When the potential energy curve is a parabola such as $U(x) = \tfrac{1}{2}kx^2$, a particle with total energy $K + U$ executes simple harmonic motion of amplitude A.

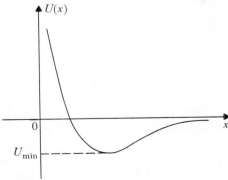

Figure 10-12   For this potential energy function, a particle is bound only when its total energy $K + U$ is negative.

Dissipation of Energy

If a force of kinetic friction opposes the motion of the particle and does a (negative) work W_f as the particle moves from A to B, we find (using the work-energy principle and rearranging terms) that

$$K_A + U_A = K_B + U_B + |W_f| \qquad (10.29)$$

where U is the sum of the potential energies associated with all the other forces acting on the particle and $|W_f|$ is the magnitude of the work done by the force of kinetic friction. The positive quantity $|W_f|$ is often called the work done against friction. Equation 10.29 shows that the total energy $K + U$ decreases by an amount $|W_f|$. Anticipating the idea that this energy has not actually disappeared, we say that energy has been dissipated by friction and that friction is a dissipative force. As we shall discuss in the following chapter, when dissipative forces do work, energy associated with the ordered motion of the entire body (that we have considered as a "particle") is transformed into disordered energy randomly distributed among the molecules of the body, and the body becomes warmer.

Example 8   A block of mass m starts from rest and slides a distance s down a plane inclined at an angle θ with the horizontal. Its motion is opposed by a force of kinetic friction f. Find the block's speed v in terms of s, f, θ, m, and g.

Solution   We may select the horizontal level of the block's final position as the level of zero gravitational potential energy. The magnitude $|W_f|$ of the work done by friction is fs. Since this is the amount of energy which has been dissipated we have

$$0 + mgs \sin \theta = \tfrac{1}{2}mv^2 + 0 + fs$$

$$v = \sqrt{2s\left(g \sin \theta - \frac{f}{m}\right)}$$

Summary

☐   The work done by a force **F** on a particle which moves from A to B along a specified path is

$$W_{AB} = \int_A^B F_s \, ds$$

where F_s is the component of **F** in the direction of motion.

☐   A particle of mass m traveling at speed v has a kinetic energy given by

$$K = \tfrac{1}{2}mv^2$$

☐   The work-energy principle for a particle states that the work W_{AB} done by the *resultant* force acting on a particle is equal to the change $K_B - K_A$

of the particle's kinetic energy:

$$W_{AB} = K_B - K_A$$

☐ A potential energy function U associated with a force $\mathbf{F}$ is a function of position such that, at any point,

$$F_s = -\frac{dU}{ds}$$

where F_s is the component of the force in the direction of a displacement of magnitude ds away from the point. The potential energy function associated with the Hooke's law force, $F_x = -kx$, is $U = \frac{1}{2}kx^2$. In a uniform gravitational field $\mathbf{g}$ directed vertically downward, the gravitational potential energy of a particle of mass m at altitude y is $U = mgy$.

☐ A force $\mathbf{F}$ that can be derived from a potential energy function U (by $F_s = -dU/ds$) is called a conservative force. The work done by a conservative force is the negative of the change in the associated potential energy function

$$W_{AB} = \int_A^B F_s\, ds = \int_A^B -\frac{dU}{ds}\, ds = -(U_B - U_A)$$

This work is independent of the path joining the end points A and B and, for a closed path, this work is zero. The work done by a nonconservative force (such as the force of kinetic friction) depends on the path taken between the end points.

☐ The law of conservation of energy for a particle is: For any two points A and B on the particle's path,

$$K_A + U_A = K_B + U_B$$

provided no work is done on the particle by nonconservative forces.

☐ When a force of kinetic friction does work of magnitude $|W_f|$ as a particle moves from A to B, we have

$$K_A + U_A = K_B + U_B + |W_f|$$

and we say that energy has been dissipated by friction.

Questions

1 A girl exerts a 10.0-N horizontal force as she pushes a box across a horizontal table top through a distance of 0.90 m. How much work is done on the box by this force?

2 A man pushing vertically upward on a stone with a constant force of 200 N moves from point A to B to C to D. Find the work he does for each portion of the trip, given that:
(a) B is 3.0 m above A.
(b) C is 13 m east of B at the same height.
(c) D is 1.0 m below C.

3 A constant horizontal force of 900 N pushes on a crate as it slides up a plane inclined at an angle of 60° to the horizontal. What is the work done by this force on the crate during a 3.0-m displacement of the crate along the inclined plane?

4 The crate in the preceding question weighs 400 N. What is the work done by this weight during the 3.0-m displacement along the inclined plane?

5 Find the work done by a force given in SI units by

$$F_x = 5.0x - 4.0$$

when this force acts on a particle that moves from $x = 2.0$ m to $x = 6.0$ m.

6 The force acting on a suspended particle is given by

$$F_y = -ky + mg$$

Find the work done by this force on a particle which moves from y_1 to y_2.

7 The 2.0×10^3-kg hammer of a pile driver is lifted 2.0 m in 3.0 s. What power does the engine furnish to the hammer?

8 At 4.0 cents per kilowatt hour, what is the cost of operating a 5.0 hp motor for 2.0 h?

9 While a boat is being towed at a speed of 10 m/s, the tension in the towline is 3000 N. What is the power supplied to the boat by the towline?

10 A jet aircraft engine develops a thrust of 12×10^3 N. What power does it supply when the aircraft is traveling at 300 m/s?

11 What power must a motor supply to lift a 70-kg man at a constant speed of 0.50 m/s?

12 A balloon lifts a 200-kg mass with a constant speed to a height of 1500 m in 10.0 min. Find the work done by the force exerted on the mass by the balloon in this ascent. What is the power supplied by this force to the mass?

13 A railroad car is accelerated from rest until its kinetic energy is 5.0×10^5 J. What is the work done by the resultant force acting on the car? If the force has constant magnitude of 1000 N and is in the direction of the motion, find the distance covered by the car before it attains this kinetic energy.

14 Find the force derived from the potential energy function $U = 5x^3$.

15 Show that the force derived from the potential energy function

$$U = -xy^3 + 3xyz^2$$

has components

$$F_x = y(y^2 - 3z^2) \qquad F_y = 3x(y^2 - z^2) \qquad F_z = -6xyz$$

16 Find the components of the force derived from the following potential energy functions:

(a) $U = \frac{1}{2}kr^2$, where k is a constant and $r^2 = x^2 + y^2 + z^2$.
(b) $U = \frac{1}{2}k_x x^2 + \frac{1}{2}k_y y^2 + \frac{1}{2}k_z z^2$, where k_x, k_y, and k_z are constants.

17 Find the potential energy function associated with the force given in SI units by

$$F_x = 27x^2$$

18 Find the potential energy function $U(x)$ associated with the force

$$F_x = \frac{k}{x^2}$$

where k is a constant. Choose $U(x)$ so that $U(\infty) = 0$.

19 A puck slides down a toboggan slide which dips and flattens out, twists and turns. Relate the puck's speed v_1 at altitude y_1 to its speed v_0 at altitude y_0 assuming that the work done by frictional forces is negligible. Show that

$$v_1^2 + 2gy_1 = v_0^2 + 2gy_0$$

20 In Figure 10-13 a 2.0-kg particle slides along a smooth curved surface. The level of the point C is selected as the reference level for the measurement of gravitational potential energy. At each of the points A, B, C, and D find the particle's:

(a) Potential energy.
(b) Total energy $K + U$.
(c) Kinetic energy.
(d) Speed.

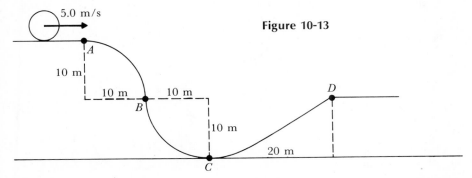

Figure 10-13

21 Use the law of conservation of energy to show that, when a particle executes simple harmonic motion of amplitude A and angular frequency ω, the maximum speed of the particle is

$$v_{\text{max}} = \omega A$$

22 An 8.0-kg mass is suspended from one spring and set into oscillation with an amplitude of 0.15 m. A second identical spring suspends a 2.0-kg mass which oscillates with an amplitude of 0.60 m. What is the ratio of the total energies of these masses?

23 A particle executes simple harmonic motion along the axis OX, where O is the equilibrium position. Specify, in terms of the amplitude A, the values of the position coordinate at which:

(a) The kinetic energy is zero.
(b) The potential energy is zero.
(c) The potential energy is one-fourth the total energy $K + U$.
(d) The potential energy and the kinetic energy are equal.

24 A 0.20-kg cardboard box acquires a speed of 5.0 m/s in falling from rest through a distance of 2.5 m. How much of the box's total energy $K + U$ has been dissipated?

Supplementary Questions

S-1 Show that the force derived from the potential energy function

$$U = \frac{k}{r}$$

where k is a constant and $r = \sqrt{x^2 + y^2 + z^2}$, has components

$$F_x = \frac{kx}{r^3} \qquad F_y = \frac{ky}{r^3} \qquad F_z = \frac{kz}{r^3}$$

and consequently that the force has a magnitude $F = \sqrt{F_x^2 + F_y^2 + F_z^2}$ given by

$$F = \frac{k}{r^2}$$

S-2 A particle of mass m is fastened to one end of a string which is fixed at the other end. The particle moves with varying speed in a vertical circle of radius L (Figure 10-14). At its uppermost position the particle has a speed v_1:
 (a) What is the tension in the string when the string is horizontal?
 (b) Find the tension in the string when the par icle reaches its lowest position.

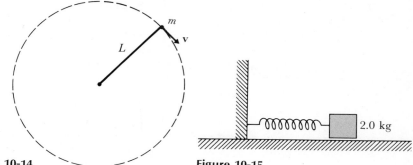

Figure 10-14   **Figure 10-15**

S-3 A 2.0-kg mass is fastened to a spring of force constant 200 N/m and placed on a smooth table as in Figure 10-15. The mass is pulled 0.100 m to the right of its equilibrium position and then projected to the right at a speed of 0.20 m/s:
 (a) What is the amplitude of the motion?
 (b) If the mass were projected to the left from the same position at the same speed, what would be the amplitude?

S-4 A particle's potential energy is given in SI units by

$$U(x) = \frac{8.0}{4.0 + x^2}$$

where x is the position coordinate:
 (a) Sketch a graph of this potential energy function. Locate any positions of equilibrium and classify these as positions of stable or unstable equilibrium.
 (b) The particle is released at rest 2.0 m to the right of the origin. If the mass of the particle is 0.50 kg, what is its initial acceleration? Give a qualitative description of the particle's motion. What is its final speed?
 (c) If the particle is projected from $x = -\infty$ toward the origin, what is the critical projection speed that must be exceeded in order for the particle to continue past the origin and go off to $x = +\infty$?

S-5 A particle's potential energy is given in SI units by

$$U(x) = -\frac{8.0}{4.0 + x^2}$$

where x is the position coordinate:

(a) Sketch a graph of this function. Show that the origin is a position of stable equilibrium.

(b) Find the boundaries of the motion when the particle's total energy $K + U$ is -1.6 J. What is the maximum kinetic energy attained by the particle?

(c) How much energy would have to be supplied to the particle of part (b) to allow it to move off to infinity?

S-6 The Lennard–Jones potential energy function for the interaction of two atoms in a diatomic molecule is

$$U = 4\epsilon \left[\left(\frac{\sigma}{x} \right)^{12} - \left(\frac{\sigma}{x} \right)^6 \right]$$

where σ and ϵ are positive constants and x is the distance between the atoms. Assume that one of the atoms is very heavy and remains at rest at the origin while the other moves along a straight line:

(a) Find the equilibrium position and show that it is a position of stable equilibrium.

(b) What is the minimum value U_{min} of the potential energy?

(c) If the light atom has negligible kinetic energy at $x = \infty$ and moves toward the origin, what will be its kinetic energy at the equilibrium position? How close will this atom come to the origin?

S-7 A block is projected at a speed v_0 along a rough horizontal plane. The coefficient of sliding friction between the block and the plane is μ_k. Using the fact that the decrease of the block's total energy $K + U$ must be equal to the magnitude of the work done by friction, show that the block comes to rest after moving a distance $v_0^2/2\mu_k g$.

S-8 A 10-kg block slides down a rough plane inclined at an angle of 53.1° to the horizontal. The block's speed increases from 4.0 m/s to 8.0 m/s in a distance of 5.0 m measured along the plane:

(a) Find the change in the block's total energy $K + U$.

(b) Find the magnitude of the work done by the frictional force acting on the block.

(c) Compute the coefficient of kinetic friction between the plane and the block.

S-9 In the Atwood's machine in Figure 10-16, the masses are released at rest in the

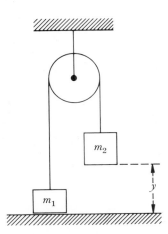

Figure 10-16

positions shown. The mass m_2 is greater than m_1. Show that the speed of m_2 just before it strikes the floor is

$$v = \left[\frac{2(m_2 - m_1)gy}{m_1 + m_2} \right]^{1/2}$$

if no energy is dissipated and the kinetic energy of the pulley is negligible.

S-10 A triangle is formed by the vectors **A**, **B**, and **C** = **A** − **B**. The tails of **A** and **B** coincide at one vertex and enclose an angle θ. By evaluating the scalar products in the equation

$$(\mathbf{A} - \mathbf{B}) \cdot (\mathbf{A} - \mathbf{B}) = \mathbf{C} \cdot \mathbf{C}$$

prove the trigonometric relation known as the *law of cosines:*

$$A^2 + B^2 - 2AB \cos \theta = C^2$$

S-11 Vectors **i**, **j**, and **k** are unit vectors in the positive $x, y,$ and z directions, respectively (see Question S-8 of Chapter 3):
(a) Show that

$$\mathbf{i} \cdot \mathbf{i} = 1 \qquad \mathbf{j} \cdot \mathbf{j} = 1 \qquad \mathbf{k} \cdot \mathbf{k} = 1$$

$$\mathbf{i} \cdot \mathbf{j} = 0 \qquad \mathbf{j} \cdot \mathbf{k} = 0 \qquad \mathbf{k} \cdot \mathbf{i} = 0$$

(b) Use the expressions

$$\mathbf{A} = A_x\mathbf{i} + A_y\mathbf{j} + A_z\mathbf{k}$$

$$\mathbf{B} = B_x\mathbf{i} + B_y\mathbf{j} + B_z\mathbf{k}$$

to evaluate the scalar product **A** · **B** in terms of the components of **A** and **B**. Make use of the simplifications provided by the results of part (a) and thereby show that

$$\mathbf{A} \cdot \mathbf{B} = A_x B_x + A_y B_y + A_z B_z$$

11.

energy of
systems of particles

In the previous chapter we considered the energy of a single particle acted upon by external forces. This discussion will now be extended to systems comprising any number of particles or larger objects. The important idea of *internal energy* is introduced and the general law of conservation of energy is stated. The laws of conservation of energy and of momentum are applied to collisions.

11.1 Work-Energy Principle

During the motion of a system of particles, the work done by the forces acting on the particles of the system includes not only *external work* W_{ext} done by the forces exerted on the system by the particles outside the system but also *internal work* W_{int} done by the forces exerted by two particles of the system on each other when their separation changes. For a *system* of particles the *work-energy principle* becomes

$$W_{\text{ext}} + W_{\text{int}} = \Delta K$$

where ΔK is the change in the system's kinetic energy. (K is the sum of the kinetic energies of all the particles of the system.)

In differential form, the work-energy principle is

$$dW_{ext} + dW_{int} = dK$$

and leads to the power equation

$$\frac{dW_{ext}}{dt} + \frac{dW_{int}}{dt} = \frac{dK}{dt}$$

which states that the algebraic sum of powers supplied by all the forces acting on the particles of the system is equal to the time rate of change of the system's kinetic energy.

The internal work is zero if the distance separating the particles of the system remains constant. Thus, for any motion of a perfectly rigid body, there is no internal work. But for some systems the internal work is all important, as is illustrated in the following example.

Example 1   A car, initially at rest, attains a kinetic energy K by accelerating without skidding along a horizontal road. Neglecting the retarding force exerted on the car by the air, find:
(*a*) The work done by the external forces which accelerate the car.
(*b*) The work done inside the car.

Solution   The car is selected as the system:
(*a*) The external forces acting on the car are shown in Figure 11-1. The resultant external force is the sum $\mathbf{f}_1 + \mathbf{f}_2$ of the forces of *static* friction exerted on the tires by the road. This resultant force accelerates the car. (According to the law of motion of the center of mass, $\mathbf{f}_1 + \mathbf{f}_2 = M\mathbf{a}_c$.) However, since the portions of the tires instantaneously in contact with the road are at rest relative to the road (no skidding), the forces $\mathbf{f}_1$ and $\mathbf{f}_2$ acting on these portions do no work. Therefore, $W_{ext} = 0$.
(*b*) With $W_{ext} = 0$, the work-energy principle implies

$$W_{int} = K$$

The increase of kinetic energy of the car is due entirely to *internal work*.

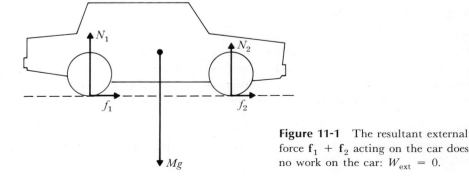

Figure 11-1   The resultant external force $\mathbf{f}_1 + \mathbf{f}_2$ acting on the car does no work on the car: $W_{ext} = 0$.

11.2 External and Internal Potential Energy

If the external forces are conservative, the system will have an associated external potential energy which is the sum of the corresponding potential energies of the particles of the system. For instance, in a uniform external gravitational field g, a system of particles with masses $m_1, m_2, \ldots, m_n$, at altitudes $y_1, y_2, \ldots, y_n$, respectively, has a potential energy U which is the sum

$$U = m_1 g y_1 + m_2 g y_2 + \cdots + m_n g y_n$$

This potential energy U can be calculated as if the entire mass M of the system were concentrated at the center of mass (Figure 11-2). To prove this we note that

$$U = (m_1 y_1 + m_2 y_2 + \cdots + m_n y_n)g$$

and the sum in the parentheses is equal to $M y_c$ from the definition (Eq. 4.3) of the center of mass. Therefore

$$U = Mg y_c$$

When the internal forces are conservative, the system has an *internal potential energy* associated with each pair of particles due to their interaction. Such an internal potential energy belongs to the *pair* and cannot be assigned to one specific particle of the pair.

The total potential energy of the system is the sum of its external and internal potential energies and can be characterized as the energy that the system possesses because of the positions of its parts. The system's total energy, the sum of its kinetic and potential energies, is *conserved* during its motion, provided no work is done by nonconservative external or internal forces.

Example 2   A block of mass m_1 is connected to another block of mass m_2 by a light spring with a force constant k. The spring has been stretched from a

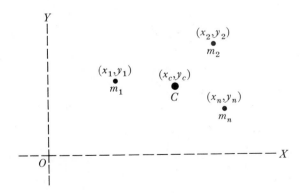

Figure 11-2   In a uniform gravitational field g, the gravitational potential energy of the system of particles can be calculated as if the entire mass were concentrated at the center of mass C.

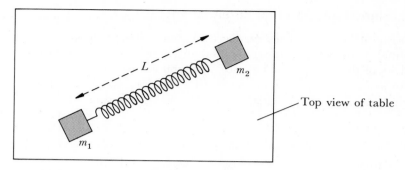

Figure 11-3   Because of the spring, this system of particles of masses m_1 and m_2 has an internal potential energy $U_{12} = \frac{1}{2}k(L - L_0)^2$.

length L_0 to a length L (Figure 11-3). This assembly is placed on a smooth horizontal table and released at rest. Find the speed v_1 of the mass m_1 when the separation of the blocks has been reduced to L_0.

Solution   Consider the system comprising both blocks. This system has an *internal* potential energy $\frac{1}{2}k(L - L_0)^2$. The law of conservation of energy gives

$$\tfrac{1}{2}m_1v_1^2 + \tfrac{1}{2}m_2v_2^2 + 0 = 0 + 0 + \tfrac{1}{2}k(L - L_0)^2$$

where v_2 is the speed of the block of mass m_2 when the separation is L_0. The speeds v_1 and v_2 can be related using the law of conservation of momentum (there are no horizontal *external* forces):

$$m_1\mathbf{v}_1 + m_2\mathbf{v}_2 = 0 + 0$$

$$v_2 = \frac{m_1}{m_2}v_1$$

Substitution of this result in the energy equation gives

$$\tfrac{1}{2}m_1\left(1 + \frac{m_1}{m_2}\right)v_1^2 = \tfrac{1}{2}k(L - L_0)^2$$

$$v_1 = \sqrt{\frac{k/m_1}{1 + m_1/m_2}}(L - L_0)$$

11.3 Kinetic Energy of a System

The kinetic energy K of a system of particles is the sum of the kinetic energies $K_1, K_2, \ldots, K_n$ of the individual particles of the system:

$$K = K_1 + K_2 + \cdots + K_n$$

$$= \tfrac{1}{2}m_1v_1^2 + \tfrac{1}{2}m_2v_2^2 + \cdots + \tfrac{1}{2}m_nv_n^2$$

An important result, known as *König's theorem,* states that the kinetic energy K of any system can be expressed as a sum of two kinetic energies:

$$K = K_c + K' \qquad (11.1)$$

where K_c, called the *translational kinetic energy,* is equal to $\frac{1}{2}MV_c^2$ (M is the mass of the system and $\mathbf{V}_c$ is the velocity of the center of mass), and K' is the kinetic energy of the motion relative to the CM-frame of the system. This will be proved first for a one-dimensional motion along the X axis. Interpreting each velocity as an x component and using primes to denote velocities relative to the CM-frame, we have

$$
\begin{aligned}
K &= \tfrac{1}{2}m_1 v_1^2 + \cdots + \tfrac{1}{2}m_n v_n^2 \\
&= \tfrac{1}{2}m_1(V_c + v_1')^2 + \cdots + \tfrac{1}{2}m_n(V_c + v_n')^2 \\
&= \tfrac{1}{2}(m_1 + \cdots + m_n)V_c^2 + (\tfrac{1}{2}m_1 v_1'^2 + \cdots + \tfrac{1}{2}m_n v_n'^2) \\
&\quad + V_c(m_1 v_1' + \cdots + m_n v_n') \\
&= \tfrac{1}{2}MV_c^2 + K' + V_c P' \qquad (11.2)
\end{aligned}
$$

where $P' = m_1 v_1' + \cdots + m_n v_n' = 0$, from the definition of the CM-frame. Therefore, König's theorem is valid for a one-dimensional motion. The proof can be extended without difficulty to apply to a general three-dimensional motion by an analogous treatment of y and z components (see Question S-1).

11.4 Internal Energy and the Law of Conservation of Energy

The *internal energy* E_{int} of a system includes the kinetic energy K' of the motion of its particles relative to their CM-frame, the internal potential energy U_{int} of their interaction, and the internal energies of the individual particles.

The *total energy* of an isolated system with a translational kinetic energy K_c is

$$E = K_c + E_{\text{int}} \qquad (11.3)$$

The total energy E of the isolated system thus includes the translational and internal energy of each object of the system as well as the potential energy associated with the interactions of the objects within the system. For example, the total energy E of an isolated system comprising two objects is

$$E = (K_{1c} + E_{1,\text{int}}) + (K_{2c} + E_{2,\text{int}}) + U_{12} \qquad (11.4)$$

where the first object has an internal energy $E_{1,\text{int}}$ and a translational kinetic energy K_{1c}, and a similar notation is used for the second object. The term U_{12} denotes the potential energy associated with the interaction of the two objects.

An object is said to be *macroscopic* if it is large enough so that its gross properties are directly measurable. A macroscopic object (such as a stone) is composed of many billions of molecules. Suppose that a macroscopic object is at rest in its CM-frame. Although no motion is apparent, the molecules of the object have an incessant *random* motion and the energy of these molecules contributes to the internal energy E_{int} of the object.

In many of the systems considered in Newtonian mechanics, the "particles" are actually macroscopic objects (such as stones). If, during a process, work is done by frictional forces, the energy of these macroscopic objects is dissipated. Energy is not lost but merely transformed into internal energy of these objects which then become warmer.

Generalizing from the results of all experimental evidence from every domain of science, physicists have postulated the general *law of conservation of energy* which for an *isolated* system can be stated in the form

$$E_i = E_f \tag{11.5}$$

where E_i and E_f refer to the values of the total energy E of the system at any two instants.

In Chapter 10 we discussed the law of conservation of kinetic plus potential energy for a system consisting of a single particle. A better understanding of the terms involved and the conditions which must be fulfilled for this law to be valid can be obtained by considering an enlarged system comprising the particle as the first object (with translational kinetic energy K_{1c} and internal energy $E_{1,int}$) and everything with which this particle interacts as the second object. Then from Eq. 11.4 we see that since the total energy E of this large isolated system is conserved, the sum $K_{1c} + U_{12}$ will be conserved provided the internal energies $E_{1,int}$ and $E_{2,int}$ do not change and K_{2c} is constant. It is the sum $K_{1c} + U_{12}$ that is referred to as the particle's total energy $K + U$ in Chapter 10. We can now see that the true total energy of the particle must include the particle's internal energy, but if there is no change in this internal energy, we need not consider it in applying the law of conservation of energy. And in Chapter 10, U is called the potential energy of the particle. When we consider the enlarged system that includes everything with which the particle interacts (object 2), we recognize that this potential energy is the potential energy U_{12} associated with the interaction of object 1 and object 2.

Although in Section 10.2 we defined energy as a measure of the ability to do work, it turns out that in many ways energy is the more fundamental concept. We can characterize work in terms of energy by noting that *work is a measure of the energy transferred by a force.*

11.5 Collisions

The laws of conservation of momentum and energy alone can be used to obtain important results concerning collisions, results that are independent of the details of the interaction between the particles involved.

We consider an isolated system of particles that interact only during the collision. Since the total momentum of an isolated system is conserved, we can equate the *vector sum* of the momenta before collision to the vector sum of the momenta after the collision. The total energy of an isolated system also is conserved.

The word collision is used in the broadest sense and does not necessarily imply actual contact during the interaction. We consider interactions that give rise to various processes: the disintegration of an object initially present, the combination of objects, or simply an *elastic* collision in which the objects separate after their approach without any change in their internal energy. Collisions that involve changes in internal energies are said to be *inelastic*.

The law of conservation of energy gives

$$K_i + E_{i,\text{int}} = K_f + E_{f,\text{int}} \tag{11.6}$$

where K_i is the sum of the kinetic energies of the objects of the system before the collision and K_f is the sum of the kinetic energies of the (possibly different) objects after the collision. The energy released during the collision is

$$Q = E_{i,\text{int}} - E_{f,\text{int}} \tag{11.7}$$

where $E_{i,\text{int}}$ is the sum of the internal energies of the objects that are present before the collision and $E_{f,\text{int}}$ is the sum of the internal energies of the objects that exist after the collision. In collisions of atoms, molecules, and nuclei the changes in internal energy can be such that Q can be positive (an exoergic collision), negative (an endoergic collision), or zero (an elastic collision). Collisions of inert macroscopic bodies always involve work done by dissipative forces. The internal energies of the colliding bodies increase (Q is negative) and the bodies become warmer.

Equations 11.6 and 11.7 imply

$$Q = K_f - K_i \tag{11.8}$$

which shows that the energy Q released in a collision appears as kinetic energy of the objects present after the collision. Part of the kinetic energy of an isolated system cannot be changed by collisions within the system. This is a consequence of Eq. 11.1:

$$K = K_c + K'$$

Since collisions within the system cannot change the velocity $\mathbf{V}_c$ of its center of mass, the translational kinetic energy K_c remains constant. The only part of an isolated system's kinetic energy that can change is the kinetic energy K' arising from the motion of the particles of the system relative to their CM-frame:

$$Q = K_f - K_i = K_f' - K_i' \tag{11.9}$$

Example 3   Particles of masses m_1 and m_2 move along the X axis with x components of velocity v_{1i} and v_{2i}, respectively, relative to the laboratory (L-frame). After an *elastic* collision these particles move along the X axis with x components of velocity given by v_{1f} and v_{2f}, respectively. Find these final velocities in terms of the initial velocities and the masses.

Solution by Analysis in L-Frame   The law of conservation of momentum gives, for x components,

$$m_1 v_{1i} + m_2 v_{2i} = m_1 v_{1f} + m_2 v_{2f}$$

Because the collision is *elastic*, $Q = 0$ and kinetic energy is conserved ($K_i = K_f$):

$$\tfrac{1}{2} m_1 v_{1i}^2 + \tfrac{1}{2} m_2 v_{2i}^2 = \tfrac{1}{2} m_1 v_{1f}^2 + \tfrac{1}{2} m_2 v_{2f}^2$$

These two equations can be solved for the two unknowns, v_{1f} and v_{2f}. From the momentum equation

$$m_1(v_{1i} - v_{1f}) = m_2(v_{2f} - v_{2i}) \tag{11.10}$$

and the equation for the kinetic energy can be rearranged to give

$$m_1(v_{1i}^2 - v_{1f}^2) = m_2(v_{2f}^2 - v_{2i}^2)$$

$$m_1(v_{1i} + v_{1f})(v_{1i} - v_{1f}) = m_2(v_{2f} + v_{2i})(v_{2f} - v_{2i})$$

Dividing this equation by Eq. 11.10, assuming that the denominators are not zero, we obtain $v_{1i} + v_{1f} = v_{2f} + v_{2i}$, which can be rearranged to give

$$v_{1i} - v_{2i} = v_{2f} - v_{1f} \tag{11.11}$$

which states that the relative velocity of approach is equal to the relative velocity of separation in this one-dimensional elastic collision. Solving Eq. 11.11 for v_{2f} and substituting this expression in Eq. 11.10, we find

$$v_{1f} = \frac{m_1 - m_2}{m_1 + m_2} v_{1i} + \frac{2m_2}{m_1 + m_2} v_{2i} \tag{11.12}$$

The expression for v_{2f} can be obtained from Eq. 11.11 and 11.10 by a similar procedure which yields

$$v_{2f} = \frac{m_2 - m_1}{m_2 + m_1} v_{2i} + \frac{2m_1}{m_2 + m_1} v_{1i} \tag{11.13}$$

(Notice that the expression for v_{2f} can be obtained from Eq. 11.12 simply by exchanging the labels 1 and 2.)

Solution by First Analyzing Collision in CM-Frame   We use primes to denote values measured relative to the CM-frame. Since $\mathbf{P}' = 0$, the two initial momenta must have the same magnitude $|p_i'|$ and the two final momenta must have the same magnitude $|p_f'|$. A particle's kinetic energy can be expressed in terms of its momentum using

$$K = \tfrac{1}{2} m v^2 = \frac{p^2}{2m}$$

Since the collision is elastic, $Q = 0$ and $K_i' = K_f'$:

$$\frac{p_i'^2}{2m_1} + \frac{p_i'^2}{2m_2} = \frac{p_f'^2}{2m_1} + \frac{p_f'^2}{2m_2}$$

This shows that $|p_i'| = |p_f'|$. Therefore each particle emerges from the collision with its speed restored to its initial value but with its direction of motion reversed: for a one-dimensional motion $v_{1f}' = -v_{1i}'$ and $v_{2f}' = -v_{2i}'$. We transform back to the L-frame using the velocity transformation law $\mathbf{v} = \mathbf{V}_c + \mathbf{v}'$ with

$$\mathbf{V}_c = \frac{m_1 \mathbf{v}_{1i} + m_2 \mathbf{v}_{2i}}{m_1 + m_2}$$

This yields

$$\mathbf{v}_{1f} = \mathbf{V}_c + \mathbf{v}_{1f}' = \mathbf{V}_c - \mathbf{v}_{1i}'$$
$$= \mathbf{V}_c - (\mathbf{v}_{1i} - \mathbf{V}_c) = 2\mathbf{V}_c - \mathbf{v}_{1i}$$
$$= \frac{m_1 - m_2}{m_1 + m_2}\mathbf{v}_{1i} + \frac{2m_2}{m_1 + m_2}\mathbf{v}_{2i}$$

with a similar result for $\mathbf{v}_{2f}$. This example illustrates that both the pictorial and the mathematical descriptions of collisions are simplified by referring the motion to the CM-frame.

Example 4   A cannon of mass m_1 recoils with a momentum $\mathbf{p}_1$ when it fires a cannonball of mass m_2 with momentum $\mathbf{p}_2$. Find the kinetic energies of the cannon (K_1) and of the ball (K_2) in terms of the masses and the Q value for this process.

Solution   Because the cannon is initially at rest, the system comprising both the cannon and the ball has a total momentum $\mathbf{P} = 0$. The momentum of this system is conserved during the process of firing: $\mathbf{p}_1 + \mathbf{p}_2 = 0$ or $\mathbf{p}_1 = -\mathbf{p}_2$.

$$Q = K_f - K_i = K_1 + K_2 = \frac{p_1^2}{2m_1} + \frac{p_2^2}{2m_2} = \frac{p_1^2}{2}\left(\frac{1}{m_1} + \frac{1}{m_2}\right)$$

Therefore

$$p_1^2 = \frac{2m_1 m_2 Q}{m_1 + m_2}$$

and

$$K_1 = \frac{p_1^2}{2m_1} = \frac{m_2 Q}{m_1 + m_2} = Q\left(1 - \frac{m_1}{m_1 + m_2}\right)$$

Similarly

$$K_2 = Q\left(1 - \frac{m_2}{m_1 + m_2}\right)$$

Example 5   Show how the speed v_1 of a bullet of known mass m_1 can be determined by measurement of the maximum vertical rise y of the block of mass m_2 in the "ballistic pendulum" shown in Figure 11-4.

Solution   The system we consider comprises both the block and the bullet. The problem will be analyzed in two separate stages:

1   *The collision*   Because the collision is of such a short duration that the block does not have time to move an appreciable distance, the suspension cords remain essentially vertical and exert only vertical forces on the block. Since no external horizontal forces act on the system, its horizontal component of momentum will be conserved during the collision:

$$m_1\mathbf{v}_1 + 0 = (m_1 + m_2)\mathbf{v}$$

where $\mathbf{v}$ is the common (horizontal) velocity of the block and bullet after the collision.

2   *The swing of the pendulum*   There are no appreciable frictional effects to dissipate the system's energy as the pendulum swings to its maximum height. The law of conservation of energy yields

$$\tfrac{1}{2}(m_1 + m_2)v^2 + 0 = 0 + (m_1 + m_2)gy$$

$$v = \sqrt{2gy}$$

Therefore

$$v_1 = \frac{m_1 + m_2}{m_1} v = \frac{m_1 + m_2}{m_1} \sqrt{2gy}$$

Notice that although the system's translational kinetic energy is *not* conserved in the completely inelastic collision and the system's momentum is *not* conserved during the swing of the pendulum, this problem has been solved by *appropriate* use of the conservation laws for momentum and energy.

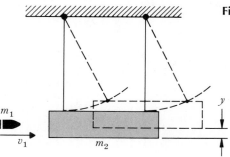

Figure 11-4   Ballistic pendulum.

11.6 Internal Energy and Mass

The internal energy of a system is related to the system's mass M by Einstein's equation*

$$E_{\text{int}} = Mc^2 \tag{11.14}$$

where c denotes the speed of light in a vacuum (3.00×10^8 m/s). This remarkable relationship is suggested by the special theory of relativity and has been verified in experiments involving many different physical systems.

According to Einstein's equation, the very fact of the existence of a mass M implies the presence of an internal energy Mc^2, so we can regard any particle which has mass as a localized bundle of energy. And for any complex system of mass M, whether or not its internal energy can be analyzed in terms of the parts of the system, the system's internal energy is Mc^2.

When an isolated system has a translational kinetic energy K_c, its total energy is†

$$E = K_c + Mc^2 \tag{11.15}$$

The internal energy Mc^2 is called the *rest energy* since it is the energy of an isolated system when it has no translational motion.

Familiar macroscopic objects have rest energies that are enormous compared to their kinetic energies. For instance, for a bullet traveling at 10^3 m/s, we have

$$\frac{Mc^2}{\frac{1}{2}MV^2} = 2\left(\frac{3 \times 10^8}{10^3}\right)^2 = 1.8 \times 10^{11}$$

The energy Q released in a collision can be calculated using the Einstein equation for the internal energy of each object. This gives

$$Q = E_{i,\text{int}} - E_{f,\text{int}} = M_i c^2 - M_f c^2 \tag{11.16}$$

where M_i is the sum of the masses of the separated objects that are present before the collision and M_f is the sum of the masses of the separated objects that exist after the collision. Notice that for inelastic collisions, the total mass of the system changes. The chemists' famous law of conservation of mass is not true, although the changes of mass that presumably do occur are so small relative to the masses involved (about one part in a billion) that they cannot be detected by direct measurement of masses before and after a chemical reaction. Changes of masses are evident in processes studied in nuclear and particle physics. It is now recog-

---

*In the early literature of special relativity the quantity $M/\sqrt{1 - v^2/c^2}$ is called the *relativistic mass* of an object moving with a speed v, and M is called the *rest mass* of the object. In this book the only mass to which we refer is the *rest mass* and we therefore adopt the convention that the term *mass*, used without an adjective, is understood to mean *rest mass*. This is now the practice in almost all the literature of high-energy physics. The reader is cautioned that, in many introductory physics and chemistry textbooks, the term mass is used to mean relativistic mass.

†In Chapter 40 we show that, in the special theory of relativity, the translational kinetic energy is $K_c = Mc^2(1 - 1/\sqrt{1 - v^2/c^2})$. If v is small compared to c, this is approximately $\frac{1}{2}Mv^2$ (see Question S-11).

nized that instead of having a law of conservation of mass and another law of conservation of energy we have only one law, the law of *conservation of energy, in which the energy includes the rest energy.*

Binding Energy

The internal energy Mc^2 of a composite object includes not only the rest energy of each of the constituent particles but also the kinetic energy of these particles in their CM-frame and their energies of interaction with each other. For example, if an object of mass M is composed of two particles with masses m_1 and m_2, the internal energy of the composite object is

$$Mc^2 = m_1c^2 + m_2c^2 + K'_1 + K'_2 + U_{12} \qquad (11.17)$$

where U_{12} is the energy of interaction of particles 1 and 2.*

Equation 11.17 illustrates that, in general, the mass of a composite object is not equal to the sum of the masses of its constituent particles. The mass difference determines the binding energy

$$E_B = [(sum\ of\ masses\ of\ separated\ parts) - M]c^2$$

The binding energy is the minimum energy which must be furnished to a composite object in order to separate it into specified parts, when this energy is evaluated in the CM-frame of the separated parts.

Example 6   To show that the changes of mass that occur in most mundane situations are too small to detect, we consider a collision of two lumps of putty, each of mass m, moving in opposite directions with the same speed v of 3.0 m/s. The lumps collide and stick together forming a single stationary lump of mass M. Since the kinetic energy of the original lumps disappears, M must be greater than $2m$. Evaluate the ratio of the mass increase to the original mass.

Solution   Equating the sum of the total energies before collision to the total energy after collision we have

$$(mc^2 + \tfrac{1}{2}mv^2) + (mc^2 + \tfrac{1}{2}mv^2) = Mc^2 + 0$$

This can be rearranged to give the change in mass,

$$M - 2m = \frac{mv^2}{c^2}$$

Therefore the fractional mass increase is

$$\frac{M - 2m}{2m} = \frac{1}{2}\frac{v^2}{c^2} = \frac{1}{2}\left(\frac{3.0}{3.0 \times 10^8}\right)^2 = 0.50 \times 10^{-16}$$

---

*In Newtonian mechanics, the value of the interaction energy U in a given configuration is defined only to within an arbitrary additive constant. But Einstein's equation requires that an object's internal energy is always positive and is equal to Mc^2. This implies that there is no longer an arbitrariness in value of the interaction energy. Infinite separation corresponds to zero interaction energy.

This fractional change is far too small to be detected by direct measurements of mass.

Example 7   In a certain fusion reaction, a deuteron ($^2_1$H) of mass m_d interacts with a triton ($^3_1$H) of mass m_t and produces an α particle ($^4_2$He) of mass m_α and a neutron ($^1_0$n) of mass m_n. This reaction is represented by

$$^2_1\text{H} + {}^3_1\text{H} \longrightarrow {}^4_2\text{He} + {}^1_0\text{n}$$

Given the masses of the nuclei, find the energy released by this fusion reaction.

Solution   It is convenient to do the arithmetic using the *atomic mass unit* (u) which is defined as one-twelfth of the mass of the carbon-12 atom. (1 u = $1.6605655 \times 10^{-27}$ kg.) The masses of the nuclei are known:

$$
\begin{array}{ll}
m_d = 2.01355 \text{ u} & m_\alpha = 4.00151 \text{ u} \\
m_t = 3.01550 \text{ u} & m_n = 1.00866 \text{ u} \\
\hline
M_i = 5.02905 \text{ u} & M_f = 5.01017 \text{ u}
\end{array}
$$

The decrease in mass because of the reaction is $M_i - M_f = 0.01888$ u, and the energy released in the reaction is

$$Q = (M_i - M_f)c^2$$

$$= (0.01888 \text{ u})(1.66 \times 10^{-27} \text{ kg/u})(3.00 \times 10^8 \text{ m/s})^2$$

$$= 2.82 \times 10^{-12} \text{ J}$$

In the physics of atoms, nuclei, and elementary particles, the unit of energy that is most often used is the *electronvolt* (eV) which is related to the joule by

$$1 \text{ eV} = 1.6021892 \times 10^{-19} \text{ J}$$

Commonly used multiples are 1 MeV = 10^6 eV and 1 GeV = 10^9 eV. Our result is

$$Q = \frac{2.82 \times 10^{-12} \text{ J}}{1.60 \times 10^{-13} \text{ J/MeV}} = 17.6 \text{ MeV}$$

Example 8   What is the minimum energy that must be supplied to a deuteron in order to separate it into a neutron and a proton?

Solution   The energy required is the binding energy

$$E_B = (m_n + m_p - m_d)c^2$$

The masses of the nuclei are known:

$$
\begin{array}{rl}
m_n = & 1.00866 \text{ u} \\
m_p = & 1.00728 \text{ u} \\
\hline
m_n + m_p = & 2.01594 \text{ u} \\
m_d = & 2.01355 \text{ u} \\
\hline
m_n + m_p - m_d = & 0.00239 \text{ u}
\end{array}
$$

The binding energy is

$$E_B = \frac{(0.00239 \text{ u})(1.66 \times 10^{-27} \text{ kg/u})(3.00 \times 10^8 \text{ m/s})^2}{1.60 \times 10^{-13} \text{ J/MeV}} = 2.23 \text{ MeV}$$

Example 9 *Radioactivity of radium* We consider the spontaneous disintegration (called a radioactive decay) of the radium nucleus $^{226}_{88}\text{Ra}$ into an α particle (^4_2He) and a "daughter" radon nucleus, $^{222}_{86}\text{Rn}$, according to the equation

$$^{226}_{88}\text{Ra} \longrightarrow {}^{222}_{86}\text{Rn} + {}^4_2\text{He}$$

(a) Find the Q for this process.
(b) Evaluate the kinetic energies of the decay products.

Solution

(a) The *atomic** masses are

$$
\begin{array}{lll}
M(\text{Ra}) & = & 226.02536 \text{ u} \\
M(\text{Rn}) & = 222.01753 \text{ u} & \\
M(\text{He}) & = 4.00260 \text{ u} & \\
M_f & = \overline{226.02013 \text{ u}} & 226.02013 \text{ u} \\
M_i - M_f & = & 0.00523 \text{ u}
\end{array}
$$

There is enough energy for this decay to occur because the radium isotope has a greater mass and therefore a greater rest energy than that of the decay products. Units can be converted using $1 \text{ u} = 931.5 \text{ MeV}/c^2$. The energy released is

$$Q = (M_i - M_f)c^2 = (0.00523 \text{ u})(931.5 \text{ MeV}/\text{u}c^2)c^2 = 4.87 \text{ MeV}$$

(b) The sum of the kinetic energies of the decay products is

$$K_{\text{Rn}} + K_\alpha = Q = 4.87 \text{ MeV}$$

The law of conservation of momentum implies that these products emerge in opposite directions with momentum vectors of equal magnitude. Since all speeds are small compared to the speed of light, the expressions of Newtonian mechanics can be used for momentum and kinetic energy, and the kinetic energies can be calculated as in Example 4. This yields

$$K_\alpha = (4.87 \text{ MeV})\left(1 - \frac{4}{226}\right) = 4.78 \text{ MeV}$$

Experimental measurement of the energies of the α particles emitted by radium gives $K_\alpha = 4.8 \text{ MeV}$, in good agreement with the prediction obtained using mass measurements and Einstein's equation.

---

*Tabulated masses are given in Appendix D for neutral *atoms*, rather than for bare *nuclei*. However, when atomic numbers balance in the equation for a nuclear reaction, the numbers of orbital electrons also balance, with the consequence that the error introduced by the inclusion of the masses of each atom's orbital electrons cancels out.

$$\tfrac{1}{2}mv_i^2 + \tfrac{1}{2}m(0)^2 = \tfrac{1}{2}m(v_1')^2 + \tfrac{1}{2}m(v_2')^2$$

$$v_i^2 = (v_1')^2 + (v_2')^2 \qquad \text{through}$$

Summary

☐ In a process involving a system of particles, if internal work W_{int} is done, the work-energy principle becomes

$$W_{ext} + W_{int} = \Delta K$$

☐ König's theorem states

$$K = K_c + K'$$

where K_c is the system's translational kinetic energy and K' is the kinetic energy of the motion relative to the system's CM-frame.

☐ The internal energy E_{int} of a system includes the kinetic energies of its particles in their CM-frame, their interaction energy, and their individual internal energies.

☐ The total energy E of an isolated system with translational kinetic energy K_c is

$$E = K_c + E_{int}$$

An expression of the general law of conservation of energy for an isolated system is

$$E_i = E_f$$

where E_i and E_f are the values of the system's total energy at any two instants.

☐ The energy released in a collision (or any process) within an isolated system is

$$Q = E_{i,int} - E_{f,int} = K_f - K_i = K_f' - K_i'$$

☐ The internal energy of a system of mass M is

$$E_{int} = Mc^2$$

This implies that the value of Q for a process is

$$Q = M_i c^2 - M_f c^2$$

The binding energy of a composite object of mass M against separation into specified parts is

$$E_B = [(\text{sum of masses of separated parts}) - M]c^2$$

Questions

1   A 60-kg skater, initially at rest, pushes against a wall of the rink and leaves the wall with a speed of 5.0 m/s. Taking the skater as the system, find the external work and the internal work in this motion.

2   A car of mass M, initially at rest, attains a kinetic energy K by accelerating down an incline that descends through a vertical distance h. Taking the car as the system, find the external work and the internal work during this motion.

3   A long chain of length L rests on a smooth horizontal table. One link is allowed to dangle over the edge. The chain then slides over the edge. Find the speed of the chain as the last link leaves the table.

4   When a system at rest breaks up into two fragments of masses m and M, show that the law of conservation of momentum implies that the ratio of kinetic energies is given by

$$\frac{kinetic\ energy\ of\ M}{kinetic\ energy\ of\ m} = \frac{m}{M}$$

5   A bullet of mass m traveling at a speed v strikes a block of mass M and is embedded in it. Find how far the block (with the bullet embedded in it) will slide along a rough horizontal board. The coefficient of kinetic friction between the board and the block is μ_k.

6   Show that in a head-on elastic collision between two particles of equal mass moving along the X axis, the particles simply exchange velocities.

7   From the results of Example 3 for the case where m_2 is initially at rest, show that:
   (a) If m_2 is very much greater than m_1, v_{1f} is approximately $-v_{1i}$ and v_{2f} is approximately zero.
   (b) If m_2 is very much smaller than m_1, v_{1f} is approximately v_{1i} and v_{2f} is approximately $2v_{1i}$.

8   Find the results in the CM-frame for the collisions examined in parts (a) and (b) of the preceding question.

9   A sled of mass m_1 is sliding horizontally with a constant speed v_1 when a crate of mass m_2 is dropped vertically into the sled. Find the subsequent speed of the sled.

10   An elevator is descending at a constant speed of 9.8 m/s. A ball is dropped from the top of the elevator shaft at the instant the elevator is 14.7 m from the top. If the ball rebounds elastically, what height will it reach?

11   Show that, in the ballistic pendulum in Figure 11-4, the ratio of the kinetic energy of the bullet and of the pendulum after collision, to the original kinetic energy of the bullet, is given by $m_1/(m_1 + m_2)$.

12   **(a)** What is the total energy of an isolated object with translational kinetic energy K and mass M?
   (b) What is a formula for its kinetic energy that can be used when its speed is much less than c?

13   **(a)** Why is Mc^2 called the *rest* energy?
   (b) Show that Mc^2 has the dimensions of energy.

14   A convenient unit for measuring masses of atoms and nuclei is the atomic mass unit (u) which is exactly $\frac{1}{12}$ the mass of an atom of carbon whose nucleus has 6 protons and 6 neutrons. The masses of a neutron, proton, and hydrogen atom are only about 1% greater than 1 u. The rest energy associated with a mass of 1 u is therefore a useful number. Calculate this rest energy in MeV using the fact that 1 u = 1.6606×10^{-27} kg; 1 eV = 1.6022×10^{-19} J; c = 2.9979×10^8 m/s.

15   The energy transformed into kinetic energy of the exploding fragments in the chemical explosion of one ton of TNT is 4.2×10^9 J. This amount of energy is now called a "ton," and the energy released in fusion and fission bombs is measured in kilotons and megatons. Show that the rest energy of a 1.0-kg mass is 21 megatons.

16   How many cars, each of mass 10^3 kg, traveling at 30 m/s are required in order that their kinetic energies add up to the rest energy of a 1.0-kg mass?

17   Are changes of mass confined to nuclear phenomena, or do they occur whenever the kinetic energy of any isolated system changes? Explain the fact that mass changes were not observed by Newton.

18   When an α particle interacts with the nitrogen nucleus $^{14}_{7}$N, the oxygen nucleus $^{17}_{8}$O and a proton may be formed:
     (a) Write this nuclear reaction in the symbolic form illustrated in Example 7.
     (b) What is the Q (in MeV) for the reaction?

19   Find the Q for the fusion reaction

$$^{2}_{1}\text{H} + ^{2}_{1}\text{H} \longrightarrow ^{3}_{1}\text{H} + ^{1}_{1}\text{H}$$

20   **(a)** Is it energetically possible for a neutron (n) to decay into a proton (p), an electron (e), and an antineutrino ($\bar{\nu}$)? Explain.

$m_n = 1.00866$ u
$m_p = 1.00728$ u
$m_e = 5.4860 \times 10^{-4}$ u
$m_{\bar{\nu}} = 0$

   (b) What is the sum of the kinetic energies (in MeV) of the particles produced if the neutron decays according to the reaction examined in part (a)?

21   It is observed that an H_2 *molecule* can be dissociated into two separated hydrogen atoms if it absorbs at least 4.5 eV in some process. What is the binding energy of the H_2 molecules against this type of breakup? (Chemists call this energy the *dissociation* energy of the molecule.)

22   What is the energy required to separate a neutron from the nucleus $^{3}_{1}$H?

23   What energy is required to separate $^{3}_{1}$H into three particles, a proton and two neutrons?

24   Determine the energy released in the fission reaction

$$^{1}_{0}\text{n} + ^{235}_{92}\text{U} \longrightarrow ^{140}_{56}\text{Ba} + ^{89}_{36}\text{Kr} + 7^{1}_{0}\text{n}$$

25   In the radioactive decay

$$^{238}_{92}\text{U} \longrightarrow ^{234}_{90}\text{Th} + ^{4}_{2}\text{He}$$

the α particle emitted has an energy of 4.2 MeV. What is the value of Q for this process?

26   In the radioactive decay

$$^{235}_{92}\text{U} \longrightarrow ^{231}_{90}\text{Th} + ^{4}_{2}\text{He}$$

the α particle emitted has an energy of 4.58 MeV. Find the mass of a neutral atom of the thorium daughter.

Supplementary Questions

S-1 Prove König's theorem for a general three-dimensional motion by:
(a) Following the method given in Section 11.3 for each component and using the fact that

$$v^2 = v_x^2 + v_y^2 + v_z^2$$

(b) Employing vector algebra and using the fact that

$$v^2 = \mathbf{v} \cdot \mathbf{v} = (\mathbf{V}_c + \mathbf{v}') \cdot (\mathbf{V}_c + \mathbf{v}')$$
$$= V_c^2 + v'^2 + 2\mathbf{v}' \cdot \mathbf{V}_c$$

S-2 Consider an elastic collision between two particles of equal mass, one of which is initially at rest. Show that after the collision the particles move at right angles to each other (unless the collision is head-on).

S-3 Show that, in an elastic one-dimensional collision when a particle of mass m_1 strikes a stationary particle of mass m_2, the fractional decrease in the kinetic energy of the mass m_1 is given by

$$\frac{K_{1i} - K_{1f}}{K_{1i}} = \frac{4m_1 m_2}{(m_1 + m_2)^2}$$

S-4 A ball of mass m_1 is fastened to a cord of length L (Figure 11-5) and released when the cord is horizontal. When the cord is vertical the ball makes an elastic collision with a block of mass m_2 which is resting on a smooth table. Find the speed of the ball after collision and the speed of the block in terms of m_1, m_2, L, and g.

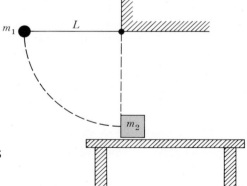

Figure 11-5

S-5 Consider a two-particle elastic collision which is not head-on. Show that, in the CM-frame, the collision simply rotates velocities, which remain opposite in direction and unchanged in magnitude.

S-6 The Lennard–Jones potential energy function is

$$U = 4\epsilon\left[\left(\frac{\sigma}{r}\right)^{12} - \left(\frac{\sigma}{r}\right)^{6} \right]$$

where σ and ϵ are positive constants. Assume that this is the potential energy associated with the interaction of two atoms of masses m_1 and m_2 when separated by a distance r. The two atoms are initially at rest with an infinite separation, and then they move toward one another under the influence of the attractive forces that they exert on each other. Find:

(a) The speed of each atom at the instant that the separation of the atoms is equal to σ.

(b) The maximum speed of each atom, in terms of ϵ, σ, m_1, and m_2.

S-7 The first study of nuclear reactions produced by energetic particles from a particle accelerator was published by John D. Cockcroft and Ernest T. S. Walton in England in 1932. They had bombarded nuclei of lithium-7 with protons accelerated in their new machine. It was then observed that in some instances there was a nuclear reaction in which these initial particles (lithium nucleus and proton) were transformed into two α particles. Moreover, Cockcroft and Walton noted that the two α particles produced in this manner had more kinetic energy than the incident proton:

(a) Write the nuclear reaction in the symbolic form illustrated in Example 7.

(b) Evaluate Q, first from the measured masses given in Appendix D, and then from the following kinetic energies measured in this experiment:

kinetic energy of each ^{4}He nucleus   8.6   MeV
kinetic energy of proton   0.25 MeV

(c) Explain how these data provide experimental evidence of the validity of Einstein's equation, $E_{int} = Mc^2$.

S-8 (a) Show that a reaction cannot occur unless

$$Q + K'_i \geq 0$$

where K'_i is the sum of the kinetic energies of the particles present before the reaction, evaluated in their CM-frame.

(b) In the L-frame, a particle of mass m_1 and kinetic energy K_1 strikes a stationary target of mass m_2. Using Newtonian mechanics, show that

$$\frac{K'_i}{K_1} = \frac{m_2}{m_1 + m_2}$$

(c) From parts (a) and (b) show that the reaction cannot occur unless the incident projectile m_1 has a kinetic energy greater than the so-called threshold value,

$$K_{1,threshold} = -Q\left(1 + \frac{m_1}{m_2}\right)$$

S-9 To obtain a beam of monoenergetic neutrons, we bombard a foil containing $^7_3$Li nuclei by protons. The reaction in which a neutron and $^7_4$Be are produced has a Q value of -1.64 MeV. What is the threshold energy for this reaction?

S-10 A proton is incident on a stationary tritium nucleus ($^3_1$H). What is the threshold energy for the production of a neutron and $^3_2$He?

S-11 Using the binomial theorem, show that

$$\left(1 - \frac{v^2}{c^2}\right)^{-1/2} = 1 + \frac{1}{2}\left(\frac{v^2}{c^2}\right) + \frac{3}{8}\left(\frac{v^4}{c^4}\right) + \cdots$$

Use this to show that the relativistic expression for the translational kinetic energy of an object of mass M

$$K_c = Mc^2\left(1 - \frac{1}{\sqrt{1 - v^2/c^2}}\right)$$

is approximately equal to $\frac{1}{2}Mv^2$ when v is small compared to c.

12.

motion of
a rigid body

When the system of particles constitutes a rigid body, simple expressions can be found for physical quantities, such as the kinetic energy of the system, and it is then worthwhile to give the laws of mechanics a special formulation appropriate for such a system. Since most machinery has rotating parts that are almost rigid, the application of Newtonian mechanics to the rotational motion of rigid bodies is of particular interest to mechanical engineers.

12.1 Kinematics of Translation and Rotation

Rather than consider the most general rigid body motion in three dimensions, we shall assume in this chapter that the motion is always parallel to the XY plane. We can describe such a motion as a combination of a *translational* motion and a *rotational* motion.

The rotational motion is described by giving the angular displacement, the angular velocity* ω, and the angular acceleration α of the rigid body. Since all lines painted on a rigid body experience the same angular displacement $\Delta\theta$ in any motion, the quantities $\Delta\theta$, ω, and α are characteristic of the body as a whole, and their values do not depend on the specification of some particular axis as the axis of rotation.

The translational motion is described by specifying the motion of some arbitrary "base point" on the rigid body. Often the center of mass of a rigid body is selected as the base point. If the rigid body happens to be rotating about a fixed axis perpendicular to the XY plane, a base point on this axis is convenient because it remains at rest.

---

*In these next two chapters it will be convenient to adopt the convention that the counterclockwise sense of rotation is positive for all angular quantities (θ, ω, α) and the clockwise sense negative. This accords with the sign convention adopted for torque in Chapter 5. When following this convention, the magnitude of ω is the angular speed and ω itself will be called the angular velocity.

12.2 Kinetic Energy of a Rigid Body Rotating about a Fixed Axis

When a rigid body is rotating about a fixed axis, the particles of the body travel in circular paths about the axis (Figure 12-1). The particle of mass m_i at a distance r_i from the axis has a speed v_i given from Eq. 8.3 by

$$v_i = r_i \omega$$

where ω is the angular speed of the rotating rigid body. The kinetic energy of this particle is

$$\tfrac{1}{2} m_i v_i^2 = \tfrac{1}{2} m_i r_i^2 \omega^2$$

The kinetic energy K of the rigid body is the sum of the kinetic energies of its constituent particles:

$$K = \tfrac{1}{2} m_1 r_1^2 \omega^2 + \tfrac{1}{2} m_2 r_2^2 \omega^2 + \cdots + \tfrac{1}{2} m_n r_n^2 \omega^2$$
$$= \tfrac{1}{2}(m_1 r_1^2 + m_2 r_2^2 + \cdots + m_n r_n^2) \omega^2$$

This can be written as

$$K = \tfrac{1}{2} I \omega^2 \qquad \text{(fixed axis)} \quad (12.1)$$

where the quantity I, called the *moment of inertia* of the body about the axis, is defined by

$$I = m_1 r_1^2 + m_2 r_2^2 + \cdots + m_n r_n^2 \qquad (12.2)$$

We shall find that, as is suggested by the equation $K = \tfrac{1}{2} I \omega^2$, the moment of inertia I is a measure of the *rotational inertia* of a rigid body.

The kinetic energy $\tfrac{1}{2} I \omega^2$ is called the rotational kinetic energy of the body. This is *not* a new type of kinetic energy; it is merely a useful expression for the sum of the kinetic energies of the particles of the rigid body. The problem of calculating the kinetic energy of a rigid body rotating at a given angular speed is solved if the moment of inertia I can be determined.

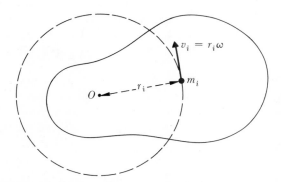

Figure 12-1   Rigid body rotating about a fixed axis through O and perpendicular to diagram.

12.3 Moment of Inertia

To find the moment of inertia I of a body about a given axis we must evaluate the sum $m_1 r_1^2 + m_2 r_2^2 + \cdots + m_n r_n^2$. Since the distances $(r_1, r_2, \ldots, r_n)$ of the particles from the axis do not change as a rigid body rotates, its moment of inertia about the given axis is *constant*. The value of I depends not only on the

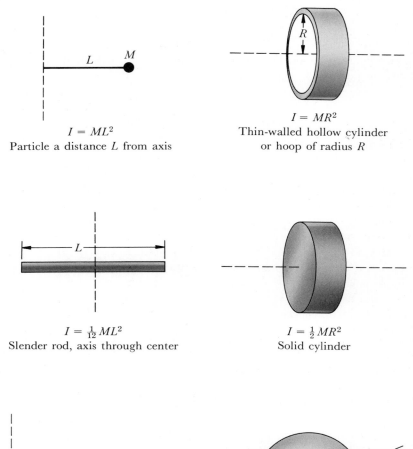

$I = ML^2$
Particle a distance L from axis

$I = MR^2$
Thin-walled hollow cylinder
or hoop of radius R

$I = \frac{1}{12} ML^2$
Slender rod, axis through center

$I = \frac{1}{2} MR^2$
Solid cylinder

$I = \frac{1}{3} ML^2$
Slender rod, axis through end

$I = \frac{2}{5} MR^2$
Solid sphere, axis through center

Figure 12-2   Moments of inertia about indicated axes.

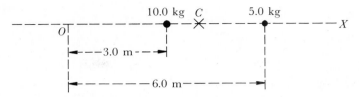

Figure 12-3   System of particles of Example 1.

masses of the particles but also on the distances of the particles from the axis. The further a particle is from the axis, the greater its contribution to the moment of inertia. A simple formula expressing the moment of inertia of a body about an axis can usually be found if the body is geometrically simple. Results for several objects are given in Figure 12-2. In every case the moment of inertia depends, not only on the total mass M, but also on geometrical factors which reflect the fact that a moment of inertia depends on the way that the mass is distributed about the axis. The axis must always be specified because a given body generally has a different moment of inertia about a different axis.

Example 1   Find the moment of inertia of the system shown in Figure 12-3:
   (a) About an axis OZ perpendicular to the plane of the figure and passing through the origin.
   (b) About an axis CZ' passing through the center of mass and parallel to the axis OZ.

Solution
   (a) The moment of inertia about the axis OZ is

$$I = (10.0 \text{ kg})(3.0 \text{ m})^2 + (5.0 \text{ kg})(6.0 \text{ m})^2 = 270 \text{ kg} \cdot \text{m}^2$$

   (b) The center of mass is between the two particles and is 1.0 m from the 10.0-kg particle (see Example 2 of Chapter 4). The moment of inertia of this system about the axis CZ' is

$$I_c = (10.0 \text{ kg})(1.0 \text{ m})^2 + (5.0 \text{ kg})(2.0 \text{ m})^2 = 30 \text{ kg} \cdot \text{m}^2$$

Continuous Distributions of Mass

For a rigid body with a continuous distribution of mass, we can regard the body as subdivided into elements of mass dm. A mass dm at a distance r from an axis makes a contribution $r^2 \, dm$ to the moment of inertia I of the body about the axis. Then

$$I = \int r^2 \, dm \tag{12.3}$$

where the integral denotes the sum of the contributions $r^2 \, dm$ of all the elements of mass that constitute the body.

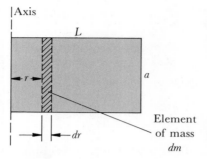

Element of mass dm

Figure 12-4   The flat plate of Example 2.

Example 2   Find the moment of inertia of a uniform rectangular flat plate about an axis along the edge of width a (Figure 12-4). The plate has a mass M and a length L.

Solution   The important step in calculating a moment of inertia is the selection of a typical element dm. All parts of dm must be at essentially the *same distance* r from the axis. Since the plate is uniform, the mass of any part is proportional to its area. The mass dm of the typical element shown in Figure 12-4 is therefore given by

$$\frac{dm}{M} = \frac{a\,dr}{aL}$$

$$dm = \frac{M}{L}\,dr$$

The moment of inertia of this element about the axis at a distance r is

$$r^2\,dm = \frac{M}{L}r^2\,dr$$

The sum of the contributions from all the elements that comprise the plate is

$$I = \int_0^L \frac{M}{L}r^2\,dr = \tfrac{1}{3}ML^2 \tag{12.4}$$

Theorems on Moments of Inertia

Several useful theorems relate different moments of inertia:

1   *Method of decomposition*   If a system is composed of two parts with moments of inertia I_1 and I_2, the moment of inertia of the complete system is

$$I = I_1 + I_2 \tag{12.5}$$

The proof consists of simply splitting the sum $I = \Sigma\, m_i r_i^2$ into the sum which gives I_1 and the sum which gives I_2.

2   *Parallel axis theorem*   A system's moment of inertia I_c about an axis through its center of mass, and its moment of inertia I through a parallel axis, are related by

$$I = I_c + MD^2 \tag{12.6}$$

where D is the distance between the two parallel axes. A simple proof of this valuable theorem is given in the next section.

3   *Perpendicular axis theorem*   If a flat plate in the XY plane has moments of inertia I_{OX}, I_{OY}, and I_{OZ} about the mutually perpendicular axes OX, OY, and OZ, then

$$I_{OZ} = I_{OX} + I_{OY} \tag{12.7}$$

The method of proof is indicated in Question 8.

Example 3   Find the moment of inertia I_c of the plate of Example 2 about an axis passing through its center of mass and parallel to the side of width a.

Solution   From Example 2, the moment of inertia I of the plate about an axis along the edge is

$$I = \tfrac{1}{3}ML^2$$

Using the parallel axis theorem with $D = L/2$, we find

$$\tfrac{1}{3}ML^2 = I_c + M(\tfrac{1}{2}L)^2$$

This yields

$$I_c = \tfrac{1}{12}ML^2 \tag{12.8}$$

Radius of Gyration

It is sometimes convenient to express the moment of inertia I of a body of mass M in terms of a length k, called the *radius of gyration*, which is defined by

$$Mk^2 = I \tag{12.9}$$

This implies that a single particle of mass M at a distance k from the axis has a moment of inertia equal to that of the body in question.

Another way of stating the result of Example 2, $I = \tfrac{1}{3}ML^2$, is to say that the radius of gyration of the plate about an axis along its edge is

$$k = \frac{L}{\sqrt{3}}$$

In terms of radii of gyration, the parallel axis theorem states

$$k^2 = k_c^2 + D^2 \tag{12.10}$$

12.4 Rotation and the Law of Conservation of Energy

König's theorem (Section 11.3) states that the kinetic energy K of any system of particles can be written as a sum of two kinetic energies

$$K = \tfrac{1}{2}MV_c^2 + K'$$

where K' is the kinetic energy of the system evaluated in its CM-frame. When the system is a rigid body rotating with an angular speed ω, Eq. 12.1 gives

$$K' = \tfrac{1}{2}I_c\omega^2 \tag{12.11}$$

since, in the CM-frame, the system rotates about a fixed axis through the center of mass and perpendicular to the XY plane. Therefore, the kinetic energy of a rigid body rotating with an angular speed ω is

$$K = \tfrac{1}{2}MV_c^2 + \tfrac{1}{2}I_c\omega^2 \tag{12.12}$$

The first term $\tfrac{1}{2}MV_c^2$ is the translational kinetic energy associated with the motion of the center of mass. The second term $\tfrac{1}{2}I_c\omega^2$ is the rotational kinetic energy due to motion relative to the center of mass. Using the expression for K given by Eq. 12.12 we can analyze a variety of interesting motions by using the law of conservation of energy.

Suppose a rigid body moves in a uniform gravitational field (Figure 12-5). Then the body has a gravitational potential energy Mgy. If the body does not slip, no work is done by friction. Then the law of conservation of energy yields

$$\tfrac{1}{2}MV_c^2 + \tfrac{1}{2}I_c\omega^2 + Mgy = constant \tag{12.13}$$

If in Figure 12-5 the uniform sphere of radius R rolls without slipping, the velocity $\mathbf{v}_p$ of the point of contact with the plane is zero. This constraint implies that V_c and ω must be related. In the CM-frame, the point P has a velocity of magnitude $v_p' = \omega R$ directed up the plane. Substitution of these values in the velocity transformation equation (Eq. 6.10)

$$\mathbf{v}_p = \mathbf{V}_c + \mathbf{v}_p'$$

gives the *condition of rolling,*

$$V_c = \omega R \tag{12.14}$$

With $\omega = V_c/R$ in Eq. 12.13, we can easily relate speeds V_c at different heights y. For instance, if the sphere starts from rest, its speed V_c after descending a vertical distance y is found from

$$\tfrac{1}{2}MV_c^2 + \tfrac{1}{2}I_c\left(\frac{V_c}{R}\right)^2 + 0 = 0 + 0 + Mgy$$

This gives

$$V_c = \sqrt{\frac{2gy}{(1 + I_c/MR^2)}} \tag{12.15}$$

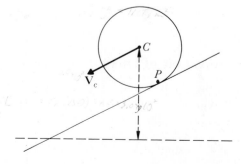

Figure 12-5   Sphere rolling down an inclined plane.

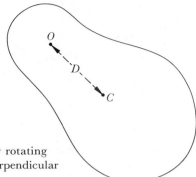

Figure 12-6   Rigid body rotating about a fixed axis OZ perpendicular to diagram.

Proof of the Parallel Axis Theorem

Consider a rigid body rotating about a fixed axis OZ (Figure 12-6). The kinetic energy of this motion is

$$K = \tfrac{1}{2}I\omega^2$$

where ω is the angular velocity of the body and I is its moment of inertia about OZ. We may also evaluate this kinetic energy using Eq. 12.12:

$$K = \tfrac{1}{2}MV_c^2 + \tfrac{1}{2}I_c\omega^2$$

where I_c is the moment of inertia of the rigid body about an axis CZ' which is parallel to OZ. If the distance from OZ to CZ' is D, then $V_c = D\omega$. We have

$$\tfrac{1}{2}I\omega^2 = K = \tfrac{1}{2}M(D\omega)^2 + \tfrac{1}{2}I_c\omega^2$$

This yields the *parallel axis theorem,*

$$I = MD^2 + I_c \tag{12.6}$$

12.5 Torque and Rotation about a Fixed Axis

Figure 12-7 shows an angular displacement $\Delta\theta$ of a rigid body rotating about a fixed axis. The work W done by a force $\mathbf{F}$ acting on a portion of the body that moves a distance $\Delta s = r\Delta\theta$ is

$$W = F_s\Delta s = F_s r\Delta\theta \tag{12.16}$$

provided the tangential component F_s is constant. The torque of $\mathbf{F}$ about the axis of rotation is the sum of the torques of its tangential component F_s and its normal component F_N. Since the line of action of F_N passes through the axis, its moment arm is zero and consequently the torque of F_N is zero. The moment arm of F_s is the distance r from the axis to the point of application of $\mathbf{F}$. The conclusion is that the torque of $\mathbf{F}$ about the axis of rotation is

$$\tau = F_s r \tag{12.17}$$

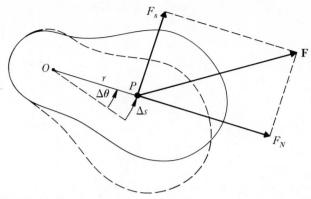

Figure 12-7 Rigid body rotating about a fixed axis through O and perpendicular to diagram.

Then Eq. 12.16 gives

$$W = \tau \Delta\theta \qquad (\tau \text{ constant}) \quad (12.18)$$

for the work done by the torque τ acting on a rigid body which undergoes an angular displacement $\Delta\theta$. From this result it can be inferred that the general expression for the work done by a variable torque τ acting on a rigid body is, in differential form,

$$dW = \tau \, d\theta \qquad (12.19)$$

and

$$W = \int_{\theta_i}^{\theta_f} \tau \, d\theta \qquad (12.20)$$

when the angular displacement of the rigid body is $\theta_f - \theta_i$. From Eq. 12.19, the power supplied by the torque τ acting on a rigid body that is rotating with an angular velocity ω is

$$power = \frac{dW}{dt} = \tau \frac{d\theta}{dt} = \tau\omega \qquad (12.21)$$

From the power equation of Section 11.1,

$$\frac{dW_{\text{ext}}}{dt} + \frac{dW_{\text{int}}}{dt} = \frac{dK}{dt}$$

we obtain for a rigid body

$$\frac{dW_{\text{ext}}}{dt} = \frac{dK}{dt}$$

since for a *rigid* body there is no internal work. Consequently

$$\tau_{\text{ext}}\omega = \frac{dW_{\text{ext}}}{dt} = \frac{dK}{dt} = \frac{d}{dt}(\tfrac{1}{2}I\omega^2) = I\omega \frac{d\omega}{dt}$$

or

$$\tau_{\text{ext}} = I\alpha \qquad (12.22)$$

Table 12-1

Linear motion along X axis		Rotation about a fixed axis (OZ)	
Displacement	Δx	Angular displacement	$\Delta\theta$
Velocity	v_x	Angular velocity	ω
Acceleration	a_x	Angular acceleration	α
Mass	m	Moment of inertia	I
Force	F_x	Torque	τ
$F_x = ma_x$		$\tau_{\text{ext}} = I\alpha$	
Kinetic energy	$K = \frac{1}{2}mv_x^2$	Kinetic energy	$K = \frac{1}{2}I\omega^2$
Work	$dW = F_x\,dx$	Work	$dW = \tau\,d\theta$
Power	$power = F_x v_x$	Power	$power = \tau\omega$

where α is the angular acceleration of the rigid body, I is its moment of inertia about the fixed axis of rotation, and τ_{ext}, called the *total external torque*, is the algebraic sum of the torques of the *external* forces about the axis of rotation.

This important dynamical equation for rotational motion should be compared to the dynamical equation which determines the linear acceleration a of a particle of mass m, $F = ma$. Analogous quantities for rotational and linear motion are seen to be *angular acceleration* and *linear acceleration, torque* and *force,* and *moment of inertia* and *mass* (see Table 12-1).

The role of I as a measure of rotational inertia is evident from the equation $\tau = I\alpha$. The larger the value of I, the greater the torque required to produce a given angular acceleration.

Example 4 A uniform solid flywheel with mass M of 10.0 kg and a radius R of 0.160 m is mounted as shown in Figure 12-8 on a horizontal axle with bearings that can be assumed to be frictionless. A light rope is wrapped around the rim of the flywheel, and a steady pull T of 20.0 N is exerted on the free end of the rope. Find the angular acceleration α of the flywheel and the time it takes to make 2.00 rev starting from rest.

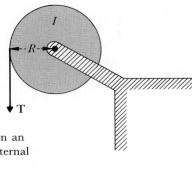

Figure 12-8 Flywheel is given an angular acceleration by an external torque.

Solution   The moment of inertia of this cylindrical flywheel about the axis of rotation, from Figure 12-2, is

$$I = \tfrac{1}{2}MR^2 = \tfrac{1}{2}(10.0 \text{ kg})(0.160 \text{ m})^2 = 0.128 \text{ kg} \cdot \text{m}^2$$

To determine the resultant external torque τ_{ext} about the axis of rotation, we must consider all the external forces acting on the flywheel. The weight Mg of the flywheel, and the upward force exerted on the axle by the bearings, have zero torque about the axis of rotation. (They have zero moment arm because their directions pass through the axis.) We assume that there is no frictional torque at the bearings. There remains only the torque of the force T which has the moment arm R. Therefore

$$\tau_{\text{ext}} = TR = (20.0 \text{ N})(0.160 \text{ m}) = 3.20 \text{ N} \cdot \text{m}$$

From $\tau_{\text{ext}} = I\alpha$ we obtain

$$\alpha = \frac{\tau_{\text{ext}}}{I} = \frac{3.20 \text{ N} \cdot \text{m}}{0.128 \text{ kg} \cdot \text{m}^2} = 25.0 \text{ rad/s}^2$$

The angular displacement is $\theta - \theta_0 = \tfrac{1}{2}\alpha t^2$ which yields

$$t = \sqrt{\frac{2 \times 4\pi \text{ rad}}{25.0 \text{ rad/s}^2}} = 1.00 \text{ s}$$

Example 5   A mass m is suspended from the rope of the preceding example. Find the acceleration of this suspended mass in terms of m, M, R, and g.

Solution   When the system under consideration is the flywheel alone, the free-body diagram excludes m and shows the force T as an external force as in the

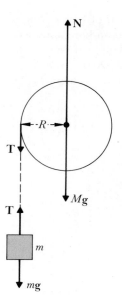

Figure 12-9   Free-body diagram for the suspended mass m is shown in lower portion of figure. The upper portion is the free-body diagram for the flywheel. The force **N** is exerted on the axle by the bearings.

upper portion of Figure 12-9. The equation $\tau_{ext} = I\alpha$, applied to the flywheel alone, gives

$$(Mg \times 0) + (N \times 0) + TR = \tfrac{1}{2}MR^2\alpha \qquad (12.23)$$

From the free-body diagram for the suspended mass (Figure 12-9), we see that Newton's second law yields

$$mg - T = ma \qquad (12.24)$$

where the positive direction is downward. The relationship between α and a is found by considering a point on the rope above m but in contact with the flywheel. This point has the same acceleration as m and also has an angular acceleration α about the flywheel axis. From Eq. 8.16 we obtain

$$\alpha = \frac{a}{R}$$

Therefore Eq. 12.23 yields

$$T = \tfrac{1}{2}MR\alpha = \tfrac{1}{2}Ma$$

Substituting this result for T in Eq. 12.24, we find

$$mg - \tfrac{1}{2}Ma = ma$$

$$a = \frac{mg}{m + M/2} = \frac{g}{1 + M/2m}$$

and

$$T = \tfrac{1}{2}Ma = \frac{mg}{1 + 2m/M}$$

The tension is less than the weight mg of the suspended mass when the mass has a downward acceleration.

Example 6 *Physical pendulum* The rigid body in Figure 12-10 is free to oscillate about a fixed horizontal axis through the point O. Find the period T of small oscillations in terms of the mass M, the moment of inertia I about the fixed axis, and the distance D from this axis to the center of mass C.

Solution The only external force which exerts a torque about the fixed axis is the weight Mg. In the position shown in Figure 12-10, the weight has a moment arm $D \sin \theta$ and the torque is $-MgD \sin \theta$ (negative because it is clockwise). Therefore $\tau_{ext} = I\alpha$ gives

$$-MgD \sin \theta = I\frac{d^2\theta}{dt^2}$$

For small angles $\sin \theta$ is approximately equal to θ. With this approximation we obtain

$$\frac{d^2\theta}{dt^2} = -\frac{MgD}{I}\theta$$

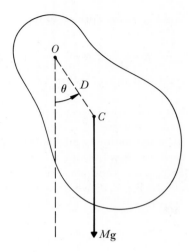

Figure 12-10   Physical pendulum. A rigid body oscillates about a fixed horizontal axis through O.

This is the differential equation for simple harmonic motion (Eq. 9.17) with an angular frequency $\sqrt{MgD/I}$ and a period

$$T = 2\pi \sqrt{\frac{I}{MgD}} \qquad (12.25)$$

A simple and accurate experimental measurement of a rigid body's moment of inertia can be made by measuring the period of small oscillations. The simple pendulum of Section 9.6 is an idealized physical pendulum in which all the mass is concentrated at one point.

Summary

☐   The moment of inertia of a system of particles about a given axis is

$$I = m_1 r_1^2 + m_2 r_2^2 + \cdot \; \cdot \; \cdot + m_n r_n^2$$

where the particles of the system have masses $m_1, m_2, \ldots , m_n$ and are located at distances from the axis $r_1, r_2, \ldots , r_n$, respectively. The moment of inertia of a body with a continuous distribution of mass is

$$I = \int r^2 \, dm$$

where the integral is the sum of the contributions $r^2 \, dm$ of all the elements of mass that constitute the body.

☐   The parallel axis theorem states that

$$I = I_c + MD^2$$

where I_c is the moment of inertia about an axis through the center of mass and I is the moment of inertia about a parallel axis a distance D from the center of mass.

☐ The kinetic energy of a rigid body rotating about a fixed axis is

$$K = \tfrac{1}{2}I\omega^2$$

where I is the moment of inertia of the body about the fixed axis and ω is the angular speed of the body.

☐ For any motion parallel to the XY plane, the kinetic energy of a rigid body can be expressed as the sum of the translational kinetic energy $\tfrac{1}{2}MV_c^2$ and the rotational kinetic energy $\tfrac{1}{2}I_c\omega^2$ associated with the motion relative to the center of mass:

$$K = \tfrac{1}{2}MV_c^2 + \tfrac{1}{2}I_c\omega^2$$

☐ A torque τ acting on a rigid body that is rotating with an angular velocity ω supplies a power given by

$$power = \tau\omega$$

☐ When a rigid body rotates about a fixed axis

$$\tau_{\text{ext}} = I\alpha$$

where α is the angular acceleration of the body, I is its moment of inertia about the axis, and τ_{ext} is the algebraic sum of the torques of the *external* forces about the axis.

Questions

1   A dumbbell consists of a 1.20-kg particle and a 0.80-kg particle placed at opposite ends of a light rod which is 0.40 m long. Calculate the moment of inertia of this dumbbell about an axis perpendicular to the rod and passing through the rod at a point 0.10 m from the 0.80-kg particle.

2   A rigid right triangle ABC is formed with sides of lengths $AB = 0.50$ m, $AC = 0.40$ m, and $BC = 0.30$ m with particles of masses $m_A = 2.0$ kg, $m_B = 3.0$ kg, and $m_C = 4.0$ kg placed at the vertices A, B, and C, respectively. Find the moment of inertia of this rigid body about each of the three sides.

3   A particle of mass m is fixed at one end of a uniform bar which has length L and mass M. Find the moment of inertia I about an axis perpendicular to the bar and passing through the end of the bar remote from the mass m.

4   A thin hoop has a mass M and a radius r. Deduce an expression for the moment of inertia I about an axis perpendicular to the plane of the hoop and passing through the center of the hoop.

5   Show that the moment of inertia of an annular cylinder about the axis of the cylinder is given by

$$I = \frac{M}{2}(R_i^2 + R_o^2)$$

where M is the cylinder's mass, and R_i and R_o are the inner and outer radii, respectively.

6 Verify that the moments of inertia calculated in Example 1 satisfy the parallel axis theorem.

7 A sphere of mass M and radius R has a moment of inertia about a diameter given by

$$I = \tfrac{2}{5}MR^2$$

Find the moment of inertia of this sphere about an axis that is tangent to the sphere.

8 Prove the perpendicular axis theorem. [A flat plate in the XY plane has $I_{OX} = \Sigma m_i y_i^2$, $I_{OY} = \Sigma m_i x_i^2$, and $I_{OZ} = \Sigma m_i (x_i^2 + y_i^2)$.]

9 The moment of inertia of a flat disk of radius R about an axis through its center and perpendicular to the plane of the disk is $\tfrac{1}{2}MR^2$, where M is the mass of the disk. Use the perpendicular axis theorem to find the moment of inertia of the disk about a diameter.

10 Starting from the result of Example 3, use the perpendicular axis theorem to determine the moment of inertia of a flat rectangular plate about an axis through its center and perpendicular to the plane of the plate.

11 A 20-kg pail of water is suspended by a rope wrapped around a windlass which is a solid cylinder 0.40 m in diameter with a mass of 50 kg. The pail is released at rest at the top of a well and descends 20 m before it strikes the water. At this instant:
(a) What is the sum of the kinetic energies of the pail and of the rotating windlass?
(b) What is the ratio of the kinetic energies of these objects?
(c) Find the speed of the pail and the angular speed of the windlass.

12 A 40-kg flywheel with a moment of inertia of 0.30 kg·m² rotates with an angular speed of 50 rad/s:
(a) Find the kinetic energy of the flywheel if it is rotating about a fixed central axis.
(b) Find the kinetic energy of this flywheel if it is rolling without slipping. The radius of its rim is 0.10 m.

13 A homogeneous 2.0-kg cylinder with a radius of 5.0×10^{-2} m starts from rest and rolls down a plane inclined at an angle of 30° with the horizontal. When it has traveled 1.20 m along the inclined plane, what is its kinetic energy and what is the speed of its center of mass?

14 A homogeneous sphere and a homogeneous cylinder of the same radius start from rest and race down an incline. Which one wins? What is the ratio of the speeds they have attained at the finish line?

15 A uniform rod can rotate in a vertical plane about a fixed horizontal axis at a distance D from its center. The rod is released in a horizontal position. Find the angular speed of the rod when it passes through the vertical position, in terms of D, the mass M, and the length L of the rod.

16 A wheel has a thin rim of mass M and eight spokes, each of length R and mass m:
(a) Find the moment of inertia of this wheel about its central axis.
(b) What is its kinetic energy when it is rolling with an angular speed ω?

17 An automobile engine develops 500 kW while the crankshaft has an angular speed of 400 rad/s. What is the torque exerted on the crankshaft?

18 A seesaw, which is 4.0 m long and pivoted about a horizontal axis through its center, makes an angle of 60° with the vertical. A child pushes vertically downward at one end with a force of 100 N. Find the component of this force which is perpendicular to the seesaw and use this component to calculate the torque of the 100-N force about the axis of the seesaw.

19 In spinning the wheel of a stationary bicycle, a girl exerts a force which has a component of 40 N parallel to the spokes and a component of 60 N perpendicular to the spokes and to the axis of the wheel. The radius of the wheel is 0.33 m. Find the torque exerted by the girl about the wheel axis.

20 Specify the type of motion and the type of physical system to which the equation, $\tau_{ext} = I\alpha$, can be applied. Define each quantity in this equation and give its dimensions. Verify that the equation is dimensionally correct.

21 A resultant external torque of 80 N·m gives a flywheel an angular acceleration of 1.6 rad/s². What is the moment of inertia of the flywheel about its axis?

22 The moment of inertia of a grindstone about its axis is 0.20 kg·m². What is the angular acceleration of this grindstone when a torque about the axis of 4.0 N·m is applied?

23 A flywheel with a moment of inertia of 2.00 kg·m² about its axis has its angular speed decreased from 15.0 rad/s to 10.0 rad/s in 50 s because of frictional torque at the bearings. Find the magnitude of this torque, assuming it to be constant.

24 The rotation of the flywheel in Example 4 is opposed by a frictional torque of 0.20 N·m at the bearings. In this circumstance, what will be the angular acceleration of the flywheel when a steady pull of 20 N is exerted on the free end of the rope?

25 A flywheel with a moment of inertia of 40.0 kg·m² is rotating with an angular speed of 5.0 rad/s. The flywheel axle has a radius of 0.030 m. What tangential braking force must be applied to the surface of the flywheel axle in order to stop the flywheel after 150 revolutions?

26 A flywheel 0.80 m in diameter is mounted on a horizontal axis. When a steady pull of 40 N is applied to the free end of a rope which has been wrapped around the outside of the flywheel, the flywheel completes 3.0 revolutions in 5.0 s:
  (a) What is the angular acceleration of the flywheel?
  (b) What is the moment of inertia of the flywheel?
  (c) What is the final angular speed?
  (d) What is its final kinetic energy?
  (e) Compute the work done by the applied torque τ during the displacement $\Delta\theta$ by evaluating $\tau\Delta\theta$. Compare this result to the kinetic energy found in part (d).

Supplementary Questions

S-1 Give an alternative proof of the parallel axis theorem, proceeding directly from the definition of the moment of inertia and the definition of the center of mass.

S-2 Show that the moment of inertia of a uniform solid cylinder about a central axis perpendicular to the axis of the cylinder is

$$I = \frac{MR^2}{4} + \frac{ML^2}{12}$$

where the cylinder has mass M, length L, and radius R.

S-3 A heavy chain is wrapped around a cylindrical drum with a short portion of the chain left hanging free. The chain unwraps itself as the drum rotates about a fixed horizontal axis. Find the final angular speed of the drum in terms of the length L of the chain, the drum's radius R and mass M, and the moment of inertia I of the drum about its axis.

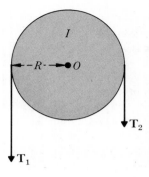

$\downarrow \mathbf{T}_2$ **Figure 12-11** In general, the tension
in a rope does change when the rope
passes over a real pulley.

$\downarrow \mathbf{T}_1$

S-4 The rope in Figure 12-11 passes over a rough pulley surface. The pulley has a radius
R and a moment of inertia I about its axis. At the pulley axis a frictional torque
of magnitude τ_f opposes rotation:
  (a) Show that the difference between the tensions in the rope on opposite sides
  of the pulley is given by

$$(T_1 - T_2)R = \tau_f + I\alpha$$

  when the pulley is rotating counterclockwise in Figure 12-11 with an angular
  acceleration α.
  (b) In mechanics problems with "light frictionless" pulleys we assume that
  $T_1 = T_2$. In making this assumption about a pulley, what physical quantities
  do we assume to have negligible values?

S-5 A uniform pole of mass M and length L starts at rest in an upright position and
falls by pivoting about its fixed base. The moment of inertia of the pole about its
base is $\frac{1}{3}ML^2$:
  (a) Show that the torque of the pole's weight about its base increases as the pole
  falls.
  (b) Is the angular acceleration of the falling pole constant? If not, why not?
  (c) Denote by ω the angular speed attained by the pole just before striking the
  ground. Deduce an expression for ω in terms of M, g, and L by using the law
  of conservation of energy.
  (d) Find an expression for the speed of the top of the pole in terms of M, g, and L
  for the instant referred to in part (c).

S-6 Compute the moment of inertia about the axis of suspension of an irregular 2.00-kg
iron bar from the following observations: the center of gravity of the bar is 0.200 m
from the axis of suspension; the bar executes 200 cycles in 250 s when set into
oscillation.

S-7 The period of oscillation of the physical pendulum of Example 6 was found to be
$T = 2\pi\sqrt{I/MgD}$:
  (a) Show that this expression can be rewritten to give

$$T = \frac{2\pi}{\sqrt{g}}\sqrt{D + \frac{k_c^2}{D}}$$

  where k_c is the radius of gyration of the rigid body about its center of mass.
  (b) The period of oscillation is observed using different locations of the axis of
  suspension. Show that the period will be a minimum when the axis of suspension is located so that $k_c = D$.

13.

angular momentum

The most significant dynamical quantity associated with rotational motion turns out to be angular momentum. Its importance as a quantity that is conserved in any isolated system is revealed in situations as varied as the rotation of galaxies, gyroscopic motion, and the most recent experiments with atoms, nuclei, and elementary particles.

The relationships between position, force, torque, and angular momentum are most easily understood by defining a new mathematical entity, the **vector product a × b** of two vectors **a** and **b**.

13.1 Vector Product

The vector product of two vectors **a** and **b** is denoted by **a × b** and is defined to be a **vector** with:

1 A *magnitude ab* sin ϕ, where ϕ is the angle between the vectors **a** and **b** (Figure 13-1).

2 A *direction* perpendicular to **a** and to **b** with a sense given by the screw rule illustrated in Figure 13-1. With the tails of **a** and **b** together, a right-handed screw is placed perpendicular to both **a** and **b** and then rotated from the first factor **a** to the second factor **b** through the smaller angle ϕ between them. The direction of advance of the screw is defined to be the direction of the vector product **a × b**.

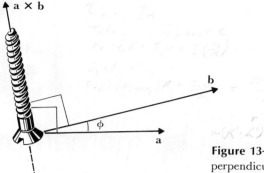

Figure 13-1 The vector **a** × **b** is perpendicular to both **a** and **b** with a sense given by the screw rule.

The cross × between the factors **a** and **b** of the vector product **a** × **b** must not be omitted because it distinguishes the vector product from other products (such as **a**·**b** and q**a**) that are used in vector algebra. We read **a** × **b** as "**a** cross **b**" and often call this product the **cross product** of **a** and **b**.

The order of the factors in the vector product **a** × **b** is important because interchanging the order changes the sign:

$$\mathbf{b} \times \mathbf{a} = -\mathbf{a} \times \mathbf{b} \tag{13.1}$$

The proof follows from the observation that, although **a** × **b** and **b** × **a** have the same magnitude $ab \sin \phi$, their *directions* are opposite because, when a screw rotates from the first factor to the second, the sense of rotation and the consequent direction of advance of the screw for **b** × **a** is opposite to that for **a** × **b**.

If the vectors **a** and **b** are parallel, ϕ and $\sin \phi$ (Figure 13-1) are zero, and **a** × **b** = 0.

The following useful rules can be derived from the definition of the vector product:

$$\mathbf{a} \times (\mathbf{b} + \mathbf{c}) = \mathbf{a} \times \mathbf{b} + \mathbf{a} \times \mathbf{c}$$

$$\frac{d}{dt} (\mathbf{a} \times \mathbf{b}) = \frac{d\mathbf{a}}{dt} \times \mathbf{b} + \mathbf{a} \times \frac{d\mathbf{b}}{dt}$$

A scalar factor q may be shifted to any position without altering the value of the product:

$$(q\mathbf{a}) \times \mathbf{b} = \mathbf{a} \times (q\mathbf{b}) = q(\mathbf{a} \times \mathbf{b})$$

13.2 Torque and Angular Momentum Vectors

Torque

The torque of a force about an *axis* was defined in Section 5.1 and used throughout Chapters 5 and 12. The torque τ_{OZ} of a force about the axis OZ is

$$\tau_{OZ} = \pm FD \tag{13.2}$$

where $F = \sqrt{F_x^2 + F_y^2}$ and the moment arm D is the distance from the line of action of $\mathbf{F}$ to the axis OZ. The torque about an axis is positive if it has a counterclockwise sense, negative if it has a clockwise sense.

The torque about a *point* is a generalization of the familiar torque about an *axis*. Consider a force $\mathbf{F}$ acting on a particle with a position vector $\mathbf{r}$ (Figure 13-2). The torque $\boldsymbol{\tau}_0$ of the force $\mathbf{F}$ about the origin O is defined to be the vector product of $\mathbf{r}$ and $\mathbf{F}$:

$$\boldsymbol{\tau}_0 = \mathbf{r} \times \mathbf{F} \tag{13.3}$$

According to this definition, the *torque about a point is a vector quantity* with a direction that is perpendicular to the plane of $\mathbf{r}$ and $\mathbf{F}$ with a sense given by the screw rule of Figure 13-1. The magnitude of $\boldsymbol{\tau}_0$ is

$$|\boldsymbol{\tau}_0| = rF \sin \phi$$

where ϕ is the smaller angle between $\mathbf{r}$ and $\mathbf{F}$. The distance $r \sin \phi$ is equal to the moment arm of the force $\mathbf{F}$ about an axis through O and perpendicular to the plane of $\mathbf{r}$ and $\mathbf{F}$.

The torque about the axis OZ is the z component of the torque vector $\boldsymbol{\tau}_0$ about the point O. This is easily seen in the special case when $\mathbf{r}$ and $\mathbf{F}$ both lie in the XY plane so that $\boldsymbol{\tau}_0$ has only a z component (Figure 13-2). In the general case, $\boldsymbol{\tau}_0$ has x and y components which are torques about the OX and OY axes, respectively, and which are given by equations analogous to Eq. 13.2. General theorems are most easily derived using the torque vector, but for applications to specific problems involving motion parallel to the XY plane, it is generally more convenient to proceed as in Chapters 5 and 12, using torques about an axis (which now can be recognized as z components of torque vectors).

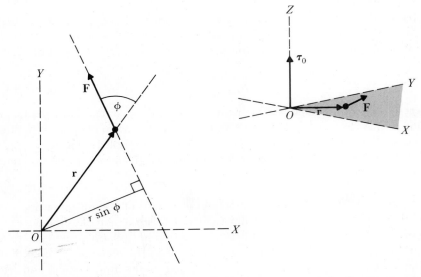

Figure 13-2   The torque *vector* $\boldsymbol{\tau}_0$ of $\mathbf{F}$ about the *point* O is defined to be $\mathbf{r} \times \mathbf{F}$. Here $\boldsymbol{\tau}_0$ is in the direction of the axis OZ.

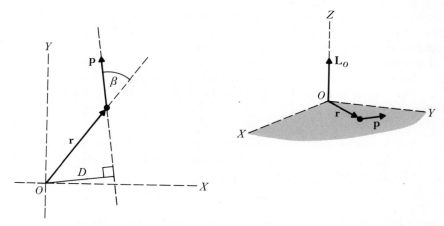

Figure 13-3   Angular momentum of the particle about the origin is $L_O = r \times p$. Here L_O is in the direction of the axis OZ.

Angular Momentum of a Particle

The angular momentum about the origin O of a particle with position vector r and momentum p is the **vector** L_O defined by

$$L_O = r \times p \tag{13.4}$$

This definition implies that the particle's angular momentum vector about the origin has a direction perpendicular to the plane containing r and p with a sense given by the screw rule (Figure 13-3). Also $|L_O| = rp \sin \beta$ where β is the angle between r and p.

   Angular momentum about an axis is defined in terms of *momentum* and its *moment arm* in exactly the same way that *torque about an axis* is defined in terms of *force* and its *moment arm*. The angular momentum L_{OZ} of the particle about the Z axis is

$$L_{OZ} = \pm pD \tag{13.5}$$

where $p = \sqrt{p_x^2 + p_y^2}$ and the moment arm D is the distance from the line of action of p to the axis OZ (Figure 13-3). L_{OZ} is positive when the particle motion is counterclockwise about OZ, negative when this motion is clockwise. L_{OZ} is the z component of L_O; the x and y components are given by expressions analogous to Eq. 13.5.

Example 1   A 0.0020-kg bullet, traveling at a speed of 500 m/s in a direction perpendicular to an upright door (Figure 13-4), strikes a nail in the door at a point 0.80 m from the axis of the door's hinges. The bullet ricochets, emerging with a speed of 300 m/s at an angle of 30° with the plane of the door. Find the bullet's angular momentum about the axis of the hinges before and after impact.

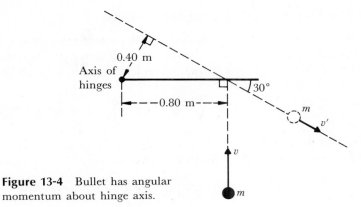

Figure 13-4 Bullet has angular momentum about hinge axis.

Solution Before impact, the bullet's momentum vector has a moment arm D of 0.80 m. Therefore the bullet's angular momentum about the axis of the hinges OZ is

$$L_{OZ} = pD = (0.0020 \text{ kg})(500 \text{ m/s})(0.80 \text{ m}) = 0.80 \text{ kg} \cdot \text{m}^2/\text{s}$$

After impact, the bullet's momentum vector has a moment arm $D = (0.80 \text{ m})$ sin $30° = 0.40$ m. The motion is clockwise about the axis of the hinges, and the bullet's angular momentum about this axis is

$$L_{OZ} = -(0.0020 \text{ kg})(300 \text{ m/s})(0.40 \text{ m}) = -0.24 \text{ kg} \cdot \text{m}^2/\text{s}$$

The angular momentum acquired by the door as a consequence of this impact is examined in Example 3.

Angular Momentum of a Rigid Body

The angular momentum of a system of particles is defined to be the vector sum of the angular momenta of its particles. A rigid body rotating with an angular velocity ω about a fixed axis OZ has an angular momentum about OZ given by

$$L_{OZ} = I\omega \tag{13.6}$$

This is proved by noting that the angular momentum about OZ of the particle of mass m_i at a distance r_i from the axis is

$$m_i v_i r_i = m_i r_i^2 \omega$$

Therefore

$$L_{OZ} = m_1 r_1^2 \omega + m_2 r_2^2 \omega + \cdots + m_n r_n^2 \omega$$

$$= (m_1 r_1^2 + m_2 r_2^2 + \cdots + m_n r_n^2)\omega$$

$$= I\omega$$

where I is the moment of the inertia of the body about the axis OZ.

13.3 The Principle of Angular Momentum

Newton's second law, $\mathbf{F} = d\mathbf{p}/dt$, yields an important relationship between the torque τ_O about the origin of the resultant force $\mathbf{F}$ acting on a particle of mass m, and the rate of change of the particle's angular momentum $\mathbf{L}_O$ about the origin:

$$\frac{d\mathbf{L}_O}{dt} = \frac{d}{dt}(\mathbf{r} \times \mathbf{p}) = \frac{d\mathbf{r}}{dt} \times \mathbf{p} + \mathbf{r} \times \frac{d\mathbf{p}}{dt} = \mathbf{v} \times m\mathbf{v} + \mathbf{r} \times \mathbf{F}$$

Noting that the vector product $\mathbf{v} \times \mathbf{v}$ of parallel vectors is zero and, substituting $\mathbf{r} \times \mathbf{F} = \tau_O$, we have the *principle of angular momentum* for a particle:

$$\tau_O = \frac{d\mathbf{L}_O}{dt} \tag{13.7}$$

which states that *the torque of the resultant force acting on the particle is equal to the rate of change of a particle's angular momentum.*

This principle can be extended to apply to any system of particles. For each particle of a system, there is an equation of the form of Eq. 13.7. When these equations are summed we obtain the *principle of angular momentum* for a system:

$$\tau_{O,\text{ext}} = \frac{d\mathbf{L}_O}{dt} \tag{13.8}$$

where $\mathbf{L}_O$ is the total angular momentum of the system and $\tau_{O,\text{ext}}$, called the resultant external torque, is the vector sum of the torques of the *external* forces acting on the particles of the system. (Torques of internal forces do not appear in the sum on the left of Eq. 13.8 because Newton's third law requires that such torques cancel out in pairs, as is illustrated in Figure 13-5.)

For z components, Eq. 13.8 implies

$$\tau_{OZ,\text{ext}} = \frac{d}{dt} L_{OZ} \tag{13.9}$$

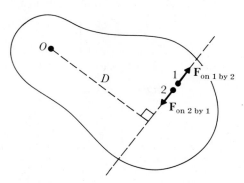

Figure 13-5   The algebraic sum of the torques of an action-reaction pair is zero. This leads to the result that the total *internal* torque is zero.

For a rigid body rotating about the axis OZ, $L_{OZ} = I\omega$ and Eq. 13.9 gives

$$\tau_{OZ,\text{ext}} = I\frac{d\omega}{dt} = I\alpha \qquad (13.10)$$

which is Eq. 12.22, the fundamental dynamical equation for rigid body rotation about a fixed axis.

The *principle of angular momentum*, $\tau_{\text{ext}} = d\mathbf{L}/dt$, holds when the torques and the angular momentum are taken about a reference point that is *fixed in an inertial frame of reference*. It can be shown that the principle of angular momentum is valid also when the reference point is the center of mass, even when the center of mass has an arbitrary motion relative to an inertial frame. For an axis CZ' through the center of mass and parallel to an axis OZ fixed in an inertial frame, this implies

$$\tau_{CZ',\text{ext}} = \frac{d}{dt}L_{CZ'} \qquad (13.11)$$

and for a rigid body,

$$\tau_{CZ',\text{ext}} = I_c\alpha \qquad (13.12)$$

where I_c is the body's moment of inertia about the axis CZ'.

Example 2 Find the acceleration a_c of the center of mass of a sphere which rolls down a rough plane inclined at an angle θ with the horizontal (Figure 12-5).

Solution Figure 13-6 is the free-body diagram for the sphere. The motion relative to the center of mass is governed by $\tau_{CZ',\text{ext}} = I_c\alpha$ which gives

$$fR = \tfrac{2}{5}MR^2\alpha$$

The law of motion of the center of mass, $F_{\text{ext}} = Ma_c$, yields

$$Mg\sin\theta - f = Ma_c$$

The condition of rolling (Eq. 12.14), requires that $V_c = R\omega$ and $a_c = R\alpha$. Substituting $\alpha = a_c/R$ in the torque equation, we find

$$f = \tfrac{2}{5}Ma_c$$

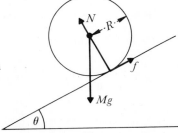

Figure 13-6 Force of static friction prevents slipping.

The acceleration of the center of mass can then be determined from

$$Mg \sin \theta - \tfrac{2}{5}Ma_c = Ma_c$$

This gives

$$a_c = \tfrac{5}{7}g \sin \theta$$

13.4 Conservation of Angular Momentum

The principle of angular momentum has a most important consequence known as the *law of conservation of angular momentum: If the resultant external torque* $\boldsymbol{\tau}_{\text{ext}}$ *is zero, the total angular momentum of a system is constant.* The reference point for angular momentum must be either the center of mass or a point that is fixed in an inertial frame. Even though $\boldsymbol{\tau}_{\text{ext}}$ is not zero, it may happen that a component is zero and then the corresponding component of angular momentum is conserved. For instance, Eq. 13.9 implies that if $\tau_{OZ,\text{ext}}$ is zero, $L_{O\hat{z}}$ is constant.

Table 13-1 summarizes laws involving angular momentum and shows the analogous laws for (linear) momentum.

A striking example of the conservation of angular momentum is provided in the experiment indicated in Figure 13-7. A man with outstretched arms, holding weights in each hand, stands on a turntable which is rotating about its vertical axis *OZ*. The only torque about the axis *OZ* is the frictional torque acting on the rotating central axis of the turntable, and if this torque is small, the angular momentum about this axis of the entire rotating system will not change appreciably over short time intervals. That is

$$I\omega = constant$$

where I is the moment of inertia about the axis *OZ* of the system comprising the man, the weights, and the rotating portion of the turntable. Now, if the man pulls the weights in toward his body, there is a dramatic increase in his angular speed. Why? The contribution mr^2 of the weights to the moment of inertia I of this system decreases as these weights are forced closer to the axis of rotation. Conservation of angular momentum requires that the decrease in I be accompanied by an increase in angular speed ω in such a way that the angular momentum $I\omega$ remains constant.

Table 13-1 Analogous Quantities and Laws

Angular momentum **L**	(Linear) momentum **P**
Resultant external torque $\boldsymbol{\tau}_{\text{ext}}$	Resultant external force $\mathbf{F}_{\text{ext}}$
$\boldsymbol{\tau}_{\text{ext}}$ = rate of change of total angular momentum	$\mathbf{F}_{\text{ext}}$ = rate of change of total (linear) momentum
If $\boldsymbol{\tau}_{\text{ext}} = 0$, total angular momentum is conserved	If $\mathbf{F}_{\text{ext}} = 0$, total (linear) momentum is conserved

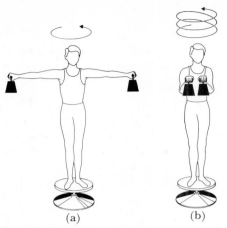

Figure 13-7 (a) Large moment of inertia, small angular velocity. (b) Small moment of inertia, large angular velocity.

A figure skater or a ballet dancer exploits conservation of angular momentum $I\omega$ in the same way. The angular speed of spin about a vertical axis can be increased simply by decreasing the body's moment of inertia about this axis. This is accomplished by decreasing the distance of the arms and legs from the axis of rotation; because I depends upon the square of the distance of the parts from the axis, a large variation is possible.

The contortions a diver goes through while executing a gainer somersault have a similar motivation. After a diver has left the diving board, there is no resultant external torque about any axis through his center of mass except for rather small torques arising from air resistance. In particular, the torque of the diver's weight about an axis through his center of mass is precisely zero since his weight can be considered to be acting at his center of mass. Consequently, from the time the diver leaves the board until he strikes the water, he travels with a constant value of angular momentum $I\omega$ about an axis through his center of mass. His angular speed about a horizontal axis CZ' through his center of mass can be increased only by decreasing his moment of inertia about this axis, that is, by going into the tuck position. Then, by straightening out again, he reduces his angular speed as his moment of inertia increases.

When a system is isolated so that there are no external forces, the external torque is obviously zero. Consequently, the resultant angular momentum of an isolated system is conserved. Changes in the angular momentum of one part of an isolated system are always accompanied by compensating changes in another part in such a way that the resultant angular momentum for the entire system remains constant. All the experimental evidence in every realm of physics is in accord with the law of conservation of angular momentum. This law, originally derived from Newton's second and third laws, is apparently valid even in atomic and nuclear physics where Newtonian mechanics fails.

All the mechanical variables that are conserved in an isolated system have now been identified. There are seven independent quantities: the three components of total momentum, the three components of total angular momentum, and the total energy of the system.

Example 3   Find the angular velocity ω imparted to the door in Example 1 when it is struck by the bullet. The door has a moment of inertia about the hinge axis of 4.0 kg·m².

Solution   Consider the system comprising the door and the bullet. For this system there is no external torque about the axis of the hinges (OZ). (The force exerted on the door by the bullet and the force exerted on the bullet by the door are internal forces for this system and contribute only internal torques, which cancel.) The total angular momentum about the hinge axis for the composite system is therefore conserved:

$$L_{OZ,\text{before impact}} = L_{OZ,\text{after impact}}$$

This gives, using the values of the bullet's angular momentum found in Example 1,

$$0.80 \text{ kg·m}^2/\text{s} + 0 = -0.24 \text{ kg·m}^2/\text{s} + (4.0 \text{ kg·m}^2)\omega$$

Solving for ω, we obtain

$$\omega = 0.26 \text{ rad/s}$$

Example 4   In a piece of machinery, two disks of moments of inertia I_1 and I_2 are mounted on the same shaft and may be connected or disconnected by a clutch. Initially the disk with moment of inertia I_1 is rotating with an angular speed ω_0, and the other disk is stationary. The disks are then coupled by the clutch. Find their common angular speed ω in terms of ω_0 and their moments of inertia.

Solution   Consider the system comprising both disks. For this system there is no external torque about the shaft. (As they are being brought to a common angular speed each disk does exert a torque on the other, but these are *internal* torques for the system which includes both disks.) The angular momentum of this system about the shaft (axis OZ) is therefore conserved:

$$L_{OZ,\text{initial}} = L_{OZ,\text{final}}$$

That is,

$$I_1\omega_0 + 0 = I_1\omega + I_2\omega$$

Therefore

$$\omega = \frac{I_1\omega_0}{I_1 + I_2}$$

13.5 Nature's Unit of Angular Momentum

One of the real surprises that emerges from a study of the behavior of atoms is that the angular momentum of a system cannot assume any one of a continuous range of values. There is a natural unit of angular momentum, universally denoted by $\hbar$, which is very small:

$$\hbar = 1.0545887 \times 10^{-34} \text{ J·s}$$

The component of the angular momentum of any system along a given direction is restricted to be an integral or half-integral multiple of $\hbar$. This restriction, a typical quantum mechanical phenomenon, is inexplicable from the point of view of Newtonian mechanics.

Each type of particle possesses a characteristic angular momentum in its CM-frame. This intrinsic angular momentum $\sqrt{S(S+1)}\,\hbar$, called *spin*, is conventionally given by specifying the value of the so-called *spin quantum number S*. (See Table 43-4. Electrons, protons, and neutrons all have $S = \frac{1}{2}$. Photons have $S = 1$. Certain particles, such as the pion, have $S = 0$.)

Summary

☐   The vector product $\mathbf{a} \times \mathbf{b}$ is a vector perpendicular to the plane containing $\mathbf{a}$ and $\mathbf{b}$ and in the direction of advance of a screw which rotates from $\mathbf{a}$ to $\mathbf{b}$ through the smaller angle. The magnitude of $\mathbf{a} \times \mathbf{b}$ is $ab \sin \phi$ where ϕ is the angle between $\mathbf{a}$ and $\mathbf{b}$.

☐   The torque of a force $\mathbf{F}$ about the origin is the vector

$$\boldsymbol{\tau}_O = \mathbf{r} \times \mathbf{F}$$

where $\mathbf{r}$ is the position vector of the point of application of $\mathbf{F}$. The z component of $\boldsymbol{\tau}_O$ is the torque about the OZ axis,

$$\tau_{OZ} = \pm FD$$

where $F = \sqrt{F_x^2 + F_y^2}$ and D is the moment arm of $\mathbf{F}$ about OZ. Counterclockwise torque about OZ is positive; clockwise torque is negative.

☐   The angular momentum about the origin of a particle with position vector $\mathbf{r}$ and (linear) momentum $\mathbf{p}$ is

$$\mathbf{L}_O = \mathbf{r} \times \mathbf{p}$$

The angular momentum about the OZ axis is the z component of $\mathbf{L}_O$,

$$L_{OZ} = \pm pD$$

where $p = \sqrt{p_x^2 + p_y^2}$ and D is the moment arm of $\mathbf{p}$ about OZ. Counterclockwise motion about OZ corresponds to positive L_{OZ}, clockwise to negative L_{OZ}.

☐   The principle of angular momentum for a system of particles with total angular momentum $\mathbf{L}$ is

$$\boldsymbol{\tau}_{\text{ext}} = \frac{d\mathbf{L}}{dt}$$

where $\boldsymbol{\tau}_{\text{ext}}$ is the resultant external torque. This principle is valid when the reference point for $\mathbf{L}$ and $\boldsymbol{\tau}$ is either the center of mass C or a point O fixed in an inertial frame of reference. About the axis OZ this implies

$$\tau_{OZ,\text{ext}} = \frac{d}{dt} L_{OZ}$$

About the axis CZ' through C but parallel to OZ,

$$\tau_{CZ',\text{ext}} = \frac{d}{dt} L_{CZ'}$$

☐ The law of conservation of angular momentum states that, if τ_{ext} is zero, a system's total angular momentum $\mathbf{L}$ is constant. If $\tau_{OZ,\text{ext}}$ is zero, L_{OZ} is constant. If $\tau_{CZ',\text{ext}}$ is zero, $L_{CZ'}$ is constant.

Questions

1   A 1.20×10^3-kg car, heading east along a straight highway at a constant speed of 30 m/s, passes 40 m south of a particular telephone pole. Viewed from above, what is the angular momentum of the car with respect to the pole:
   (a) When it is closest to the pole?
   (b) When it is 60 m from the pole?

2   Evaluate the orbital angular momentum of the earth as it moves in an approximately circular orbit about the sun. (Use the data from Appendix C.)

3   A 30-kg child, riding on a carrousel in a circle of 2.4 m radius, travels at a speed of 1.2 m/s. What is the child's angular momentum about the vertical central axis of the carrousel?

4   The carrousel in the preceding question has a moment of inertia about its central axis of 100 kg·m²:
   (a) What is its angular momentum about its axis?
   (b) What is the angular momentum of the system comprising both the carrousel and the child?

5   Evaluate the "spin" angular momentum of the earth as it revolves about a polar axis. (Use the data from Appendix C and assume that the earth's moment of inertia is given in terms of its mass M and radius R by $\frac{2}{5}MR^2$. Here the fact that the earth's density is greater within its inner core is ignored.)

6   A 2.0-kg stick which is 3.0 m long lies in a north-south direction on a sheet of ice. It is pushed by a 30-N force applied at one end and directed east. Find the acceleration of the center of mass and the initial angular acceleration of the stick.

7   The carrousel and the child in Question 4 are initially both at rest, and the carrousel is free to rotate without friction about its central axis. The child now runs around (counterclockwise looking downward) on the rim of the carrousel in a circle of 2.4 m radius and travels at a speed of 1.2 m/s relative to the ground. Find the angular speed of the carrousel.

8   The moment of inertia of the system comprising the rotating turntable, the man, and the weights in the configuration shown in Figure 13-7(a) is 8.0 kg·m², and the turntable is rotating with an angular speed of 2.0 rad/s. When the man pulls in the weights to the configuration shown in Figure 13-7(b), the moment of inertia of the rotating system is reduced to 2.0 kg·m²:
   (a) Find the angular speed corresponding to Figure 13-7(b).
   (b) Find the total kinetic energy of the system before and after the weights are pulled inward and thereby determine how much work the man must have done in pulling in the weights.

9   Assume that the bullet in Examples 1 and 3, instead of ricocheting, becomes embedded in the door. Find the angular velocity of the door for this case.

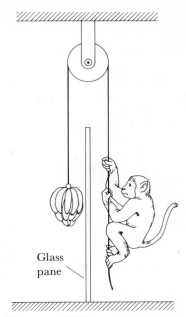

Figure 13-8 Will the monkey get the bananas?

Glass
pane

10 Two disks with moments of inertia I_1 and I_2 are rotating about a common axis with angular velocities of the same magnitude ω; but the first rotates clockwise and the second counterclockwise. They are then coupled so that they reach a common angular velocity ω':

(a) Find ω' in terms of ω, I_1, and I_2.

(b) Deduce an expression for the energy which is dissipated as the disks are brought to a common angular velocity.

11 A carrousel of radius 2.4 m is initially at rest but is free to rotate about its central axis. Its moment of inertia about its axis is 100 kg·m². A 30-kg child runs along the ground at a speed of 6.0 m/s in a direction tangent to the rim of the carrousel and then jumps on. Find the consequent angular speed ω of the carrousel.

12 Assume that the bullet in Examples 1 and 3, instead of ricocheting, emerges from the door with its direction unchanged but with its speed reduced to 300 m/s. Find the door's angular velocity after this impact.

13 A monkey and a bunch of bananas of equal mass m are supported at rest by a rope which passes over a pulley (Figure 13-8). The rope and the pulley have negligible mass, and the pulley is frictionless. The monkey attempts to reach the bananas by climbing the rope:

(a) Consider the system comprising the monkey, the bananas, the rope, and the pulley. What are the external forces for this system? Evaluate the resultant external torque about the pulley axis.

(b) By considering the angular momentum of both the monkey and the bananas about the pulley axis, determine what will be the motion of the monkey, the bananas, and the rope.

Supplementary Questions

S-1 Prove that if $F_1 + F_2 = 0$, the sum of the torques of these forces has the same value about different reference points O and O'.

S-2 Prove that the angular momentum of a system of particles has the same value for different reference points O and O' that are fixed in the CM-frame.

S-3 As a bowling ball of mass M and radius R starts down an alley, its center of mass has a speed v_0 and its angular speed is zero. The ball slides until its angular speed has reached a value such that the condition of rolling (Eq. 12.14) is satisfied. At the instant when sliding stops, what is the speed of the center of mass of the ball? How much energy has been dissipated?

S-4 A spool of thread rests on a rough horizontal table. The thread is pulled in the manner shown in Figure 13-9 and the spool rolls without slipping. What will be the direction of motion? Express the acceleration of the center of mass of the spool in terms of the applied force F, the spool's mass M, outside radius R, inside radius r, and radius of gyration k about the central axis.

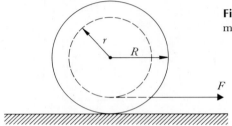

Figure 13-9 What is the direction of motion of the spool?

S-5 The vectors $\mathbf{i}$, $\mathbf{j}$, and $\mathbf{k}$ are unit vectors in the positive x, y, and z directions, respectively (see Question S-8 of Chapter 3):

(a) Show that

$$\mathbf{i} \times \mathbf{i} = 0 \qquad \mathbf{j} \times \mathbf{j} = 0 \qquad \mathbf{k} \times \mathbf{k} = 0$$

$$\mathbf{i} \times \mathbf{j} = \mathbf{k} \qquad \mathbf{k} \times \mathbf{i} = \mathbf{j} \qquad \mathbf{j} \times \mathbf{k} = \mathbf{i}$$

(b) Use the expressions

$$\mathbf{A} = A_x\mathbf{i} + A_y\mathbf{j} + A_z\mathbf{k}$$

$$\mathbf{B} = B_x\mathbf{i} + B_y\mathbf{j} + B_z\mathbf{k}$$

to evaluate the vector product $\mathbf{A} \times \mathbf{B}$ in terms of the components of $\mathbf{A}$ and $\mathbf{B}$. Make use of the simplifications provided by the results of part (a) and thereby show that

$$\mathbf{A} \times \mathbf{B} = (A_yB_z - A_zB_y)\mathbf{i} + (A_zB_x - A_xB_z)\mathbf{j} + (A_xB_y - A_yB_x)\mathbf{k}$$

(c) The determinant

$$\begin{vmatrix} A_x & A_y \\ B_x & B_y \end{vmatrix}$$

is equal to $A_xB_y - A_yB_x$. Verify that if the unit vectors are manipulated as numbers

$$\mathbf{A} \times \mathbf{B} = \begin{vmatrix} \mathbf{i} & \mathbf{j} & \mathbf{k} \\ A_x & A_y & A_z \\ B_x & B_y & B_z \end{vmatrix}$$

S-6 Show that $|\mathbf{A} \times \mathbf{B}|$ is the area of the parallelogram with sides A and B.

14.

gravitation

Gravitational forces govern the motions of the planets about the sun and the motions of the moon and various man-made space capsules about the earth. In this chapter we investigate the significant characteristics of the orbits of such satellites using the idea of a gravitational field to describe gravitational interaction.

14.1 Gravitational Fields and Gravitational Force

If a mass m at a given point in space experiences a gravitational force $\mathbf{F}$, the gravitational field $\mathbf{g}$ at that point is

$$\mathbf{g} = \frac{\mathbf{F}}{m} \tag{14.1}$$

In Chapter 2 we discussed only what the gravitational field at a certain point in space does. The field $\mathbf{g}$ exerts a force $m\mathbf{g}$ on an object of mass m placed at that point. We now ask: What is the *source* of this gravitational field? The answer is that other objects produce this field $\mathbf{g}$ according to the following rule: A particle of mass M produces at a distance r a gravitational field $\mathbf{g}$ of magnitude

$$g = \frac{GM}{r^2} \tag{14.2}$$

The vector $\mathbf{g}$ points toward the source M, as shown in Figure 14-1. The constant G is called the *gravitational constant*. Its value has been measured to be approximately $6.6720 \times 10^{-11} \, \text{N} \cdot \text{m}^2/\text{kg}^2$ in very difficult and delicate experiments that will be outlined later in this section.

Figure 14-1 The gravitational field vector at several points in space due to a particle of mass M.

According to Eq. 14.2, the field is large (or strong) at points near the source, and small (or weak) at points far from the source. At a given distance, the larger the mass M of the source, the stronger the field **g**.

Gravitational forces obey the superposition principle (Section 2.2). When forces $m\mathbf{g}_1, m\mathbf{g}_2, \ldots, m\mathbf{g}_n$ act on a particle of mass m, they are equivalent to a single force $m\mathbf{g}$ which is their vector sum:

$$m\mathbf{g} = m\mathbf{g}_1 + m\mathbf{g}_2 + \cdot \cdot \cdot + m\mathbf{g}_n$$

Division of this equation by m yields the *superposition principle for gravitational fields:*

$$\mathbf{g} = \mathbf{g}_1 + \mathbf{g}_2 + \cdot \cdot \cdot + \mathbf{g}_n$$

which states that the gravitational field produced at a given point by several sources is the vector sum of fields that each source would individually contribute, if it alone were present (Figure 14-2). The vector sum **g** is often called the *superposition* of the fields contributed by the individual particles.

The superposition principle provides a method of predicting the gravitational field that will be produced by any distribution of particles, for instance, all the particles that comprise the planet earth. In Chapter 22 we shall show that the gravitational field produced at external points by mass distributed uniformly over a *spherical* shell is the same as if the entire mass of this shell were concentrated at its center. Therefore, to the approximation that the earth can be

Figure 14-2 The superposition principle. At point P, mass M_1 alone would produce field $\mathbf{g}_1$; M_2 alone would produce $\mathbf{g}_2$. When both sources are present the field is the vector sum of $\mathbf{g}_1$ and $\mathbf{g}_2$.

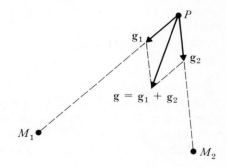

regarded as an assembly of concentric uniform spherical shells, the gravitational field outside the earth can be computed as if its entire mass M_e were located at its center. Then, from Eq. 14.2, the earth's field at a distance r from the earth's center is

$$g = \frac{GM_e}{r^2} \qquad (r \text{ greater than } r_e) \quad (14.3)$$

for r not less than the radius r_e of the earth. This equation describes the gravitational field that is shown in Figure 2-4. Equation 14.3 can be rewritten in a form that is often more convenient for calculations:

$$g = g_s \left(\frac{r_e}{r}\right)^2 \qquad (r \text{ greater than } r_e) \quad (14.4)$$

where g_s is the gravitational field at the earth's surface.

Newton's Law of Gravitation

To determine the gravitational force exerted by a particle of mass M on another particle of mass m we think about the problem in two distinct stages. First, we think of M as a source of a gravitational field at all points in space and determine the field that M produces at the location of the other mass m which is a distance r away. According to Eq. 14.2 this field has a magnitude $g = GM/r^2$ and a direction toward the source M. This field $\mathbf{g}$ exerts a force of magnitude $F = mg$ on the mass m. Substituting for g the value given by Eq. 14.2 we get

$$F = G\frac{Mm}{r^2} \tag{14.5}$$

This is the magnitude of the force exerted on m by M; it acts on m and is directed toward M, so it is an attractive force.

To find the force exerted on M by m we compute the field produced by m at the location of M, $g = Gm/r^2$. This field exerts a force $\mathbf{F} = M\mathbf{g}$ on M so that the magnitude of the force on M is $F = GMm/r^2$; the direction is toward the source m. Notice that gravitational forces obey Newton's third law: the force exerted on M by m is equal in magnitude and opposite in direction to the force exerted on m by M (Figure 14-3). The force law, $F = GMm/r^2$, where F is the magnitude of the attractive force exerted by one particle on another, is called *Newton's law*

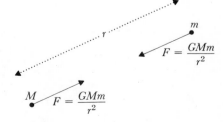

Figure 14-3   Newton's law of gravitation.

of gravitation, since this was the gravitational force law assumed by Newton in his successful attempt to understand planetary motion.

The law need not be restricted to particles. It turns out to be true if M and m are the masses of any two spherically symmetric distributions of mass with a distance r between their centers. The gravitational force exerted by one object on another is extremely weak unless at least one of the objects has a huge mass, like the mass of our planet.

Example 1   What is the gravitational force exerted on a 0.0100-kg marble by a 10.0-kg cannonball which is 0.100 m away?

Solution   We use Newton's law of gravitation and substitute the appropriate numbers:

$$F = \frac{(6.67 \times 10^{-11} \text{ N} \cdot \text{m}^2/\text{kg}^2)(10.0 \text{ kg})(0.0100 \text{ kg})}{(0.100 \text{ m})^2}$$

$$= 6.67 \times 10^{-10} \text{ N}$$

Measurement of G

Example 1 illustrates that the gravitational force exerted by one laboratory-size object on another is so small that it is completely negligible in engineering practice. Yet, to determine the value of the gravitational constant G, we must measure such a minute gravitational force between known masses. This was done by Lord Cavendish (1731–1810) in 1798, using a very sensitive torsion balance (Figure 14-4). When the masses M are brought into position, the attractive forces F that they exert on the suspended masses m can be determined by measuring the angle through which the suspension fiber is twisted. Then G is calculated from $G = Fr^2/Mm$. Results of modern experiments give

$$G = 6.6720 \times 10^{-11} \text{ N} \cdot \text{m}^2/\text{kg}^2$$

With the value of G determined, a remarkable feat can be accomplished: the mass of our planet can be calculated. Equation 14.3 gives

$$M_e = \frac{g_s r_e^2}{G}$$

We know, from measurement of the gravitational force on a unit mass at the surface of the earth, that $g_s = 9.8$ N/kg. From surveying data applied to the

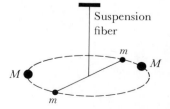
Suspension fiber

m

M

M

m

Figure 14-4   Schematic diagram of torsion balance used to measure the gravitational forces exerted on suspended masses m by large fixed masses M.

geometry of a sphere, we find that the radius of the earth r_e is 6.37×10^6 m. Therefore the mass M_e of the earth is given by

$$M_e = \frac{(9.8 \text{ N/kg})(6.37 \times 10^6 \text{ m})^2}{(6.67 \times 10^{-11} \text{ N} \cdot \text{m}^2/\text{kg}^2)} = 6.0 \times 10^{24} \text{ kg}$$

The Field Concept

The attractive gravitational forces, each of magnitude $F = GMm/r^2$, that two masses exert on one another were postulated by Newton without any mention of the idea of a gravitational field. It was simply assumed that one body can act directly on a second body which is distant from the first; that is, it was assumed that *action-at-a-distance* is possible. The predictions of gravitational phenomena obtained from this viewpoint are the same as those which have been developed using the concept of a gravitational field. We introduced the idea of a field before stating Newton's law of gravitation because there are advantages to acquiring the habit of thinking about physical phenomena in terms of fields.

Using the field concept, we discuss gravitational forces in a two-stage description with the gravitational field playing an intermediate role:

1   Each source produces a field at all points in space.
2   The field at a given location exerts a force on any particle which happens to be at that location.

A key assertion of a field theory of any type of force, whether gravitational, electric, or magnetic, is that *the force acting on an object is determined by the field at the location of the object.* Many different arrangements of sources could have produced this field, but if the field is known, the discussion of the force on an object can proceed without explicit reference to the field source.

The great value of the field concept was first made apparent by Michael Faraday (1791–1867) in his studies of electricity and magnetism. The notion of electric and magnetic fields is central to the unification of the theories of light, electricity, and magnetism, and the consequent discovery of radio waves, the topics of later chapters. To this very day, the concept of a field is fundamental to the most successful theories that physicists have created; the field idea has staying power.

14.2 Gravitational Potential Energy

Particle in an External Field

Consider a system consisting of a single particle of mass m in an external gravitational field $g = GM/r^2$, where r is the distance from M to m. The external force acting on the mass m is $F = mg = GMm/r^2$. Using the terminology of Section 10.3, the potential energy function we associate with this force is a function U such that $F_s = -dU/ds$. Let us investigate the function defined by

$U = -GMm/r$. For $ds = dr$, we find that the force derived from this potential energy function has a component in the radial direction (the direction of increasing r) given by

$$F_r = -\frac{d}{dr}\left(-\frac{GMm}{r}\right) = -\frac{GMm}{r^2}$$

For a displacement ds perpendicular to the radial direction, r is constant and we find $F_s = dU/ds = 0$. Therefore the force derived from this potential energy function is directed toward M and has a magnitude GMm/r^2. Since this is the gravitational force acting on m, the associated potential energy function is called the *gravitational potential energy*.

We have found that *in an external gravitational field, g* $= GMm/r^2$, *a particle of mass m has a gravitational potential energy*

$$U = -\frac{GMm}{r} \tag{14.6}$$

This function (Figure 14-5) is negative for all values of r and increases from $-\infty$ to 0 as r increases from 0 to ∞.

Example 2   Use Eq. 10.26 to find the potential energy U of a mass m in the field produced by M, by integration of the expression for the force exerted on m, $F_r = -GMm/r^2$.

Solution   Equation 10.26 gives

$$U(x,y,z) = -\int_a^{x,y,z} F_s\, ds + U_a$$

From Figure 14-6 we see that $ds = dr'/\cos\theta$, where θ is the angle between ds and the radial direction. Also $F_s = F_{r'}\cos\theta$. Therefore

$$F_s\, ds = F_{r'}\, dr'$$

and

$$U = -\int_{r_a}^{r} F_{r'}\, dr' + U_a = \int_{r_a}^{r} \frac{GMm}{r'^2}\, dr' + U_a$$

This gives

$$U = -\frac{GMm}{r} + \frac{GMm}{r_a} + U_a$$

If we select the arbitrary reference point a to be at infinity, the second term is zero. This expression is further simplified by choosing the arbitrary value U_a of the potential energy at the reference point to be zero. Then

$$U = -\frac{GMm}{r}$$

Since the integral in Eq. 10.26 is a work, we can think of the potential energy $U = -GMm/r$ as the negative of the work that is done by the gravitational force acting on m as this mass is brought from infinity to a distance r from M.

Or, equivalently, U is the work that would have to be done by an external applied force to bring m from infinity to its position at a distance r from M.

Example 3 A particle is released at rest at an altitude above the earth's surface equal to the radius r_e of the earth. Find its speed v_s as it strikes the earth.

Solution The gravitational force $GM_e m/r^2$ increases by a factor of four as the particle's distance from the earth's center changes from $2r_e$ to r_e. There is a corresponding change in the particle's acceleration. Therefore we can use *neither* formulas for motion with constant acceleration *nor* the expression $mg_s y$ for the gravitational potential energy in a *uniform* field $\mathbf{g}_s$. However, the problem is easily solved using the general expression $U = -GM_e m/r$. We neglect the air resistance encountered as the particle falls through the earth's atmosphere. Then the law of conservation of the particle's total energy $K + U$ gives

$$\tfrac{1}{2}mv_s^2 - \frac{GM_e m}{r_e} = 0 - \frac{GM_e m}{2r_e}$$

Therefore, $v_s = \sqrt{GM_e/r_e}$.

Figure 14-5 Graph of the gravitational potential energy function $U(r) = -GMm/r$.

Figure 14-6 Mass m is taken from point a a distance r_a from M to a point that is a distance r from M. The element of path length is $ds = dr'/\cos\theta$.

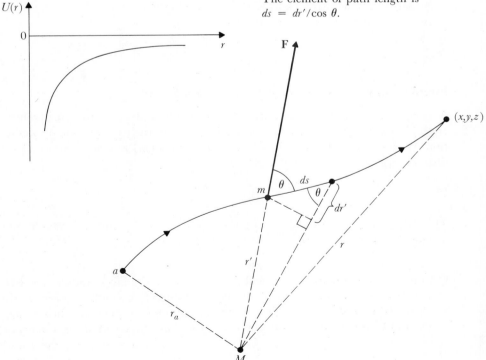

Example 4   Show that for altitudes y that are small compared to the radius r_e of the earth, the gravitational potential energy can be approximated by the function $mg_s y$.

Solution   We have

$$U = -\frac{GM_e m}{r} = -\frac{GM_e m}{r_e + y} = -\frac{GM_e m}{r_e}\left(1 + \frac{y}{r_e}\right)^{-1}$$

The binomial theorem gives

$$\left(1 + \frac{y}{r_e}\right)^{-1} = 1 - \frac{y}{r_e} + \left(\frac{y}{r_e}\right)^2 - \left(\frac{y}{r_e}\right)^3 + \cdots$$

When $y/r_e \ll 1$, $(y/r_e)^2 \ll (y/r_e)$, $(y/r_e)^3 \ll (y/r_e)^2$, etc. Then $(1 + y/r_e)^{-1}$ can be approximated by $1 - y/r_e$ and the expression for U becomes

$$U = -\left(\frac{GM_e m}{r_e}\right)\left(1 - \frac{y}{r_e}\right) = mg_s y + constant$$

where $g_s = GM_e/r_e^2$ and the constant is $-GM_e m/r_e$. If we choose the location of zero potential energy to be at $y = 0$ rather than at an infinite value of r, then we obtain

$$U = mg_s y \qquad \text{(approximately, for } y \ll r_e)$$

in agreement with the expression used in Chapter 10 for the gravitational potential energy near the earth's surface.

Internal Gravitational Potential Energy of a System

A system comprising several particles has an internal gravitational potential energy associated with each pair of particles due to their gravitational interaction. Within the system, a pair of particles of masses m_1 and m_2 separated by a distance r_{12} has a gravitational potential energy

$$U_{12} = -\frac{Gm_1 m_2}{r_{12}}$$

This internal gravitational potential energy belongs to the pair and is not assigned to one specific particle of the pair.

Example 5   Two identical spherical spaceships, each of mass m and diameter D, are initially at rest (relative to an inertial frame of reference) in outer space separated by a distance r. Their gravitational interaction causes these spaceships to be accelerated toward one another. Find the speed of each ship just before they collide.

Solution Consider the system consisting of the two spaceships. Just before collision, the ships have velocities $\mathbf{v}_1$ and $\mathbf{v}_2$, respectively. Since the system is isolated, its total momentum is conserved:

$$m\mathbf{v}_1 + m\mathbf{v}_2 = 0 + 0$$

where we have used initial values (zero) on the right-hand side of this equation. Therefore $|\mathbf{v}_1| = |\mathbf{v}_2|$ and the ships have a common speed v at impact.

Throughout the motion of these spaceships, no energy is dissipated. Equating the initial total energy of the system to its total energy just before impact, we have

$$0 - \frac{Gm^2}{r} = \tfrac{1}{2}mv^2 + \tfrac{1}{2}mv^2 - \frac{Gm^2}{D}$$

When solved for v this yields

$$v = \sqrt{Gm\left(\frac{1}{D} - \frac{1}{r}\right)}$$

14.3 Planetary Motion

Kepler's Laws

Observations of planetary motions have been made for thousands of years. The description of a planet's trajectory relative to an earth-centered frame is complex. A simple description of roughly circular orbits about a fixed sun, known in ancient times, was rediscovered by Copernicus (1473–1543). This heretical idea of demoting mankind's home from the center of the universe to just one of many minor planets led to fierce disputes which were ultimately settled, not by the decisions of authorities, but by the lonely efforts of the few who were willing to make precise experimental measurements and who did their best to apply unfettered human reason to their results. Years of precise observations of planetary positions were tabulated by Tycho Brahe (1546–1601), and his data were given neat mathematical description by Johannes Kepler (1571–1630) in the form of the following three laws:

1 The planets move around the sun in orbits which are ellipses, with the sun at one focus (the law of orbits).
2 A position vector from sun to planet sweeps over equal areas in equal times (the law of areas).
3 The squares of the periods of revolution of the planets around the sun are proportional to the cubes of the average radii of their respective orbits (the law of periods).

In an astounding synthesis of knowledge, Newton showed that Kepler's empirical laws were a consequence of $\mathbf{F} = m\mathbf{a}$ with $F = GMm/r^2$. We shall see that Newton's laws are sufficient to account for the motion of a planet about the sun, the orbit of an earth satellite, or simply the familiar trajectory of a baseball.

In the following applications of Newtonian mechanics to the motion of planets, we shall use the *astronomical frame* of reference, a frame with origin at the center of mass of the solar system and with axes not rotating relative to distant stars. This is an *inertial frame* to an excellent approximation. We shall assume that the sun of mass M is fixed at the origin. This is a good approximation because the mass of the sun is much greater than the masses of its planets.

Angular Momentum and Planetary Motion

In Figure 14-7 a planet is represented by a particle of mass m with a position vector $\mathbf{r}$ and the sun is represented by a particle of mass M at the origin. Since the line of action of the gravitational force exerted on the planet by the sun passes through the sun, the torque of this force about the sun is zero. Therefore, if this is the only force acting on the planet, the planet's angular momentum $\mathbf{L}$ about the sun is conserved:

$$\mathbf{L} = \mathbf{r} \times \mathbf{p} = \text{constant vector}$$

Since $\mathbf{L}$ is perpendicular to $\mathbf{r}$, the constancy of $\mathbf{L}$ implies that $\mathbf{r}$ remains in a plane perpendicular to $\mathbf{L}$ throughout the motion: the path of the particle is confined to one plane.

We can describe the particle's motion most conveniently by using polar coordinates (r, θ) in the plane of motion. The component of the particle's velocity in the radial direction is

$$v_r = \frac{dr}{dt} \tag{14.7}$$

Perpendicular to the radial direction in the direction of increasing θ, the particle has a velocity component v_θ given by

$$v_\theta = r\omega = r\frac{d\theta}{dt} \tag{14.8}$$

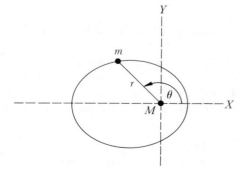

Figure 14-7   Planet of mass m moves in an orbit about the sun of mass M.

The particle's angular momentum L about the origin is found by noting that the moment arm for v_θ is r and for v_r is zero. Therefore

$$L = rmv_\theta = mr^2 \frac{d\theta}{dt} \qquad (14.9)$$

From this relationship we can give a geometric interpretation of the conservation of a planet's angular momentum about the sun (Figure 14-8). The area dA of the sector bounded by two radius vectors and the element of a path is approximated by the area of a triangle of base r and height $r\,d\theta$:

$$dA = (\tfrac{1}{2}r)(r\,d\theta)$$

Therefore

$$\frac{dA}{dt} = \tfrac{1}{2}r^2 \frac{d\theta}{dt} = \frac{L}{2m} \qquad (14.10)$$

Evidently the rate dA/dt at which the radius vector sweeps out the area is proportional to the planet's angular momentum. Kepler's law of areas, which states that dA/dt is constant, is seen to be equivalent to the statement that a planet has a constant angular momentum about the sun.

From Kepler's law of areas, Newton could learn only that the force acting on a planet was a *central* force, a force with a line of action always passing through a fixed center of force (the sun). All other characteristics of gravitational force, such as its dependence on the masses and their separation, had to be discovered or tested by considering other features of planetary motion.

Energy, Angular Momentum, and Orbits

The equation of the orbit and a general understanding of the motion is most simply obtained by using the laws of conservation of energy and of angular

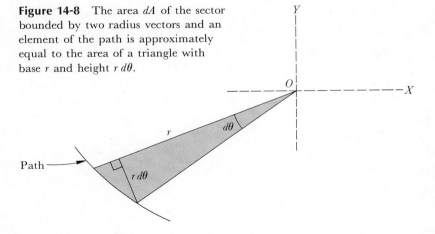

Figure 14-8 The area dA of the sector bounded by two radius vectors and an element of the path is approximately equal to the area of a triangle with base r and height $r\,d\theta$.

momentum. Substituting $d\theta/dt = L/mr^2$ in the expression for the planet's total energy

$$E = K + U = \tfrac{1}{2}m\left[\left(\frac{dr}{dt}\right)^2 + r^2\left(\frac{d\theta}{dt}\right)^2\right] - \frac{GMm}{r}$$

we obtain

$$E = \tfrac{1}{2}m\left(\frac{dr}{dt}\right)^2 + \frac{1}{2}\frac{L^2}{mr^2} - \frac{GMm}{r} \qquad (14.11)$$

This expression shows that the radial part of the motion can be regarded as taking place in one dimension with an "effective potential energy"

$$U_{\text{eff}} = \frac{1}{2}\frac{L^2}{mr^2} - \frac{GMm}{r} \qquad (14.12)$$

Figure 14-9 shows how the *turning points,* r_{min} and r_{max}, can be determined graphically for a given value of the energy E and a given nonzero value of L. At a turning point, the radial velocity dr/dt is zero. Because the angular velocity $d\theta/dt = L/mr^2$ is never zero, a planet does not come to rest at a turning point, but its distance r from the sun begins to increase instead of decreasing, or vice versa.

It is clear from Figure 14-9 that if $E < 0$, the motion is confined to values of r between r_{min} and r_{max}. A value $E > 0$ corresponds to a motion in which the mass m comes from and returns to infinity.

The relationship between r and θ will give the equation of the path. From Eq. 14.11 we find

$$\frac{dr}{dt} = \sqrt{\frac{2}{m}\left(E + \frac{GMm}{r}\right) - \frac{L^2}{m^2 r^2}}$$

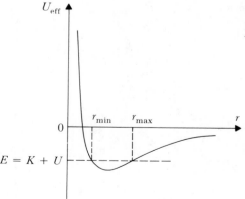

Figure 14-9   Graph of the effective potential energy. Turning points occur where $E = U_{\text{eff}}$.

From this result and Eq. 14.9, we obtain

$$d\theta = \frac{L}{mr^2} dt = \frac{L \, dr/r^2}{\sqrt{2m(E + GMm/r) - L^2/r^2}}$$

This implies

$$\theta = \int \frac{-L \, du}{\sqrt{2m(E + GMmu) - L^2 u^2}} + constant$$

where $u = 1/r$ and $du = -dr/r^2$. Completing the square in the denominator, we have

$$\theta = \int \frac{-L \, du}{\sqrt{(2mE + G^2 M^2 m^4/L^2) - (Lu - GMm^2/L)^2}} + constant$$

$$= \cos^{-1} \frac{(L/r) - (GMm^2/L)}{\sqrt{2mE + G^2 M^2 m^4/L^2}} + constant$$

Selecting the direction from which θ is measured so that the constant of integration is zero, and putting

$$R = \frac{L^2}{GMm^2} \quad and \quad \epsilon = \sqrt{1 + \frac{2EL^2}{G^2 M^2 m^3}} \tag{14.13}$$

we obtain the equation of the path

$$\frac{R}{r} = 1 + \epsilon \cos \theta \tag{14.14}$$

This is the equation of a conic section with one focus at the origin; R is the *semi-latus rectum* of the orbit and ϵ the *eccentricity*. The point of closest approach to M, called perihelion, occurs at $\theta = 0$.

The type of orbit depends on the eccentricity as follows:

$\epsilon > 1, E > 0$:   hyperbola   (Figure 14-10)

$\epsilon = 1, E = 0$:   parabola   (Figure 14-11)

$\epsilon < 1, E < 0$:   ellipse   (Figure 14-12)

(which becomes a circle for $\epsilon = 0$).

The only orbit corresponding to a bound motion is an ellipse, in agreement with Kepler's law of orbits. The major axis of an elliptical orbit is

$$2a = r_{min} + r_{max} = \frac{R}{1 + \epsilon} + \frac{R}{1 - \epsilon} = \frac{2R}{1 - \epsilon^2}$$

With R and ϵ given by Eq. 14.13, the semi-major axis is

$$a = \frac{GMm}{2|E|} \tag{14.15}$$

Notice that the major axis depends only on the energy and not on the angular

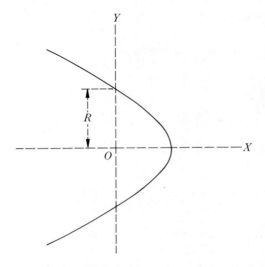

Figure 14-10 Hyperbola with semi-latus rectum R.

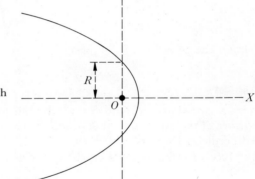

Figure 14-11 Parabola with semi-latus rectum R.

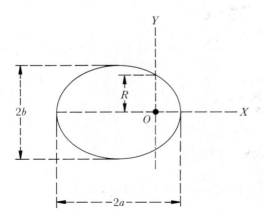

Figure 14-12 Ellipse with semi-latus rectum R, major axis $2a$, and minor axis $2b$.

momentum. From the geometry of an ellipse it can be shown that the semi-minor axis is

$$b = a\sqrt{1 - \epsilon^2}$$

After substitution for a and ϵ from Eq. 14.13, this yields

$$b = \frac{L}{\sqrt{2m|E|}} \qquad (14.16)$$

To find the period T of revolution in an elliptical orbit, we integrate Eq. 14.10

$$\int_0^T \frac{dA}{dt} \, dt = \frac{L}{2m} \int_0^T dt$$

This gives $A = LT/2m$. Since the area A of an ellipse is πab, we have

$$T = \frac{2m\pi ab}{L} = \frac{2m\pi a}{\sqrt{2m|E|}}$$

after substituting for b from Eq. 14.16. From Eq. 14.15, this can be written as

$$T = \frac{2\pi}{\sqrt{GM}} a^{3/2} \qquad (14.17)$$

This shows that *the squares of the periods of the various planets are proportional to the cubes of their major axes,* which can be taken as a precise statement of Kepler's law of periods. This result is useful in various ways in astronomy. The mass of the sun can be calculated from measurement of a planet's period and the major axis of its orbit. When the mass of the sun is known, the distance a for a planet can be calculated from measurement of its period.

All the results of this section can be used to describe orbits of objects about the earth (Figure 14-13). The center of the earth is then taken as the origin and

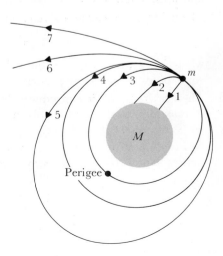

Figure 14-13 Orbits of the mass m when projected perpendicular to **g** with different speeds.

coordinate axes have directions that are fixed relative to distant stars. Such a frame is approximately an inertial frame of reference.

Example 6   Starting from Newton's second law, find the speed v and the period T of a satellite traveling in a *circular* orbit of radius r about the earth of mass M_e.

Solution   We use the frame of reference described in the preceding paragraph. For a satellite of mass m, Newton's second law for normal components $F_N = ma_N$, with $F_N = GM_em/r^2$ and $a_N = v^2/r$, gives

$$\frac{GM_em}{r^2} = \frac{mv^2}{r}$$

Therefore the satellite's speed is

$$v = \sqrt{\frac{GM_e}{r}}$$

The satellite's period is

$$T = \frac{2\pi r}{v} = \frac{2\pi r^{3/2}}{\sqrt{GM_e}}$$

which is Kepler's law of periods for circular orbits of radius r.

Summary

☐   A particle of mass M is the source of a gravitational field which, at a distance r, has a magnitude

$$g = \frac{GM}{r^2}$$

and a direction toward M. (G is called the gravitational constant.)

☐   The field $\mathbf{g}$ produced at a given point by several sources is the superposition

$$\mathbf{g} = \mathbf{g}_1 + \mathbf{g}_2 + \cdots + \mathbf{g}_n$$

where $\mathbf{g}_1, \mathbf{g}_2, \ldots, \mathbf{g}_n$ are the fields that the sources would individually contribute, if each were present alone.

☐   The gravitational field produced by a spherically symmetric distribution of mass at a point outside this distribution is the same as if all the mass were concentrated at the center.

☐   Newton's law of universal gravitation states that any two particles of masses m and M exert attractive forces on each other of magnitude

$$F = \frac{GMm}{r^2}$$

□ A particle of mass m, in an external gravitational field $g = GM/r^2$, has a gravitational potential energy

$$U = -\frac{GMm}{r}$$

The gravitational potential energy of any pair of particles within a system is

$$U = -\frac{Gm_1m_2}{r_{12}}$$

where m_1 and m_2 are the masses of the particles and r_{12} is their separation.

□ The orbit of a particle of mass m, acted on by an attractive force GMm/r^2 directed toward a mass M fixed at the origin of an inertial frame, is a conic section

$$\frac{R}{r} = 1 + \epsilon \cos \theta$$

with one focus at the origin, semi-latus rectum R proportional to L^2, and eccentricity $\epsilon = \sqrt{1 + 2EL^2/G^2M^2m^3}$, where L is the particle's angular momentum about the origin and $E = K + U$. The orbit is an ellipse for $E < 0$, a parabola for $E = 0$, and a hyperbola for $E > 0$. The period in an elliptical orbit with semi-major axis a is

$$T = \frac{2\pi}{\sqrt{GM}} a^{3/2}$$

Questions

1 Find the magnitude and direction of the gravitational field produced by a 1.0-kg particle at a point which is 1.0 m from this particle. What force will this field exert on a 10-kg particle?

2 (a) The gravitational field produced by a particle has a value g_1 at a distance r_1 and a value g_2 at a distance r_2. Express the ratio g_1/g_2 in terms of r_1 and r_2.
    (b) A particle produces a gravitational field of magnitude 1.0×10^{-11} N/kg at a distance of 2.0 m. What field does it produce at a distance of 6.0 m?

3 At what distance above the earth's surface is the earth's gravitational field $\frac{1}{25}$ of its value at the earth's surface?

4 Using the data in Appendix C, calculate the gravitational field produced by the moon's mass:
    (a) At the surface of the moon.
    (b) At the location of the earth.

5 Using the data in Appendix C, calculate the gravitational field produced by the sun's mass:
    (a) At the surface of the sun.
    (b) At the location of the earth.

6 The sources of a gravitational field are *two* particles, one of mass 1.0 kg and the other of mass 2.0 kg separated by 1.0 m:
 (a) Find the gravitational field at the point midway between the particles.
 (b) Find the gravitational field at a point in space which is 2.0 m from the 1.0-kg mass and 1.0 m from the 2.0-kg mass.

7 Locate the point between the earth and the sun at which the superposition of their gravitational fields is zero.

8 It sometimes happens that the gravitational fields contributed by the earth, the moon, and the sun are all in the same direction at a point P on the earth's surface:
 (a) Sketch a diagram showing the relative positions of P, the earth's center, the moon, and the sun.
 (b) Using the results of Questions 4 and 5, calculate the superposition of these three fields at the point P.
 (c) Calculate the superposition of these three fields at a point Q on the earth's surface such that PQ is a diameter of the earth. (Neglect the change in magnitude of the individual fields due to the change PQ in the distance from the moon and the sun.)

9 **(a)** An 80.0-kg mass and a 60.0-kg mass are 3.00 m apart. Find the superposition of their gravitational fields at a point P which is 5.00 m from the 80.0-kg mass and 4.00 m from the 60.0-kg mass.
 (b) Calculate the gravitational force that would be exerted on a 2.0-kg mass placed at the point P.

10 Masses of 6.00 kg and 4.00 kg are placed at two of the vertices of an equilateral triangle whose sides are 2.00 m long. Find the gravitational field produced at the location of the third vertex.

11 Show that Newton's law of gravitation, $F = GMm/r^2$, together with the definition of a gravitational field ($\mathbf{g} = \mathbf{F}/m$), imply that the field produced at a distance r from a source M is given by

$$g = \frac{GM}{r^2}$$

as stated in Eq. 14.2.

12 Use Newton's law of gravitation to compute the force exerted by a 1.0-kg particle on a 10-kg particle which is 1.0 m away. Check your answer by comparing it with the result obtained by the two-stage process used in Question 1.

13 A 100-kg man and a 50-kg woman are approximately 20 m apart. Estimate the attractive force exerted on the man by the woman. Compare this force to the man's weight and decide whether or not the man would be aware of this attraction.

14 **(a)** Calculate the magnitude of the force exerted on the earth by the sun. (Use the data in Appendix C.)
 (b) What is the ratio of the force exerted on the moon by the sun to the force exerted on the moon by the earth?

15 Explain how Cavendish was able to determine the mass of the earth.

16 How much work is done by the gravitational force acting on a 1000-kg spaceship which moves from an altitude of 10×10^6 m to the surface of the earth?

17 A spaceship returning home out of fuel is traveling slowly when it is a few million kilometres from the earth so that its kinetic energy is negligible and the potential energy of the ship-earth system is also negligible. The ship is attracted by the earth and comes in on a parabolic orbit which unfortunately intercepts the earth. What is its velocity on impact? Justify your answer.

18 The escape speed of an object on the earth is the speed that is just sufficient to enable the object to keep moving away from the earth forever and reach, with negligible kinetic energy, remote regions where r is so large that $GM_e m/r$ is negligible. Deduce an expression for the escape speed v_e in terms of G and the mass M_e and radius r_e of the earth. What is the numerical value of v_e?

19 The mass of Mars is 0.107 times the mass of the earth and the radius of Mars is 0.53 times the radius of the earth:
   (a) What is the value of g on Mars?
   (b) What is the escape speed for an object on Mars?

20 Give an expression for the total gravitational potential energy of the sun-earth-moon system.

21 Spaceships are at rest a distance r_0 apart. Under the influence of their gravitational attraction, they move toward each other. Find their speeds v_1 and v_2 in terms of their separation r, their initial separation r_0, and their masses m_1 and m_2, respectively.

22 Show that Eq. 14.18, $v = \sqrt{GM/r}$, can be rewritten in the convenient form

$$v = \sqrt{gr}$$

and that for an earth satellite we have

$$v = \sqrt{g_s r_e \frac{r_e}{r}}$$

where g_s is the gravitational field at the earth's surface and r_e is the earth's radius:
   (a) From this result, evaluate the speed of a satellite for an orbit near the earth's surface ($r = r_e$).
   (b) Calculate the orbital speed of the moon ($r/r_e = 60$).
   (c) Show that the moon's period is nearly one month.

23 At what distance from the earth's center will a satellite moving in the plane of the equator always be vertically above the same point on the earth's surface?

24 Explain how the mass of a planet with a moon can be determined.

25 A satellite of Jupiter has a period of 1.77 days and an orbital radius of 4.22×10^8 m. Find the mass of Jupiter.

26 A 20×10^3-kg spaceship is drifting with its engines turned off in a remote region of space where the gravitational field due to other objects is very small. A small satellite, moving almost solely under the influence of the gravitational force exerted by the spaceship, has an orbit about the spaceship of radius 10^2 m. What is the period of revolution and the speed of this satellite?

27 Determine the mass of the sun from the value of G and the following data concerning the earth's orbital motion:

$$1 \text{ yr} = 3.15 \times 10^7 \text{ s}$$

$$\text{radius of earth's orbit} = 1.5 \times 10^{11} \text{ m}$$

28 Calculate the mass of the earth from the data given in Appendix C concerning the moon's orbit.

29   The orbit of Mars about the sun has an average radius which is 1.52 times the average radius of the earth's orbit. What is the period (in years) of the orbital motion of Mars?

30   Find the speed v_A of a planet when it is at aphelion (the position when it is farthest from the sun) in terms of the distances r_{max}, r_{min}, and the speed v_P at perihelion (Figure 14-14).

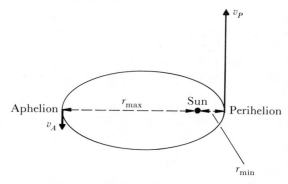

Figure 14-14   Perihelion and aphelion positions in a planet's orbit about the sun.

Supplementary Questions

S-1   An earth satellite travels through the atmosphere in an orbit which is approximately circular but has a slowly diminishing radius r, decreasing by an amount Δr in one revolution. Assuming that Δr is much less than r, show that the retarding force F' exerted on the satellite by the atmosphere is

$$F' = \frac{F}{4\pi} \left(\frac{\Delta r}{r} \right)$$

where F is the gravitational force exerted on the satellite.

S-2   Consider the motion of a mass m acted on by a central attractive force given by

$$F_r = -kr$$

Which of Kepler's laws must be modified to apply to this case? Explain.

S-3   Express the geometric constants ϵ, R, a, and b of a planet's orbit in terms of the angular momentum per unit mass, $L' = L/m$, and the energy per unit mass $E' = E/m$.

S-4   A meteor of mass m is approaching the earth (of mass M_e) from remote regions with a speed $v_0 = \sqrt{GM_e/2D}$ where the constant D is the moment arm of the velocity vector $\mathbf{v}_0$ about the earth:

(a)   What type of conic section is the orbit of m?

(b)   Find the semi-latus rectum and the eccentricity of this orbit.

(c)   What is the distance of closest approach to the earth?

S-5 A satellite, at a distance r_0 from the earth's center, is projected with a speed

$$v_0 = \sqrt{\frac{\beta GM_e}{r_0}}$$

in a direction perpendicular to its radius vector from the earth's center (Figure 14-13). Identify the type of orbit and find the distance of closest approach to the earth and the maximum distance from the earth (for the elliptical orbits) for the following values of β: $\frac{1}{2}$, 1, $\frac{3}{2}$, 2, and 4.

S-6 In Example 6, Kepler's law of periods is derived for a circular orbit, making the approximation that M is fixed at the center of mass which is the origin. In fact, if M and m comprise an isolated system, M moves in a circular orbit of radius R about the center of mass, where $MR = mr$. Show that the exact calculation of the period of the mass m in a circular orbit of radius r, gives

$$T = \frac{2\pi(r + R)^{3/2}}{\sqrt{G(M + m)}}$$

15.

fluid statics

A fluid is a substance that can flow. The term therefore includes both *liquids* and *gases*. An understanding of the behavior of fluids is necessary for the comprehension of many practical devices.

It is most convenient to introduce the new terms, pressure and density, for a system that is a fluid, in order to work out the consequences of the laws of Newtonian mechanics. But we shall find that the so-called "principles" of fluid statics, such as Pascal's principle and Archimedes' principle, rather than being new fundamental laws of physics, are direct consequences of one law of Newtonian mechanics, the first condition of equilibrium (Section 5.3).

15.1 Pressure

In fluid statics we assume that the fluid and any other relevant objects, such as containers, are at rest. Then the fluid in contact with any surface exerts a force which is a *push perpendicular to the surface.** If we measure the force exerted by the fluid on a small plane surface, we find that this force has the same magnitude no matter what the orientation of the surface, provided the small surface is

*If there are tangential forces (or shearing stresses) on any surface of a fluid, adjacent layers of the fluid slide over one another as long as these forces are maintained. Hence, in a fluid at rest, the tangential force on any surface is zero.

Figure 15-1 The force exerted by a fluid on a surface is a push of magnitude $F = PA$, where P is the pressure in the fluid at the surface of area A.

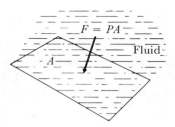

kept at the same height in the fluid. The ratio of the magnitude dF of the normal force exerted by the fluid on a small surface to the area dA of the surface is called the *pressure* P in the fluid:

$$P = \frac{dF}{dA} \qquad \text{or} \qquad dF = P\,dA \qquad (15.1)$$

We thus associate a number called the pressure with each point in the fluid. This number, as all divers know, increases with increasing depth.

If the pressure P has the same value at all points on a plane surface of area A, then the fluid exerts a force F (Figure 15-1) which is a push perpendicular to the surface with a magnitude

$$F = PA \qquad (15.2)$$

The plane surface A may be the surface of a solid within the fluid or it may be any mathematical surface we care to imagine within the fluid. When the pressure varies over a plane surface, as it does on the vertical surface of a dam, we must use Eq. 15.1 to obtain

$$F = \int P\,dA \qquad (15.3)$$

where the integral denotes the sum of the forces $P\,dA$ on all the elements of area dA that constitute the surface.

The unit in which pressure is expressed is a force unit divided by an area unit. In the SI system, this unit is the N/m^2 which is called the *pascal* (Pa):

$$1\ Pa = 1\ N/m^2$$

The pressure of the atmosphere at the earth's surface is approximately 100 kPa. At a given location the atmospheric pressure varies from day to day with fluctuations occasionally as large as 5%.

A commonly used unit of pressure is the *standard atmosphere* (atm) which is defined by

$$1\ atm = 101.325\ kPa$$

Example 1 A horizontal glass pane with an area of 0.20 m^2 is exposed to air pressure of 1.00×10^5 Pa on its lower surface. The upper surface is exposed to a vacuum. What is the force exerted on the glass by the air?

Solution   The air exerts an upward force on the lower surface given by

$$F = PA = (1.00 \times 10^5 \text{ Pa})(0.20 \text{ m}^2) = 2.0 \times 10^4 \text{ N}$$

(A window pane does not have to withstand such a large unbalanced force because there is usually about the same air pressure on *both sides* and therefore the *resultant* force exerted on the pane by the air is more or less zero when there is no wind.)

15.2 Density

The *density* ρ of any homogeneous substance, be it a solid or a fluid, is defined as its mass M divided by its volume V; that is,

$$\rho = \frac{M}{V}$$

The SI unit of density is the kg/m³. Water has a density of 1.00×10^3 kg/m³; the density of gold is almost twenty times greater; while air at 0°C at a pressure of 1.00 atm has a density of only 1.29 kg/m³, which is approximately 0.1% of the density of water. Mercury, with a density of 13.6×10^3 kg/m³, has the greatest density of any substance that is a *liquid* at room temperature and normal pressures.

The density of gases changes markedly when the temperature and pressure are changed, as will be discussed in Chapter 17. The densities of solids and liquids change relatively little for modest temperature and pressure changes, so it is usually an adequate approximation in fluid statics to consider such densities as constant. Typical densities of gases, liquids, and solids are given in Table 15-1.

Table 15-1   Densities (kg/m³) at 0°C and 1 atm Pressure

Hydrogen	0.0899	Wood, white pine	0.42×10^3
Helium	0.179	Aluminum	2.70×10^3
Air	1.29	Iron	$7.8 \ \times 10^3$
Oxygen	1.43	Silver	$10.5 \ \times 10^3$
Olive oil	0.92×10^3	Lead	$11.3 \ \times 10^3$
Water	1.00×10^3	Gold	$19.3 \ \times 10^3$
Mercury	$13.6 \ \times 10^3$	Platinum	$21.4 \ \times 10^3$

15.3 Change of Pressure with Depth

The first condition of equilibrium enables us to understand the pressures encountered at different points in a fluid at rest. In fluid statics every portion of the fluid is in equilibrium. We select as the system to which we apply the equilibrium condition the portion of the fluid within an imaginary cylindrical surface of height dy and cross-sectional area A (Figure 15-2). The free-body diagram is

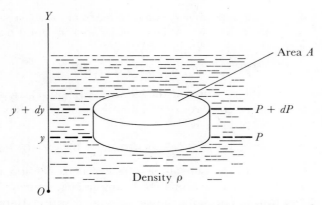

Figure 15-2 A system in equilibrium is the fluid within a cylinder of cross section A and height dy.

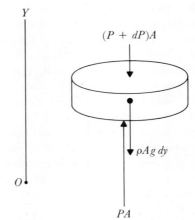

Figure 15-3 Free-body diagram for system of Figure 15-2. Only the vertical forces are shown.

shown in Figure 15-3. The system has a volume equal to $A\, dy$. From $M = \rho V$ its mass is given by $\rho A\, dy$. Therefore its weight Mg is equal to $(\rho A\, dy)g$. The fluid outside this cylindrical region exerts:

1 Horizontal forces on the vertical sides of the cylinder.
2 An upward force PA on the lower surface.
3 A downward force $(P + dP)A$ on the upper surface.

The first condition of equilibrium for this system is

algebraic sum of y components of external forces $= 0$

$$PA - (P + dP)A - \rho gA\, dy = 0$$

Therefore

$$dP = -\rho g\, dy \tag{15.4}$$

According to this equation, positive dy (an increase in elevation) corresponds to negative dP (a decrease in pressure). When ρ and $\mathbf{g}$ are constant, Eq. 15.4 can be integrated to give

$$P_2 - P_1 = -\rho g(y_2 - y_1) \tag{15.5}$$

where P_1 and P_2 are the pressures at the elevations y_1 and y_2, respectively.

Thus, in descending a distance of 10.0 m through water of density 1.00×10^3 kg/m^3, the increase in pressure is

$$(1.00 \times 10^3 \text{ kg/m}^3)(9.8 \text{ N/kg})(10.0 \text{ m}) = 0.98 \times 10^5 \text{ Pa}$$

which is about 1 atm. If the pressure at the surface of a lake is 1 atm, the pressure at a depth of 10 m is thus approximately 1 atm greater, namely 2 atm.

By way of contrast, the pressure change associated with a 10-m change in altitude in *air* of density 1.29 kg/m^3 is merely

$$(1.29 \text{ kg/m}^3)(9.8 \text{ N/kg})(10.0 \text{ m}) = 126 \text{ Pa}$$

which is only about 0.1% of typical atmospheric pressures. This result illustrates the fact that, for *gases* in laboratory containers, the difference in pressure at two points is usually negligible and the pressure can be taken to be the same throughout the container. If, however, we are concerned with an increase in altitude of a few kilometres or more, there is an appreciable drop in the pressure of the air. (But because the air density changes when the pressure changes, Eq. 15.5 can be applied only to small changes of altitude.)

We can extend the proof of Eq. 15.5, which was concerned only with the pressures at points like A and B on the same vertical line (Figure 15-4), to apply to a much more general case. First we can show that equilibrium of horizontal forces acting on a horizontal cylindrical portion of the fluid requires that the pressures at points like B and C on the same horizontal line are the same. It then follows that Eq. 15.5 applies to points A and C since it applies to A and B and $P_B = P_C$.

In this way we obtain the desired generalization, *the fundamental law of fluid statics*. The equation

$$P_2 - P_1 = -\rho g(y_2 - y_1) \tag{15.5}$$

gives the pressure difference between two points (such as A and D in Figure 15-4) with a vertical separation $y_2 - y_1$, regardless of their horizontal separation, provided that the two points can be connected by a path lying in the fluid of constant density. Note that this law applies for static fluids. In general, it is not true if the fluid inside the container is moving.

According to this law, the shape of the containing vessel does not affect the pressure, and the pressure is the same at all points at the same horizontal level in a connected static fluid. This is in accord with the readily observable fact that the levels of the liquid in the left- and right-hand portions of a container such as that of Figure 15-4 are the same.*

*We are assuming that the adhesive forces exerted by the container on the fluid are negligible.

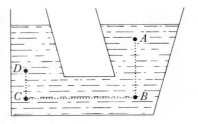

Figure 15-4 The fundamental law of fluid statics is the statement that the pressure change between two points such as A and D is equal to $\rho g \Delta y$ where Δy is the difference in elevation of these two points.

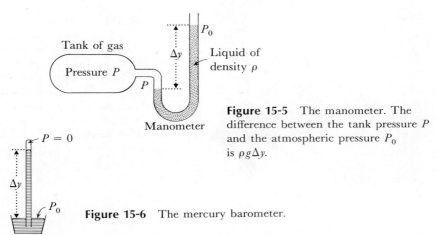

Tank of gas

Pressure P

P

P_0

Δy

Liquid of density ρ

Manometer

Figure 15-5 The manometer. The difference between the tank pressure P and the atmospheric pressure P_0 is $\rho g \Delta y$.

$P = 0$

Δy

P_0

Figure 15-6 The mercury barometer.

15.4 Pressure Gauges

The *open-tube manometer* shown in Figure 15-5 is used to determine the pressure P of a gas in the tank. The liquid in the manometer has a pressure P at the liquid surface exposed to the gas in the tank, while at the other surface the pressure is the atmospheric pressure P_0. With the vertical separation of these two liquid surfaces denoted by Δy, the fundamental law of fluid statics gives

$$P - P_0 = \rho g \Delta y$$

The pressure P is called the *absolute pressure* to distinguish it from the pressure difference $P - P_0$ which is named the *gauge pressure*. Knowledge of ρ and a measurement of the difference in height Δy of the liquid columns allows calculation of the gauge pressure.

The *mercury barometer* (Figure 15-6) is a glass tube, some 80 cm long, which has been filled with mercury and then inverted in a dish of mercury. The space above the mercury column contains only mercury vapor whose pressure at room temperature is so small that it can be neglected. This instrument was invented in 1643 by Galileo's pupil, Evangelista Torricelli (1608–1647). His barometer showed that the air of our atmosphere exerts a pressure, and it also provided a means for measuring changes in this pressure.

The "Torricellian vacuum" above the mercury column of this barometer created a sensation because it violated the Aristotelian tenet that "nature abhors a vacuum." With this sterile idea out of the way, the development of vacuum

Figure 15-7 The Magdeburg hemispheres (bronze, approximately 60 cm in diameter) and pump devised by Otto von Guericke (1602–1686). Using these hemispheres, von Guericke demonstrated in a most striking manner the reality of atmospheric pressure. *Once the air had been evacuated from the bronze sphere by an air pump, apparently two teams of eight horses each were unable to pull the two halves of the sphere apart.* This public scientific drama was staged before the Reichstag at Regensburg on May 8, 1654, during von Guericke's tenure as burgomaster of Magdeburg in Saxony. (Photo courtesy Deutsches Museum Bildstelle.)

pumps proceeded and knowledge of gases consequently increased rapidly, as we shall see in Chapter 17. The first practical vacuum pump, invented by Otto von Guericke in Germany around 1650, was used to stage the Magdeburg hemisphere demonstration (Figure 15-7), a masterpiece of scientific showmanship.

The theory underlying Torricelli's barometer is provided by the fundamental law of fluid statics. If we denote by Δy the height of the mercury column above the mercury level in the dish, Eq. 15.5 (with $P_2 - P_1 = 0 - P_0$) gives the pressure of the atmosphere at the mercury surface in the dish:

$$P_0 = \rho g \Delta y \tag{15.6}$$

Laboratory workers using mercury barometers and manometers find it convenient to specify pressures directly in terms of the height Δy of the column of mercury. The standard atmosphere is, within experimental accuracy, the pressure difference in a vertical mercury (Hg) column which is exactly 76 cm long under standard conditions (0°C and $g = 9.80665$ N/kg); that is

$$76 \text{ cm of Hg} = 1 \text{ atm} = 101.325 \text{ kPa}$$

Example 2   On a day when the barometer reads 77 cm of Hg, a mercury manometer (Figure 15-5) connected to a tank of compressed air has the top of the mercury column which is exposed to the tank at a level 33 cm below the top of the column which is open to the atmosphere. What is the absolute pressure in the tank in pascals?

Solution The gauge pressure $(P - P_0)$ of the air in the tank is given by

$$P - P_0 = 33 \text{ cm of Hg}$$

The barometer gives $P_0 = 77$ cm of Hg, so $P = 110$ cm of Hg. This is the absolute pressure in the tank. We can express this pressure in pascals by using the previous result that 76 cm of Hg represent a pressure of 1.01×10^5 Pa:

$$P = (110 \text{ cm of Hg}) \left(\frac{1.01 \times 10^5 \text{ Pa}}{76 \text{ cm of Hg}} \right) = 1.46 \times 10^5 \text{ Pa}$$

15.5 Pascal's Principle

An immediate consequence of the fundamental law of fluid statics is a fact discovered in the seventeenth century by the French philosopher, mathematician, and physicist, Blaise Pascal (1623–1662). This fact is now known as *Pascal's principle: If added pressure is applied anywhere to a confined fluid, this added pressure is transmitted undiminished to every portion of the fluid and to the walls of the confining vessel.*

To understand why this is so, we need only to notice that any increase in pressure at one point in the fluid must result in the same increase at all points because, according to the fundamental law of fluid statics, pressure *differences* do not change. The pressure *difference* between two points in a connected fluid is determined only by the difference in vertical height, the fluid density ρ, and the gravitational field **g**.

Pascal's principle is exploited in the functioning of a *hydraulic press,* a machine employing a confined liquid. This device is used (in car lifts, hospital operating tables, barber and dentist chairs, as well as in hydraulic automobile brakes) to obtain in a convenient way a very large ratio of output to input force (mechanical advantage). A small input force f applied to a piston of small area a results in a pressure $P = f/a$ which is transmitted through the liquid and gives rise to a large output force $F = PA$ exerted by the liquid on a piston of large area A (Figure 15-8). The pressure P is the same under both pistons since these points are at the same horizontal level in the same liquid. We have

$$\frac{F}{f} = \frac{PA}{Pa} = \frac{A}{a} \tag{15.7}$$

which shows that the mechanical advantage F/f is given by the ratio of the areas of the pistons A/a.

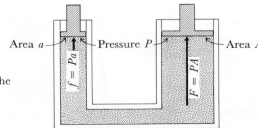

Area a — Pressure P — Area A

$f = Pa$ $F = PA$

Figure 15-8 Principle of the hydraulic press.

In hydraulic brakes, when the brake pedal pushes on a piston in the master cylinder, the pressure increase is transmitted through the brake fluid to the four wheel cylinders. At each wheel a small piston then moves to activate a brake shoe or a disk.

15.6 Archimedes' Principle

The Buoyant Force

A fluid exerts a buoyant force on a body which is totally or partially immersed in it. Why this is so is not hard to discover. The fluid exerts pressure on the body's surface, causing an upward force on the bottom and a downward force on the top of an immersed object. Since the pressure is greater at greater depths, the upward force will be the greater and the resultant force will be an upward or *buoyant* force.

A very simple expression for the precise magnitude of this buoyant force was discovered over 2000 years ago by Archimedes. His result can be deduced from the laws of fluid statics as follows.

First consider a fluid in equilibrium without any immersed object. Focus attention on a certain portion of this fluid occupying a volume V with a surface S (Figure 15-9). This volume V is held in equilibrium by the forces exerted by the remainder of the fluid on the surface S. Therefore these forces must have a resultant force which is directed vertically upward through the center of gravity of the fluid in V and has a magnitude equal to the weight ($Mg = \rho Vg$) of the fluid in V. If we now replace the fluid inside V by some other object, since the pressures in the outside fluid are unchanged at each point on the surface S, the forces exerted by the outside fluid are unchanged. *That is, the fluid of density ρ exerts a buoyant force B given by*

$$B = \rho Vg \tag{15.8}$$

on an object which displaces a volume V of the fluid. The force **B** passes through

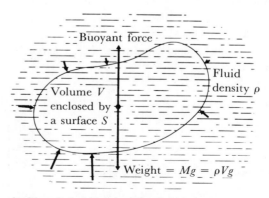

Figure 15-9   Archimedes' principle. The fluid outside the surface S exerts a buoyant force ρVg on this surface.

the center of gravity (which is named the center of buoyancy) of this volume V of the displaced fluid. This law for ascertaining the buoyant force is known as *Archimedes' principle.*

A balloon of volume V which has an average density ρ_b less than the density ρ of the surrounding air will experience a resultant upward force

$$B - Mg = \rho Vg - \rho_b Vg = (\rho - \rho_b)Vg$$

and therefore will be accelerated upward. The hydrogen or helium inside the balloon does not lift the balloon; it merely serves to keep the balloon distended.

A body whose average density is less than that of a liquid's density can float partially submerged in the liquid. Such a body sinks until its weight is balanced by the buoyant force. According to Archimedes' principle, this buoyant force is equal to the weight of the displaced liquid. Hence the weight of the displaced liquid is equal to the weight of a body floating in it. Saying that a ship displaces a given weight of water is just another way of specifying the weight of the loaded ship.

Example 3   When a submarine of total volume V is floating on the surface of water of density ρ, the part of the submarine that is above the water level has a volume V':

(*a*) What is the weight $M_f g$ of the floating submarine?

(*b*) Water is admitted to the submarine's ballast tanks until the submarine is hovering just below the surface. What is the weight $M_s g$ of the submerged submarine?

(*c*) What volume of water has been admitted to the ballast tanks?

Solution

(*a*) When the submarine is floating, the volume of water displaced by the submarine is $V - V'$. Therefore, according to Archimedes' principle, the buoyant force B_f exerted on the floating submarine by the water is

$$B_f = \rho(V - V')g$$

The vertical forces acting on the submarine are in equilibrium. Consequently when the submarine is floating, its weight is

$$M_f g = \rho(V - V')g$$

(*b*) The buoyant force exerted on the submerged submarine is

$$B_s = \rho V g$$

Equilibrium of vertical forces acting on the submarine implies that the weight $M_s g$ of the submarine has been increased to be equal in magnitude to this buoyant force:

$$M_s g = \rho V g$$

(*c*) The weight of water admitted to the ballast tanks is

$$M_s g - M_f g = \rho V g - \rho(V - V')g = \rho V' g$$

The volume occupied by this water is V'.

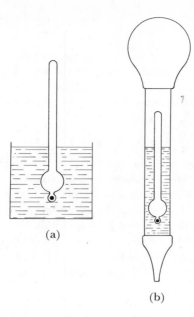

(a)

(b)

Figure 15-10 (a) Hydrometer. (b) Hydrometer used as a battery tester.

Hydrometer

The density of a liquid is easily measured by floating a *hydrometer* [Figure 15-10(a)] in it and noting the depth to which the hydrometer sinks. The hydrometer is usually a glass cylinder with a weight at the bottom and a thin calibrated tube at the top. The denser the liquid, the less the depth to which the hydrometer must sink to achieve a buoyant force equal to its weight.

The glass stem is usually calibrated in terms of the *specific gravity* rather than the density of the liquid. The specific gravity of a substance is defined as the ratio of the density of the substance to the density of water. Pure water, therefore, has a specific gravity of exactly 1. From the data of Table 15-1, we see that mercury has a specific gravity of 13.6.

The most familiar hydrometer is the battery tester [Figure 15-10(b)] used in every garage to ascertain the state of charge of a car's storage battery. When the lead plates of the battery have reacted with much of the battery acid, the specific gravity of the liquid drops to around 1.14, the "no charge" condition. After a charging current has reversed these chemical reactions to produce again sufficient acid to raise the liquid's specific gravity to 1.28, "full charge" is indicated.

Summary

☐ The pressure P at a point in a fluid is the ratio of the magnitude dF of the normal force exerted by the fluid to the area dA of a small plane surface passing through the point:

$$P = \frac{dF}{dA} \qquad \text{or} \qquad dF = P \, dA$$

☐ For a sizable plane surface

$$F = \int P \, dA$$

and when P is constant over the surface of area A,

$$F = PA$$

☐ The SI unit of pressure is the pascal (Pa):

$$1 \text{ Pa} = 1 \text{ N/m}^2$$

☐ The density ρ of a homogeneous substance of mass M and volume V is

$$\rho = \frac{M}{V}$$

☐ The fundamental law of fluid statics is

$$P_2 - P_1 = -\rho g (y_2 - y_1)$$

where $P_2 - P_1$ is the pressure difference between any two points with a difference in elevations $y_2 - y_1$, provided that the two points can be connected by a path lying within a fluid of constant density ρ and that there is a constant gravitational field $\mathbf{g}$ directed vertically downward. If ρ or $\mathbf{g}$ vary, we must use the differential form of this law:

$$dP = -\rho g \, dy$$

☐ Archimedes' principle states that a fluid exerts a buoyant (upward) force $\mathbf{B}$ on a body equal to the weight of the fluid displaced by the body. The line of action of $\mathbf{B}$ passes through the center of gravity of this volume V of the displaced fluid. A fluid of density ρ exerts a buoyant force

$$B = \rho V g$$

Questions

1 The pressure in the water next to a 9.0×10^{-3} m^2 flat tile is 1.2×10^5 Pa. Find the magnitude of the force exerted on the tile by the water.

2 While skating, a hockey player applies a force of 1000 N distributed over the 1.00×10^{-3} m^2 area of the bottom of a skate:
   (a) What is the pressure on the surface of the ice underneath this skate?
   (b) If the player were to replace the skate by a shoe, would this pressure decrease? Explain.

3 Assume that the diameters of the Magdeburg hemispheres (Figure 15-7) are 0.60 m. With an atmospheric pressure of 101 kPa, what would be the force pressing the one hemisphere against the other if, inside the sphere, there were a perfect vacuum? [By first considering the equilibrium of a single closed hemispherical surface in air, we can deduce that the force exerted by the air on the curved surface has the same magnitude as the force exerted on the flat surface of area $\pi(0.30 \text{ m})^2$.]

4 The piston of a hydraulic lift for automobiles has a radius of 0.15 m. What pressure is required to lift an automobile weighing 12×10^3 N?

5 What is the mass in kilograms, and the weight in newtons, of a volume of 1.0 cm^3 (10^{-6} m^3) of water?

6 On a certain day the pressure at the surface of a lake is 1.00×10^5 Pa. How far must one descend to encounter a pressure of 3.00×10^5 Pa?

7 A tube contains a 2.00-m layer of olive oil floating above a 1.50-m layer of water which in turn floats above a 0.500-m layer of mercury. The upper surface of the olive oil is exposed to the pressure of the atmosphere which is 1.00×10^5 Pa. What is the absolute pressure at the bottom of the mercury layer?

8 When a man who is 2.00 m tall is standing, what is the hydrostatic difference between the blood pressure in his feet and the blood pressure in his brain? Assume that his blood has a density of 1.06×10^3 kg/m^3.

9 A long, thin vertical tube passes through a rubber stopper into a flask. When 100 cm^3 of water are poured into this assembly, the flask and tube are filled so that the height from the base of the flask to the free water surface is 1.50 m. The area of the flask base is 2.00×10^{-2} m^2. Find the force exerted by the water on the flask base. (Ignore the effect of the air pressure which contributes equal forces to the top and bottom surfaces of the base of the flask.) Show that the force exerted by the water on the base is much greater than the weight of 100 cm^3 of water. Explain this "hydrostatic paradox."

10 The manometer in Figure 15-5 contains mercury. The level difference is as shown in the figure with $\Delta y = 19$ cm. The barometric pressure is 76 cm of Hg:
    (a) Give the gauge pressure and the absolute pressure of the gas in the tank in centimetres of mercury.
    (b) What is the absolute pressure in pascals?

11 After hearing of Torricelli's mercury barometer, Pascal made a water barometer. What is the height of the water column in such a barometer on a day when the atmospheric pressure is 1.00×10^5 Pa?

12 A mercury barometer at the top of a building reads 1.00 cm less than a barometer on the ground floor. How tall is the building?

13 The gauge pressure of the air within the tires of a 2000-kg automobile is 2.0×10^5 Pa. How much total surface area of the tires is in contact with the road?

14 **(a)** If the hydraulic press of Figure 15-8 were constructed with pistons of areas 0.10 m^2 and 2.0×10^{-4} m^2, what would be the mechanical advantage?
    (b) What force would have to be applied to the small piston to lift an automobile weighing 15×10^3 N?
    (c) What would be the pressure underneath the pistons?

15 **(a)** Will a solid iron ball float in mercury? Will a solid gold ball float in mercury? Explain.
    (b) An ocean-going ship is loaded at a port in Lake Ontario. Will the ship sink lower or rise higher when it reaches salt water? Explain. (The density of salt water is greater than that of fresh water.)

16 A block of aluminum with a volume of 0.100 m^3 is completely immersed in water. The block is suspended by a rope. Find:
    (a) The mass and the weight of the aluminum block.
    (b) The buoyant force exerted on the block by the water.

   (c) The force exerted on the block by the rope. (This force is equal in magnitude to what is called the *apparent weight* of the submerged block.)

17 A barge, loaded with coal, has a total mass of 2.40×10^5 kg. The barge is 15.0 m long and 8.0 m wide:

   (a) Using Archimedes' principle, calculate the depth D of the barge below the water line.

   (b) Using Eq. 15.5 calculate the pressure increase at this depth.

   (c) From $F = PA$, calculate the force exerted on the barge by the water, and compare this with the weight of the barge.

18 **(a)** The density of ice is 0.92×10^3 kg/m^3. What fraction of the volume of a floating ice cube is above water?

   (b) What volume of an ice floe in a river is just large enough to carry a child who weighs 400 N?

19 **(a)** What is the volume of a helium-filled balloon that will support a weight of 1000 N?

   (b) How great a weight would the same balloon support if the helium were replaced by hydrogen?

20 **(a)** From the data of Table 15-1, compute the specific gravity of gold.

   (b) Successive readings of a battery tester are taken as a storage battery is being recharged. Will the hydrometer float higher or lower? Explain.

Supplementary Questions

S-1 A stone has an apparent weight of 5.40 N when it is completely immersed in water, and 6.00 N when it is completely immersed in oil of density 0.80×10^3 kg/m^3. What is the stone's specific gravity?

S-2 The apparent weight of an ore sample is 206 N when the ore is completely immersed in water, and 230 N when the ore is in air. What is the specific gravity of the ore?

S-3 Archimedes devised a nondestructive test to determine whether his king's crown was pure gold. A balance was used to select a quantity of pure gold which had a mass equal to that of the crown. He then immersed the crown in a container which had been completely filled with water and measured the amount that overflowed. After refilling the container, the pure gold was immersed in the water and again the overflow was measured. From these measurements he determined that the crown was not pure gold. Explain how this was possible. If the ratio of the mass of water which overflowed when the pure gold was immersed to the mass which overflowed when the crown was immersed was 0.900, what was the density of the crown?

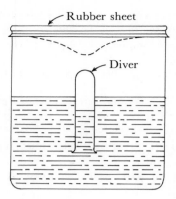

Rubber sheet

Diver

Figure 15-11

S-4 An ingenious toy called a cartesian diver is represented in Figure 15-11. The diver is an inverted test tube which contains just enough air to float in a large tank of water. The tank is closed by a rubber sheet stretched over the top. When this sheet is pushed down slightly, the diver sinks to the bottom. When the rubber sheet returns to its original position, the diver rises to the surface. Explain this behavior and account for the fact that the diver cannot find a stable position between the surface and the bottom of the tank.

S-5 A beaker of water on one pan of a balance and weights on the other pan are in equilibrium. Without touching the beaker, a student places a finger tip in the water. Explain what happens.

S-6 An ice cube floats in a glass filled to the brim with ice water. Will the water overflow when the ice melts? Do the experiment and explain your observation.

S-7 A bottle of water containing air bubbles is falling freely with an acceleration **g**. Do the bubbles rise through the water?

S-8 A balloon filled with helium is fastened to the floor of a car by a length of string. While the car is going around a bend, which way will the balloon move relative to the car?

16.

fluid dynamics

Fluid dynamics, the study of the motion of a fluid and the related forces, forms one of the foundations of aeronautics, mechanical engineering, meteorology, marine engineering, civil engineering, and bioengineering. In fluid dynamics, the research scientist can still find fascinating experimental phenomena that are far from being understood. Although the underlying physical principles are the well-known laws of Newtonian mechanics, the application of these laws to describe a general motion of a real fluid often leads to problems of such complexity that attempts to gain a theoretical understanding are defeated. Difficulties in extracting general answers from the Newtonian mechanics of fluid motion arise from the fact that, except for the gravitational force, the forces acting on different portions of the fluid are not known in advance but instead are determined by the fluid motion. And the motions of all portions of the fluid have to be determined in a single calculation. Nevertheless, we shall see that with several simplifying assumptions it is quite easy to obtain a result known as *Bernoulli's equation*, which proves to be very useful in analyzing many situations.

16.1 Streamlines, Equation of Continuity

To describe fluid motion, we consider first a small volume ΔV of the fluid, called a volume element. The velocity $\mathbf{v}$ of the center of mass of this volume element is termed the *fluid velocity* at the location of the volume element. In this way, at

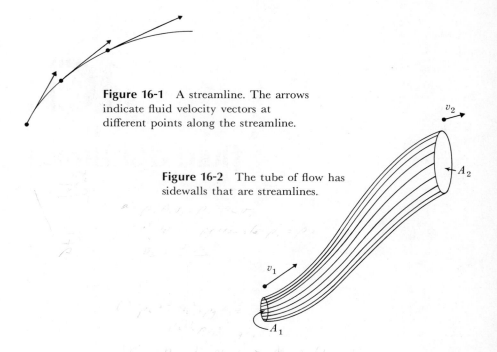

Figure 16-1   A streamline. The arrows indicate fluid velocity vectors at different points along the streamline.

Figure 16-2   The tube of flow has sidewalls that are streamlines.

any instant, a velocity vector is associated with each point in a region occupied by a fluid. A *streamline* is a line drawn so that at each point its direction is the direction of the fluid velocity at that point (Figure 16-1).

A fluid flow is said to be *steady* if the fluid velocity at every point does not change as time goes on. Steady flow occurs at low speeds in pipes and streams. A given volume element may have different velocities at different points, but in a steady flow all volume elements have the same velocity as they pass a given point. Streamlines then are stationary, and the path of a given volume element is a streamline.

A tubular region of a fluid, with sidewalls that are streamlines, is called a *tube of flow* (Figure 16-2). Since the fluid velocity is parallel to these sidewalls, no fluid flows through the sidewalls. A tube of flow is therefore like a pipe in that the fluid that enters at one end must leave at the other.

This simple observation leads to a relationship between the cross-sectional area A_1, the speed v_1, and the density ρ_1 at one end of a thin tube of flow and the corresponding quantities A_2, v_2, and ρ_2 at the other end of the tube. In a small time interval Δt the fluid at the input moves a distance $v_1\Delta t$. The volume of fluid entering through A_1 is therefore $A_1 v_1 \Delta t$, and the mass of this fluid is

$$\Delta M = \rho_1 A_1 v_1 \Delta t \tag{16.1}$$

In the same time interval, the fluid at the output moves a distance $v_2\Delta t$. The volume of fluid leaving through A_2 is therefore $A_2 v_2 \Delta t$, and the output mass is

$$\Delta M = \rho_2 A_2 v_2 \Delta t \tag{16.2}$$

Since mass is conserved (in Newtonian mechanics), the mass which enters through A_1 in the interval Δt must equal the mass which leaves through A_2 in the same interval. Therefore

$$\rho_1 A_1 v_1 = \rho_2 A_2 v_2 \tag{16.3}$$

a result which is known as the *equation of continuity.*

The variations in density that occur in the flowing fluid are often relatively small, and it is a good approximation to assume that $\rho = constant$; that is, to assume that the fluid is *incompressible.* This is usually an excellent approximation for flowing liquids. And even though gases are highly compressible, it often happens that the pressure changes encountered by a flowing gas are rather small and that the consequent changes in gas density are unimportant. For instance, at speeds well below the speed of sound in air, the flow of air past an airfoil does not involve large changes in air density. Consequently, a useful theory of sub-sonic aerodynamics can be developed by assuming that the air is incompressible.

For an incompressible fluid, $\rho_1 = \rho_2$, and the equation of continuity simplifies to

$$v_1 A_1 = v_2 A_2 \tag{16.4}$$

This shows that where the tube area is large, the fluid speed is small, and vice versa. And in a stream of constant width, the water flows rapidly where the stream is shallow, but we observe that "still waters run deep."

16.2 Bernoulli's Equation

A particularly useful theorem of fluid dynamics was discovered in 1738 by Daniel Bernoulli. This theorem is now recognized as a statement of the law of conservation of energy. It can be derived by examining the energy transforma-tions that occur as a portion of a fluid moves through a tube of flow.

We shall assume that the fluid is *nonviscous.* When one layer of a real fluid slides over another layer, this slipping is opposed by a frictional force called a *viscous* force which results in the dissipation of energy. Although viscosity is of great importance in some circumstances, there are many situations in which viscous effects are small enough so that a good approximation is obtained by assuming that the flow is *nonviscous.*

The situation is further simplified by assuming that the fluid is incompressible ($\rho = constant$), so that we do not need to take into account the volume changes of a portion of the fluid or the associated work done on this portion as it com-presses or expands.

We focus now on the system shown in Figure 16-3(a) which comprises not only the mass of the fluid within the flow tube between the ends A_1 and A_2, but also the mass ΔM that is about to enter the tube through A_1. After a time interval Δt, the fluid of this system has moved to the position shown in Figure 16-3(b).

In the configuration shown in Figure 16-3(a), this system has a total energy

$$E_1 = \tfrac{1}{2}(\Delta M)v_1^2 + (\Delta M)gy_1 + \textit{energy of fluid within tube}$$

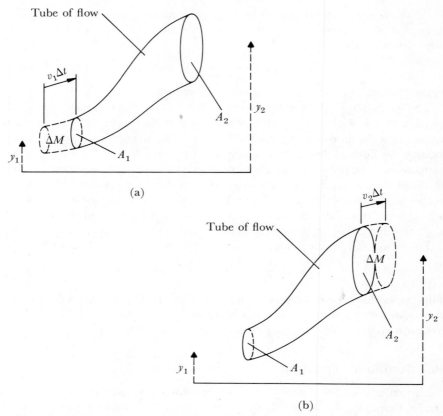

Figure 16-3   (a) Position at the beginning of the time interval Δt of the system comprising the fluid within the tube and the mass ΔM. (b) Position at the end of the time interval Δt of the system in (a).

The ends of the tube have vertical coordinates y_1 and y_2 measured from some convenient reference level. After the time interval Δt, the total energy of this system is

$$E_2 = \tfrac{1}{2}(\Delta M)v_2^2 + (\Delta M)gy_2 + \textit{energy of fluid within tube}$$

We assume that the flow is *steady*. Then the fluid velocity at a given location remains unchanged, and the tube is stationary. Consequently the energy of the fluid within the tube is constant; the change in the system's energy, from the preceding equations, is therefore

$$E_2 - E_1 = \tfrac{1}{2}(\Delta M)v_2^2 + (\Delta M)gy_2 - \tfrac{1}{2}(\Delta M)v_1^2 - (\Delta M)gy_1$$

Energy is transferred to the system by forces exerted on the system by the surrounding fluid. The work done by these forces measures the energy transferred to the system. The surrounding fluid exerts a force P_1A_1 on the mass ΔM of Figure 16-3(a) in the direction of motion of this mass. During the time interval

Δt, this mass moves a distance $v_1 \Delta t$, and the work done on the system by the force $P_1 A_1$ is

$$W_1 = P_1 A_1 v_1 \Delta t = \frac{P_1 \Delta M}{\rho}$$

where we have used $A_1 v_1 \Delta t = \Delta M / \rho$ from Eq. 16.1. At the other end of the tube, the surrounding fluid exerts a force $P_2 A_2$ in the direction opposite to that of the motion of the system, which moves a distance $v_2 \Delta t$ in the time interval Δt. The corresponding work done by this force on the system is therefore the negative quantity

$$W_2 = -P_2 A_2 v_2 \Delta t = -\frac{P_2 \Delta M}{\rho}$$

since $A_2 v_2 \Delta t = \Delta M / \rho$ from Eq. 16.2. No work is done by the forces exerted on the sidewalls by the surrounding fluid, because these forces are perpendicular to the sidewalls (assuming no viscosity) and are therefore perpendicular to the direction of motion of the fluid on which they act.

Conservation of energy now implies that the energy $W_1 + W_2$ transferred to the system by the forces exerted by the surrounding fluid is equal to the change $E_2 - E_1$ in the system's total energy; that is

$$W_1 + W_2 = E_2 - E_1 \tag{16.5}$$

which gives

$$P_1 \Delta M / \rho - P_2 \Delta M / \rho = \tfrac{1}{2}(\Delta M) v_2^2 + (\Delta M) g y_2 - \tfrac{1}{2}(\Delta M) v_1^2 - (\Delta M) g y_1$$

Canceling ΔM from each term, multiplying by ρ, and then rearranging, we obtain the famous *Bernoulli equation,*

$$P_1 + \tfrac{1}{2} \rho v_1^2 + \rho g y_1 = P_2 + \tfrac{1}{2} \rho v_2^2 + \rho g y_2 \tag{16.6}$$

for any two points (1 and 2) on the same streamline in a steady flow of a nonviscous incompressible fluid.

Notice that this fundamental result of fluid dynamics includes the fundamental law of fluid statics as the special case that obtains when v_1 and v_2 are equal:

$$P_1 + \rho g y_1 = P_2 + \rho g y_2$$

which yields

$$P_1 - P_2 = \rho g (y_2 - y_1)$$

in agreement with Eq. 15.5.

A fluid flow is called *irrotational* if there is no rotation of a small paddle wheel immersed anywhere in the fluid. This excludes not only whirlpools but also a flow which involves a variation of the velocity vector in a transverse direction. It can be shown that, for irrotational flow, Bernoulli's equation applies to any two points within the fluid whether or not they lie on the same streamline.

16.3 Applications of Bernoulli's Equation

Figure 16-4 represents a steady flow of water through a horizontal pipe which has a varying cross section. The equation of continuity, $v_1 A_1 = v_2 A_2$, requires that the speed of the water be greatest at the constriction. The pipe very nearly constitutes a tube of flow, so Bernoulli's equation may be applied to the pipe if we disregard viscous effects. For points 1 and 2 at the same horizontal level, $y_1 = y_2$, and Bernoulli's equation gives

$$P_1 + \tfrac{1}{2}\rho v_1^2 = P_2 + \tfrac{1}{2}\rho v_2^2$$

This implies that, at the constriction where the speed is the greatest, the pressure is a minimum, in accord with the pressure measurements indicated by the heights of the vertical columns of water in Figure 16-4.

This effect is exploited in a *Venturi meter* which is used to measure the speed of flow of a fluid. The meter introduces a constriction in the flow tube. A manometer measures the amount $P_1 - P_2$ by which the pressure P_2 at the constriction drops below the pressure P_1 where the pipe has its normal cross section A_1. Manometer readings can be calibrated to give the corresponding speed of flow v_1 within the pipe. The calculation of v_1 from knowledge of $P_1 - P_2$ and A_1/A_2 can be effected using both the equation of continuity and Bernoulli's equation (see Question S-1).

In the *aspirator pump* shown in Figure 16-5, the pressure within the flowing stream of water drops at the throat (the constriction) to a value below atmospheric pressure, and suction occurs through a side tube. This simple pump can reduce the pressure to values as low as a few centimetres of mercury.

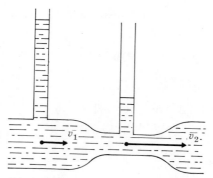

Figure 16-4   Venturi effect. Where the fluid speed is greatest, the pressure is least.

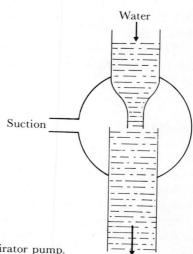

Figure 16-5   Aspirator pump.

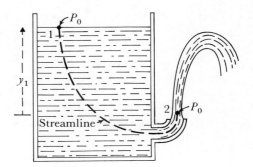

Figure 16-6 Flow from an orifice in a large container.

The speed with which a liquid flows out of an orifice in a large container was observed by Torricelli to be such that the emerging stream, when it is directed vertically upward, can rise almost to the level of the liquid in the container. Bernoulli's equation permits a simple analysis of this phenomenon. Figure 16-6 shows a streamline which connects the upper surface at point 1 to a point 2 just outside the container. If the container's cross section is large compared to that of the outlet, the speed v_1 of water at the upper surface is approximately zero. The pressure at the upper surface and on the sides of the emerging jet is the atmospheric pressure P_0. Bernoulli's equation therefore yields

$$P_0 + 0 + \rho g y_1 = P_0 + \tfrac{1}{2}\rho v_2^2 + 0$$

where y_1 is the height of the surface of water above the level of the output. Solving for the flow speed just outside the outlet, we obtain Torricelli's law

$$v_2 = \sqrt{2gy_1} \qquad (16.7)$$

which is exactly the speed of vertical projection necessary for a projectile to rise to a height y_1.

Example 1 Water flows in glass tubing (with a 2.0 mm inner diameter) located in the basement of a laboratory. The tubing rises a height of 7.0 m to the second floor, with the diameter of the tubing tapering to 1.0 mm. Find the water pressure P_2 and the speed v_2 in the pipe at the second floor when, within the pipe in the basement, the water pressure is $P_1 = 2.00 \times 10^5$ Pa and the water speed is $v_1 = 1.0$ m/s. Assume that Bernoulli's equation applies in this example.

Solution The equation of continuity, $v_2 A_2 = v_1 A_1$, allows us to determine v_2 from v_1 and the ratio of the pipe cross section, A_1/A_2.

$$\frac{A_1}{A_2} = \frac{(2.0 \text{ mm})^2}{(1.0 \text{ mm})^2} = 4.0$$

so

$$v_2 = \left(\frac{A_1}{A_2}\right)v_1 = 4.0 \times (1.0 \text{ m/s}) = 4.0 \text{ m/s}$$

Disregarding the effects of viscosity, and assuming that the glass tubing constitutes a tube of flow, we may apply Bernoulli's equation which gives

$$P_2 + \tfrac{1}{2}\rho v_2^2 + \rho g y_2 = P_1 + \tfrac{1}{2}\rho v_1^2$$

where y_2 is the height of the second floor above the basement. Therefore

$$P_2 = P_1 + \tfrac{1}{2}\rho v_1^2 - \tfrac{1}{2}\rho v_2^2 - \rho g y_2$$
$$= (2.00 + 0.005 - 0.08 - 0.69) \times 10^5 \text{ Pa}$$
$$= 1.24 \times 10^5 \text{ Pa}$$

16.4 Viscosity and Bernoulli's Equation

All the effects that have been discussed in this chapter are modified by the presence of viscous forces. Because of work done by viscous forces in a flowing fluid, energy is dissipated in heating the fluid. Then Eq. 16.5 is replaced by

$$W_1 + W_2 = E_2 - E_1 + energy \ dissipated$$

which leads to the result that, for a point 2 downstream from the point 1 and on the same streamline,

$$P_1 + \tfrac{1}{2}\rho v_1^2 + \rho g y_1 > P_2 + \tfrac{1}{2}\rho v_2^2 + \rho g y_2 \qquad (16.8)$$

an expression that we will call Bernoulli's *inequality*.

In viscous flow through a horizontal pipe of constant cross section, the pressure is observed to drop as one moves downstream (Figure 16-7). This accords with the Bernoulli inequality, which for this case, with $y_1 = y_2$ and $v_1 = v_2$, gives

$$P_1 > P_2$$

And because of viscosity, the downstream pressure P_2 in Example 1 will actually be less than the value we computed using Bernoulli's equation.

The reason that the emerging jet of water in Figure 16-6 does not quite attain the level in the container is that there is a dissipation of energy associated with viscosity. And when the analysis that led to the Torricelli law is repeated, using the Bernoulli inequality instead of the Bernoulli equation, we find that v_2 is less than $\sqrt{2gy_2}$.

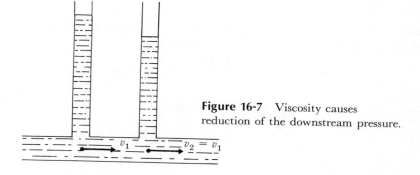

Figure 16-7 Viscosity causes reduction of the downstream pressure.

These examples indicate that in many real situations the results obtained using Bernoulli's equation, when uncorrected for energy dissipation, should not be regarded as anything more than estimates. In this respect, we must mention *turbulence*—one of the most important phenomena of fluid dynamics. When a fluid moves sufficiently rapidly down a pipe or past an obstacle, a churning of the fluid occurs, eddies are formed, and the detailed motion of a volume element of the fluid generally becomes chaotic. This is *turbulent flow*. Considerable energy is dissipated within a fluid where the flow is turbulent. In such turbulent regions there can be pronounced departures from the predictions of Bernoulli's equation.

Summary

☐ A streamline is a line with a direction at each point in the direction of the fluid velocity at that point.

☐ In steady flow, the fluid velocity at every point is time independent.

☐ The fluid speed v and density ρ at two points within a tube of flow of cross section A are related by the equation of continuity, $\rho_1 A_1 v_1 = \rho_2 A_2 v_2$.

☐ Bernoulli's equation relates the pressure P, the fluid speed v, and the altitude y at any two points on the same streamline in a steady flow of a nonviscous incompressible fluid:

$$P_1 + \tfrac{1}{2}\rho v_1^2 + \rho g y_1 = P_2 + \tfrac{1}{2}\rho v_2^2 + \rho g y_2$$

☐ Because of energy dissipation due to viscosity, the sum $P + \tfrac{1}{2}\rho v^2 + \rho g y$ decreases as we move downstream along a streamline.

Questions

1 What is *steady* flow? Is a gently flowing stream an example of steady flow? Is a babbling brook?

2 Show that the fluid volume per unit time passing through a cross section A of a pipe is given by

$$V = Av$$

where v is the fluid speed in the pipe.

3 The water entering a house flows with a speed of 0.10 m/s through a pipe of 21 mm inside diameter:
   (a) What is the speed of the water at a point where the pipe tapers to a diameter of 7 mm?
   (b) What mass and what volume of water enter the house in 60 s?

4 (a) Show that the work done during the time interval Δt by the external fluid forces on the fluid of the system shown in Figure 16-3 is given by

$$W = (P_1 - P_2)\Delta V$$

where ΔV is the volume of the mass ΔM (the volume that flows through the tube in the time interval Δt).

(b) How much work is done by the pressure in forcing 5.0 m³ of water through a pipe when the pressure difference between the ends of the pipe is 1.2×10^5 Pa?

5   Give the dimensions of each quantity appearing in Bernoulli's equation, and verify that the equation is dimensionally correct.

6   Water flows upward through a tapered vertical tube. Where the cross section is 4.0×10^{-5} m², the pressure is 2.0×10^5 Pa, and the fluid speed is 1.5 m/s. At a point which is 3.0 m higher, the cross section is 12.0×10^{-5} m²:
(a) Find the fluid speed at the higher point.
(b) Use Bernoulli's equation to estimate the pressure at the higher point.

7   The constriction in Figure 16-4 has an inner diameter of 0.5 cm, and the normal diameter of the pipe is 1.0 cm. When the volume of water per second flowing through the pipe is 0.20×10^{-4} m³/s, find:
(a) The fluid speed at the constriction and also at a point far from the constriction.
(b) The pressure difference between the normal fluid pressure P_1 and the pressure at the constriction, P_2.

8   Water enters the aspirator pump in Figure 16-5 at a rate of 0.16 m³/min at a pressure of 2.0×10^5 Pa through a pipe with a cross section of 4.0×10^{-4} m². In order to use this pump to achieve pressures as low as 3.0×10^4 Pa, what should be the cross-sectional area of the constriction?

9   **(a)** What is the speed at which water flows from a small hole 3.0 m below the water level in a full tank which is 5.0 m high?
(b) How far will the jet of water be projected horizontally before striking the ground at the level of the bottom of the tank?

10   In a closed pressure tank, the air pressure above the water surface is 2.0×10^5 Pa. A jet of water is squirted vertically upward from an aperture 4.0 m below the water surface. How high will this jet rise?

Supplementary Questions

S-1   **(a)** A Venturi flowmeter introduces a constriction of cross section A_2 in a pipe of normal cross section A_1. The meter determines the pressure difference $P_1 - P_2$ between the normal fluid pressure P_1 and the pressure at the constriction P_2. Using Bernoulli's equation and the equation of continuity, show that where the cross section has the normal value A_1, the fluid speed v_1 is given by

$$v_1 = A_2 \sqrt{\frac{2(P_1 - P_2)}{\rho(A_1^2 - A_2^2)}}$$

where ρ is the fluid density.
(b) Show that the fluid volume per unit time passing through any cross section of the pipe is given by

$$V = A_1 A_2 \sqrt{\frac{2(P_1 - P_2)}{\rho(A_1^2 - A_2^2)}}$$

S-2   The siphon shown in Figure 16-8 transfers water from one container to a lower container but first raises the liquid to a height h_2 above the level in the first container. Consider a siphon of cross section 4.0×10^{-4} m² with $h_2 = 3.0$ m and the

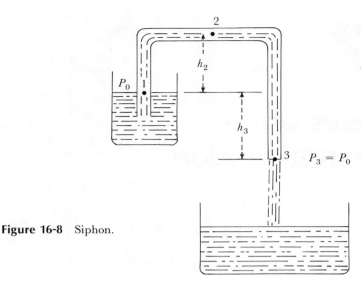

Figure 16-8   Siphon.

lower end of the siphon (point 3) such that $h_3 = 5.0$ m. At point 3 the pressure is the atmospheric pressure, $P_0 = 1.0 \times 10^5$ Pa. Find:

(a) The pressure P_1 at point 1 within the siphon.

(b) The pressure P_2.

(c) The speed v of the water within the siphon.

(d) The volume per second of water flowing through the siphon.

S-3 Show that, if the flow of water in a siphon is assumed to be governed by Bernoulli's equation, one can make the following predictions about siphon operation:

(a) The flow rate is proportional to $\sqrt{h_3}$. (The distance h_3 is shown in Figure 16-8.)

(b) The siphon can operate only if the sum $h_2 + h_3$ is less than the height h of a column of water in a water barometer: that is, we must have

$$h_2 + h_3 < h$$

where

$$\rho g h = P_0$$

(*Hint*: Minimum fluid pressure, which occurs at point 2, must always remain a positive quantity.)

S-4 The drop in pressure encountered in a 1.0-km length of a certain oil pipeline is 1.0×10^5 Pa:

(a) How much work is done by the pressure in forcing 1.0 m^3 of oil through 1.0 km of this pipeline?

(b) The pipeline is horizontal and has a constant cross section. What drop in pressure is predicted by Bernoulli's equation? Comment.

(c) What is the energy dissipated per cubic metre of oil transported per kilometre of pipeline length?

17.

gases and
thermal motion

The next four chapters treat the phenomena of heat and deal with *macroscopic* systems. Such systems are large compared to a single molecule—large enough so that their gross characteristics can be determined by direct experimental measurement. A typical macroscopic system is a gas of some 10^{23} molecules contained in a cylinder fitted with a piston (Figure 17-1).

Within any macroscopic system, the molecules are in incessant motion. The characteristic feature of this molecular motion is *randomness* which is always present to some extent. This random molecular motion, called *thermal motion*, is the underlying cause of the phenomena of heat. In this chapter we examine thermal motion in the simplest macroscopic system—an ideal gas.

17.1 Equation of State of an Ideal Gas

The number N of molecules in a mass M of a sample of gas containing identical molecules can be calculated if the mass m of an individual molecule has been determined. Molecular masses can be measured using a mass spectrometer (Section 27.2). Evidently $M = Nm$ and

$$N = \frac{M}{m} \tag{17.1}$$

Now if we measure the volume occupied by different amounts of a given type of gas at constant pressure and temperature, we discover that, as we expect, this volume is proportional to the number N of gas molecules present.

Suppose that, with a definite amount of a gas contained in the cylinder of Figure 17-1, we measure the gas pressure P corresponding to various values of

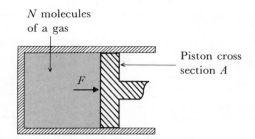

N molecules
of a gas

Piston cross
section A

F

Figure 17-1   The gas pressure
$P = F/A$.

the volume V occupied by the gas. We find that if the volume changes, the pressure changes in such a way that the product PV is a constant if the temperature remains constant. This result, discovered in 1660 by Robert Boyle (1627–1691), is known as Boyle's law.

Continuing the experimental investigation, we observe that when the gas is heated, the product PV increases. A most convenient temperature scale, called the Kelvin temperature scale, can be defined in such a way that the Kelvin temperature T of the gas (at low pressures and sufficiently high temperatures) is proportional to this product PV.

All these statements are summarized in the one equation

$$PV = NkT \qquad (17.2)$$

where k, known as Boltzmann's constant, will shortly be discussed in detail. This equation is called the *equation of state for an ideal gas* and a fictitious substance that would behave this way for all temperatures and pressures is named an *ideal gas*. The *equation of state* of a system is the equation relating pressure, volume, and temperature. For real gases, the experimental results displayed in Figure 17-2 show that Eq. 17.2 is an excellent approximation when the pressure is not too high and the temperature is well above the condensation temperature, but deviations from this behavior do occur under other conditions.

Various Temperature Scales

Before further discussion of ideal gases, we shall digress briefly to give the relationship between the Kelvin temperature T and other commonly used temperature scales. Household and medical thermometers in the United States are calibrated using the Fahrenheit scale. On this scale, when the pressure is 1 atm, the temperature of melting ice is $32.0°F$ (32.0 degrees Fahrenheit) and the temperature of boiling water is $212.0°F$. The corresponding temperatures on the Celsius temperature scale, commonly used in scientific work, are $0.00°C$ (0.00 degrees Celsius) and $100.00°C$. The relationship between the Celsius temperature t_C and the Fahrenheit temperature t_F is

$$t_C = \tfrac{5}{9}(t_F - 32°F)$$

For example, what is often referred to as normal room temperature, $68°F$, corresponds to $t_C = \tfrac{5}{9}(68 - 32)°C = 20°C$.

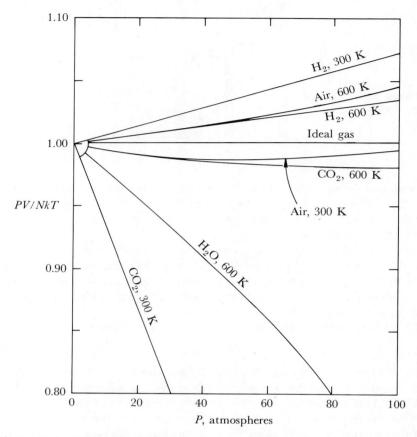

Figure 17-2 Comparison of the behavior of several actual gases with that of an ideal gas.

The Celsius temperature scale, and, indirectly, the Fahrenheit temperature scale, are defined in terms of the Kelvin temperature scale by the equation

$$t_C = T - 273.15 \text{ K}$$

where T is the Kelvin temperature corresponding to the Celsius temperature t_C. Thus 0°C corresponds to 273.15 K (273.15 kelvins), and normal room temperature is 293.15 K. The kelvin is the SI unit of temperature. The Celsius scale is not in itself part of the SI system, but a difference of one degree on the Celsius scale is equal to one kelvin. According to the SI convention the degree symbol is always associated with Celsius temperature but not with the kelvin. The Fahrenheit, Celsius, and Kelvin temperature scales are compared in Figure 17-3.

A precise definition of the Kelvin temperature is given later (Sections 19.3 and 20.5). The interesting relationship between the Kelvin temperature and molecular motion is examined in Section 17.3.

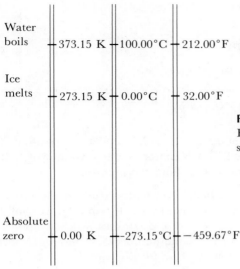

Water
boils ⊢ 373.15 K ⊢ 100.00°C ⊢ 212.00°F

Ice
melts ⊢ 273.15 K ⊢ 0.00°C ⊢ 32.00°F

Figure 17-3 Comparison of the Kelvin, Celsius, and Fahrenheit scales of temperature.

Absolute
zero ⊢ 0.00 K ⊢ -273.15°C ⊢ -459.67°F

Boltzmann's Constant k

We return to the equation of state of an ideal gas, $PV = NkT$, and consider different types of gases. In 1811, some hundred years before the number of molecules in a sample of gas could be determined, A. Avogadro (1776–1856), after considering evidence accumulated from study of chemical reactions between gases, advanced the following hypothesis: equal volumes of different gases at the same temperature and pressure contain the same number of molecules. Experiments confirm Avogadro's hypothesis; at a pressure of 1.013×10^5 Pa and a temperature of 273 K, a cubic metre of any gas contains 2.68×10^{25} molecules. From this experimental result we can evaluate Boltzmann's constant. Solving Eq. 17.2 for k we obtain

$$k = \frac{PV}{NT}$$

Therefore

$$k = \frac{(1.013 \times 10^5 \text{ Pa})(1.00 \text{ m}^3)}{(2.68 \times 10^{25} \text{ molecules})(273 \text{ K})} = 1.38 \times 10^{-23} \text{ J/K}$$

Boltzmann's constant, according to Avogadro's hypothesis, has this same value for all gases. We say that k is a universal constant.

The use of the equation of state of an ideal gas is illustrated in the following examples.

Example 1 A mass of 32.0×10^{-3} kg of oxygen is contained in a cylinder at a pressure of 1.013×10^5 Pa and a temperature of 273 K. What volume does the gas occupy? (The mass of an oxygen molecule is 5.32×10^{-26} kg.)

Solution   The number of oxygen molecules in the cylinder is

$$N = \frac{M}{m} = \frac{32.0 \times 10^{-3} \text{ kg}}{5.32 \times 10^{-26} \text{ kg}} = 6.02 \times 10^{23} \text{ molecules}$$

The equation of state of an ideal gas gives

$$V = \frac{NkT}{P} = \frac{(6.02 \times 10^{23})(1.38 \times 10^{-23} \text{ J/K})(273 \text{ K})}{1.013 \times 10^5 \text{ Pa}} = 22.4 \times 10^{-3} \text{ m}^3$$

Example 2   A gas which occupies a volume of 2.24 m³ at a pressure of 1.01 × 10⁵ Pa and a temperature of 273 K is compressed to a volume of 1.12 m³ and heated to a temperature of 819 K. Find the pressure.

Solution   Denoting the pressure, volume, and temperature in the first state by P_1, V_1, and T_1, and in the second state by P_2, V_2, and T_2, we shall first deduce a simple relationship between these quantities which is applicable whenever we are concerned with the same number N of molecules in two different states. The equation of state of an ideal gas gives

$$kN = \frac{P_1 V_1}{T_1} = \frac{P_2 V_2}{T_2}$$

The desired relationship is

$$\frac{P_1 V_1}{T_1} = \frac{P_2 V_2}{T_2} \tag{17.3}$$

To find the pressure P_2, we use Eq. 17.3 and obtain

$$P_2 = \left(\frac{V_1}{V_2}\right)\left(\frac{T_2}{T_1}\right)P_1$$

$$= \left(\frac{2.24 \text{ m}^3}{1.12 \text{ m}^3}\right)\left(\frac{819 \text{ K}}{273 \text{ K}}\right)(1.01 \times 10^5 \text{ Pa})$$

$$= 2.00 \times 3.00 \times (1.01 \times 10^5 \text{ Pa})$$

$$= 6.06 \times 10^5 \text{ Pa}$$

Gas Constant R

Instead of expressing the equation of state of an ideal gas directly in terms of the number N of the gas molecules, we can specify (this is common practice, particularly in chemistry) the quantity of the gas in terms of the *mole* (mol). This is a measure of quantity of matter that is convenient for macroscopic amounts. *The mole is defined as the amount of substance of a system containing as many elementary entities as there are atoms in exactly* 0.012 kg *of carbon-12.** The elementary entities

*A particular isotope is defined as an atom with a uniquely specified nucleus; the number written after the name of the element gives the total number of nucleons (protons plus neutrons) in the nucleus.

must be specified, and they may be molecules, atoms, ions, electrons, or other particles.

The number of atoms in exactly 0.012 kg of carbon-12 is called Avogadro's number and is denoted by N_A. A mole of any specified kind of molecule (or atom) consists of N_A molecules (or atoms) of this kind. The experimental value of Avogadro's number N_A is 6.022045×10^{23} mol^{-1}. This truly enormous number is of the order of magnitude of the number of atoms in a handful of carbon or of the number of molecules in a test tube full of water. Avogadro's number is therefore typical of the number of molecules involved in chemical laboratory experiments.

To determine the number of moles of a measured mass of a pure substance we must know the molecular weight of the molecules of the substance (or the atomic weight of the atoms). This is the ratio of the mass m of a molecule (or atom) of the substance to one-twelfth the mass of an atom of carbon-12. In other words, a molecular or atomic weight* is a dimensionless ratio of masses given by

$$\mu = \frac{m}{\frac{1}{12}mass \ of \ carbon\text{-}12 \ atom}$$

From this definition, the atomic weight of carbon-12 is exactly 12. By experimental measurement of mass ratios we obtain results such as $\mu = 1.007825$ for hydrogen-1 and $\mu = 15.994915$ for oxygen-16. These definitions imply that the number of moles n of a substance composed of molecules of molecular weight μ (or of atoms of atomic weight μ) is given by

$$n = \frac{1000M}{\mu}$$

where M is the mass of the substance in kilograms.

The number N of molecules in n moles of a substance is

$$N = nN_A$$

Making the substitution $N = nN_A$ in Eq. 17.2, we can express the equation of state of an ideal gas in terms of the number of moles n of the gas:

$$PV = nRT \qquad (17.4)$$

where R, called the *gas constant*, is given by

$$R = N_A k$$

The value of R can be determined from measurements of P, V, and T when n is known from measurements of M and μ. Modern experiments give

$$R = 8.31441 \text{ J/mol} \cdot \text{K}$$

The corresponding value of the Boltzmann constant is

$$k = \frac{R}{N_A} = 1.380662 \times 10^{-23} \text{ J/K}$$

*This name is misleading since a molecular (or atomic) weight is a mass ratio, not a weight.

To illustrate the use of the equation of state of an ideal gas in the alternative form, $PV = nRT$, we shall reconsider Example 1. Suppose that, instead of knowing the mass of the oxygen molecules, we are given $\mu = 32.0$ for the molecular weight of oxygen. (The molecules of gaseous oxygen O_2 are formed from two oxygen atoms, and the molecular weight is approximately $2 \times 16 = 32$.) A mass of 0.032 kg of oxygen comprises

$$n = \frac{1000 \times 0.032}{32.0} \text{ mol} = 1.00 \text{ mol}$$

The volume occupied by this gas at a pressure of 1.013×10^5 Pa and a temperature of 273 K is

$$V = \frac{nRT}{P} = \frac{(1.00 \text{ mol})(8.31 \text{ J/mol} \cdot \text{K})(273 \text{ K})}{1.013 \times 10^5 \text{ Pa}} = 22.4 \times 10^{-3} \text{ m}^3$$

It is worth noting that, as this calculation illustrates, a mole of *any* ideal gas occupies 22.4 litres under standard conditions of pressure and temperature (that is, 1 atm and 0°C).

17.2 Kinetic Theory of Gases

We now attempt to understand the behavior of gases in terms of molecular motion. A gas is a collection of billions of molecules engaged in random motion. The impact of these molecules with the container walls gives rise to the force exerted on the walls by the gas. This force and the associated gas pressure can be calculated from the average of the translational kinetic energy $\frac{1}{2}mv^2$ possessed by the gas molecules. Then comparison with the equation of state of an ideal gas reveals that temperature is a measure of the average translational kinetic energy of the molecules.

Temperature and Molecular Kinetic Energy

Let us first examine a very simple situation in which one molecule travels with a constant speed v in an empty box of length L with ends of cross-sectional area A. In Figure 17-4, we take the direction of the X axis to be perpendicular to the right-hand wall. The molecule exerts a force on this end of the box as it collides with the box and rebounds. The average force arising from such molecular impacts, divided by the area of the box surface, is the pressure P contributed by this molecule. We assume that collisions with a wall are such that the molecule's velocity component perpendicular to the wall is reversed, but that other components are not altered; then the molecule's speed and kinetic energy are not changed. In a collision with the right-hand wall the x component of the molecule's momentum vector changes from mv_x to $-mv_x$, which implies that the *change* of momentum Δp_x is $2mv_x$. The time interval Δt between collisions is $2L/v_x$

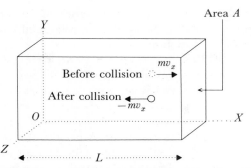

Figure 17-4 Molecule exerts a force on right-hand wall as it collides and rebounds. Molecule's x component of momentum changes from mv_x to $-mv_x$.

since the molecule must travel twice the length of the box before returning to the same wall. The average force exerted on the molecule by the wall is given by

$$\text{average force} = \text{average rate of change of momentum}$$

$$= \frac{\Delta p_x}{\Delta t} = \frac{2mv_x}{2L/v_x} = \frac{mv_x^2}{L} \tag{17.5}$$

From Newton's third law, this is also the magnitude of the average force exerted on the wall by the molecule.

When there are N molecules in the box, the average total force exerted on the wall is

$$\text{average force} = \frac{1}{L} \sum mv_x^2 = \frac{N}{L} \overline{mv_x^2}$$

where Σ denotes summation over all the gas molecules and this sum has been expressed as N times the average value $\overline{mv_x^2}$. The average pressure P on the wall of area A is

$$P = \frac{\text{average force}}{A} = \frac{N}{V} \overline{mv_x^2}$$

where $V = AL$ is the volume of the box. Because the x, y, and z directions are equivalent, we have $\overline{mv_x^2} = \overline{mv_y^2} = \overline{mv_z^2}$ and, since $v_x^2 + v_y^2 + v_z^2 = v^2$,

$$\overline{mv_x^2} = \tfrac{1}{3}\overline{mv^2}$$

Substitution of this result in $PV = N\overline{mv_x^2}$ gives

$$PV = \tfrac{2}{3}N(\overline{\tfrac{1}{2}mv^2}) \tag{17.6}$$

Interactions of the gas molecules do not affect this result, provided that the range of the intermolecular forces is small compared to the average distance between the molecules. A comparison of this kinetic theory equation, $PV = \tfrac{2}{3}N(\overline{\tfrac{1}{2}mv^2})$, with the equation of state of an ideal gas, $PV = NkT$, shows that

$$\overline{\tfrac{1}{2}mv^2} = \tfrac{3}{2}kT \tag{17.7}$$

This equation gives the kinetic theory interpretation of temperature: the Kelvin temperature T is proportional to the average translational kinetic energy of the molecules' random thermal motion.

The validity of Eq. 17.7 is not restricted to an ideal gas. Newtonian mechanics leads to this result for any substance. *From the point of view of Newtonian mechanics, the temperature of any macroscopic object is a measure of the average translational kinetic energy of its molecules, evaluated in the object's CM-frame.*

When two macroscopic objects are placed in contact, the atoms in them collide and transfer energy to one another. A net transfer of energy from the hotter object to the colder object will take place until a state of *thermal equilibrium* is reached with both objects at the same temperature. This fact, fundamental to the temperature concept, is emphasized in Section 19.3. When the two objects are in thermal equilibrium, although individual particles transfer energy in both directions, there is no net transfer of energy from either object to the other. Although the calculation is too involved to be undertaken in this book, it is possible to confirm the kinetic theory interpretation of temperature by showing that the condition of balanced energy transfers between two objects implies that the average translational kinetic energies of their molecules are equal.

At room temperature (20°C or about 293 K), the average translational kinetic energy of a molecule $\frac{3}{2}kT$ has the value 6.1×10^{-21} J. The joule is obviously an inconveniently large unit for such energies. In Section 11.6 we introduced the energy unit generally used in atomic physics, the electronvolt (eV), which is related to the joule by

$$1 \text{ eV} = 1.6021892 \times 10^{-19} \text{ J}$$

It is useful to note that at room temperature kT is about $\frac{1}{40}$ eV. The temperature corresponding to $kT = 1$ eV is 11.6×10^3 K.

Molecular Speeds

The molecules of a substance undergo incessant thermal motion. At any instant, different molecules are generally moving with different speeds. It can be shown that at temperature T, in any substance with molecules of mass m whose motion is described by Newtonian mechanics, the number of molecules with speeds in the interval between v and $v + dv$ is

$$f(v) \, dv = \frac{4}{\sqrt{\pi}} \left(\frac{m}{2kT} \right)^{3/2} e^{-mv^2/2kT} v^2 \, dv \tag{17.8}$$

This result (Figure 17-5), known as the *Maxwellian distribution of speeds*, was deduced by James Clerk Maxwell (1831–1879) and has been experimentally verified by direct measurement of the speeds of molecules emerging from a very small hole in a vessel.

A typical molecular speed encountered at temperature T is given by the thermal speed v_T defined by

$$\tfrac{1}{2}mv_T^2 = \overline{\tfrac{1}{2}mv^2} = \tfrac{3}{2}kT$$

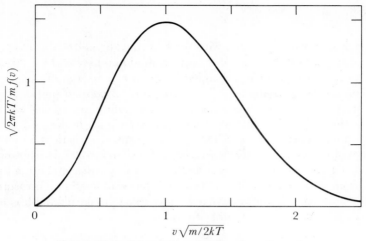

Figure 17-5 Maxwellian distribution of speeds.

Therefore

$$v_T = \sqrt{\overline{v^2}} = \sqrt{\frac{3kT}{m}} \qquad (17.9)$$

For hydrogen molecules at room temperature, the thermal speed is 1.9 $\times$ 10^3 m/s. The more massive the particle, the smaller its thermal speed. Thermal motion is very violent for molecules and is still appreciable in the Brownian motion of visible particles discussed next, but it is entirely negligible for massive bodies.

Brownian Motion

In 1827, an English botanist, Robert Brown, peering through a microscope, noticed that tiny grains of plant pollen suspended in a liquid were in continual motion, zigzagging about in a random fashion (Figure 17-6). Brown, suspecting

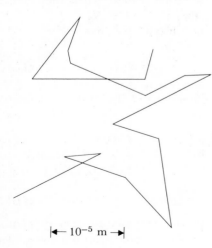

Figure 17-6 Brownian motion. The line joins positions of a small particle (mass 10^{-16} kg) observed at 2-minute intervals, as it is buffeted about by water molecules.

$\vdash$ 10^{-5} m $\dashv$

that this motion was connected with the fact that the pollen was living matter, was surprised to find that a similar random motion is executed by a small speck of dust or a smoke particle suspended in still air or in a liquid.

Kinetic theory gives the explanation for this "Brownian motion" of visible specks. Such particles (each composed of some million molecules) are buffeted about by molecular impacts and, just like individual molecules, acquire an average kinetic energy of $\frac{3}{2}kT$. This was the central idea in the quantitative treatment of Brownian motion advanced by Einstein in 1905. He showed how to calculate the average kinetic energy $\frac{3}{2}kT$ of a visible particle executing Brownian motion from measurement of the distance it has wandered in a measured time. Once $\frac{3}{2}kT$ and hence k was known, it was possible to make a reliable calculation (20% error) of Avogadro's number from

$$N_A = \frac{R}{k}$$

and thereby determine the mass of the individual molecules in a sample of measured mass.

Consequently, before the invention of mass spectrometers (Section 27.2) in the second decade of this century, Brownian motion served as an important source of molecular data. And it was the recognition of Brownian motion as a visible manifestation of molecular impacts that convinced even the most skeptical scientists of the value of a molecular theory of matter.

Mixtures of Gases

In the kinetic theory derivation of $PV = \frac{2}{3}N(\overline{\frac{1}{2}mv^2})$, we did not assume that the molecules were all identical. With the kinetic theory interpretation of the temperature, $\overline{\frac{1}{2}mv^2} = \frac{3}{2}kT$, we find that $PV = NkT$ is a *universal* equation, involving no quantities dependent on the nature of the gas—a result which is a consequence of neglecting the interaction between molecules. When the gas is a mixture of several different ideal gases, so that $N = N_1 + N_2 + \cdots + N_n$ where N_i is the number of molecules of the ith type, we have

$$PV = N_1kT + N_2kT + \cdots + N_nkT \qquad (17.10)$$

If the entire volume V were occupied by molecules of the ith kind alone, the pressure would be

$$P_i = \frac{N_ikT}{V} \qquad (17.11)$$

Equation 17.10 implies that

$$P = P_1 + P_2 + \cdots + P_n \qquad (17.12)$$

This is *Dalton's law of partial pressures:* The pressure of a mixture of gases is equal to the sum of the pressures that each individual gas would exert if it alone filled the volume. The pressures $P_1, P_2, \ldots, P_n$ are called the *partial pressures* of the respective gases. Experimental results with mixtures of real gases such as air are in accord with Dalton's law under conditions such that the components behave as ideal gases.

17.3 Mean Free Path

The interactions of gas molecules can be treated as collisions. Between collisions each molecule moves with a constant speed in a straight line. Since collisions occur randomly, the distance traveled by a molecule between collisions may have any value.

The average distance traveled between successive collisions is called the *mean free path* l. A related quantity is the average time interval between successive collisions, called the *mean free time* τ. In order of magnitude

$$\tau \approx \frac{l}{v_T} \tag{17.13}$$

where v_T is the thermal speed of a molecule.

To estimate the magnitude of l we examine molecular collisions in more detail. When we consider the possibility of collision of two given molecules, it is convenient to regard one molecule as a target fixed in a certain plane and the other molecule as a projectile that crosses the plane. The interaction of the two molecules will be counted as a collision only if the encounter is so close that the projectile trajectory is appreciably altered. This trajectory then meets the plane somewhere within a small region around the fixed molecule. The target area on the plane which the projectile must strike in order for a collision to occur is called the *collision cross section* σ.

For example, if the molecules are regarded as solid spheres of radius r_0, this target area is circular with a radius equal to $2r_0$, and

$$\sigma = 4\pi r_0^2$$

Although real molecules are not solid spheres, the interaction between molecules does decrease very rapidly with increasing distance between them and the collision cross section can be regarded as the order of magnitude of a molecule's cross-sectional "area."

When a molecule travels a distance L, it collides with all the molecules in a volume of cross section σ and length L. Therefore, if there are N_v molecules per unit volume, the number of collisions experienced by the given molecule in a length L is $N_v \sigma L$ and the average distance l between collisions is $L/N_v \sigma L$; that is

$$l \approx \frac{1}{N_v \sigma} \tag{17.14}$$

This shows that if σ is constant,* the mean free path l depends only on the number of molecules per unit volume.

For air at $0°C$ and at a pressure of 1 atm, $N_v \approx 3 \times 10^{25}$ molecules/m^3. Using as a typical molecular radius $r_0 \approx 10^{-10}$ m, we find $\sigma \approx 12 \times 10^{-20}$ m^2 and

$$l \approx 3 \times 10^{-7} \text{ m}$$

Since $l \gg r_0$, a molecule generally travels a relatively long distance before encountering another molecule. The thermal speed of the molecule is given by $v_T \approx 5 \times 10^2$ m/s and the mean free time of a molecule is

$$\tau \approx \frac{l}{v_T} \approx 6 \times 10^{-10} \text{ s}$$

Because $N_v \propto P$, the mean free path is inversely proportional to the pressure. In a high vacuum of the order of 10^{-8} cm of Hg, $\lambda \approx 400$ m. In this circumstance, nearly all the collisions are with the walls of the gas container rather than with the other molecules.

17.4 Internal Energy and Temperature

The internal energy of many substances at room temperature comprises important contributions from:

1   The translational kinetic energy of molecules associated with their random thermal motion.
2   The rotational and vibrational kinetic energy arising from the motion of the atoms of a molecule relative to the center of mass of the molecule.
3   The potential energy associated with the interaction of the atoms within a molecule.
4   The potential energy arising from the interaction of one molecule with another. When molecules are packed tightly together, as in a liquid or a solid, they continually exert appreciable forces on each other. The associated potential energy is then by no means the negligible quantity that it is assumed to be for an ideal gas.

There is an important distinction between the temperature of an object and its internal energy. The temperature, according to the result

$$\overline{\tfrac{1}{2}mv^2} = \tfrac{3}{2}kT \tag{17.7}$$

is a measure of the average translational kinetic energy *only*. The internal energy generally includes not only these translational kinetic energies but also the internal energies of the molecules and the intermolecular potential energies.

*In fact, σ increases slightly as the temperature is decreased. At lower temperatures the speeds of the gas molecules are reduced. Then the average duration of the collisions is increased and this leads to an increased effect of a relatively distant encounter between two molecules. Consequently the collision cross section is larger. For example, in N_2 and O_2 there is a 30% increase in σ when the temperature drops from 373 K to 173 K.

The zero of temperature on the Kelvin scale, called *absolute zero*, corresponds to zero translational kinetic energy of the molecules. Thermal motion ceases entirely at absolute zero, at least according to Eq. 17.7. This is the prediction of classical mechanics. However, at sufficiently low temperatures and correspondingly low energies, the conditions for the validity of Newtonian mechanics are no longer satisfied and quantum phenomena become evident. According to quantum mechanics, the motion of the particles in a body never ceases completely. Even at absolute zero, there remains a vibrational motion of atoms within the molecules, or vibrations of atoms about crystal lattice sites. These vibrations are called *zero-point vibrations*, and the energy associated with these vibrations is called the *zero-point energy*.

Energy levels are important in quantum phenomena. Any bound system, be it composed of atoms rotating or vibrating within a molecule or of electrons bound to an atom, has as possible values of its internal energy only certain discrete values which are named the *energy levels* of the system. As the molecules of a gas are banged about by the thermal agitation at a Kelvin temperature T, some inelastic molecular collisions occur. During a collision, translational kinetic energy may be converted into "excitation energy" of one of the molecules. In absorbing an energy ΔE, a molecule makes a transition from a lower energy level E_ℓ to an upper energy level E_u, where

$$\Delta E = E_u - E_\ell$$

In subsequent collisions, this molecule may lose this energy passing back from level E_u to E_ℓ; or alternatively it may be promoted to a still higher energy level.

Eventually, after many collisions, an equilibrium distribution is obtained, with the number of molecules in level E_ℓ fluctuating about an equilibrium population N_ℓ with a lesser equilibrium population N_u in the upper level E_u. The ratio of equilibrium populations can be shown to be

$$\frac{N_u}{N_\ell} = \frac{e^{-E_u/kT}}{e^{-E_\ell/kT}} = e^{-(E_u - E_\ell)/kT} = e^{-\Delta E/kT} \tag{17.15}$$

According to this result, the ratio of equilibrium populations N_u/N_ℓ will be small unless the temperature is high enough so that the average translational kinetic energy $\frac{3}{2}kT$ is large enough to be comparable to the energy difference, $\Delta E = E_u - E_\ell$. For instance, if we let E_ℓ represent the *ground* state energy (the lowest energy level) and E_u the first *excited* state, and if the temperature is so low that $\frac{3}{2}kT$ is much less than ΔE, then the population of molecules in the ground state is very much greater than the population in the excited state. At a temperature this low, molecules hardly ever collide with enough kinetic energy to make excitation possible. We say the possibility of excitation to the energy level E_u is *frozen out*.

The way that the internal energy of any substance is partitioned among the various possibilities is, therefore, dependent on the size of $\frac{3}{2}kT$ relative to the energy difference ΔE between different energy levels of the particles. As the temperature is increased, an increasing variety of excitations becomes possible. The situation at various temperatures is indicated in Table 17-1 and Figure 17-7.

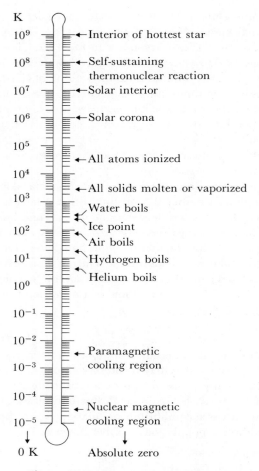

K

10^9 ←Interior of hottest star

10^8 ←Self-sustaining
thermonuclear reaction

10^7 ←Solar interior

10^6 ←Solar corona

10^5

10^4 ← All atoms ionized

10^3 ← All solids molten or vaporized

Water boils

10^2 Ice point

Air boils

10^1 Hydrogen boils

10^0 Helium boils

10^{-1}

10^{-2}

10^{-3} ← Paramagnetic
cooling region

10^{-4}

10^{-5} ← Nuclear magnetic
cooling region

0 K   Absolute zero

Figure 17-7 Phenomena occurring at various Kelvin temperatures. The temperature scale is logarithmic.

Table 17-1 Typical Phenomena at Various Temperatures

Temperature (K)	$\frac{3}{2}kT$		Phenomena
0	0	eV	Zero-point vibration
273	0.035	eV	Ice melts
2000	0.26	eV	Most substances gaseous
6000	0.78	eV	Temperature at sun's surface
10^6	130	eV	Light elements largely ionized
10^7	1.3	KeV	Temperature at sun's center
10^{10}	1.3	MeV	Nuclei undergo thermal excitation and disintegration

Summary

☐ The equation of state of an ideal gas containing N molecules or n moles is

$$PV = NkT \qquad \text{or} \qquad PV = nRT$$

where T is the Kelvin temperature. From experimental measurement of P, V, and N (or n) at an assigned value of T, the value of k (or R) is calculated:

$$k = 1.38 \times 10^{-23} \text{ J/K} \qquad R = 8.31 \text{ J/mol} \cdot \text{K}$$

We have $N = nN_A$ and $R = N_A k$, where N_A is Avogadro's number ($6.02 \times 10^{23} \text{ mol}^{-1}$).

☐ According to Newtonian mechanics, the temperature of any macroscopic object is a measure of the average translational kinetic energy $\overline{\frac{1}{2}mv^2}$ of its molecules, evaluated in the object's CM-frame:

$$\overline{\tfrac{1}{2}mv^2} = \tfrac{3}{2}kT$$

☐ In a gas with N_v molecules per unit volume, the mean free path is

$$\ell \approx \frac{1}{N_v \sigma}$$

where σ is the collision cross section.

Questions

1 The mass of a hydrogen molecule determined by mass spectrometer measurement is 3.35×10^{-27} kg. How many hydrogen molecules are there in 2.0 kg of hydrogen?

2 The temperature of the human body is normally 37.0°C. Find the corresponding temperatures on the Fahrenheit and the Kelvin scales.

3 By pushing in the plunger, we compress the volume of air within a syringe to one-fourth of its original volume. Initially the pressure is 1.01×10^5 Pa. Find the final pressure for the case when:
   (a) The compression is performed so slowly that the temperature of the air within the syringe remains about the same as that of the surroundings.
   (b) The temperature of the air within the syringe increases from an original value of 300 K to a final value of 330 K.

4 Nitrogen is contained in a steel cylinder at a pressure of 2.0×10^5 Pa when the temperature is 300 K. What is the pressure when the temperature is raised to 500 K?

5 By what percentage will the pressure in a car's tire increase if the temperature of the air in the tire is raised from 300 K to 400 K? Assume that the change in the volume of the tire is negligible.

6 A hydrogen balloon at the earth's surface has a volume of 5.0 m³ on a day when the temperature is 300 K and the pressure is 1.00×10^5 Pa. The balloon rises and expands as the pressure drops. What is the volume of the balloon (at an altitude of about 40 km) when the pressure is merely 0.33×10^3 Pa and the temperature is 260 K?

7   During the compression stroke of the piston, air within a cylinder of a diesel engine is compressed to $\frac{1}{16}$ of its initial volume and the pressure of this air increases from 1.0 atm to 42 atm. The initial temperature of the air is 300 K. Find its temperature at the end of the compression stroke.

8   Gas in a cylinder is heated until both the pressure and the volume have tripled. If the initial temperature is 300 K, what is the final temperature?

9   Low pressures achieved in a vacuum system are often given in terms of a unit called the *torr* (in honor of Torricelli), which is related to the pascal by

$$1 \text{ Torr} = 1.33 \times 10^2 \text{ Pa}$$

At an ultra-high vacuum of 10^{-10} Torr, at a temperature of 300 K, how many molecules still remain in a volume of 1.00 cm^3?

10 (a) How many H_2O molecules are there in a test tube containing 3.0 mol of water?
    (b) A test tube contains 54×10^{-3} kg of water. The molecular weight of water is approximately $2 + 16 = 18$. How many moles of water are there in the test tube?

11 What is the pressure in a 50-litre tank containing 0.44 kg of carbon dioxide (CO_2) at a temperature of 20°C? (1 litre $= 10^{-3}$ m^3.)

12 What is the mass of the carbon dioxide contained in a 50-litre tank at a pressure of 3.0 atm and a temperature of 30°C?

13 What is the mass of helium in a helium-filled balloon which has a volume of 6.00 m^3 at a pressure of 0.50×10^5 Pa and a temperature of 250 K?

14 What pressure is exerted by 0.600 kg of oxygen at 300 K within a 50-litre tank?

15 What interpretation of temperature is given by the kinetic theory?

16 What is the temperature of a substance whose molecules in random motion have an average translational kinetic energy of 0.10 eV?

17 (a) What is the thermal speed of oxygen molecules (mass 5.32×10^{-26} kg) at 300 K?
    (b) At what temperature will this thermal speed be doubled?

18 What is the relationship between the thermal speed v_T and $\overline{v^2}$? The thermal speed is often called the root mean square speed. Explain.

19 Explain how the mass of an individual molecule in a gas can be calculated once the value of Boltzmann's constant has been determined.

20 Calculate the thermal speed of a smoke particle, of mass 1.0×10^{-16} kg, which executes Brownian motion in air at a temperature of 300 K.

21 Explain how Avogadro's number can be calculated once the value of Boltzmann's constant has been determined.

22 From the data given at the end of Section 17.3, determine the number of collisions per second experienced by a molecule in air at 0°C and a pressure of 1.0 atm.

23 In an ideal gas, show that the mean free path is

$$\ell \approx \frac{kT}{\sigma P}$$

24 Given that the collision cross section of H_2 is 14×10^{-20} m^2, calculate the mean free path of a hydrogen molecule in a container of H_2 under the following conditions:
    (a) $T = 300$ K, $P = 1.00$ atm.
    (b) $T = 600$ K, $P = 1.00$ atm.
    (c) $T = 300$ K, $P = 1.00 \times 10^{-6}$ atm.

25   Using the data of the preceding question, find the total number of collisions occurring in 1.0 s in a volume of 1.0 cm^3 of H_2 at $T = 300$ K and $P = 1.00$ atm.

26   What is the distinction, as far as kinetic theory is concerned, between the temperature and the internal energy of a collection of molecules?

27   **(a)** Calculate the thermal speed of argon molecules ($\mu = 40$) at $0°C$.
   (b) What is the kinetic energy of a mole of argon gas at this temperature?
   (c) What is the total momentum relative to the laboratory of this amount of argon in a container which is at rest in the laboratory?

28   **(a)** Find the temperature at which the population of molecules in an excited state E_u will be equal to $1/e$ times the population in the ground state E_ℓ (use Eq. 17.15).
   (b) If $E_u - E_\ell = 1.0$ eV, what is the value of the temperature found in part (a)?
   (c) At room temperature will there be an appreciable fraction of these molecules excited to level E_u? ($E_u - E_\ell = 1.0$ eV.)

29   At 300 K, what is the internal energy in joules within a balloon containing 6.0 $\times$ 10^{23} molecules of neon (0.020 kg)? Consider only the part of the internal energy that depends on the temperature; that is, do not include the rest energies of the neon molecules.

Supplementary Questions

S-1   Air is trapped in a barometer tube above the mercury column. On a day when the pressure of the atmosphere is 76.0 cm of Hg, the height of the mercury column is merely 72.0 cm and the air space above the column is 12.0 cm long. On a day when this mercury column is 70 cm long, what is the pressure of the atmosphere?

S-2   Calculate the ratio of the final to the initial volume of an air bubble that rises from the bottom of a lake which is 30 m deep. Assume that the temperature change of the bubble is negligible and that the atmospheric pressure is 1.00×10^5 Pa.

S-3   **(a)** For one state of a given type of ideal gas, denote the density, pressure, and temperature by ρ_1, P_1, and T_1, respectively. In a second state denote them by ρ_2, P_2, and T_2. Show that

$$\frac{\rho_1 T_1}{P_1} = \frac{\rho_2 T_2}{P_2}$$

   (b) What is the density of air at the top of a mountain on a day when the temperature is $17°C$ and the pressure is 0.90 atm? (At $0°C$ and 1 atm, air has a density of 1.29 kg/m^3.)

S-4   As oxygen is withdrawn from a 50-litre tank, the pressure within the tank drops from 21 atm to 7 atm, and the temperature drops from 300 K to 280 K. How many kilograms of oxygen were withdrawn from the tank?

S-5   **(a)** Assume that the temperature in the upper portions of our atmosphere is $0°C$ and calculate the thermal speed of hydrogen molecules (H_2) in this region.
   (b) Compare this thermal speed to the speed required to escape from the earth. Can the absence of hydrogen in our atmosphere be attributed to the ease with which hydrogen molecules can escape from the earth? Comment.
   (c) Will the rate of escape be greater for hydrogen than for nitrogen and oxygen? Explain.

S-6 **(a)** Calculate the thermal speed of hydrogen *atoms* at the sun's surface where the temperature is 6000 K.

(b) Compute the escape speed for projectiles at the sun's surface and, by comparing this with the answer to part (a), show that these facts are consistent with the observation that there is an abundance of hydrogen in the sun's atmosphere.

S-7 After calculating relevant thermal speeds and escape speeds, explain why the moon cannot have an atmosphere similar to the earth's atmosphere.

S-8 **(a)** In a gas of uranium hexafluoride, calculate the ratio of the thermal speeds of the two different isotopes, $^{235}UF_6$ and $^{238}UF_6$, which have molecular weights 349 and 352, respectively.

(b) When uranium hexafluoride is allowed to diffuse through a porous barrier into an evacuated space, which constituent will diffuse through the barrier at the greater rate? Explain.

18.

heat and the first law of thermodynamics

18.1 Microscopic and Macroscopic Viewpoints: Equilibrium States

A macroscopic system in mechanical equilibrium, say some apparently motionless water in a test tube, represents an enormously complicated system from a microscopic point of view. There are some 10^{24} interacting molecules in motion. A description which involves some 10^{24} details is far beyond the capacity of the largest computers and would hardly be very illuminating, even if it were available.

Fortunately, as we have seen, the gross characteristics of a macroscopic system can often be described in terms of a few *macroscopic quantities,* such as *temperature, pressure,* and *volume,* that can be determined by direct experimental measurement. We find that if a system is left undisturbed by interaction with anything else, that is, if the system is *isolated,* it will eventually reach a situation in which no more changes are apparent. This special situation, obtained simply by waiting long enough, is called an *equilibrium state.* Except for minute fluctuations, the pressure, volume, temperature, and all other macroscopic quantities have constant values in an equilibrium state.

When a system attains an equilibrium state, three types of equilibrium are obtained:

1   *Mechanical equilibrium*   There is no acceleration of any macroscopic portion of the system. Mechanical wave motion, turbulence, and eddies do not occur.
2   *Thermal equilibrium*   All portions of the system are at the same temperature, which is constant.
3   *Chemical equilibrium*   The chemical composition of a macroscopic portion of the system at any location does not change as time elapses.

From the microscopic viewpoint, equilibrium requires that the change of a macroscopic property, caused by the motion of some molecules, must be counter-balanced by the change caused by other molecules. Equilibrium evidently is a very special situation as far as molecular motions are concerned. It is reasonable to hope that, from the condition of exact balancing, microscopic theory can deduce the special macroscopic properties of substances in equilibrium states. This has been done with considerable success using what is called *statistical mechanics*.

The science that is concerned with a macroscopic description only is named *thermodynamics*. It deals with systems in equilibrium states and processes in which a system changes from one equilibrium state to another. Thermodynamics is a practical subject, to a large extent developed and first used with attention riveted upon applications in engineering and chemistry, but the laws of thermodynam-ics are so general that the subject merits the attention of all scientists.

When thermodynamics is developed without reference to the underlying atomic phenomena, it is a rather abstract subject, and its many logical inter-connections make it difficult for us to understand one part before understanding the whole. In this book we do not attempt to give a formal axiomatic presenta-tion of thermodynamics. Although emphasis is placed on the macroscopic description provided by thermodynamics, we adopt from the beginning the point of view that the thermodynamic variables are a manifestation of atomic behavior.

18.2 Work for a Thermodynamic Process

We consider a system that is a fluid, and for simplicity, we assume that this fluid is contained in a cylinder fitted with a piston. In an equilibrium state, the fluid occupies the entire volume V within the cylinder and exerts a uniform pressure P on the cylinder walls and on the piston. If the fluid expands, it moves the piston and does work.

The work done by this system can be calculated in terms of macroscopically measurable quantities for a process that is *quasi-static,* that is, a process that is carried out sufficiently slowly so that the system is at all times arbitrarily close to equilibrium. For a quasi-static process in which the piston moves out a distance ds (Figure 18-1), the work done by the system on the piston is

$$\dlap{d}W = F\,ds = PA\,ds$$

(The differential symbol with a line drawn through it will be explained in Section 18.3.) Since $A\,ds$ is the change in volume of the system, the expression for the work done by the system on the surroundings for a quasi-static process involving a volume change dV is*

$$\dlap{d}W = P\,dV \qquad (18.1)$$

---

*It is not difficult to show that this expression for $\dlap{d}W$ is valid for a fluid enclosed in a container with an arbitrary shape.

The work done *by* the system is positive if the system expands ($dV > 0$) against external forces and negative if the system is compressed ($dV < 0$). No work is done by the system during a quasi-static process in which the volume is constant.

In Figure 18-2, called a *PV diagram,* the initial state is represented by the point *i* with coordinates P_i, V_i and the final state by the point *f* with coordinates P_f, V_f. The succession of intermediate states of the quasi-static process are represented by the points of the curve joining *i* and *f.* The work, $dW = P\,dV$, is represented by the hatched area of the narrow rectangle of height P and width dV. The work for the finite quasi-static process in which the volume changes from V_i to V_f is

$$W = \int_{V_i}^{V_f} P\,dV \tag{18.2}$$

and this work is represented on the *PV* diagram by the area under the curve which represents the quasi-static process. To evaluate W, P must be expressed as a function of V; that is, the path of integration must be defined.

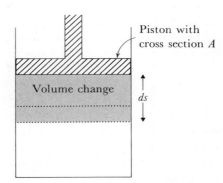

Figure 18-1 When the piston moves out a distance *ds,* the cylinder volume changes by an amount equal to *A ds.*

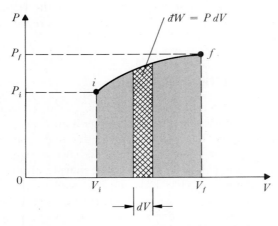

Figure 18-2 A quasi-static process is represented by the curve from the initial state (P_i, V_i) to the final state (P_f, V_f). The work for this process is represented by the area under the curve.

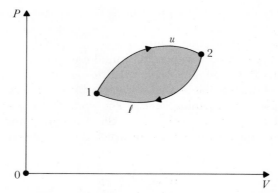

Figure 18-3   A quasi-static cyclic process is represented by a closed curve. The area enclosed by the curve represents the work for this process.

When the pressure is constant throughout the process (an *isobaric* process), Eq. 18.2 gives a particularly simple expression for the work:

$$W = P(V_f - V_i) \qquad (P \text{ constant}) \quad (18.3)$$

In a quasi-static isothermal (constant temperature) expansion or compression of n moles of an ideal gas at temperature T,

$$W = \int_{V_i}^{V_f} P \, dV = nRT \int_{V_i}^{V_f} \frac{dV}{V} = nRT(\ln V_f - \ln V_i) = nRT \ln \frac{V_f}{V_i} \quad (18.4)$$

In a *cyclic process,* the system returns to its initial state and the process is represented by a closed curve (Figure 18-3). On the upper portion $1u2$ of the curve the system expands and does work represented by the area under that curve; on the lower portion $1\ell2$ of the curve the system is compressed and the work done by the system is therefore negative and equal in magnitude to the area under the lower curve. The total work done by the system in the cyclic process is therefore the difference of these areas which is represented by the *area enclosed by the curve.*

18.3 The First Law of Thermodynamics

In thermodynamics the symbol $\overline{E}$ (rather than E_{int}) will be used to denote the internal energy of a macroscopic system. Consider a macroscopic system such as a bouncing ball: the system has a kinetic energy due to its bulk motion and also a potential energy because of its altitude. This energy associated with bulk motion or position of the ball as a whole is to be sharply distinguished from the internal energy $\overline{E}$ of the ball. The bulk motion involves an *ordered* motion of the molecules of the system, but as has been emphasized in Section 17.4, the internal energy $\overline{E}$ comprises contributions arising from randomly directed molecular motions and randomly distributed internal excitations.* The *internal*

*In many discussions of thermal phenomena, internal energy is referred to as *thermal energy.*

energy $\overline{E}$ is disordered energy. The frame of reference used in thermodynamics is generally the system's CM-frame. Then for an isolated system in a thermodynamic equilibrium state, $\overline{E}$ is the system's total energy.

Conservation of Energy for a Thermal Process

For a thermodynamic process in which a system is initially in an equilibrium state with internal energy $\overline{E}_i$ and finally is in an equilibrium state with internal energy $\overline{E}_f$, the law of conservation of energy gives

$$\overline{E}_f - \overline{E}_i = \textit{energy transferred to system during process} \qquad (18.5)$$

As in Section 18.2, W denotes the macroscopically measurable work done *by* the system on its surroundings during the process. Then the macroscopic work done *on* the system is $-W$, and this is the energy transferred to the system by macroscopically measurable external forces. The system may also gain or lose energy by direct transfer without doing macroscopic work. This energy transfer to the system is denoted by the symbol Q and is called the *heat for the process.* The energy transferred to the system during the process thus consists of two parts, Q and $-W$:

$$\overline{E}_f - \overline{E}_i = Q - W \qquad (18.6)$$

This expression of the law of conservation of energy for a thermal process is called the *first law of thermodynamics.*

The heat Q for a process is defined by Eq. 18.6 as the energy which is transferred to the system without the performance of macroscopically measurable work; that is, Q is the energy transferred because of a temperature difference between the system and its surroundings. In Eq. 18.6, Q is counted positive for energy transferred to the system and negative for energy taken from the system. Since the heat Q is an energy, the SI unit for heat is of course the SI energy unit, the joule.*

The Internal Energy Function

The change of internal energy $\overline{E}_f - \overline{E}_i$, depends only on the initial and final states and has the same value for all processes connecting these states. The internal energy $\overline{E}$ is what is called a *state function:* in any equilibrium state the system's internal energy has a definite value. If the equilibrium state is specified by the temperature T and the pressure P, then $\overline{E}$ is some definite mathematical function of T and P that we denote by $\overline{E}(T,P)$. This state function can be specified by a formula in T and P, by graphs, or by a table such as Table 18-1 giving the value of $\overline{E}$ at various temperatures and pressures. In thermodynamics only

---

*In the older literature of science and engineering, we find heat measured in terms of the *calorie,* a unit that was introduced in the eighteenth century before it was recognized that heat is a form of energy. The calorie is now defined in terms of the joule. By international agreement the thermochemical kilocarlorie (kcal) is defined by

$$1 \text{ kcal} = 4184 \text{ J}$$

changes in internal energy are measured. Consequently internal energy functions are specified simply by giving the change in internal energy from any convenient reference condition (97°C and 1 atm in Table 18-1).

With given initial and final states, the values of W and Q for a process depend upon the way the process is carried out. The symbols W and Q have a meaning only with reference to a specified process. In particular, W and Q are *not* state functions and we cannot speak of the "heat contained by the system" or the "work contained by the system." The first phrase is an unfortunate remnant of the old caloric theory of heat (in which the phenomena of heat were attributed to a light indestructible fluid called caloric). Although the phrase "the heat in a body" still crops up frequently, particularly in popular literature, the word "heat" is then being used colloquially in place of the correct scientific term, which is "the internal energy of the body."

The First Law in Differential Form

For an infinitesimal process, the first law of thermodynamics is written in the differential form

$$d\overline{E} = dQ - dW \tag{18.7}$$

The differential symbols for the heat for the process dQ and the work for the process dW have lines drawn through them to emphasize that these quantities are *not* the differentials of state functions, there being no state functions Q or W.

For a quasi-static change of volume dV, the work dW is equal to $P\,dV$, and the first law of thermodynamics gives

$$d\overline{E} = dQ - P\,dV \tag{18.8}$$

Experimental Measurement of Internal Energy

A system can be thermally isolated by enclosing it within thermal insulators such as glass wool or asbestos. Then, unless a very long time is involved, the heat for any process occurring within the system is negligible. An idealized

Table 18-1 Internal Energy per Kilogram of H_2O Molecules at 1.00 atm When Water at 97°C Is Assigned Zero Internal Energy

	Temperature (°C)	Internal energy per kilogram (kJ/kg)
Water	97	0.0
Water	98	4.2
Water	99	8.4
Water	100	12.6
Steam	100	2099

process in which the heat for each portion of the process is zero is called an *adiabatic* process. For an adiabatic process, $Q = 0$ and the first law of thermodynamics reduces to

$$\overline{E}_f - \overline{E}_i = -W \qquad \text{(adiabatic process)} \qquad (18.9)$$

Consequently data for the internal energy tables (such as Table 18-1) for a system can be obtained experimentally by measuring the macroscopic work W done during an adiabatic process and measuring the system's initial and final temperatures and pressures.

In the years between 1840 and 1868, J. P. Joule (1818–1889) performed many such experiments, one of which is depicted schematically in Figure 18-4. While the suspended mass m falls slowly through a distance Δy, it loses potential energy $mg\Delta y$. The falling weight causes a paddle wheel to rotate and churn the water. Since the macroscopic work done *on* the water is $mg\Delta y$, the macroscopic work W done by the water in this process is $-mg\Delta y$. The initial and final temperatures are noted and the pressure is recorded. Assuming that the process is adiabatic, we can calculate the change in the water's internal energy from

$$\overline{E}(T_f,P) - \overline{E}(T_i,P) = mg\Delta y$$

where P is the pressure of the atmosphere. Joule's experiments give the result that for 1.00 kg of water at a pressure of 1.00 atm, there is an increase in internal energy of about 4.2 kJ for each 1 K (1 C°) rise in temperature for temperatures from 0°C to 100°C; that is for a mass M of water

$$\overline{E}_f - \overline{E}_i = (4.2 \text{ kJ/kg·K})M(T_f - T_i)$$

Experiments such as the foregoing indicate how changes in the internal energy of any system can be measured. But the great historical significance of Joule's experiments lies in the fact that these experiments also provide strong quantitative evidence for our modern notions of energy; namely:

1 Energy is conserved.
2 In heating a system we are simply increasing its internal energy.

These two ideas gained widespread acceptance after Joule's demonstration that the amount of macroscopic mechanical work required to change the state of a

Figure 18-4 The falling weight causes the paddle wheel to rotate and churn the water. The temperature rise $T_f - T_i$, associated with the loss of potential energy $mg\Delta y$, is measured by the thermometer.

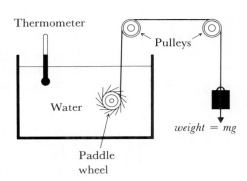

thermally insulated system is independent of how the work is done and depends only on the initial and final states of the system.

Example 1   One kilogram of H_2O molecules is contained in a large cylinder fitted with a piston. The water temperature is 100°C and the pressure is 1.000 atm (1.013×10^5 Pa). Heat is applied until the water is entirely converted into steam at the same temperature and pressure. During this process the change in the volume occupied by the H_2O molecules is 1.672 m³. Find the heat absorbed by the H_2O molecules in this process.

Solution   We select as the system the 1.00 kg of H_2O molecules within the cylinder. During the expansion at constant pressure, this system does work on its surroundings given by

$$W = P(V_f - V_i) = (1.013 \times 10^5 \text{ Pa})(1.672 \text{ m}^3) = 169.4 \text{ kJ}$$

From Table 18-1, the change in this system's internal energy is

$$\overline{E}_f - \overline{E}_i = (2099 - 12.6) \text{ kJ} = 2086 \text{ kJ}$$

The first law of thermodynamics gives

$$Q = (\overline{E}_f - \overline{E}_i) + W = 2256 \text{ kJ}$$

Since the sign is positive, the system absorbs heat from its surroundings. (The quantity that we have evaluated is called the *latent heat of vaporization* of water; it is discussed in Section 19.4.)

Example 2   The piston of Example 1 is clamped so that the volume of the cylinder's contents remains constant. Now heat is applied and the pressure of the steam and water within the cylinder rises to 10.0 atm. The piston is now unclamped and allowed to move out slowly with the pressure maintained at 10.0 atm as heat continues to be applied. When the steam has expanded to the same volume attained in Example 1, the cylinder is again clamped. Now heat is given off by the steam until its temperature falls to 100°C and the pressure is again 1.000 atm. Find the work and the heat for this entire process.

Solution   In the initial state there is 100% water at 100°C and at a pressure of 1.000 atm. In the final state there is 100% steam at 100°C and at a pressure of 1.000 atm. The system therefore has the same initial and final states as in Example 1, so the initial and final internal energies are the same as in Example 1. Consequently, for the entire process,

$$\overline{E}_f - \overline{E}_i = 2086 \text{ kJ}$$

as in Example 1.

No work is done during the processes in which the volume is constant. During the expansion at a pressure of 10.0 atm we obtain, from Eq. 18.3, a work which has a magnitude ten times that of Example 1. Thus for the entire three-stage process the work done by the system of H_2O molecules is

$$W = 0 + 1694 \text{ kJ} + 0 = 1694 \text{ kJ}$$

The first law of thermodynamics gives

$$Q = \overline{E}_f - \overline{E}_i + W = 2086 \text{ kJ} + 1694 \text{ kJ} = 3780 \text{ kJ}$$

The system absorbs almost twice as much heat in this process as it did in the process of Example 1.

Examples 1 and 2 illustrate the important difference between a *state function* such as $\overline{E}$ and the process quantities Q and W. When the transformation of a system from a given initial state (such as water, 100°C, 1 atm) to a given final state (steam, 100°C, 1 atm) is accomplished by *different* processes, *different* amounts of heat and work are involved. But the change in the system's internal energy is the same for both processes because $\overline{E}_f - \overline{E}_i$ depends only on the initial and final states.

18.4 Heat Capacities

Knowledge of certain aspects of the thermal behavior of a system, gained by experiment or by theory, is conveniently expressed by specifying what is called the *heat capacity* of the system. The system's heat capacity C for a specified process is defined by

$$dQ = C \, dT \qquad \text{or} \qquad C = \frac{dQ}{dT} \qquad (18.10)$$

where dQ is the heat for the process and dT is the temperature change of the system. When the system is a homogeneous substance, dQ is proportional to the amount of the substance which can be given by specifying the number of moles n, the mass m, or the number of particles N. Quantities independent of the amount and thus characteristic of the type of substance are:

1   The heat capacity per mole or the molar heat capacity, $c = C/n$.
2   The heat capacity per unit mass or the specific heat capacity, $c' = C/m$.
3   The heat capacity per particle, $c'' = C/N$.

A given system has different heat capacities for processes performed in different ways. Of particular interest is the heat capacity for a process in which the system is confined to a constant volume. This heat capacity is called the heat capacity at constant volume and is denoted by C_V. If the volume remains constant, $dV = 0$ and the first law of thermodynamics, $d\overline{E} = dQ + P \, dV$, gives $d\overline{E} = dQ$; the heat for this process serves only to change the internal energy since no work is done. Using $dQ = d\overline{E}$ in Eq. 18.10, we obtain

$$C_V = \left(\frac{\partial \overline{E}}{\partial T} \right)_V \qquad (18.11)$$

The subscript V signifies that the partial derivative is to be evaluated for a constant value of V. It is necessary to show explicitly the quantity that is assumed to be constant because the internal energy of a system usually depends not only on temperature but also on other quantities such as V or P that are used to specify the state of the system, and the value of the derivative $\partial \overline{E}/\partial T$ depends on which of these quantities is maintained constant; that is, in general $(\partial \overline{E}/\partial T)_V$ is not the same as $(\partial \overline{E}/\partial T)_P$.

In many experiments, it is the pressure rather than the volume that is kept constant. If the pressure is constant throughout a process, the work is

$$dW = P \, dV = d(PV) \qquad (P \text{ constant})$$

and the first law of thermodynamics then gives

$$dQ = d(\overline{E} + PV) \qquad (P \text{ constant})$$

The heat for a process at constant pressure is seen to be the change in the quantity

$$H = \overline{E} + PV$$

This state function H is called the *enthalpy* of the system. Since, at constant pressure, $dQ = dH$, the heat capacity at constant pressure is the partial derivative

$$C_P = \left(\frac{\partial H}{\partial T}\right)_P \qquad (18.12)$$

Several specific heat capacities that are often required in laboratory work or engineering practice are given in Table 18-2.* Average heat capacities in the temperature range from 0°C to 100°C can be measured using a water calorimeter (Figure 18-5) and following a procedure called the method of mixtures which is illustrated in the following example.†

Example 3   The inner can in the calorimeter (Figure 18-5) is made of aluminum (specific heat capacity = 0.91 kJ/kg·K) and has a mass of 0.400 kg. The can contains 0.140 kg of water. This water is thoroughly stirred and its temperature is measured to be 10.0°C. A 0.200-kg block of a new alloy, the substance under investigation, is heated to 100.0°C in a steam bath and then

---

*Historically the average specific heat capacity of water in the temperature range from 14.5°C to 15.5°C at a pressure of 1 atm was assigned the value of exactly one kcal/kg·C°. This determined the size of the kilocalorie. In the modern definition of the thermochemical kilocalorie (1 kcal = 4184 J) the size of the kilocalorie has been defined to yield values that are in agreement to better than 0.1% with values obtained using the older definition.

†Modern calorimetry employs the Joule heating (Chapter 25) by an electric current to measure specific heat capacities at temperatures ranging from 0.1 K to the melting point. A heater coil of resistance wire is wound around the sample under investigation. A current I is established in the heater coil of resistance R for a short time t and thereby supplies an energy I^2Rt to the sample. The temperature rise ΔT of the sample is measured. The specific heat capacity is then calculated from

$$I^2Rt = mc'\Delta T$$

quickly transferred to the calorimeter. The water is again stirred and the common temperature of the alloy and water is found to be 30.0°C. The average value of the specific heat capacity of water between 10°C and 30°C is 4.18 kJ/kg·K. Find the specific heat capacity c'_a of this alloy.

Solution   We assume that the heat insulating jacket surrounding the inner calorimeter container reduces the heat loss to the surroundings to a negligible amount. Then

$$heat\ given\ off\ by\ alloy = heat\ absorbed\ by\ water$$

$$+\ heat\ absorbed\ by\ calorimeter$$

We evaluate each term of this equation using $Q = mc'\Delta T$, which gives the

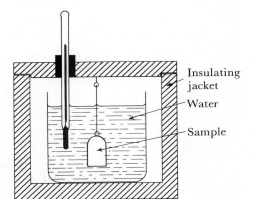

Figure 18-5   A water calorimeter. The sample whose specific heat capacity is under investigation is immersed in the water.

Table 18-2 Specific Heat Capacities of Common Substances at a Pressure of 1 atm and Ordinary Temperatures

Substance	Specific heat capacity (kJ/kg·K)
Hydrogen	20
Water*	4.2
Ice	2.1
Window glass	0.84
Iron	0.50
Copper	0.39
Mercury	0.14

*The specific heat capacity of water is 0.5% lower near 35°C and rises to values 0.4% higher at 0°C and at 100°C, but except where a more precise value is specified, we will use 4.2 kJ/kg·K in worked examples and questions.

heat absorbed in a temperature rise or the heat given off in a temperature drop:

$$c_a'(0.200 \text{ kg})(70.0 \text{ K}) = (0.140 \text{ kg})(4.18 \text{ kJ/kg·K})(20.0 \text{ K})$$
$$+ (0.400 \text{ kg})(0.91 \text{ kJ/kg·K})(20.0 \text{ K})$$

This yields

$$c_a' = 1.36 \text{ kJ/kg·K}$$

as the average value of the alloy's specific heat capacity in the temperature range from 30°C to 100°C.

Heat Capacities of Ideal Gases

The discussion in Section 17.4 shows that each molecule of an ideal gas has an average translational kinetic energy $\frac{3}{2}kT$ and an average internal energy $\bar{\epsilon}_{int}(T)$ which depends on the temperature. The average total energy of a molecule of an ideal gas is therefore

$$\bar{\epsilon} = \tfrac{3}{2}kT + \bar{\epsilon}_{int}(T)$$

The internal energy of an ideal gas of N such molecules is

$$\overline{E} = \tfrac{3}{2}NkT + N\bar{\epsilon}_{int}(T) \qquad (18.13)$$

According to this expression, internal energy of an ideal gas is independent of the volume of the container and is a function of the temperature only: that is,

$$\overline{E} = \overline{E}(T) \qquad (18.14)$$

To understand this result, consider the effect of changing the volume of the container. When the volume changes, the average separation of the gas molecules changes and, if the molecules interact with each other, the potential energy associated with this interaction will change. Consequently, for a sufficiently dense real gas, the internal energy will depend not only on the temperature but also on the volume: $\overline{E} = \overline{E}(T,V)$. But for an ideal gas, because the potential energy of molecular interaction is assumed to be negligible, $\overline{E}$ is not affected by changes in V.

For an ideal gas, since $\overline{E}$ depends only on T and is independent of V, the partial derivative $(\partial \overline{E}/\partial T)_V$ is the same as the derivative $d\overline{E}/dT$, and

$$C_V = \frac{d\overline{E}}{dT} \qquad \text{(ideal gas)} \quad (18.15)$$

There is a simple relationship between C_P and C_V for an ideal gas. The enthalpy of an ideal gas is

$$H = \overline{E} + PV = \overline{E}(T) + NkT$$

which is a function of temperature only. Therefore

$$C_P = \left(\frac{\partial H}{\partial T}\right)_P = \frac{d\overline{E}}{dT} + Nk$$

Since $C_V = d\overline{E}/dT$, we have

$$C_P = C_V + Nk = C_V + nR \qquad \text{(ideal gas)} \quad (18.16)$$

When the internal energy function is known, C_V can be calculated and then C_P determined from Eq. 18.16.

Heat Capacities of Monatomic Ideal Gases

For molecules consisting of a single atom (the monatomic gases: helium, neon, argon, krypton, xenon, radon) the only part of the internal energy function $\overline{E}$ that is affected by the gas temperature is the translational kinetic energy. At usual terrestrial temperatures, $\frac{3}{2}kT$ is much less than the approximately 10-eV energy required to allow one of these atoms to make a transition from its ground state to its first excited state. Therefore each of the atoms of a monatomic gas acquires an average translational kinetic energy $\frac{3}{2}kT$, but no internal excitation of the atom occurs. Consequently for a monatomic ideal gas, Eq. 18.13 becomes

$$\overline{E} = \tfrac{3}{2}NkT + constant \qquad \text{(monatomic ideal gas)}$$

Then

$$C_V = \frac{d\overline{E}}{dT} = \tfrac{3}{2}Nk = \tfrac{3}{2}nR \qquad (18.17)$$

$$C_P = C_V + nR = \tfrac{5}{2}Nk = \tfrac{5}{2}nR \qquad (18.18)$$

The ratio of these heat capacities is

$$\gamma = \frac{C_P}{C_V} = \frac{5}{3} \qquad (18.19)$$

Experiments with inert gases give results in good agreement with these predictions.

Example 4   Find the relationship between the pressure and volume in an ideal gas during a quasi-static adiabatic process. Assume that $\gamma = c_P/c_V$ is constant throughout the process.

Solution   For this process we have $dQ = 0$, $dW = P\,dV$, and $d\overline{E} = nc_V\,dT$ (from Eq. 18.15). The first law of thermodynamics therefore gives

$$nc_V\,dT = -P\,dV$$

From the equation of state of an ideal gas we have

$$nR\,dT = P\,dV + V\,dP$$

Eliminating dT between these two equations we obtain

$$-\frac{R}{c_V}P\,dV = P\,dV + V\,dP$$

Using $R/c_V = (c_P - c_V)/c_V = \gamma - 1$, we find

$$\frac{dP}{P} + \gamma \frac{dV}{V} = 0$$

Integration of this equation gives

$$\ln P + \gamma \ln V = constant$$

Therefore the equation for an adiabatic process of an ideal gas is

$$PV^\gamma = constant \tag{18.20}$$

For example, if the initial state is specified by (P_i, V_i) and the final state by (P_f, V_f), then

$$P_i V_i^\gamma = P_f V_f^\gamma$$

18.5 Equipartition of Energy

A system's internal energy can be computed in a simple way using the *equipartition theorem* which states: *When a system obeying Newtonian mechanics is in equilibrium at the Kelvin temperature T, every independent quadratic term in the expression for the system's energy has an average value equal to $\frac{1}{2}kT$.*

For example, in a system of molecules of mass m, each molecule has a translational kinetic energy

$$\tfrac{1}{2}mv^2 = \tfrac{1}{2}mv_x^2 + \tfrac{1}{2}mv_y^2 + \tfrac{1}{2}mv_z^2$$

and, according to the equipartition theorem, in equilibrium each of these three independent quadratic terms has an average value equal to $\frac{1}{2}kT$:

$$\overline{\tfrac{1}{2}mv_x^2} = \overline{\tfrac{1}{2}mv_y^2} = \overline{\tfrac{1}{2}mv_z^2} = \tfrac{1}{2}kT$$

Consequently

$$\overline{\tfrac{1}{2}mv^2} = \tfrac{3}{2}kT \tag{18.21}$$

and the equipartition theorem asserts that this result, which we had deduced for an ideal gas in Section 17.2, is valid for the particles of *any* system in equilibrium at temperature T.

The proof of the equipartition theorem is beyond the scope of this book, but we must emphasize that the condition for the validity of the theorem is that the particles obey Newtonian mechanics. Consequently, we should expect that the excitation of a given type of energy to its full value $\frac{1}{2}kT$ will occur only when quantum phenomena such as the existence of discrete energy levels are relatively unimportant. This is the case when kT is large enough to be approximately equal to or greater than the spacing of the energy levels associated with the excitation in question.

Heat Capacities of Solids

In a simple solid (for example: copper, gold, aluminum, or diamond) the interatomic forces are such that the situation of stable mechanical equilibrium of the solid is achieved when the atoms are located at regular positions in a crystal lattice. For small displacements, the resultant force acting on an atom is proportional to its displacement from its equilibrium position. Consequently, an atom executes simple harmonic oscillations about its equilibrium position.* With origin at this equilibrium position and proper choice of the coordinate axes, the energy associated with these vibrations has the form

$$\epsilon = \epsilon_x + \epsilon_y + \epsilon_z$$

with

$$\epsilon_x = \tfrac{1}{2}mv_x^2 + \tfrac{1}{2}k_x x^2$$

where k_x is the force constant for displacement of an atom of mass m along the X axis; there are analogous expressions for ϵ_y and ϵ_z.

If a solid is in equilibrium at temperature T, the equipartition theorem implies that, for each atom

$$\bar{\epsilon}_x = \overline{\tfrac{1}{2}mv_x^2} + \overline{\tfrac{1}{2}k_x x^2} = \tfrac{1}{2}kT + \tfrac{1}{2}kT = kT$$

and similarly $\bar{\epsilon}_y = \bar{\epsilon}_z = kT$. Each atom therefore has an average vibrational energy $3kT$. The internal energy of a solid consisting of N atoms then is

$$\overline{E} = 3NkT = 3nRT$$

The molar heat capacity at constant volume is given by

$$c_V = \frac{1}{n}\left(\frac{\partial \overline{E}}{\partial T}\right)_V = 3R \tag{18.22}$$

a result which is remarkably simple and independent of the nature of the atoms and the geometry of the crystal.

This prediction of the equipartition theorem is in agreement with the empirical *law of Dulong and Petit*: *At sufficiently high temperatures, all solids have a molar heat capacity c_V equal to $3R$.* Experimental values of c_V for various substances at $T = 298$ K are given in Table 18-3. Figure 18-6 shows the temperature dependence of c_V which is typical of crystalline solids. At sufficiently low temperatures c_V is less than $3R$ and approaches zero as T approaches zero.

The "freezing out" of various excitations as the temperature is lowered was a great mystery from the point of view of Newtonian mechanics. In 1907 Einstein recognized that this could be understood if the possible changes in the energy of an oscillating atom could not be arbitrarily small. Heat capacities thus provided early evidence of the quantum mechanical behavior of matter. In quantum mechanics any bound system such as an oscillating particle has certain

---

*In a rigorous analysis, the simultaneous oscillations of all the atoms must be considered. However the final result, Eq. 18.22 is valid since it is independent of the masses and of the force constants.

Table 18-3 Molar Heat Capacities c_V at $T = 298$ K

Solid	c_V J/mol·K
Aluminum	23.4
Carbon (diamond)	6.1
Copper	23.8
Gold	24.5
Lead	24.8
Silver	24.4
Tungsten	24.4

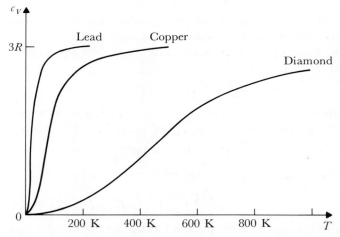

Figure 18-6 Temperature dependence of the molar heat capacity at constant volume c_V. Curves for all metals and nonmetals lie within the curves given for lead and diamond. At high temperatures, these curves generally approach the Dulong and Petit value $3R$, lead attaining this value at about 200 K and diamond at temperatures well above 2000 K.

discrete energy levels as the only possible values of its energy. At temperatures low enough so that kT is much less than the energy required for transitions from one energy level to another, excitation of oscillations is a rare event.

Summary

☐   If a macroscopic system is isolated it eventually reaches an equilibrium state that can be described in terms of a few macroscopic quantities, such as temperature, pressure, and volume, that can be determined by direct experimental measurement.

☐ In an equilibrium state there is mechanical, thermal, and chemical equilibrium.

☐ The work dW done by a fluid on its surroundings during a quasi-static volume change dV at pressure P is

$$dW = P\,dV$$

☐ The internal energy $\overline{E}$ of a system is a state function. In a process in which the internal energy of the system changes from an initial value $\overline{E}_i$ to a final value $\overline{E}_f$, the first law of thermodynamics states

$$\overline{E}_f - \overline{E}_i = Q - W$$

where W is the work done by the system on the surroundings during the process. This equation defines the heat Q for the process. For an infinitesimal process

$$d\overline{E} = dQ - dW$$

The values of the heat and the work depend on the process. Q and W are not state functions; therefore dQ and dW are not differentials of functions.

☐ A system's heat capacity C for a specified process is

$$C = \frac{dQ}{dT}$$

where dQ is the heat for the process and dT is the temperature change of the system. For n moles or a mass m of a homogeneous substance, the molar heat capacity is $c = C/n$ and the specific heat capacity is $c' = C/m$.

$$C_V = \left(\frac{\partial \overline{E}}{\partial T}\right)_V \quad \text{and} \quad C_P = \left(\frac{\partial H}{\partial T}\right)_P$$

where $H = \overline{E} + PV$. The state function H is called the enthalpy.

☐ For an ideal gas the internal energy is a function of the temperature only, $\overline{E} = \overline{E}(T)$. It follows that $C_V = d\overline{E}/dT$ and $C_P = C_V + nR$.

☐ The theorem of equipartition of energy states that when a system obeying Newtonian mechanics is in equilibrium at Kelvin temperature T, every independent quadratic term in the expression for the system's energy has an average value equal to $\frac{1}{2}kT$.

☐ According to the empirical law of Dulong and Petit, at sufficiently high temperatures, all solids have a molar heat capacity

$$c_V = 3R$$

Questions

1   Some gas is confined to the left half of a rigid box by a thin partition (Figure 18-7). Within the right half there is a vacuum. The partition is suddenly ruptured and the gas undergoes what is called a *free expansion:*
    (a) While it is expanding, is the gas in an equilibrium state? Explain.
    (b) Describe the approach to equilibrium and the equilibrium attained.

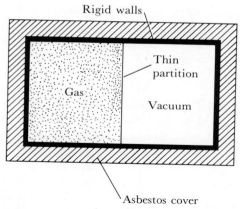

Figure 18-7   When the partition is ruptured, the gas undergoes a free expansion.

2   In the free expansion of Question 1, do the contents of the box (the gas) perform any work on objects outside the *rigid* box?

3   Gas at a pressure of 8.0×10^5 Pa is confined within a cylinder fitted with a piston of radius 3.0×10^{-2} m. As the gas is warmed, the piston moves out in such a way that the pressure within the cylinder is kept constant. How much work is done on the piston by the gas during an expansion in which the piston moves 0.12 m?

4   One kilogram of ice at 0°C is placed in a cylinder. A piston maintains a constant pressure of 1.00 atm (1.01×10^5 Pa) on the cylinder contents. Heat is applied and the ice melts, turning into water at 0°C. As the ice melts, the piston moves inward. When all the ice has melted, the volume of the cylinder contents has decreased by 8.6×10^{-5} m³. Find the work done by the piston on the contents of the cylinder.

5   In Figure 18-8, points *A*, *B*, *C*, and *D* represent different states of a fluid. Find the work done by this fluid in a quasi-static process which is represented by:
   (a) Path $A \to B \to D$.
   (b) Path $A \to C \to D$.
   (c) Path $A \to D$ (straight line).

Figure 18-8

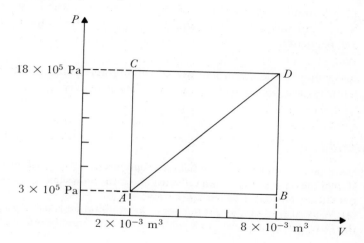

6 In Figure 18-8, what is the work done by the fluid in a quasi-static process represented by:

  (a) The closed triangular path $A \rightarrow C \rightarrow D \rightarrow A$.

  (b) The closed triangular path $A \rightarrow D \rightarrow B \rightarrow A$.

  (c) The closed rectangular path $A \rightarrow C \rightarrow D \rightarrow B \rightarrow A$.

7 A cylinder fitted with a piston contains 0.10 mol of air at room temperature (20°C). The piston is pushed in so slowly that the air within the cylinder remains essentially in thermal equilibrium with the surroundings. Find the work done by the air within the cylinder if the final volume is one-half the initial volume.

8 What is the increase in the internal energy of a system in a process in which 8.0 kJ of work are done *on* the system and 12.0 kJ of heat are absorbed by the system?

9 The internal energy of a system increases by 3.0 kJ in a process in which 5.0 kJ of heat are absorbed by the system. What is the work done *by* the system in this process?

10 For the process described in Question 4, the internal energy increase is given by

$$\overline{E}_f - \overline{E}_i = 333 \text{ kJ/kg}$$

Find the heat for the process.

11 For the system of Question 4, assume the initial and final temperature and pressure are still 0°C and 1 atm, but we arrange to have the compression take place with a high constant pressure of 1.0×10^4 atm:

  (a) Find the change in internal energy of the cylinder contents.

  (b) Find the work done by the cylinder contents in this process.

  (c) Find the heat for the entire process.

12 Use the words *work, heat,* or *internal energy* to correctly fill in the blanks:

  (a) Water has a greater _____ at 20°C than at 0°C at the same pressure.

  (b) The _____ of two stones is increased by rubbing them together.

13 A system is slowly compressed by a constant pressure of 5.0×10^5 Pa in a process in which the system gives off 2.0 kJ of heat and the internal energy of the system remains unchanged. Find the decrease in the volume of the system.

14 **(a)** How much heat is required to raise the temperature of 3.0 kg of water from 15°C to 20°C in a process for which the specific heat capacity of water is 4.18 kJ/kg·K?

  (b) What temperature rise would the same amount of heat produce in the same mass of mercury? (The specific heat capacity of mercury is 0.14 kJ/kg·K.)

15 A lead bullet traveling at 500 m/s strikes a board and comes to rest. Calculate the temperature rise of the bullet, assuming that the heat loss to the board is negligible. (The specific heat capacity of lead is 0.13 kJ/kg·K.)

16 What is the temperature rise in water that passes over Niagara Falls, tumbles 50 m down, and splashes into the river below?

17 A 10.0-kg copper block at 100°C is placed in 10.0 kg of water at 20°C. What is the final temperature of the copper and water? (The specific heat capacity of copper is 0.39 kJ/kg·K.)

18 The calorimeter of Example 3, with the calorimeter containing 0.120 kg of water at 18.0°C, is used to determine the specific heat capacity of a metal whose mass is 0.300 kg. The metal is heated to 98.0°C and placed in the calorimeter. The final temperature of the water and metal is 28.0°C. Find the specific heat capacity of the metal.

19   A 0.600-kg sample of an alloy at a temperature of 100°C is dropped into 0.500 kg of water at 15.1°C which is contained in a 0.200-kg calorimeter made of the same alloy. The water, the calorimeter, and the sample reach a temperature of 20.1°C. Find the specific heat capacity of the alloy.

20   **(a)** Compare the heat capacities of equal masses of water and lead. (Specific heat capacity of lead is 0.13 kJ/kg·K.)
     (b) Compare the heat capacities of equal volumes of water and lead. (Specific gravity of lead is 11.3.)

21   The temperature of 0.20 mol of helium is raised from 73 K to 273 K. Find the heat absorbed by the helium if the process is performed quasi-statically at:
     (a) Constant volume.
     (b) Constant pressure.

22   A cylinder fitted with a piston contains 1.00×10^{22} molecules of argon. This gas is heated from 27°C to 77°C with the pressure maintained constant at the value of 2.0×10^5 Pa:
     (a) Calculate the work done by the gas in this expansion.
     (b) Find the change in the internal energy of the argon.
     (c) Find the heat for this process.

23   Assume that the heat capacity C_V of an ideal gas is constant. Show that the work done by this gas during a quasi-static adiabatic expansion is given by

$$W = C_V(T_i - T_f)$$

24   A monatomic ideal gas, initially at 0°C, undergoes a quasi-static adiabatic expansion to a volume which is three times its initial volume. Find the final temperature.

25   Show that the equation for a quasi-static adiabatic process of an ideal gas,

$$PV^\gamma = constant$$

     implies that

$$TV^{(R/c_V)} = constant$$

26   The fireball, about 100 ms after the detonation of a uranium fission bomb, consists of a sphere of gas with a radius of about 15 m at a temperature of 3×10^5 K. Estimate the radius at which the temperature of the expanding fireball will have dropped to 3×10^3 K. (Assume that $\gamma = 1.4$.)

27   A mole of lead has a mass of 0.207 kg. At 298 K the specific heat capacity of lead at constant volume is 0.120 kJ/kg·K. Calculate the molar heat capacity and compare the answer to the Dulong and Petit value.

Supplementary Questions

S-1   Use Einstein's equation (Eq. 11.14),

$$E_{int} = Mc^2$$

     to calculate the relative change in mass $\Delta M/M$ that occurs when a mass M of water at 100°C is converted into a steam at 100°C. (Assume that the pressure is 1 atm and use the data in Table 18-1.) Notice that the change in mass is too small to be detected by direct measurements of mass.

S-2 The Saint Lawrence River drops 75 m as it flows from Lake Ontario to the sea. A letter published in a Montreal newspaper suggested that freezing of the river be prevented by establishing a chain of hydroelectric plants along the river and by using their output to power electric heaters immersed in the river. Evaluate the temperature rise that could be produced in this way and comment on the merits of the proposal.

S-3 The free expansion of Question 1 is, to an excellent approximation, an adiabatic process:

(a) Is it a quasi-static adiabatic process? Can the relationship found in Example 4,

$$PV^\gamma = constant$$

be applied to this process?

(b) Show that the initial and final internal energies of the gas are equal.

(c) If the gas is an ideal gas, so that $\overline{E}$ is independent of the volume V and is an increasing function of the temperature, show that the initial and final gas temperatures are equal.

S-4 Suppose that the PV diagram of Figure 18-8 represents the states of n moles of a monatomic ideal gas. Find the heat absorbed by the gas for the quasi-static process represented by:

(a) Path $A \rightarrow B \rightarrow D$.

(b) Path $A \rightarrow C \rightarrow D$.

(c) Path $A \rightarrow D$ (straight line).

S-5 Assume that the equation of state for a gas is van der Waals' equation,

$$\left(P + \frac{a}{V^2}\right)(V - b) = RT$$

where a and b are constants. Find an expression for the work done by this gas in a quasi-static isothermal expansion, in terms of the initial volume V_i, the final volume V_f, and the constants T, a, and b.

S-6 A vertical cylinder, fitted at its upper end with a heavy piston of weight Mg and cross section A, contains n moles of a monatomic ideal gas. Initially the piston is clamped; the Kelvin temperature of the gas is T_i and its volume is V_i. Then the piston is released. It falls a certain distance and then executes oscillations of diminishing amplitude. Eventually equilibrium is attained with the gas at a Kelvin temperature T_f and occupying a volume V_f. Assume that this compression of the gas is adiabatic (but *not* quasi-static):

(a) Find the final pressure of the gas, assuming that the pressure on the outside surface of the piston is negligible.

(b) Find the work W done by the gas in this compression, in terms of Mg, A, V_f, and V_i.

(c) Calculate V_f and T_f in terms of T_i, V_i, n, Mg, and A.

S-7 In a quasi-static adiabatic process, the pressure and volume of a gas change from (P_i, V_i) to (P_f, V_f). Show that the work done by the gas on its surroundings in this process is

$$W = \frac{P_i V_i - P_f V_f}{\gamma - 1}$$

where $\gamma \ (= C_P / C_V)$ is assumed to be constant.

S-8 A system of molecules, each of mass m, is in equilibrium at temperature T. In the CM-frame of this system, a molecule has a velocity vector $\mathbf{v}$ with components v_x, v_y, v_z. Use symmetry arguments and the equipartition theorem to evaluate the following average values:

(a) $\overline{v_x}$ **(b)** $\overline{v_x^2}$ **(c)** $\overline{v_x^3}$ **(d)** $\overline{v_x^2 + v_y^2}$

S-9 The surface of a solid in a reasonably good vacuum is covered by a single layer of molecules which are said to be *adsorbed* on the surface. The forces exerted by the atoms of the solid keep these adsorbed molecules from escaping. However, the adsorbed molecules are free to move around on the two-dimensional surface. Treating the adsorbed layer as a two-dimensional gas of molecules obeying classical mechanics, find the internal energy at temperature T of a mole of adsorbed molecules. What is the molar heat capacity of the adsorbed layer?

19.

thermal phenomena

This chapter surveys several thermal phenomena that have an immediate relevance in our daily lives. Applications in thermometry are given. The important fact that a macroscopic system, left undisturbed, will eventually reach an *equilibrium state* is emphasized by considering thermal equilibrium and phase equilibrium, which are attained, for instance, when ice and water coexist indefinitely.

19.1 Thermal Expansion

Solids

The atoms of a crystalline solid are arranged in a regular array whose geometry is determined by the interatomic forces. Each atom oscillates back and forth with an amplitude of about 0.1 Å and a frequency of about 10^{13} Hz. When the temperature of the crystal is raised, there is an increase in the average translational kinetic energy of the atoms and in the amplitude of their oscillations. The motion with larger amplitude brings about a change in the interaction between adjacent molecules, and this usually leads to a slightly greater average distance between atoms. Therefore the entire solid expands as its temperature is raised.

Experiments show that the increase ΔL in the length L of a solid when its temperature is raised by an amount ΔT is given by

$$\Delta L = \alpha L \Delta T \qquad (19.1)$$

where α, named the *coefficient of linear expansion*, has a value characteristic of the type of solid (Figure 19-1). To an approximation adequate for most applications, α can be assumed to be a constant independent of the temperature interval or the reference length appearing in Eq. 19.1. Values of α for several solids are shown in Table 19-1. The order of magnitude of the expansion of most solids is easy to remember, being about one millimetre per metre length per hundred kelvins.

Example 1 A steel railway track is 20.000 m long on a winter day when the temperature is $-15\,°C$. What is its length on a summer day when the temperature is $35\,°C$?

Solution The temperature increase is

$$\Delta T = 35\,°C - (-15\,°C) = 50\ C° = 50\ K$$

Therefore

$$\Delta L = \alpha L \Delta T = (12 \times 10^{-6}/K)(20.000\ m)(50\ K) = 0.012\ m$$

The new length is

$$L + \Delta L = 20.000\ m + 0.012\ m = 20.012\ m$$

This example illustrates the fact that engineers must take thermal expansion into consideration. Railway tracks would buckle if not enough space were left between sections to permit their length to increase by an amount $\alpha L \Delta T$ where

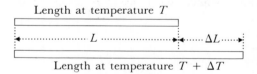

Length at temperature T

L ···· ΔL

Length at temperature $T + \Delta T$

Figure 19-1 Thermal expansion. When the temperature is raised by an amount ΔT, the length increases by an amount $\Delta L = \alpha L \Delta T$. In this figure the expansion has been exaggerated for clarity.

Table 19-1 Approximate Coefficients of Linear Expansion (α) Near 293 K

Substance	α (per kelvin)	Substance	α (per kelvin)
Steel	12×10^{-6}	Granite	8×10^{-6}
Brass	19×10^{-6}	Glass (Pyrex)	3×10^{-6}
Aluminum	24×10^{-6}	Glass (ordinary)	9×10^{-6}
Lead	29×10^{-6}	Brick	9×10^{-6}
Invar	0.9×10^{-6}	Hard rubber	80×10^{-6}

ΔT is the largest temperature increase that is anticipated. For the same reason, spaces must be left between the concrete slabs used in highway construction and between adjacent sections of large buildings.

Thermostats are made from thin strips of different metals which are welded or riveted together. When the temperature changes, one metal expands more than the other; thus the combination must bend in an arc (Figure 19-2).

Uneven heating and the consequent uneven expansion give rise to large forces within a solid. Ordinary glass shatters when heated unless it is warmed so slowly that appreciable temperature differences between different regions of the glass do not occur. Pyrex has the merit of having a lower coefficient of linear expansion than ordinary glass (Table 19-1) and thus is not subjected to such markedly different expansions at different temperatures.

Materials which are composed of millions of minute crystals oriented at random will expand equally in all directions. For such materials, a temperature increase ΔT will produce increases of any area A and any volume V given by

$$\Delta A = 2\alpha A \Delta T \tag{19.2}$$

and

$$\Delta V = 3\alpha V \Delta T \tag{19.3}$$

These equations can be deduced from $\Delta L = \alpha L \Delta T$ and are good approximations only when the fractional expansion ($\Delta A/A$ or $\Delta V/V$) is small. Area expansion is exploited when one loosens the metal lid of a glass jam jar by heating it under hot water.

Liquids

Liquids generally expand when heated. The approximate increase in a volume V associated with a temperature rise ΔT is given by

$$\Delta V = \beta V \Delta T \tag{19.4}$$

where β is called the *coefficient of volume expansion* of the liquid (Table 19-2).

Example 2   On a summer day when the temperature is 31°C, a steel gasoline tank at 20°C is filled with cold gasoline pumped from underground storage

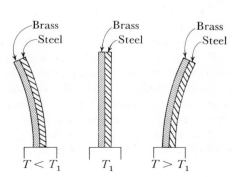

Figure 19-2   Principle of a thermostat element. Each metal strip expands a different amount when the temperature changes from T_1 to $T > T_1$.

Table 19-2 Coefficients of Volume
Expansion (β) at 293 K

Liquid	β (per kelvin)
Mercury	18×10^{-5}
Gasoline	95×10^{-5}
Benzene	124×10^{-5}
Ether, ethyl	166×10^{-5}

tanks where the temperature is also 20°C. What fraction of the gasoline will have overflowed when the gasoline and the tank temperatures reach 31°C?

Solution   From Eq. 19.3, the expansion of the volume V of the steel tank is

$$\Delta V_{\text{tank}} = (3 \times 12 \times 10^{-6}/\text{K})(11 \text{ K})V = 4.0 \times 10^{-4}V$$

The expansion of the volume V of the gasoline, from Eq. 19.4, is

$$\Delta V_{\text{gasoline}} = (95 \times 10^{-5}/\text{K})(11 \text{ K})V = 105 \times 10^{-4}V$$

Therefore

$$overflow\ volume = \Delta V_{\text{gasoline}} - \Delta V_{\text{tank}}$$

$$= (105 - 4) \times 10^{-4}V$$

$$= 0.0101V$$

About 1% of the gasoline overflows.

Anomalous Expansion

After having been exposed to numerous examples of thermal expansion, we may be led to believe that materials always expand when heated. This is not true. At some temperatures water contracts when heated, and so do certain other substances. Water is particularly interesting in this regard since its coefficient of volume expansion is negative at very low temperatures (about 0 K to 70 K), as well as in the region from 0°C to 4°C.

As for other substances, a certain type of iron expands as its temperature increases from 0 K up to about 1100 K, at which point it contracts slightly. (Subsequently it continues expansion with further temperature increase.) At 1100 K certain molecular rearrangements take place in the iron crystal, resulting in contraction. Other materials which at certain temperatures have negative expansion coefficients are silicon, selenium, tellurium, and a certain cobalt-iron-chromium alloy. Though relatively rare, there *is* such a phenomenon as contraction with increased temperature.

An interesting and simple demonstration of the negative thermal expansion coefficient of ordinary elastic bands can be done this way: suspend a small mass by means of a rubber band; then hold a flaming match near the rubber. The small mass will be seen to rise noticeably as the rubber contracts on being heated.

That water displays this anomalous thermal expansion has profound consequences for life on our planet. The volume of a given mass of water does not increase regularly as its temperature is increased. Instead the volume *decreases* as the water is warmed from 0°C to 4°C. Above 4°C the water expands when heated. The volume occupied by a given mass of water is therefore a minimum at 4°C.

The reason for the contraction of water as it is heated from 0°C to 4°C involves the structure of ice. Ice crystals have a rather open structure with empty spaces. When a chunk of ice melts, ice crystals collapse and the volume is reduced, but many ice crystals remain in water at 0°C. If the temperature is raised, more and more ice crystals collapse. This is the dominant effect up to 4°C.

These properties determine the sequence of events that occur in the cooling and freezing of lakes. The implications of Archimedes' principle (Section 15.6) are most easily understood in terms of water and ice *densities* rather than volumes occupied by a given mass. Two important facts are:

1   Ice is less dense than water. The buoyant force exerted by water on ice is therefore sufficient to cause the ice to float.
2   Water has a maximum density at 4°C. Below 4°C, cooler portions of water will be forced upward by the buoyant forces exerted by warmer portions. Therefore when the surface of a lake is cooled below 4°C, mechanical equilibrium will be attained with the coldest water at the surface and the temperature rising as the depth increases.

Consequently ice forms first at the surface and, because ice floats, it remains there. With continued cooling the ice thickens but, unless all the water freezes, the water temperature near the bottom will remain at 4°C. Most solids expand when melted and liquids expand when heated so that the solid and the coldest liquid sink. If water and ice behaved in the usual manner lakes would freeze from the bottom up and easily become a solid mass of ice in colder climates. The effects on marine life and on our weather would be drastic.

19.2 Heat Transfer

Thermal energy can be transferred from one place to another by three different mechanisms: conduction, convection, and radiation. *Conduction* is a relatively slow process in which thermal energy is transferred by electron motion or molecular interaction. *Convection* is a more rapid process which occurs when a portion of a heated fluid moves from one place to another. *Radiation* involves the emission and absorption of electromagnetic waves which travel at the speed of light.

Conduction

Conduction involves a transport of thermal energy without the conducting material itself being moved. Consider a thin slab of material (Figure 19-3) of thickness dx with faces of cross section A. One face is maintained at temperature T and the other at temperature $T + dT$, producing a *temperature gradient* dT/dx within the slab. After steady-state conditions have been reached we find, from measurement of the heat dQ that passes through such a slab in a time interval dt, that dQ is proportional to both dt and A, and it is also proportional to the temperature gradient dT/dx provided that this gradient is not too large. These experimental results are summarized in the fundamental law of heat conduction,

$$\frac{dQ}{dt} = -KA\frac{dT}{dx} \tag{19.5}$$

The negative sign corresponds to the fact that thermal energy flows in the direction in which the temperature decreases. K is a constant of proportionality called the thermal conductivity of the material.

Table 19-3 gives values of K for various substances. Metals, good conductors of electricity (Chapter 25), are also good thermal conductors. The electrons which account for a metal's conduction of electric charge (Section 43.4) effect

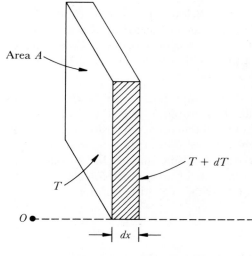

Area A

$T + dT$

T

O — — — — — — — — — — — X

dx

Figure 19-3 Slab of thickness dx and cross section A. Within the slab the temperature gradient is dT/dx.

Table 19-3 Thermal Conductivities at Ordinary Temperatures

Substance	kW/m·K	Substance	kW/m·K
Silver	0.42	Aluminum	0.020
Copper	0.38	Ice	2.2×10^{-3}
Gold	0.29	Brick	7×10^{-4}
Brass	0.11	Asbestos	8×10^{-5}
Steel	0.046	Air	2.3×10^{-5}

a rapid transfer of thermal energy from a hotter to a colder region. In solid insulators and semiconductors a transfer of thermal energy is accomplished by the interactions of neighboring atoms. Relatively slow thermal conduction, due to molecular collisions, occurs in nonmetallic liquids and gases.

Example 3   A metal pipe of length L carries steam at temperature T_1. The outer surface of the pipe is maintained at temperature T_2. The inner and outer radii of the pipe are R_1 and R_2, respectively, and the thermal conductivity of the metal is K. In the steady-state, find the rate dQ/dt at which heat flows radially outward through the pipe.

Solution   The rate at which heat flows outward through a cylindrical shell with inner radius R and outer radius $R + dR$ (Figure 19-4), from the fundamental law of heat conduction, is

$$\frac{dQ}{dt} = -K2\pi RL \frac{dT}{dR} \qquad (19.6)$$

In the steady-state, for any given volume element of a conducting material, there is no time variation of the volume or internal energy, and consequently, in any time interval, the heat entering the volume element has the same magnitude as the heat leaving the volume element. It follows that in Eq. 19.6 dQ/dt is a constant (independent of the value of R). Equation 19.6 yields

$$\int_{T_1}^{T_2} dT = -\frac{dQ/dt}{2\pi LK} \int_{R_1}^{R_2} \frac{dR}{R}$$

which implies

$$T_2 - T_1 = -\frac{dQ/dt}{2\pi LK} \ln \frac{R_2}{R_1}$$

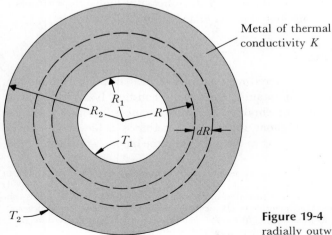

Figure 19-4   Heat is conducted radially outward through the metal.

Therefore the rate at which heat flows radially outward through the pipe is

$$\frac{dQ}{dt} = \frac{2\pi LK(T_1 - T_2)}{\ln (R_2/R_1)}$$

Convection

Convection consists of the actual motion of a portion of a heated fluid. Natural convection currents are set up if a heated region of fluid expands, becoming less dense than the surrounding fluid. Then, according to Archimedes' principle, the buoyant force exerted by the surrounding cooler fluid is greater than the weight of the heated portion. The heated fluid thus rises. Similarly, a cooler portion sinks. The thermal energy from a hot-water radiator is carried to other parts of a room by such natural convection currents. In forced convection, the fluid motion is either caused or increased by a blower or pump.

Thermal Radiation

All objects, such as a glowing tungsten wire in a light bulb, the surface of the sun, and even human skin, continuously radiate energy as electromagnetic waves.* The rate at which an object emits this thermal radiation increases rapidly as its temperature increases, being approximately proportional to T^4, where T is the Kelvin temperature of the object. At high temperatures, radiation is frequently the principal mechanism of heat loss.

The distribution of energy among the different wavelengths is characteristic of the temperature of the emitting surface. The predominant wavelengths of thermal radiation decrease when the temperature of the radiating object is increased. Below about 900 K, these wavelengths are too long to be detected by the eye, but such *infrared* electromagnetic waves do cause a sensation of heat in the skin. At 1000 K, an object appears dull red; evidently some radiation is visible. At still higher temperatures, the radiating object emits relatively more of the shorter wavelength blue light, and at sufficiently high temperatures it appears bluish-white. White-hot is hotter than red-hot.†

Heat Transfer in Humans

The human body provides interesting examples of heat transfer. Conduction, convection, and radiation maintain a remarkably constant body temperature in spite of extreme environmental temperature changes. A fairly uniform temperature of 37°C throughout the body's interior is achieved by forced convection, with the heart serving as the pump and the blood as the heater fluid.

*A detailed discussion of electromagnetic waves and of the electromagnetic spectrum is given in Chapter 35.
†The study of thermal radiation had a profound impact on the history of physics. Planck's constant h and the first evidence of the quantum mechanical nature of the world made their appearance in 1900 in the efforts made by Max Planck (1858–1947) to understand the way that electromagnetic energy is distributed among the different wavelengths in thermal radiation.

Heat is brought to the skin by convection and conduction within the body. The skin then loses heat to the surroundings. Some 8×10^3 kJ a day is a typical heat loss of a sedentary man.

The relative importance of the different mechanisms of heat loss varies greatly according to circumstances. Experiments show that over 50% of the heat loss from a dry nude in still air is in the form of infrared radiation; this radiation loss, however, amounts to only about 5% of the total heat loss by a person properly attired for outdoor activity in the winter. (The presence of perspiration and the temperature and dryness of the surrounding air greatly affect the rate at which heat is lost, because, for each kilogram of perspiration evaporated nearly 2.5×10^3 kJ of heat are absorbed from the body. Evaporation and the "latent heat" required to change water to vapor are discussed in Section 19.4.)

Dewar Flask

An instructive practical example of efforts to circumvent convection, conduction, and radiation is provided by the Dewar flask, or Thermos bottle (Figure 19-5). The flask is a double-walled glass vessel which is silvered on the inside. The space between the walls is evacuated to prevent convection and conduction. Because glass is a poor conductor, little heat is conducted through the glass walls over the neck. The silvered surfaces reflect most of the thermal radiation that would leave from the inside or enter from the outside.

19.3 Thermometers

Several of the phenomena involved in the operation of different practical thermometers have now been discussed. Before examining the details of these useful devices we pause to consider the familiar but fundamental concept of temperature.

Figure 19-5   Dewar flask.

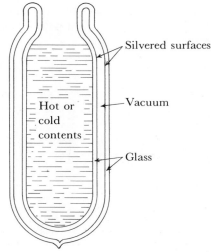

Silvered surfaces

Hot or
cold
contents

Vacuum

Glass

We have already discussed a microscopic interpretation of temperature as a measure of random translational kinetic energy of molecular motion. In Chapter 20 we give a very general, though somewhat abstract, definition of temperature in terms of quantities related to the microscopic description of matter. But it is interesting to put aside briefly all microscopic notions and consider temperature as a purely macroscopic quantity, a fundamental concept in the macroscopic science of thermal phenomena, thermodynamics.

The Concept of Temperature

Our sense of touch involves a temperature sense which enables us to distinguish hot bodies from cold bodies and even to decide on the order of hotness of several different objects. This life-saving temperature sense, although it is too subjective and unreliable to provide a basis for scientific work, does give us the useful intuitive notion that a temperature, which is a measure of the hotness of an object, can somehow be assigned.

The macroscopic temperature concept can be made precise by considering *thermal equilibrium*. Any object such as a copper bar which is insulated from its surroundings by an asbestos sheet soon reaches a state of thermal equilibrium in which its macroscopically observable properties, such as its length or volume, remain unchanged as time elapses. If an iron plate, also in equilibrium, is placed in contact with the copper bar, equilibrium is usually disturbed. Lengths, volumes, and other properties alter as time goes on. In this case we say the two objects are not in thermal equilibrium with each other. After a sufficiently long time the copper bar and the iron plate will reach a state of thermal equilibrium in which the observable properties of the objects are constant. We then say that these objects [A and B in Figure 19-6(a)] are in thermal equilibrium with each other.

We now consider several different macroscopic objects and examine thermal equilibrium between different pairs of these objects. [Figures 19-6 (b) and 19-6 (c)]. All our experience with such experiments is expressed concisely in the *zeroth law of thermodynamics*: *Two systems in thermal equilibrium with a third are in thermal equilibrium with each other.*

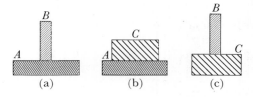

Figure 19-6   Thermal equilibrium, the zeroth law of thermodynamics, and temperature. (a) Objects A and B are in thermal equilibrium. (b) Objects A and C are in thermal equilibrium with the state of A the same in both cases. (c) Experiment shows that objects B and C then will be found to be in thermal equilibrium with each other. Objects A, B, and C have the same *temperature*.

The *temperature* of an object is identified in thermodynamics *as the property which determines whether or not the object will be in thermal equilibrium with other objects.* Objects in thermal equilibrium have the same temperature. When objects with different temperatures are placed in an enclosure and allowed to exchange thermal energy by any means—conduction, convection, or radiation—thermal equilibrium eventually will be reached with all objects at the same temperature.

Thermometers and the Kelvin Temperature Scale

A *thermometer* is simply a relatively small macroscopic system which is arranged so that, when the system absorbs or gives off heat, one property of the system changes in an evident and readily measurable manner. The reading of a thermometer is the temperature of all systems in thermal equilibrium with it.

In different types of thermometers, different measurable properties are selected to indicate the temperature. Each of the thermometers to be described has features that make it particularly useful in certain applications.

To assign numbers to the various temperatures indicated by a thermometer, we must establish a temperature scale. In setting up the most important scale, the Kelvin temperature scale, the temperature at which pure water, ice, and water vapor coexist in equilibrium (Figure 19-7) is arbitrarily assigned the value 273.16 K. Water in this state is said to be at its *triple point*. The reason for choosing the triple point of water as a standard state is that there is only one definite value of temperature and pressure (4.58 mm of Hg) at which water will remain in equilibrium with ice and water vapor. Changes in the relative amounts of the solid, liquid, and gaseous forms do not affect the temperature or pressure. This state of water therefore furnishes a reproducible standard of temperature which, by definition, is

$$T_{\text{triple point}} = 273.16 \text{ K } exactly*$$

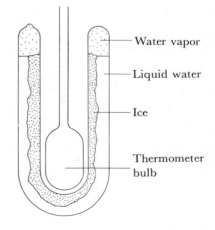

Figure 19-7   Triple-point cell containing ice, liquid water, and water vapor.

Water vapor

Liquid water

Ice

Thermometer bulb

*This value, adopted in 1954 by international convention, has been selected so that the modern scale will agree as closely as possible with the less accurate values obtained using other procedures.

Numbers are now assigned to other temperatures on the Kelvin temperature scale in accordance with the definition of temperature given in Section 20.5. Over a wide range of temperatures, these Kelvin temperatures can be obtained directly from readings of the pressure of helium contained in the constant-volume gas thermometer that we shall now describe. The other thermometers can then be calibrated.

Constant-Volume Gas Thermometer

A constant-volume gas thermometer can be used to determine temperatures as low as 1 K. A small amount of helium is put into a glass bulb and the pressure of the gas is measured at various temperatures. Provision is made to keep the gas volume constant as both its temperature and pressure change (Figure 19-8). If the helium gas obeys the ideal gas law,* then

$$PV = NkT \tag{17.2}$$

where P is the pressure reading of the thermometer at the temperature T. The pressure reading P_t of the thermometer at the triple-point temperature of water (273.16 K) therefore satisfies

$$\frac{P_t}{273.16 \text{ K}} = \frac{Nk}{V} = \frac{P}{T}$$

where we have used the fact that N and V are maintained constant in this thermometer. Therefore

$$\frac{T}{273.16 \text{ K}} = \frac{P}{P_t}$$

In this way the Kelvin temperature T can be calculated after measurement of the ratio of the pressures P and P_t. We can check that the helium density is low enough to ensure that the ideal gas law is obeyed by verifying that the same pressure ratios are obtained with a similar thermometer containing less helium.

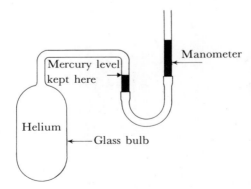

Mercury level kept here →

Manometer

Helium

← Glass bulb

Figure 19-8 Constant-volume gas thermometer. The manometer is adjusted so that the mercury level in the left tube is maintained at a fixed mark. In this way the volume of the helium is kept constant. The manometer pressure reading determines the temperature of the helium in the bulb.

*With temperature defined as in Section 20.5, we can deduce that gases of sufficiently low density must obey this law.

Gas thermometers have the disadvantage of being cumbersome and slow in reaching thermal equilibrium. Their use is therefore confined to research and standards laboratories.

Liquid-in-Glass Thermometers

The thermal expansion of a liquid, usually mercury or alcohol, is exploited in the most common type of thermometer. The liquid fills a thin-walled bulb at the end of a capillary tube. A very small expansion of the liquid produces an appreciable increase in the height of the liquid column in the capillary tube.

The *clinical thermometer* is a mercury-in-glass thermometer with a constriction (Figure 19-9) between the bulb and the capillary tube. This constriction inhibits the return flow of mercury from the capillary to the bulb, so the thermometer indicates the highest temperature that the bulb has attained.

Liquid-in-glass thermometers are portable, relatively inexpensive, and easy to read. For accurate work, however, many corrections must be applied, and for this reason this type of thermometer has been almost entirely replaced in research laboratories by resistance thermometers or thermocouples.

Bimetallic Thermometer

A rugged portable thermometer (Figure 19-10) can be made using a strip formed of two metals which expand by different amounts when their temperature is raised. As the strip bends, it moves a pointer over a scale.

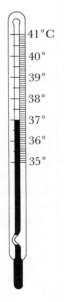

Figure 19-9   Clinical thermometer.

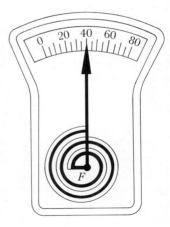

Figure 19-10   A bimetallic thermometer. The bimetallic strip, which employs the principle of differential expansion to activate the pointer, is in the form of a spring centered at point *F*.

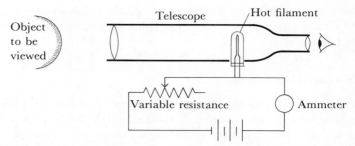

Figure 19-11   Optical pyrometer. A glowing object is viewed through the telescope. The filament current is adjusted until the filament brightness matches that of the glowing object.

Optical Pyrometer

When an object is so hot that it emits visible thermal radiation, its temperature can be measured using the *Disappearing Filament Optical Pyrometer* (Figure 19-11). The glowing object is viewed through the pyrometer telescope and the observer compares the brightness of the object with the brightness of an electric lamp filament. The filament current* is increased until the filament brightness matches that of the glowing object. The unknown temperature then can be determined from the ammeter reading of the filament current if the instrument has been previously calibrated at known temperatures.

Thermography

Modern infrared scanners can detect small variations in the temperature of radiating surfaces and convert this information into a picture called a thermograph in which each region is displayed with a shading characteristic of its temperature. An obvious military application is the nighttime detection of hostile troops. In industry, one of many applications of thermography involves the testing of electronic apparatus. Any flaw which leads to a change in an electric current distribution will also alter the thermal pattern.

Thermography also finds use in medicine. The human body, with a skin temperature of about 31°C, radiates approximately 50 W/m² of skin surface. This energy is largely in the infrared region. An instrument known as a thermograph can scan small segments of the body, record the emitted energy, and thus pinpoint local temperature differences. Since present thermography instrumentation is capable of a thermal sensitivity of less than 0.07 K, the slight temperature increase generally associated with breast cancers (a few tenths of a kelvin) can be readily detected (Figure 19-12).

---

*The electrical terms current, resistance, EMF, and voltage that are used in this section are discussed in detail in Chapters 25 and 26.

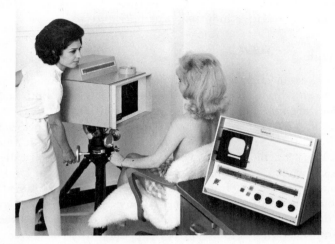

Figure 19-12   A very promising application of thermography is the routine scanning of women to detect breast cancers while they are still in what is called the subclinical stage—a time when the growth is of the order of a centimeter or less in diameter and when therapeutic measures are most effective. Since the scan simply involves reading the infrared radiation emitted by the body this method of diagnosis avoids hazards such as those involved in x-ray examinations.

Resistance Thermometers

The electrical resistance of a conductor changes when the temperature changes. Extremely precise measurements of resistance can be made by standard electrical methods, so resistance thermometers furnish some of the most precise temperature measurements. A thermometer to measure high temperatures employs a coil of platinum wire (melting point 2047 K) wound on a silica spool (melting point 1693 K). At extremely low temperatures the platinum wire is replaced by a germanium crystal or a carbon cylinder.

Thermistor*

A resistance thermometer which can measure temperature changes as small as 0.001 K is obtained by measuring the resistance of a *thermistor*—a small bead of semiconductor material placed between two wire leads. A small temperature increase leads to a very large increase in the number of electrons in the conduction band of a semiconductor (Section 43.5). This greatly increases the electrical

*The physical mechanisms responsible for the functioning of a thermistor are not discussed until Section 43.5. This useful device is mentioned here to complete the survey of thermometers commonly used in modern science and engineering.

conductivity of the semiconductor. The electrical resistance of a semiconductor is therefore a very sensitive indicator of temperature.

Thermocouple

Physicists and engineers generally rate the *thermocouple* as the most useful thermometer. When wires of two different metals are joined at both ends and the two junctions are maintained at different temperatures, an EMF is set up in the circuit. This EMF depends on the temperature difference between the junctions. Thus the EMF can be used to measure this temperature difference.

In operation as a thermometer, one thermocouple junction called the test junction is embedded in the material whose temperature is to be determined. At the other end of each thermocouple wire, copper wires are joined and these two junctions (Figure 19-13), maintained at any desired reference temperature, constitute the reference junction. The temperature difference between the test junction and the reference junction is determined from measurement of the EMF between the two copper wires. Previous calibration with known temperatures is obviously necessary.

The thermocouple has many merits. It can follow temperature changes rapidly because the test junction has such a small mass that it comes quickly into thermal equilibrium with the material under investigation. The reading of the instrument, the EMF measurement, can be performed at a location remote from the test junction.

Thermocouples with one test junction wire of copper and the other wire of an alloy called constantan are used to measure temperatures ranging from 80 K to 600 K. This *copper–constantan* thermocouple gives the comparatively large EMF of 40 microvolts per kelvin temperature difference between the junctions. Since this EMF can be measured with an accuracy of about 1 microvolt using an

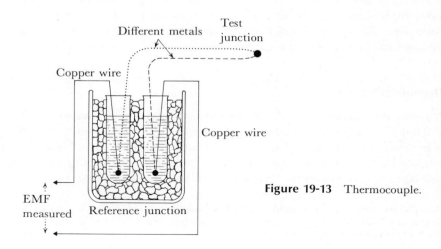

Figure 19-13   Thermocouple.

instrument called a *potentiometer,* temperature differences can be determined with an accuracy of about 0.02 K. Thermocouples of platinum and a platinum–rhodium alloy are used for temperatures up to 1900 K.

19.4 Changes of Phase and Phase Equilibrium

Latent Heats

Substances can exist in several different *phases* or forms. Thus a collection of H_2O molecules can exist in the *solid phase* as ice, the *liquid phase* as water, or the *gaseous phase* as steam.* For a given substance a change from one phase to another occurs at a well-defined temperature and involves a volume change as well as the absorption or liberation of heat. The magnitude of the heat absorbed in melting ice is impressive and has pronounced effects on our climate.

The main features of phase changes are brought out by considering the sequence of events that occurs when heat is continuously supplied to a mass m of ice contained in a cylinder fitted with a piston, with the pressure on the piston maintained at 1 atm. We start with ice, fresh from the refrigerator, at a temperature of $-20°C$. The H_2O molecules of the ice vibrate about definite equilibrium positions with an average kinetic energy of $\frac{3}{2}kT$. This energy is insufficient to allow a molecule to break away from its equilibrium position which is determined by strong intermolecular forces. We now apply heat. The ice does not immediately begin to melt but rather warms up in accordance with

$$Q = mc'_p \Delta T$$

where c'_p is the specific heat capacity of ice at a pressure of 1 atm in this range of temperatures. As the ice warms, its molecules jiggle more violently.

At $0°C$, the melting point for ice (at a pressure of 1 atm), the temperature ceases to rise, even though heat continues to be supplied to the ice. Instead, the most energetic molecules enter the liquid phase which is characterized by the molecules still being closely packed together but no longer with fixed relative positions. We emphasize that, as heat is applied, the temperature remains at $0°C$ until all the ice has melted.

The heat absorbed by the mass m which changes from the solid to the liquid phase without undergoing a change of temperature or pressure is given by

$$Q = mL_f \tag{19.7}$$

where L_f is called the *latent heat of fusion.* For an ice–water transition at $0°C$ and 1 atm,

$$L_f = 333 \text{ kJ/kg}$$

---

*Some authors make a distinction between water vapor and steam by defining steam as vapor condensed on dust particles in the air (like fog). However, we use the terms steam and water vapor synonymously.

When this process is reversed so that water freezes, each kilogram of water at 0°C *gives off* 333 kJ of heat in changing to ice at 0°C. Equation 19.7 yields the heat given off in freezing *or* the heat absorbed in melting.

Now with all the ice melted so that the cylinder contains only water at 0°C, we continue to apply heat. The water molecules execute a still more violent random thermal motion and the temperature increases in accordance with

$$Q = mc'_P \Delta T$$

where c'_P is now the specific heat capacity of water at a pressure of 1 atm.

At 100°C and 1 atm, the most energetic molecules near the surface of the water have enough energy to escape and form water vapor or steam. Evaporation takes place. Again, since only the most energetic molecules escape from the liquid phase, energy, named a *latent heat*, must be supplied to accomplish the phase transition.

The heat absorbed by the mass m which changes from the liquid phase to the gaseous phase without change in temperature or pressure is given by

$$Q = mL_v \tag{19.8}$$

where L_v is called the *latent heat of vaporization* of the substance. For water L_v is large, 2256 kJ/kg at 100°C and 1 atm. For the reverse process, condensation, Eq. 19.8 still applies, but now the heat is given off by the condensing vapor.

Example 4 How much heat is required to change 1.00 kg of ice at -8.0°C into water at 20°C in a process which takes place at a pressure of 1 atm? Assume that between -8.0°C and 0°C, ice has an average specific heat capacity $c'_P = 2.1$ kJ/kg·K and that between 0°C and 20°C, water has an average specific heat capacity $c'_P = 4.2$ kJ/kg·K.

Solution To warm the ice from -8.0°C to 0.0°C requires a heat

$$Q = mc'_P \Delta T = (1.00 \text{ kg})(2.1 \text{ kJ/kg·K})(8.0 \text{ K}) = 17 \text{ kJ}$$

The phase transition from ice at 0°C into water at 0°C requires a heat

$$Q = mL_f = (1.00 \text{ kg})(333 \text{ kJ/kg}) = 333 \text{ kJ}$$

The heat required to raise the temperature of the water from 0°C to 20°C is

$$Q = (1.00 \text{ kg})(4.2 \text{ kJ/kg·K})(20 \text{ K}) = 84 \text{ kJ}$$

The heat required for the entire process is therefore

$$(17 + 333 + 84) \text{ kJ} = 434 \text{ kJ}$$

The fact that the temperature of an ice–water system does not drop below 0°C until all the water has frozen can be used to provide a handy method of temperature control. An old trick, known to farmers (see Question 21), has been revived by the United States Air Force in sending fresh fruits and vegetables to Arctic outposts. Freezing and spoilage of most of these foods starts at about -1°C. They find that, by using a water-soaked layer of material sealed between inner and outer insulating layers, they can maintain the interior at a safe $0°C$ for about six hours, until the water has completely frozen, even when the outside temperature is as low as $-50°C$.

Phase Equilibrium

Consider again the ice and water in the cylinder at a temperature of $0°C$ and a pressure of 1 atm. If the heat transfer to or from the cylinder is stopped, the ice and water can coexist indefinitely. The escape of energetic molecules from the solid phase is offset by the arrival, from the liquid, of molecules which become bound to the solid. At the melting or fusion temperature ($0°C$ for ice at a pressure of 1 atm), these two competing processes proceed at equal rates. We say that *phase equilibrium* is obtained and that the solid phase is in equilibrium with the liquid phase.

Similarly, we can obtain equilibrium between liquid and vapor phases. At a pressure of 1 atm, water and steam confined to a cylinder and isolated will coexist indefinitely at a temperature of $100°C$. Of course a liquid tends to evaporate at any temperature, and in most circumstances the vapor is not in equilibrium with the liquid. However, if the liquid and vapor are confined, the vapor pressure alters until the evaporation of the liquid is balanced by the condensation of the vapor, and equilibrium is attained. The pressure of the vapor at which the vapor and the liquid coexist in equilibrium is called the *saturation vapor pressure*. For reasons that we shall now investigate, *the saturation vapor pressure depends only on the temperature,* not on the volume of vapor or liquid.

Evaporation and Condensation

The approach to equilibrium between water and its vapor is illustrated in Figure 19-14. Some water is placed in a cylinder. A vacuum pump is used to remove the air and the vapor from the region above the water [Figure 19-14 (a)].

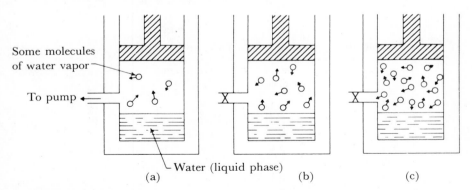

Some molecules of water vapor

To pump

Water (liquid phase)

(a)   (b)   (c)

Figure 19-14   (a) Molecules of water vapor are pumped away as fast as they are evaporated from the liquid. (b) The valve is closed. The number of molecules of water vapor increases. (c) Phase equilibrium is attained. The pressure now exerted by the vapor is called the *saturation vapor pressure*.

The valve is then closed so the pumping stops [Figure 19-14 (b)]. The collection of H_2O molecules within the cylinder is not in an equilibrium state and will not remain entirely in the liquid phase. Within the liquid, the most energetic of the molecules which approach the surface will have sufficient energy to cross the surface and break away from the attractive forces exerted by the molecules remaining in the liquid phase. Some of the most energetic molecules thus enter the vapor phase [Figure 19-14(b)]. They have been *evaporated* or *vaporized*.

The rate of evaporation per unit liquid surface area will be determined entirely by the translational kinetic energies (average value $\frac{3}{2}kT$) of the molecules within the liquid, and these energies are determined by the temperature. Consequently the *rate of evaporation* per unit liquid surface area is determined by the liquid *temperature* only.

As the molecules in the vapor phase bounce around, some will strike the liquid surface and re-enter the liquid phase. We say such molecules undergo *condensation* from the vapor phase to the liquid phase. The number of vapor molecules striking the surface is proportional to the pressure of the vapor. Consequently the *rate of condensation* per unit area of liquid surface is proportional to the *pressure of the vapor.*

Evaporation and condensation compete. Evaporation proceeds at a rate determined by the liquid temperature. In Figure 19-14 (b) evaporation is the dominant effect because the pressure of the vapor is so low that the condensation rate is low. The pressure of the vapor thus increases until, at a value called the *saturation vapor pressure* [Figure 19-14 (c)], *phase equilibrium* is attained with

$$evaporation\ rate\ =\ condensation\ rate$$

Since the evaporation rate depends only on the temperature, and the condensation rate depends on the pressure of the vapor, this phase equilibrium equation implies that the *saturation vapor pressure depends only on the temperature.* Experiments confirm this expectation; results are given in Table 19-4 and the vaporization curve is shown in Figure 19-15.

Table 19-4 Saturation Vapor Pressure of Water

Temperature °C	Pressure mm of Hg	Temperature °C	Pressure atm
0.0	4.58	100	1.00
5.0	6.54	120	1.96
10.0	9.21	150	4.70
15.0	12.8	200	15.4
20.0	17.5	250	39.3
40.0	55.3	300	84.8
60.0	149	350	163.2
80.0	355	374.15*	218.4*
100.0	760		

*critical point

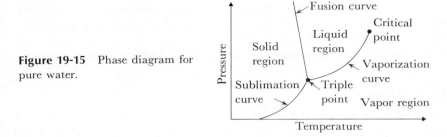

Figure 19-15   Phase diagram for pure water.

If the piston is suddenly pushed down a certain distance the pressure of the vapor will momentarily rise, but condensation now occurs at a greater rate. Equilibrium will soon be reached with less vapor and more liquid but with the *same saturation vapor pressure*, provided the temperature is maintained constant.

If the vapor is removed as fast as it evaporates, phase equilibrium is never attained. Continued evaporation takes place. This happens when pumping as in Figure 19-14 (a) or when dry air blows over a lake or past wet laundry. Evaporation is a very efficient cooling process, as we all discover when we stand soaking wet in a breeze. A large *latent heat of vaporization* is absorbed from a wet object as water is vaporized. The energetic molecules that escape into the vapor phase deplete the liquid's thermal energy.

Sublimation

Solids also evaporate. A direct transition between the solid and vapor phase, without passing through the liquid phase, is called *sublimation.* The evaporation of ice, which accounts for the drying of wet clothes hung outside in below freezing temperatures, is a familiar example. As with other phase transitions, a latent heat is involved in sublimation. There is also a very definite temperature at which equilibrium is obtained between the solid and the vapor at a specified pressure. For instance, solid carbon dioxide (*Dry Ice*) reaches equilibrium with its vapor at a pressure of 1 atm when the temperature falls to $-79°C$.

Sublimation is exploited in the process of *freeze-drying* foods. A solidly frozen food is exposed to a low pressure (approximately 0.1 mm of Hg), and heat is supplied. The ice in the food continues to sublime as the water vapor is pumped away. By this process the moisture in the food is removed without appreciably altering the shape, color, or taste of the product. Vacuum packed, the food will keep for long periods at room temperature. Reconstitution is accomplished simply by adding water.

Amorphous Solids

The phase changes that we have been discussing for ice are typical of the behavior of many solids which have a definite crystal structure. However, there are certain materials, called *amorphous solids,* such as glass and various resins, that lack a crystal structure. These amorphous substances do not possess a definite melting point. As glass is heated it gradually softens. For this process there is no phase transition and no latent heat of fusion. An amorphous solid is more closely related to a very viscous liquid than to a crystalline solid.

Phase Diagram

Many of the foregoing facts about the behavior of a substance that relate to fusion, vaporization, and sublimation can be depicted on a phase diagram (Figure 19-15). This diagram shows the phase or phases of a *pure* substance that are present in equilibrium at every possible combination of pressure and temperature.

Each point on a phase diagram refers to the substance in an *equilibrium state*, the condition reached by the substance when the external conditions have been left unchanged for a sufficiently long time. An *equilibrium state* of a typical system, such as a collection of 10^{24} H_2O molecules confined within a cylinder, is characterized by:

1   A uniform temperature T throughout the cylinder.
2   A uniform pressure P throughout the cylinder.
3   A definite mass of H_2O molecules in each possible phase.

The interpretation of the phase diagram of Figure 19-15 is clarified by singling out certain particular states, as shown in Figure 19-16. Thus the equilibrium state of H_2O molecules at a temperature of 110°C and a pressure of 1.00 atm is represented by point A. In this state there is only water vapor or steam. Point B corresponds to an equilibrium state at a temperature of 100°C and a pressure of 1.00 atm. Under these conditions, water and its vapor can coexist in equilibrium, and the 1.00 atm pressure is called the *saturation vapor pressure*. At point C (50°C, 1.00 atm) there is only water. At point D (0.0°C, 1.00 atm) ice and water can coexist in equilibrium, while at E (-20°C, 1.00 atm) there is only ice. The line $EDCB$ represents the process described at the beginning of this section: ice is warmed and then melted to water, which is warmed and then vaporized.

We see that there is a region of the phase diagram corresponding to equilibrium states in which the substance is entirely solid, a second region corresponding to the liquid phase, and a third region corresponding to the vapor phase.

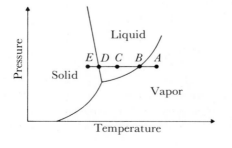

Figure 19-16 Various equilibrium states of H_2O.

The liquid and vapor regions are separated by the *vaporization curve* which gives all the values of temperature and pressure for which the liquid and vapor can coexist in equilibrium. In other words, the vaporization curve gives the saturation vapor pressure at different temperatures. The data from which this curve is plotted are given in Table 19-4. The *sublimation curve* separates the solid and vapor regions and gives the values of temperature and pressure at which solid and vapor can coexist in equilibrium. States with solid and liquid in equilibrium are given by the *fusion curve* which separates the solid and liquid regions. A phase change involves passing from one region to another by crossing one of these three curves.

The vaporization, sublimation, and fusion curves intersect in a point called the *triple point*. At the triple-point temperature and pressure the three phases can coexist in equilibrium (Figure 19-7). This state is easy to achieve and gives an accurately reproducible temperature which is now the basis for modern thermometry.

The vaporization curve of Figure 19-15 does not extend indefinitely. With a liquid and its vapor in equilibrium we find that, as the temperature increases, the distinction between the two phases decreases and finally disappears at a certain temperature and pressure (and volume per unit mass) that we call the *critical point*.

Boiling

Boiling occurs when bubbles of vapor form throughout the volume of a liquid. Since incipient bubbles do not collapse, the vapor is in phase equilibrium with the liquid. Therefore at the location of the bubble the pressure must be the saturation vapor pressure. That is, *the pressure in a boiling liquid is the saturation vapor pressure at the temperature of the liquid.* The vaporization curve thus acquires an added significance. The vaporization curve gives the corresponding temperatures and pressures of a boiling liquid. From the vaporization curve (Figure 19-15 or Table 19-4) we can conclude that at room temperature water will boil at a pressure of 17.5 mm of Hg, while at 100°C boiling occurs at 1.00 atm. One can control the temperature of boiling by adjusting the pressure. The valve in a pressure cooker is usually set to maintain the cooker pressure at about 1.0 atm above the kitchen pressure of 1.0 atm. The pressure inside the cooker will then be about 2.0 atm and the water inside the cooker will boil at the corresponding temperature on the vaporization curve. This is about 120°C. The cooking time is thereby reduced by as much as a factor of ten.

19.5 Water Vapor in Air

In our atmosphere, water vapor is mixed with air. The presence of the air has practically no effect on the behavior of the water vapor. The total pressure exerted by the mixture is the sum of the pressures that each gas would exert if it alone were present [in accord with Dalton's Law of partial pressures (Section 17.2)].

The presence of air above the surface of water does not change the equilibrium conditions shown by Table 19-4 and the corresponding phase diagram (Figure 19-15) for pure water. The pressure of water vapor *in equilibrium with water* is still the *saturation vapor pressure* given by Table 19-4, the same as if the air were not there at all.

In our atmosphere, however, the water vapor is usually *not* in equilibrium with water. We are confronted with a variety of different situations. Water vapor at a given temperature is said to be *unsaturated, saturated,* or *supersaturated* according to whether the pressure of the water vapor is less than, equal to, or greater than the saturation vapor pressure at the temperature in question.

Example 5   The pressure of the water vapor in the atmosphere on a certain day is 17.5 mm of Hg. Classify this vapor as unsaturated, saturated, or supersaturated when the temperature is (a) 10°C, (b) 20°C, (c) 40°C.

Solution

Temperature	Saturation vapor pressure from Table 19-4	Condition of vapor with pressure of 17.5 mm of Hg
10°C	9.21 mm of Hg	Supersaturated
20°C	17.5 mm of Hg	Saturated
40°C	55.3 mm of Hg	Unsaturated

The supersaturated state of a vapor is not an equilibrium state. The excess water vapor eventually condenses into water, usually forming tiny drops on minute particles in the air, until the pressure of the water vapor has been reduced to the saturation vapor pressure.

Relative Humidity

Unsaturated vapor is most common in our homes and in the atmosphere. The drier the air, the more rapid the evaporation from any moist objects. A useful characterization of the degree of saturation is furnished by the *relative humidity* defined by the equation

$$percentage\ relative\ humidity = 100 \times \frac{\left\{ \begin{array}{c} pressure\ of\ vapor\ at \\ temperature\ in\ question \end{array} \right\}}{\left\{ \begin{array}{c} saturation\ vapor\ pressure \\ at\ same\ temperature \end{array} \right\}}$$

The relative humidity is 100% when the vapor is saturated and 0% if no water vapor is present.

Example 6   Find the relative humidity within a house when the temperature is 20°C and the pressure of the water vapor is 5.0 mm of Hg.

Solution   From Table 19-4 the saturation vapor pressure at 20°C is 17.5 mm of Hg. Therefore

$$relative\ humidity = 100 \times \frac{5.0\ \text{mm of Hg}}{17.5\ \text{mm of Hg}}$$

$$= 29\%$$

This is an uncomfortably dry environment. Both for comfort and health a humidifier should be used to introduce more water vapor and to maintain the relative humidity around 49–50%.

When air with unsaturated vapor is cooled sufficiently, a temperature called the *dew point* is reached. At this temperature the vapor is saturated. Further cooling will cause condensation. Clouds, fog, and rain result from the cooling of a portion of the atmosphere below its dew point. On a clear night, the earth's surface is cooled by emitting thermal radiation to space without receiving a balancing radiation from clouds. If this cooling brings the water vapor near the earth below its dew point, grass is covered with dew.

The most accurate method of measuring relative humidity is simply to determine the dew point by observing the temperature at which dew forms on the polished surface of a cool metal container.

Example 7   The air temperature inside a house is 20°C. There is dew on the surface of a metal can containing beer at 10°C. No dew forms on warmer surfaces. Find the relative humidity within the house.

Solution   The dew point is 10°C. A thin film of air next to the metal surface is at 10°C and the water vapor in that layer is saturated. Therefore the vapor pressure in this film is the saturation vapor pressure at 10°C, 9.21 mm of Hg, from Table 19-4. The vapor pressure of the warm air elsewhere in the house, being the same as the vapor pressure in the cool film, is also 9.21 mm of Hg. But at 20°C, the vapor pressure necessary *for saturation*, again from Table 19-4, is 17.5 mm of Hg. Therefore

$$relative\ humidity = 100 \times \frac{9.21\ \text{mm of Hg}}{17.5\ \text{mm of Hg}}$$

$$= 53\%$$

Cloud Chamber

If saturated vapor is cooled rapidly, it becomes supersaturated. This is a non-equilibrium state, and during the next few seconds a gradual condensation into a uniform fog occurs. The formation and growth of minute water drops is enhanced if the drops carry an electric charge (Chapter 21). This fact is exploited

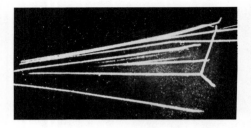

Figure 19-17   Cloud-chamber photograph of the collision between an electrically charged particle and the nucleus of an atom. (From A. B. Arons, *Development of Concepts of Physics*, Addison-Wesley, 1965.)

in that important instrument of particle physics, the *Wilson cloud chamber*. The chamber is filled with saturated vapor and then suddenly expanded. The expansion is accompanied by a rapid cooling, so the vapor becomes supersaturated. For the next few seconds growing water drops will form on the ions left in the wake of any high-energy charged particle that passes through the chamber. Such cloud-chamber tracks (Figure 19-17) are a prime source of information about the subatomic world.

Summary

☐   The increase ΔL in the length L of a solid when its temperature is raised an amount ΔT is

$$\Delta L = \alpha L \Delta T$$

where α is the coefficient of linear expansion of the solid. The corresponding area and volume expansions are given by

$$\Delta A = 2\alpha A \Delta T \qquad \text{and} \qquad \Delta V = 3\alpha V \Delta T$$

respectively. For a liquid with a coefficient of volume expansion β,

$$\Delta V = \beta V \Delta T$$

☐   The fundamental law for the rate dQ/dt at which heat is conducted (in the steady-state) through a slab of cross section A and thickness dx is

$$\frac{dQ}{dt} = -KA \frac{dT}{dx}$$

where dT/dx is the temperature gradient within the slab and K is its thermal conductivity.

☐   The zeroth law of thermodynamics states that two systems in thermal equilibrium with a third are in thermal equilibrium with each other.

☐   The temperature of an object is the property which determines whether or not the object will be in thermal equilibrium with other objects.

☐   Temperatures above about 1 K can be measured using a constant-volume gas thermometer containing helium. When the helium pressure is sufficiently low, the Kelvin temperature T of the helium at pressure P can be calculated from

$$\frac{T}{273.16 \text{ K}} = \frac{P}{P_t}$$

where P_t is the helium pressure at a temperature equal to that of water at its triple point.

☐ Equilibrium states corresponding to different phases of a pure substance can be represented by points on a pressure-versus-temperature diagram called a phase diagram. Regions corresponding to the solid, liquid, and vapor phases are separated by the following curves representing states with different phases in equilibrium: vaporization curve (liquid-vapor), fusion curve (solid-liquid), sublimation curve (solid-vapor). At the triple point, the solid, liquid, and vapor phases coexist in equilibrium. For these phases, a phase transition at constant pressure involves a volume change and an amount of heat called a latent heat.

Questions

1 (a) What is the increase in length in a steel metrestick when its temperature is raised 10 K?
   (b) In what way is an invar measuring tape superior to a steel measuring tape?

2 A steel cable supporting a suspension bridge has a length of 500 m at 30°C. What is the change in its length when the temperature is decreased to $-10°$C?

3 A locomotive wheel is 1.30 m in diameter at 20°C. A steel tire with a diameter 0.50 mm under size is to be shrunk on. To what temperature must the tire be heated to make its diameter 0.50 mm over size?

4 The control element in a thermostat is a bimetallic strip. Explain how it works.

5 A steel steampipe has a cross-sectional area of 8.00×10^{-3} m^2 at 20°C. What will be its area when filled with superheated steam at a temperature of 170°C?

6 Explain why the column of mercury first descends and then rises when a mercury-in-glass thermometer is thrust into boiling water.

7 A 0.100 m^3 bottle made of ordinary glass is completely filled with benzene at a temperature of 15.0°C. What volume of benzene overflows when the temperature is raised to 25.0°C?

8 What is unusual about the thermal expansion of water? Comment on some of the practical implications of this peculiarity.

9 What water temperature would you expect to find near the bottom of a lake which has ice on its surface? Why?

10 Why is a thick-walled bottle made of ordinary glass prone to shatter when heated?

11 A rod of length L has a cross-sectional area A. One end is maintained at temperature T_1 and the other end at a higher temperature T_2. If the heat loss through the sides

is negligible, show that the rate at which heat is conducted down the rod is

$$\frac{dQ}{dt} = KA\frac{T_2 - T_1}{L}$$

12 A household refrigerator has walls made of corkboard with a thermal conductivity of 0.04 W/m·K. These walls are 6.0 cm thick and have a total area of 5.0 m². The temperature difference between the inside and outside surfaces of the walls is 20 K. What quantity of heat is conducted through the walls in 1.0 h?

13 What is the principal means of heat transfer from a hot object to a cold object when they are separated by:
(a) A vacuum?
(b) A solid metal?
(c) A gas when the warmer object is beneath the cooler object?

14 (a) What method of heat transfer is employed to heat the upstairs of a house with a furnace in the basement?
(b) How is heat transfer from the sun to the earth accomplished?
(c) Name at least one household situation which relies on conduction to transfer heat.

15 A certain man loses 5.0×10^3 kJ in a day by evaporation of his perspiration:
(a) Approximately how much water does his body lose in this process?
(b) If this energy could have been retained and used with 100% efficiency to climb a mountain, how high would this energy permit a 70 kg man to climb?

16 Assume that the gas in the bulb of a constant-volume gas thermometer obeys the ideal gas law. If T_1 and T_2 are the Kelvin temperatures of two objects for which the thermometer registers pressures P_1 and P_2, respectively, show that

$$\frac{T_1}{T_2} = \frac{P_1}{P_2}$$

17 A constant-volume gas thermometer registers a pressure of 0.200 atm when its bulb is in thermal equilibrium with water at the triple point. What is the Kelvin temperature of a bath if this thermometer registers a pressure of 0.225 atm when its bulb is in thermal equilibrium with the bath water?

18 You wish to monitor the temperature within a factory chimney. Which type of thermometer would you select and why?

19 A 12.0-kg block of ice at 0°C absorbs 1665 kJ of heat in a certain process which takes place at a pressure of 1.00 atm. What is the final temperature? Explain.

20 How much heat must be supplied to 3.0 kg of water at 20°C in order to convert it to steam at 100°C? Assume that the process takes place at a constant pressure of 1.00 atm.

21 A farmer places a barrel containing 200 kg of water in a storage room to prevent his produce from freezing:
(a) Find the heat given off by the water in cooling from 20°C down to 0°C.
(b) Find the heat given off while half the water in the barrel freezes, changing from water at 0°C to ice at 0°C?
(c) If the barrel of water were to be replaced by a 1.0 kW electric heater, for how long would the heater have to operate to furnish the same energy as that given off in part (b) by the freezing of 100 kg of water?

22 How much heat is given off in a process in which 5.0 kg of steam at 100°C are converted into ice at $-10°C$? (The pressure is maintained constant at 1 atm.)

23 A mass of 0.100 kg of ice at 0.0°C is introduced into a calorimeter can containing 0.800 kg of water at 30.0°C. At what temperature will thermal equilibrium be attained? (Neglect the heat given off by the calorimeter can.)

24 Pure water is introduced into a closed container from which all the air has been removed. The temperature is maintained at 20°C. The pressure of the vapor is measured as time goes on. Describe what you will observe.

25 Draw a line on a phase diagram to show a process, taking place at constant pressure, in which a solid is warmed and then undergoes sublimation, whereupon the vapor is warmed.

26 Draw a line on a phase diagram to show a process in which a vapor is liquefied by increasing the pressure while the temperature is maintained constant.

27 The phase change from water to ice can lead to the cracking of rocks and the destruction of highways. Explain.

28 The temperature of phase equilibrium between ice and water decreases 0.0075 K for each atmosphere increase in pressure. A large sample of ice at 0.0°C is insulated and the pressure is increased from 1.00 atm to 1.00×10^3 atm (typical of the pressure underneath the blade of a skate). Some of the ice melts. At what temperature will phase equilibrium be attained between the ice and water?

29 At the top of a mountain which is 6 km above sea level, we observe that water boils at 80°C. What is the pressure of the atmosphere at this location?

30 What information about boiling is provided by the vaporization curve? Explain.

31 What purpose does the water fulfill in the cooking of a soft-boiled egg?

32 Heat is applied to a pressure cooker which is set to allow the pressure within the cooker to rise to 2 atm. When the pressure reaches 2 atm what will be the temperature if:
(a) There is water in the cooker?
(b) There is air but no water in the cooker? (Before heat was applied this air was at a pressure of 1 atm and a temperature of 27°C.)

33 On a day when the temperature is 15°C, the dew point is found to be 10°C. Calculate the relative humidity. Explain your reasoning at each step of the calculation.

34 On a certain winter day, the outside temperature is 0°C, and the water vapor is saturated at this temperature. Some of this air is taken inside and warmed to 20°C without changing the pressure of the water vapor. What is the relative humidity? (This value will be typical of the environment inside buildings that do not use humidifiers.)

Supplementary Questions

S-1 Consider the thermal expansion of a rectangular area A with sides of lengths L_1 and L_2. When the fractional expansions $\Delta L_1 / L_1$ and $\Delta L_2 / L_2$ are small, the approximate increase in area is given by

$$\Delta A = 2\alpha A \Delta T$$

Prove this algebraically from Eq. 19.1 and draw a figure showing the relevant quantities.

S-2   Consider the thermal expansion of a cube with sides of length L and show, from Eq. 19.1, that the approximate increase in volume is given by

$$\Delta V = 3\alpha V \Delta T$$

S-3   A thermostat, which is based on the fact that mercury is a good electrical conductor, is constructed as illustrated in Figure 19-18. Two wires from an electrical circuit are inserted in the tube, one at the junction with the bulb and the other 15 cm away. The volume inside the bulb is 1.00 cm^3 and at 0.00°C the mercury just fills it. The area of the cross section of the glass tube is 3.00×10^{-4} cm^2. The mean value (for the temperature in question here) of the thermal coefficient of volume expansion of mercury is 18×10^{-5} per K. Disregarding the expansion of the glass, calculate the temperature at which the circuit will be completed and the current will commence.

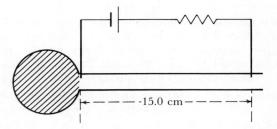

Figure 19-18

S-4   Two slabs of thicknesses L_1 and L_2 and thermal conductivities K_1 and K_2, respectively, are placed in contact (Figure 19-19). The slabs have the same cross-sectional area. The outer surfaces are maintained at temperatures T_1 and T_2. Show that in the steady-state the temperature of the surfaces in contact is

$$T = \frac{K_1 T_1 L_2 + K_2 T_2 L_1}{K_1 L_2 + K_2 L_1}$$

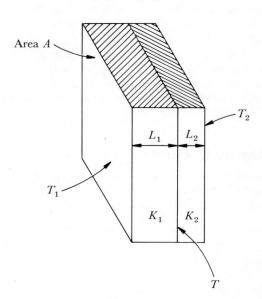

Figure 19-19

S-5   In the preceding question, show that the rate at which heat is conducted from the surface at the higher temperature T_2 to the surface at the lower temperature T_1 is

$$\frac{dQ}{dt} = \frac{A(T_2 - T_1)}{(L_1/K_1) + (L_2/K_2)}$$

S-6   A layer of ice on a lake is 3.0 cm thick. The temperature at the upper surface of the ice is $-20°C$ and at the lower surface is $0°C$. What is the rate of increase of thickness of the ice?

S-7   Show that the rate at which heat is conducted radially outward through a spherical shell of material which has a thermal conductivity K, an inner radius r_1, and an outer radius r_2, is

$$\frac{dQ}{dt} = \frac{4\pi K(T_1 - T_2)}{(1/r_1) - (1/r_2)}$$

where T_1 and T_2 are the temperatures of the inner and outer surfaces, respectively.

S-8   Burns caused by steam are often severe. To understand why this is so, consider the heat that would be given off to a person's skin during the transformation of 0.010 kg of steam at 100°C to water at 37°C, and compare this heat to that which would be given off in cooling 0.010 kg of hot water from 100°C to 37°C.

S-9   The specific heat capacity of a metal is to be determined in an ice calorimeter experiment. A 0.600-kg block of the metal at a temperature of 100°C is introduced into a calorimeter which contains a large quantity of ice at 0°C. When contents of the calorimeter have reached thermal equilibrium, 0.082 kg of ice has melted and the final temperature is 0°C. Find the specific heat capacity of the metal and state the assumptions that are made in this calculation.

S-10 In an experiment to measure the latent heat of vaporization L_v of water, the water is boiled at atmospheric pressure, and steam is led into a calorimeter and allowed to condense. The calorimeter is made of aluminum ($c' = 0.91$ kJ/kg·K) and has a mass of 0.400 kg. Initially the calorimeter contains 0.300 kg of water at 15°C. When the temperature within the calorimeter has risen to 65°C, the total mass of the calorimeter and its contents is found to be 0.733 kg. Calculate L_v from these data.

S-11 One type of solar heating system exploits the fact that sodium sulphate has a convenient melting point (31°C) and a sizable latent heat of fusion (215 kJ/kg):
   (a) Explain how sodium sulphate could be used to store energy received during periods of sunshine and to give off heat to a house at other times.
   (b) If it takes about 10^6 kJ per day to heat a house, what mass of sodium sulphate would be required to supply heat for one day?

S-12 Explain why it is that the saturation vapor pressure depends only on the temperature and is independent of the volume of the vapor or the liquid.

S-13 It is possible to start with water in the liquid phase and to transform it into a vapor without ever reaching a state where the liquid and the vapor phases are present simultaneously. Indicate such a transformation by drawing on a phase diagram a path from an initial state in the liquid region, through the intermediate states, to a final state in the vapor region. Then describe in words how such a transformation could be accomplished.

S-14 A closed bathroom with a volume of 30 m³ is maintained at a temperature of 20.0°C. The relative humidity in the room is 30%. A basin is then filled with water at 20.0°C. What mass of this water will evaporate?

20.

entropy

Thermal energy is disordered energy. The amount of energy of a system is measured by the state function $\overline{E}$ and the amount of disorder by a new state function S called the *entropy*. In any process which occurs in nature, the total energy is conserved and the *total entropy increases*. The quantitative formulation of these ideas and the study of their implications is the topic of this chapter.

The *principle of entropy increase* is one of the most profound laws of physics. Its ramifications extend throughout chemistry and biology as well as physics and engineering.

The historical roots of this subtle subject lie in a study of that all-important device of the Industrial Revolution, the steam engine. In 1824 a young French engineer, Nicholas Leonard Sadi Carnot (1796–1832), published a brochure titled *Reflections on the Motive-Power of Heat, and on Machines Fitted to Develop That Power.* Near the beginning of his paper he questioned ". . . whether the motive power of heat is unbounded, whether the possible improvements in steam engines have an assignable limit—a limit which the nature of things will not allow to be passed by any means whatever." He asked the right question. Penetrating analyses and plausible assumptions led Carnot to the result that all "heat engines" have an efficiency limit determined entirely by the temperature at which they take in heat and the temperature at which they exhaust or reject heat.

The significance of Carnot's discovery was not recognized until the middle of the nineteenth century. It was then that William Thomson (1824–1907), later to become Lord Kelvin, and R. J. E. Clausius (1822–1888) realized that Carnot had pointed the way to a new law whose generality ranged far beyond the confines of steam engines. The law they formulated, called the *second law of thermodynamics,* has been stated in many different but logically equivalent ways. Two phrasings of the second law are:

Kelvin statement It is impossible by means of inanimate material agency to derive mechanical effect from any portion of matter by cooling it below the temperature of the coldest of the surrounding objects.

Clausius statement No process is possible whose *sole* result is the transfer of heat from a cooler to a hotter body.

Clausius also stressed the utility of entropy and showed that the second law was the same as saying that the total entropy increases in any process. The microscopic interpretation of entropy as a measure of molecular disorder was discovered by Boltzmann.

We approach the concept of entropy following a route in keeping with William Thomson's frequently quoted general view: "I often say that when you can measure what you are speaking about, and express it in numbers, you know something about it; but when you cannot express it in numbers, your knowledge is of a meagre and unsatisfactory kind; it may be the beginning of knowledge, but you have scarcely, in your thoughts, advanced to the stage of Science, whatever the matter may be."

20.1 Calculation of Entropy Changes

Entropy, A State Function That Measures Disorder

When a macroscopic system is in equilibrium, its thermodynamic state is specified by the values of a few macroscopic quantities. Any *state function* that the system possesses is a definite mathematical function of the quantities that specify the equilibrium state. Table 20-1 gives examples of state functions for a system consisting of n moles of an ideal gas, when the state is specified by the temperature T and the volume V. Included in this table is a new *state function, the entropy S.*

When the mathematical form of a state function is complicated, it is often convenient to represent the state function graphically or simply to give a table of values. The latter procedure is illustrated by Table 20-2, which gives values of the entropy of 1.00 kg of H_2O molecules in several different states. Here the temperature T and the pressure P have been selected as the thermodynamic variables which specify the equilibrium state.

Table 20-1 Examples of State Functions for an Ideal Gas with Constant Molar Heat Capacity c_V

Name	State function, state specified by (T, V)
Internal energy	$\overline{E} = \overline{E}_0 + nc_V(T - T_0)$
Enthalpy	$H = \overline{E} + nRT$
Entropy	$S = S_0 + nc_V \ln(T/T_0) + nR \ln(V/V_0)$

Table 20-2 Entropy per Kilogram of H_2O Molecules at 1.00 atm When Ice at $-10°C$ Is Assigned Zero Entropy

Phase	Temperature °C	Specific entropy kJ/kg·K
Ice	-10.0	0
Ice	0.0	0.078
Water	0.0	1.30
Water	20.0	1.59
Water	40.0	1.87
Water	60.0	2.13
Water	80.0	2.38
Water	100.0	2.61
Steam	100.0	8.65
Steam	250.0	9.3

Entropy is a measure of molecular disorder. Thus a system consisting of 1 kg of H_2O molecules has little entropy when the molecules are arranged in an orderly fashion, as in a crystal at low temperature. As the crystal is warmed, there is an increased disorder associated with the random thermal motion of its molecules, so the entropy increases. In a liquid the molecules are free to roam about. The liquid phase is considerably more disordered than the solid phase. Consequently, melting involves an appreciable increase in entropy. In the gaseous phase there is practically no correlation between the motions of different molecules. This is an even more disordered situation than when the molecules are in the liquid phase. A large entropy increase therefore is associated with vaporization.

Clausius Equation

By considering a system of a large number of molecules from the point of view of Newtonian or of quantum mechanics, we can establish the numerical measure of the system's disorder that we call the *entropy of the system.* This procedure is discussed later. But decades before entropy was understood from a molecular viewpoint, Clausius learned that substances can be assigned a rather mysterious state function that he named the entropy S. The change dS in a system's entropy during a process in which the system absorbs heat dQ is given by the *Clausius equation*

$$dS = \frac{dQ}{T} \qquad \text{(reversible process)} \quad (20.1)$$

where T is the Kelvin temperature of the system during the process. (As is discussed in Section 20.2, a process is reversible if it is quasi-static and if there are no dissipative effects. Although the validity of the Clausius equation is restricted to such idealized processes, we can always use it to determine the entropy function of a substance and to construct entropy tables such as Table 20-2.)

Notice that the Clausius equation is at least roughly in accord with the interpretation of entropy as a measure of molecular disorder. Heating a substance increases the random molecular motion and therefore increases the disorder; the Clausius equation states that heating a substance increases its entropy. It is also reasonable that the increase in disorder accompanying the absorption of a given quantity of heat should depend on the relative magnitude of this heat to the internal energy of the body, and therefore decrease with increasing temperature.

Clausius showed that the existence of a state function S that satisfies this equation is a consequence of the postulate that no process is possible whose sole result is the transfer of heat from a cooler to a hotter body. To simplify the discussion, we shall simply adopt as a postulate the existence of a state function S satisfying Eq. 20.1.

The plausibility of this postulate is enhanced by the discovery that it is certainly true in the simple case when the system is an ideal gas. Then the equation of state is $PV = nRT$, and for an infinitesimal reversible process, the first law of thermodynamics, $d\bar{E} = dQ - dW$, with $d\bar{E} = nc_V dt$ and $dW = P\,dV = nRT\,dV/V$, gives

$$dQ = nc_V\,dT + \frac{nRT}{V}\,dV \qquad (20.2)$$

Now Q is *not* a state function. There is no function of T and V whose differential is dQ. *But by multiplying dQ by the factor $1/T$ we obtain the differential dS of a state function S*:

$$dS = \frac{dQ}{T} = nc_V\frac{dT}{T} + nR\frac{dV}{V}$$

The equation can now be integrated. Assuming that c_V is constant in the temperature range of interest, we find that for a reversible transformation from state (T_0,V_0) to state (T,V),

$$S - S_0 = nc_V \ln\frac{T}{T_0} + nR \ln\frac{V}{V_0} \qquad (20.3)$$

where S_0 is the value of S for the state (T_0,V_0). This gives the state function S displayed in Table 20-1.

In this derivation the factor $1/T$ serves as an *integrating factor* for dQ; that is, after multiplication by $1/T$ we obtain an expression which can be integrated to yield a *state function*. The Clausius equation, $dS = dQ/T$, asserts that this is true, not just for an ideal gas, but for any system.

Example 1   What is the change in the entropy of a system consisting of 1.00 kg of H_2O molecules when it is transformed (at a constant pressure of 1 atm) from water at 100°C to steam at the same temperature?

Solution   From the Clausius equation, the change in entropy of the system is

$$S_f - S_i = \int \frac{dQ}{T} = \frac{1}{T} \int dQ$$

since the temperature is constant throughout the process. The heat for the process is

$$\int dQ = mL_v = (1.00 \text{ kg})(2256 \text{ kJ/kg}) = 2256 \text{ kJ}$$

Therefore

$$S_f - S_i = \frac{2256 \text{ kJ}}{373.15 \text{ K}} = 6.046 \text{ kJ/K}$$

Example 2   A mass of 1.00 kg of water is warmed (at a constant pressure of 1 atm) from 0.0°C to 100.0°C:
(a) What is the entropy change $S_f - S_i$ of the water during this process?
(b) What is the entropy change of this water when it is cooled from 100.0°C to return to its initial state at 0.0°C and a pressure of 1 atm?

Solution
(a) The entropy change of a system of mass m with a constant heat capacity mc'_P, during a quasi-static isobaric transformation from temperature T_i to T_f, is

$$S_f - S_i = \int_{T_i}^{T_f} \frac{dQ}{T} = \int_{T_i}^{T_f} \frac{mc'_P\, dT}{T} = mc'_P \ln \frac{T_f}{T_i}$$

This gives

$$S_f - S_i = (1.00 \text{ kg})(4.2 \text{ kJ/kg·K}) \ln (373.15/273.15) = 1.31 \text{ kJ/K}$$

(b) When the system returns to its initial state, the system's entropy returns to its initial value S_i. Consequently in the cooling process the entropy of the water *decreases* by an amount equal to the increase found in part (a), 1.31 kJ/K.

These examples illustrate how we can calculate entropy changes from measurements of heat capacities and latent heats. Such calculations provide the data from which entropy tables like Table 20-2 are constructed. Notice that since we are concerned only with *changes* of entropy, we can arbitrarily select some state as a reference state and tabulate only the differences in entropy between other states and this reference state.

The entropy of a composite system is the sum of the entropies of its parts. This is implied by the Clausius equation. Consequently the entropy of a mass m of a substance is simply m times the entropy per unit mass (the specific entropy) of the substance.

Heat Reservoirs

A heat reservoir is a system so large compared to the systems with which it interacts that its temperature remains essentially unchanged when it absorbs or gives off heat. The Clausius equation allows the calculation of the entropy change of the reservoir to be performed very simply. For a reservoir at Kelvin temperature T we obtain

$$entropy\ increase\ of\ reservoir = \frac{heat\ absorbed\ by\ reservoir}{T} \tag{20.4}$$

or

$$entropy\ decrease\ of\ reservoir = \frac{heat\ given\ off\ by\ reservoir}{T} \tag{20.5}$$

These equations give the entropy changes of a heat reservoir correctly for *any* process without restriction. We shall use these equations frequently in the following sections.

Example 3   A hot stone is placed in a lake which has a temperature of 290 K. As the stone cools, it gives off 300 kJ. What is the change in entropy of the water of the lake?

Solution   To an excellent approximation the lake can be considered to be a heat reservoir. The initial and final water temperature is 290 K. Therefore the entropy increase of the lake is given by

$$\Delta S = \frac{heat\ absorbed\ by\ reservoir}{T} = \frac{300\ kJ}{290\ K} = 1.03\ kJ/K$$

20.2 Second Law of Thermodynamics

Principle of Entropy Increase

Many processes that would seem possible according to the first law of thermodynamics, because they do conserve energy, in fact never occur. For example, a brick resting on a horizontal road never spontaneously accelerates along the road by gathering up thermal energy from the road (Figure 20-1). If this process occurred, the energy which is randomly distributed over many molecules of the road would be converted into an ordered motion of the brick as a whole. This transformation from disorder to order is quite compatible with the first law, but for some reason it does not happen.

Figure 20-1   The spontaneous acceleration of the brick along a horizontal road does *not* occur. This would involve a transformation of *disordered* molecular motion into *ordered* motion of the brick as a whole.

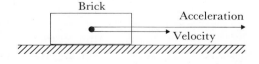

The reverse process (Figure 20-2) does occur and it illustrates a feature present in all our experience. When a brick is projected along the road, it slows down as the kinetic energy due to the *ordered* motion of the brick as a whole is transformed into thermal energy associated with *disordered* molecular motion within the road and the brick.

Examination of this and many other processes reveals that there is a general rule which determines the direction in which energy transformations occur. Ordered energy is transformed into disordered energy; that is, using an arrow to denote the direction of the process,

order → disorder

But the reverse transformation is not observed. The disorder in one object can be reduced in a process only if there is a more than compensating increase in the disorder of other objects that participate in the process.

This law of nature which determines the direction in which processes occur can be given a quantitative formulation in terms of *entropy*. The entropy of an object is a measure of its molecular disorder. Increasing disorder implies increasing entropy. The rule is that processes take place only in the direction that increases the *total entropy* of the participating objects. Stated formally, this is the *principle of entropy increase*: *In any process, the total entropy of the participating objects is increased or* (in an idealized "reversible process") *unchanged. The total entropy never decreases.*

Reversible and Irreversible Processes

Life is a one-way street. We are born, we grow old, and we die. The physical law that distinguishes the past from the future, that puts the arrow on time in

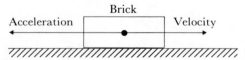

Figure 20-2   The reverse of the process shown in Figure 20-1. The brick slows down as the kinetic energy associated with the *ordered* motion of the brick as a whole is transformed into *disordered* molecular motion (thermal energy) of the brick and the road. This process occurs.

Arrow of time

Figure 20-3   The situation with less disorder comes before the situation with greater disorder.

Figure 20-3 is the *principle of entropy increase*. The situation with lower entropy (less disorder) must come before that with higher entropy (more disorder). This is how we can tell that, if a movie shows a broken egg reassemble into a perfect egg, the movie must be running backward.

Any real process involves an increase in the total entropy. A reversal of such a process, leading to the restoration of the initial entropies of all the participating objects, would therefore involve a decrease in entropy. The entropy principle thus implies that entropy-increasing processes are *irreversible* processes.

A detailed examination shows that *irreversible* entropy increases occur when:

1   Macroscopic kinetic or potential energy is dissipated into thermal energy. This happens when work is done by a frictional force [or when Joule heating occurs in a resistor (Chapter 25)].
2   Heat is transferred from an object at a higher temperature to an object at a lower temperature.
3   Mechanical equilibrium of all portions of a system is not maintained; therefore accelerations, eddies, or turbulence develop.
4   Phase equilibrium (Section 19.4) is not maintained. (We shall not consider chemical reactions and the irreversibility arising from failure to maintain chemical equilibrium.)

The principle of entropy increase leads to the classification of processes displayed in Table 20-3. Processes for which there is no change in the total entropy can occur in *either* direction. They are called *reversible processes*.

A reversible process is an idealization never observed in macroscopic experiments; however it can be closely approximated by (1) minimizing dissipative effects resulting from frictional forces or electrical resistance, (2) transferring heat only through very small temperature differences, (3) avoiding accelerations and turbulence, and (4) avoiding appreciable departures from phase or chemical equilibrium.

The Clausius equation, $dS = dQ/T$, is valid only for reversible processes. This is why reversible processes are so important in thermodynamic theory. To determine the entropy function for a substance we always consider reversible processes and employ the Clausius equation. Once we know the entropy of a substance in different states, we can use this information to calculate the substance's change in entropy, no matter what process brings it from one state to another.

Table 20-3 Classification of Processes According to Principle of Entropy Increase

ΔS_{total} *increases*	Process is possible; it will be irreversible.
ΔS_{total} *decreases*	Process is impossible (statistical mechanics says improbable).
$\Delta S_{total} = 0$	Process in reversible; it may occur in either direction.

Example 4   A kettle contains 1.00 kg of hot water at 80.0°C. This is poured into a pot containing 1.00 kg of cold water at 0.0°C. After thorough mixing, the water has a uniform temperature of 40.0°C throughout. Find the change in the total entropy which occurred in this process.

Solution   We use the entropies of water given in Table 20-2:

$$\text{final total entropy} = (2.00 \text{ kg})(1.87 \text{ kJ/kg·K}) = 3.74 \text{ kJ/K}$$

$$\text{initial entropy of hot water} = (1.00 \text{ kg})(2.38 \text{ kJ/kg·K})$$

$$\text{initial entropy of cold water} = (1.00 \text{ kg})(1.30 \text{ kJ/kg·K})$$

$$\text{initial total entropy} = 3.68 \text{ kJ/K}$$

$$\Delta S_{\text{total}} = 3.74 \text{ kJ/K} - 3.68 \text{ kJ/K} = 0.06 \text{ kJ/K}$$

The total entropy increases so the process is irreversible. This is no surprise; we know we cannot "unmix" the hot and cold water.

Notice that the Clausius equation was used only in the construction of the entropy table, as outlined in Section 20.1. Having once done this job, we can use these data to compute entropy changes in any process, no matter how irreversible the process is.

Example 5   Calculate the total entropy change ΔS_{total} associated with the free expansion (Figure 18-7) of an ideal gas from an initial equilibrium state (volume V_i, temperature T) to a final equilibrium state (volume $V_f = 2V_i$, temperature T).

Solution   During the free expansion, the entropy change of the surroundings is zero. From Eq. 20.3, the change in entropy of the ideal gas is

$$S_f - S_i = nR \ln \frac{V_f}{V_i} = nR \ln 2$$

Therefore the total entropy change associated with this process is

$$\Delta S_{\text{total}} = nR \ln 2$$

Since this is positive, the free expansion is irreversible. This accords with the fact that we do not observe the gas to pass spontaneously from a state where it fills the entire box of Figure 18-7 to a state where it is confined to the left half of the box.

In this example, the Clausius equation is used only in the determination of the entropy function of an ideal gas. After this function is found, it can be used to determine the difference of entropies for any two equilibrium states of the gas.

Entropy and the Second Law of Thermodynamics

The most important properties of entropy for macroscopic theory are gathered together in a summary that constitutes a statement of the *second law of thermo-*

dynamics: To each equilibrium state of a system we can assign a definite value of a quantity S called *entropy* which has the following properties:

1   *Clausius equation*   In any infinitesimal reversible process in which the system absorbs heat dQ, the change in entropy of the system is

$$dS = \frac{dQ}{T} \qquad (20.1)$$

where T is the Kelvin temperature of the system.

2   *Principle of entropy increase*   In any process, the total entropy of the participating objects is increased or (in an idealized reversible process) unchanged. The total entropy of an isolated system never decreases.

The second law is relevant not only in engineering and physics, but also in chemistry and biology. The direction in which chemical reactions proceed is governed by the principle of entropy increase. Consequently, the entropy of different chemicals is of vital interest to chemists.

In biological processes, reduction of disorder is often obvious as random collections of molecules are exquisitely organized into living matter. Do such processes violate the second law? Apparently not. If we look at *all* the objects participating in the process, we find that the obvious entropy decreases have been more than matched by entropy increases elsewhere.

Entropy Increase Due to Heat Conduction

The quantitative implications of the second law are brought out by studies of certain simple physical processes. Heat conduction provides an instructive example.

We consider a process in which a finite amount of heat Q flows spontaneously from a reservoir at temperature T_1 to a reservoir at temperature T_2 (Figure 20-4). The process takes place by heat conduction through a copper wire joining the reservoirs. The entropy changes associated with this process are the following:

1   *Entropy increase of cold reservoir* $= Q/T_2$.
2   *Entropy decrease of hot reservoir* $= Q/T_1$.

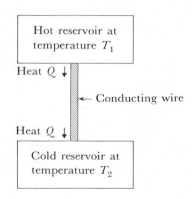

Figure 20-4   Conduction. A heat Q is given off by the hot reservoir and is absorbed by the cold reservoir. The resulting entropy change is $(Q/T_2) - (Q/T_1)$.

Hot reservoir at temperature T_1

Heat Q ↓

← Conducting wire

Heat Q ↓

Cold reservoir at temperature T_2

We neglect the entropy change of the wire. This entropy change is proportional to the mass of the wire and will be small if this mass is small. In fact, if each portion of the wire is maintained in the same thermodynamic state (constant temperature and pressure), its entropy will not change at all.

The change in the total entropy because of this process is therefore given by

$$\Delta S_{total} = \frac{Q}{T_2} - \frac{Q}{T_1} \tag{20.6}$$

Therefore

$$\Delta S_{total} = Q \left(\frac{T_1 - T_2}{T_2 T_1} \right)$$

This result shows that if T_1 is greater than T_2, that is, if the heat flow is from higher to lower temperature, ΔS_{total} is positive. Then the total of the entropies of the participating objects increases. This verifies that our everyday experience with the direction of heat flow agrees with the principle of entropy increase. It is worth emphasizing that when heat flows from a higher to a lower temperature, the total entropy always is increased; it never merely remains unchanged.

This process is irreversible. A heat flow from the lower temperature T_2 to the higher temperature T_1 would result in a decrease in the total entropy in violation of the second law of thermodynamics. The *Clausius statement* of the second law ("No process is possible whose *sole* result is the transfer of heat from a cooler to a hotter body.") has emerged as a consequence of our formulation of the second law in terms of the principle of entropy increase. But some interesting speculations on ways and means to circumvent this stubborn one-way process have been made from time to time. Consider the following.

A well-informed little athlete with lightning reflexes, called a Maxwell demon, is pictured in Figure 20-5. This figment of Maxwell's imagination devotes his life to confounding Clausius by opening the door at the correct times to permit the fastest molecules to move to the right half of the box, and the slowest molecules to move to the left half. The right half gets hotter and the left half cooler.

Maxwell pointed out that if this demon or his mechanized equivalent existed and could continue indefinitely without intervention from outside the box,

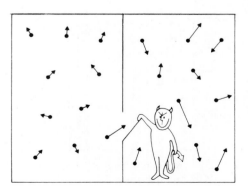

Figure 20-5   Maxwell demon.

Figure 20-6 Schematic diagram of a perpetual-motion engine of the second kind which extracts heat from one reservoir and converts this heat entirely into work. Such a device is an impossibility according to the second law of thermodynamics.

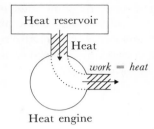

then the second law of thermodynamics could be violated. But so far no demons have been detected and none are expected.

Perpetual-Motion Engines of the Second Kind

Another immediate and important consequence of the second law is that *no process is possible whose sole effect is to extract heat from one reservoir and convert this heat entirely into work.** The key word in this statement is *sole*. If, at the end of a process, all objects except the heat reservoir (Figure 20-6) are returned to the equilibrium states in which they started, the entropy of each object will return to its initial value. Therefore the entropy *change* of each of these objects will be zero for the process. The only entropy change will be the *entropy decrease* of the heat reservoir. This process would therefore cause a decrease in the total entropy of the participating objects. According to the second law, this is impossible.

Inventors who have given up trying to beat the first law by creating energy with a "perpetual-motion engine of the first kind" will be equally frustrated in their efforts to devise a "perpetual-motion engine of the second kind" which conserves energy but violates the second law. An engine which can run forever solely by extracting heat from the ocean is unfortunately impossible. In the following section we shall find that to make a "heat engine" that works, a temperature difference is required and the engine which would use heat from the ocean must give off some heat at a temperature lower than that of the ocean. To persistent seekers of perpetual-motion engines, the first law says that you can't win; then the second law adds that you can't even break even!

20.3 Heat Engines and Refrigerators

Heat Engines

A heat engine is a device that absorbs a heat Q and performs a work W. Practical examples are the steam, gasoline, and diesel engines. Consider an engine that runs for an indefinite period and therefore operates in *cycles*, returning after each

---

*The impossibility of such a process is often taken as the (Kelvin) statement of the second law of thermodynamics. It is then possible, by following a long and interesting chain of reasoning, to deduce the statements about entropy that we have adopted as the second law.

cycle to exactly the same state as that in which it started. The entropy and the internal energy of the engine and its contents therefore return to their initial values after each cycle. In other words, the change of the entropy and the change of the internal energy of the engine and its contents are zero after one cycle. That is, for one cycle

$$\Delta S_{\text{engine}} = 0$$

and

$$\Delta \overline{E}_{\text{engine}} = 0$$

The *thermal efficiency* of a heat engine is defined in terms of the heat Q furnished to the engine in one cycle and the work output W in one cycle:

$$thermal\ efficiency = \frac{W}{Q}$$

This is the ratio of the output desired to the input furnished.

The second law of thermodynamics is such a sweeping generalization that it is possible to discover interesting limitations on the efficiency of a heat engine without any consideration of detailed mechanisms. An examination of entropy changes is necessary. The extraction, during each cycle, of a heat Q_{hot} from a hot reservoir at Kelvin temperature T_{hot} decreases the entropy of this reservoir by an amount $Q_{\text{hot}}/T_{\text{hot}}$. Therefore a compensating entropy increase must be provided. This is not to be found in the engine itself nor in its contents, because after one cycle both return to the equilibrium state in which they started and their entropies consequently return to their initial values ($\Delta S_{\text{engine}} = 0$). Therefore some heat Q_{cold} must be given off by the engine in each cycle to increase the entropy of some other object (Figure 20-7). If this heat Q_{cold} is absorbed by a second heat reservoir at Kelvin temperature T_{cold}, the entropy of the reservoir will increase by an amount $Q_{\text{cold}}/T_{\text{cold}}$. To achieve the largest possible compensating entropy increase for a given quantity of heat Q_{cold}, the temperature T_{cold} should be as low as possible. Entropy considerations thus show that the reservoir at temperature T_{cold} should be the coldest available. As far as the reservoir at temperature T_{hot} is concerned, the hotter the better.

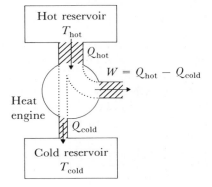

Figure 20-7 Schematic diagram of a heat engine which, in one cycle, extracts heat Q_{hot} and rejects heat Q_{cold}, performing a work W equal to $Q_{\text{hot}} - Q_{\text{cold}}$. The entropy increase of the cold reservoir ($Q_{\text{cold}}/T_{\text{cold}}$) more than compensates for the entropy decrease of the hot reservoir ($Q_{\text{hot}}/T_{\text{hot}}$).

The principle of entropy increase requires that the entropy decrease of the hot reservoir be not greater than the entropy increase of the cold reservoir. That is,

$$\frac{Q_{hot}}{T_{hot}} \leq \frac{Q_{cold}}{T_{cold}}$$

which implies

$$\frac{T_{cold}}{T_{hot}} \leq \frac{Q_{cold}}{Q_{hot}} \tag{20.7}$$

It is important that we examine the thermal efficiency of this engine. Let us apply the first law of thermodynamics to the system consisting of the engine and its contents. For one cycle, since $\Delta \overline{E} = 0$, the first law gives

$$0 = (Q_{hot} - Q_{cold}) - W$$

where W is the work output of the engine in one cycle. Solving this equation for W/Q_{hot}, we obtain

$$thermal\ efficiency = \frac{W}{Q_{hot}} = 1 - \frac{Q_{cold}}{Q_{hot}}$$

Together with Eq. 20.7, this implies

$$thermal\ efficiency \leq 1 - \frac{T_{cold}}{T_{hot}} \tag{20.8}$$

The maximum possible thermal efficiency of an engine operating between two given heat reservoirs, called the *Carnot efficiency*, is then seen to be given by

$$Carnot\ efficiency = 1 - \frac{T_{cold}}{T_{hot}} \tag{20.9}$$

Sadi Carnot obtained this important result in the early 1820s before the nature of heat was understood. His ideas were taken up decades later by Kelvin and Clausius and thus contributed to the modern formulation of the second law of thermodynamics.

The foregoing analysis provides guidelines in the design of practical heat engines. At some portion in a cycle, an engine must exhaust some heat to a cool body, usually to the atmosphere or to a nearby river. With the "cold reservoir" temperature determined by nature, the only way an engine designer can increase the Carnot efficiency is to have the heat intake occur at as high a temperature as possible. For this reason steam engines utilize high-pressure boilers to obtain steam superheated to temperatures well above 100°C.

Example 6 What is the Carnot efficiency of a steam engine with a boiler temperature of 177°C which exhausts heat to a condenser at a temperature of 27°C?

Solution

$$Carnot\ efficiency = 1 - \frac{300\ K}{450\ K} = 0.33$$

This value (33%) is considerably greater than the overall efficiency of about 20% that is typically achieved in practice.

Only when there is no increase in the total entropy does the engine's thermal efficiency attain its maximum possible value—the Carnot efficiency. (The equality signs then hold in Eq. 20.7 and 20.8.) When $\Delta S_{total} = 0$, the process is *reversible*. For this to be true, each process in the engine's operation must be reversible and therefore subject to all the limitations mentioned in Section 20.2. An engine capable of operating in this idealized way is called a *Carnot engine*. The efficiency of any real engine is reduced below the Carnot efficiency because of work done by frictional forces, heat transfer through rather large temperature differences, turbulence and acceleration of gases in the cylinders, and failure to maintain phase equilibrium or chemical equilibrium.

Thermal Pollution

When a river, lake, or ocean is used as a cold reservoir, the heat Q_{cold} that it absorbs may cause damage called "thermal pollution." For example, when electric power is generated using oil-fueled installations, the discharge of cooling water to the ocean causes the ocean temperature in the vicinity to rise appreciably, a typical increase being 10 K.

The increased water temperature has two major effects. There is a reduction in the amount of dissolved oxygen in the water and an increase in the rate of chemical reactions occurring within an organism. Fish can be adversely affected. Since their metabolism is speeded up they require more oxygen, but there is less dissolved oxygen in the water. This difficulty coupled with the evidence that reproduction is often inhibited in thermally stressed environments can result in a sharp decline or elimination of certain fish species.

More subtle changes also occur. At elevated temperatures there is a shift from an edible algae (from the point of view of small aquatic invertebrates and fish) to a less desirable or even toxic blue-green algal species. There is then danger of a general disruption in the entire food web. Already, cries of concern from sports fishermen and bird enthusiasts have gained public attention.

Our study of heat engines has shown that although the rejection of the polluting heat Q_{cold} seems a senseless waste of energy as far as the first law of thermodynamics is concerned, this rejection is absolutely necessary according to the second law. The compensating entropy increase of the cold reservoir Q_{cold}/T_{cold} must be provided for the engine to operate. Thermal pollution is an unpleasant consequence of the second law of thermodynamics.

Refrigerators

Significant general features of the operation of a refrigerator can be obtained from an analysis similar to that given for heat engines. In fact a refrigerator can be regarded as a heat engine run in reverse (Figure 20-8).

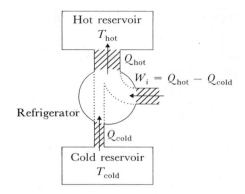

Figure 20-8 Schematic diagram of a refrigerator (a heat engine run in reverse). In one cycle there is a work input W_i equal to the difference between the heat rejected to the hot reservoir and the heat extracted from the cold reservoir.

Consider a refrigerator which has a *work input* W_i in one cycle during which it extracts a heat Q_{cold} from a cold reservoir at Kelvin temperature T_{cold} and gives off a heat Q_{hot} to a hot reservoir at Kelvin temperature T_{hot}. The purpose of the refrigerator is to extract as much heat Q_{cold} as is possible for a given work input W_i. In a home refrigerator the work is done by an electrically driven compressor. The ice and food inside the refrigerator constitute the cold reservoir; the kitchen air is the hot reservoir.

The lower limit to the work required to operate the refrigerator is easy to determine. The first law of thermodynamics is applied to one cycle of operation of the device represented by the circle in Figure 20-8. For one cycle, $\Delta \overline{E} = 0$, and the first law gives

$$0 = (Q_{cold} - Q_{hot}) - (-W_i)$$

Therefore

$$W_i = Q_{hot} - Q_{cold} = Q_{cold}\left(\frac{Q_{hot}}{Q_{cold}} - 1\right) \qquad (20.10)$$

The entropy of the hot reservoir increases and that of the cold reservoir decreases. For the compressor and contents after one cycle $\Delta S_{engine} = 0$. From the second law of thermodynamics we conclude that the entropy increase Q_{hot}/T_{hot} of the hot reservoir must be greater than the entropy decrease Q_{cold}/T_{cold} of the cold reservoir. That is

$$\frac{Q_{hot}}{T_{hot}} \geq \frac{Q_{cold}}{T_{cold}}$$

which implies

$$\frac{Q_{hot}}{Q_{cold}} \geq \frac{T_{hot}}{T_{cold}}$$

This result, together with Eq. 20.10, gives

$$W_i \geq Q_{cold}\left(\frac{T_{hot}}{T_{cold}} - 1\right) \qquad (20.11)$$

Example 7   What is the minimum amount of power that must be supplied to a refrigerator that freezes 2.00 kg of water at 0°C into ice at 0°C in a time interval of 5.00 min? The room temperature is 20°C.

Solution   The heat Q_{cold} which is extracted from the water is given by

$$Q_{cold} = mL_f = (2.00 \text{ kg})(333 \text{ kJ}) = 666 \text{ kJ}$$

From Eq. 20.11, the minimum work which must be supplied is given by

$$W_i = Q_{cold}\left(\frac{T_{hot}}{T_{cold}} - 1\right) = (666 \text{ kJ})\left(\frac{293 \text{ K}}{273 \text{ K}} - 1\right) = 48.8 \text{ kJ}$$

$$power = \frac{48.8 \text{ kJ}}{300 \text{ s}} = 163 \text{ W}$$

Those who are interested in the ingenious methods used to make practical engines and refrigerators may be disappointed to find that we have not examined how particular devices work. But the beauty of thermodynamics is that we can make useful statements about *all* such devices no matter what the details.

20.4 Degradation of Energy

The entropy increases that occur in natural processes measure the increased disorder. For engineers and philosophers, another interpretation of the significance of entropy increases is perhaps more useful. A total entropy increase is associated with a lost opportunity to obtain useful mechanical work. When a process leads to an increase in the total entropy, some energy which could have been used to perform macroscopic mechanical work has been *degraded to a form in which it is unavailable for work*. Natural processes occur in a direction such that, on a scale of availability for mankind's use, energy runs downhill.

Heat conduction (Figure 20-4) provides a clear example. We found in Section 20.2 that when heat Q passes from a hot reservoir at temperature T_1 to a cold reservoir at temperature T_2 there is an increase of total entropy given by

$$\Delta S_{total} = \frac{Q}{T_2} - \frac{Q}{T_1} \tag{20.6}$$

If, instead, this heat Q had been used as the input to a heat engine operating with the Carnot efficiency $(1 - T_2/T_1)$, the work output that would have been obtained is

$$W = Q\left(1 - \frac{T_2}{T_1}\right) = T_2\left(\frac{Q}{T_2} - \frac{Q}{T_1}\right)$$

Comparison with Eq. 20.6 shows that

$$W = T_2\Delta S_{total} \tag{20.12}$$

We see that when a heat conduction process transfers all the heat Q into the cold reservoir, the opportunity to achieve a work $T_2\Delta S_{total}$ is lost. An amount of energy $T_2 \Delta S_{\text{total}}$ proportional to the entropy increase ΔS_{total} is made unavailable for work by the entropy-increasing process.

The mixing of hot and cold water which results in lukewarm water is an everyday example of the lost opportunity associated with a total entropy increase (see Example 4). Before mixing, a heat engine could have performed useful work by taking in heat from the hot water and rejecting heat to the cold water. This cannot be done with the lukewarm water at a uniform temperature.

The entropy-increasing direction of all natural processes tends to bring about uniformity of temperature, pressure, and composition. A gloomy topic for natural philosophers is the possibility of an ultimate "heat death" of the universe when a situation of complete uniformity is reached. Since no process could then increase the total entropy, physical, chemical, and biological change would cease.

Instead of pursuing this perhaps unjustified extrapolation of our knowledge, we shall survey our present situation. Our modern society has an ever-increasing requirement for macroscopic work. To fulfill this requirement, we take energy that has been stored in an available form and ultimately degrade it into thermal energy of our environment. Apart from nuclear fuel, our planet's main source of available energy is solar radiation. Most of the solar energy arriving at our planet is immediately degraded into thermal energy. Fortunately, there are natural processes which store a significant amount of the incident solar energy in very available forms. Short-term storage as potential energy of water is found in elevated lakes and rivers. Long-term storage of available energy occurs as complex molecules are formed and become constituents of trees and eventually of fossil fuels.

20.5 Entropy in Statistical Mechanics

Microscopic Definition of Entropy

Preceding sections illustrate the utility of entropy calculations in situations of practical interest. We now examine the significance of entropy from a molecular point of view and show how the entropy of a collection of molecules is defined.

Consider a typical macroscopic system such as 0.0180 kg of water in a test tube. A specification of the temperature and volume determines the macroscopic equilibrium state of this water. When this same system, a collection of 6.02 $\times$ 10^{23} H_2O molecules, is described from a microscopic viewpoint using Newtonian mechanics or quantum mechanics (Chapter 42), we find that there are many different microscopic states* which correspond to essentially the same macroscopic equilibrium state. Roughly speaking, there are many different molecular motions that produce the same behavior on the macroscopic scale.

---

*In Newtonian mechanics, a microscopic state of a system of N particles is specified by giving the position and velocity of each particle. In quantum mechanics a microscopic state is specified by giving the wave function (Chapter 42) of the system.

When we know only the macroscopic equilibrium state, the system can be in any one of many microscopic states. These are called the *accessible* states for this particular macroscopic state.

The disorder of a macroscopic system is measured by the *number Ω of accessible states for the macroscopic state in question.* The number Ω is just the number of microscopically different situations that correspond to the given macroscopic equilibrium state. As Feynman puts it, "we measure 'disorder' by the number of ways that the inside can be arranged, so that from the outside it looks the same."[*]

The entropy S of a macroscopic system in a macroscopic equilibrium state is defined in terms of the number Ω of its accessible states by the following equation:

$$S = k \ln \Omega$$

where k is Boltzmann's constant. Using this definition, the entropy of a system can be calculated from knowledge of the microscopic states of the system.

The definition of entropy implies that a system composed of parts has an entropy which is the sum of the entropies of each part. This can be seen by considering a macroscopic state of a composite system with two parts. One part of the system can be found in Ω_1 microscopically different states corresponding to an entropy S_1. The other part can be found in Ω_2 microscopically different states corresponding to an entropy S_2. The number of microscopically different states for the composite system is given by

$$\Omega = \Omega_1 \Omega_2$$

since each of the Ω_1 states of one part can occur in conjunction with the Ω_2 different states of the other part. Therefore the entropy S of the composite system is

$$S = k \ln \Omega = k \ln \Omega_1\Omega_2 = k \ln \Omega_1 + k \ln \Omega_2 = S_1 + S_2$$

which shows that the entropy of the composite system is the sum of the entropies of its parts. This is the advantage of working with the entropy S instead of with the related number Ω of microscopically different states.

Entropy and Probability

The development of the microscopic theory of the behavior of large collections of molecules constitutes the subject of *statistical mechanics.* In this subject, the second law of thermodynamics acquires a new and interesting interpretation.

Statistical mechanics focuses attention on the number Ω of accessible states. As time goes on, a macroscopic system makes transitions from one microscopic state to another. Ultimately the system will have spent some time in each of its accessible states. The situation achieved after many transitions is analogous to

[*]*The Feynman Lectures on Physics,* Vol. 1, pp. 46–7, Addison-Wesley, 1963.

that produced by repeated shuffling of a pack of cards. No matter how the deck is stacked initially, after sufficient shuffling no one arrangement of the cards is particularly favored. All possible arrangements become equally likely. Similarly, after sufficient time has elapsed, a collection of many molecules reaches a situation where the system is found with *equal probability* in each one of its Ω accessible states. This situation characterizes an *equilibrium* state of the macroscopic system.

Suppose a system which had been confined to only a portion Ω_1 of Ω microscopic states has new possibilities opened up so that all of these Ω states become accessible. Transitions occur and after sufficient time the system is equally likely to be in each of the Ω accessible states. This is a spontaneous change from an entropy $S_1 = k \ln \Omega_1$ to a larger entropy $S = k \ln \Omega$. The direction of the spontaneous change in an isolated system is seen to be the direction which increases the entropy.

After an isolated system has reached equilibrium with Ω accessible states, the probability of its being found in some situation corresponding to a restriction to only Ω_1 of these Ω states is the fraction Ω_1/Ω. From this fraction one can compute the probability of fluctuations of a given size in macroscopic quantities such as the pressure and the density. Fluctuations occur continually, but if the system contains a large number of molecules, the fluctuations will be small.

In general, we find that in statistical mechanics the second law of thermodynamics and all its consequences are statements about probabilities, not certainties. Entropy-decreasing processes are not impossible, merely improbable. The principle of entropy increase is toned down to say "it is highly improbable that entropy-decreasing processes will be observed." But computations of the probabilities involved show that the measurable entropy decreases of isolated macroscopic systems considered in Section 20.2 are indeed so highly improbable that the second law can be regarded as stating engineering certainties. Eddington knew how to give us a feeling for these small probabilities. Comtemplating a group of wild monkeys punching randomly on a set of typewriters, he deduced that they could be expected to reel off without error Shakespeare's complete works many quadrillion times in succession with about the same frequency as an observer would see a stone leap into the air by converting one joule of thermal energy into macroscopic kinetic energy.

Temperature Defined as $d\overline{E}/dS$

When we describe thermal phenomena using statistical mechanics, there is no need to introduce temperature by means of separate postulates.* Temperature can be *defined* in terms of the entropy change associated with a change of internal energy.

*It is not necessary to resort to statistical mechanics to define temperature. In the purely *macroscopic* science of thermodynamics, the Kelvin temperature scale can be defined using the zeroth law and the Clausius statement of the second law. Such an analysis is appropriate material for more advanced courses.

The entropy of a system is $S = k \ln \Omega$. This definition determines the value of S/k. Notice that S can be calculated only when the value of Boltzmann's constant k has been fixed.

Consider a system whose entropy is determined by its internal energy $\overline{E}$ and volume V. The temperature of this system is *defined* by

$$\frac{1}{T} = \left(\frac{\partial S}{\partial \overline{E}}\right)_V \tag{20.13}$$

This means that in a process in which the volume is maintained constant, if a small flow of heat produces a change $d\overline{E}$ in the system's internal energy and a change dS in the system's entropy, the Kelvin temperature of the system is

$$T = \frac{d\overline{E}}{dS} \qquad (V \text{ constant}) \quad (20.14)$$

This definition implies that the "hotness" of an object is determined not by its internal energy nor by its entropy but rather by the amount that its entropy *changes* when its internal energy is changed. For a given change $d\overline{E}$ in internal energy, the entropy change of a hot object is smaller than the entropy change of a cold object. In other words, the hot object has a larger value of $d\overline{E}/dS$ than the cold object. We burn our fingers if we touch something whose entropy changes too little when its internal energy changes.

The familiar properties of temperature that were outlined in Section 19.3 can now be deduced from the definition $T = d\overline{E}/dS$. Consider the transfer of thermal energy $d\overline{E}$ from an object at temperature $T_1 = d\overline{E}/dS_1$ to another object at temperature $T_2 = d\overline{E}/dS_2$. If the first object is warmer ($T_1 > T_2$), the entropy decrease of the first object has a smaller magnitude than the entropy increase of the second. The transfer of thermal energy then increases the total entropy. Thus our new definition of temperature, together with the rule that entropy-increasing processes will occur spontaneously, implies this: When two objects are placed in thermal contact, thermal energy will be transferred from the warmer object to the cooler until both reach equilibrium states at the same temperature, that is, the same value of $d\overline{E}/dS$.

In the process that we have been considering, the changes in internal energy were entirely due to a thermal energy transfer (heat). Then for a given system, $d\overline{E} = đQ$. For such a process the definition, $T = d\overline{E}/dS$, implies that $dS = đQ/T$. This is the familiar Clausius equation. With some labor we can deduce that the Clausius equation remains valid for a reversible process involving both heat and work so that the internal energy change is

$$d\overline{E} = đQ - đW$$

It turns out that we still have

$$dS = \frac{đQ}{T}$$

Both the work and the heat change a system's internal energy, but in a reversible process only the heat changes a system's entropy.

Kelvin Temperature Scale and Boltzmann's Constant

The *Kelvin* temperature T is defined by arbitrarily assigning the value 273.16 K as the Kelvin temperature of water at its triple point. This fixes the value of Boltzmann's constant k. We take any substance whose thermal behavior is theoretically understood [so that $(1/k)(\partial S/\partial \overline{E})_V$ can be expressed in terms of macroscopically measurable properties] and make experimental measurements when the substance is in thermal equilibrium with water at its triple point. Then k can be calculated from

$$\frac{1}{kT} = \frac{1}{k}\left(\frac{\partial S}{\partial \overline{E}}\right)_V \tag{20.15}$$

One simple substance is any collection of N weakly interacting particles. Theory shows that for this substance the right-hand side of Eq. 20.15 is N/PV. This equation then becomes the ideal gas law:

$$kT = \frac{PV}{N}$$

Therefore Boltzmann's constant can be determined by measurement of N, the pressure P, and the volume V, for any sufficiently rarefied gas in equilibrium with water at its triple point. (The experimental data were given and calculation of k performed in Section 17.1.) Furthermore, the measurements of pressures P_1 and P_2 of a sufficiently rarefied gas in a constant-volume gas thermometer at different Kelvin temperatures T_1 and T_2 can be used to determine the ratio of these temperatures. This follows from the ideal gas law, which implies that

$$\frac{T_1}{T_2} = \frac{P_1}{P_2}$$

when N and V are constant.

Summary

☐ *A statement of the second law of thermodynamics is*: Any thermodynamic system has a state function S, called its entropy, with the following properties:

1 *Clausius equation* In any infinitesimal reversible process in which the system absorbs a heat dQ, the change in entropy of the system is

$$dS = \frac{dQ}{T}$$

where T is the Kelvin temperature.

2 *Principle of entropy increase* In any process, the total energy of the participating objects is increased or (in a reversible process) unchanged. The total entropy of an isolated system never decreases. (A reversible process is an idealization obtained if the process is quasi-static and if there are no dissipative effects.)

☐ The maximum possible thermal efficiency of an engine operating between two given heat reservoirs is the Carnot efficiency given by

$$\text{Carnot efficiency} = 1 - \frac{T_{\text{cold}}}{T_{\text{hot}}}$$

where the Kelvin temperatures of the hot and cold reservoirs are T_{hot} and T_{cold}, respectively.

☐ When a refrigerator in one cycle extracts a heat Q_{cold} from a cold reservoir, the work input W_i to the refrigerator satisfies

$$W_i \geq Q\left(\frac{T_{\text{hot}}}{T_{\text{cold}}} - 1\right)$$

☐ When an irreversible process occurs producing an increase ΔS_{total} in the entropy of all the participating objects, the energy that is thereby made unavailable for work is $T_2 \Delta S_{\text{total}}$, where T_2 is the lowest available temperature.

☐ The entropy of a system is a measure of its molecular disorder. In a given macroscopic equilibrium state which corresponds to Ω microscopically different states, the entropy is

$$S = k \ln \Omega$$

where k is Boltzmann's constant.

☐ The Kelvin temperature T, of a system which has an entropy function S determined by the system's internal energy $\overline{E}$ and volume V, is defined by

$$\frac{1}{T} = \left(\frac{\partial S}{\partial \overline{E}}\right)_V$$

Questions

1 **(a)** What does entropy measure?
  (b) Would you expect a kilogram of solid lead to have more or less entropy than a kilogram of molten lead? Explain on the basis of your answer to part (a).

2 Using the data of Table 20-2, calculate the change in entropy of 10.0 kg of H_2O molecules when they are transformed from water at 20.0°C to steam at 250°C. (Pressure = 1 atm.)

3 Using the data of Table 20-2, calculate the change in entropy of 5.00 kg of H_2O molecules when they are transformed from water at 20.0°C to ice at -10.0°C. (Pressure = 1 atm.)

4 Find the change in entropy of 2.00 kg of carbon dioxide during sublimation from "Dry Ice" to gas at a pressure of 1.00 atm and a temperature of -78.5°C. The latent heat of sublimation under these conditions is 577 kJ/kg.

5 Find the change in entropy of 3.00 kg of H_2O molecules at a pressure of 1.00 atm when the temperature is raised from 26.0°C to 28.0°C.

6 What is the entropy change of 50.0 kg of copper sulphate when it is melted at 31°C? (Latent heat of fusion is 215 kJ/kg.)

7   A mass of 5.0 kg of water at 37°C is vaporized (latent heat of vaporization at this temperature is 2.4×10^3 kJ/kg). Find the entropy change of the system in this process.

8   Evaluate the entropy change of 2.0 mol of a monatomic ideal gas when the temperature is increased from 300 K to 600 K without changing the volume.

9   A heat reservoir at 100°C gives off 2.00 kJ of heat. What is the entropy change of this reservoir?

10 (a) What entropy change will be produced in a heat reservoir at 0°C if it absorbs the 2.00 kJ of heat given off by the reservoir of the preceding question?
   (b) What is the total of the entropy changes in the two reservoirs because of this process?

11 Consider the possibility of 3.00 kg of water in a pail suddenly spontaneously cooling itself from 28.0°C to 26.0°C and springing up into the air:
   (a) If in this way thermal energy could be converted into kinetic energy of ordered motion upward, how high would the water rise?
   (b) What is the entropy change of the water in this process? (See Question 5.)
   (c) Are there other entropy changes due to this process?
   (d) Is the process possible according to the first law of thermodynamics?
   (e) Is the process possible according to the principle of entropy increase?

12 Describe three different processes that you have observed which are irreversible, as far as your experience would indicate.

13 During a certain process a system absorbs 600 kJ of heat from a heat reservoir at 400 K. The consequent increase in the entropy of the system is 1.80 kJ/K. Is this process reversible? Explain.

14 During a certain process the entropy of a system decreases by 1.20 kJ/K and the system gives off 360 kJ of heat to a reservoir at a temperature of 300 K. Is this process reversible? Explain.

15 A mass of 2.00 kg of hot water at a temperature of 100.0°C is mixed with a mass of 3.00 kg of cold water at a temperature of 0.0°C. After stirring, the temperature of the mixture is 40.0°C:
   (a) Find the change in the total entropy for this process.
   (b) Is the process irreversible?

16 Specify a system which interacts with other systems in such a way that its entropy decreases. Then show that a more than compensating increase of entropy occurs in another system.

17 (a) What is the entropy change of a substance that undergoes a reversible adiabatic process?
   (b) The answer to part (a) does not give the entropy change of the ideal gas for the adiabatic process considered in Example 5. Explain.

18 Show that if a Maxwell demon could operate without increasing his own entropy, the demon would produce a decrease in the total entropy.

19 Calculate the changes in the total entropy that occur when 2.0 kJ of heat are conducted from a hot reservoir at 100.0°C to a cold reservoir at 0.0°C. The conduction takes place through a copper wire of negligible mass. Is the process reversible?

20 Prove that the second law of thermodynamics, as stated in Section 20.6, implies the Kelvin statement: "No process is possible whose sole effect is to extract heat from one reservoir and convert this heat entirely into work."

21 What is a perpetual-motion engine of the second kind?

22 In one cycle, a heat engine extracts 4.8 kJ from a hot reservoir and rejects 3.6 kJ to a cold reservoir:
(a) What is the engine's work output in one cycle?
(b) What is the efficiency of this engine?

23 If a heat engine gives off heat to the atmosphere which is at 27°C, how high must be the temperature of the hot reservoir in order to achieve a Carnot efficiency of 50%?

24 Prove that any heat engine, operating between a hot reservoir at temperature T_{hot} and a cold reservoir at temperature T_{cold}, has a thermal efficiency less than or equal to the Carnot efficiency $[1 - (T_{cold}/T_{hot})]$.

25 A power plant uses heat engines which exhaust heat to the environment. If thermal efficiency were the only consideration, when should the heat engines exhaust to the air and when to the nearby ocean? Assume that average temperatures in the winter are 0°C for the air and 6°C for the ocean, while in the summer these are 20°C for the air and 15°C for the ocean.

26 Why are elevated boiler temperatures desirable in steam engines? Contrast the Carnot efficiency of a steam engine having a boiler temperature of 100°C with that of the engine in Example 6. Assume the condenser temperature is 27°C.

27 Compare the Carnot efficiency of a steam engine which has a boiler temperature of 170°C and a condenser temperature of 40°C to the Carnot efficiency of a gasoline engine which has a combustion temperature of 1500°C and an exhaust temperature of 400°C.

28 Identify several processes which occur in practical heat engines that result in the engine's efficiency being less than the Carnot efficiency.

29 Is it theoretically possible to devise a heat engine which will create no thermal pollution? Explain.

30 The refrigerator of Example 7 has a work input of 48.8 kJ. Calculate the heat given off by this refrigerator to the room which is at 20°C. Compare this heat to the work input.

31 On a hot day a housewife leaves the refrigerator door open in an effort to cool the kitchen. Will this work? Explain.

32 A refrigerator absorbs heat from water at 0°C and gives off heat to the room at a temperature of 27°C. When 20 kg of water at 0°C have been converted to ice at 0°C, what is the minimum amount of energy that has been supplied to operate the refrigerator? What is the minimum amount of heat that has been given off to the room?

33 Consider the entropy changes of the cold and the hot reservoirs and show that the principle of entropy increase requires that a refrigerator give off more heat to the hot reservoir than it takes in from the cold reservoir.

34 How much energy becomes unavailable for work because of the heat conduction process described in Question 19?

35 Energy which is carried to the earth by electromagnetic radiation from the sun is ultimately used to heat an element in a kitchen stove. Examine the energy transformations involved and point out the degradations of energy that occur.

36 A system, initially in a macroscopic state which can occur in Ω_1 microscopically different ways, undergoes a process which brings it into a different macroscopic state which can occur in Ω_2 microscopically different ways. If the initial entropy is S_1 and the final entropy is S_2, show that

$$\frac{\Omega_2}{\Omega_1} = e^{(S_2 - S_1)/k}$$

37   Contrast the entropy changes that occur in hot objects with those that occur in cold objects when they experience a given small change of thermal energy.

38   If water at the triple point were assigned the value of exactly one degree "Jones," what would be the value of Boltzmann's constant in joules per kilogram Jones degree?

Supplementary Questions

S-1   A 4.0-kg block of ice at $0°C$ is placed in 8.0 kg of water at $20°C$. This system is insulated and maintained at a pressure of 1.00 atm. Find the final temperature and the change in entropy for this process.

S-2   If the heat engine of Figure 20-6 could be built, in violation of the Kelvin statement of the second law of thermodynamics, show how this could be used in conjunction with a refrigerator to produce a device which violates the Clausius statement of the second law of thermodynamics.

S-3   An engine runs at 600 rev/min and in each revolution takes in 50 kJ of heat from a boiler which is at a temperature of $167°C$. The exhaust temperature is $107°C$. The efficiency of the engine is 50% of the Carnot efficiency. What is the power output of this engine?

S-4   The electric power input to a certain refrigerator is 1.5 kW. Heat is taken from the cold reservoir at a rate of 0.60 kW. At what rate is heat given to the hot reservoir?

S-5   The coefficient of performance K of a refrigerator (Figure 20-8) is defined by

$$K = \frac{Q_{cold}}{W_i}$$

Show that the maximum possible value of K is

$$K_{Carnot} = \frac{T_{cold}}{T_{hot} - T_{cold}}$$

S-6   What is the relationship between the coefficient of performance K_{Carnot} of an ideal refrigerator and the Carnot efficiency of the same device run backwards as a heat engine?

S-7   The heat leakage in a certain refrigerator is 600 kJ/h. The cooling element operates at $-13°C$ and the condenser at $37°C$. The coefficient of performance of the refrigerator is 40% of the Carnot value K_{Carnot} discussed in Question S-5. What is the average power consumption?

S-8   The best way of heating any building, from the point of view of operating cost, is provided by a "heat pump." A heat pump works in cycles and absorbs a heat Q_o from a cool outside environment at a temperature T_o and gives off a heat Q_r to a warm room at temperature T_r.

(a)   Show that the heat pump requires a work input

$$W_i = Q_r - Q_o$$

(b) Assume that all processes are reversible, that is $\Delta S_{\text{total}} = 0$. Then show that

$$Q_r = W_i \left(\frac{T_r}{T_r - T_o} \right)$$

(c) Assume that ocean water is available at a temperature of $2°C$ in the winter and that the room temperature is $22°C$. Calculate how much heat the reversible heat pump will furnish if $W_i = 2.0$ kJ. How much heat would an electric heater furnish when supplied with 2.0 kJ?

S-9 Explain how a refrigerator could be employed as a heat pump to heat a house in winter.

S-10 In a house heated by hot water radiators, the radiator temperature is maintained at $70°C$ by supplying 5.0 kW to electrical heaters immersed in the water within the radiators. If instead, an ideal (reversible) heat pump were used with the heat intake at $0°C$, what power would be needed to operate the heat pump?

S-11 An energy of 2.3 eV [equal to the energy of a single photon (Chapter 41) in a beam of green light] is absorbed as heat by a system at room temperature such that $kT = \frac{1}{40}$ eV. The entropy of the system thus increases by an amount $S_2 - S_1 = \Delta S = 2.3$ eV/T. Use the result of Question 36 to determine that the number of microscopically different states increases by a factor (Ω_2/Ω_1) of 10^{40}.

S-12 (a) Show that when N atoms of a monatomic ideal gas are confined to a constant volume and heated ($C_V = \frac{3}{2}Nk$) so that the temperature rises a small amount ΔT, the factor Ω_2/Ω_1 defined in Question 36 is given by

$$\frac{\Omega_2}{\Omega_1} = e^{1.5N\Delta T/T}$$

(b) Evaluate this ratio when $T = 325$ K, $\Delta T = 1$ K, and the gas contains a mere 500 atoms.

S-13 The number Ω of states accessible to N atoms of a monatomic ideal gas with a volume V, when the energy of the gas is between $\overline{E}$ and $\overline{E} + d\overline{E}$ (where $d\overline{E}$ is small compared to $\overline{E}$), can be shown to be given by

$$\Omega = (constant) V^N \overline{E}^{3N/2}$$

(a) Determine the entropy S as a function of V and $\overline{E}$.

(b) Using the entropy function of part (a) and the definition of the Kelvin temperature T,

$$\frac{1}{T} = \left(\frac{\partial S}{\partial \overline{E}} \right)_V$$

show that $\overline{E} = \frac{3}{2}NkT$.

S-14 (a) For an ideal gas with constant heat capacities per mole, c_P and c_V, show that the entropy of n moles can be expressed in the form

$$S = nc_V \ln PV^\gamma + constant$$

where $\gamma = c_P/c_V$.

(b) From the preceding result, show that during a reversible adiabatic process the pressure and volume of an ideal gas are related by

$$PV^\gamma = constant$$

21.

electric charge
and electric fields

The electromagnetic interaction is one of the four fundamental interactions known in nature. This is the interaction that is involved in the study of electricity, magnetism, and electromagnetic waves. Since the configuration of electrons in an atom is determined by the electromagnetic interaction of these electrons with each other and with the atom's nucleus, the chemical behavior of atoms and the internal structure of various bodies is ultimately governed by this interaction.

The electromagnetic interaction between two particles depends on the existence of a fundamental physical characteristic of these particles, their *electric charge*. In the general case, when two charged particles are in motion, each charge exerts on the other an electric force and a magnetic force. The simplest situation is obtained when both charges remain at rest; then there are no magnetic forces and the electromagnetic interaction is determined entirely by electric forces.

We begin the study of electricity and magnetism by first considering the forces exerted by charges at rest, a subject called *electrostatics*.

21.1 Electric Charge

Positive and Negative Charges

When the charges remain at rest, the electric force exerted by one charged particle on another charged particle acts along the line joining the charges (Figure 21-1). For certain pairs of charged particles this electric force is attractive while for other pairs it is repulsive. Experiments show that there are two different types of electric charge: one type is called *positive charge* and the other type *negative charge.** Two particles carrying electric charges of the same type, both positive or both negative, exert repulsive forces on each other. Attractive forces are exerted when one of a pair of particles is positive and the other negative. In summary: *Like charges repel, unlike attract.*

Operational Definition of Charge

The charge of a particle can be defined by an operational procedure illustrated in Figure 21-2. Measurement is made of the force F exerted on a stationary particle with a charge q which is placed a distance r from another stationary particle with a charge Q. Then the particle with charge q is replaced by a particle with charge q', and the force F' acting on q' is measured. The values of the charges q and q' are defined to be proportional to the forces F and F'; that is

$$\frac{q}{q'} = \frac{F}{F'} \tag{21.1}$$

Consequently two electric charges can be compared by measuring forces, and an unknown charge can be assigned a numerical value by comparing it with whatever quantity of charge is adopted as unit charge.

The Coulomb

The electric current I in a wire is the rate at which charge flows through a given cross section of the wire. If a charge dQ flows through a cross section in a time interval dt, then the current is

$$I = \frac{dQ}{dt} \tag{21.2}$$

and the charge that flows through a cross section in a time interval between t_1 and t_2 is

$$Q = \int_{t_1}^{t_2} I \, dt \tag{21.3}$$

---

*The choice as to which type of charge is called positive and which negative follows a historical convention. There would be no change in the laws of physics if all positive charges were to be called negative and vice versa.

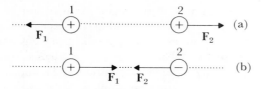

Figure 21-1   Electric forces. (a) Like charges repel each other. (b) Unlike charges attract. In each case $\mathbf{F}_1$ is the force exerted on charge 1 by charge 2, and $\mathbf{F}_2$ is the force exerted on charge 2 by charge 1.

Figure 21-2   The ratio q/q' of the charges is defined to be equal to the ratio F/F' of the forces exerted by the particle with a charge Q at a distance r.

Practical laboratory procedures for the comparison of electric currents are more accurate than those for the direct comparison of charges. For this reason, in the SI system the unit of electric charge is defined in terms of the unit of electric current. The definition of the SI unit of current, the *ampere* (A), is given in Section 28.2.

The SI unit of charge, named the *coulomb* (C), *is defined as the charge that flows through a given cross section of a wire in* 1 s *when there is a steady current of* 1 A *in the wire.* Equations 21.2 and 21.3 give the charge in coulombs when the current is measured in amperes and the time in seconds.

Charges Found in Nature

The electron is the particle of primary interest in much of physics, chemistry, and electrical technology. The charge of an electron is negative and is conventionally denoted by the symbol $-e$. Experimental measurement gives

$$e = 1.6021892 \times 10^{-19} \text{ C}$$

This particular amount of charge is apparently of fundamental significance because every sufficiently accurate measurement of charge indicates that the only charges that occur in nature are integral multiples of the charge e. For instance, in a gas through which an electric spark is passing, we find particles with charges $+3e$, $+2e$, $+e$, and 0 (uncharged or electrically neutral) and $-e$, but never anything with a charge of something like $\frac{3}{11}e$ or $2.7e$.

Conservation of Charge

A composite object has a charge that is found by experiment to be the algebraic sum of the charges of the various particles of the object. An object with equal amounts of positive and negative charges has zero net charge and is said to be electrically *neutral*. For example, a hydrogen atom, whose constituents are a proton of charge $+e$ and an electron of charge $-e$, is electrically neutral. The nuclei of all other types of atoms are composed of protons and neutrons. The neutron is electrically neutral. The charge of a nucleus containing Z protons is Ze. (The number Z is called the atomic number.) The nucleus together with Z orbital electrons constitutes a neutral atom.

An atom or molecule which is not electrically neutral is called an ion. The process of *ionization* generally consists of the removal of one or more electrons from a neutral atom or molecule which then becomes a positive ion with a charge of magnitude equal to that of the electrons which have been removed. Certain types of neutral atoms and molecules can bind an extra electron and form a negative ion with a charge $-e$.

In electrification by friction, such as occurs when a glass rod is rubbed with a silk cloth, a very small fraction of the electrons in the glass are transferred to the silk. The original neutrality of the glass and of the silk is thus disturbed with the glass acquiring a positive charge and the silk acquiring a negative charge of equal magnitude.

In the reactions studied in chemistry and in nuclear and particle physics, there are often radical changes in the structure of composite objects and sometimes even creation or annihilation of particles such as electrons, protons, and neutrons. For instance, a free neutron ($_0$n) experiences a sudden spontaneous transformation into a proton ($_1$H), an electron ($_{-1}$e), and an antineutrino ($_0\bar{\nu}$) in a process symbolized by

$$_0\text{n} \longrightarrow {}_1\text{H} + {}_{-1}\text{e} + {}_0\bar{\nu}$$

where subscripts on each particle symbol indicate the charge in units of e. Notice that the total electric charge has the same value before and after this process. In all experiments that have kept track of electric charge during various processes, a similar result has been obtained. These experimental observations are summarized by stating the *law of conservation of electric charge: Electric charge is conserved; throughout any process occurring within a system, the algebraic sum of the electric charges within the system does not change.*

21.2 Electric Field

It is advantageous to treat electric forces using a two-stage description analogous to that given for gravitational forces:

1   A source produces a field at all points in space.
2   The field at any location exerts a force on a charged particle at that location.

A useful analogy between electrical and gravitational quantities is displayed in Table 21-1. The electric charge possessed by an object plays a role in determining electrostatic forces analogous to that of mass in determining gravitational forces. With the substitution of *charge* for *mass,* the formalism developed for gravitation can be carried over to electrostatics. Nevertheless, we should

Table 21-1 Analogous Quantities

Electrical	Gravitational
Electric charge q	Mass m
Electric field **E**	Gravitational field **g**
Electric force **F** exerted by **E** on q:	Gravitational force **F** exerted by **g** on m:
$\mathbf{F} = q\mathbf{E}$	$\mathbf{F} = m\mathbf{g}$
Source of electric field: charge Q	Source of gravitational field: mass M
Particle of charge Q produces at distance r an electric field of magnitude	Particle of mass M produces at distance r a gravitational field of magnitude
$E = \dfrac{kQ}{r^2}$	$g = \dfrac{GM}{r^2}$
Coulomb's law:	Newton's law of gravitation:
$F_r = \dfrac{kQq}{r^2}$	$F_r = -\dfrac{GMm}{r^2}$

emphasize that there are tremendous differences between electrical and gravitational phenomena. Some of the reasons for these differences are:

1   The electric force exerted by one charged particle within an atom on another is much stronger than the feeble gravitational force with which one particle presumably attracts the other.
2   There are two types of electric charge, positive and negative. Consequently, cancellation occurs and large objects can have a small net charge or be electrically neutral (zero net charge). Electric forces can be repulsive or attractive. On the other hand, only one type of mass occurs and gravitational forces are always attractive.
3   The interaction of charges in motion generally involves magnetic forces as well as electric forces.

Definition of the Electric Field

The electric field at a point in space is a vector **E** defined by the equation

$$\mathbf{E} = \frac{\mathbf{F}}{q} \tag{21.4}$$

where **F** is the electric force that a particle with a charge q experiences when placed at the point in question. This "test charge" q should be small enough so that the force it exerts has a negligible influence on the positions of the charges that are producing the field. The SI unit of **E** is seen to be the N/C.

When the electric field at a given location is known, the electric force on a charge q at that location can be calculated from

$$\mathbf{F} = q\mathbf{E} \tag{21.5}$$

This electric force is in the direction of **E** if the charge is positive, but it is in the opposite direction if the charge is negative. For example, in a region of space where E = 200 N/C north, the electric force exerted on a proton is (1.60 $\times$ 10^{-19} C)(200 N/C) = 3.2 $\times$ 10^{-17} N north, while the electric force exerted on an electron in this region is 3.2 $\times$ 10^{-17} N south.

Again we emphasize that the fundamental assumption of a *field* theory of forces is that the force acting on an object is determined entirely by the field at the location of the object. It is not necessary for us to know what configuration of sources produced this field in order to know the force exerted by the field on an object. For instance, the force on an electric charge in a home television antenna is determined by the electric field at the location of the electron. Knowing this field, we do not need to keep in mind the pertinent source charges in the television broadcasting station and in the adjacent earth, trees, and roofs.

Electrostatic Field Produced by a Point Charge

The source of an electric field is electric charge. A stationary particle possessing a positive charge Q produces at a distance r an electric field E of magnitude

$$E = \frac{kQ}{r^2} \tag{21.6}$$

where k is a constant. This result can be verified, at least in principle, by measuring the force F on a test charge q at a distance r from Q and applying the definition E = F/q. To an accuracy sufficient for most purposes in this book, the measured value of k is

$$k = 9.0 \times 10^9 \text{ N} \cdot \text{m}^2/\text{C}^2$$

It is sometimes convenient to express this constant in the form

$$k = \frac{1}{4\pi\epsilon_0} \tag{21.7}$$

where ϵ_0, called the *permittivity constant,* has the measured value 8.854187818 $\times$ 10^{-12} $\text{C}^2/\text{N} \cdot \text{m}^2$.

The electric field vector E points away from the stationary positive charge that produced it (Figure 21-3). When the source is a negative charge, the field points toward the source. These rules for the direction of E are correctly given for positive or negative Q by the statement that the direction of E is radial and that the radial component of E (the component of E in the direction of increasing r) is given by

$$E_r = \frac{kQ}{r^2} \tag{21.8}$$

The electric field created by the charge Q is strong at points near the source and weak far from the source.

Superposition Principle for Electric Fields

When several different charges contribute to the electric field at a given point in space, this electric field vector is the *vector sum* of the fields that each charge would individually contribute, if each were the only source of the electric field. In other words, the electric fields created by different charges are superposed without affecting one another. This simple and fundamental law of nature is known as the superposition principle for electric fields.

In the next three chapters we shall study the electrostatic field in a *vacuum* populated only by point charges or by a continuous distribution of charge. In practice, to have these charges held in place they would have to be located in or on an insulator (Section 24.1) such as glass or a plastic. The modifications of an electric field due to charges within neutral molecules of an insulator are considered only in Chapter 24, where we also examine the effects of the presence of mobile charges within metals. In the meantime, we shall assume that no metals are present.

Example 1   Figure 21-4 shows two point charges, Q and $-Q$, a distance s apart. Find the electric field **E** produced by this configuration of charges at the point P which is a distance r along the perpendicular bisector of the line joining the charges.

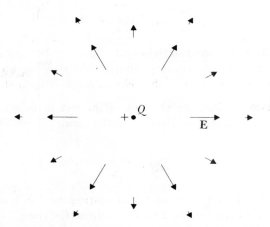

Figure 21-3   Representation of the electric field due to a positive point charge. If the point charge were negative the arrows representing the electric field at the various points would be directed inward. (Compare this to Figure 14-1 in which a gravitational field is shown.)

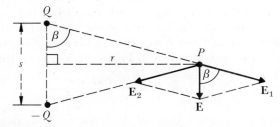

Figure 21-4   The superposition principle gives $\mathbf{E} = \mathbf{E}_1 + \mathbf{E}_2$.

Solution   The superposition principle gives

$$\mathbf{E} = \mathbf{E}_1 + \mathbf{E}_2$$

where $\mathbf{E}_1$ is the field produced by Q at P and $\mathbf{E}_2$ is the field produced by $-Q$ at P. From Eq. 21.6, the vectors $\mathbf{E}_1$ and $\mathbf{E}_2$ have magnitudes

$$E_1 = E_2 = \frac{kQ}{(s/2)^2 + r^2}$$

From Figure 21-4, we see that the resultant of $\mathbf{E}_1$ and $\mathbf{E}_2$ points downward and has a magnitude

$$E = 2E_1 \cos \beta$$

where

$$\cos \beta = \frac{s/2}{\sqrt{(s/2)^2 + r^2}}$$

Therefore the magnitude of the field at P is

$$E = \frac{kQs}{(s^2/4 + r^2)^{3/2}}$$

Example 2   An infinitely long straight rod of negligible thickness has charge distributed along it with a constant charge density λ per unit length. Find the electric field $\mathbf{E}$ at a point P that is a distance R from the rod.

Solution   In Figure 21-5(a) the element of the rod of length dx contains a charge $dQ = \lambda \, dx$ and contributes, at the point P, an electric field of magnitude

$$dE = \frac{k\lambda \, dx}{r^2}$$

The resultant field $\mathbf{E}$ at P is found by calculating the vector sum of the contributions $d\mathbf{E}$ of all the elements that constitute the rod:

$$\mathbf{E} = \int d\mathbf{E}$$

Different elements of the charged rod produce contributions $d\mathbf{E}$ that have not only different magnitudes but also different directions. We evaluate the vector sum by summing components. From the symmetry of the charge distribution it is clear that components of $d\mathbf{E}$ parallel to the rod sum to zero [Figure 21-5(b)]. Therefore the direction of the resultant field at P is radial (normal to the rod). The radial component of the contribution $d\mathbf{E}$ furnished by the element of length dx shown in Figure 21-5(a) is $dE \sin \beta$. Therefore

$$E_R = \int \sin \beta \, dE = \int \frac{k\lambda \sin \beta \, dx}{r^2}$$

where the integration ranges over all the elements that constitute the rod.

The quantities β, r, and x are variables and, to evaluate the integral, we must express two of these variables in terms of the third. Simplification is obtained if β is chosen as the independent variable. From Figure 21-5(a),

$$r = R \csc \beta \qquad \text{and} \qquad x = R \cot \beta$$

Therefore

$$dx = -R \csc^2 \beta \, d\beta$$

Substituting these expressions into the integral for E_R, we find

$$E_R = -\frac{k\lambda}{R} \int_\pi^0 \sin \beta \, d\beta = \frac{k\lambda}{R}(\cos 0 - \cos \pi) = \frac{2k\lambda}{R}$$

The direction of **E** is radially outward if the rod has a positive charge, inward if negative.

Example 3   Charge is distributed uniformly over an infinite plane with a charge per unit area or *surface charge density* σ. Find the electric field at the point P in Figure 21-6.

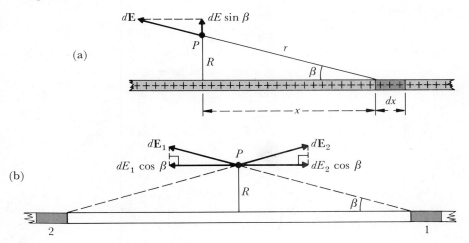

Figure 21-5   (a) The field $d\,\mathbf{E}$ produced by an element of an infinitely long charged rod. (b) The components parallel to the rod of the field produced by symmetrically placed elements, 1 and 2, sum to zero.

Figure 21-6   Charged strip of width dx is a portion of the charged plane.

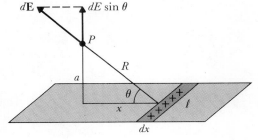

Solution   The plane can be regarded as composed of narrow strips. Since each narrow strip is a line charge, the result of Example 2 can be used to give the contribution of the charges on a strip to the field at P. In Figure 21-6, a length l of a strip of width dx has an area $l\,dx$ and consequently a charge of $\sigma l\,dx$. Therefore the charge per unit length of this strip is

$$\lambda = \frac{\sigma l\,dx}{l} = \sigma\,dx$$

From Example 2, this infinitely long strip will contribute at P an electric field of magnitude

$$dE = \frac{2k\sigma\,dx}{R}$$

The resultant field at P is the vector sum of the contributions of all the strips which comprise the plane. The symmetry of charge distribution implies that the components of $d\mathbf{E}$ parallel to the plane sum to zero. The resultant field at P is therefore perpendicular to the charged plane. The component of $d\mathbf{E}$ perpendicular to the plane is $\sin\theta\,dE$ and the resultant field at P has a magnitude

$$E = \int \sin\theta\,dE = 2k\sigma \int_{x=-\infty}^{x=+\infty} \frac{\sin\theta\,dx}{R}$$

Since $\sin\theta = a/R$ and $R^2 = a^2 + x^2$, we have

$$E = 2k\sigma a \int_{-\infty}^{\infty} \frac{dx}{a^2 + x^2} = 2k\sigma a \left[\frac{1}{a}\tan^{-1}\frac{x}{a}\right]_{-\infty}^{+\infty} = 2\pi k\sigma$$

Using Eq. 21.7, $k = 1/4\pi\epsilon_0$, this result can be rewritten as

$$E = \frac{\sigma}{2\epsilon_0}$$

Notice that the field is independent of the distance from the charged plane. The direction of $\mathbf{E}$ is away from the plane if the charges are positive, toward the plane if negative.

Electrical apparatus often contains parallel plates with adjacent surfaces carrying charges equal in magnitude and opposite in sign (Figure 21-7). The

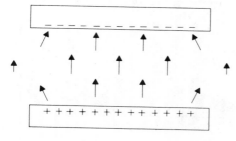

Figure 21-7   Between charged parallel plates, $\mathbf{E}$ is uniform except in regions near the edges.

surface charge density on such a plate is approximately constant except near the edge of the plate. As is suggested by the preceding example, the charges on the plates set up an electric field **E** which has the same value at all points between the plates; that is, **E** is *uniform*, except in regions near the edges of the plates. Notice that **E** is directed away from the positive charges toward the negative charges.

Electric Flux Lines

To visualize the electric field produced by a given distribution of charges, we must associate with each point in space a vector **E** of the appropriate magnitude and direction. Convenient graphical representations of electric fields can be given using what are called *electric flux lines* or *lines of force,* as shown in Figures 21-8, 21-9, and 21-10. *An electric flux line is a curve whose tangent at any point is in the direction of* **E** *at that point.*

Figure 21-8   Electric flux lines (or lines of force) for the field set up by a positive charge.

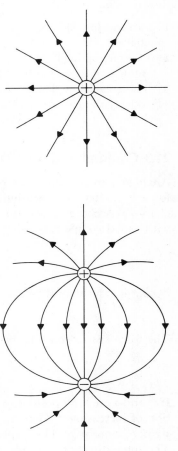

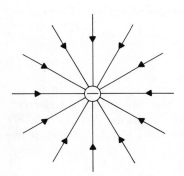

Figure 21-9   Electric flux lines for the field produced by a negative charge.

Figure 21-10   Electric flux lines for the field produced by the equal and opposite charges of Example 1.

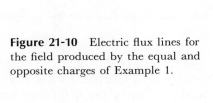

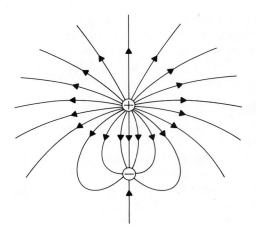

Figure 21-11 Electric flux lines for the field that is the superposition of fields produced by charges of unequal magnitudes and opposite signs.

Figure 21-11 illustrates the following properties of flux lines that can be shown to be true for any electrostatic field:

1   Electric flux lines do not intersect at points in space where there are no charges.
2   One end of an electric flux line is at a charge; the other end may be either at infinity or at a charge of the opposite sign.
3   The greater the density of flux lines in a given region, the larger the magnitude of the electric field in that region.

21.3 Coulomb's Law

Analogous to Newton's law of gravitation is the force law giving the electric force **F** exerted by a stationary particle of charge Q on another particle of charge q. A charge Q sets up at all points at a distance r a radial electric field with radial component

$$E_r = \frac{kQ}{r^2} \tag{21.8}$$

A charge q located at such a point experiences an electric force

$$\mathbf{F} = q\mathbf{E}$$

The force exerted on q is therefore radial and has a radial component

$$F_r = \frac{kQq}{r^2}$$

If Q and q are both positive or both negative, F_r is positive and **F** is in the direction of increasing r. That is, q is repelled by Q. But if Q and q have opposite signs, F_r is negative. This implies that **F** is directed toward Q (an attractive force). The force directions for any sign of the charges are correctly summarized by the statement: *Like charges repel and unlike attract.*

By considering the field established by a stationary charge q at the location of Q, and then calculating the force exerted on Q by this field, we find that the force exerted on Q by q is equal in magnitude but opposite in direction to the force exerted on q by Q; that is, electrostatic forces obey Newton's third law. These rules for the magnitude and the direction of the electric force exerted by a stationary charged particle on another charged particle are known as *Coulomb's law.*

The validity of this law for a rather limited range of values of the charge separation r was demonstrated in experiments performed in 1784 by Charles Augustin de Coulomb (1736–1806) using a torsion balance analogous to that described in Section 14.1 in connection with the Cavendish experiment. Modern experiments and their theoretical interpretation suggest that Coulomb's law is valid for charge separations over the enormous range from 10^{-15} m to at least several kilometres.

Coulomb's law is the fundamental experimental law of electrostatics. Instead of first introducing the idea of an electric field and the expression $E_r = kQ/r^2$ for the field produced by a stationary charge Q, Coulomb's law can be taken as the starting point for the development of electrostatics. It then follows that, with **E** defined as **F**$/q$, a stationary charge Q sets up at a distance r a radial electric field with a radial component

$$E_r = \frac{F_r}{q} = \frac{kQ}{r^2}$$

21.4 Charges in Electric Fields

Millikan's Measurement of e

As small drops of oil are squirted out of an atomizer, they acquire various electric charges by friction. Wandering around in air, even a neutral drop will occasionally encounter ions and acquire an electric charge. Consider such a charged oil drop, of mass M and net positive charge q, which has been squirted into the air in the region between charged plates (Figure 21-7) where there is a uniform electric field **E** pointing vertically upward. The forces acting on the oil drop are then:

1 An electric force

$$\mathbf{F}_{\text{electric}} = q\,\mathbf{E}$$

   acting vertically upward.

2 A gravitational force (the weight of the drop)

$$\mathbf{F}_{\text{gravitational}} = M\mathbf{g}$$

   acting vertically downward.

3 A viscous drag exerted on the drop by the air when the drop moves through the air.

By adjusting the charge on the plates, we can vary the electric field $\mathbf{E}$ until the electric force on the drop just balances the drop's weight, that is, $qE = Mg$. The drop will then neither rise nor fall.

The mass M of the drop can be determined from a detailed study of its fall through the viscous air. The field $\mathbf{E}$ is determined by routine electrical measurements. Hence the charge on the drop can be calculated from $q = Mg/E$.

This is the basic idea behind a long series of experiments carried out by R. A. Millikan (1868–1953) during the first decades of this century. He found that the measured charge on the oil drop was always

$$\pm 1.60 \times 10^{-19} \text{ C}, \ \pm 3.20 \times 10^{-19} \text{ C}, \ \pm 4.80 \times 10^{-19} \text{ C}, \ldots$$

That is,

$$q = integer \times e$$

where (to three significant figures)

$$e = 1.60 \times 10^{-19} \text{ C}$$

This experimental result quoted in Section 21.1 has been confirmed in particle track recorder observations of thousands of nuclear reactions.

The presence of positive ions, that is, atoms or molecules from which an electron has been stripped, accounts for the positively charged oil drops. Drops with a negative charge have acquired an excess of electrons.

Motion of an Electron in a Uniform Electric Field

If we enclose the metal plates of Figure 21-7 in a glass tube and pump out the air inside the tube we have what is called a *vacuum tube*. The metal plates are called *electrodes*, and since there are just two of them this vacuum tube is called a *diode*. If we heat the negative electrode (named the *cathode*), electrons will be emitted from the hot surface.

Between the plates an electron (charge $q = -e$) experiences only an electric force

$$\mathbf{F} = q\mathbf{E} = -e\mathbf{E}$$

The gravitational force on an electron is entirely negligible compared to even the most feeble electric forces encountered in practical work. If a good vacuum is maintained, an electron usually gets a free run from the cathode to the positive electrode (the *anode*), with little chance of a collision with the remaining air molecules. Newton's second law, $\mathbf{F} = m\mathbf{a}$, applied to an electron gives

$$eE = m_e a$$

where m_e is the mass of an electron. The electron's acceleration consequently has a magnitude

$$a = \frac{e}{m_e} E$$

The electron is of course accelerated in the direction of the electric force, and this direction is opposite to that of **E**.

Notice that in a given field **E**, the acceleration of any charged particle is proportional to its *"charge-to-mass ratio"* (e/m_e for the electron). Measurement of a particle's acceleration in a known electric field suffices to determine its charge-to-mass ratio.

The electron has a charge-to-mass ratio which is the largest found in nature, 1836 times that of the proton. Agility, a large acceleration in a given field, is therefore a characteristic of an electron. It is this characteristic that is exploited in many electronic devices, and in fact it led to the discovery of the electron in 1897 by J. J. Thomson (1856–1940). Using electric and magnetic forces (Section 27.2), he was able to measure the electron's charge-to-mass ratio. The modern value is

$$\frac{e}{m_e} = 1.7588047 \times 10^{11} \text{ C/kg}$$

From this knowledge and from measurement of e, the electron's mass can be computed. Modern data yield

$$m_e = 9.109534 \times 10^{-31} \text{ kg}$$

The motion of the electron in a uniform electric field is a motion with a constant acceleration. It is therefore mathematically identical to that of a baseball in a uniform gravitational field. If the electron starts from rest near the cathode it will travel in a straight line to the anode, speeding up as it goes, with its distance from the cathode after an elapsed time t given by

$$x = \tfrac{1}{2}a_x t^2$$

and its velocity by

$$v_x = a_x t$$

The only difference between the electron's motion and that of a falling baseball lies in the numerical value of the acceleration (and the subsequent speeds attained). Electric fields of 10^6 N/C are common. The acceleration imparted by such a field to an electron is

$$a = \frac{e}{m_e} E$$

$$= (1.76 \times 10^{11} \text{ C/kg})(10^6 \text{ N/C})$$

$$= 1.76 \times 10^{17} \text{ m/s}^2$$

This is about 2×10^{16} times the baseball's acceleration (9.8 m/s^2).

Clearly it does not take very long (about 10^{-8} s) for this electron to be accelerated to a speed approaching the speed of light (3.00×10^8 m/s). As these high speeds are reached, Newtonian mechanics no longer gives the right answers. The modifications required are given by Einstein's special theory of relativity (Chapter 40).

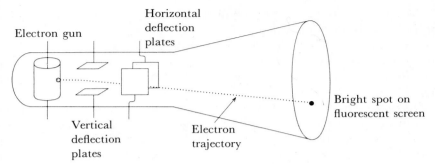

Figure 21-12   Cathode ray oscilloscope.

If an electron is projected horizontally and enters the region between the plates of Figure 21-7, it will follow a parabolic trajectory similar to a segment of the projectile's path in Figure 7-3. After it has emerged from the region between the plates and moves through the region where the electric field E is zero, its velocity vector will be constant. The net result is that the electron's trajectory has been bent by passage between the plates. The electron has been deflected.

Such deflection of a beam of electrons by passage through "deflection plates" is the key to the functioning of a television picture tube* or a *cathode ray oscilloscope* (Figure 21-12). A beam of electrons is fired from an electron gun, proceeds down the picture tube, and makes a bright spot where it strikes the fluorescent screen. On route the beam passes between plates arranged to give a vertical deflection and then between plates which give a horizontal deflection. The charge on the deflection plates is periodically altered; the electric field then changes accordingly, and the deflection of the beam passing between the plates is changed. The spot on the screen where the beam strikes is thus moved back and forth and up and down. When the beam intensity is altered according to the incoming signal, the spot brightness changes as the spot moves and a picture is "painted" on the screen.

Summary

- [ ] Electric charge is a fundamental physical characteristic of a particle. There are two types of electric charge—positive and negative. Like charges repel and unlike attract.

- [ ] The ratio of two charges q and q' is defined by

$$\frac{q}{q'} = \frac{F}{F'}$$

where F and F' are the forces exerted on q and q', respectively, by a stationary charge Q which is a distance r away. The coulomb, the SI unit of

*In practice, modern television picture tubes employ a magnetic deflecting force (Section 27.2) instead of the electric forces described in this chapter.

charge, is the charge that flows through a given cross section of a wire in 1 s when there is a steady current of 1 A in the wire.

☐ All experiments are in accord with the law of conservation of electric charge which states that, throughout any process occurring within a system, the algebraic sum of the electric charges within the system does not change.

☐ The electric field **E** at a given point in space will exert a force $\mathbf{F} = q\mathbf{E}$ on a charge q placed at the given point.

☐ A stationary particle with a charge Q is the source of an electric field which is radial, with radial component

$$E_r = \frac{kQ}{r^2}$$

The experimental value of k is approximately 9.0×10^9 N·m^2/C^2. The permittivity constant ϵ_0 is related to k by $k = 1/4\pi\epsilon_0$.

☐ Electric fields obey the superposition principle: The electric field produced at a given point in space by several charges is the vector sum of the fields that each charge would individually contribute, if each were the only source.

☐ A line of force or an electric flux line is a curve whose tangent at any point is in the direction of **E** at that point.

☐ Coulomb's law states that the force exerted by a stationary charge Q on a stationary charge q a distance r away is in the radial direction from Q and has a radial component

$$F_r = \frac{kQq}{r^2}$$

This Coulomb force obeys Newton's third law.

☐ All experimental evidence indicates that the only electric charges that occur in nature are integral multiples of $e = 1.60 \times 10^{-19}$ C.

Questions

1 What is the charge of a system of 6.2×10^{18} electrons?

2 Compare and contrast gravitational forces and electrostatic forces.

3 Define the term *electric field*. What is the SI unit for electric field?

4 A charge of 8.0×10^{-8} C is 0.40 m from a charge of -2.0×10^{-8} C:
   (a) Determine the electric field produced by the negative charge at the location of the positive charge.
   (b) What force does this electric field exert on the positive charge?

5 **(a)** Determine the electric field produced by the charges in the preceding question at a point P which is 0.30 m from the positive charge and 0.10 m from the negative charge.
   (b) What force would be exerted by this field on an electron placed at P?

6 A charge of $+8e$ is a distance D to the left of a charge $-2e$. Find the point where the electric field is zero.

7) Two charges are located on the X axis: a 300×10^{-12} C charge at $x = 2.0 \times 10^{-2}$ m, and a -400×10^{-12} C charge at $x = -2.0 \times 10^{-2}$ m:

   (a) Find the x and y components of the electric field at the point on the Y axis with coordinates $(0, 1.0 \times 10^{-2}$ m$)$.

   (b) Where is the electric field zero?

8) A thin lucite rod which is 6.0 m long has a positive charge of 8.0×10^{-9} C uniformly distributed along its length. Find the electric field produced by this charge distribution at a point which is 4.0 m from the midpoint of the rod.

9 Two infinite plane sheets of charge, with surface charge densities of 0.80×10^{-6} C/m² and -0.60×10^{-6} C/m², are 0.40 m apart and parallel. Find the electric field in each of the three regions into which space is divided by these sheets.

10 The plane sheets of charge of the preceding question are placed at right angles to each other. Find the electric field in each of the four regions into which space is divided by these sheets.

11 Check the result of Question 4 by computing the force exerted on the positive charge directly from Coulomb's law.

12 Consider the gravitational force and the electric force exerted on an electron by a proton. Show that the ratio of these forces is given by

$$\frac{\text{gravitational force}}{\text{electric force}} = \frac{1}{2.3 \times 10^{39}}$$

13) Figure 21-13 shows two identical particles of mass M, each carrying a charge q. Find the angle θ between the strings and the vertical in terms of M, q, and the length l of the strings.

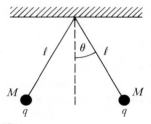

Figure 21-13

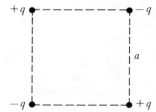

Figure 21-14

14 Charges $+q$ and $-q$ are located at the corners of a square of side a (Figure 21-14). Find the force on each charge.

15 A positively charged oil drop is injected into the region between two oppositely charged, horizontal plates like those of Figure 21-7:

   (a) If the charge on the drop is $+2e$, what is the magnitude of the electric field when an electric force of 1.6×10^{-15} N is exerted on the drop?

   (b) Sketch the orientation of the plates and indicate the kind of charge on each in order that the electric field between them exerts a force which might balance the gravitational force on the oil drop.

16 In the Millikan oil drop experiment, an oil drop with a charge of -3.2×10^{-19} C is held in equilibrium by an electric field of magnitude 8.0×10^{3} N/C directed downward. What is the mass of the oil drop?

17 Show that when the only force acting on an electron is the force exerted by an electric field $\mathbf{E}$, the electron's acceleration is given by $\mathbf{a} = -\mathbf{E}e/m_e$, where m_e is the electron's mass.

18 The plates of Figure 21-7 are 4.0×10^{-2} m apart. The plates are charged as shown and produce a uniform field of 5.0×10^{4} N/C in the region between the plates:
   (a) Find the acceleration of an electron in this region.
   (b) Find the time required for an electron, which starts at rest at the upper plate, to reach the lower plate.

19 An electron emerges from the hot cathode of a vacuum tube with a speed of 6.0 $\times 10^{5}$ m/s. The electron then encounters an electric field of 20 N/C in a direction opposite to the velocity vector of the electron:
   (a) Find the magnitude and direction of the force acting on the electron.
   (b) Determine the acceleration of the electron.
   (c) Find the distance the electron will travel in 3.0×10^{-8} s and its speed after this time interval.

Supplementary Questions

S-1 A total charge Q is uniformly distributed over a thin ring of radius a. Show that the electric field produced by this charge distribution at a point on the axis of the ring a distance z from its center is directed along the axis and has a magnitude

$$E = \frac{kQz}{(a^2 + z^2)^{3/2}}$$

S-2 Charge is uniformly distributed with a surface charge density of 0.30×10^{-6} C/m^2 over an infinitely long broad strip which is 0.12 m wide. Find the electric field at a point which is 0.050 m from one edge of the strip and 0.13 m from the other edge.

S-3 A total charge Q is uniformly distributed over a thin ring of radius a. Then a very short segment of length s is removed from the ring. Find the electric field at the center of the ring by:
   (a) Summing the contributions $d\mathbf{E}$ furnished by each element of the charge distribution.
   (b) Considering the situation with the segment of length s restored to the ring, and then exploiting symmetry and the superposition principle.

S-4 An electron emerges from an electron gun in a cathode ray oscilloscope at a speed of 2.0×10^{7} m/s and then travels horizontally with a constant velocity until, while passing between deflection plates, it encounters an electric field of 2.0×10^{4} N/C directed vertically downward. This deflecting field is established over a region which has a horizontal length of 2.0×10^{-2} m:
   (a) For how long a time interval will the electron be subjected to a deflecting force?
   (b) What is the vertical acceleration of the electron during the time interval determined in part (a)?
   (c) Find the vertical component of velocity and the vertical displacement at the end of this time interval.

22.

gauss's law

From Coulomb's law and its consequences, $E_r = kQ/r^2$ where $\mathbf{F} = q\mathbf{E}$, we can deduce a remarkable theorem, known as Gauss's law, which shows how to calculate the amount of charge in any region from a knowledge of the electric field on the surface bounding that region. Certain problems that would be difficult using the methods of the preceding chapter have solutions that are transparent when Gauss's law is considered.

22.1 Electric Flux

Gauss's law involves a quantity called electric *flux*. For the particular case of a uniform field $\mathbf{E}$ and a plane surface of area A, the electric flux ψ through the surface is defined by the equation

$$\psi = E_N A = EA \cos \theta \tag{22.1}$$

where θ is the angle between $\mathbf{E}$ and the outward normal to the surface (Figure 22-1) and $E_N = E \cos \theta$ is the component of $\mathbf{E}$ normal to the surface. The outward direction is determined only when the surface in question is a portion of a closed surface, such as the surface of the prism in Figure 22-2. The flux through a surface can be visualized as a quantity proportional to the number of flux

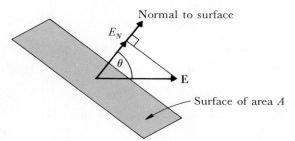

Figure 22-1   The flux through the plane surface of area A is $\psi = E_N A = EA \cos \theta$.

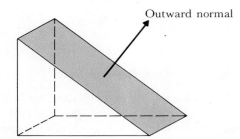

Figure 22-2   A closed surface and an outward normal.

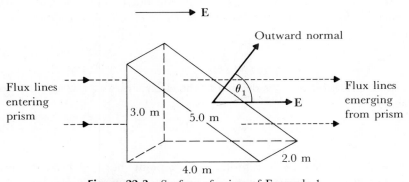

Figure 22-3   Surface of prism of Example 1.

lines passing through the surface, the flux being positive where the flux lines emerge from a closed surface and negative where the flux lines enter the surface (Figure 22-3). (This useful relationship between flux and the number of flux lines, where flux lines have the properties enumerated in Section 21.2, can be established after Gauss's law has been deduced.)

Example 1   In the uniform horizontal electric field of Figure 22-3, with $E = 2.0 \times 10^4$ N/C, find the electric flux through each side of the surface of the prism. What is the total flux through this closed surface?

Solution   For the surface on the right, $\cos \theta_1 = \frac{3}{5}$ and the flux is

$$\psi_1 = (2.0 \times 10^4 \text{ N/C})(2.0 \times 5.0 \text{ m}^2)(\tfrac{3}{5}) = 12 \times 10^4 \text{ N} \cdot \text{m}^2/\text{C}$$

For the surface on the left, $\cos \theta_2 = \cos 180° = -1$ and the flux is

$$\psi_2 = (2.0 \times 10^4 \text{ N/C})(2.0 \times 3.0 \text{ m}^2)(-1) = -12 \times 10^4 \text{ N} \cdot \text{m}^2/\text{C}$$

Notice that the negative flux corresponds to flux lines entering the closed surface. The flux through all other surfaces of the prism is zero. This follows from the fact that $\cos \theta = 0$ for each of these surfaces, but it can be seen most easily if we note that no flux lines go through these surfaces which are parallel to **E**. The net flux through the entire prism is

$$\psi = \psi_1 + \psi_2 = 0$$

Definition of Electric Flux for the General Case

In the general case of a curved surface and a nonuniform electric field, we subdivide the surface into many small surfaces and, for each small surface, we evaluate the product $E_N \times area$ where E_N is a value of the normal component of the electric field that occurs on the small surface in question. The sum of such products for all the small surfaces that constitute the original closed surface is called an approximating sum. The limit of the approximating sums, as the number of subdivisions approaches infinity and the area of each small surface approaches zero, is called the integral of E_N over the closed surface and is denoted by

$$\oint_S E_N \, dA$$

This surface integral of E_N is defined as the electric flux ψ_S through the closed surface S:

$$\psi_S = \oint_S E_N \, dA \tag{22.2}$$

Example 2   An electric field is produced by a point charge Q. Find the electric flux ψ_S through a spherical surface S of radius r centered at Q (Figure 22-4).

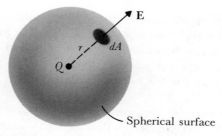

Figure 22-4   Spherical surface of radius r centered at point charge Q.

Solution For any element dA of this surface, since $\mathbf{E}$ is directed away from Q, the normal component of the electric field is

$$E_N = E_r = \frac{kQ}{r^2}$$

Because E_N has the same value at all points of the surface over which the integration is to be performed, it can be factored out through the integration sign, giving

$$\psi_S = \oint_S E_N \, dA = \oint_S \frac{kQ}{r^2} \, dA = \frac{kQ}{r^2} \oint_S dA$$

Noting that

$$\oint_S dA = \text{total area of spherical surface} = 4\pi r^2$$

we obtain

$$\psi_S = 4\pi kQ$$

Using $k = 1/4\pi\epsilon_0$, we can write

$$\psi_S = \frac{Q}{\epsilon_0}$$

(The motivation for the introduction of the factor 4π in the denominator of the relationship $k = 1/4\pi\epsilon_0$ is to simplify expressions such as the foregoing which involve the factor 4π because of the geometry of the sphere.)

Solid Angle

In the evaluation of certain surface integrals we encounter a quantity known as a *solid angle*. This is the extension to three dimensions of the familiar two-dimensional quantity, the angle $d\theta$ between two lines (Figure 22-5). For a given pair of intersecting lines, the ratio ds/r of the subtending arc ds to the radius r is a constant independent of r and characteristic of the pair of lines. This ratio is defined as the measure of the angle $d\theta$ in *radians*:

$$d\theta = \frac{ds}{r}$$

Figure 22-5 The measure of the angle in radians is defined by $d\theta = ds/r$.

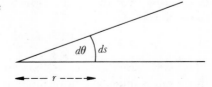

For a pair of lines opened out so that the subtending arc is the circumference of a circle, the angle in radians is given by

$$\theta = \frac{2\pi r}{r} = 2\pi$$

Analogously, in three dimensions, we consider a cone and a spherical surface of radius r centered at the vertex of the cone (Figure 22-6). The area of the spherical surface included within the cone is dA_s. The ratio dA_s/r^2 is a constant independent of the radius of the spherical surface and therefore characteristic of the cone. The cone is said to enclose a *solid angle* $d\omega$ and the ratio dA_s/r^2 is defined as the measure of this solid angle in *steradians*:

$$d\omega = \frac{dA_s}{r^2} \tag{22.3}$$

For a cone opened out so that the entire spherical surface is enclosed, the solid angle enclosed by the cone is given in steradians by

$$\omega = \frac{4\pi r^2}{r^2} = 4\pi \tag{22.4}$$

Example 3   An electric field is produced by a point charge Q. Find the electric flux ψ through an arbitrary closed surface which encloses Q [Figure 22-7(a)].

Solution   Figure 22-7(a) shows an element of the surface of area dA at a distance r from the charge Q. The flux through this element is

$$d\psi = E \, dA \, \cos \theta = kQ \frac{dA \, \cos \theta}{r^2}$$

From Figure 22-7(b) we see that $dA \cos \theta$ is the projection of the area dA on a surface normal to the radial direction. That is,

$$dA \, \cos \theta = dA_s$$

where dA_s is the area of a spherical surface of radius r which is enclosed by the cone with vertex at Q and sides touching the boundaries of the element dA. This cone encloses a solid angle

$$d\omega = \frac{dA_s}{r^2} = \frac{dA \, \cos \theta}{r^2}$$

Using this result in the expression for the flux through dA, we obtain

$$d\psi = kQ \, d\omega$$

Integrating over the entire surface, we find that the total flux through the closed surface is

$$\psi_S = kQ \oint_S d\omega = 4\pi kQ = \frac{Q}{\epsilon_0}$$

This is the same answer as that found in Example 2 but we now see that this result is independent of the shape of the closed surface S through which

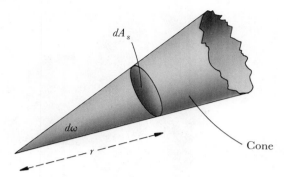

Figure 22-6 Cone includes an area dA_s of a spherical surface of radius r. The solid angle enclosed by the cone is $d\omega = dA_s/r^2$.

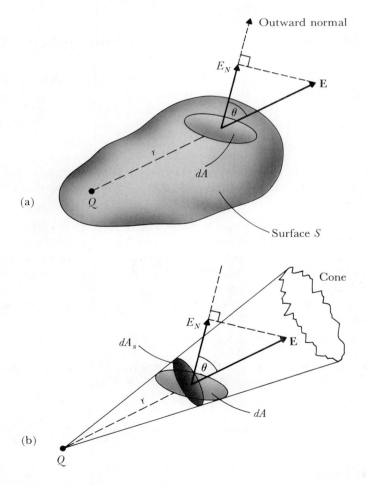

Figure 22-7 The element of area dA is a portion of an arbitrary closed surface S that encloses the point charge Q.

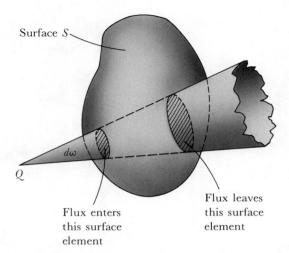

Surface S

$d\omega$

Q

Flux enters
this surface
element

Flux leaves
this surface
element

Figure 22-8   Positive point charge Q is outside an arbitrary closed surface S.

the flux is computed. In Figure 22-7 we consider the charge that set up the electric field to be positive. However, the result, $\psi_S = Q/\epsilon_0$, is valid also in the case where Q is negative. Then ψ_S is negative and flux lines, which are directed from infinity toward the negative charge, enter the surface which encloses Q.

Example 4   An electric field is set up by a positive point charge Q. Find the flux through an arbitrary closed surface S located so that the charge Q is *outside* the surface.

Solution   Figure 22-8 shows a narrow cone with vertex at Q. This cone encloses a solid angle $d\omega$ and intercepts the surface S an even number of times. From the arguments of Example 3, at each interception the flux through the enclosed surface has a magnitude $kQ\,d\omega$. This flux is *negative* at the first interception where flux lines enter the surface, and *positive* at the next interception where these flux lines leave the surface. The net flux for this pair of surface elements is

$$-kQ\,d\omega + kQ\,d\omega = 0$$

Considering in this way the flux through every pair of surface elements, we conclude that the flux through the entire surface is zero:

$$\psi_S = 0$$

It is clear that we obtain the same result when Q is negative.

22.2 Gauss's Law

Examples 3 and 4 of the preceding section show that, if an electric field $\mathbf{E}_i$ is produced by a single point source Q_i, the flux ψ_{Si} through any closed surface S

is given by

$$\epsilon_0 \psi_{Si} = Q_i \qquad \text{(if } Q_i \text{ is inside } S\text{)}$$

$$\epsilon_0 \psi_{Si} = 0 \qquad \text{(if } Q_i \text{ is outside } S\text{)}$$

(22.5)

In the general case where the field **E** is the superposition

$$\mathbf{E} = \mathbf{E}_1 + \mathbf{E}_2 + \cdot \ \cdot \ \cdot + \mathbf{E}_n$$

of the fields $\mathbf{E}_1, \mathbf{E}_2, \ldots, \mathbf{E}_n$ contributed by the point sources $Q_1, Q_2, \ldots, Q_n$, the flux through S is

$$\oint_S E_N \, dA = \oint_S (E_{1N} + E_{2N} + \cdot \ \cdot \ \cdot + E_{nN}) \, dA$$

This equation states the *superposition principle for flux*:

$$\psi_S = \psi_{S1} + \psi_{S2} + \cdot \ \cdot \ \cdot + \psi_{Sn} \tag{22.6}$$

The flux when several sources are present is the sum of the fluxes that the individual sources would contribute if each were present alone.

Evaluating each term ψ_{Si} using Eq. 22.5, we see that this superposition principle gives

$$\epsilon_0 \psi_S = Q_S \tag{22.7}$$

where Q_S is the algebraic sum of the charges enclosed by the surface S. This is *Gauss's law: The electric flux through any closed surface is proportional to the total charge enclosed by the surface.*

Often it is convenient to regard the charge as continuously distributed throughout space with a charge per unit volume (or charge density) ρ defined so that the charge Q_V in any volume V is the integral of ρ throughout that volume

$$Q_V = \int_V \rho \, dV \tag{22.8}$$

For a continuous distribution of charge with charge density ρ, Gauss's law becomes

$$\epsilon_0 \oint_S E_N \, dA = \int_V \rho \, dV \tag{22.9}$$

where V is the volume enclosed by the surface S.

A review of the preceding derivation will confirm that the validity of Gauss's law depends upon:

1   The inverse square law for the force exerted by one particle on another.
2   The fact that the direction of this force lies along the line joining the particles.
3   The superposition principle for such forces.

Since gravitational forces, as well as electrostatic forces, have these characteristics, Gauss's law also holds for gravitational fields with mass replacing charge as the source of the field.

Gauss's law has been presented as a theorem of electrostatics, a consequence of Coulomb's law and the definitions of charge and electric field. But Gauss's law itself can be taken as the basic law of electrostatics. Then Coulomb's law can be deduced, as will be apparent after the following section (see Question 5).

22.3 Applications of Gauss's Law

The surface referred to in Gauss's law, called a *Gaussian surface*, may be any closed surface that we care to consider. In certain cases the *symmetry of the charge distribution* is such that we can identify a surface on which the normal component E_N of the electric field is constant. When such a surface is selected as the Gaussian surface (or as part of the Gaussian surface), because E_N then can be factored out of the flux integral, Gauss's law immediately yields the relationship between E and the source charges. The electric field produced by a given charge distribution can be found by direct use of Gauss's law only when the symmetry of the charge distribution can be exploited in this way.

Field Produced by a Spherically Symmetric Distribution of Charge

Figure 22-9 represents a distribution of charge which is spherically symmetric, that is, the charge density ρ depends only on the distance r from a central point O. The graph of Figure 22-10 shows ρ as a function of r. We wish to find the electric field produced by this charge distribution at a point P a distance r from the center O.

Figure 22-9   A spherically symmetric charge distribution.

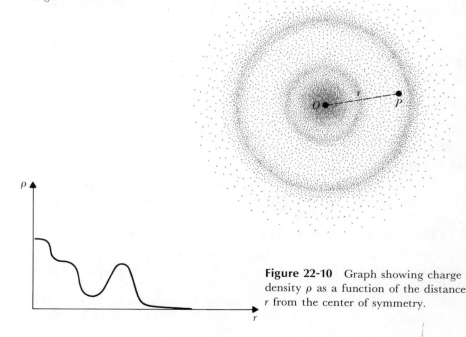

Figure 22-10   Graph showing charge density ρ as a function of the distance r from the center of symmetry.

One method is to proceed as in the examples of Section 21.2 and to attempt to calculate the vector sum of the fields $d\mathbf{E}$ contributed by the charge in each volume element of the distribution. A much simpler method is to first determine as much as possible about the electric field from the symmetry of the charge distribution. A Gaussian surface that exploits this symmetry can then be selected and Gauss's law applied.

The spherical symmetry of the charge distribution implies that:

1 The direction of $\mathbf{E}$ is radial; no other direction is unique.
2 The radial component E_r has the same value at all points on any spherical surface centered at O.

To find the radial component E_r at the point P a distance r from O, we select the Gaussian surface S to be a spherical surface of radius r centered at O. Then $E_N = E_r$, and the electric flux ψ_S through this surface is

$$\psi_S = \oint_S E_r \, dA = E_r \oint_S dA = 4\pi r^2 E_r$$

Gauss's law gives

$$\epsilon_0 4\pi r^2 E_r = Q_S$$

where Q_S is the charge inside the surface S. Therefore

$$E_r = \frac{1}{4\pi\epsilon_0} \frac{Q_S}{r^2} = \frac{kQ_S}{r^2} \qquad (22.10)$$

Comparison with the expression given by Eq. 21.8 for the field produced by a point charge Q, $E_r = kQ/r^2$, shows that *the field produced by a spherically symmetric distribution of charge at a point P a distance r from the center of symmetry O is the same as the field that would be produced by a point charge located at O and equal to the charge inside a spherical surface centered at O and passing through P.*

For example, Eq. 22.10 implies that in the case where a total charge Q is uniformly distributed over a spherical shell (Figure 22-11), the field produced by this charge is zero at all points within the shell and is given by $E_r = kQ/r^2$ at points outside the shell at a distance r from its center.

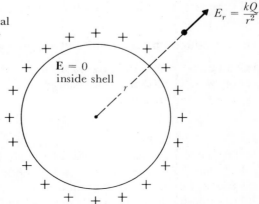

Figure 22-11 Total charge Q is uniformly distributed over a spherical shell. The resultant $\mathbf{E}$ is zero at any point inside this shell of charge.

The same line of reasoning leads to the analogous results for gravitational fields that were quoted in Section 14.1. In particular, the assumption that the earth is spherically symmetric implies that, at points outside the earth, the earth's gravitational field is the same as if all its mass were concentrated at its center (Figure 2-4). A simple rule of this nature is *not* valid for arbitrary distributions of mass. For instance, unless a mass distribution is spherically symmetric, it is *not* true that mass distribution sets up a gravitational field that is the same as if all its mass were concentrated at its center of mass.

Field Produced by a Line Charge

To illustrate the relative ease with which fields of symmetric charge distributions can be found using Gauss's law, we reconsider Example 2 of Chapter 21—the field of a charged infinitely long straight rod with a constant charge per unit length λ. The symmetry of this charge distribution implies that:

1   At every point in space the direction of **E** is radial, that is, perpendicular to the rod. (The field can have no component along the rod since the two directions parallel to the rod are entirely equivalent.)
2   The radial component E_R has the same value at all points on any cylindrical surface with axis along the rod.

The radial component E_R of the field at the point P a distance R from the rod can be found by selecting a Gaussian surface S that is cylindrical with axis along the rod, radius R, and length ℓ (Figure 22-12). The flux through the flat ends of this surface is zero since the flux lines are everywhere perpendicular to the rod. The flux through the curved cylindrical surface of area $2\pi R\ell$ is $2\pi R\ell E_R$. This Gaussian surface encloses a charge $\lambda\ell$. Gauss's law therefore gives

$$\epsilon_0 2\pi R\ell E_R = \lambda\ell$$

Consequently

$$E_R = \frac{\lambda}{2\pi\epsilon_0 R} = \frac{2k\lambda}{R}$$

in agreement with the result found in Section 21.2 by integration of the contributions $d\mathbf{E}$ furnished by each element of the charge distribution.

Field Produced by an Infinite Plane Sheet of Charge

In Example 3 of Section 21.2 we showed that an infinite plane sheet of positive charge with a constant surface charge density σ sets up at any point in space an electric field of magnitude

$$E = \frac{\sigma}{2\epsilon_0}$$

The same result can be derived from Gauss's law without any elaborate calculation. From the symmetry of the charge distribution it is evident that the electric field is perpendicular to the charged plane and has the same magnitude E but opposite direction at two points P and P' on opposite sides of the plane

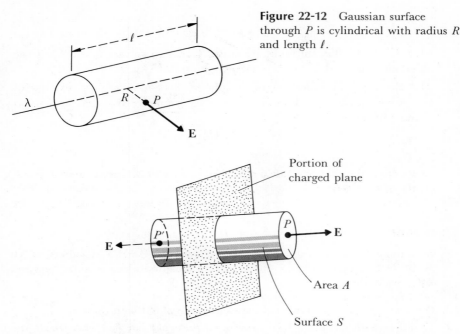

Figure 22-12 Gaussian surface through P is cylindrical with radius R and length ℓ.

Portion of charged plane

Area A

Surface S

Figure 22-13 Gaussian surface S in the field produced by an infinite plane sheet of charge.

and equidistant from it. We therefore select as the Gaussian surface a cylinder (Figure 22-13) with sides perpendicular to the charged plane and with flat ends of area A passing through P and P' and parallel to the charged plane. No flux passes through the sides of this Gaussian surface. A flux EA emerges through each end of area A. Since the charge within the closed surface is σA, Gauss's law gives

$$\epsilon_0 2EA = \sigma A$$

which confirms that

$$E = \frac{\sigma}{2\epsilon_0}$$

Summary

☐   The electric flux ψ_S through a closed surface S is defined by

$$\psi_S = \oint_S E_N\, dA$$

where E_N is the component of **E** normal to the element of area dA of the surface S. In the particular case of a uniform field **E**, the flux through a plane surface of area A is

$$\psi = E_N A$$

☐ Gauss's law is

$$\epsilon_0 \psi_S = Q_S$$

where ψ_S is the flux through any closed Gaussian surface S and Q_S is the algebraic sum of the electric charges within S.

☐ The electric field produced by a spherically symmetric distribution of charge at a distance r from the center of symmetry is radial and has a radial component

$$E_r = \frac{kQ_S}{r^2}$$

where Q_S is the charge inside a spherical surface of radius r centered at the center of symmetry.

Questions

1   In a uniform field, **E** is directed vertically upward and has a magnitude of 3.0×10^4 N/C. Find the electric flux through a table top which is 3.0 m long and 2.0 m wide when:
   (a) The table top is horizontal.
   (b) The normal to the table top makes an angle of 53° with the vertical.

2   Find the electric flux through each side of the prism of Example 1 when the electric field is directed vertically downward and has a magnitude of 2.0×10^4 N/C.

3   Use Gauss's law to determine the algebraic sum of the electric charges within the prism of Example 1.

4   An electric field is produced by a point charge Q. Find the electric flux through each side of a closed surface which is cubical with a side of length l when:
   (a) Q is at the center of the cube.
   (b) Q is at a corner of the cube.

5   Derive the expression for the field produced at a distance r by a point charge Q

$$E_r = \frac{kQ}{r^2}$$

using Gauss's law and assuming that the field is spherically symmetrical with the center of symmetry at Q.

6   A total charge Q is distributed uniformly throughout a spherical region of radius r_0. Show that the electric field at a distance r from the center of this distribution is radial with a radial component given by

$$E_r = \frac{kQr}{r_0^3}$$

for $r < r_0$, and by

$$E_r = \frac{kQ}{r^2}$$

for $r > r_0$. Sketch a graph of E_r as a function of r from $r = 0$ to $r = 3r_0$.

7   A total charge Q is uniformly distributed throughout a sphere of radius r_0. This sphere is placed inside a spherical shell of radius $2r_0$. Find the electric field in all regions of space when the charge uniformly distributed over the shell has the value:
   (a) $-Q$ **(b)** $-\frac{1}{3}Q$

8 Gauss's law for gravitational fields is

$$\frac{1}{4\pi G} \oint_S g_N \, dA = -M_S$$

where g_N is the normal component of **g** on the Gaussian surface S and M_S is the mass enclosed by S. Use this law to find the earth's gravitational field at all distances r from its center, making the simplifying assumption that the earth has a uniform density ρ.

9 Charge is distributed with a uniform surface density σ on an infinitely long cylindrical shell of radius R_0. Find expressions for the electric field inside and outside this shell.

10 Charge is distributed uniformly with a density ρ throughout an infinitely long cylinder of radius R_0. Show that the field at a distance R from the axis of the cylinder is radial with a radial component given by

$$E_R = \frac{\rho R}{2\epsilon_0}$$

for $R < R_0$, and by

$$E_R = \frac{\rho R_0^2}{2\epsilon_0 R}$$

for $R > R_0$. Sketch a graph of E_R as a function of R from $R = 0$ to $R = 3R_0$.

Supplementary Questions

S-1 An infinite region between the plane at $z = 0$ (the XY plane) and the plane at $z = D$ is filled with electric charge with a constant density ρ (Figure 22-14). Find the electric field produced by this charge distribution in all regions of space.

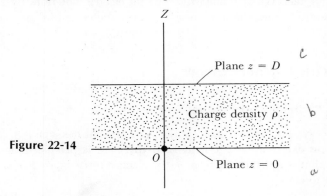

Figure 22-14

S-2 A total charge Q_1 is uniformly distributed throughout a sphere with radius r_1 and center C_1. Then, without disturbing the remaining charge, the charge is removed from a spherical region which lies entirely within the first sphere and has its center at C_2. Show that the electric field at C_2 is in the radial direction from C_1 with

$$E_r = \frac{kQ_1 r'}{r_1^3}$$

where r' is the distance from C_1 to C_2.

23.

electric potential

The description of a static electric field is enhanced and calculations are simplified by the introduction of the scalar function of position $V(x, y, z,)$ known as the electric potential. The potential drop $V_a - V_b$ in going from point a to point b is a fundamental concept required for the treatment of electric circuits given in later chapters. We approach the topic of electric potential by considering a related quantity, the potential energy of a point charge in a given electric field.

23.1 Electric Potential Energy

Charged Particle in an External Electric Field

Consider a system consisting of a single particle that has a charge q and is acted upon by an external electric field $\mathbf{E}$. If this field is produced by a stationary charge Q that is a distance r from q, the electric force exerted on q is the radial Coulomb force with radial component

$$F_r = \frac{kQq}{r^2}$$

The potential energy function (Section 10.3) associated with this force is a function U such that

$$F_s = -\frac{dU}{ds}$$

With a little foresight (or remembering the analogous gravitational problem of Section 14.2) we are led to investigate the potential energy function defined by $U = kQq/r$. For $ds = dr$, we find that the force derived from this potential energy function has a radial component

$$F_r = -\frac{dU}{dr} = \frac{kQq}{r^2}$$

For a displacement ds perpendicular to the radial direction, r is constant and $F_s = dU/ds = 0$. These results show that the force derived from

$$U = \frac{kQq}{r} \tag{23.1}$$

is the Coulomb force, and that U is therefore the potential energy of the charge q in the field produced by Q. U is called the *electric potential energy*.

When Q and q have the same sign, the potential energy U is everywhere positive but the slope dU/dr of the graph of U as a function of r [Figure 23-1(a)] is everywhere negative, corresponding to the repulsive Coulomb force, $F_r = -dU/dr$. If Q and q have opposite signs, U is negative but dU/dr is positive [Figure 23-1(b)] corresponding to an attractive Coulomb force. As r approaches infinity, both U and $F_r = -dU/dr$ approach zero.

Example 1   Use Eq. 10.26 to find the potential energy U of the charge q in the field produced by Q by integration of the expression for the force exerted on q, $F_r = kQq/r^2$.

Solution   Equation 10.26 gives

$$U(x,y,z) = -\int_a^{x,y,z} F_s \, ds + U_a$$

From Figure 23-2 we see that $ds = dr'/\cos\theta$, where θ is the angle between

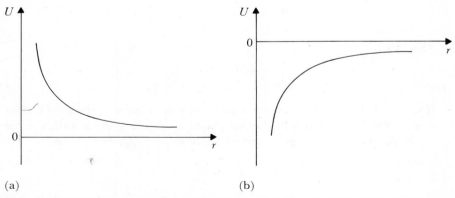

(a)   (b)

Figure 23-1   Graph of the potential energy U of charge q in field produced by Q.
(a) Q and q have same sign. (b) Q and q have opposite signs.

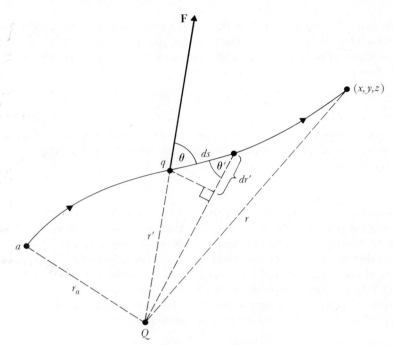

Figure 23-2   Charge q is taken from point a a distance r_a from Q to a point that is a distance r from Q. The element of path length is $ds = dr'/\cos \theta'$, which approaches the value $dr'/\cos \theta$ as ds approaches zero.

ds and the radial direction. Also $F_s = F_{r'} \cos \theta$. Therefore

$$F_s \, ds = F_{r'} \, dr'$$

and

$$U = -\int_{r_a}^{r} F_{r'} \, dr' + U_a = -\int_{r_a}^{r} \frac{kQq}{r'^2} \, dr' + U_a$$

This gives

$$U = \frac{kQq}{r} - \frac{kQq}{r_a} + U_a$$

If we select the arbitrary reference point a to be at infinity, the second term is zero. This expression is further simplified by choosing the arbitrary value U_a of the potential energy at the reference point to be zero. Then

$$U = \frac{kQq}{r}$$

Since the integral in Eq. 10.26 is a work, we can think of the potential energy $U = kQq/r$ as the negative of the work that is done by the electric force acting on q as it is brought from infinity to a distance r from Q. Or, equivalently, U is the work that would have to be done by an external applied force to bring q from infinity to its position at a distance r from Q.

System of Several Charged Particles

A system comprising several charged particles has an internal electric potential energy associated with each pair of particles due to their Coulomb interaction. Within the system, a pair of particles with charges Q_1 and Q_2 separated by a distance r_{12} has an electric potential energy

$$U_{12} = \frac{kQ_1Q_2}{r_{12}} \tag{23.2}$$

This internal electric potential energy belongs to the pair and is not assigned to one specific particle of the pair. (Potential energy is assigned to a single particle only when the source of the associated force is not included in the system.)

The superposition principle for Coulomb forces leads to additivity of the associated potential energies. For example, the system of three charges shown in Figure 23-3 has a total electric potential energy

$$U = \frac{kQ_1Q_2}{r_{12}} + \frac{kQ_1Q_3}{r_{13}} + \frac{kQ_2Q_3}{r_{23}}$$

This is the amount of work that would have to be done by external applied forces to bring these charges into their present configuration from positions where they were separated by infinite distances.

Example 2 Find the electric potential energy of the system of charges shown in Figure 23-4.

Solution

$$U = \frac{k(-2e^2)}{r_0} + \frac{ke^2}{2r_0} + \frac{k(-2e^2)}{r_0} = -\frac{7}{2}\frac{ke^2}{r_0}$$

Figure 23-3 System of three point charges.

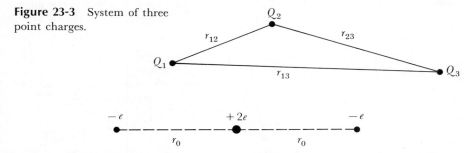

Figure 23-4 System of three point charges of Example 2.

23.2 Electric Potential Function

Definition of Electric Potential

An electrostatic field $\mathbf{E}$ is produced by some distribution of stationary charges $Q_1, Q_2, \ldots, Q_n$. If a test charge q is placed in this field at the location shown in Figure 23-5, the system consisting of this single test charge has an electric potential energy

$$U = \frac{kqQ_1}{r_1} + \frac{kqQ_2}{r_2} + \cdots + \frac{kqQ_n}{r_n} \tag{23.3}$$

We see that the electric potential energy of the test charge is proportional to q; that is,

$$U = qV \tag{23.4}$$

where V is a function of position and is independent of q. The quantity $V = U/q$, the potential energy per unit test charge, is called the *electric potential* of the electric field. The potential V is a scalar quantity and may be a positive or a negative number.

The SI unit of potential, the joule per coulomb, is called the *volt* (V):

$$1 \text{ V} = 1 \text{ J/C}$$

Potential in the Field Produced by a Single Point Charge

When the electric field is produced by a single stationary point charge Q, a test charge q placed at a point P that is a distance r from Q has a potential energy $U = kQq/r$. The potential at the point P is $V = U/q$ which gives

$$V = \frac{kQ}{r} \tag{23.5}$$

Figure 23-6 shows the potential at various locations in the field produced by a positive point charge of 1.0×10^{-8} C. In any electric field the points in space at which the potential has a given value lie on a surface called an *equipotential* surface. From Eq. 23.5 it is evident that, in the field produced by a single point charge, the equation of an equipotential surface is $r = \textit{constant}$. The equi-

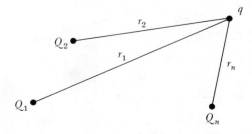

Figure 23-5   Point charge q placed in the electrostatic field established by sources $Q_1, Q_2, \ldots, Q_n$.

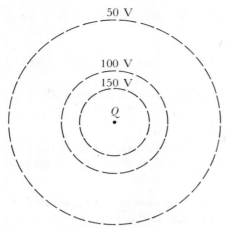

Figure 23-6 Equipotentials at 50 V, 100 V, and 150 V in the field produced by a positive point charge $Q = 1.0 \times 10^{-8}$ C.

potential surfaces of Figure 23-6 are therefore spherical surfaces centered at Q. For example, all points where $V = 100$ V are at a distance from Q given by

$$r = \frac{kQ}{V} = \frac{(9.0 \times 10^9)(1.0 \times 10^{-8})}{100} \text{ m} = 0.90 \text{ m}$$

In other words, the equipotential surface corresponding to a potential of 100 V is a spherical surface of radius 0.90 m.

Potential in the Field Produced by Several Point Charges

In the general case where the electrostatic field is produced by the distribution of charges $Q_1, Q_2, \ldots, Q_n$ of Figure 23-5, Eq. 23.3 together with the definition $V = U/q$, gives

$$V = \frac{kQ_1}{r_1} + \frac{kQ_2}{r_2} + \cdots + \frac{kQ_n}{r_n} \tag{23.6}$$

where V is the potential at the point which is a distance r_1 from Q_1, r_2 from $Q_2, \ldots, r_n$ from Q_n.

This equation states the superposition principle for potentials: The potential at a given point in a field produced by several charges is the *algebraic sum* of the potentials that each charge alone would contribute. The advantage of working with potentials rather than field vectors is that algebraic sums of scalar quantities are easier to evaluate than vector sums.

When the charges that produce the field are continuously distributed throughout some region of space, the potential is the sum of contributions $k\,dQ/r$ where dQ is the charge in an element of the charge distribution and r

is the distance from dQ to the point in question. The potential due to the entire distribution is

$$V = \int \frac{k\,dQ}{r} \tag{23.7}$$

where the integration ranges over all elements dQ that comprise the charge distribution.

Example 3   A uniformly charged plastic disk of radius a has a surface charge density σ. Find the potential V at the point P on the axis of the disk with coordinates $(0,0,z)$ (Figure 23-7).

Solution   All charges in the shaded ring of Figure 23-7 are at the same distance r from P. The total charge in this ring of area $2\pi s\,ds$ is

$$dQ = \sigma 2\pi s\,ds$$

The potential contributed by this element of charge is

$$dV = \frac{k\,dQ}{r} = \frac{k\sigma 2\pi s\,ds}{r} = \frac{2\pi\sigma ks\,ds}{\sqrt{z^2 + s^2}}$$

The potential at P due to all the charge on the disk is

$$V = \int \frac{k\,dQ}{r} = 2\pi\sigma k \int_{s=0}^{s=a} \frac{s\,ds}{\sqrt{z^2 + s^2}} = \left[2\pi\sigma k \ \sqrt{z^2 + s^2} \ \right]_{s=0}^{s=a}$$

This gives

$$V = 2\pi\sigma k (\sqrt{z^2 + a^2} - |z|)$$

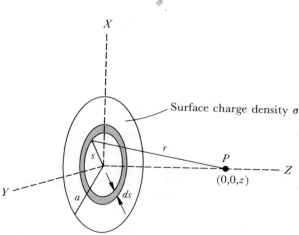

Figure 23-7   Charged disk of Example 3.

23.3 Calculation of **E** from V

The relationship between the electric field vector **E** at a point in space and the potential function V can be found by considering the force **F** exerted by **E** on a test charge q placed at the point in question. Since $\mathbf{F} = q\mathbf{E}$ and $F_s = -dU/ds$ where $U = qV$, we have

$$qE_s = \frac{d}{ds}(qV)$$

which gives

$$E_s = -\frac{dV}{ds} \tag{23.8}$$

This relationship states that, at any point in space, the component E_s of the electric field in a given direction is the negative of the spatial rate of change of the potential in that direction. By considering a displacement ds parallel to the X axis, we obtain

$$E_x = -\frac{\partial V}{\partial x} \tag{23.9}$$

Similarly

$$E_y = -\frac{\partial V}{\partial y} \quad \text{and} \quad E_z = -\frac{\partial V}{\partial z} \tag{23.10}$$

The gradient of V at a given point is defined to be a vector in the direction in which dV/ds is a maximum and with a magnitude equal to this maximum value. Equation 23.8 implies that

$$\mathbf{E} = -\operatorname{grad} V$$

where grad V denotes the gradient of V. The minus sign corresponds to the fact that **E** points from regions of higher to lower potential. From Eq. 23.9 and 23.10 we see that the components of the vector grad V are given by

$$(\operatorname{grad} V)_x = \frac{\partial V}{\partial x} \quad (\operatorname{grad} V)_y = \frac{\partial V}{\partial y} \quad (\operatorname{grad} V)_z = \frac{\partial V}{\partial z} \tag{23.11}$$

If the displacement ds lies on an equipotential surface, $dV = 0$. Then $E_s = -dV/ds = 0$. This shows that, at any point, the electric field **E** *is normal to the equipotential surface* through the point in question. It follows that electric flux lines are everywhere perpendicular to equipotential surfaces, as is illustrated in Figure 23-8.

The unit of **E** that has been used up to this point is the newton per coulomb (N/C). The relationship $E_s = -dV/ds$ shows that the unit of **E** can equally well be the *volt per metre* (V/m) and we shall now adopt this unit to follow common scientific practice.

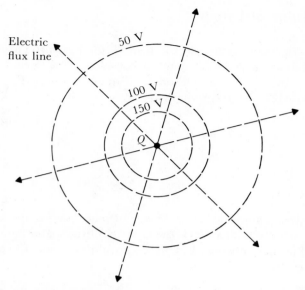

Figure 23-8   Flux lines are everywhere perpendicular to equipotential surfaces.

Example 4   Find the electric field **E** at the point P on the axis of the charged disk of Example 3 (Figure 23-7).

Solution   The symmetry of the charge distribution implies that, at P, **E** has only a z component. E_z can be most easily determined by differentiating the potential function found in Example 3. At the point with coordinates $(0,0,z)$ with $z > 0$,

$$E_z = -\frac{\partial V}{\partial z} = -\frac{\partial}{\partial z}\left[2\pi\sigma k(\sqrt{z^2 + a^2} - z)\right]$$

$$= 2\pi\sigma k\left(1 - \frac{z}{\sqrt{z^2 + a^2}}\right)$$

23.4 Calculation of V from E

When **E** is known, the change in potential can be calculated by integration. Equation 23.8 implies

$$dV = -E_s\, ds$$

Integrating both sides of this equation, we have

$$-\int_a^b dV = \int_a^b E_s\, ds$$

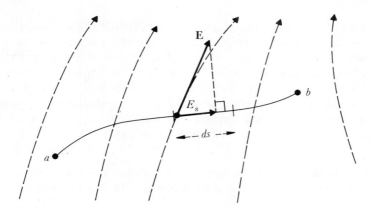

Figure 23-9   Solid line represents the path of integration for the line integral $\int_a^b E_s \, ds$. Dashed lines represent electric flux lines.

Therefore

$$V_a - V_b = \int_a^b E_s \, ds \tag{23.12}$$

The quantity $V_a - V_b$ is called the *potential drop in going from* point a to point b and is denoted by the single symbol V_{ab}.* The potential drop in going from one point to another is often spoken of as the *voltage* between the two points. Equation 23.12 states that V_{ab} is the line integral of the electric field along *any* path from a to b (Figure 23-9).

The reason that it is possible to express the line integral of the electric field as the change in a scalar function of position $V(x,y,z)$ stems ultimately from the fact that Coulomb forces are *conservative*. Equation 23.12 implies that an electrostatic field has the following equivalent properties:

1   The line integral of **E** between any two given points in space is independent of the path of integration joining these points.
2   The line integral of **E** around any closed curve is zero

$$\oint E_s \, ds = 0 \tag{23.13}$$

(This follows from Eq. 23.12 for the case where points a and b coincide.)

Instead of defining potential as potential energy per unit test charge, the line integral $\int_a^b E_s \, ds$ is often taken as the definition of the potential drop $V_a - V_b$ in going from point a to point b. The latter approach emphasizes that

*The *difference of potential* between point a and point b is defined by some authors as $V_a - V_b$ and by others as $V_b - V_a$. In this book where the sign is important we avoid the term "difference of potential" by referring instead to an unambiguous quantity, the potential drop in going from one point to another. If a potential drop V_{ab} has a negative value, the potential actually rises as one goes from a to b.

the physically significant quantities are potential drops, rather than the values assigned to the potential function $V(x,y,z)$ at the various points of space. These values depend on the arbitrary choice of the value of the potential at an arbitrarily selected reference point. When using the expression kQ/r for the potential a distance r from Q, the reference point is at infinity and there the potential is zero. But if the field is produced by a charge distribution that itself extends to infinity, as in the following examples, then it is generally no longer possible to assign zero potential to points at infinity.

Example 5   A uniform electric field has a constant magnitude E and points in the positive x direction. Find an expression for the potential function V.

Solution   The potential drop in going from the origin to the point (x,y,z) is

$$V(0,0,0) - V(x,y,z) = \int_{0,0,0}^{x,y,z} E_s \, ds$$

Selecting the path of integration shown in Figure 23-10, we find

$$V(0,0,0) - V(x,y,z) = \int_{0,0,0}^{x,0,0} E \, dx + \int_{x,0,0}^{x,y,z} E_s \, ds$$

The second integral is zero because along this portion of the path **E** is perpendicular to ds and consequently the component E_s is zero. Therefore

$$V(0,0,0) - V(x,y,z) = E \int_0^x dx = Ex$$

The potential function is

$$V(x,y,z) = -Ex + V(0,0,0)$$

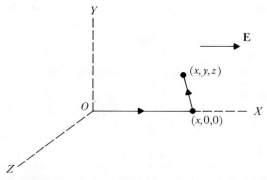

Figure 23-10   Uniform field **E** in the x direction. The path of integration goes from the origin along the X axis to $(x,0,0)$ and then perpendicular to the field until the point (x,y,z) is reached.

Notice that as x approaches infinity, this potential becomes infinite rather than zero. If the potential at the origin is assigned the value zero, this potential function simplifies to

$$V(x,y,z) = -Ex$$

The answer can be checked by differentiation of the proposed potential function:

$$E_x = -\frac{\partial V}{\partial x} = -\frac{\partial}{\partial x}(-Ex) = E \qquad E_y = -\frac{\partial V}{\partial y} = 0 \qquad E_z = -\frac{\partial V}{\partial z} = 0$$

This shows that the field **E** associated with the potential $V(x,y,z) = -Ex$ is indeed a uniform field in the x direction.

Example 6 Find the potential function in the radial field $E_R = 2k\lambda/R$ produced by the infinitely long charged rod of Example 2, Chapter 21.

Solution The potential drop in going from point P_0 to point P in Figure 23-11 is

$$V_0 - V(R) = \int_{P_0}^{P} E_s \, ds = 2k\lambda \int_{R_0}^{R} \frac{dR'}{R'}$$

since for a radial path of integration, $ds = dR'$ and $E_s = 2k\lambda/R'$. Therefore

$$V_0 - V(R) = 2k\lambda \left[\ln R' \right]_{R'=R_0}^{R'=R} = 2k\lambda \ln R - 2k\lambda \ln R_0$$

The potential function is

$$V(R) = -2k\lambda \ln R + \text{constant}$$

Notice that in this example, as in Example 5, the potential function is infinite at infinity.

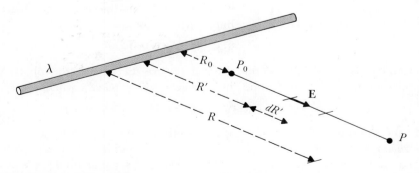

Figure 23-11 Charged rod sets up a radial field $E_R = 2k\lambda/R'$. The potential drop in going from P_0 to P is found by evaluating the line integral of this field between these points.

23.5 Energy of a Particle in an Electric Field. The Electronvolt

If a particle of charge q in an electrostatic field $\mathbf{E}$ (established by other charges) is moved from point a to point b, the work W_{ab} done by the electric field on the charge q is equal to q multiplied by the potential drop V_{ab}:

$$W_{ab} = \int_a^b F_s\, ds = q \int_a^b E_s\, ds = q(V_a - V_b) = qV_{ab} \qquad (23.14)$$

where we have used Eq. 23.12. In other words, the work done by the electric field on q is the negative of the change $qV_b - qV_a$ of the particle's electric potential energy.

If a particle with charge q is acted upon only by the electric force $\mathbf{F} = q\mathbf{E}$, the particle's potential energy is qV and the law of conservation of total energy $K + U$ gives

$$K_a + qV_a = K_b + qV_b$$

Therefore a charged particle accelerated through a potential drop V_{ab} experiences a change in kinetic energy $\Delta K = K_b - K_a$ given by

$$\Delta K = q(V_a - V_b) = qV_{ab}$$

This is the relationship that is used to determine the energy of the charged particles in a beam emerging from the electrostatic particle accelerators used in nuclear physics. In a typical accelerator of the Van de Graff type, protons are accelerated from rest through a potential drop of several million volts and acquire sufficient kinetic energy to produce nuclear reactions in the target material.

In expressions such as $W_{ab} = qV_{ab}$, $U = qV$, or $\Delta K = qV_{ab}$, we must multiply a potential drop by a charge to obtain work or energy. The SI unit of work or energy is the joule, which is equal to the coulomb volt. Since the charges of molecules, atoms, and subatomic particles are always a small integral multiple of the magnitude of the electronic charge e, a useful unit of energy is the *electronvolt* (eV) which is defined as the *work done by the electric field on a particle with charge e when the particle moves through a potential drop of one volt*. This definition implies that the conversion factor between the joule and the electronvolt has the numerical value of the electronic charge in coulombs:

$$1 \text{ eV} = 1.6021892 \times 10^{-19} \text{ J}$$

Commonly used multiples of the electronvolt are the kiloelectronvolt (keV), the megaelectronvolt (MeV), and the gigaelectronvolt (GeV), where

$$1 \text{ keV} = 10^3 \text{ eV} \qquad 1 \text{ MeV} = 10^6 \text{ eV} \qquad 1 \text{ GeV} = 10^9 \text{ eV}$$

In the equations of this chapter, if the electronic charge e is adopted as unit charge and a charge is expressed as a multiple q of this unit, then a term such as qV is an energy in electronvolts if V is expressed in volts. For example, at a point where $V = 100$ V a proton (charge $= e$) has a potential energy

$$qV = 100 \text{ eV}$$

and an α particle (a doubly ionized helium nucleus with charge $= 2e$) accelerated from rest through a potential drop of 100 V acquires a kinetic energy

$$K = qV_{ab} = 200 \text{ eV}$$

The electronvolt is widely used in atomic, nuclear, and particle physics not
only because of its convenience in the calculation of the energy of charged
particles but also because this energy unit has a convenient size, 1 eV being
typical of the amount of energy change involved in an atomic process such as a
chemical reaction.

Summary

☐   The potential energy of a charge q in an electrostatic field produced by
a charge Q is

$$U = \frac{kQq}{r}$$

☐   The internal potential energy of a system consisting of stationary charges
Q_1, Q_2, Q_3 is the electric potential energy

$$U = \frac{kQ_1Q_2}{r_{12}} + \frac{kQ_1Q_3}{r_{13}} + \frac{kQ_2Q_3}{r_{23}}$$

☐   In an electrostatic field, the electric potential at any point is defined by

$$V = \frac{U}{q}$$

where U is the potential energy that a test charge q would have at the
point in question. The SI unit of potential is the volt (V): 1 V $=$ 1 J/C.

☐   The potential in the field produced by the point sources $Q_1, Q_2, \ldots ,$
Q_n is

$$V = \frac{kQ_1}{r_1} + \frac{kQ_2}{r_2} + \cdot \;\; \cdot \;\; \cdot + \frac{kQ_n}{r_n}$$

at the point which is a distance r_i from Q_i $(i = 1, 2, \ldots , n)$. The potential
at a point P produced by a continuous distribution of charges is

$$V = \int \frac{k\,dQ}{r}$$

where r is the distance from the element dQ to point P and the integration
ranges over all elements dQ that comprise the distribution.

☐   All points on an equipotential surface are at the same potential. Electric
flux lines (or lines of force) are everywhere perpendicular to equipotential
surfaces.

☐ At any point in space, the component E_s of the electric field in a given direction is the negative of the spatial rate of change of potential in that direction:

$$E_s = -\frac{dV}{ds}$$

In particular,

$$E_x = -\frac{\partial V}{\partial x} \qquad E_y = -\frac{\partial V}{\partial y} \qquad E_z = -\frac{\partial V}{\partial z}$$

These three equations are equivalent to the one vector equation

$$\mathbf{E} = -\text{grad } V$$

☐ The potential drop, $V_{ab} = V_a - V_b$, in going from point a to point b is the line integral of the electric field along any path joining these points:

$$V_{ab} = V_a - V_b = \int_a^b E_s \, ds$$

It is the potential drops, rather than the values of the potential function itself, that have physical significance.

☐ In any electrostatic field

$$\oint E_s \, ds = 0$$

☐ The electronvolt is an energy unit related to the joule by

$$1 \text{ eV} = 1.602 \times 10^{-19} \text{ J}$$

A term qV is an energy in electronvolts if V is measured in volts and the charge is equal to the magnitude of the electronic charge multiplied by q.

Questions

1   What is the electric potential energy (in joules) of an electron which is 0.53×10^{-10} m from a nucleus with a charge e?

2   What is the electric potential energy of the system of the five particles in Figure 23-12? The charge $+4e$ is at the center of the square and charges $-e$ are located at each corner. A side has the length a.

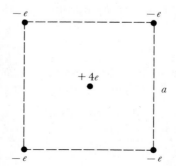

Figure 23-12

3 For the system of Figure 23-12, how much work must be done to completely separate the positively charged particle from this structure?

4 Find the electric potential energy of the system of charges in Figure 23-13. The charge $+2e$ is at the center of the cube and charges $-e$ are located at each corner. A side has the length a.

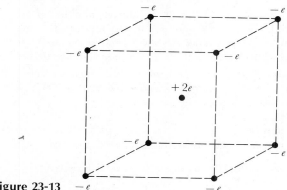

Figure 23-13

5 In the field produced by a point charge of -5.0×10^{-9} C, find the equipotential surface corresponding to $V = -200$ V.

6 An electric field is produced by the system of four charges shown in Figure 23-14. Find the electric potential at the center of the square. What is the potential energy of a charge $q = +2e$ if it is placed at the center of the square?

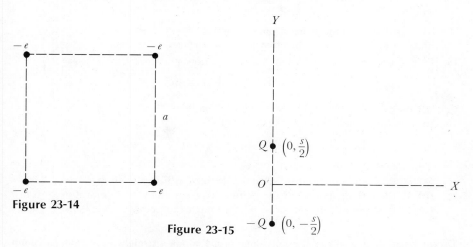

Figure 23-14

Figure 23-15

7 An electric field is produced by a pair of charges Q and $-Q$ separated by a distance s (Figure 23-15). Describe the equipotential surface corresponding to $V = 0$. Deduce an expression for the potential at any point $(0,y)$ on the line through the two charges.

8 A thin ring of radius a has a uniform charge per unit length λ. Find the potential at the center of the ring.

9 If the potential function is constant throughout a given region of space, what can be said about **E** in that region?

10   Starting from a given point P in space, we find that, in a displacement of 2.0 $\times 10^{-3}$ m north, the potential decreases by 80 V; in a displacement of 1.0 $\times 10^{-3}$ m west, the potential increases by 30 V; and in any small vertical displacement there is no change in potential. Estimate the magnitude and direction of the electric field at P. (Since $E_s = -dV/ds$, E_s is approximately equal to $-\Delta V/\Delta s$ for a displacement Δs small enough so that little change in $\mathbf{E}$ is encountered.)

11   The potential at a point P is 500 V. The potential is also 500 V at points near P on a vertical line through P and on a north-south line through P. The potential at a point 3.0 $\times 10^{-4}$ m west of P is 490 V; and at a point 3.0 $\times 10^{-4}$ m east of P, the potential is 510 V. Estimate the magnitude and direction of the electric field at P.

12   The electric field in a certain region which includes a point P is 5.0 $\times 10^3$ V/m directed vertically upward. The potential at P is 200 V. Estimate the potential at points with the following locations:
(a) 3.0 mm above P.
(b) 4.0 mm below P.
(c) 2.0 mm north of P in the same horizontal plane.

13   The potential function in an electrostatic field is given in SI units by

$$V = 20x + 10$$

(a) Describe the equipotential surfaces.
(b) Find the magnitude and direction of the electric field everywhere.

14   The potential function in an electrostatic field is

$$V = \frac{kQ}{r}$$

where r is the distance from a point charge Q. Use the equation $E_s = -dV/ds$ to find the magnitude and direction of $\mathbf{E}$ everywhere.

15   Find $\mathbf{E}$ corresponding to the potential function given in SI units by

$$V = 100 \ln R + 50$$

where R is the distance from an infinite straight line.

16   In a uniform field, $\mathbf{E}$ has the components

$$E_x = 300 \text{ V/m} \qquad E_y = 0 \qquad E_z = 0$$

The potential at the origin is 100 V:
(a) What is the potential at the point with coordinates (0.40 m,0,0)?
(b) What is the potential at the point with coordinates (0.40 m,0.30 m,0.80 m)?
(c) Describe the equipotential surface through the point (0.40 m,0,0).

17   In Example 5, evaluate the line integral of $\mathbf{E}$ along a straight path from the point (x,y,z) to the origin (Figure 23-10). Then verify that $\oint E_s \, ds = 0$ for the closed triangular path from (0,0,0) to (x,0,0) to (x,y,z) to (0,0,0).

18   Show that in a uniform field $\mathbf{E}$ directed from an equipotential surface at potential V_a to an equipotential surface at potential V_b,

$$V_a - V_b = E\ell$$

where ℓ is the distance between the equipotential surfaces.

19 A point charge Q sets up a radial field with radial component

$$E_r = \frac{kQ}{r^2}$$

By calculating the line integral of **E**, show that

$$V(r_a) - V(r_b) = \frac{kQ}{r_a} - \frac{kQ}{r_b}$$

20 Concentric spherical surfaces of radii r_a and r_b ($r_b > r_a$) carry charges Q_a and Q_b, respectively, spread uniformly over these surfaces. Calculate the potential drop in going from the inner to the outer surface.

21 Find a potential function for all points in the electric field produced by the charged cylinder of Question 10, Chapter 22.

22 Concentric infinitely long cylindrical surfaces of radii R_a and R_b ($R_b > R_a$) carry charges per unit length λ_a and λ_b, respectively, spread uniformly over these surfaces. Calculate the potential drop in going from the inner to the outer surface.

23 There is a uniform surface charge density σ on each of two infinite planes, one at $x = -l$ and the other at $x = l$. Find the electric field and the potential everywhere.

24 In a region in which there is an electric field, the electric forces do a work of 3.00 J on a positive charge of 1.50 C when the charge moves from point a to point b:
(a) What is the potential drop in going from point a to point b?
(b) Which point, a or b, is at the higher potential?

25 What is the electric potential energy in electronvolts of an electron that is 0.53×10^{-10} m from a nucleus with charge e?

26 Point a is at a potential of 300 V, and point b is at a potential of 100 V:
(a) Evaluate, in joules and in electronvolts, the kinetic energy of an α particle which starts at rest at point a and is accelerated to point b.
(b) Calculate the kinetic energy at b, in joules and in electronvolts, of an electron which has a kinetic energy of 600 eV at a.

Supplementary Questions

S-1 Find the potential energy per particle of an infinite row of equally spaced charges of alternating sign, each with a magnitude e (Figure 23-16).

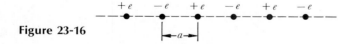

Figure 23-16

S-2 A thin rod of length l lies on the X axis with its center at the origin. A total charge Q is uniformly distributed along the length of this rod. Find the potential at a point $(0,y,0)$ where $y > 0$.

S-3   A total charge Q is uniformly distributed throughout a sphere of radius r_0 producing a radial electric field with radial component given by

$$E_r = \frac{kQr}{r_0^3}$$

for $r < r_0$, and by

$$E_r = \frac{kQ}{r^2}$$

for $r > r_0$. If the zero of potential is at infinity, show that

$$V = \frac{kQ}{r}$$

for $r > r_0$, and that

$$V = \frac{3kQ}{2r_0} - \frac{kQr^2}{2r_0^3}$$

for $r < r_0$. Sketch a graph of V as a function of r from $r = 0$ to $r = 3r_0$.

S-4   In a region where there is a uniform electric field, an electron starts from rest and moves northeast, acquiring a kinetic energy of 60 eV after traveling 2.0 mm. Find the magnitude and direction of the electric field in this region.

S-5   A particle accelerator accelerates α particles from rest through a potential drop of 2.5×10^6 V:

   (a) What is the kinetic energy of an α particle in the beam emerging from this accelerator?

   (b) Assume that one of these α particles approaches a gold nucleus head-on. This nucleus has a charge of $79e$ and a radius of 5.0×10^{-15} m. What is the distance of closest approach to the center of the nucleus? Will this α particle reach the surface of the nucleus?

S-6   The particle accelerator of the preceding question is used to accelerate protons instead of α particles. What is the energy of the emerging proton beam? How close can one of these protons come to the center of a gold nucleus?

S-7   Find the potential energy function for the field produced by the charge distribution in Figure 22-14 described in Question S-1, Chapter 22.

24.

conductors
and dielectrics
in electrostatics

In the preceding three chapters we have been studying electrostatic fields in empty space populated only by point charges or by continuous distributions of charge. In this chapter we consider modifications of the electric field caused by the presence of matter in bulk. This investigation leads to a discussion of a most useful element of electric circuits, the capacitor.

24.1 Conductors in an Electrostatic Field

Within any material there is a certain density of charged particles that are mobile and drift through the material if a nonzero average electric field is established within the material. The electrical *conductivity* of a material measures the rate at which charge is conducted through the material per unit cross-sectional area per unit electric field. Materials with the highest conductivities are classified as *conductors*, those with the lowest as *insulators* or *dielectrics*. The ratio of the conductivity of good conductors such as metals to the conductivity of a good insulator like glass is tremendous, of the order of 10^{20}. Within a metal there is an appreciable density of highly mobile charged particles since, when atoms combine to form a metallic crystal, the valence electrons of each atom become free to roam throughout the entire crystal. But within an insulator an appreciable density of mobile charges is lacking.

If, for the duration of an experiment, there is no significant drift of charges through a material, the material may be treated as if it were a perfect insulator with zero conductivity. On the other hand, if the mobility of the charges that can drift through the material is so large that mobile charges will have moved in a very short time interval to positions where they experience zero average resultant force, and then remain there for the duration of the experiment, the material behaves as if it were a perfect conductor with infinite conductivity. For many practical problems in electrostatics, the common insulators may be assumed to be perfect insulators and the common conductors to be perfect conductors.

In this chapter we shall be concerned only with the *average* position and the average motion of charges within a material. The charge density ρ and the electric field **E** will be assumed to be average values for macroscopic volume elements which, although small compared to the entire macroscopic object, are large enough to contain billions of atoms. The potential V is the potential associated with this average electric field.

E, V, and ρ within a Conductor

Within a homogeneous isotropic conductor, a mobile charge q is free of all forces except the electric force $q\mathbf{E}$. In *electrostatics*, these charges remain at rest, and we conclude that the *electric field is zero*:

$$\mathbf{E} = 0 \qquad \text{(inside a conductor, in electrostatics)} \quad (24.1)$$

The potential drop in going from a to b, for any two points within the conductor, is

$$V_a - V_b = \int_a^b E_s \, ds = 0$$

since, for a path of integration lying entirely within the conductor, $\mathbf{E} = 0$ at every point of the path. We see that all points of a conductor are at the same potential. *In electrostatics, the entire conductor is an equipotential region:*

$$V = constant \qquad \text{(all points of a conductor, in electrostatics)} \quad (24.2)$$

The total charge Q_S within any volume inside the conductor (Figure 24-1) is given by Gauss's law as

$$Q_S = \int_S E_N \, dA = 0$$

since **E** is zero at all points of the Gaussian surface S lying within a conductor.

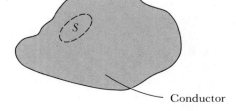

Figure 24-1 Gaussian surface S *within a conductor. In electrostatics, the electric field is zero at all points of this surface.*

Conductor

It follows that the *average charge density ρ is zero inside a conductor:*

$$\rho = 0 \qquad \text{(inside a conductor, in electrostatics)} \quad (24.3)$$

E and σ at the Surface of a Conductor

In electrostatics, all points of a conductor are at the same potential and there-fore the *surface of a conductor is an equipotential surface.* **E** is normal to an equipoten-tial. This implies that *in electrostatics the direction of the electric field at any point on the surface of a conductor is normal to the surface.*

We have discovered that the charge density is zero throughout the volume of a conductor. This means that in electrostatics, a *net charge can reside only on the surface of a conductor.* The relationship between the surface charge density σ and the electric field **E** at the surface of a conductor is found by applying Gauss's law to the cylindrical Gaussian surface S of Figure 24-2. Because **E** is normal to the surface, no flux emerges through the cylindrical sidewalls of S.

Outside the conductor, S is closed by a flat end of area A parallel and arbi-trarily close to the conductor's surface. The flux through this end is $E_N A$, where E_N is the component of **E** along the outward normal to the surface. (If **E** is out-ward, $E_N = |\mathbf{E}|$; if **E** is inward, $E_N = -|\mathbf{E}|$.) Since **E** $= 0$ inside a conductor, there is no flux through the part of S lying within the conductor. The charge inside S is σA. Gauss's law therefore gives

$$\epsilon_0 E_N A = \sigma A$$

Consequently, the special relationship that must hold at every surface of a con-ductor in electrostatics is

$$E_N = \frac{1}{\epsilon_0}\sigma \qquad \text{(conducting surface in electrostatics)} \quad (24.4)$$

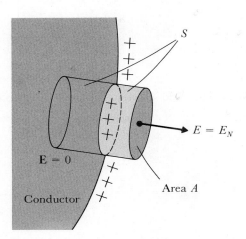

Figure 24-2   Gaussian surface S is partially within the conductor. Cylindrical sidewalls of S are perpendicular to the conductor's surface.

From a microscopic point of view, the transition from zero electric field within the conductor to the value $E_N = \sigma/\epsilon_0$ takes place over a thickness of several atomic layers and the surface charge is actually distributed throughout these layers (Figure 24-3).

Hollow Conductors

Figure 24-4 shows an *empty* hollow conductor carrying a net charge Q. We know that $\mathbf{E} = 0$ inside the conductor and that the charge must reside on the surface, but is there any charge on the *inside* surface and is there an electric field within the cavity? The answer is that the field in the empty cavity is zero and this implies (from $E_N = \sigma/\epsilon_0$) that the surface charge density is zero everywhere on the inside surface. The charge resides entirely on the *outside* surface of an empty hollow conductor.

One way to prove that $\mathbf{E} = 0$ inside the cavity is to consider the contrary situation, as indicated in Figure 24-5. If there is a nonzero field in the cavity, then there will be electric flux lines that must begin on positive charges on one part of the inner surface and end on negative charges on another part of this surface. (This is required by Gauss's law.) Then $\int_a^b E_s \, ds$, evaluated for a path

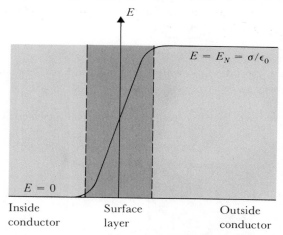

| Inside conductor | Surface layer | Outside conductor |

Figure 24-3 The electric field variation crossing the surface of a conductor.

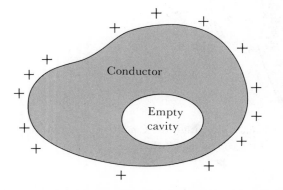

Figure 24-4 Hollow conductor.

which is the flux line shown in Figure 24-5, will be a sum of positive contributions and therefore will be positive, implying that $V_a \neq V_b$. Since we know that $V_a = V_b$, we can conclude that the situation shown in Figure 24-5 cannot exist in electrostatics. Because there is no electric field within an empty cavity in a conductor, the cavity is said to be electrostatically *shielded* from the effects of source charges on or outside this conductor.

Experiments confirm that static equilibrium is obtained with all the charge on the outside of any closed conducting surface. Modern experimental results verify this consequence of Gauss's law with such precision that it can be asserted that the exponent in Coulomb's law [*force* $\propto 1/(distance)^2$] does not deviate from 2 by more than about one part in ten million.

Conductors and the Superposition Principle for Electric Fields

The somewhat surprising facts about electric fields in the presence of conductors should not create the impression that a conductor somehow prevents the electric field contributions that are to be expected from the superposition principle. In every case, the field at any point is the superposition of the contributions from every charge that is present. For example, at a point P inside an empty cavity in a conductor (Figure 24-6), the field contributed by the external point

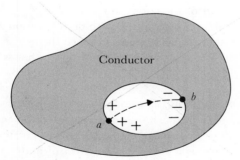

Figure 24-5   A flux line from point a to point b implies V_a is not equal to V_b. Therefore this situation does not occur in electrostatics.

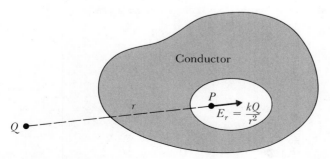

Figure 24-6   Charge Q furnishes a contribution $E_r = kQ/r^2$ to the field at point P within the cavity. Resultant field at P is zero because of contributions of other charges.

charge Q is the same as if Q were the only charge present. But other charges on the outer surface of the conductor move into positions such that their contributions at P cancel the contribution of Q, giving a superposition that is zero.

The situation is similar at any point in the interior of a conductor. Each charge furnishes a contribution, but the configuration of these charges in electrostatics must be such that the superposition of all these contributions is zero.

The relationship at a conducting surface, $E_N = \sigma/\epsilon_0$, does not imply that the local surface charge density σ is the only source of the field **E** at the surface. The field is the superposition of the contributions of *all* the charges, charges on all portions of the surface and any other charges that happen to be present.

24.2 Capacitors

Two conductors separated by a dielectric constitute a device called a *capacitor* (Figure 24-7). In most applications the conductors are given charges of equal magnitude and opposite sign. These charges produce an electric field and there is a potential drop V_{ab} between the conductors given by

$$V_{ab} = V_a - V_b = \int_a^b E_s \, ds$$

where V_a is the potential of the conductor with charge Q, V_b the potential of the conductor with charge $-Q$, and the path of integration is any path joining the two conductors. Since, at every point in space the field **E** is proportional to Q, the potential drop V_{ab} is also proportional to Q; that is

$$Q = CV_{ab} \qquad (24.5)$$

where C is a positive constant of proportionality called the *capacitance* of the capacitor. The value of the capacitance depends upon the geometry of the pair of conductors and upon the nature of the dielectric. For the present we shall assume that the dielectric is a vacuum—the effect of material dielectrics is discussed in Section 24.5.

From the definition of capacitance it follows that the SI unit of capacitance is the coulomb per volt which is named the *farad* (F) in recognition of the pioneering studies of capacitors conducted by Faraday:

$$1 \text{ F} = 1 \text{ C/V}$$

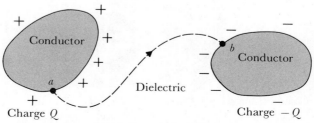

Figure 24-7   A capacitor with a capacitance $C = Q/V_{ab}$.

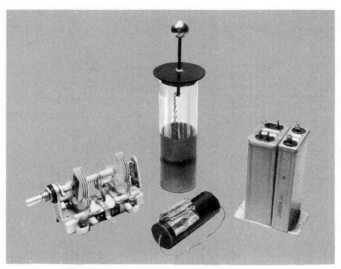

Figure 24-8   Capacitors. The glass Leyden jar is shown in the center. To the left is a radio tuning capacitor with a capacitance that can be varied from 20 pF to 150 pF. To the right are 2.0-μF and 10-μF oil-filled capacitors and in front center (partially cut open) a 0.50-μF metal-foil capacitor.

Figure 24-9   Circuit symbol for a capacitor.

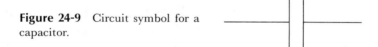

Because this is an inconveniently large unit for the capacitance of most of the capacitors that it is practical to construct (Figure 24-8), submultiples such as the *microfarad* (1 μF $=$ 10^{-6} F) and the *picofarad* (1 pF $=$ 10^{-12} F) are often used.

In an electric circuit, a capacitor is represented by the symbol in Figure 24-9. Electric charge is stored in a capacitor, and it is this feature that is exploited in typical circuit applications such as the use of capacitors to "smooth" the rectified current furnished by a power supply or to tune radio circuits. Details of the reasons for the utility of capacitors will become apparent as we continue the study of electrical phenomena.

24.3 Capacitance Calculation

Parallel-Plate Capacitor

Figure 24-10 represents a capacitor formed by two large parallel metal plates separated by a vacuum. If the plate separation l is small compared to the linear

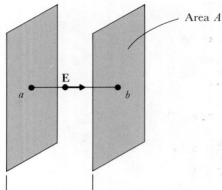

Figure 24-10   Parallel-plate capacitor.

dimensions of the plates, then, to a good approximation, the field is uniform between the plates and zero elsewhere. In Figure 24-10,

$$V_{ab} = \int_a^b E_s \, ds = E \int_a^b ds = E\ell$$

Equation 24.4 gives

$$E = \frac{\sigma}{\epsilon_0} = \frac{Q}{\epsilon_0 A}$$

where $\sigma = Q/A$ is the surface charge density on the inner surface of the left plate in Figure 24-10. Therefore $V_{ab} = Q\ell/\epsilon_0 A$ and

$$C = \frac{Q}{V_{ab}} = \frac{\epsilon_0 A}{\ell} \tag{24.6}$$

Practical electrostatic methods of measurement of the permittivity constant ϵ_0 (or the constant of proportionality in Coulomb's law, $k = 1/4\pi\epsilon_0$) are based on measurements of capacitance. As Eq. 24.6 illustrates, measurement of C together with the geometrical data A and ℓ permit evaluation of ϵ_0. For example, with a capacitor having plates of area 0.1000 m^2 and separation 1.000 mm, if measurement of Q and V_{ab} yields $C = Q/V_{ab} = 885$ pF, these data imply that the value of the permittivity constant is

$$\epsilon_0 = \frac{C\ell}{A} = \frac{(885 \times 10^{-12} \text{ F})(1.000 \times 10^{-3} \text{ m})}{0.1000 \text{ m}^2} = 8.85 \times 10^{-12} \text{ F/m}$$

Spherical Capacitor

The spherical capacitor in Figure 24-11 consists of an inner metal sphere of radius r_a which is concentric with a hollow metal sphere of inner radius r_b. With this geometry, a straightforward calculation gives the exact value of the capacitance.

Since the charges are distributed with spherical symmetry, the electric field between the spheres is radial with a radial component given by Eq. 22.10:

$$E_r = \frac{kQ}{r^2}$$

where Q is the charge on the inner sphere. The potential drop between the spheres is

$$V_{ab} = V_a - V_b = \int_a^b E_r \, dr = \frac{kQ}{r_a} - \frac{kQ}{r_b} \tag{24.7}$$

Therefore

$$C = \frac{Q}{V_{ab}} = \left(\frac{k}{r_a} - \frac{k}{r_b}\right)^{-1} \tag{24.8}$$

An isolated sphere can be regarded as the special case of the spherical capacitor that is obtained when the radius r_b of the hollow metal sphere is infinite. Then the isolated sphere with radius r_a and charge Q is at a potential

$$V_a = \frac{kQ}{r_a}$$

and has a capacitance

$$C = \frac{r_a}{k} \tag{24.9}$$

Capacitors in Series

It is often useful to know the capacitance of the single capacitor C_s that is equivalent, as far as external connections are concerned, to a combination of capacitors connected in *series*, as shown in Figure 24-12(a).

Figure 24-11   Spherical capacitor.

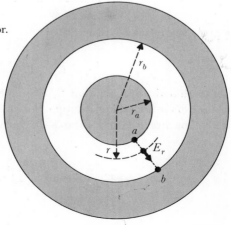

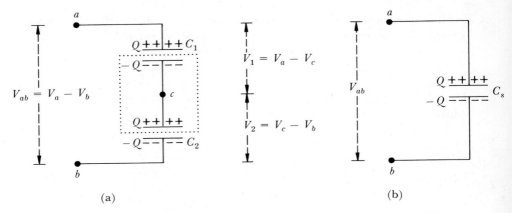

Figure 24-12   (a) Two capacitors connected in series, and (b) their equivalent.

If all the conductors are uncharged, then when a voltage V_{ab} is established across the series combination, each plate acquires a charge of the same magnitude. This follows from the requirement that the charges on the central portion enclosed by the dotted line in Figure 24-12(a) sum to zero. The voltage V_{ab} across the series combination is the sum of the voltages V_1 and V_2 across the individual capacitors:

$$V_{ab} = V_1 + V_2$$

Using $V_1 = Q/C_1$ and $V_2 = Q/C_2$, we find

$$V_{ab} = \frac{Q}{C_1} + \frac{Q}{C_2}$$

The equivalent single capacitor C_s [Figure 24-12(b)] will store the same charge Q when the voltage V_{ab} is the same:

$$C_s = \frac{Q}{V_{ab}} = \frac{Q}{Q/C_1 + Q/C_2}$$

This gives the relationship required:

$$\frac{1}{C_s} = \frac{1}{C_1} + \frac{1}{C_2} \tag{24.10}$$

Notice that this implies that C_s is smaller than the smallest capacitance of the capacitors in series. For example, a 12-μF capacitor connected in series with a 4.0-μF capacitor is equivalent to a single capacitor with a capacitance C_s given by

$$\frac{1}{C_s} = \frac{1}{4.0\ \mu\text{F}} + \frac{1}{12\ \mu\text{F}} = \frac{4}{12\ \mu\text{F}}$$

$$C_s = 3\ \mu\text{F}$$

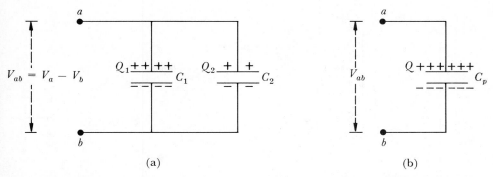

Figure 24-13   (a) Two capacitors connected in parallel, and (b) their equivalent.

Capacitors in Parallel

Capacitors connected as in Figure 24-13(a) are said to be connected in parallel. With this connection, there is the same voltage V_{ab} across both capacitors. The first capacitor has a charge $Q_1 = C_1 V_{ab}$ and the second, $Q_2 = C_2 V_{ab}$. The total charge on the upper conductor is

$$Q = Q_1 + Q_2 = C_1 V_{ab} + C_2 V_{ab}$$

The single capacitor C_p [Figure 24-13(b)] equivalent to this parallel combination will store this same charge Q when the voltage is V_{ab}. Therefore

$$C_p = \frac{Q}{V_{ab}} = \frac{C_1 V_{ab} + C_2 V_{ab}}{V_{ab}}$$

which gives

$$C_p = C_1 + C_2 \tag{24.11}$$

The equivalent capacitance is simply the sum of the capacitances for the parallel combination.*

24.4 Energy Stored in a Capacitor

Figure 24-14 represents a capacitor of capacitance C at a time when the voltage between the plates is V', the charge on one plate is Q', and the charge on the other is $-Q'$. We now increase the charge to $Q' + dQ'$ by transporting a positive charge dQ' from the negative to the positive plate. The work that is done against the electric forces exerted on dQ' in moving this charge through a voltage V' is (from Eq. 23.14)

$$dW' = V' \, dQ' = \frac{Q'}{C} \, dQ'$$

---

*Readers may be more familiar with the rules for series and parallel combinations of resistors (Section 26.2). Notice that the rules for capacitors in series are analogous to the rules for resistors in parallel and vice versa. Many a design error has arisen from erroneously adding the capacitances of a series combination.

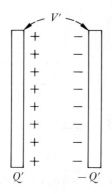

Figure 24-14   Capacitor with charge Q' and voltage V'.

Therefore the work required to charge the capacitor, starting with the capacitor uncharged ($Q' = 0$) and reaching a final charge $Q' = Q$, is given by

$$W = \frac{1}{C} \int_{Q'=0}^{Q'=Q} Q' \, dQ' = \frac{1}{2} \frac{Q^2}{C}$$

This is the energy U stored in the capacitor. Since $Q = CV$, three equivalent expressions for the stored energy are

$$U = \frac{1}{2} \frac{Q^2}{C} = \tfrac{1}{2}CV^2 = \tfrac{1}{2}QV \tag{24.12}$$

The energy associated with any configuration of charges can be regarded as stored in the electric field created by these charges. The general relationship between the energy per unit volume u and the electric field $\mathbf{E}$ can be found by considering a parallel-plate capacitor with plate area A and plate separation l. The capacitance is $C = \epsilon_0 A/l$ and the electric field between the plates is $E = V/l$. The energy stored in this capacitor is

$$U = \tfrac{1}{2}CV^2 = \frac{1}{2}\left(\frac{\epsilon_0 A}{l}\right)(El)^2 = \tfrac{1}{2}\epsilon_0 E^2 Al$$

The volume Al is the volume throughout which there is the uniform field $\mathbf{E}$. We have

$$U = (\tfrac{1}{2}\epsilon_0 E^2)(volume)$$

which shows that the energy per unit volume or *energy density* is

$$u = \tfrac{1}{2}\epsilon_0 E^2 \tag{24.13}$$

Although we have considered only the very special case of the uniform field within a parallel-plate capacitor, the result $u = \tfrac{1}{2}\epsilon_0 E^2$ turns out to be the correct expression for the energy density in any electric field $\mathbf{E}$ in a vacuum.

Example 1   The capacitor C_1 in Figure 24-15 initially has a charge Q_0, and the voltage between its plates is V_0. After the switch S is closed, the capacitor C_2 becomes charged. Find the ratio V/V_0 where V is the final voltage

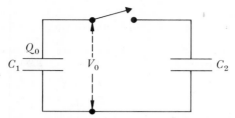

Figure 24-15   Initially the voltage across C_1 is V_0 and the charge of this capacitor is Q_0. There is no charge on C_2.

across the capacitors, and evaluate the ratio U/U_0 of the final to the initial stored energy.

Solution   Since charge is conserved, the sum of the charges Q_1 and Q_2 on the upper conductor in the final state must be equal to the initial charge Q_0:

$$Q_0 = Q_1 + Q_2$$

Therefore

$$C_1 V_0 = C_1 V + C_2 V$$

and

$$\frac{V}{V_0} = \frac{C_1}{C_1 + C_2}$$

The ratio of the final to the initial stored energy is

$$\frac{U}{U_0} = \frac{C_1 V^2 + C_2 V^2}{C_1 V_0^2} = \frac{C_1 + C_2}{C_1}\left(\frac{V}{V_0}\right)^2 = \frac{C_1}{C_1 + C_2}$$

This shows that the final stored energy is less than the initial energy. The difference $U - U_0$ appears, for the most part, as increased internal energy of the connecting wires which become warmer.

24.5 Dielectrics

In the preceding sections it is assumed that there is a vacuum in the region separating the conductors of a capacitor. If, instead of a vacuum, this space is filled with a uniform dielectric, we observe that a capacitor which had the capacitance C_{vacuum} now has a larger capacitance

$$C = \kappa C_{\text{vacuum}}$$

The constant $\kappa = C/C_{\text{vacuum}}$ is called the *dielectric constant* of the dielectric. From data in Table 24-1 we see that the capacitance can be increased significantly by choice of an appropriate dielectric.*

---

*A quantity that is not related to κ but also must be considered in selecting a dielectric for some particular application is the *dielectric strength* given in Table 24-1. The dielectric strength is the maximum electric field that can exist in the dielectric without causing electrical breakdown.

Table 24-1 Dielectric Constants and Dielectric Strengths

Material	Dielectric constant κ	Dielectric strength MV/m
Vacuum	1 (exactly)	∞
Air (1 atm, 20°C)	1.0006	3
Water (20°C)	80	–
Transformer oils	2.2	5–15
Paper (paraffined)	2	40–60
Rubber (vulcanized)	3	16–50
Glass	5–10	20–40
Mica	4.5–7.5	25–200
Porcelain	6–8	10–20
Nitrocellulose plastics	6–12	10–40

For given charges Q and $-Q$ on the plates of a capacitor, the effect of the dielectric is to reduce the voltage across the capacitor from a value V_{vacuum} to the value

$$V = \frac{Q}{C} = \frac{Q}{\kappa C_{\text{vacuum}}} = \frac{V_{\text{vacuum}}}{\kappa}$$

This implies that the electric field between the plates is reduced from the value E_{vacuum} to

$$E = \frac{V}{l} = \frac{1}{\kappa}\frac{V_{\text{vacuum}}}{l} = \frac{E_{\text{vacuum}}}{\kappa} \tag{24.14}$$

This reduction of the field can be understood in terms of *polarization*—the alignment of atomic or molecular dipoles in the dielectric when an electric field is applied. An electric *dipole* is a configuration of equal amounts of positive and negative charge with the positive charge displaced relative to the negative charge (Figure 24-16). When an electric field is established within a dielectric, the positive nucleus of an atom is displaced slightly in the direction of the field while the electrons are displaced in the opposite direction. The atomic dipoles formed in this way are called induced dipoles (Figure 24-17). Certain types of molecules, H_2O for example, are permanent dipoles. Thermal agitation tends to randomize the dipole alignment in a dielectric, but if an electric field is established, partial alignment occurs (Figure 24-18).

When a uniform dielectric is polarized by a uniform electric field, the interior of the dielectric contains equal numbers of positive and negative ends of dipoles, but at one face of the dielectric there will be a net positive charge arising from a preponderance of positive ends of dipoles, while at the opposite face of the dielectric there will be a net negative charge from the negative ends of dipoles (Figure 24-19). These "polarization charges" produce an electric field within the dielectric that is in opposition to the field produced by the "free charges" Q_{free} and $-Q_{\text{free}}$ on the conducting plates (Figure 24-20). Consequently the resultant electric field between the plates is reduced, in agreement with Eq. 24.14.

Figure 24-16 An electric dipole. The total charge is zero but the positive charge is displaced relative to the negative charge.

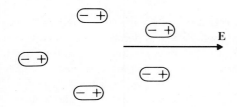

Figure 24-17 Induced electric dipoles. When an electric field is applied the positive nucleus of an atom is displaced in the direction of the field while the negative electrons are displaced in the opposite direction.

Figure 24-18 Molecules such as H_2O that are permanent dipoles become partially aligned when an electric field is applied. Alignment is incomplete because of thermal agitation.

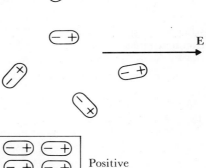

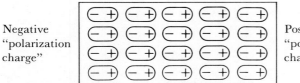

Negative "polarization charge"

Positive "polarization charge"

Figure 24-19 Ends of dipoles give "polarization charge" which is positive at one face of the dielectric and negative at the other face.

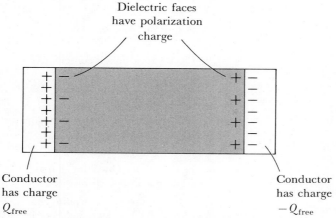

Dielectric faces have polarization charge

Conductor has charge Q_{free}

Conductor has charge $-Q_{\text{free}}$

Figure 24-20 Polarization charges produce a field within the dielectric that is in opposition to the field produced by the "free charges," Q_{free} and $-Q_{\text{free}}$, on the conducting plates.

Reformulation of Gauss's Law

In Gauss's law,

$$\epsilon_0 \oint_S E_N \, dA = Q_S$$

the charge Q_S is the total of the free charges and the polarization charges. Since it is awkward to keep track of polarization charges, it is desirable to reformulate this law so that the polarization charges do not appear explicitly and the entire effect of the dielectric is represented by its dielectric constant κ. Equation 24.14 suggests (and a detailed analysis verifies) that this can be accomplished by simply replacing $\mathbf{E}$ in Gauss's law by $\kappa\mathbf{E}$; that is, the general statement of Gauss's law that is convenient when dielectrics are present is

$$\epsilon_0 \oint_S \kappa E_N \, dA = Q_{\text{free},S} \tag{24.15}$$

where $Q_{\text{free},S}$ is the algebraic sum of the free charges (polarization charges are excluded) inside a closed surface S, and κ is the value of the dielectric constant at the location of the surface element of area dA.

The permittivity of a material is defined as

$$\epsilon = \kappa \epsilon_0$$

Then Gauss's law can be written

$$\oint_S \epsilon E_N \, dA = Q_{\text{free},S} \tag{24.16}$$

Table 24-2 Equations of Electrostatics in a Medium with Dielectric Constant κ and Permittivity ϵ

$\oint_S \epsilon E_N \, dA = Q_{\text{free},S}$	Gauss's law
$V_{ab} = \int_a^b E_s \, ds$	Definition of potential drop
$\oint E_s \, ds = 0$	Electrostatic field is conservative
$E_N = \dfrac{\sigma}{\epsilon}$	Normal component of $\mathbf{E}$ at conductor-dielectric interface; surface charge density σ on conductor
$u = \tfrac{1}{2}\epsilon E^2$	Energy density in dielectric where electric field is $\mathbf{E}$
$E_r = \dfrac{1}{4\pi\epsilon}\dfrac{Q}{r^2} = \dfrac{k}{\kappa}\dfrac{Q}{r^2}$	Field produced by point charge Q surrounded by a *uniform* dielectric
$V = \dfrac{1}{4\pi\epsilon}\dfrac{Q}{r} = \dfrac{k}{\kappa}\dfrac{Q}{r}$	Potential produced by point charge Q surrounded by a *uniform* dielectric
$C = \kappa C_{\text{vacuum}}$	Capacitance of a pair of conductors surrounded by a *uniform* dielectric

Using this formulation of Gauss's law, we can verify that, when a dielectric is present, equations of electrostatics are modified as indicated in the lower portion of Table 24-2.

Summary

☐ In electrostatics:

1 Within a conductor, $\mathbf{E} = 0$, $\rho = 0$, and $V = constant.$
2 The surface of a conductor is an equipotential; $\mathbf{E}$ is normal to the surface with $E_N = \sigma/\epsilon_0$ in a vacuum just outside the surface.
3 Charge resides entirely on the outer surface of an empty hollow conductor and the empty cavity is a shielded region where $\mathbf{E} = 0$ for any distribution of charges on and outside the hollow conductor.

☐ Two conductors separated by a dielectric constitute a capacitor which has a capacitance defined by

$$C = \frac{Q}{V}$$

where one conductor has a charge Q and the other a charge $-Q$, and V is the voltage across the capacitor. The SI unit of capacitance is the farad (F): $1 \text{ F} = 1 \text{ C/V}$.

☐ Parallel conducting plates in a vacuum have a capacitance

$$C = \frac{\epsilon_0 A}{\ell}$$

where the plates have area A and separation ℓ.

☐ A combination of two capacitors with capacitances C_1 and C_2 are equivalent to a single capacitor with a capacitance given by

$$\frac{1}{C_s} = \frac{1}{C_1} + \frac{1}{C_2} \qquad (C_1 \text{ and } C_2 \text{ connected in series})$$

$$C_p = C_1 + C_2 \qquad (C_1 \text{ and } C_2 \text{ connected in parallel})$$

☐ The energy U stored in a capacitor with a charge $Q = CV$ is

$$U = \frac{1}{2}\frac{Q^2}{C} = \tfrac{1}{2}CV^2 = \tfrac{1}{2}QV$$

☐ In a vacuum in a region where the electric field is $\mathbf{E}$, the energy density associated with this field is

$$u = \tfrac{1}{2}\epsilon_0 E^2$$

☐ The dielectric constant κ of a dielectric is defined by

$$\kappa = \frac{C}{C_{vacuum}}$$

where C_{vacuum} is the capacitance of a pair of conductors in a vacuum and

C is their capacitance when the vacuum is replaced by a uniform dielectric. The permittivity ϵ of the dielectric is defined by $\epsilon = \kappa\epsilon_0$.

☐ Gauss's law can be expressed as

$$\oint_S \epsilon E_N \, dA = Q_{\text{free},S}$$

where ϵ is the value of the permittivity of the dielectric at the location of the surface element of area dA. The algebraic sum of the free charges inside the closed surface S is denoted by $Q_{\text{free},S}$. The polarization charges (charges at the ends of aligned dipoles within a dielectric) are excluded from this sum.

Questions

1   What is the surface charge density on a horizontal copper roof when the electric field just above the roof is directed vertically downward and has a magnitude of 20 V/m?

2   Show that the field produced by a total charge Q on an isolated spherical conductor of radius r_0 is:
(a) Zero for $r < r_0$.
(b) Radial with a radial component $E_r = kQ/r^2$ for $r \geq r_0$.

3   Verify that, at the surface of the spherical conductor of the preceding question, the expression $E_r = kQ/r^2$ agrees with the field calculated from $E_N = \sigma/\epsilon_0$.

4   Show that the potential in the field produced by a total charge Q on an isolated spherical conductor of radius r_0 is given by:
(a) $V = kQ/r$ for $r > r_0$.
(b) $V = kQ/r_0$ on and within the conducting sphere.

5   Find expressions for the electric field and the potential when an isolated cylindrical conductor of radius R_0 has a charge per unit length λ.

6   Figure 24-21 shows a small brass sphere with a total charge Q. This sphere has been placed within a cavity in a copper conductor. The net charge on the copper is zero. Prove that the charge on the copper walls of the cavity is $-Q$. What is the charge on the outer surface of the copper conductor?

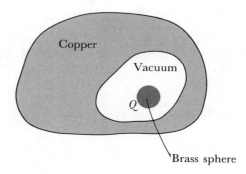

Figure 24-21

7   Find expressions for the electric field and the potential in the preceding question, for the case where the inner and outer copper surfaces are spherical and concentric with radii r_i and r_o, respectively; the brass sphere has a radius r_b and is concentric with the copper shell.

8   Find the electric field and the potential when the inner conductor of Figure 24-22 carries a charge per unit length λ and the net charge on the outer conductor is zero.

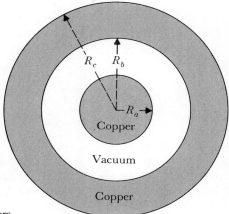

Figure 24-22   Cross section of infinitely long coaxial conductors.

9   What plate area of a parallel-plate capacitor is required to achieve a capacitance of 1.0 F if the plate separation is 1.0×10^{-3} m and the plates are separated by a vacuum?

10 When the knob of a certain radio tuning capacitor is turned, the effective area of the plates in the interleaved position is reduced from 40 cm^2 to 10 cm^2. What is the ratio of the initial to the final capacitance?

11 What is the capacitance of the coaxial cylinders of Figure 24-22? Use this result to estimate the capacitance of a vacuum diode with concentric cylindrical electrodes 2.0 cm long with radii $R_a = 1.0$ mm and $R_b = 20$ mm.

12 Considering the earth as an isolated spherical conductor of radius 6.4×10^6 m, find its capacitance.

13 Compute the equivalent capacitance between the points a and b of Figure 24-23. Find the charge on each capacitor when the voltage between points a and b is 600 V.

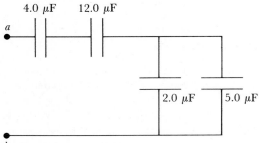

Figure 24-23 b

14 Compute the equivalent capacitance between the points a and b of Figure 24-24. Find the voltage between points a and b if the charge on one of the 5.0-μF capacitors is 15×10^{-6} C.

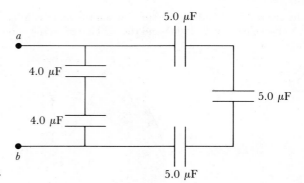

Figure 24-24

15 What is the energy density in a vacuum just outside the surface of a conducting sphere of 0.100-m radius when the sphere is at a potential of 5.00 kV?

16 A certain type of glass has a dielectric constant of 6.0 and a dielectric strength of 30 MV/m. Using this glass as the dielectric, find the minimum plate area of a parallel-plate capacitor that has a capacitance of $4.0 \times 10^{-2} \mu$F and can withstand a voltage of 5.0 kV.

17 A parallel-plate capacitor, with a dielectric which is a slab of material with dielectric constant κ, has a capacitance C. The plates are connected to the terminals of a battery which establishes a voltage V_0 between the plates of the capacitor. Then the battery is disconnected. Find the work that must be done to remove the slab by evaluating the energy stored in the capacitor before and after the slab is removed.

18 Use Gauss's law in the form

$$\oint_S \epsilon E_N \, dA = Q_{\text{free},S}$$

to show that in electrostatics, at a surface between a conductor and a dielectric with permittivity ϵ, the normal component of the electric field is

$$E_N = \frac{\sigma}{\epsilon}$$

where σ is the charge density on the conducting surface.

Supplementary Questions

S-1 Figure 24-25 shows three large parallel conducting plates. The isolated inner plate carries a total charge of 200×10^{-9} C. The two outer plates are connected by a copper wire. Find the charge on each surface of the inner plate.

S-2 The total charge on the copper sphere of Figure 24-26 is zero:

(a) Assume that r is much greater than r_0. What force acts on each of the three objects, Q_a, Q_b, and the copper conductor?

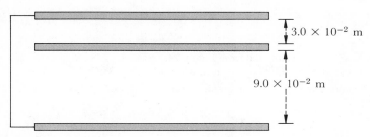

Figure 24-25

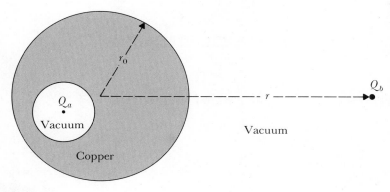

Figure 24-26

(b) When r is not large compared to r_0, what can be said about these three forces without an elaborate calculation?

S-3 The voltage applied across a series combination of a 4.0-μF capacitor and a 6.0-μF capacitor is 500 V:

(a) Find the charge and the voltage for each capacitor.

(b) The 500-V source is removed and the charged capacitors are disconnected and then reassembled with their positive plates connected and their negative plates connected. Find the charge and the voltage for each capacitor.

S-4 A charging battery establishes a voltage V_0 between the plates of a parallel-plate capacitor with plate area A and plate separation l. The battery is then disconnected and one plate is pulled away from the other until the plate separation is $2l$:

(a) Express the initial and final energy stored in the capacitor in terms of A, l, and V_0.

(b) Find the work that is done in pulling apart the plates.

S-5 For the situation described in the preceding question, evaluate the work $dW = F\,dl$ that must be done in increasing the plate separation from l to $l + dl$. Then show that the force of attraction exerted by one capacitor plate on the other is

$$F = \frac{Q^2}{2\epsilon_0 A}$$

where the charge on one plate is Q and on the other $-Q$.

S-6 Find the force exerted on a capacitor plate of area A carrying a charge Q, by using

$$F = QE_{\text{average}}$$

where E_{average} is estimated from Figure 24-3 as $\frac{1}{2}(E_{\text{inside}} + E_{\text{outside}}) = \frac{1}{2}(0 + \sigma/\epsilon_0)$. Compare this result with that found in Question S-5.

S-7 Use Gauss's law (Eq. 24.15) to show that in Figure 24-27,

$$\kappa_1 E_1 = \kappa_2 E_2$$

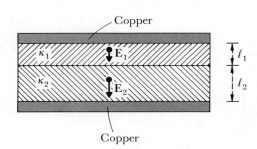

Figure 24-27

S-8 Find the capacitance of the capacitor in Figure 24-27.

25.

electric current
and resistance

An electric field established in a region where there are mobile charged particles produces a streaming motion of these particles that is called an *electric current* (Figure 25-1). For example, an applied electric field causes (1) a current of electrons within a metallic conductor, (2) a current of positive and negative ions within an ionized gas or an electrolytic solution, (3) a current of electrons emerging from a heated cathode in a vacuum tube, or (4) a current of protons when these particles are accelerated within an electrostatic particle accelerator.

In this chapter we define the terms necessary for a quantitative description of currents and investigate the relationship between the current and the electric field in a conductor.

25.1 Current and Current Density

Electric Current I

The rate at which charge flows through a surface is called the *electric current* through the surface: thus *if a net charge dQ is transported through a surface in a time interval dt, the current I through the surface is the scalar quantity defined by*

$$I = \frac{dQ}{dt} \qquad (25.1)$$

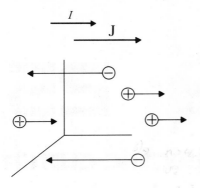

Figure 25-1 The streaming motion of positive and negative charges that constitutes an electric current. The direction of the current density **J** and the current I is the direction of motion of the positive charges.

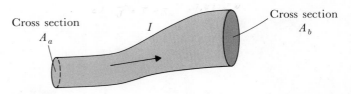

Figure 25-2 When charge flows through a tubular region and does not accumulate within the tube, the current I is the same at any cross section of the tube.

The SI unit of current, the coulomb per second, is called the *ampere* (A):

$$1 \text{ A} = 1 \text{ C/s}$$

Commonly used submultiples are the milliampere (1 mA $= 10^{-3}$ A) and the microampere (1 μA $= 10^{-6}$ A).

A standard convention for labeling the direction of a current has been adopted by the overwhelming majority of engineers and scientists. According to this convention, the direction of the current through a surface is the direction of motion of positively charged particles and is opposite to the direction of motion of negatively charged particles (such as the mobile electrons in a wire).

Consider charge flowing within a tubular region (Figure 25-2) and any two cross sections A_a and A_b. Suppose that the amount of charge between these cross sections remains constant. Then, since charge is conserved, the amount of charge that enters at one cross section must be the same as the amount of charge that leaves at the other cross section in any time interval. It follows that the *current is the same at each cross section.* In this circumstance it is no longer necessary to explicitly refer to a specific surface through which the charge flows and we speak simply of "the current." For example, we refer to "the current in a wire" meaning the current through any cross section of the wire.

Electric Current Density

At any point in a region where there is an electric current there is a current density **J** defined as a vector with:

1   A *direction* in the direction of the average velocity vector of the positive charges (or equivalently, a direction *opposite* to that of the average velocity vector of the negative charges).
2   A *magnitude* equal to the current per unit area normal to **J**.

Then if there is a constant current density **J** over a surface normal to **J** of area A, the current through the surface is $I = JA$. (In general, the current I through any surface S is the integral over the surface of the normal component J_N of the current density: $I = \int_S J_N \, dA$.)

It is useful to relate the current density to quantities referring to the mobile charged particles. In Figure 25-3 there are n particles per unit volume, each with the same charge q and moving with the same velocity **v**. After a time interval dt has elapsed, all the particles within the volume with cross section dA and length $v \, dt$ will have passed through the surface dA. The number of such particles is $n \times volume = nv \, dt \, dA$ and the charge transported through dA in the time interval dt is $nqv \, dt \, dA$. The current through dA is $nqv \, dA$ and the magnitude and direction of the current density is correctly given by

$$\mathbf{J} = nq\mathbf{v} \tag{25.2}$$

This expression can be used also when the particles with charge q have different velocities, if **v** is replaced by the average value $\mathbf{v}_d$ of these velocity vectors:

$$\mathbf{J} = nq\mathbf{v}_d \tag{25.3}$$

Since $\mathbf{v}_d$, called the *drift velocity*, is a vector average, it will be zero for a distribution of velocities in which all directions are equally likely, no matter how great the particle speeds may be. For example, in a metal the mobile electrons are continually in motion with speeds of the order of 10^6 m/s, but if the average

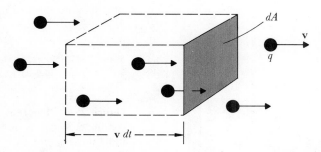

Figure 25-3   All the particles in the volume $v \, dt \, dA$ pass through dA in a time interval dt.

electric field within the metal is zero, this motion is completely random and the drift velocity is zero. When an electric field is applied, a slight systematic drift is superimposed on the random motion of the electrons.

Example 1   A copper wire with a cross-sectional area of 2.0×10^{-6} m² carries a current of 6.0 A. If there are approximately 10^{29} mobile electrons per cubic metre of copper, what is the drift speed of the electrons in the wire?

Solution   Assuming that the current density **J** is constant over the cross-sectional area A of the wire, we have

$$J = \frac{I}{A} = \frac{6.0 \text{ A}}{2.0 \times 10^{-6} \text{ m}^2} = 3.0 \times 10^6 \text{ A/m}^2$$

From Eq. 25.3 the electron drift velocity $\mathbf{v}_d$ is related to the current density by $\mathbf{J} = n(-e)\mathbf{v}_d$, where n is the number of mobile electrons per unit volume. Therefore the electron drift speed is

$$v_d = \frac{J}{ne} = \frac{3.0 \times 10^6 \text{ A/m}^2}{(10^{29} \text{ m}^{-3})(1.60 \times 10^{-19} \text{ C})} = 2 \times 10^{-4} \text{ m/s}$$

These electrons drift less than a metre in an hour. It should be emphasized that the drift speed of the electrons is not the speed with which electric field disturbances travel along the wire. This latter speed is close to the speed of light in a vacuum, and it is for this reason that we obtain almost instantaneous response when we switch electrical appliances on and off.

25.2 Resistance

Definition of Resistance

The current I through a conductor depends on the potential drop V between the ends of the conductor (Figure 25-4). The resistance R of a conductor between surfaces at potentials V_a and V_b is defined by the equation

$$R = \frac{V}{I} \tag{25.4}$$

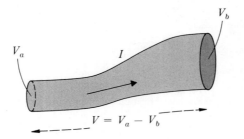

Figure 25-4   The resistance R of a conductor between surfaces at potentials V_a and V_b is defined by $R = V/I$, where $V = V_a - V_b$.

where $V = V_a - V_b$. The name *resistance* for the quantity R is appropriate because, for a given potential drop, the larger the value of R, the smaller the current.

The SI unit of resistance is the volt per ampere which is named the *ohm* (Ω):

$$1 \ \Omega = 1 \ V/A$$

When considered as an element of a circuit, a conductor with resistance is called a *resistor* and is represented diagramatically as shown in Figure 25-5. The current in the resistor is directed from the terminal at the higher potential toward the terminal at the lower potential. In other words, as a resistor is traversed in the direction of the current, the potential always drops. The amount of the drop is given by $V = IR$. This potential drop therefore is called an *IR drop*.

Ohm's Law

The graph in Figure 25-6 represents the experimental results obtained by measurement of the corresponding values of the current I through a metallic conductor and the potential drop V between its terminals. The temperature of the conductor is maintained constant. The straight line graph implies that V is proportional to I. This means that the *resistance $R = V/I$ has the same value no matter what the value of the current or of the potential drop*. This is *Ohm's law:*

$$R = \frac{V}{I} = a \ constant \ independent \ of \ I \ or \ V \tag{25.5}$$

Metallic conductors at constant temperature obey Ohm's law with great accuracy up to the highest current densities that have been investigated experimentally.

We must stress, however, that Ohm's law is merely a description of the observed behavior of some conductors over a certain range of conditions. Many conductors do *not* obey Ohm's law. Their current-voltage graph is not a straight

Figure 25-5   Circuit symbol for a resistor.

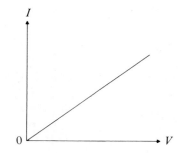

Figure 25-6   Graph of I versus V for a metallic conductor at constant temperature.

line (Figure 25-7). Such a *nonlinear* conductor has a resistance that is significantly different for different values of the voltage. For example, a selenium rectifier [Figure 25-7(b)] has a low resistance for one polarity of the applied voltage but a very high resistance for the opposite polarity. (Reversing the polarity of the voltage applied to an *ohmic* conductor changes the direction but not the magnitude of the current.)

Conductivity and Resistivity

At any point within a conducting material, the current density **J** is determined by the electric field **E** and the nature of the material. For an isotropic material, **J** is in the direction of **E**, and we can write

$$\mathbf{J} = \sigma \mathbf{E} \tag{25.6}$$

where the positive quantity σ defined by this equation is called the *conductivity* of the material. The reciprocal of the conductivity is defined as the *resistivity* ρ:

$$\rho = \frac{1}{\sigma} \tag{25.7}$$

The resistance R of a uniform conductor is proportional to the resistivity ρ of the conducting material:

$$R = (constant)\rho \tag{25.8}$$

where the constant of proportionality has the dimensions of a reciprocal length and is determined, for a well-defined distribution of current within the conductor, by the shape and size of the conductor. This is illustrated in the following example (and also in Questions 13 and S-2).

Example 2   Find the resistance of a uniform cylindrical conductor of length ℓ and cross section A, carrying a current I with a uniform current density $J = I/A$ directed parallel to the axis of the cylinder (Figure 25-8).

Solution   Within the conductor there is a uniform electric field

$$\mathbf{E} = \frac{\mathbf{J}}{\sigma} = \rho \mathbf{J}$$

The voltage between the ends of the cylinder is $V = E\ell$. The resistance of this conductor is

$$R = \frac{V}{I} = \frac{E\ell}{JA} = \left(\frac{\ell}{A}\right)\rho \tag{25.9}$$

Equation (25.9) shows how resistivities can be determined from resistance measurements. Representative values of resistivities are given in Table 25-1. The SI unit of resistivity is the *ohm metre* ($\Omega \cdot m$). Notice the tremendous range of values between the resistivities of good conductors ($\rho \approx 10^{-8}\ \Omega \cdot m$) and those of good insulators ($\rho \approx 10^{15}\ \Omega \cdot m$).

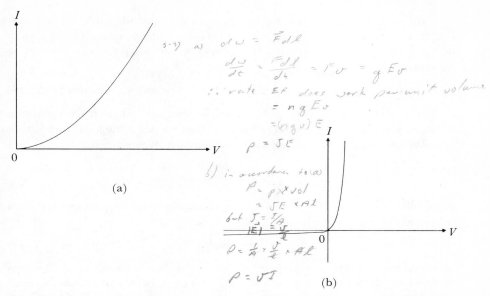

(a)

(b)

Figure 25-7   Nonlinear circuit elements. (a) Vacuum tube. (b) Selenium rectifier.

Figure 25-8   Uniform cylindrical conductor with resistivity ρ.

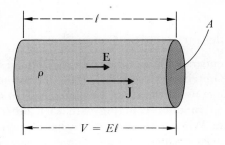

Table 25-1 Resistivities and Their Temperature Coefficients at Room Temperature (20°C)

Material	Resistivity ρ $\Omega \cdot$m	Temperature coefficient of resistivity α (per kelvin)
Silver	1.6×10^{-8}	3.8×10^{-3}
Copper	1.7×10^{-8}	3.9×10^{-3}
Aluminum	2.8×10^{-8}	3.9×10^{-3}
Nickel	7.8×10^{-8}	6×10^{-3}
Iron	1.0×10^{-7}	5.0×10^{-3}
Graphite	3.5×10^{-5}	-5×10^{-4}
Glass	$\approx 10^{11}$	
Rubber	$\approx 10^{15}$	
Quartz	$> 10^{16}$	

If the resistance of a conductor has a constant value independent of the current then the resistivity of the conducting material is also a constant independent of the value of the current density, and vice versa. Therefore, an alternative statement of *Ohm's law* is

$$\mathbf{J} = \sigma \mathbf{E} \qquad (\sigma = 1/\rho = constant) \quad (25.10)$$

The resistivity of a conductor generally changes when its temperature changes. The fractional change in resistivity $d\rho/\rho$ per unit change in temperature is called the conductor's temperature coefficient of resistivity, α:

$$\alpha = \frac{1}{\rho}\frac{d\rho}{dT}$$

For a limited temperature change ΔT, the change in resistivity can be calculated from the approximate expression

$$\Delta\rho = \rho\alpha\Delta T$$

using values of α and ρ corresponding to an appropriate temperature. (Room temperature values are given in Table 25-1.) The resistivity of metals increases with increasing temperature (α is positive). Some conductors, carbon for example, have resistivities that decrease with increasing temperature, and their temperature coefficients of resistivity are negative.

A Model for Electrical Conduction

Some understanding of the processes that are significant in electrical conduction can be obtained by considering the motion of mobile charged particles within a conducting medium. In the absence of an electric field, these particles move about at random, colliding with each other and with other particles of the medium. When a uniform electric field $\mathbf{E}$ is applied, a mobile particle of charge q and mass m experiences a constant acceleration $\mathbf{a}$ between collisions, given by

$$\mathbf{a} = \frac{q\mathbf{E}}{m}$$

In the time interval t since the last collision the particle's velocity changes from an initial value $\mathbf{v}_0$ to the value

$$\mathbf{v} = \mathbf{v}_0 + \frac{q\mathbf{E}t}{m}$$

The drift velocity $\mathbf{v}_d$ is the average value of $\mathbf{v}$ taken over all possible times t since the last collision and over all possible velocities $\mathbf{v}_0$. If we assume that these velocities $\mathbf{v}_0$ are randomly distributed, their vector average is zero and the drift velocity is

$$\mathbf{v}_d = \frac{q\mathbf{E}\tau}{m}$$

where τ is the average time since the last collision. For collisions that are independent random events, τ is also the *average time between collisions*. The systematic drift of n such particles per unit volume gives rise to a current density

$$\mathbf{J} = nq\mathbf{v}_d = \frac{nq^2\tau\mathbf{E}}{m}$$

The conductivity of the medium is therefore

$$\sigma = \frac{J}{E} = \frac{nq^2\tau}{m} \tag{25.11}$$

A conductor obeys Ohm's law if its conductivity is a constant independent of **E**. The model adopted for the preceding calculation of σ suggests that any material should obey Ohm's law if the electric field is not strong enough to affect the concentration n of charge carriers and the average time τ between collisions. Departures from Ohm's law certainly are to be expected in fields strong enough to change the carrier concentration. This is what happens in an electric spark.

The theory that we have considered is useful in understanding conduction through gases, liquids, and solids and "explains" the fact that metals obey Ohm's law. In a metal, the distribution of speeds of the mobile electrons is affected only very slightly by an applied field **E**. (The relatively large field in Example 1 produces a drift speed of only about 10^{-4} m/s, much less than the average speed of 10^6 m/s of the mobile electrons.) Consequently the average time τ between collisions of a mobile electron in a metal is essentially independent of **E**. It follows that σ is independent of **E**; a metallic conductor obeys Ohm's law.

In spite of some successes, this theory fails to explain important features of conduction in metals. The motion of an electron through a metallic crystal is quite different from that of a particle obeying Newtonian mechanics. Experimental evidence indicates, and quantum mechanics predicts, that an electron would move freely without collision through an idealized crystal with a *perfectly* periodic array of atoms. The electron "collisions" within a metal crystal that account for the metal's resistivity arise because the periodicity of the crystal is less than perfect; there are various sorts of imperfections in the crystal. And the crystal atoms, instead of remaining at rest in a regular array, execute thermal oscillations about equilibrium positions.

25.3 Current and Power

Consider a current I directed from an equipotential at potential V_a to an equipotential at potential V_b (Figure 25-9). In a time interval dt, a charge dQ passes the V_a equipotential while the same amount of charge passes the V_b equipotential. The charge passing through the region between these equipotentials therefore experiences a change in potential energy given by

$$dW = V_a\,dQ - V_b\,dQ = V_{ab}I\,dt \tag{25.12}$$

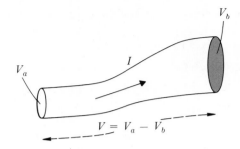

Figure 25-9 Current I confined to a tubular region between equipotentials at V_a and V_b.

This is the work done by the electric forces on the charges between the V_a and V_b equipotentials. The *power* supplied by the electric field to the charges moving between the V_a and V_b equipotentials is

$$P = \frac{dW}{dt} = V_{ab}I \tag{25.13}$$

This equation will be applied to many different physical situations. We consider first an example in which the charges move through a vacuum.

Example 3 In an electrostatic accelerator, protons emerge with negligible kinetic energy from an ion source maintained at a potential of 2.00×10^6 V. These protons are accelerated through a vacuum toward an electrode at zero potential. The proton beam emerging from the accelerator contains 10^{13} protons per second. What is the power supplied to the protons by the electric field?

Solution Since the charge of each proton is 1.60×10^{-19} C, the beam current is

$$I = 1.60 \times 10^{-19} \times 10^{13} \text{ A} = 1.60 \times 10^{-6} \text{ A}$$

The power supplied by the electric field to the proton beam is

$$P = V_{ab}I = (2.00 \times 10^6 \text{ V})(1.60 \times 10^{-6} \text{ A}) = 3.2 \text{ W}$$

[Consistent use of SI units must give the answer in the SI unit of power, the *watt*: $1 \text{ V} \cdot \text{A} = (1 \text{ J/C})(1 \text{ C/s}) = 1 \text{ J/s} = 1 \text{ W}$.] In this accelerator, and whenever a charged particle is accelerated through a vacuum by an electrostatic field, electric potential energy of the particle is converted into *kinetic energy* of this particle. At the ion source, a proton has a potential energy of 2.00 MeV and no kinetic energy. Emerging from the accelerator, the same proton has zero potential energy but has acquired a kinetic energy of 2.00 MeV.

Power in Electric Circuits

The interpretation of the power equation

$$P = V_{ab}I \tag{25.13}$$

is of particular interest for the case when Figure 25-9 represents a portion of an electric circuit. The electric potential energy of the circulating charges changes at the rate $V_{ab}I$ as they drift through this portion of the circuit with a kinetic energy that remains negligibly small. In this circumstance, $P = V_{ab}I$ is a general expression for the power input or output of this portion of the circuit:

1   If $V_a > V_b$, the circulating charges give up energy and there is a power input

$$P_{\text{in}} = V_{ab}I$$

2   If $V_a < V_b$, the circulating charges gain energy and there is a power output

$$P_{\text{out}} = V_{ab}I$$

Power Dissipation in a Resistor, Joule's Law

If the portion of the circuit is a pure resistance R, the potential drop $V_{ab} = V_a - V_b$ is always positive (an IR drop in the direction of the current), so there is a power *input* to the resistor, $P = V_{ab}I$.

Mobile charged particles are accelerated by the electric field within the conductor, but the kinetic energy gained by the charge carriers is transferred by collisions to the atoms of the conductor. The net result is that electric potential energy of the mobile charged particles is converted into internal energy (or thermal energy) of the conductor. As the internal energy of the conductor increases, its temperature rises until there is an outward flow of heat at the same rate as the energy input. In this process, called *Joule heating,* the input power is dissipated within the conductor. Various expressions for the *power dissipated* are:

$$P = VI = I^2R = \frac{V^2}{R} \tag{25.14}$$

For an ohmic resistor, this equation is called *Joule's law.*

Example 4   A wire-wound resistor with a resistance of 250 Ω will overheat if the power dissipated exceeds 10 W. What is the maximum constant voltage V that can be applied across the terminals of this resistor?

Solution

$$V = \sqrt{PR} = \sqrt{(10 \text{ W})(250 \ \Omega)} = 50 \text{ V}$$

Summary

☐   If a net charge dQ is transported through a surface in a time interval dt, the current through the surface is

$$I = \frac{dQ}{dt}$$

☐   The current density $\mathbf{J}$ is a vector in the direction of motion of positive charges with a magnitude that is the current per unit area normal to $\mathbf{J}$.

☐ In a region where there are n particles per unit volume, each with charge q and moving with velocities whose vector average is the drift velocity $\mathbf{v}_d$, the current density is

$$\mathbf{J} = nq\mathbf{v}_d$$

☐ The resistance R of a conductor carrying a current I with a potential drop V between its ends is defined by

$$R = \frac{V}{I}$$

Ohm's law is

$$R = \frac{V}{I} = a\ constant\ independent\ of\ V\ or\ I$$

Metallic conductors at constant temperature obey Ohm's law. Many other materials do not.

☐ The conductivity σ and resistivity $\rho = 1/\sigma$ of an isotropic conductor are defined by

$$\mathbf{J} = \sigma\mathbf{E}$$

An alternative statement of Ohm's law is

$$\sigma = \frac{1}{\rho} = \frac{J}{E} = a\ constant\ independent\ of\ \mathbf{J}\ or\ \mathbf{E}$$

☐ The resistance of a uniform cylinder with a constant current density $\mathbf{J}$ parallel to its axis is

$$R = \frac{\rho\ell}{A}$$

☐ If, in an electric field $\mathbf{E}$, particles of charge q and mass m move through a material with an average time τ between collisions, the conductivity of the material is

$$\sigma = \frac{nq^2\tau}{m}$$

☐ A portion of a circuit with a potential drop $V_{ab} = V_a - V_b$, carrying a current I directed from a to b, has:

1   A power input

$$P_{in} = V_{ab}I$$

if V_{ab} is positive.

2   A power output

$$P_{out} = V_{ab}I$$

if V_{ab} is negative.

□ In a conductor with resistance $R = V/I$, the power dissipated is

$$P = VI = I^2R = \frac{V^2}{R}$$

Questions

1 **(a)** If the electric current in a wire is 7.00 A, what is the charge which drifts past a given point in the wire in 1.00 s?
   (b) In a two-cell flashlight about 1.08 C of charge pass by any given point in the flashlight circuit in 2.00 s. What is the circuit current?

2 Calculate the charge passing through an x-ray tube during a 0.20-s exposure at 100 mA.

3 Find the beam current in a synchrotron that has 10^{11} electrons traveling almost at the speed of light $(3.0 \times 10^8$ m/s) in a circular path with a circumference of 240 m.

4 The total charge that a certain automobile storage battery should circulate in normal operation is 70 ampere hours:
   (a) How many coulombs is a charge of 70 ampere hours?
   (b) How long could a starter be driven, if the starter current is 500 A?

5 Find the magnitude of the current density in a wire with a square cross section 1 mm on a side, when the wire carries a current of 20 A.

6 Find the magnitude and direction of the current density vector in a region where there are:
   (a) 4.0×10^{16} electrons per cubic metre, with a drift velocity of 2.0×10^6 m/s west.
   (b) 4.0×10^{16} singly charged positive ions per cubic metre, with a drift velocity of 5.0×10^5 m/s east.
   (c) Particles of parts (a) and (b) simultaneously present.

7 A potential drop of 100 V is maintained across each of three resistors which have resistances of 5.0 Ω, 50 Ω, and 500 Ω, respectively. Find the current through each resistor.

8 What is the resistance of a toaster that operates with a current of 6.0 A when the voltage across it is 120 V?

9 The electrical wiring in a home is ordinarily copper wire with a cross section of approximately 2.0×10^{-6} m². At room temperature, what is the resistance of a 30-m length of this wire?

10 If there is a 0.50-V IR drop along a 10.0-m length of a copper wire, find:
   (a) The electric field within the wire.
   (b) The current density within the wire.
   (c) The current in the wire, if the cross section is 20×10^{-7} m².

11 An aluminum wire and a copper wire of the same length have the same resistance:
   (a) What is the ratio of their radii?
   (b) What is the ratio of their masses?
   (The density of copper is 8.9×10^3 kg/m³; the density of aluminum is 2.7×10^3 kg/m³.)

12 If copper and aluminum wires of the same length and resistance were to cost the same amount, what would be the ratio of their costs per kilogram?

13   The dimensions of a carbon bar with a square cross section are $(2.0 \times 10^{-2} \text{ m}) \times (2.0 \times 10^{-2} \text{ m}) \times (30 \times 10^{-2} \text{ m})$:

   (a) The bar is placed in a circuit in such a way that the current is directed parallel to the long dimension from one square face to the other. What is the resistance of the bar?

   (b) When the current is directed from one rectangular face to the opposing rectangular face, what is the bar's resistance?

14   What is the fractional change in the resistance of copper wire when the temperature increases from $20°C$ to $40°C$?

15   Estimate the average time between collisions for a mobile electron in copper, making the assumption that copper contains about 10^{29} mobile electrons per cubic metre.

16   An electric light bulb dissipates 100 W when the voltage between its terminals is 120 V. Find the current. What is the resistance of the bulb under these conditions?

17   An immersion heater coil dissipates 5.0 kW when 236 V are applied across it. If the voltage is reduced to 118 V and the coil resistance remains unchanged, what power will be dissipated by the coil?

18   During a one-month period a certain electric refrigerator, connected to a 120-V supply, runs for a total of 150 h:

   (a) If the current required by the refrigerator were 3.2 A, how much energy would be supplied by the electric power company?

   (b) If the electrical energy were supplied at 1.50 cents per kilowatt hour, what would be the monthly cost of operating this refrigerator?

   (c) Compare this cost with that of a television set which requires 2.4 A at 120 V and which runs for 90 h during the month.

Supplementary Questions

S-1   The cathode and anode of a vacuum diode are parallel plates with a separation l. Electrons emitted from the hot cathode provide a current density J. Find the charge density ρ as a function of the distance x from the cathode. Known parameters are J, l, and the potential drop V between the anode and the cathode. Assume that ρ is so small that the electric field between the plates remains at the value $E = V/l$ everywhere between the electrodes.

S-2   Graphite of resistivity ρ fills the space between an inner copper cylinder of radius R_1 and an outer concentric copper tube of radius R_2. Show that for a radial current through the graphite, the resistance between the copper electrodes is

$$R = \frac{\rho}{2\pi l} \ln \frac{R_2}{R_1}$$

where l is the length of the electrodes.

S-3   Show that in a conductor with a current density

$$\mathbf{J} = nq\mathbf{v}$$

arising from n particles per unit volume, each with charge q and moving with the same velocity **v**, the power per unit volume p delivered by the electric field **E** to these particles is given by

$$p = JE$$

Verify that this is consistent with

$$P = VI$$

when P is the power delivered to a uniform cylindrical conductor carrying a current I and having a potential drop V between its ends.

26.

electric circuits

The law of conservation of electric charge and the law of conservation of energy are applied in this chapter to a very practical study of immediate utility—the behavior of electric circuits.

26.1 Electromotive Force

To maintain for an indefinite period of time a current I through a resistor, we must have a *complete circuit* (Figure 26-1) so that charges circulate around a closed conducting path. Since electric potential energy is continually dissipated in the resistor at the rate I^2R, the circuit must include some device that will increase the electric potential energy of circulating charges. Such a device is called a *source of EMF.**

A localized source of EMF such as a storage battery has a low potential terminal indicated by a minus sign and a high potential terminal indicated by a plus sign. A source of EMF functions as a "charge pump" which can move charges in a direction opposite to that of the forces exerted by the electrostatic

---

*The name is not well chosen. A source of EMF is neither a source of charge nor a source of energy. Instead, a source of EMF serves as an energy converter as charge passes through it. The label EMF, pronounced ee-em-eff, is an abbreviation for "electromotive force." This is a misleading term since an electromotive force is a work per unit charge rather than a force.

field. When the current I *within* a localized source is directed from the low to the high potential terminal, energy is converted from some nonelectrostatic form into electric potential energy of mobile charges at a rate proportional to the current: that is, the *power converted* is

$$P = \mathcal{E} I \tag{26.1}$$

where $\mathcal{E}$ is a constant that characterizes the source. The constant $\mathcal{E}$ defined by this equation is called the *EMF of the localized source*. The SI unit of $\mathcal{E}$ is the volt.

With $I = dQ/dt$ and $P = dW/dt$, the definition of $\mathcal{E}$ may be rewritten as

$$\mathcal{E} = \frac{dW}{dQ} \tag{26.2}$$

which shows that the EMF of a source is the nonelectrostatic work per unit charge passing through the source. In Chapter 31 we shall study EMFs arising from work done by electric forces associated with time-varying magnetic fields and EMFs due to magnetic forces. But even when the only macroscopic field present is an *electrostatic* field, we can still have nonelectrostatic forces which can do work on charges and establish an EMF. Such forces arise where there are differences in chemical composition or where there are temperature gradients. These "effective forces," when examined on a microscopic scale, are attributable to the electromagnetic interaction, but they are not determined by the macroscopic electric field which is an average value for volume elements containing billions of molecules.

Figure 26-2 is a schematic representation of a source of EMF in normal usage with the conventional current directed from low to high potential *within* the source. Such a current is said to be *in the direction of the EMF*. Energy is converted at the rate $\mathcal{E} I$ into electric potential energy from internal energy of the chemical constituents if the source is a storage battery or a dry cell, or from macroscopic

Figure 26-1 Storage battery is the source of EMF in the circuit.

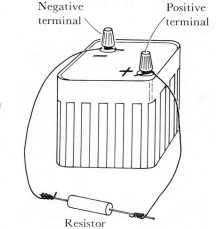

Negative terminal

Positive terminal

Resistor

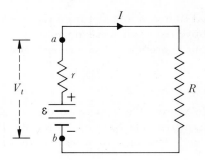

Figure 26-2 Schematic diagram showing source of EMF with terminals at a and b. The source has an internal resistance r and an EMF ε. The direction of the EMF is from b to a which is in the direction of the current for this circuit. If the source is a storage battery, the battery is "discharging." The terminal voltage is $V_t = V_a - V_b$.

kinetic energy if the source is a dynamo. Any source has some internal resistance r and therefore a power I^2r is dissipated within the source. The electrical power output of the source is $V_t I$ where V_t, called the *terminal voltage*, denotes the voltage $V_a - V_b$ between the positive and the negative terminals of the source (Figure 26-2). The power output of the source is supplied to the external circuit. The law of conservation of energy for this source leads to the power equation,

power output = power converted − power dissipated

or

$$V_t I = \varepsilon I - I^2r$$

Therefore, when I is in the direction of ε so that the source has a power output, its terminal voltage is

$$V_t = \varepsilon - Ir \tag{26.3}$$

Notice that when $I = 0$, $V_t = \varepsilon$; *the EMF of a source is its open-circuit terminal voltage.*

The expression for the power converted

$$P = \varepsilon I$$

is valid no matter what the direction of the current through the source of EMF. However, if the current is opposite to the direction of the EMF (Figure 26-3), then energy is converted from electric potential energy into internal energy of the chemical constituents of a storage battery* or into macroscopic kinetic energy of a dynamo. (In the latter case we call the dynamo an electric motor.) With this current direction, the electrical power input to the source of EMF is $V_t I$, a portion εI is converted and a portion I^2r is dissipated:

$$V_t I = \varepsilon I + I^2r$$

---

*In daily usage, a storage battery is said to be "charging" or "recharging" when the current direction is opposite to the direction of its EMF, and "discharging" when the current is in the direction of its EMF. These terms are misnomers since in each case it is the internal energy of the battery, rather than its charge, that changes. A storage battery stores internal energy, not charge.

Therefore, when I is in the direction opposite to that of ε so that the source is receiving electrical power from the external circuit, the terminal voltage of the source is

$$V_t = \varepsilon + Ir \qquad (26.4)$$

If the direction of the flow of charge in a circuit does not change, the current is said to be a *direct current* (DC). A constant EMF in a circuit produces a direct current which, except for *transients* that occur when switches are opened or closed (Section 21.4), is also a constant current.

Example 1   A storage battery with an EMF of 12.0 V has an internal resistance of $2 \times 10^{-3}\ \Omega$. What is the terminal voltage of the battery at an instant when it is supplying power to the starter of a car by a current of 500 A?

Solution   The terminal voltage is

$$V_t = \varepsilon - Ir = 12.0\ \text{V} - (500\ \text{A})(2 \times 10^{-3}\ \Omega)$$

$$= 11.0\ \text{V}$$

26.2 Simple Electric Circuits

Single-Loop Circuits

Figure 26-4 is a schematic diagram of a circuit which provides only one conducting path for the current. Such a circuit is called a single-loop circuit. The current I is the same throughout the circuit. All such circuits can be analyzed using the *Kirchhoff loop rule: The algebraic sum of the EMFs in a loop is equal to the sum of the IR drops in the loop.* In applying this rule an EMF is considered positive if it is in the direction of the current, and negative if it is in the opposite direction.

Figure 26-3   Current direction when ε_1 is greater than ε. This current is in a direction opposite to the direction of the EMF ε. If the source at the left is a storage battery, this is the current direction for "recharging" the battery.

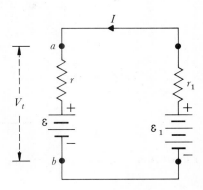

For the circuit of Figure 26-4, the Kirchhoff loop rule gives

$$\mathcal{E}_1 + (-\mathcal{E}_2) = Ir_1 + IR + Ir_2 \tag{26.5}$$

where $\mathcal{E}_1$ and $\mathcal{E}_2$ are *positive* numbers representing the magnitudes of the EMFs.

The method we shall use to prove Eq. 26.5 can be extended easily to provide a general proof of the Kirchhoff loop rule. The potentials at the points a, b, and c are denoted by V_a, V_b, and V_c, respectively. Since

$$(V_a - V_b) + (V_b - V_c) + (V_c - V_a) = 0$$

the algebraic sum of the potential drops around the loop is zero. Noting that $V_a - V_b$ is the potential drop IR and using Eq. 26.4 for the terminal voltage $V_b - V_c$ and Eq. 26.3 for the terminal voltage $V_a - V_c$, we obtain

$$(IR) + (\mathcal{E}_2 + Ir_2) - (\mathcal{E}_1 - Ir_1) = 0$$

Rearrangement of terms gives the Kirchhoff loop rule, as expressed by Eq. 26.5.

Example 2   A storage battery with an EMF of 12.0 V and an internal resistance of 55×10^{-3} Ω (commercial batteries in good condition have an internal

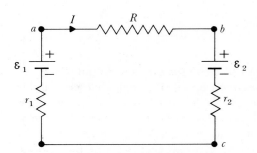

Figure 26-4   A single-loop circuit.

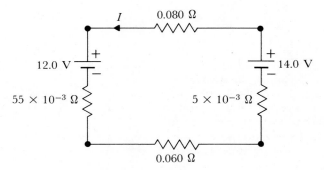

Figure 26-5   Circuit for "recharging" the 12.0-V storage battery of Example 2.

resistance of only a few thousandths of an ohm) is being "recharged" using the circuit of Figure 26-5. Find the circuit current I.

Solution   The Kirchhoff loop rule gives

$$-12.0 \text{ V} + 14.0 \text{ V} = (0.080 \ \Omega)I + (0.055 \ \Omega)I + (0.060 \ \Omega)I + (0.005 \ \Omega)I$$

$$I = \frac{2.0 \text{ V}}{0.200 \ \Omega} = 10 \text{ A}$$

Example 3   Find the power delivered to a load with resistance R by a source with EMF $\mathcal{E}$ and internal resistance r (Figure 26-6).

Solution   Kirchhoff's loop rule gives

$$\mathcal{E} = IR + Ir$$

Therefore

$$I = \frac{\mathcal{E}}{R + r}$$

and the power delivered to the load is

$$P = I^2R = \frac{\mathcal{E}^2R}{(R + r)^2}$$

The power delivered, considered as a function of the load resistance R, reaches a maximum value $\mathcal{E}^2/4r$ when $R = r$ (Question S-1). In this condition the resistances of the load and of the power source are said to be *matched*.

Series and Parallel Circuits

Analysis is often facilitated by replacing a network of resistors between two points in a circuit by a single resistor that is equivalent in that it draws the same current as the network and has the same voltage across it.

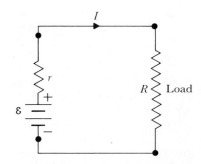

Figure 26-6   Source delivers power to a load with a resistance R.

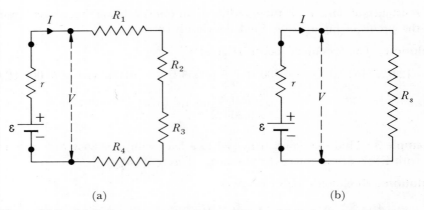

Figure 26-7 (a) Series connection of four resistors. (b) R_s is the single resistance equivalent to the series combination of four resistors.

Figures 26-7(a) and 26-8(a) show two different connections of four resistors and a source of EMF. Resistors in *series* provide only one conduction path and therefore each resistor carries the same current. Resistors in *parallel* have common terminals and therefore the voltage is the same across each resistor.

The resistance R of the single resistor equivalent to a parallel combination is defined by $R_p = V/I$ where V is the voltage across each resistor of the parallel combination and I is the total current [Figure 26-8(a)]. Since charge is conserved and does not accumulate at a junction, the total current I is the sum of the currents in the individual resistors of the parallel combination:

$$I = I_1 + I_2 + I_3 + I_4$$

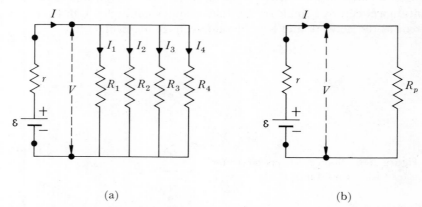

Figure 26-8 (a) Parallel connection of four resistors. (b) R_p is the single resistor equivalent to the parallel combination of four resistors.

This gives

$$\frac{V}{R_p} = \frac{V}{R_1} + \frac{V}{R_2} + \frac{V}{R_3} + \frac{V}{R_4}$$

Therefore the *reciprocal* of the resistance of the equivalent single resistor is the sum of the *reciprocals* of the resistances of the resistors connected in parallel:

$$\frac{1}{R_p} = \frac{1}{R_1} + \frac{1}{R_2} + \frac{1}{R_3} + \frac{1}{R_4} \qquad (26.6)$$

With more than one conduction path, the equivalent resistance is less than the smallest of the individual resistances.

The student is asked in Question 10 to show that the single resistor equivalent to the *series* combination of resistors in Figure 26-7 has a resistance given by

$$R_s = R_1 + R_2 + R_3 + R_4 \qquad (26.7)$$

A complex network that consists only of series and parallel combinations can be reduced to a single equivalent resistor by successive application of Eq. 26.6 and 26.7, as illustrated in Figure 26-9.

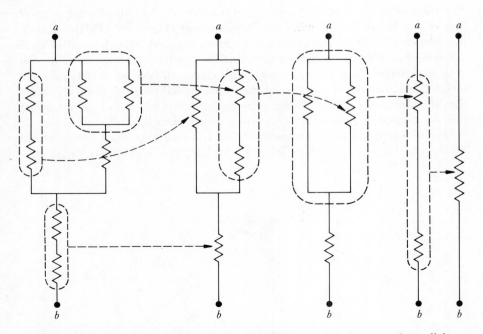

Figure 26-9 Reduction of a network that consists only of series and parallel combinations.

Short Circuit

A short circuit is a current path which bypasses the load (the principal circuit resistance R in Figure 26-10). For example, a source is "shorted" when its terminals are connected directly. The total resistance R_p in the external circuit is then given by

$$\frac{1}{R_p} = \frac{1}{r_w} + \frac{1}{R}$$

where r_w is the very small resistance of the wire providing the short circuit. Usually r_w is much smaller than R, so R_p is essentially the same as r_w. If r_w is also small compared to the internal resistance r, the total circuit resistance is approximately r. Then the current through the source is limited only by the internal resistance of the source: $I = \varepsilon/r$. The internal resistance is usually so small that short circuit current is tremendous. Overheating occurs causing damage and perhaps fire. To prevent this, *fuses* or *circuit breakers* are installed in domestic and industrial circuits. These safety devices are designed to open the circuit if the current is too large.

Galvanometers, Ammeters, and Voltmeters

The basic electrical detecting instrument used within several different types of meters is a D'Arsonval moving-coil galvanometer (Figure 26-11). In a galvanometer, a constant current under investigation is passed through a coil which is suspended in a magnetic field established by a permanent magnet. The coil then experiences current-dependent magnetic forces (a topic of Chapter 27) which cause the coil to rotate until their effect is balanced by restoring forces furnished by a hair spring. The amount of coil rotation, which is a measure of the coil current, is indicated by a pointer attached to the coil.

The sensitivity of a galvanometer with a coil resistance R_G is characterized by the coil current I_G or the voltage $V_G = I_G R_G$ that is required for *full-scale*

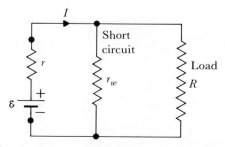

Figure 26-10   A load resistor short-circuited by a wire of resistance r_w.

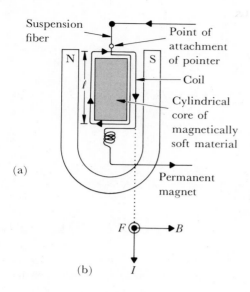

(a)

(b)

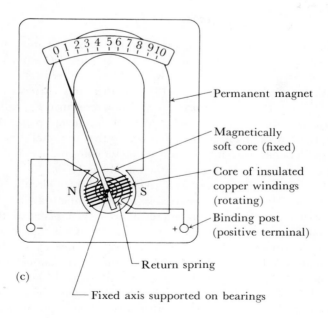

(c)

Figure 26-11   D'Arsonval moving-coil galvanometer. (a) Diagram of the meter.
(b) The force on the longitudinal part of the winding on the right, employing the
directions associated with Eq. 27.8, is outward. (c) Galvanometer with the moving
coil attached to an axis supported on bearings.

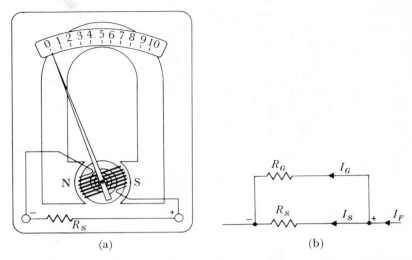

Figure 26-12   Ammeter. (a) Diagram showing the moving-coil galvanometer connected in parallel with the very low resistance shunt R_S. (b) Circuit diagram showing currents that produce full-scale deflection. The circuit current being measured is $I_F = I_S + I_G$.

deflection. Typical values for a cheap galvanometer used in panel meters are $R_G = 50\ \Omega$, $I_G = 2.0$ mA, and $V_G = 0.10$ V. Most DC ammeters and voltmeters are simply galvanometers used with appropriate resistors.

A DC ammeter (Figure 26-12) that measures currents ranging from, say, 0 to 10 A, is obtained by using a very low resistance R_S, called a *shunt* resistor, in *parallel* with the galvanometer. Most of the current passes through this shunt. The resistances involved can be chosen so that the small fraction of the current which does pass through the galvanometer coil produces full-scale deflection of the galvanometer needle when the current is 10 A.

To measure the current in a given branch of a circuit, the circuit must be broken and the ammeter inserted in *series* with the branch.* After the ammeter is inserted, since the total branch resistance is then increased, the branch current, although correctly indicated by the ammeter, is less than it was before the ammeter was inserted. However, the current change caused by insertion of an ammeter will be small if the total resistance of the ammeter is small compared to the branch resistance. An ideal ammeter would have zero resistance.

A DC *voltmeter* (Figure 26-13) can be constructed by connecting a very high resistance R_M, called a multiplier, in series with a galvanometer. For example,

---

*A mistake frequently made in student laboratories is the connection of the ammeter in parallel with the branch. Then the current through the ammeter is *not* the current through the branch in question and no useful information is obtained. Even worse, the voltage across the branch may produce a current through the low resistance ammeter that is so large that the galvanometer element is destroyed.

to construct a voltmeter reading from 0 to 150 V, we select the high resistance so that the galvanometer current attains the value sufficient to produce full-scale galvanometer deflection when 150 V are applied across the voltmeter terminals.

To measure the voltage between two points in a circuit, we connect a voltmeter terminal to each point. Some current then passes through the voltmeter, and the voltage, although correctly indicated by the voltmeter, will be less than it was before the voltmeter was connected. But if the voltmeter resistance is large compared to the circuit resistance between the points in question, the voltage change caused by connecting the voltmeter will be small. An ideal voltmeter has an infinite resistance.

Ground Connections

Often a point in a circuit is connected to the earth by a conductor called a *ground wire*. Since a closed conducting path with a source of EMF is required for the maintenance of a steady current, there will be no such current in the ground wire in the circuit shown in Figure 26-14. Analysis of such a circuit is in no way affected by the presence of the ground wire. When absolute potentials are assigned to different points of an electric circuit, it is conventional to assign zero potential to the earth.

Often one terminal of a source of EMF is grounded. Then, if one terminal of the load is grounded, the circuit is completed through the conducting earth and

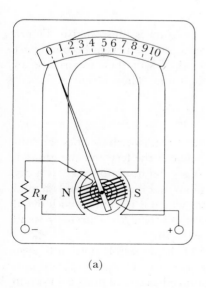

(a)

$$\underset{-}{\bullet}\!\!-\!\!\!\overset{R_M}{\wedge\!\!\wedge\!\!\wedge}\!\!-\!\!\!\overset{R_G}{\wedge\!\!\wedge\!\!\wedge}\!\!-\!\!\underset{+}{\bullet}$$

(b)

Figure 26-13   (a) Voltmeter composed of a moving-coil galvanometer connected in series with a high resistance R_M. (b) Equivalent circuit diagram of this voltmeter.

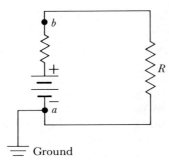

Figure 26-14 Circuit with a ground connection. There is no steady current from point *a* to ground. The potential of the point *a* is zero.

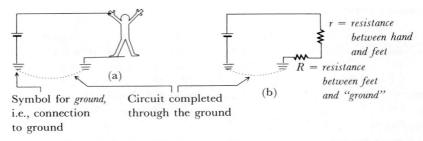

Figure 26-15 Electric shock. (a) The physical situation. (b) The associated circuit diagram.

no return wire is required. A hazard in this situation is illustrated in Figure 26-15. By touching the conductor that is not grounded, a person becomes part of a complete circuit. This unfortunate individual will then experience an *electric shock* caused by a current passing through him. This current is

$$I = \frac{\varepsilon}{r + R}$$

where *r* is the resistance of the body between the hand (which touches the conductor) and the feet, and *R* is the resistance between the feet and the ground.

If the person is standing on a surface that is a good insulator, such as a dry wooden floor, then *R* is large and *I* may be so small that it will cause only a mild electric shock. But, if the feet are on a wet floor which is in contact with metal pipes that ultimately serve as a ground connection, then *R* is extremely small and the current can be large enough to cause death.

Figure 26-16 illustrates that the grounding of the accessible portion of electrical appliances can be an important safety measure. The ground wire provided for this purpose, together with a three-pronged electrical plug, is shown in Figure 26-17.

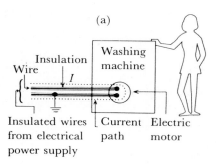

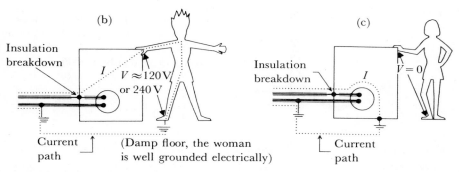

Figure 26-16   The use of an independent third wire to ground the frame or casing of an electrical appliance: a safety measure. (a) Normal operation without the safety ground wire. (b) Electric shock: insulation breakdown and casing not properly grounded. Current I through the woman is $I = V/r$ where r is the resistance between her hand and foot. (c) Protection against electrical fault: casing is grounded. The voltage V across the woman is zero; therefore the current through the woman is zero. Note that in both cases (b) and (c) the current passes through the ground which provides the completion of the circuit.

26.3 Multiloop Circuits

Many circuits are not merely series or parallel combinations. The method we have used for a single-loop circuit can be extended to provide a systematic procedure for the analysis of complex circuits. We now use *two Kirchhoff rules*:

Junction rule   The sum of the currents entering any junction is equal to the sum of the currents leaving that junction. This rule follows from the law of conservation of electric charge and from the fact that electric charge does not accumulate at the junction.

Loop rule   The algebraic sum of the EMFs in any loop is equal to the algebraic sum of the IR products in the same loop. (A loop is any closed conducting path that we care to consider in the circuit.) This rule is a generalization of Eq. 26.5 and is a consequence of the law of conservation of energy.

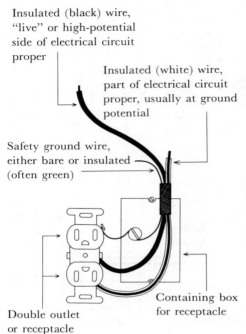

Insulated (black) wire,
"live" or high-potential
side of electrical circuit
proper

Insulated (white) wire,
part of electrical circuit
proper, usually at ground
potential

Safety ground wire,
either bare or insulated
(often green)

Figure 26-17   Double electrical outlet showing how both the actual receptacle and its containing box are connected directly to ground by the third wire which is not part of the regular circuit.

Containing box
for receptacle

Double outlet
or receptacle

To apply the loop rule we arbitrarily choose one direction around the loop (clockwise or counterclockwise) as the positive direction. Then, in the equation for this loop,

$$algebraic\ sum\ of\ EMFs = algebraic\ sum\ of\ IR\ products$$

the EMFs and the currents in this positive direction are considered positive; EMFs and currents in the opposite direction are negative.

A general procedure which applies to the most complex circuits is outlined in the following example.

Example 4   Find the current in every branch of the circuit of Figure 26-18.

Solution   Unknown branch currents are assigned symbols and *directions*; the junction rule is used as these assignments are made. For example, having made the assignments of I_1 and I_2 indicated in Figure 26-18, we apply the junction rule to junction A and find that the current in the remaining branch is $I_1 + I_2$ directed away from the junction.

We choose a sufficient number of loops to traverse every branch of the network at least once. One way of doing this is to consider the two loops selected in Figure 26-18.

We now apply the loop rule first to loop 1 and then to loop 2, having indicated that the clockwise direction has been selected as the positive direction

Figure 26-18 The 2-loop circuit with the branch currents I_1, I_2, and $I_1 + I_2$ that have been assumed for the application of Kirchhoff's loop rule. Solution gives $I_2 = -4.0$ A, which implies that the actual current in the 5.0-Ω resistor is opposite to the indicated direction.

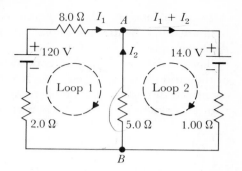

for each loop:

$$+120 \text{ V} = (8.0 \ \Omega + 2.0 \ \Omega)I_1 - (5.0 \ \Omega)I_2$$

$$-14.0 \text{ V} = (1.00 \ \Omega)(I_1 + I_2) + (5.0 \ \Omega)I_2$$

which can be rewritten as

$$120 \text{ V} = (10.0 \ \Omega)I_1 - (5.0 \ \Omega)I_2$$

$$-14.0 \text{ V} = (1.00 \ \Omega)I_1 + (6.0 \ \Omega)I_2$$

The solution of this pair of equations is $I_1 = 10.0$ A and $I_2 = -4.0$ A. The significance of the minus sign in the answer for I_2 is that the actual current in the 5.0 Ω resistor is in the direction opposite to that assumed for I_2. Therefore, the current in this branch is 4.0 A directed from junction A toward junction B. The current in the 1.00-Ω resistor is $I_1 + I_2 = (10.0 - 4.0)$ A $= 6.0$ A, directed toward junction B.

Electrical Measurements Using Null Methods

The Wheatstone bridge circuit in Figure 26-19 provides a precise method for the measurement of an unknown resistance R_x, when R_1, R_2, and R_3 are adjustable known resistances. The known resistances are adjusted until there is no galvanometer deflection when the switch K is closed. Then the bridge is said to be *balanced*, the potentials at the points a and b are equal, the current in R_1 is I_x, and the current in R_2 is I_3. Consequently

$$I_x R_x = V_d - V_a = V_d - V_b = I_3 R_3$$

and

$$I_x R_1 = V_a - V_c = V_b - V_c = I_3 R_2$$

Therefore

$$\frac{R_x}{R_3} = \frac{R_1}{R_2} \tag{26.8}$$

and the unknown resistance R_x can be calculated.

In actual practice, the galvanometer is protected by a shunt during pre-liminary adjustments, and full galvanometer sensitivity is exploited only when the balance is almost perfect.

The *potentiometer* circuit in Figure 26-20 is used for the precision measurement of the ratio of two voltages. A source of EMF ε, usually a storage battery, is connected through an adjustable resistance R_a to a fixed resistance R. A source with an unknown EMF ε_x is connected between terminals a and b through a

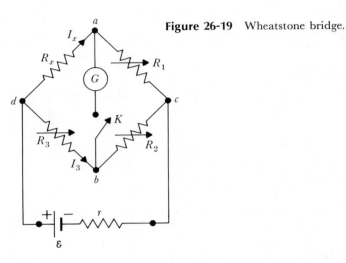

Figure 26-19   Wheatstone bridge.

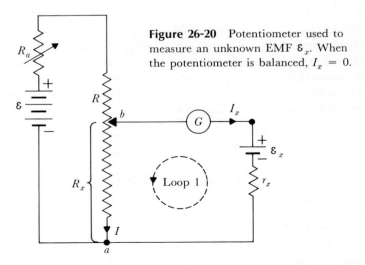

Figure 26-20   Potentiometer used to measure an unknown EMF ε_x. When the potentiometer is balanced, $I_x = 0$.

galvanometer. The potentiometer is balanced by moving the sliding contact b until the galvanometer indicates zero current ($I_x = 0$). Kirchhoff's loop rule applied to loop 1 in Figure 26-20 gives

$$\mathcal{E}_x = IR_x - I_x r_x$$

where R_x is the potentiometer resistance between terminal a and the sliding contact b. When the potentiometer is balanced, $I_x = 0$ and we have

$$\mathcal{E}_x = IR_x$$

If the unknown EMF is replaced by a standard source with a known EMF $\mathcal{E}_s$,

$$\mathcal{E}_s = IR_s$$

where R_s is the potentiometer resistance between terminal a and the position of the sliding contact b when balance is obtained. Therefore

$$\frac{\mathcal{E}_x}{\mathcal{E}_s} = \frac{R_x}{R_s}$$

and the unknown EMF can be calculated if the resistance ratio R_x/R_s is known.

In a type of potentiometer called a *slide-wire* potentiometer, the fixed resistor R is a wire that is extremely uniform and

$$\frac{R_x}{R_s} = \frac{\ell_x}{\ell_s}$$

where ℓ_x and ℓ_s are the values of the lengths of the slide-wire (between a and the sliding contact b) that have resistances R_x and R_s, respectively. Using a slide-wire potentiometer, the ratio of two EMFs can be determined by measuring a ratio of lengths:

$$\frac{\mathcal{E}_x}{\mathcal{E}_s} = \frac{\ell_x}{\ell_s}$$

In both the Wheatstone bridge and the potentiometer, the basic detecting instrument, the galvanometer, serves only as an indicator of balance. Since the calibration of this detector is not involved in the measurement, an important source of error is eliminated. In any field of science or engineering, when a precision measurement is desired, it is usually advantageous to devise a null method rather than a direct method which requires reading a calibrated scale.

26.4 Transients in *R-C* Circuits

When a sudden change is made in a physical system there is usually a certain interval of time during which the system adjusts itself to the new conditions. The temporary effects are called *transients*. In this section we investigate the transient currents that follow the opening or closing of a switch in a circuit containing a capacitor.

Charging a Capacitor

Figure 26-21(a) shows a source with a constant EMF ε and an uncharged capacitor. At $t = 0$ the switch S is thrown to position 1 and a current I is established in the circuit. The capacitor charge Q and voltage V_C increase until $V_C = \varepsilon$ which corresponds to a steady-state with no current. To discover exactly what happens we must determine I and Q as functions of the time t.

A sign convention for Q, I, and V_C is required. The charge on the upper plate of the capacitor is denoted by Q. Then $V_C = Q/C$ is positive when Q is positive and the current $I = dQ/dt$ is positive when directed toward the upper plate, as indicated in Figure 26-21(b).

At any given instant, the algebraic sum of the potential drop across the capacitor V_C and the potential drop across the resistor V_R must be equal to the terminal voltage ε of the source (for simplicity, the internal resistance of the source has been assumed to be negligible):

$$\varepsilon = V_C + V_R \tag{26.9}$$

This gives

$$\varepsilon = \frac{Q}{C} + IR \tag{26.10}$$

Differentiation of this equation yields

$$0 = \frac{1}{C}\frac{dQ}{dt} + R\frac{dI}{dt}$$

Since $I = dQ/dt$, the differential equation that I must satisfy is

$$\frac{dI}{dt} = -\frac{I}{RC}$$

The variables can be separated giving

$$\frac{dI}{I} = -\frac{dt}{RC}$$

Integration of both sides yields

$$\ln I = -\frac{t}{RC} + constant$$

or

$$I = I_0 e^{-t/RC} \tag{26.11}$$

where I_0 is a constant of integration. As t approaches zero through positive values, Eq. 26.11 gives $I = I_0$. Therefore I_0 is the current just after the switch is thrown to position 1. At this instant the charge on the capacitor is still zero so Eq. 26.10 reduces to

$$\varepsilon = 0 + I_0 R$$

Therefore $I_0 = \varepsilon/R$.

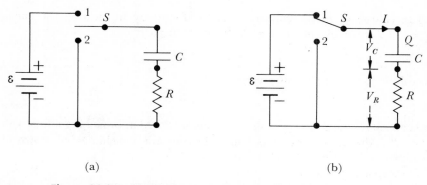

(a)   (b)

Figure 26-21   Circuit for charging or discharging a capacitor.

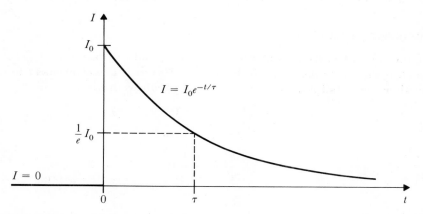

Figure 26-22   The exponential decay $I = I_0 e^{-t/\tau}$ of the current in an *R-C* circuit while the capacitor is charging. The time constant is $\tau = RC$.

Figure 26-22 gives the graph of the current as a function of time. When the switch is thrown, applying the EMF across the *R-C* circuit, the current rises abruptly to the value $I_0 = \mathcal{E}/R$ and then "decays" exponentially to zero. The characteristic time for this decay is the quantity RC which is called the *time constant* τ of the circuit. At the instant $t = \tau = RC$, the current has the value $I = I_0 e^{-RC/RC} = I_0 e^{-1}$; that is,

$$I = \frac{1}{e} I_0 \qquad (\text{at } t = \tau = RC) \quad (26.12)$$

In any time interval equal to the time constant τ, *the current decays to a fraction* $1/e$ *(about 0.37) of its value at the beginning of the interval.* After an interval long compared to the time constant, the current has a negligible value.

The charge Q on the capacitor can be found by integrating the current ($Q = \int_0^t I\,dt$) or by solving Eq. 26.10 for Q to yield

$$Q = C\varepsilon - CRI$$

Since $I = (\varepsilon/R)e^{-t/RC}$, this gives

$$Q = C\varepsilon(1 - e^{-t/RC})$$

The capacitor charge grows as shown in Figure 26-23 approaching a final value $Q_f = C\varepsilon$. After a time interval of one time constant, the charge has grown to a fraction $1 - 1/e$ (about 0.63) of its final value.

The capacitor voltage is

$$V_C = \frac{Q}{C} = \varepsilon(1 - e^{-t/RC})$$

This is a growth function with a time constant RC and a final value ε.

Discharging a Capacitor

Suppose that when the capacitor has a charge Q_0 and a voltage $V_0 = Q_0/C$, the switch is thrown to position 2 in Figure 26-21(a). It is now convenient to call this time the initial instant, $t = 0$. For this circuit we have

$$V_R + V_C = 0$$

or

$$R\frac{dQ}{dt} + \frac{Q}{C} = 0$$

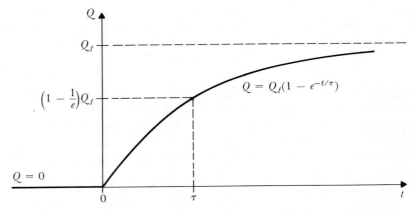

Figure 26-23 The growth function $Q = Q_f(1 - e^{-t/\tau})$ for the charge on a capacitor which is being charged through a resistance R. The time constant is $\tau = RC$.

Following the same steps as those that led to Eq. 26.11, we find that the solution of this differential equation for Q is

$$Q = Q_0 e^{-t/RC}$$

which is an exponential decay with time constant $\tau = RC$.

Since $V_C = Q/C$, the capacitor voltage also decays to zero. The current function is

$$I = \frac{dQ}{dt} = -\frac{Q_0}{RC} e^{-t/RC} = -\frac{V_0}{R} e^{-t/RC}$$

The minus sign shows that the current is directed away from the positively charged plate of the capacitor. The magnitude of the current *decays* exponentially for both the charge and the discharge of the capacitor.

Notice that when a switch is thrown in an *R-C* circuit, the current changes abruptly. But the capacitor charge and the capacitor voltage do not change immediately. These quantities require a time interval of the order of $\tau = RC$ for an appreciable change in value.

Summary

☐   If the current I is in the direction of a source with EMF ε, energy is converted within the source from some nonelectrostatic form into electric potential energy at the rate

$$P = \varepsilon I$$

and a source with internal resistance r has a terminal voltage

$$V_t = \varepsilon - Ir$$

☐   If the current direction is reversed, electric potential energy is converted within the source into some other form at the rate

$$P = \varepsilon I$$

and the terminal voltage of the source is

$$V_t = \varepsilon + Ir$$

☐   Circuits can be analyzed using Kirchhoff's rules:

   Junction rule   The sum of the currents entering any junction is equal to the sum of the currents leaving that junction.

   Loop rule   The algebraic sum of the EMFs in any loop is equal to the algebraic sum of the *IR* products in the same loop.

☐   The duration of transient conditions when a capacitor is being charged or discharged through a resistance R is characterized by the time constant

$\tau = RC$. Transient behavior of the variable quantities in Table 26-1 is described by either a decay function of the form

$$(initial\ value)\,e^{-t/\tau}$$

or a growth function of the form

$$(final\ value)(1 - e^{-t/\tau})$$

Table 26-1

Quantity	Capacitor charging	Capacitor discharging
Current	Decay	Decay
Charge	Growth	Decay
Capacitor voltage	Growth	Decay

Questions

1 When a charge of 3.0 C passes through a certain source of EMF, the work done on this charge by nonelectrostatic forces is 24 J. What is the EMF of the source?

2 How much energy will a 70-A·h, 12-V storage battery deliver without recharging?

3 A car battery which has an EMF of 12.0 V provides a temporary 600-A current to the engine starter. How much energy is converted into electric potential energy within the battery if the driver presses on the starter for 10 s?

4 A car has a storage battery with an EMF of 12 V. If the battery furnishes 5.0 × 10^3 J to the starter and an additional 1.0 × 10^3 J are dissipated within the battery, what electric charge has passed through the battery?

5 What current is furnished by a 12-V storage battery to a starter which develops 3.0 hp? (1 hp = 746 W.)

6 The open-circuit voltage of a certain storage battery is 12.0 V. When the current in a resistor connected between the battery terminals is 150 A, the terminal voltage is 10.1 V. What is the internal resistance of the battery?

7 Find the current in the circuit of Figure 26-5 if the polarity of the 12.0-V storage battery is reversed.

8 A source with an EMF of 60 V and an internal resistance of 0.20 Ω is connected in series with a second source which has an EMF of 40 V and an internal resistance of 0.10 Ω. If this series combination is connected across a 0.70-Ω resistor, what is the current when:

(a) The negative terminal of one source is connected to the positive terminal of the other?

(b) The positive terminals of the sources are connected?

9 A storage battery with an EMF of 12.0 V has an internal resistance of 6 × 10^{-2} Ω. The battery is to be recharged by connecting 120-V DC across a series combination of the battery and a resistor whose resistance R is adjusted to limit the charging current to 10.0 A:

(a) Which terminal (+ or −) of the battery should be connected to the high-potential side of the power lines?

(b) What is the value of R?

(c) Find the terminal voltage of the battery while it is being recharged.

(d) What is the power input from the lines?

(e) What is the power dissipated in R and also within the battery?

(f) What is the power converted within the storage battery?

(g) If a charge of 40 A·h circulates through the battery, what will be the cost at 3 cents per kilowatt hour?

10 Show that R_s of Figure 26-7(b) is given by

$$R_s = R_1 + R_2 + R_3 + R_4$$

11 **(a)** Calculate the single resistance equivalent to three resistances in series if they are 8.0 Ω, 4.0 Ω, and 24 Ω.

(b) Calculate the equivalent resistance if the resistances of part (a) are connected in parallel.

12 Figure 26-24 shows a circuit element (called a *potentiometer*) used as a *voltage divider*. By moving the sliding contact, the voltage V_1 can be adjusted from 0 to V_0:

(a) Show that

$$\frac{V_1}{V_0} = \frac{R_1}{R_0}$$

(b) Find V_1 when a resistance R is connected in parallel with R_1.

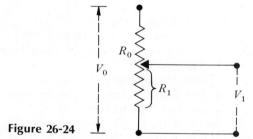

Figure 26-24

13 In Figure 26-25, what is the resistance between the terminals a and b?

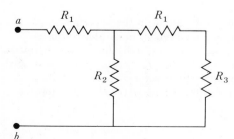

Figure 26-25

14 Consider a 1200-W heating element operated on a 120-V line:
   (a) What current is drawn by the heater?
   (b) What would be the current if a short circuit occurred because the insulation of the heater cord had become worn away? (Assume that the connecting cord wires have a resistance of 0.012 Ω.)
   (c) Would the current ever increase to the value determined in part (b)? Explain.

15 A 12-V source of EMF having an internal resistance of 0.0093 Ω is shortcircuited with a wire of resistance 7×10^{-4} Ω:
   (a) What is the magnitude of the current through this source?
   (b) How much energy is dissipated in the circuit if the short circuit is maintained for 15 s?

16 In the situation shown in Figure 26-15, what current will pass through a person who touches a "live" wire at 120 V (above ground potential) if the body resistance is 10^4 Ω and the resistance between feet and ground is 10^2 Ω?

17 A galvanometer gives full-scale deflection when a current of 1.0×10^{-4} A passes through its coil. What is the total resistance of a voltmeter constructed using this galvanometer together with a series resistance, if the voltmeter is designed to measure from 0 to 150 V?

18 Find the resistance of the shunt required to convert the galvanometer of the preceding question to an ammeter which reads 10 A full-scale. The resistance of the galvanometer coil is 30 Ω.

19 A galvanometer has an internal resistance of 1.0 Ω and gives maximum deflection for a current of 50 mA. Indicate with a sketch how this instrument can be changed into:
   (a) A voltmeter with a maximum reading of 2.5 V.
   (b) An ammeter with maximum reading of 2.5 A.
   In each case calculate the resistance of whatever resistors have to be used in conjunction with the galvanometer.

20 Consider a circuit in which the following components are connected in series: a 6.00-V battery of negligible internal resistance, a resistor of 10.0 kΩ, another resistor of 20.0 kΩ, and an ammeter of negligible resistance:
   (a) Determine the current in the circuit.
   (b) Determine the IR drop across the 20.0-kΩ resistor.
   (c) What is the current in the circuit when a voltmeter with a resistance of 30.0 kΩ is connected across the 10.0-kΩ resistor?
   (d) What is the current through the voltmeter itself?

21 A voltmeter which has a resistance of 960 Ω is connected across a 120-Ω resistor which is in turn connected in series with a 100-Ω resistor and an ammeter of negligible resistance. The source of EMF is a 3.00-V battery of negligible internal resistance:
   (a) What is the current in the circuit?
   (b) What is the current through the voltmeter?
   (c) What is the IR drop indicated by the voltmeter?
   (d) What is the (absolute) "error" in the IR drop that we can measure? That is, what is the difference between the IR drop across the 120-Ω resistor with the voltmeter connected and the IR drop across the same resistor with the voltmeter disconnected?

(e) What is the relative error (error/reading) in the IR drop introduced by the measuring device, the voltmeter?

22 Find the current in each circuit element of Figure 26-26. Show that the potential at each point of the circuit is correctly represented in Figure 26-27.

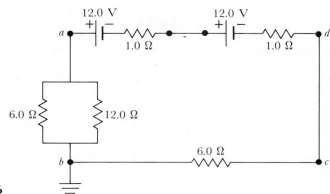

Figure 26-26

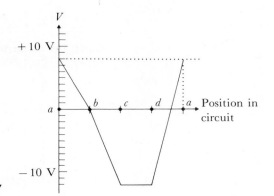

Figure 26-27

23 Redraw the circuit of the preceding question and add a resistanceless connection from point b to point d. Now, with reference to this modified circuit, find the current in each circuit element.

24 In the circuit of Figure 26-28, the voltmeter V_1 reads 12.0 V, V_2 reads 4.0 V, and the equivalent resistance of the two parallel resistors is 4.0 Ω. Assume that the meters do not affect the circuit behavior and that the source of EMF has negligible internal resistance:

(a) What is the reading of the ammeter A?
(b) What is the reading of the voltmeter V_3?
(c) How large is the EMF?
(d) How large is the resistance of the resistor R_1?

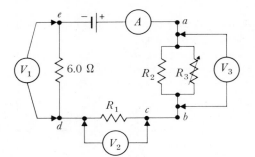

Figure 26-28

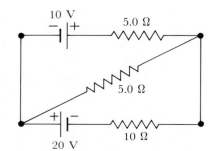

Figure 26-29

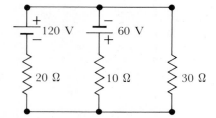

Figure 26-30

(e) If the resistor R_2 has a resistance of 8.0 Ω, what is the resistance set on the rheostat R_3?

(f) What is the potential of point d if point e is grounded?

(g) What would happen to the total current in the circuit if the rheostat R_3 were turned to a lower resistance?

25 Find the current in the diagonal resistor in Figure 26-29.

26 Find the current in each source of EMF and in the 30-Ω resistor in Figure 26-30.

27 Find the current in each branch of Figure 26-30 for the case in which the polarity of the 60-V source is reversed from that indicated in the figure.

28 A 120-V source with an internal resistance of 2.0 Ω is connected in parallel with a source which has an internal resistance of 3.0 Ω but whose EMF is unknown. When this combination is connected across a 10-Ω load, the current in the 120-V source is 40 A. Find the magnitude and polarity of the unknown EMF.

29 In the circuit of Figure 26-31, find $\mathcal{E}_1$ and $\mathcal{E}_2$.

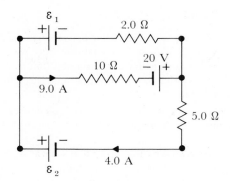

Figure 26-31

30 When the circuit elements in Figure 26-19 have the values $R_x = 10.0 \; \Omega$, $R_3 = 20.0 \; \Omega$, $R_1 = 3.0 \; \Omega$, $R_2 = 12.0 \; \Omega$, $R_G = 2.0 \; \Omega$, $r = 0$, and $\varepsilon = 10.0$ V, find the current through the galvanometer when switch K is closed.

31 A 5.0-μF capacitor is charged through a 2.0-MΩ resistor by a storage battery with an EMF of 12.0 V. The capacitor is initially uncharged and the charging starts at $t = 0$. Sketch graphs of the voltage across the resistor and the voltage across the capacitor showing initial and final values as well as values at the following instants: 10 s, 20 s, 30 s.

32 What is the final charge of the capacitor in the preceding question? If it is to be discharged through a 2.0-kΩ resistor, how long will it take for this capacitor to be discharged so completely that its predicted charge is less than the magnitude of the electronic charge?

33 In the circuit of Figure 26-32, switch S is thrown to position a. When the voltage across the 10-MΩ resistor has decreased to 30 V the switch is thrown to position b:
  (a) Find the current through each circuit element and the voltage across the capacitor just before and just after switching to position b.
  (b) Find the capacitor voltage 60 s after switching to position b.

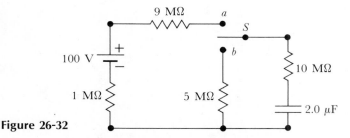

Figure 26-32

34 In the circuit of Figure 26-32, the capacitor is initially uncharged. Switch S is thrown to position a. After 80 s have elapsed, what is the power supplied to the capacitor, the power dissipated in the entire circuit, and the rate at which electric potential energy is being supplied by the source of EMF?

35 A capacitor, with initial charge Q_0 and voltage V_0, is discharged through a resistance R. Use Joule's law to evaluate the energy dissipated in the resistor and then compare this result with the amount of energy initially stored in the capacitor.

Supplementary Questions

S-1   Prove the *power transfer theorem*: the power delivered to a load with resistance R by a given source with EMF $\mathcal{E}$ and internal resistance r is a maximum when the resistance of the load matches that of the source; that is, $R = r$ (see Example 3 and Figure 26-6). Show that when the load and the source are matched, the power delivered is $\mathcal{E}^2/4r$.

S-2   A DC motor, with an internal resistance of 3.0 Ω, is connected across 120-V power lines. When its speed has built up to normal operating speed, the current is 10.0 A:

   (a) What EMF is generated by the motor and what is its polarity relative to the direction of the motor current? (Such an EMF is called a "back EMF.")
   (b) At what rate is electric potential energy converted to macroscopic kinetic energy by the motor?
   (c) What is the power dissipated within the motor?
   (d) What is the ratio of the useful power developed by the motor to the total electrical power input?

S-3   In Figure 26-33, what is the resistance between point a and point b of this combination of resistances?

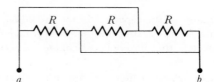

Figure 26-33 a b

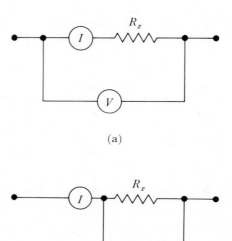

(a)

(b)

Figure 26-34

S-4   **(a)** If $R' = V/I$ is the ratio of the readings of a voltmeter and an ammeter when connected as shown in Figure 26-34(a), show that the unknown resistance R_x has the value

$$R_x = R'\left(1 - \frac{R_A}{R'}\right)$$

where R_A is the ammeter resistance.

(b) If $R' = V/I$ is the ratio of the readings of a voltmeter and an ammeter when connected as shown in Figure 26-34(b), show that the unknown resistance R_x has the value

$$R_x = \frac{R'}{1 - R'/R_V}$$

where R_V is the voltmeter resistance.

(c) Explain why one circuit is preferable for the measurement of a very low resistance and the other for the measurement of a very high resistance.

S-5   An unknown resistance R_x which is of the same order of magnitude as the known resistance R_V of a voltmeter can be measured by placing the voltmeter in *series* with the unknown resistance and known source of EMF, ε. In the circuit of Figure 26-35, $R_V = 20$ kΩ and $\varepsilon = 12.0$ V. The voltmeter reads 2.0 V. What is the unknown resistance R_x?

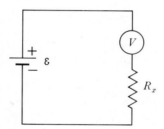

Figure 26-35

S-6   Show that when the potentiometer of Figure 26-20 is unbalanced, the galvanometer current is

$$I_x = \frac{(R_x/R)\varepsilon - \varepsilon_x}{r_x + R_x - R_x^2/R}$$

when $R_a = 0$. The galvanometer resistance, as well as the internal resistance of the working battery (of EMF ε), is neglected. R denotes the resistance of the resistor extending from the top of the figure to the point a.

S-7   Find the resistance between the terminals a and b of the combination of resistors shown in Figure 26-36. (Place a 1-V source of EMF between a and b and calculate the current through the source using Kirchhoff's rules.)

S-8   Two sources, each of EMF ε and internal resistance r, are used to produce a current in a load of resistance R:

(a) Find the load current I_s when the sources are connected in series.

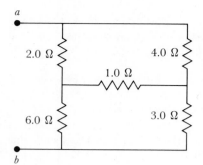

Figure 26-36

(b) Find the load current I_p when the sources are connected in parallel.

(c) Compare I_s and I_p in the following cases: $R < r$, $R = r$, $R > r$.

S-9   A hollow cube has a resistor R along each edge. Find the equivalent resistance between diagonally opposite corners of the cube.

S-10   Find the equivalent resistance R between the points a and b (the input resistance) for the infinite ladder network represented in Figure 26-37(a). Use the fact that, since the chain is infinite, its input resistance will not be changed by adding one more unit to the front of the ladder and therefore the resistance between the points a' and b' of Figure 26-37(b) also has the value R.

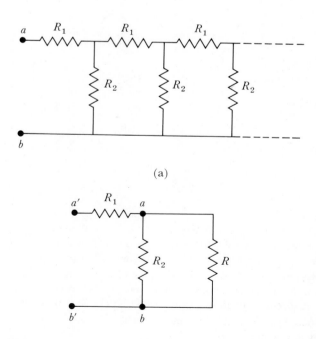

Figure 26-37

S-11 Using 1-Ω resistors, an infinite net with square meshes is formed with four of these resistors joined at each corner of every square. Show that the equivalent resistance between the terminals of one of these resistors is 0.5 Ω. (Use symmetry and superposition for an elegant solution.)

S-12 The material between the plates of a parallel-plate capacitor has a dielectric constant κ and a resistivity ρ that is not infinite. The capacitor therefore discharges itself through this material. Find the time constant for this discharge.

27.

the force exerted
by a magnetic field

When two charges ar. in *motion* there are new forces which come into play (in addition to the electric forces) that are termed magnetic forces. To describe magnetic interaction, physicists again introduce the idea of a field. With each point in space we associate, in addition to the electric field vector **E**, a second vector **B** called the *magnetic field**** at the point in question. The force exerted on a moving charged particle is determined entirely by the electric field **E** and the magnetic field **B** at the location of the particle. The experimental law for this force is used in the first section of this chapter to give a general definition of **E** and **B**. The remainder of the chapter is concerned with the effects and the uses of the force exerted by a given magnetic field in a variety of situations.

The question of the sources of a magnetic field and the relationship of a magnetic field to its sources is investigated in later chapters.

27.1 Definition of the Magnetic Field B

A charge q moving with velocity **v** in the vicinity of a current of other charges experiences not only an electric force $q\mathbf{E}$ but also an additional force proportional to both q and **v** and perpendicular to **v**. We use this experimental fact to

_____

*Other names for this vector are the *magnetic induction,* the *magnetic field strength,* and the *magnetic flux density*.

define at any given point in space two vectors, **E** and **B**, by

$$\mathbf{F} = q\mathbf{E} + q\mathbf{v} \times \mathbf{B} \tag{27.1}$$

where **F** is the force exerted on a test charge q moving with a velocity **v** at the given point. The vectors **E** and **B** are independent of q and **v**. The vector **E** is called the electric field at the given point and the new quantity, the vector **B**, is called the *magnetic field*. The resultant force given by Eq. 27.1 is named the Lorentz force.

The force exerted by the magnetic field **B** on the moving charged particle is

$$\mathbf{F} = q\mathbf{v} \times \mathbf{B} \tag{27.2}$$

This force, termed the *magnetic force*, is specified in both magnitude and direction by this vector equation involving the *cross product* (Section 13.1) of the vectors **v** and **B**. The magnitude of the magnetic force is

$$F = qvB \sin \theta$$

where θ is the angle between **v** and **B** (Figure 27-1). The direction of the magnetic force is perpendicular to both **v** and **B**. To complete the determination of the force direction we can use either the screw rule (Section 13.1) to find the direction of the cross product **v** × **B** or the right-hand rule illustrated in Figure 27-2. Notice that the direction of the magnetic force $q\mathbf{v} \times \mathbf{B}$ on a negative charge is opposite to the direction of the magnetic force on a positive charge.

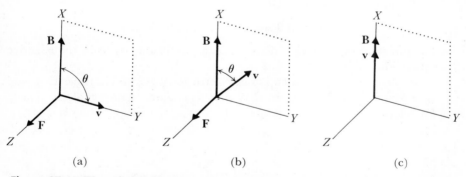

(a)   (b)   (c)

Figure 27-1 The relationship between **v**, **B**, and **F**, where **F** is the force experienced by a positive charge moving at velocity **v** while in a magnetic field **B**. (*X, Y, Z* are mutually perpendicular axes.) (a) Here $\theta = 90°$ and F has maximum value. (b) $\theta < 90°$ and F is smaller than in part (a). (c) $\theta = 0°$, **B** and **v** are parallel, and hence $F = 0$.

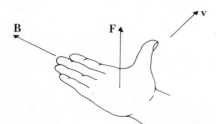

Figure 27-2 Right-hand rule to determine the direction of the magnetic force **F** exerted by a magnetic field **B** on a *positive* charge moving with a velocity **v**.

Since the relationship between the magnetic force $q\mathbf{v} \times \mathbf{B}$ and the magnetic field $\mathbf{B}$ is rather complicated and strange, it is well to emphasize certain points. First, the magnetic force is zero if the charge is at rest or moving parallel to $\mathbf{B}$. Next, this force is always a *deflecting* force perpendicular to the direction of motion of the charged particle. Finally, the direction of the magnetic force, in contrast to electric and gravitational forces, is *not* in the direction of the field. Instead $\mathbf{F}$ is perpendicular to $\mathbf{B}$.

The SI unit of the magnetic field $\mathbf{B}$ is the N·s/C·m which is named the *tesla* (T):

$$1 \text{ T} = 1 \text{ N·s/C·m}$$

A field of 1 T is a very strong magnetic field. The field between the pole pieces of common permanent magnets is often about 0.1 T. A typical value of the earth's magnetic field encountered on the earth's surface is 5×10^{-5} T. Magnetic fields are often given in terms of a much smaller unit, the *gauss*:

$$1 \text{ gauss} = 10^{-4} \text{ T}$$

Example 1　In Figure 27-3, a particle with charge q travels with velocity $\mathbf{v}$ in the positive x direction through a uniform magnetic field $\mathbf{B}$ in the z direction (out of the page).* What electric field is required so that the Lorentz force on the charged particle is zero?

Solution　Since $\mathbf{v}$ is perpendicular to $\mathbf{B}$, the magnitude of the magnetic force acting on the moving particle is

$$F = qvB \sin \theta = qvB \sin 90° = qvB$$

The direction of the magnetic force on a positive charge is in the negative y direction.

To provide an equal and opposite electric force $q\mathbf{E}$, the electric field should be directed vertically upward and have a magnitude such that $qE = qvB$. Therefore

$$E = vB$$

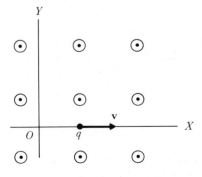

Figure 27-3　Charged particle moving perpendicular to a uniform magnetic field $\mathbf{B}$ directed out of the page.

---

*In diagrams, a magnetic field directed out of the page is indicated by ⊙ corresponding to a head-on view of the arrow representing the $\mathbf{B}$ vector. A magnetic field directed into the page is indicated by × corresponding to the tail of an arrow.

Velocity Selector

Crossed electric and magnetic fields are established in the *velocity selector* shown in Figure 27-4. A beam of charged particles is emitted from the ion source S with a wide range of velocities. The analysis of Example 1 shows that, in these crossed fields, a charged particle experiences electric and magnetic forces which have opposite directions but cancel out only if the particle speed is

$$v = \frac{E}{B}$$

Only particles with this speed can pass without deflection through the crossed fields and emerge from the output slit S_o.

27.2 Circular Motion of a Charged Particle in a Uniform Magnetic Field

Circular Orbits

From the study of the trajectories of charged particles in magnetic fields comes a rich harvest of knowledge about atoms and elementary particles. Figure 27-5 shows a particle with velocity **v** moving perpendicular to a uniform magnetic field **B**. During any very short time interval, the particle moves in the direction of **v**. Since this displacement is perpendicular to the magnetic force, *no work is done by the magnetic force*. Consequently, the magnetic force does *not* change the

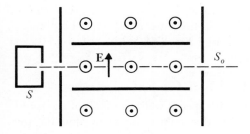

Figure 27-4 Velocity selector with crossed electric and magnetic fields.

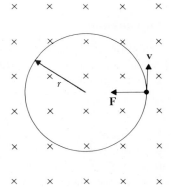

Figure 27-5 A positively charged particle moving perpendicular to a uniform magnetic field is deflected by the magnetic force **F**. The trajectory is a circle of radius r. The magnetic field is directed into the page, as indicated by the ×'s.

kinetic energy of the particle; the particle moves with constant speed. The magnetic force, $F = qvB$, therefore has a constant magnitude if the particle continues to encounter the same value of **B**. While this magnetic force does not alter the magnitude (speed) of the particle's velocity, it does uniformly change the *direction* of the particle's velocity vector. The particle's trajectory is therefore a path with a constant curvature, a circular orbit.

The acceleration vector **a**, according to Newton's second law, is parallel to the magnetic force and is therefore perpendicular to the velocity vector and points directly toward the center of the circular orbit. The acceleration is a centripetal acceleration of magnitude v^2/r where r is the orbit radius. Applied to this motion, Newton's second law,

$$\mathbf{F} = m\mathbf{a}$$

with $F = qvB$ and $a = v^2/r$, gives

$$qvB = \frac{mv^2}{r} \tag{27.3}$$

Measurement of Momentum

Solving Eq. 27.3 for the product mv, we obtain

$$momentum = qBr \tag{27.4}$$

This simple result shows that if a uniform magnetic field **B** is established over a track recorder like a cloud or bubble chamber, then a particle's momentum can be determined from a knowledge of its charge and the radius of curvature r of its track. In this way the momenta involved in elementary particle reactions are measured (Figure 27-6).

When these circular orbits in a magnetic field are analyzed according to *relativistic* dynamics (Chapter 40), we still find Eq. 27.4 holds, but

$$momentum = \frac{mv}{\sqrt{1 - v^2/c^2}}$$

Measurements of orbit radii of particles with speeds well over 99% of the speed of light are in complete agreement with this relativistic expression for the momentum.

Measurement of Charge-to-Mass Ratio, Mass Spectrometers

Equation 27.3 implies that, in a uniform magnetic field **B**, a particle with charge-to-mass ratio q/m travels in its circular orbit with an angular velocity $\omega = v/r$ given by

$$\omega = B\frac{q}{m} \tag{27.5}$$

Remarkably, the angular velocity is independent of the orbit radius. This relationship shows that any particle's charge-to-mass ratio can be determined by measurement of the time $T = 2\pi/\omega$ required to go once around a circular orbit in a known magnetic field.

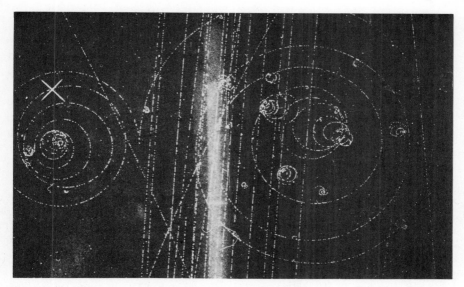

Figure 27-6 Charged particles moving in circular paths under the influence of a magnetic field. This picture shows electron-positron pair production as seen in a bubble chamber. (In the experiment in which this event occurred, the incident beam was composed of antiprotons. One of these antiprotons interacting with the liquid hydrogen in the bubble chamber has produced a neutral particle. This neutral particle, leaving no track, has decayed into an electron and a positron, which have left spiral tracks as they give up their initial energy. Courtesy Brookhaven National Laboratory.)

If an ion's charge q is known, its mass can be calculated after q/m has been measured. The apparatus used to determine the mass of an ion from observations of the ion's orbit in a magnetic field is called a *mass spectrometer*. [In certain mass spectrometers, the mass is determined from measurement of orbit radius and the ion speed (Question S-3) or kinetic energy (Question S-1).] Mass spectrometers not only provide conclusive evidence of the existence of stable and radioactive isotopes (Chapter 41) of almost all elements but also furnish precise measurements of the mass of various isotopes with an accuracy of better than one part in a million.

Example 2 Protons are circulating in a plane that is perpendicular to a uniform magnetic field of 0.400 T. The frequency of revolution is 6.10×10^6 rev/s. Determine the mass of a proton from this observation, assuming that the proton's charge is 1.60×10^{-19} C.

Solution The angular velocity of a circulating proton is

$$\omega = 2\pi(6.10 \times 10^6) \text{ rad/s} = 3.83 \times 10^7 \text{ rad/s}$$

From Eq. 27.5, a proton's mass is

$$m_p = \frac{Bq}{\omega} = \frac{(0.400 \text{ T})(1.60 \times 10^{-19} \text{ C})}{3.83 \times 10^7 \text{ rad/s}} = 1.67 \times 10^{-27} \text{ kg}$$

Particle Accelerators and Magnetic Fields

Laboratory investigation of the creation and interaction of elementary particles requires as incident projectiles a beam of particles which have been accelerated to a high kinetic energy. Consequently, enormous efforts have been directed toward the construction of particle accelerators.

All these accelerators have some common features. The type of particle to be accelerated is chosen to be an electrically *charged* particle such as an electron, a proton, or some heavier nucleus. An *electric field* is always employed to push the charged particle in order to increase its kinetic energy. The acceleration takes place in a vacuum so that the accelerated particle will not fritter away its kinetic energy in collisions with gas molecules.

Acceleration to kinetic energies much higher than a few MeV, without having to resort to extremely high voltages, is achieved by the ingenious use of alternating electric fields. In the linear (straight-line) accelerator at Stanford University, bunches of electrons are pushed along by the electric field of a confined traveling electromagnetic wave (Chapter 35). After "riding the crest of an electromagnetic wave" for about 3 km, the electrons acquire a kinetic energy of 20 GeV.

A different family of accelerators, initiated when E. O. Lawrence (1901–1958) and M. S. Livingston (1905–   ) put into operation the first *cyclotron* in 1932, uses a magnetic field to curve the path of the accelerated beam of particles into a circular or spiral trajectory. In this way a charged particle is made to pass many times between the same electrodes, and the electric field in the same region of space can be used over and over again to accelerate the particle.

The most modern machine of this type is the Fermi National Accelerator Laboratory (Fermilab) *synchrotron* near Batavia, Illinois. After an acceleration to a kinetic energy of 8 GeV in smaller machines, protons are accelerated to their final energy in the main accelerator (Figure 27-7). Here the protons travel through an evacuated tube in an orbit 6 km in circumference. An accelerating electric field increases a proton's kinetic energy by several MeV in one circuit. After 10^5 circuits during a 1.6-s acceleration interval, a proton reaches a final kinetic energy of 200 GeV. The acceleration time is approximately doubled for operation at 500 GeV.

In any accelerator that uses a guiding magnetic field, if the final orbit has a radius r_f and takes place in a magnetic field B_f, the final momentum attained by a particle in the machine is (from Eq. 27.4)

$$\text{final momentum} = q B_f r_f \tag{27.6}$$

This is the basic equation governing the design of all such accelerators. The product $B_f r_f$ determines the maximum momentum and therefore the maximum kinetic energy to which a particle can be accelerated in the machine. To achieve a high energy, a strong magnetic field must be established over an orbit of large radius. This is the reason why high-energy particle accelerators are huge.

Since the total work done by an electric field **E** on a charged particle is zero for any closed path in an *electrostatic* field ($\oint E_s \, ds = 0$), the electric field used to accelerate circulating charged particles *cannot* be an electrostatic field. This acceleration is accomplished by establishing, over certain portions of a particle's

Figure 27-7   Aerial veiw of main accelerator at Fermi National Accelerator Laboratory (Fermilab). The main accelerator is 6 km in circumference, 2 km in diameter. (Courtesy Fermilab.)

orbit, an *alternating* electric field whose direction, parallel to the particle's velocity vector, is reversed with a frequency equal to the frequency of revolution of the particles in the magnetic field. Then particles that are pushed forward by the electric field in one traversal of this region will receive a forward push each time they return after completion of a circular orbit. The accelerating electric field is therefore made to reverse in direction with an angular frequency equal to the angular velocity ω of the circulating particles. This so-called *cyclotron frequency* is

$$\omega = \sqrt{1 - \frac{v^2}{c^2}} B \frac{q}{m} \tag{27.7}$$

an expression that is the modification of Eq. 27.5 required by the special theory of relativity.

Modern synchrotrons, such as the Fermilab accelerator, keep a particle moving in the same circular orbit by increasing the magnetic field as the particle's momentum increases (*momentum* = qBr). The frequency of reversal of the accelerating electric field is correspondingly altered in accordance with Eq. 27.7.

27.3 Magnetic Force and Electric Current

Figure 27-8 shows a simple experiment which demonstrates that a magnetic field* exerts a side thrust on a wire carrying an electric current. The existence of such a force was Faraday's first great discovery in electromagnetism. Magnetic forces are exploited in all commercial electric motors and in many kinds of meters.

---

*To indicate the direction of the magnetic field produced by a magnet, we label one "pole" N (north) and the other S (south). Within the gap between the poles, **B** is directed away from the magnet's north pole toward its south pole.

The force exerted by a magnetic field **B** on a wire carrying a current is the resultant of the magnetic forces exerted on the moving charged particles within the wire. A particle of charge q, drifting along the wire with an average velocity $\mathbf{v}_d$, experiences an average magnetic force $q\mathbf{v}_d \times \mathbf{B}$. If there are n such mobile charges per unit volume, the number of mobile charges in a length dl of a wire with cross section A is $nA\,dl$ and the resultant magnetic force exerted on these charges is

$$\mathbf{F} = (nA\,dl)(q\mathbf{v}_d \times \mathbf{B}) = A\,dl\,\mathbf{J} \times \mathbf{B}$$

since the current density is $\mathbf{J} = nq\mathbf{v}_d$. The current is $I = JA$. Therefore the *magnetic force* **F** *exerted by a magnetic field* **B** *on a length* dl *of a wire carrying a current* I *is*

$$\mathbf{F} = I\,d\boldsymbol{\ell} \times \mathbf{B} \qquad\qquad (27.8)$$

where the direction of the vector $d\boldsymbol{\ell}$ is defined to be the direction of the current density vector (Figure 27-9).

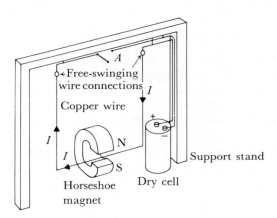

Figure 27-8 Experiment demonstrating the force on a current-carrying conductor in a magnetic field. When switch A is closed, the wire will swing out from the jaws of the magnet.

Figure 27-9 (a) A physical situation in which a wire of length dl carrying a current I would experience a force **F**. (The connecting wires to the length dl from some source of EMF are not shown.) (b) The relative directions of the current element $I\,d\boldsymbol{\ell}$, the magnetic field **B**, and the force **F** on the wire.

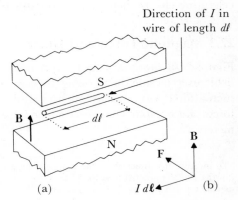

The direction of the magnetic force $I\,d\boldsymbol{\ell} \times \mathbf{B}$ is perpendicular to the wire (a side thrust) and perpendicular to the field $\mathbf{B}$ in the direction given by the right-hand rule of Figure 27-2 with $d\boldsymbol{\ell}$ in the direction of $\mathbf{v}$, the direction of motion of positive charges.

The magnitude of this magnetic force is

$$F = I\,dl\,B\,\sin\theta$$

where θ is the angle between $\mathbf{B}$ and the wire. When $\mathbf{B}$ is perpendicular to the wire, $\sin\theta = 1$ and $F = I\,dl\,B$. On the other hand, if the direction of the wire happens to be parallel to $\mathbf{B}$ so that θ and $\sin\theta$ are both zero, the magnetic force exerted by $\mathbf{B}$ on the wire is zero.

Torque on Current Loop

It is instructive to examine the magnetic forces exerted by a uniform horizontal magnetic field $\mathbf{B}$ on the rectangular loop shown in Figure 27-10. The loop is oriented so that the normal to its plane makes an angle θ with $\mathbf{B}$, and two sides of the loop are vertical. Magnetic forces of equal magnitudes are exerted upward on the top conductor and downward on the bottom conductor. The resultant force and the resultant torque of these forces is zero. The vertical conductors of length l experience magnetic forces in opposite directions (Figure 27-11), each with the magnitude

$$F = IlB$$

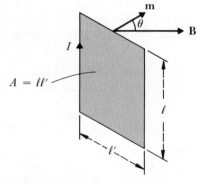

Figure 27-10   Rectangular loop in a horizontal uniform magnetic field $\mathbf{B}$. Sides of length l are vertical; sides of length l' are horizontal. Vector $\mathbf{m}$ is normal to the plane of the coil. The angle between $\mathbf{m}$ and $\mathbf{B}$ is θ.

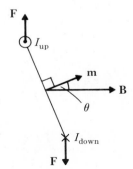

Figure 27-11   Top view of the coil of Figure 27-10. The pair of oppositely directed forces with equal magnitudes F constitutes a couple which has the same torque about any vertical axis.

These forces have components normal to the plane of the coil of magnitude $F \sin \theta$. The sum of the torques of this couple (see Question S-5 of Chapter 5) about any vertical axis is

$$\tau = l'F \sin \theta$$

where l' is the width of the loop. Substituting $F = IlB$ and noting that ll' is the area A of the loop, we find

$$\tau = IAB \sin \theta$$

For a loop with N turns, this becomes

$$\tau = NIAB \sin \theta \tag{27.9}$$

The torque vector $\boldsymbol{\tau}$ is directed vertically downward.

Magnetic Dipole Moment m

A current loop is called a *magnetic dipole* (Figure 27-12). The *magnetic moment* **m** (or the magnetic dipole moment) of a plane loop of any shape is defined to be a vector with:

1   A magnitude

$$m = NIA$$

where the loop has N turns, each carrying a current I and enclosing an area A.
2   A direction normal to the plane of the loop and such that, when the fingers of the right hand curl in the direction of the current, the thumb points in the direction of **m**.

When rewritten in terms of m, Eq. 27.9 is $\tau = mB \sin \theta$ and the magnitude and direction of the torque on the rectangular coil of Figure 27-10 are given by the vector equation

$$\boldsymbol{\tau} = \mathbf{m} \times \mathbf{B} \tag{27.10}$$

A more general analysis shows that this expression for the torque in a uniform field is valid for any magnetic moment **m**, no matter what the shape of the associated current loop.

The torque tends to twist a coil to force its magnetic moment **m** into alignment with **B**. When **m** is parallel to **B** the torque is zero. The torque exerted by

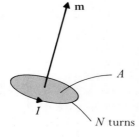

Figure 27-12   A magnetic dipole with a magnetic moment **m**.

a magnetic field on a current loop is the torque which makes the D'Arsonval galvanometer (Figure 26-11) and electric motors function. In an electric motor the direction of **m** is reversed by reversing the coil current, just as the magnetic moment **m** of a rotating coil rotates into alignment with a constant field **B**. Therefore, as the coil continues rotating, it experiences a unidirectional magnetic torque.

A current loop with magnetic moment **m** has potential energy associated with its orientation in a magnetic field **B**. During a rotation which changes the angle between **m** and **B** from θ to $\theta + d\theta$, work of magnitude $\tau\, d\theta$ (from Eq. 12.19) is done against the magnetic torque, and the associated change in potential energy is

$$dU = \tau\, d\theta = mB \sin \theta\, d\theta$$

Integration of this equation gives the potential energy function

$$U = -mB \cos \theta + constant$$

We select $\theta = \pi/2$ as the orientation corresponding to zero potential energy. Then

$$U = -mB \cos \theta \qquad (27.11)$$

As **m** is rotated from alignment with **B** to the opposite direction, the orientational potential energy changes from $-mB$ to $+mB$.

Example 3   In Figure 27-10, the coil is 0.100 m wide and 0.200 m high. The coil current is 10.0 A and there is a horizontal magnetic field of 0.50 T. An external torque is applied that slowly twists the coil so that θ increases from 0° to 180°. Find the maximum magnitude of this torque and evaluate the work done by this torque during the 180° rotation.

Solution   The magnetic moment of the coil is

$$m = IA = (10.0 \text{ A})(0.100 \text{ m} \times 0.200 \text{ m}) = 0.200 \text{ A} \cdot \text{m}^2$$

The torque exerted by the magnetic field on the coil has a magnitude $\tau = mB \sin \theta$. The maximum value of this torque, which is equal to the maximum magnitude of the applied torque, is

$$\tau_{\max} = mB = (0.200 \text{ A} \cdot \text{m}^2)(0.50 \text{ T}) = 0.10 \text{ N} \cdot \text{m}$$

The work W done by the applied torque during the 180° rotation is equal to the increase in the orientational potential energy of the coil:

$$W = mB - (-mB) = 2mB = 2(0.200 \text{ A} \cdot \text{m}^2)(0.50 \text{ T}) = 0.20 \text{ J}$$

27.4 Hall Effect

When a conducting plate with a current density **J** is placed in a magnetic field **B** perpendicular to the plate, an electric field **E** orthogonal to **B** and **J** is established giving rise to a measurable voltage between the edges of the plate.

This electric field is called the Hall field, and its appearance in an applied magnetic field is known as the transverse *Hall effect* (Figure 27-13). The magnetic field **B** deflects the carriers of the current until sufficient charge has accumulated at the edges to create a transverse electric field **E** that exerts an electric force $q\mathbf{E}$ equal in magnitude and opposite in direction to the magnetic force $q\mathbf{v}_d \times \mathbf{B}$. Then

$$qE_y = qv_xB_z$$

where v_x is the x component of the drift velocity $\mathbf{v}_d$ of a mobile particle with charge q. If there are n such mobile particles per unit volume, the x component of the current density is $J_x = nqv_x$. The Hall constant R_H is the ratio defined by

$$R_H = \frac{E_y}{J_xB_z} = \frac{v_xB_z}{nqv_xB_z} = \frac{1}{nq} \tag{27.12}$$

Figure 27-13(a) illustrates an experimental arrangement for the measurement of $E_y, J_x,$ and B_z to determine the Hall constant of a conductor. For most metals

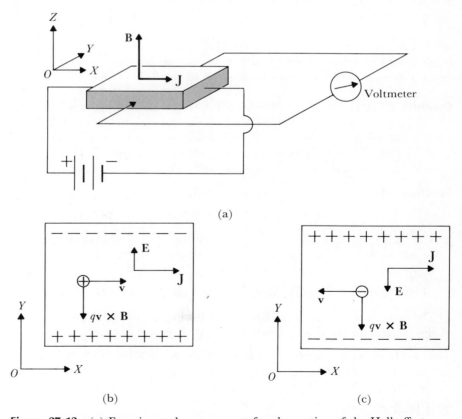

(a)

(b)   (c)

Figure 27-13   (a) Experimental arrangement for observation of the Hall effect. (b) Current carried by positive charges giving a positive Hall constant $R_H = 1/nq$. (c) Current carried by negative charges giving a negative Hall constant.

the Hall constant has a value corresponding to $q = -e$ and a carrier concentration n approximately equal to the concentration of valence electrons in the metal. But there are some surprises. Bismuth has a large Hall constant suggesting a carrier concentration of only 0.004 electrons per atom, and certain metals (e.g., beryllium, zinc, and cadmium) have *positive* Hall coefficients corresponding to current carriers with positive charges! The explanation is provided by the quantum theory of electrons in metals (Chapter 43).

Summary

☐ A charge q moving with a velocity $\mathbf{v}$ experiences a magnetic force

$$\mathbf{F} = q\mathbf{v} \times \mathbf{B}$$

at a point in space where the magnetic field is $\mathbf{B}$.

☐ The magnetic force exerted by a magnetic field $\mathbf{B}$ on a length $d\ell$ of a wire carrying a current I is

$$\mathbf{F} = I\,d\boldsymbol{\ell} \times \mathbf{B}$$

where $d\boldsymbol{\ell}$ is in the direction of the current density vector.

☐ A current loop is called a magnetic dipole. The magnetic moment $\mathbf{m}$ of a current loop is a vector normal to the plane of the loop (in the direction of the thumb when the fingers of the right hand curl in the direction of the current) with a magnitude

$$m = NIA$$

where the loop has N turns, each carrying a current I and enclosing an area A.

☐ The torque exerted by a magnetic field $\mathbf{B}$ on a current loop with a magnetic moment $\mathbf{m}$ is

$$\boldsymbol{\tau} = \mathbf{m} \times \mathbf{B}$$

The potential energy associated with the orientation of $\mathbf{m}$ is

$$U = -mB \cos \theta$$

where θ is the angle between $\mathbf{m}$ and $\mathbf{B}$.

☐ The charge carrier concentration n and the sign of the charge q carried by particles in a current of density J_x in a conductor placed in a transverse magnetic field B_z can be determined by measurement of the transverse electric field E_y, which yields the Hall constant

$$R_H = \frac{E_y}{J_x B_z} = \frac{1}{nq}$$

Questions

1 (a) Is there a magnetic force exerted by a magnetic field on a stationary electric charge?
   (b) Is the direction of the magnetic force exerted by a magnetic field on a moving charge in the same direction as the magnetic field?
   (c) What is the minimum amount of information needed to determine the direction of the magnetic force on a moving charge?

2 (a) A proton is moving horizontally north at a speed of 4.0×10^7 m/s between the pole pieces of a magnet where there is a magnetic field of 1.5 T directed vertically downward. Find the magnitude and direction of the magnetic force exerted on the proton.
   (b) If the proton moves vertically upward, what magnetic force does it experience?

3 Find the magnitude and direction of the magnetic force exerted on an electron moving vertically upward at a speed of 2.0×10^8 m/s by a horizontal magnetic field of 0.50 T which is directed west.

4 From the defining equation for **B**,

$$B = \frac{F}{qv \sin \theta}$$

determine the dimensions of **B**.

5 The equation $F = qvB \sin \theta$ involves the magnitudes of three vectors **F**, **v**, and **B**. Of these three vectors, which pairs are always at right angles? Which pairs may have any angle between them?

6 What would be the magnitude and direction of the magnetic force on the electron in Question 3 if the westerly directed magnetic field were angled at 30° above the horizontal?

7 What would be the magnitude and direction of the magnetic force on a proton moving with the same velocity and through the same magnetic field as the electron in Question 6?

8 Crossed electric and magnetic fields are established over a certain region. The magnetic field is 0.10 T vertically downward. The electric field is 2.0×10^6 V/m in a horizontal direction east. A proton, traveling horizontally northward, experiences zero resultant force from these fields and so continues in a straight line. What is the proton's speed?

9 Electrons travel through a uniform magnetic field at right angles to the field and traverse a complete circle in 10^{-9} s. What is the magnitude of the magnetic field?

10 In a magnetron (designed to generate electromagnetic waves of radar frequencies) an electron moves perpendicular to a magnetic field of 0.10 T. In a uniform field of this magnitude, what is the electron's frequency of revolution in a circular orbit?

11 Show how the mass M of an ion of charge e can be determined from a knowledge of the proton's mass m_p and from measurement of the ratio T/T_p, where T is the time required for one revolution of the ion in a uniform magnetic field, and T_p is the time required for one revolution of a proton in the same magnetic field.

12 (a) What is the time required for a proton to complete a circular orbit in a uniform magnetic field of 0.10 T?
   (b) What is the mass of a chlorine ion that takes 36.9 times as long as a proton to complete one revolution in the magnetic field of part (a)?

13 A beam consisting of a mixture of protons, deuterons, and α particles is accelerated

(essentially from rest) through a voltage V, and it emerges into a region of uniform magnetic field **B**. The particle velocities are perpendicular to **B**. Express the radii of the orbits of the deuterons and the α particles in terms of the radius r of a proton's orbit.

14 **(a)** When a magnetic field is employed in a particle accelerator, does the magnetic force increase the particle's speed?
(b) For what purpose are magnetic fields used in particle accelerators?
(c) What features of the magnet determine the maximum energy to which a particle can be accelerated in a synchroton?

15 A 5.0-A current is directed vertically upward in a straight section of a wire. The wire lies in a uniform horizontal magnetic field of 20 mT directed north. Find the magnitude and direction of the magnetic force exerted by this magnetic field on a 0.06-m section of the vertical wire.

16 A wire 0.60 m long, carrying a current of 3.0 A, is located in a region in which there is a magnetic field of 15 mT. Find the magnitude of the magnetic force on this wire if the angle between the wire and the field **B** is:
(a) $0°$ **(b)** $30°$ **(c)** $45°$ **(d)** $90°$

17 **(a)** What information besides that given in Question 16 is required in order to specify the direction of the magnetic force on the wire?
(b) Is the answer to part (a) different for the case of $30°$ between the wire and **B** than it is for the case of $45°$ between the wire and **B**? Explain.

18 The current density in a horizontal copper wire is 2.0×10^6 A/m². The current is directed from east to west. Find the magnitude and direction of the magnetic field which will exert a magnetic force on the wire sufficient to support the wire. (The density of the copper wire is 8.9×10^3 kg/m³.)

19 A coil has 20 turns each enclosing an area of 1.2×10^{-2} m²:
(a) Find the magnitude and direction of the coil's magnetic moment when the plane of the coil is horizontal and the coil current is 5.0 A directed counter-clockwise (looking downward at the coil).
(b) What is the magnitude of the torque exerted on the coil by a horizontal uniform magnetic field of 0.80 T?

20 **(a)** If the coil of Question 19 is allowed to rotate to the orientation corresponding to minimum potential energy, what is the change in its potential energy?
(b) How much work would have to be done on the coil to twist it from the orientation of minimum potential energy to the orientation of maximum potential energy?

21 Calculate the Hall constant of copper from the following observations. A copper strip 1.0 cm wide and 0.010 cm thick (Figure 27-14) carries a current of 20 A. In a vertical magnetic field of 1.2 T the Hall voltage measured between the edges of the strip is 14 μV.

Figure 27-14

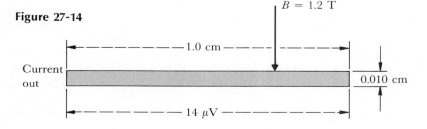

22 The Hall constant for copper is negative. From the data of the preceding question find the number of mobile electrons per unit volume of copper. If the current in Figure 27-14 is directed out of the page, what is the direction of the Hall field within the copper?

23 What would be the Hall constant for a material with equal concentrations of positive and negative mobile charge carriers if we assume that the drift speed is the same for both types of carriers? Explain.

Supplementary Questions

S-1 In a mass spectrometer designed by Dempster (Figure 27-15), ions of mass m and charge q are accelerated from rest through a voltage V and then encounter a uniform magnetic field $\mathbf{B}$. In the magnetic field, an ion moves in a semicircular path of radius r before striking a photographic plate. The ion's mass can be determined by measuring the distance D from the entry slit to the mark on the photographic plate. Show that

$$m = \frac{qB^2}{8V}D^2$$

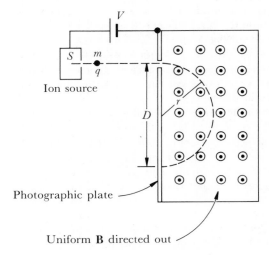

Figure 27-15 Mass spectrometer.

S-2 Consider two helium atoms, one singly ionized and the other doubly ionized, leaving the ion source of Figure 27-15. Compare the diameters of their orbits in the uniform magnetic field.

S-3 In the Bainbridge mass spectrometer (Figure 27-16), a beam of ions from an ion source enters a velocity selector where there is an electric field $\mathbf{E}'$ and an orthogonal magnetic field $\mathbf{B}'$. An ion that passes undeviated through the crossed fields of the velocity selector enters a region of uniform magnetic field $\mathbf{B}$ and traverses a

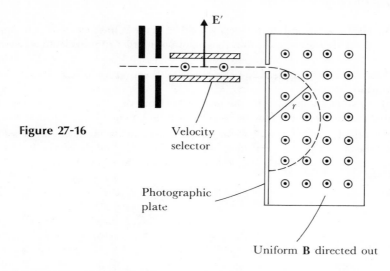

Figure 27-16

Velocity
selector

Photographic
plate

Uniform **B** directed out

circular orbit of radius r before striking a photographic plate. Show that the
charge-to-mass ratio of the ion is

$$\frac{q}{m} = \frac{E'}{rBB'}$$

S-4 In a cyclotron, a large electromagnet is used to establish an almost uniform constant magnetic field over an evacuated region between its circular pole pieces. A
particle starts from a central ion source and moves in a sequence of semicircular
orbits which increase in radius as the ion's speed is increased by the accelerating
electric field between the electrodes (Figure 27-17). In this way the ion eventually
reaches the largest possible orbit (r_1 in Figure 27-17) over which the guiding magnetic field B_1 has been established. Assume that the speed of the accelerated ion
is always small compared to the speed of light so that relativistic corrections to
Newtonian mechanics are negligible. If the accelerated ion has a mass m and a
charge q, find its angular velocity and maximum kinetic energy in the cyclotron
of Figure 27-17.

Figure 27-17 Orbit of a charged
particle in a cyclotron. A uniform
magnetic field directed out of the
page is established over a circular
region of radius r_1.

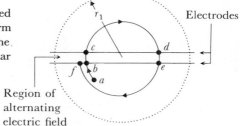

Electrodes

Region of
alternating
electric field

S-5 A cyclotron is to be designed to accelerate protons to an energy of 20 MeV using
a magnetic field of 1.5 T. Determine:
(a) The radius of the largest orbit over which the magnetic field must be
established.

(b) The period T and the frequency $f = 1/T$ of the voltage which is to be applied across the electrodes within the cyclotron in order to provide an accelerating electric field.

S-6   The cyclotron of the preceding question is to be converted to accelerate deuterons instead of protons:

(a) By what factor must the frequency be changed?

(b) What is the maximum energy that will be attained by the deuterons?

S-7   When the Hall effect is used to measure magnetic fields, bismuth is often selected because of its large Hall constant, -6×10^{-7} m^3/C. Find the sensitivity (in volts per tesla per ampere) of a *Hall probe* using a bismuth strip which is 1.0×10^{-6} m thick.

28.

magnetic fields produced by steady currents

The preceding chapter deals with the force exerted by a given magnetic field on a moving charge. To complete the description of magnetic interaction we require a specification of the magnetic field produced at each point in space by a given distribution of sources.

A moving charge is a *source* of a magnetic field at all points of surrounding space. This field is a function of both position and time.

Like other fields, magnetic fields obey a *superposition principle*: The field produced at a given instant at a certain point in space when there are several different sources is the *vector sum* of the fields that each source would individually contribute if it alone were present.

An electric current consists of a stream of moving charges, each of which contributes to the magnetic field in surrounding space. A steady electric current produces, at a given point, a magnetic field vector **B** whose value is independent of the time. Such fields are said to be constant. In this chapter we discuss the physical laws that specify the constant magnetic field produced by a distribution of steady currents.

28.1 The Biot-Savart Law

A convenient representation of magnetic fields is given by drawing *magnetic flux lines*. A magnetic flux line is a line whose tangent at any point is in the direction of the magnetic field at that point (Figure 28-1). An arrow associated with a flux line completes the specification of the direction of the magnetic field.

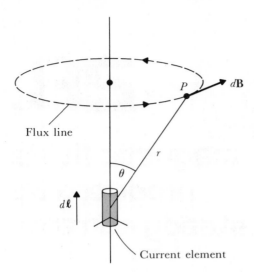

Figure 28-1   Current element $I\,d\boldsymbol{\ell}$ produces the magnetic field $d\mathbf{B}$ at the point P. The tangent to the circular flux line through P gives the direction of $d\mathbf{B}$.

The field $\mathbf{B}$ produced by a steady current in a closed loop can be regarded as a superposition of the contributions $d\mathbf{B}$ furnished by each current element $I\,d\boldsymbol{\ell}$ in the closed loop. We then write

$$\mathbf{B} = \oint d\mathbf{B}$$

The field $d\mathbf{B}$ that is produced by a current element $I\,d\boldsymbol{\ell}$ can be inferred from experimental measurement of the fields produced by closed loops of currents. A flux line in the field produced by a current element is a circle centered on the axis of the element and lying in a plane perpendicular to the element (Figure 28-1). The direction of a flux line is given by a right-hand rule: if the wire is grasped in the right hand with the *thumb* pointing in the direction of the current, the *fingers* will curl about the wire in the direction of the flux lines. The magnitude of the field produced by the current element at the point P is

$$dB = \frac{\mu_0}{4\pi}\frac{I\,d\ell\,\sin\theta}{r^2} \tag{28.1}$$

where θ is the angle between $d\boldsymbol{\ell}$ and the position vector $\mathbf{r}$ of the point P relative to the current element. This specification of the magnitude and the direction (here determined from a flux line) of the contribution $d\mathbf{B}$ produced by a current element $I\,d\boldsymbol{\ell}$ is called the *Biot-Savart* law.

The quantity μ_0 is a constant called the *permeability* constant. It has the value

$$\mu_0 = 4\pi \times 10^{-7}\ \text{T·m/A}$$

exactly, as a consequence of the definition of the ampere (Section 28.2).

Example 1   Find the magnetic field produced at a distance R from an infinitely long straight wire carrying a current I.

Solution   The wire lies along the X axis in Figure 28-2. A representative current element $I\,dx$ produces at the point P a contribution of magnitude

$$dB = \frac{\mu_0}{4\pi} \frac{I\,dx\,\sin\theta}{r^2}$$

The vector $d\mathbf{B}$ is directed into the page. Since the contributions at P of different current elements all have the same direction, the resultant field at P is simply the sum of the magnitudes dB:

$$B = \frac{\mu_0 I}{4\pi} \int \frac{\sin\theta\,dx}{r^2}$$

Two of the three variables x, r, and θ can be expressed in terms of the third. We select θ as the integration variable and use

$$x = -R\cot\theta \qquad dx = R\csc^2\theta\,d\theta \qquad r = R\csc\theta$$

Then

$$B = \frac{\mu_0 I}{4\pi R} \int_0^\pi \sin\theta\,d\theta$$

The integral has the value 2. Therefore

$$B = \frac{\mu_0 I}{2\pi R} \tag{28.2}$$

The magnetic flux lines are concentric circles around the wire. The magnitude of the field falls off as $1/R$.

Example 2   Find the field produced on the axis of a circular coil carrying a current I.

Figure 28-2   Determination of the magnetic field produced at P by a current I in an infinitely long straight wire along the X axis.

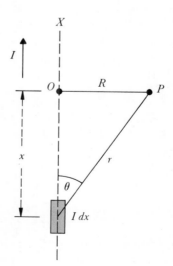

Solution   In Figure 28-3 the representative element contributes at P a field of magnitude

$$dB = \frac{\mu_0}{4\pi} \frac{I\,d\ell\,\sin\,90°}{r^2}$$

The symmetry of the current distribution implies that the resultant field at P will have only a z component. The z component of $d\mathbf{B}$ is

$$dB_z = dB \cos \alpha = \frac{a}{r} dB = \frac{\mu_0 I a\,d\ell}{4\pi r^3}$$

Therefore the magnitude of the field at P is

$$B = \oint dB_z = \frac{\mu_0 I a}{4\pi r^3} \oint d\ell$$

where, since r has the same value for all elements of the ring, it can be taken outside the integral. The total length of the ring $\oint d\ell$ is its circumference $2\pi a$. Various useful expressions for the final result are

$$B = \frac{\mu_0 I a^2}{2r^3} = \frac{\mu_0 I a^2}{2(a^2 + z^2)^{3/2}} = \frac{\mu_0 I}{2a} \sin^3 \beta \qquad (28.3)$$

the last being obtained using $\sin \beta = a/r$. Calculation of the field at points off the axis of the coil is much more difficult. Flux lines of this field are shown in Figure 28-4.

Example 3   A magnetic field stronger than that produced by a single coil can be obtained by winding wire on a cylinder to form a close-packed helix called a solenoid (Figure 28-5). Find the field on the axis of a solenoid which has n turns per unit length, each carrying a current I.

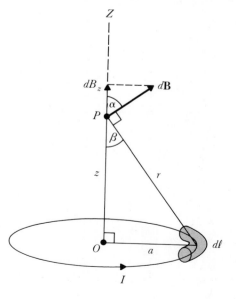

Figure 28-3   Determination of $\mathbf{B}$ at the point P on the axis of a circular coil.

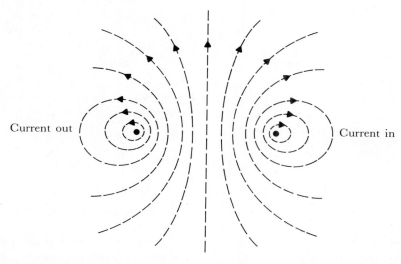

Current out

Current in

Figure 28-4   Magnetic flux lines of the magnetic field produced by a current in a circular coil. The plane of the page is perpendicular to the plane of the coil and bisects the coil.

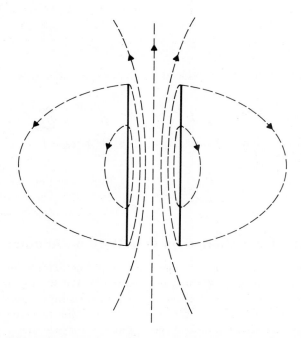

Figure 28-5   Magnetic flux lines in the field produced by a current in a solenoid.

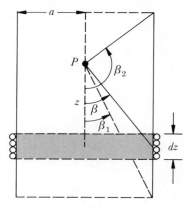

Figure 28-6   Determination of **B** at the point P on the axis of a solenoid.

Solution   To a very good approximation, a thin section of the solenoid of length dz (Figure 28-6) is equivalent to a coil carrying a current $nI\,dz$. From Example 2, the field dB contributed by this coil at the point P is given by

$$dB = \mu_0 \frac{nI\,dz}{2a} \sin^3 \beta$$

Selecting β as the variable of integration we have

$$z = -a \cot \beta \qquad \text{and} \qquad dz = a \csc^2 \beta\,d\beta$$

and therefore

$$dB = \mu_0 \frac{nI}{2} \sin \beta\,d\beta$$

The field at the point P produced by the entire solenoid is

$$B = \mu_0 \frac{nI}{2} \int_{\beta_1}^{\beta_2} \sin \beta\,d\beta = \frac{\mu_0 nI}{2}(\cos \beta_1 - \cos \beta_2) \qquad (28.4)$$

For an infinitely long solenoid, $\beta_1 = 0$, $\beta_2 = \pi$, and the expression for the magnetic field reduces to

$$B = \mu_0 nI \qquad \text{(infinitely long solenoid)} \quad (28.5)$$

28.2 Magnetic Force Exerted by One Wire on Another

The assertion that one moving charge exerts a magnetic force on another moving charge is made at the beginning of Chapter 27. We are now in a position to examine such a magnetic interaction, at least for the simple case of two distinct steady streams of moving charged particles, constituting two currents I_1 and I_2, in long straight parallel sections of wires which are a distance r apart (Figure 28-7).

Figure 28-7 Sections of two long straight parallel wires. The magnetic field **B** produced by current I_1 at the location of current I_2 exerts a force **F** on a length l of the wire carrying the current I_2.

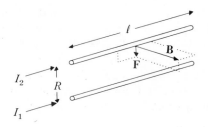

Proceeding by two distinct steps, we first evaluate the magnetic field produced by the current I_1 at the location of the second current. This field, from Eq. 28.2, has a magnitude

$$B = \frac{\mu_0 I_1}{2\pi R}$$

and is perpendicular to the second wire, as shown in Figure 28-7. Next, according to Eq. 27.8, this field exerts a force on the length l of the wire carrying the current I_2, given by

$$F = I_2 l B = \frac{\mu_0 I_1 I_2 l}{2\pi R} \tag{28.6}$$

This force, perpendicular to **B** and to the second wire, is, according to the right-hand rule of Figure 27-2, directed toward the first wire. The magnetic force is attractive when the currents are in the same direction.

Calculation of the magnetic field produced by I_2 at the location of I_1, and of the force then exerted by this field on I_1, leads again to an attractive force given by Eq. 28.6.

Definition of the Ampere

Measurement of the magnetic force between wires carrying a current gives a precise and reproducible laboratory method of establishing the size of the SI unit of current, the ampere. The size of the ampere is selected so that the constant $\mu_0/2\pi$ in Eq. 28.6 has a value of exactly 2×10^{-7} N/A^2. In other words, *one ampere is defined as the steady current in two long straight parallel wires one metre apart in empty space that causes each wire to experience a force of exactly 2×10^{-7} newton per metre of length.*

In practical procedures to standardize the ampere, the force between coils of wire a few centimetres apart is measured in an instrument called a *current balance*.

28.3 Ampere's Line Integral Law

When the electric field **E** is independent of time, the line integral of **E** around any closed curve is zero: $\oint E_s \, ds = 0$. It is interesting to examine the analogous expression for magnetic fields. The line integral of **B** around a closed path is denoted by $\oint B_s \, ds$, where B_s is the component of **B** in the direction of the path of integration.

Figure 28-8(a) shows a path of integration *abcda* in the magnetic field of magnitude $B = \mu_0 I/2\pi R$ produced by a current I in a long straight wire. The line integral of **B** along the straight path from a to b is zero because, since **B** is perpendicular to this path, the component B_s in the direction of the path is zero. Along the curved path from b to c, the field **B** has a constant magnitude $\mu_0 I/2\pi R_2$ and points in the direction opposite to the direction of the path of integration. Therefore $B_s = -\mu_0 I/2\pi R_2$ on this path and

$$\int_b^c B_s \, ds = -\frac{\mu_0 I}{2\pi R_2} \int_b^c ds = -\frac{\mu_0 I}{2\pi} \frac{\text{arc } bc}{R_2} = -\frac{\mu_0 I \theta}{2\pi}$$

where θ is the angle subtended by the arc from b to c [Figure 28-8(a)]. Proceeding in this way for each portion of the path, we obtain

$$\oint B_s \, ds = 0 + \left(-\frac{\mu_0 I \theta}{2\pi}\right) + 0 + \frac{\mu_0 I \theta}{2\pi} = 0$$

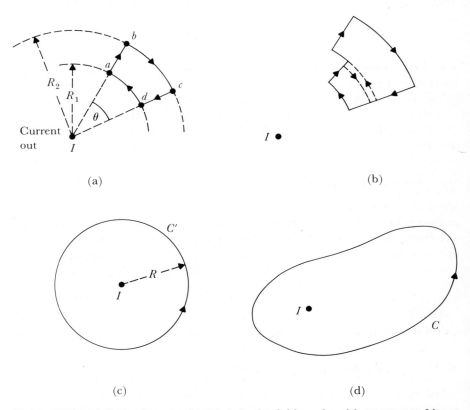

(a)   (b)

(c)   (d)

Figure 28-8 (a) Path of integration *abcda* in the field produced by a current I in an infinitely long straight wire. (b) Path of integration is composed of radial segments and arcs. (c) Path of integration is a circle with the wire at its center. (d) C is a path of integration such that the wire passes through any surface bounded by C. Then we say that C and the current I are linked.

From this result it follows that the line integral is also zero for a path that can be constructed out of radial segments and arcs, such as the path in Figure 28-8(b) (Question S-4). Since *any* path can be approximated by a sequence of radial segments and arcs, we conclude that

$$\oint B_s \, ds = 0$$

for any path that does not link the wire.

A path C' that links the wire is shown in Figure 28-8(c). On this circular path the field is everywhere in the direction of the path and has the constant magnitude $\mu_0 I / 2\pi R$. Therefore

$$\oint_{C'} B_s \, ds = \frac{\mu_0 I}{2\pi R} \oint_{C'} ds = \mu_0 I$$

since $\oint_{C'} ds$ is the circumference $2\pi R$ of this circular path of integration. The same result is obtained for any path C linking the wire (Question S-5).

The general conclusion suggested by these results is known as *Ampere's line integral law*: For any closed path of integration C in a field produced by steady currents

$$\oint_C B_s \, ds = \mu_0 I \qquad \text{(steady currents)} \qquad (28.7)$$

where I is the net current linking the path of integration.* If there are several currents $I_1, I_2, \ldots, I_n$ linking C,

$$I = I_1 + I_2 + \cdots + I_n$$

A current in this algebraic sum is positive if it passes through C in the direction of the thumb when the fingers of the right hand are directed along the path of integration; currents in the opposite direction are negative. Although we have examined Ampere's law only for the field produced by a current in a long straight wire, it can be proved valid in any field established in accordance with the Biot–Savart law.

Ampere's law is a fundamental law relating time independent fields and steady currents. When the current distribution has sufficient symmetry, the associated magnetic field can be determined most easily using this law.

Example 4   Find the magnetic field produced by a uniform steady current I parallel to the axis of a circular cylinder of radius R_0.

Solution   Figure 28-9 represents a cross section of the cylinder with the current directed outward from the plane of the paper. The symmetry of the current distribution implies that the flux lines are concentric circles centered on the cylinder axis and that, at any point, B depends only on the distance R from the axis. For a path of integration along a flux line of radius R

$$\oint B_s \, ds = \oint B \, ds = B \oint ds = B 2\pi R$$

---

*The net current linking the path of integration is the net current passing through any surface bounded by this path.

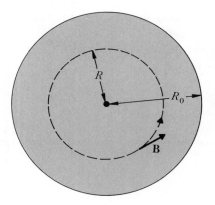

Figure 28-9 Cross section of a cylindrical conductor carrying a current outward.

For $R > R_0$, a circular path of integration of radius R links the entire current I and Ampere's law gives

$$B2\pi R = \mu_0 I \qquad \text{or} \qquad B = \frac{\mu_0 I}{2\pi R} \qquad (R > R_0)$$

This shows that at points outside the cylinder, the magnetic field is the same as if all the current were concentrated at the axis.

For $R < R_0$, a circular path of integration of radius R links only the fraction $\pi R^2 / \pi R_0^2$ of the current in the cylinder. Ampere's law yields

$$B2\pi R = \frac{\mu_0 R^2}{R_0^2} I$$

Therefore

$$B = \frac{\mu_0 I R}{2\pi R_0^2} \qquad (R < R_0)$$

Example 5　Find the magnetic field produced by a current I in a toroidal winding of N uniformly distributed closely spaced turns (Figure 28-10).

Solution　We neglect the circumferential component of the current and assume that each turn of the windings is normal to the circumference. Then the symmetry of the current distribution implies that the flux lines are concentric circles centered on the toroid axis and that the magnitude of **B** depends only on the distance R from this axis. Ampere's line integral law gives

$$2\pi R B = 0$$

for path 1 because no current links this path. The same result is obtained for path 2 since each turn carries equal currents in opposite directions through a surface bounded by this path, and the *net* current linked by the path is therefore zero. These results show that $B = 0$ everywhere outside the toroid. The magnetic field is confined to the region within the windings, and in this region, applying Ampere's line integral law to a circular path of integration of radius R, we obtain

$$2\pi RB = \mu_0 NI$$

since each turn carries a current I in the same direction through a surface bounded by this path; the return current for each coil is not included because it does not link this path. Consequently the field inside the toroid is

$$B = \frac{\mu_0 NI}{2\pi R} \qquad (28.8)$$

If variations in the length of the circumference are neglected, this reduces to

$$B = \mu_0 nI \qquad (28.9)$$

where $n = N/2\pi R$ is the number of turns per unit length. This field is the same as the field at the center of a long solenoid.

28.4 Magnetic Flux

Flux and Flux Density

The definition of magnetic flux is analogous to the definition of electric flux given in Chapter 22. The magnetic flux $d\Phi$ through an element dA of a surface is defined to be

$$d\Phi = B_N\, dA \qquad (28.10)$$

where B_N is the value of the normal component of **B** that occurs on the surface

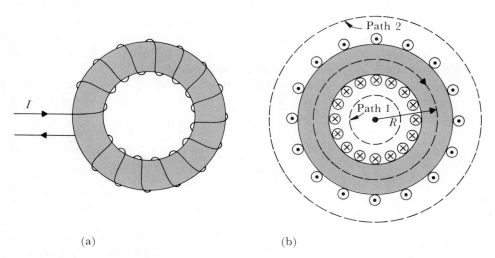

(a)   (b)

Figure 28-10   (a) Toroidal winding. (b) The net current linking paths 1 and 2 is zero. The path within the toroid links a net current NI.

element. If dA is part of a closed surface, B_N is the component in the direction of the *outward* normal to the surface. The SI unit of magnetic flux is called the *weber* (Wb).

From $d\Phi = B_N \, dA$, it follows that

$$B = \frac{d\Phi}{dA_N} \tag{28.11}$$

where $d\Phi$ is the magnetic flux through an element of area dA_N perpendicular to **B**. In keeping with this expression for B as the magnetic flux per unit normal area, the magnetic field **B** is often called the *flux density* and is measured in *webers per square metre* (1 T = 1 Wb/m^2).

The magnetic flux through a surface can be visualized as a quantity proportional to the number of magnetic flux lines that pass through the surface. The magnetic field (flux density) is strong where the flux lines are dense or crowded together; where the field is weaker the flux lines are further apart.

Gauss's Law for Magnetic Fields

The magnetic fields in the examples of this chapter have a characteristic found in all magnetic fields that have been investigated experimentally: *A flux line is a closed curve in space.* Therefore every flux line that enters a closed surface S must also leave it. The mathematical expression of this experimental fact is

$$\Phi_S = \oint_S B_N \, dA = 0 \tag{28.12}$$

for any closed surface S. This is *Gauss's law for magnetic fields.*

It can be proved that any time-independent magnetic field established by steady currents in accordance with the Biot–Savart law must satisfy Gauss's law. However, Gauss's law is not subject to the restrictions of the Biot–Savart law. Even time-dependent magnetic fields satisfy Gauss's law. Equation 28.12 is a fundamental law of electromagnetism. If we start with Gauss's law and Ampere's law as the basic equations to be satisfied by magnetic fields produced by steady currents, then we can deduce the Biot–Savart law.

Gauss's law implies that the same flux passes through any two different surfaces bounded by the same closed path C (Figure 28-11). This flux is often called the flux *enclosed by C.*

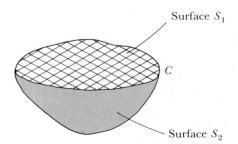

Surface S_1

C

Surface S_2

Figure 28-11   Surfaces S_1 and S_2 are both bounded by the same closed curve C. Gauss's law implies that the same flux passes through any two such surfaces. Therefore we may speak of "the flux enclosed by C" without referring to a particular surface bounded by C.

Magnetic Poles and Gauss's Law

A theory of magnetism can be developed by postulating that a magnetic field has sources called magnetic poles which are related to the magnetic field in the same way that electric charge is related to the electric field. We assume that, analogous to positive and negative electric charges, there are two types of magnetic poles, north poles and south poles. A magnetic field **B** exerts on a pole a force proportional to **B** and to the *pole strength*. This force is in the direction of **B** for a north pole, and in the opposite direction for a south pole. A needle would behave like a compass needle if it had an excess of north poles near the north-seeking end and an excess of south poles near the other. A stationary isolated north pole is the source of a magnetic field directed radially outward with a magnitude falling off inversely as the square of the distance. The field produced by a south pole is similar but directed inward. A magnetic field like that observed outside a bar magnet could be produced by an excess of north poles near one end and an excess of south poles near the other.

An isolated magnetic pole (sometimes called a monopole) is one of the most long sought-after objects in the universe. None were detected until, in a 1975 experiment using a cosmic-ray track recorder, W. Z. Osborne, L. S. Pinsky, P. B. Price, and E. K. Shirk found a track that they interpret as that of a magnetic monopole. At the moment of writing, physicists are seeking independent corroboration.

In ordinary matter, however, it appears that the only source of a magnetic field is electric charge in motion. The magnetic fields produced by moving charges satisfy Eq. 28.12. Now, analogous to Gauss's law for an electric field,

$$\oint_S E_N \, dA \propto net \; electric \; charge \; within \; S$$

Gauss's law for the magnetic field produced by magnetic poles is

$$\oint_S B_N \, dA \propto net \; magnetic \; pole \; strength \; within \; S$$

Evidently Gauss's law for magnetic fields,

$$\oint_S B_N \, dA = 0$$

summarizes the experimental evidence about the *sources* of magnetic fields in ordinary matter and states that these sources are not *isolated magnetic poles*.

Summary

☐   The Biot–Savart law states that when there is a steady current I in a closed loop, a current element $I \, d\ell$ produces at a point P a magnetic field of magnitude

$$dB = \frac{\mu_0}{4\pi} \frac{I \, d\ell \sin \theta}{r^2}$$

where θ is the angle between $d\ell$ and the position vector **r** of the point P relative to the current element. The magnetic flux lines are concentric circles about the current element, with the direction of the flux lines the same as that of the fingers of the right hand when the thumb, at the location of the current element, is in the direction of the current.

☐ The SI unit of current, the ampere, is defined (in terms of the force exerted by one current-carrying wire on another) to have a magnitude such that the permeability constant μ_0 is exactly $4\pi \times 10^{-7}$ N/A^2.

☐ Ampere's line integral law states that, for any closed path of integration C in a time-independent magnetic field produced by steady currents,

$$\oint_C B_s \, ds = \mu_0 I$$

where I is the net current linked by C.

☐ Gauss's law for magnetic fields states that the net magnetic flux through any closed surface S is zero; that is

$$\oint_S B_N \, dA = 0$$

Questions

1 In Figure 28-12, the current element produces a magnetic field of 3.0 mT directed into the page at point P_1. Find the field at the points P_2, P_3, and P_4.

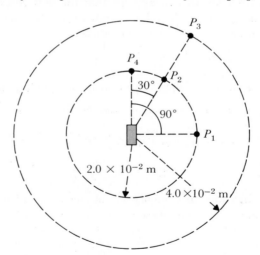

Figure 28-12

2 A current of 20 A is directed upward in a long vertical wire. Find the magnitude and direction of the magnetic field produced by this current at a point 2.0×10^{-2} m east of the wire.

3 Show that the Biot–Savart law implies that the current I in the finite length of straight wire shown in Figure 28-13 contributes at the point P a magnetic field of magnitude

$$B = \frac{\mu_0 I}{4\pi R}(\cos \theta_1 - \cos \theta_2)$$

4 Using the result of the preceding question, show that the field at the center of a square coil which has sides of length ℓ and carries a current I is

$$B = \frac{2\sqrt{2}\mu_0 I}{\pi \ell}$$

5 An infinitely long wire is bent through an angle of 90° as shown in Figure 28-14. Find the magnetic field produced by the current I at the point P.

6 Find the magnetic field produced at the center of a circular coil of 10-cm radius carrying a current of 5 A.

7 Find the magnetic field at the center P of the semicircular loop of radius a, when the long wire in Figure 28-15 carries a current I.

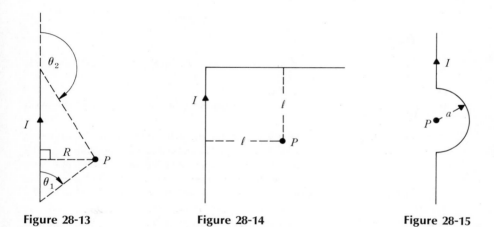

Figure 28-13　　　　　　　　**Figure 28-14**　　　　　　　　**Figure 28-15**

8 What current is required to produce a magnetic field of 10 mT at the center of a long solenoid with 10 turns per centimetre?

9 Show that, on the axis of a very long solenoid, the magnetic field at the ends is half as large as it is at the center.

10 A solenoid 16 cm long with a radius of 6.0 cm is wound uniformly with 2000 turns of wire. The coil current is 3.0 A. Find the value of B on the axis:
   (a) At the center.
   (b) At an end.

11 (a) Show that, when currents in two parallel wires have opposite directions, the wires repel one another.
   (b) Two long straight parallel wires, separated by a distance of 0.50 m, repel one another with a force of 0.10 N/m of length when they carry currents of equal magnitude in opposite directions. Find the current.

12 Find the resultant force exerted on the rectangular coil of Figure 28-16.

13 The permittivity constant has the approximate value

$$\epsilon_0 = 8.85 \times 10^{-12} \text{ C}^2/\text{N}\cdot\text{m}^2$$

and the permeability constant has the value

$$\mu_0 = 4\pi \times 10^{-7} \text{ T}\cdot\text{m/A}$$

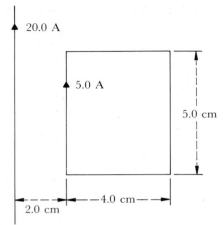

Figure 28-16

(a) Which of these values is determined by experimental measurement? Which is a consequence of definitions?

(b) What are the dimensions of $(\mu_0\epsilon_0)^{-1/2}$?

(c) Compare the value of $(\mu_0\epsilon_0)^{-1/2}$ with the value of c, where c denotes the speed of light in a vacuum.

14 A coaxial cable consists of a long solid conducting cylinder of radius R_1 supported by insulating disks on the axis of a thin-walled conducting tube of radius R_2. The inner conductor carries a current I in one direction and the outer conductor carries an equal current in the opposite direction. Find the magnetic field created by these currents at a point a distance R from the axis for $R < R_1$, $R_1 < R < R_2$, and $R > R_2$. Sketch a graph showing B as a function of R.

15 A hollow cylindrical conductor with inner radius R_i and outer radius R_o has uniform current density parallel to its axis:

(a) Show that the magnetic field due to a current I is given by

$$B = \frac{\mu_0 I (R^2 - R_i^2)}{2\pi(R_o^2 - R_i^2)} \frac{1}{R}$$

at a point a distance R from the axis such that $R_i < R < R_o$.

(b) Find B for $R < R_i$ and for $R > R_o$.

16 A thin toroid with a circumference of 0.40 m is wound uniformly with 3000 turns of wire. Find the magnetic field within the toroid when the current is 2.0 A.

17 Figure 28-17 shows a portion of an infinite plane conducting sheet with a uniform current density directed out of the page. The current within a width W is $I = I'W$,

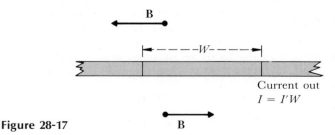

Figure 28-17

where I' is the current per unit width of the sheet. Show that the magnetic field produced by this current sheet has a magnitude

$$B = \frac{\mu_0 I'}{2}$$

18 A large circular coil of radius a carries a current I. Find the magnetic flux enclosed by a very small square coil with sides of length ℓ, when the small coil is positioned at the center of the large coil and oriented so that the angle between the planes of the two coils is:
(a) $0°$ **(b)** $90°$ **(c)** $30°$

19 The cross-sectional area of the toroid of Question 16 is 0.50 cm². Find the magnetic flux enclosed by one turn of the coil winding.

20 Find the magnetic flux through the rectangular coil of Figure 28-18 when the long straight wire carries a current I.

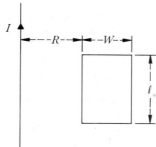

Figure 28-18

Supplementary Questions

S-1 Find the magnetic field at the center P of the circular arc of radius a, when the long wire in Figure 28-19 carries a current I.

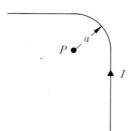

Figure 28-19

S-2 A charge Q is uniformly distributed over the surface of an insulating circular disk of radius R. The disk is rotated about its axis with an angular speed ω. Show that the magnetic field at the center of the disk is

$$B = \frac{\mu_0 \omega Q}{2\pi R}$$

S-3   Helmholtz coils shown in Figure 28-20 are two identical circular coils which carry the same current and have a separation equal to their radius a. With this arrangement, there is a large region of space centered at P throughout which the magnetic field is approximately uniform. Show that at P both dB/dz and d^2B/dz^2 are zero.

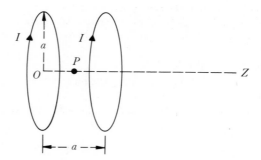

Figure 28-20

S-4   Show that $\oint B_s\, ds$ for the path of integration indicated by the solid line in Figure 28-8(b) is equal to the sum of two line integrals, each taken about a path of integration of the type indicated in Figure 28-8(a).

S-5   In Figure 28-21, the path of integration consisting of C, C', and the radial segments is a closed path which does not link the wire. Use this observation to show that

$$\oint_C B_s\, ds = -\oint_{C'} B_s\, ds = \mu_0 I$$

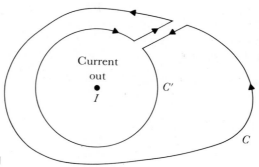

Figure 28-21

S-6   Use Ampere's line integral law to find the magnetic field at any point inside an infinite solenoid with n turns per unit length, each carrying a current I. Assume that the solenoid is equivalent to a stack of turns with planes normal to the solenoid axis. (Effects of the longitudinal component of the current are then neglected.) Show that the field inside the solenoid has a uniform value B_{in} and that the field outside has a uniform value B_{out}. By considering the path of integration $abcda$ in Figure 28-22, show that

$$B_{in} - B_{out} = \mu_0 n I$$

Compare this with the calculations of Example 3 and show that

$$B_{out} = 0 \quad \text{and} \quad B_{in} = \mu_0 n I$$

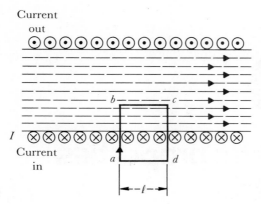

Figure 28-22

S-7   **(a)** If the current density in a wire is $\mathbf{J} = nq\mathbf{v}_d$, show that the Biot–Savart law can
be written as

$$d\mathbf{B} = \frac{\mu_0}{4\pi}\frac{q\mathbf{v}_d \times \mathbf{r}}{r^3}n\,dV$$

where $\mathbf{r}$ is the position vector of the observation point P relative to the current
element, dV is the volume of the current element, and n is the number of
mobile particles (with charge q) per unit volume.

(b) Show that the preceding result can be interpreted as stating that a particle of
charge q moving with a velocity $\mathbf{v}$ produces at the point P a magnetic field

$$\mathbf{B} = \frac{\mu_0}{4\pi}\frac{q\mathbf{v} \times \mathbf{r}}{r^3}$$

(This expression for the magnetic field produced by a moving charge is a good
approximation for speeds v small compared to the speed of light.)

S-8   At speeds small compared to the speed of light the electric field produced by a
moving charge q is given, to a good approximation, by the same expression as
that for the electric field produced by a stationary charge:

$$\mathbf{E} = \frac{q}{4\pi\epsilon_0}\frac{\mathbf{r}}{r^3}$$

Using the result of Question S-7, show that the electric field and the magnetic
field produced by a moving charge are related by

$$\mathbf{B} = \mu_0\epsilon_0\mathbf{v} \times \mathbf{E}$$

29.

magnetic properties of matter

Magnetic fields are produced by electric currents. When matter is present the magnetic field at any point is generally the superposition of the magnetic fields produced by two types of currents:

1 *Bound currents* arising from circulating charges in atoms and molecules as well as from electron spin.
2 *Free currents*, such as the familiar conduction currents in wires, that are due to the drift of mobile charges.

The general theory of the preceding chapter relates the magnetic field **B** to *all* the currents that are present. This theory will now be applied to develop an efficient method of accounting for the average magnetic effects of bound currents.

29.1 Ampere's Line Integral Law and the Magnetic Intensity H

The Magnetization M

We shall picture matter as containing a distribution of bound currents consisting of tiny current loops. Consider a piece of matter containing n current loops with the kth loop having a current I_k that encloses an area A_k (Figure 29-1).

The magnetic moment of this loop has a magnitude

$$m_k = I_k A_k$$

The magnetic moment of a system containing magnetic moments $\mathbf{m}_1$, $\mathbf{m}_2$, . . . , $\mathbf{m}_n$ is defined as the vector sum $\sum_{k=1}^{n} \mathbf{m}_k$ of all the magnetic moments of the system. A basic physical quantity for the description of the magnetic state of a piece of matter is the *magnetic moment per unit volume* **M**, called the *magnetization*. To define **M** at a given point in a material, we consider a volume element dV that includes the point and is macroscopically small but microscopically large. Then the magnetization is defined by

$$\mathbf{M} = \frac{\sum_{k=1}^{n} \mathbf{m}_k}{dV} \tag{29.1}$$

where $\mathbf{m}_1$, $\mathbf{m}_2$, . . . , $\mathbf{m}_n$ are the magnetic moments within dV (Figure 29-2).

Reformulation of Ampere's Line Integral Law

In the presence of matter, since both bound currents and free currents are generally present, Ampere's line integral law states that for any closed path C,

$$\oint_C B_s \, ds = \mu_0[(\textit{net free current linked}) + (\textit{net bound current linked})]$$

In magnetic systems the only current that can be controlled and easily measured is the free current. We therefore wish to reformulate Ampere's law so that explicit reference to bound currents is eliminated. We shall see that this can be accomplished by averaging over the violent microscopic irregularities to give a

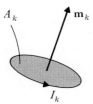

Figure 29-1 The kth current loop or magnetic dipole. The magnetic moment $\mathbf{m}_k$ has a magnitude $m_k = I_k A_k$.

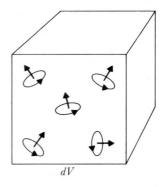

Figure 29-2 The macroscopically small volume element dV is large enough to contain millions of molecules.

macroscopic description of the magnetic state of matter. The vector **M** is a macroscopic quantity. We now associate with each point in space a *macroscopic magnetic field* **B** which is the average of the detailed microscopic field over a small but macroscopic volume. For the remainder of this chapter **B** denotes this macroscopic magnetic field.

Relationship between M and Bound Current Linkages

To investigate the relationship between **M** and the bound current linked by a path of integration, we consider for simplicity a path of integration which is a flux line. We also assume that all the magnetic moments within the matter are aligned with **B**. Then the bound current linked by the length ds of the path of integration shown in Figure 29-3 is

$$I_2 + I_4 + \cdot \cdot \cdot$$

where a given current loop within the volume element $dV = A \, ds$ contributes to this sum if and only if the path passes through the loop. The value of this sum obviously depends on the precise location of the path of integration. Let us therefore consider all paths through the small macroscopic area A which are parallel to ds. We can compute the *average* value of the net bound current linked by these paths. A fraction A_k/A of these paths will pass through a current loop of area A_k. The contribution of this current loop to the average net current linked by a path is $I_k(A_k/A)$. Therefore the average value of the net current linked by paths through A and parallel to ds is

$$I_1 \frac{A_1}{A} + I_2 \frac{A_2}{A} + \cdot \cdot \cdot + I_n \frac{A_n}{A} = \frac{\sum\limits_{k=1}^{n} m_k}{A} = \frac{M \, dV}{A} = M \, ds$$

It follows that the net bound current linked by the *closed* path C, averaged over a family of parallel paths in the vicinity of C, has the value $\oint_C M \, ds$. For an arbitrary closed path and an arbitrary orientation of the magnetic moments,

Figure 29-3   Within the volume element $dV = A \, ds$, the path of integration links the current loops I_2 and I_4.

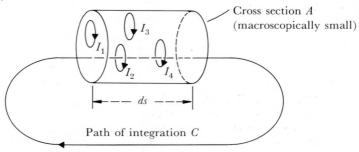

Cross section A
(macroscopically small)

Path of integration C

the same arguments lead to the general relationship,

$$average \ net \ bound \ current \ linked = \oint_C M_s \, ds \qquad (29.2)$$

where M_s is the component of **M** in the direction of the path of integration.

With this result Ampere's line integral law, expressed in terms of the macroscopic field **B** and the macroscopic quantity **M**, is

$$\oint_C B_s \, ds = \mu_0(net \ free \ current \ linked + \oint_C M_s \, ds) \qquad (29.3)$$

The Magnetic Intensity H

Rewriting the last equation, we obtain

$$\oint_C \left(\frac{B_s}{\mu_0} - M_s\right) ds = I_f$$

where I_f is the net free current linked by the path of integration. This simple relationship provides motivation for introducing a new macroscopic quantity, the vector called the *magnetic intensity* **H**, defined by

$$\mathbf{H} = \frac{\mathbf{B}}{\mu_0} - \mathbf{M} \qquad (29.4)$$

Ampere's line integral law, in terms of **H**, is

$$\oint_C H_s \, ds = I_f \qquad (29.5)$$

The line integral of **H** is determined entirely by the net *free* current I_f passing through the path of integration C.

The SI unit of **H** is the ampere per metre (A/m).

Example 1   A toroid with an iron core is formed by winding N turns of wire uniformly on a thin uniform iron ring with a circumference of $2\pi R$ (Figure 29-4). The iron is initially unmagnetized (**M** = 0). A current I is established in the wire and the magnetic field **B** within the iron is measured. Find the

Figure 29-4   Toroid with an iron core.

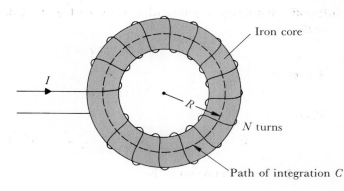

Iron core

I

R

N turns

Path of integration C

magnetic intensity **H** and the magnetization **M** within the iron core of this toroid if $N = 500$ turns, $2\pi R = 0.400$ m, $I = 800$ mA, and $B = 1.58$ T.

Solution   The symmetry of the current distribution in the wire and the symmetry of the distribution of the iron imply that the magnetic flux lines within the core will be circles concentric with the core and that **B**, **H**, and **M** will be parallel and have a constant magnitude along a flux line. Selecting a circular flux line within the core as the path of integration C, we have

$$\oint_C H_s \, ds = H \oint_C ds = H 2\pi R$$

Since each turn of the wire coil carries a free current I through the path of integration, the net free current I_f passing through the path C is given by

$$I_f = NI$$

Ampere's line integral law for **H** gives

$$H 2\pi R = NI$$

Therefore

$$H = \frac{NI}{2\pi R} = \frac{400 \text{ A}}{0.400 \text{ m}} = 1000 \text{ A/m}$$

Since $\mathbf{M} = \mathbf{B}/\mu_0 - \mathbf{H}$, and the directions of **B** and **H** are the same, the magnitude of the magnetization is

$$M = \frac{B}{\mu_0} - H = \frac{1.58 \text{ T}}{4\pi \times 10^{-7} \text{ T} \cdot \text{m/A}} - 1000 \text{ A/m}$$

$$= 1.26 \times 10^6 \text{ A/m} - 1000 \text{ A/m} = 1.26 \times 10^6 \text{ A/m}$$

M is in the direction of **B**.

Example 2   The current in the coils of the toroid in Example 1 is reduced to zero. Measurements show that the magnetic field in the iron core decreases by 1.03 T from 1.58 T to 0.55 T. Since the magnetic field **B** persists without any coil current, we say that the iron is permanently magnetized and that the iron ring is a permanent magnet. Find **H** and **M** within the iron of this permanent magnet.

Solution   Symmetry, together with Ampere's line integral law for **H**, implies that

$$H 2\pi R = NI = 0$$

Therefore

$$\mathbf{H} = 0$$

Then

$$M = \frac{B}{\mu_0} = \frac{0.55 \text{ T}}{4\pi \times 10^{-7} \text{ T} \cdot \text{m/A}} = 4.4 \times 10^5 \text{ A/m}$$

and **M** is in the direction of **B**.

29.2 Magnetic Susceptibility and Permeability

The magnetic properties of an isotropic material are characterized by a dimensionless parameter called the susceptibility χ of the material defined by

$$\mathbf{M} = \chi\mathbf{H} \tag{29.6}$$

or by the magnetic permeability μ, or the relative permeability κ_m, defined by

$$\kappa_m = \frac{\mu}{\mu_0} = 1 + \chi \tag{29.7}$$

These definitions, together with $\mathbf{H} = \mathbf{B}/\mu_0 - \mathbf{M}$, imply that

$$\mathbf{B} = \mu\mathbf{H} \tag{29.8}$$

In a vacuum, $\mathbf{M} = 0$ and $\mathbf{B} = \mu_0\mathbf{H}$. Consequently, a vacuum has a permeability μ_0, a relative permeability of exactly 1, and a susceptibility of zero.

Most materials can be classified according to their magnetic properties as diamagnetic, paramagnetic, or ferromagnetic. Representative values of χ for diamagnetic and paramagnetic materials are given in Table 29-1.

Diamagnetism

The magnetic susceptibility of a diamagnetic material is negative with a magnitude much less than 1, and the value is a constant independent of the magnetic field $\mathbf{B}$ within the material. This implies that within a diamagnetic material, the magnetic moment per unit volume $\mathbf{M}$ is proportional to $\mathbf{B}$ but in the opposite direction.

Diamagnetism arises because an external magnetic field $\mathbf{B}$ applied to any atom induces in the atom a magnetic moment $\mathbf{m}$ in the direction opposite to $\mathbf{B}$. The physical law involved is Faraday's law of electromagnetic induction, a topic of the next two chapters. Diamagnetism exists in all materials, but the induced magnetic moments are so small that their effects are masked if the atoms of the material have a permanent magnetic moment.

Paramagnetism

Paramagnetic materials have a magnetic susceptibility that is positive and much less than 1. The molecules, atoms, or ions within a paramagnetic material have

Table 29-1   Magnetic Susceptibilities at Room Temperature

Diamagnetic substance	χ	Paramagnetic substance	χ
Hydrogen (1 atm)	-2.1×10^{-9}	Oxygen (1 atm)	2.1×10^{-6}
Carbon (diamond)	-2.2×10^{-5}	Aluminum	2.3×10^{-5}
Copper	-1.0×10^{-5}	Tungsten	6.8×10^{-5}
Silver	-2.6×10^{-5}	Titanium	7.1×10^{-5}
Mercury	-3.2×10^{-5}	Platinum	3.0×10^{-4}

a permanent magnetic moment. An external field **B** exerts a torque on these magnetic moments which tends to twist them into alignment with **B**. The material acquires a magnetization **M** in the direction of **B**. Because of thermal agitation, thermodynamic equilibrium at temperature T is obtained with only partial alignment. Experiments show that the susceptibility of a paramagnetic substance in weak magnetic fields decreases with increasing temperature according to *Curie's law*,

$$\chi \propto \frac{1}{T} \tag{29.9}$$

At a given temperature, as the magnetic field is increased, the permanent magnetic moments become more perfectly aligned. For sufficiently large magnetic fields, the alignment is essentially complete and **M** attains its maximum or *saturation* value.

Although paramagnetic effects are generally stronger than diamagnetic effects, both are extremely weak. The alteration in a magnetic field produced by the presence of such materials is less than 0.01%. These materials are therefore commonly classified as "nonmagnetic" and are treated as magnetically equivalent to a vacuum for most engineering calculations.

Ferromagnetism

In the ferromagnetic materials (iron, cobalt, nickel, gadolinium, dysprosium, and certain alloys of these and other elements) atoms with permanent magnetic moments interact in such a way that tiny but microscopically visible *magnetic domains* are formed, each domain consisting of from 10^{17} to 10^{21} atoms with their magnetic moments aligned. At external points, a magnetic domain produces a strong magnetic field, since each of its aligned atoms furnishes aligned contributions. A specimen of ferromagnetic material normally contains billions of magnetic domains. When the specimen is unmagnetized, although the alignment of atomic magnetic moments within each domain is nearly perfect, different domains are aligned in different directions at random [Figure 29-5(a)]. If an external magnetic field is applied to an unmagnetized ferromagnetic material, domain walls shift and domains rotate to produce a magnetization **M** aligned with the applied field [Figure 29-5(b)].

The existence of domains is crucial for ferromagnetic behavior. The interaction responsible for domain formation is a quantum-mechanical phenomenon called the exchange interaction which favors parallel orientation of the magnetic moments of ferromagnetic atoms. At what is called the *Curie temperature* for a substance, thermal agitation is great enough to offset the exchange interaction and the domains disappear. Above its Curie temperature a substance is paramagnetic rather than ferromagnetic.

An unmagnetized ferromagnetic material, placed in a magnetic field, becomes magnetized and thereby makes a substantial alteration in the magnetic field that would otherwise be present, typically increasing the field by a factor of a thousand at points within or near the material. Permanent magnets retain the alignment of their different domains. Other "softer" ferromagnetic materials tend to revert to random domain alignment when a magnetizing field is removed.

Figure 29-6 indicates the complicated relationship between B and H in a ferromagnetic material. If the sample is initially unmagnetized and H is steadily increased from zero, a B-H graph called the *magnetization curve* is obtained. This is the graph from state 0 to state 1 in Figure 29-6. The permeability,

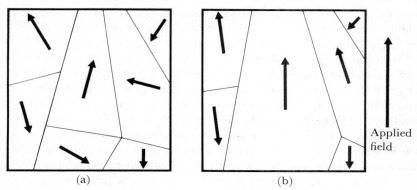

Figure 29-5 Magnetization of a ferromagnetic specimen. Domain walls shift and domains rotate to produce alignment with the applied magnetic field. If the applied field is sufficiently strong, these processes continue until complete alignment is achieved. The material is then magnetically saturated. (a) Unmagnetized. (b) Magnetized, but not to saturation.

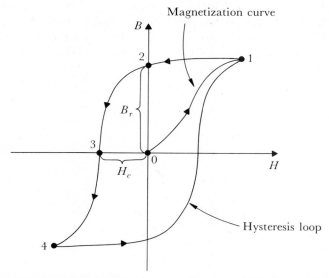

Figure 29-6 Magnetization curve (0 to 1) and hysteresis loop (1 to 2 to 3 to 4 to 1) for a ferromagnetic material.

$\mu = B/H$, is not constant. A typical ferromagnetic material, annealed iron, has a relative permeability $\kappa_m = \mu/\mu_0$ which has an initial value of 3×10^2, a maximum value of 5×10^3, and a limiting value of 1 (the value for a vacuum) as H approaches infinity.

With the sample magnetized in state 1, reduction of H does not result in B values lying on the magnetization curve. As H decreases B decreases, but along a different curve. When H has been decreased to zero (state 2), a magnetic field still remains (as in Example 2). This magnetic field B_r is called the *remanence*. Continuing with changes in the same direction, the field **H** is now established in the reverse direction, and B continues to decrease reaching the value zero at state 3. The corresponding magnetic intensity H_c is called the *coercive force*. With further change in the same direction we reach state 4, where **B** and **H** both have directions opposite to their directions in state 1. If we now reverse the direction of change of **H** we trace out the lower curve back to the original state 1. This closed *B-H* curve is called a *hysteresis loop* and the phenomenon that the magnetization curve is not retraced is referred to as *hysteresis*. The cause of hysteresis has been traced to the fact that domain boundaries, instead of shifting freely when **H** and **M** alter, tend to become stuck at crystal imperfections.

Hysteresis makes possible the existence of permanent magnetism. A good material for a permanent magnet should have both a large remanence B_r so that the magnet will be strong and a large coercive force H_c so that the field will not be greatly reduced by modest values of reverse magnetic intensity.

Because of hysteresis, the *B-H* relationship in a ferromagnetic material is always dependent on the history of the material. Such material has a memory, a fact that is exploited in magnetic tapes.

A special procedure is required to demagnetize a sample of ferromagnetic material, such as that considered in Example 1. Instead of the coil's current simply being cut off, this current is reversed several times while its magnitude is steadily reduced. The *B-H* curve for the sample is then a sequence of hysteresis loops which shrink toward the point $B = 0$, $H = 0$ (Figure 29-7).

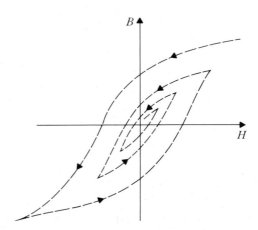

Figure 29-7 *B-H* curve for the demagnetization procedure is a sequence of shrinking hysteresis loops.

29.3 Magnetic Circuit

Figure 29-8 is the schematic diagram of an electromagnet. A current I in the N-turn coil gives rise to a strong magnetic field B_g in the air gap.

The magnetic flux lines are closed curves which link the coils. Most of the flux lines are confined to the "magnetic circuit" determined by the ferromagnetic material. However, there is some leakage flux as shown in Figure 29-8. A flux line in the air meets the surface of a ferromagnetic material almost at right angles (Question S-3).

In the design of an electromagnet, we wish to be able to calculate B_g in terms of NI and the geometry of the magnetic circuit. This formidable problem is rendered tractable if we make several approximations. We neglect leakage flux. Then there is the same flux Φ through any cross section of the magnetic circuit. If the magnetic circuit consists of several sections, with the kth section having a uniform cross section A_k, a length ℓ_k, and a permeability μ_k, the magnetic intensity within the section is

$$H_k = \frac{B_k}{\mu_k} = \frac{\Phi}{\mu_k A_k}$$

The analysis is based on Ampere's line integral law for **H**. For a path of integration C that is a flux line around the magnetic circuit,

$$\oint_C H_s \, ds = H_1 \ell_1 + H_2 \ell_2 + \cdot \cdot \cdot + H_n \ell_n$$

and the free current passing through C is given by

$$I_f = NI$$

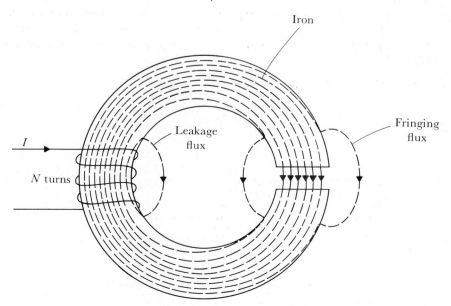

Figure 29-8 Electromagnet.

Therefore Ampere's line integral law implies

$$H_1\ell_1 + H_2\ell_2 + \cdots + H_n\ell_n = NI$$

This gives

$$\Phi\left(\frac{\ell_1}{\mu_1 A_1} + \frac{\ell_2}{\mu_2 A_2} + \cdots + \frac{\ell_n}{\mu_n A_n}\right) = NI \qquad (29.10)$$

This basic equation for a magnetic circuit determines Φ in terms of NI and the parameters of the magnetic circuit. Then the field at any cross section A can be calculated from $B = \Phi/A$.

A useful analogy between magnetic and electric circuits is shown in Table 29-2. Current in an electric circuit is analogous to magnetic flux in a magnetic circuit since, for both quantities, the net value through any closed surface is zero. The quantity $\ell/\mu A$ is called the magnetic *reluctance* $\mathcal{R}$ of a portion of the magnetic circuit. This quantity is analogous in form to an electrical resistance $\ell/\sigma A$ and it is easy to show that reluctances in series and in parallel combine like resistances. The product NI, called the *magnetomotive force* $\mathcal{M}$, plays a role in establishing the magnetic flux in a magnetic circuit analogous to that of the electromotive force in establishing a current in an electric circuit. Using this terminology the basic magnetic circuit equation can be written as

$$\mathcal{M} = \Phi\mathcal{R} \qquad (29.11)$$

which is similar to the loop equation for an electric circuit, $\mathcal{E} = IR$.

Example 3   In the electromagnet of Figure 29-8, the coil has 300 turns, the cross section of the iron is 5.0 cm^2, and the average length of the flux lines within the iron portion of the magnetic circuit is 50 cm. The gap length is 1.0 mm. Find the current required to produce a magnetic flux of 2.5×10^{-4} Wb through this magnetic circuit. The magnetization curve for the iron is known.

Solution   The flux density B in the iron is

$$B = \frac{\Phi}{A} = \frac{2.5 \times 10^{-4} \text{ Wb}}{5.0 \times 10^{-4} \text{ m}^2} = 0.50 \text{ T}$$

Table 29-2 Analogous Quantities in Magnetic and Electric Circuits

Electric circuit	Magnetic circuit
Electric current I	Magnetic flux Φ
Resistance $R = \dfrac{\ell}{\sigma A}$	Reluctance $\mathcal{R} = \dfrac{\ell}{\mu A}$
Electromotive force $\mathcal{E}$	Magnetomotive force $\mathcal{M}$
Loop equation $\mathcal{E} = IR$	Circuit equation $\mathcal{M} = \Phi\mathcal{R}$

From the magnetization curve for this iron, we determine the relative magnetic permeability κ_m when B is 0.50 T. Suppose this is $\kappa_m = 5000$. Then $\mu = \kappa_m \mu_0 = 6.28 \times 10^{-3}$ T·m/A. The reluctance of the iron portion of the circuit is

$$\mathscr{R}_{\text{iron}} = \frac{\ell}{\mu A} = \frac{0.50 \text{ m}}{(6.28 \times 10^{-3} \text{ T·m/A})(5.0 \times 10^{-4} \text{ m}^2)} = 1.6 \times 10^5 \text{ A/Wb}$$

If we neglect fringing at the air gap and assume that the effective cross section of the gap is the same as the cross section of the iron, the reluctance of the gap is

$$\mathscr{R}_{\text{gap}} = \frac{\ell_{\text{gap}}}{\mu_0 A} = \frac{1.0 \times 10^{-3} \text{ m}}{(4\pi \times 10^{-7} \text{ T·m/A})(5.0 \times 10^{-4} \text{ m}^2)} = 16 \times 10^5 \text{ A/Wb}$$

Notice that the reluctance of this short gap is ten times as great as the reluctance of the iron portion of the circuit. The total reluctance of this magnetic circuit is

$$\mathscr{R} = \mathscr{R}_{\text{gap}} + \mathscr{R}_{\text{iron}} = 18 \times 10^5 \text{ A/Wb}$$

The magnetomotive force required is

$$NI = \Phi \mathscr{R} = (2.5 \times 10^{-4} \text{ Wb})(18 \times 10^5 \text{ A/Wb}) = 4.50 \times 10^2 \text{ ampere turns}$$

Therefore the required coil current is

$$I = \frac{4.50 \times 10^2 \text{ A}}{300} = 1.5 \text{ A}$$

Summary

☐ The magnetization **M** at a point P within a material is the magnetic moment per unit volume defined by

$$\mathbf{M} = \frac{\sum\limits_{k=1}^{n} \mathbf{m}_k}{dV}$$

where $\mathbf{m}_1, \mathbf{m}_2, \ldots, \mathbf{m}_n$ are the magnetic moments of the bound current loops within a macroscopic volume element dV that includes the point P.

☐ If **B** is the average value of the magnetic field throughout dV, the magnetic intensity **H** at point P is the macroscopic quantity

$$\mathbf{H} = \frac{\mathbf{B}}{\mu_0} - \mathbf{M}$$

☐ Ampere's line integral law for **H** is

$$\oint_C H_s \, ds = I_f$$

where I_f is the net free current passing through the closed curve C.

☐   The magnetic susceptibility χ of a material is defined by

$$\mathbf{M} = \chi\mathbf{H}$$

and its permeability μ by

$$\mathbf{B} = \mu\mathbf{H}$$

The relative permeability is $\kappa_m = \mu/\mu_0 = 1 + \chi$.

☐   Magnetic matter is classified as follows:
Diamagnetic: κ_m slightly less than 1.
Paramagnetic: κ_m slightly greater than 1.
Ferromagnetic: κ_m much greater than 1 and not constant.
The *B-H* relationship is given by a magnetization curve which is not linear and by a hysteresis loop.

☐   In a magnetic circuit with flux Φ, magnetomotive force $\mathscr{M} = NI$, and total reluctance $\mathscr{R} = l_1/\mu_1 A_1 + l_2/\mu_2 A_2 + \cdots + l_n/\mu_n A_n$, the basic circuit equation is

$$\mathscr{M} = \Phi\mathscr{R}$$

Questions

1   An electron of mass m_e travels with a constant speed v in a circle of radius r:
(a) Show that the current associated with this motion is

$$I = \frac{e}{2\pi}\frac{v}{r}$$

(b) Show that the angular momentum of the electron about the center of its orbit is

$$L = m_e vr$$

(c) Show that this current loop has a magnetic moment $\mathbf{m}$ of magnitude

$$|\mathbf{m}| = \frac{e}{2m_e}L$$

(d) Evaluate the magnetic moment corresponding to an angular momentum $L = \hbar$. This magnetic moment,

$$\frac{e\hbar}{2m_e} = 9.27 \times 10^{-24} \text{ A·m}^2$$

is called the *Bohr magneton*. It is typical of the magnetic moments possessed by atoms and molecules.

2   When iron is magnetized, a typical value of the magnitude of the magnetization $\mathbf{M}$ is 1.7×10^6 A/m. If this magnetization is attributed to aligned magnetic moments of spinning electrons, each with a magnetic moment of one Bohr magneton [Question 1(d)], how many aligned electrons are there per unit volume? How many per iron atom? (Iron has an atomic weight of 56 and a density of 7.8×10^3 kg/m^3.)

3   A thin toroidal coil with N turns carrying a current I has a uniform iron core with a magnetic permeability μ. Find expressions for $\mathbf{B}$ and $\mathbf{M}$ within the core, in terms of μ, NI, and the circumference $2\pi R$ of the thin toroid.

4 A thin long straight wire carries a current I through an infinite uniform medium with a magnetic permeability μ. Find **H**, **B**, and **M** at a distance R from the wire.

5 A thick cylindrical iron wire of radius R_0 and permeability μ carries a current I. The space surrounding the wire is a vacuum. Find expressions for **H**, **B**, and **M** at a distance R from the axis of the cylinder when:
(a) $R < R_0$ **(b)** $R > R_0$

6 At what temperature will the translational kinetic energy of a molecule of a paramagnetic substance be equal to the energy required to rotate the molecule from an orientation with its magnetic dipole moment **m** aligned with **B** to an orientation with **m** in the opposite direction? Assume that $|\mathbf{m}|$ is one Bohr magneton [Question 1(d)] and that $B = 1.5$ T.

7 Find the coil current required to maintain the same flux as that in Example 3, if the length of the air gap is reduced to 0.50 mm.

8 The electromagnet of Example 3 is altered by filling the gap with a cylinder having a 2.0-cm^2 cross section and a relative permeability of 150. The iron near the gap is then tapered to this cross section and almost all the flux in the iron passes through the cylinder:
(a) Find the coil current that produces the same flux as that in Example 3.
(b) What is the flux density in the cylinder?

9 Consider a magnetic circuit that provides two parallel paths (with reluctances $\mathscr{R}_1$ and $\mathscr{R}_2$) for the magnetic flux. Show that the reluctance $\mathscr{R}_p$ of this combination of paths is given by

$$\frac{1}{\mathscr{R}_p} = \frac{1}{\mathscr{R}_1} + \frac{1}{\mathscr{R}_2}$$

Supplementary Questions

S-1 By considering the line integral of **H** around a small rectangle of arbitrarily small width (bc in Figure 29-9), show that, at the interface between two different media,

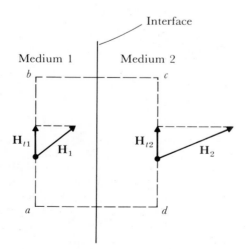

Figure 29-9

the *tangential component of* **H** *is continuous*; that is

$$\mathbf{H}_{t1} = \mathbf{H}_{t2}$$

in Figure 29-9.

S-2 By applying Gauss's law for magnetic fields to a closed cylindrical surface similar to that of Figure 24-2, show that, at the interface between two different media, the *normal component of* **B** *is continuous*; that is

$$\mathbf{B}_{N1} = \mathbf{B}_{N2}$$

in Figure 29-10.

S-3 **(a)** Using the results of Questions S-1 and S-2, show that, at the interface between two media, a flux line bends as shown in Figure 29-11, so that

$$\frac{\tan\,\theta_1}{\tan\,\theta_2} = \frac{\mu_1}{\mu_2}$$

(b) Show that at an iron–air interface, since μ_{iron} is much greater than μ_{air}, the flux line in the air near the iron is almost perpendicular to the iron surface.

S-4 When the current in the electromagnet of Example 3 is reduced to zero we find that the iron ring is permanently magnetized, the field in the gap being 0.10 T. Assume that the magnetization **M** within the iron ring is parallel to **B** and has a constant magnitude:

(a) Find **H** and **M** in the gap.

(b) Show that, within the iron, **H** has a direction opposite to that of **B**. Find H.

(c) What is the magnitude of the magnetization **M** within the iron?

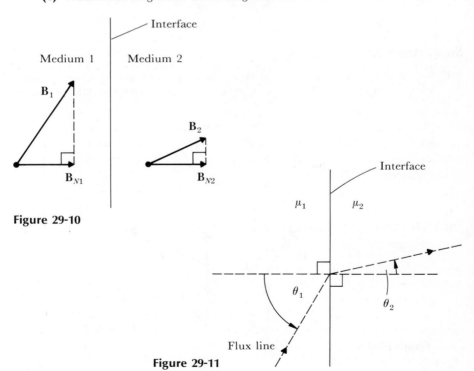

Figure 29-10

Figure 29-11

30.

time-varying fields, maxwell's equations

The laws of electricity and magnetism for *time-independent* fields are discussed in previous chapters. We now introduce two new basic physical laws which express the interconnection between time-varying electric and magnetic fields. These laws, together with Gauss's laws for electric and magnetic fields, constitute a set of equations that are the foundation of classical electromagnetic theory. This is the basic theory for much of physics (and many of the remaining chapters), particularly topics such as alternating current circuits and the study of electromagnetic waves.

30.1 The Electric Field Accompanying a Time-Varying Magnetic Field

In a region where a magnetic field is *changing*, there is an electric field which is related to the *time rate of change* of the magnetic field.

Suppose, for example, that the current in the coils of the electromagnet of Figure 29-8 is increasing. Then the magnetic field in the gap is increasing. The associated electric field in a plane perpendicular to this magnetic field is shown in Figure 30-1. This electric field at any point is proportional to the time rate of change of the magnetic field.

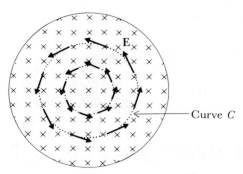

Figure 30-1 The electric field **E** (represented by arrows) associated with a *changing* magnetic field **B**. The magnetic field (represented by the X's) is directed into the page and is of *increasing* magnitude.

Curve *C*

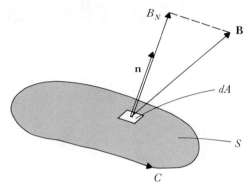

Figure 30-2 With the thumb through *C* in the direction of **n** the fingers of the right hand curl around *C* in the positive sense for the path of integration.

The fundamental relationship between the electric field **E** and the time rate of change $\partial \mathbf{B}/\partial t$ of the magnetic field* which is inferred from the results of a tremendous variety of experiments is Faraday's law

$$\oint_C E_s \, ds = - \int_S \frac{\partial B_N}{\partial t} \, dA \qquad (30.1)$$

where S is any (open) surface bounded by an arbitrary closed curve C. B_N is the component of **B** along the normal **n** to S, where the direction of **n** is related to the positive direction for the path of integration C as shown in Figure 30-2.

Notice that Faraday's law implies that the fundamental relationship of electrostatics,

$$\oint_C E_s \, ds = 0$$

is true if and only if **B** is not time-dependent.

---

*The magnetic field **B** is a function of position (x,y,z) and time t: that is, $\mathbf{B} = \mathbf{B}(x,y,z,t)$. The time rate of change of **B** at a given point (x,y,z) in space is the partial derivative $\partial \mathbf{B}/\partial t$ which is the derivative with respect to t when the values of the position coordinates are held constant.

The surface integral in Faraday's law can be interpreted in terms of the magnetic flux enclosed by C:

$$\Phi = \int_S B_N \, dA$$

The rate of change of this flux associated with time variations of **B** is

$$\frac{\partial \Phi}{\partial t} = \frac{\partial}{\partial t} \int_S B_N \, dA = \int_S \frac{\partial B_N}{\partial t} \, dA \tag{30.2}$$

Therefore Faraday's law can be written

$$\oint_C E_s \, ds = -\frac{\partial \Phi}{\partial t} \tag{30.3}$$

The direct experimental evidence for Faraday's law and its manifold practical applications are emphasized in the following chapter. In this chapter we wish to keep attention focused on the fundamental relationships between the electric and magnetic fields, one of which is Faraday's law, Eq. 30.1.

Example 1   Find the electric field associated with the changing magnetic field of Figure 30-1.

Solution   Faraday's law and the symmetry of the problem imply that the electric field is, as indicated in the figure, tangential to a centered circular path C of radius R and has a constant magnitude E on this path. Then

$$\oint_C E_s \, ds = E2\pi R$$

The average value B_{avg} of the flux density through this circular path enclosing a magnetic flux Φ is defined by

$$B_{\text{avg}} = \frac{\Phi}{area} = \frac{\Phi}{\pi R^2}$$

Faraday's law implies that the induced electric field has a magnitude given by

$$E2\pi R = \frac{\partial \Phi}{\partial t} = \pi R^2 \frac{dB_{\text{avg}}}{dt}$$

or

$$E = \tfrac{1}{2} R \frac{dB_{\text{avg}}}{dt} \tag{30.4}$$

An electron, projected so as to travel through the perpendicular magnetic field in a circular orbit C, will be continually accelerated by this electric field. Such an electric field is exploited in a *betatron*, a modern electron accelerator (Figure 30-3). An electron gains kinetic energy at a rate sufficient to maintain an orbit of constant radius in the increasing magnetic field, if the magnetic field at the orbit at any instant is equal to $\tfrac{1}{2}B_{\text{avg}}$ (Question S-1). In a typical

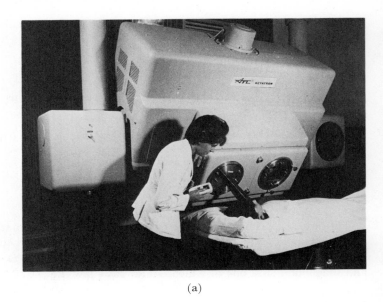

(a)

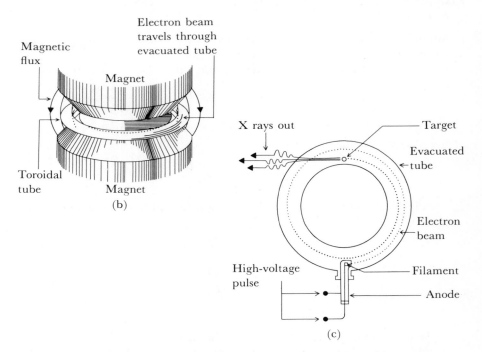

Figure 30-3 The betatron. (a) A 25-MeV betatron in use for x-ray therapy. (Courtesy ATC Betatron Corporation, formerly the Allis-Chalmers Betatron Department.) (b) Diagram of principal betatron components viewed from the side. (c) The toroidal tube showing electron beam as it might be used to produce x rays by striking a solid target.

betatron of the type used in medicine, electrons circulate within an evacuated toroidal tube many thousands of times, gaining energies up to 50 MeV. If the coil current of the betatron electromagnet alternates with a frequency of 180 Hz, the electron acceleration to its final energy is accomplished in $\frac{1}{720}$ of a second, during which time the magnetic field increases from zero to its maximum value.

Example 2   A single turn circular coil is placed at the location of path C in Figure 30-1. The coil encloses an area of 0.40 m². The average value of the magnetic field within the coil increases at a uniform rate from 0.10 T to 0.30 T in 5.0×10^{-2} s. Find the line integral of **E** about the path C, $\oint_C E_s \, ds$.

Solution   Taking the counterclockwise direction as the positive direction for the path C, and using the right-hand rule of Figure 30-2, we find that the direction of **n** in Figure 30-1 is out of the page. B_N and the magnetic flux Φ enclosed by C are consequently negative. Initially

$$\Phi = -0.10 \text{ T} \times 0.40 \text{ m}^2 = -4.0 \times 10^{-2} \text{ Wb}$$

and finally

$$\Phi = -0.30 \text{ T} \times 0.40 \text{ m}^2 = -12.0 \times 10^{-2} \text{ Wb}$$

Therefore

$$\frac{\partial \Phi}{\partial t} = -\frac{8.0 \times 10^{-2} \text{ Wb}}{5.0 \times 10^{-2} \text{ s}} = -1.6 \text{ Wb/s}$$

Faraday's law gives

$$\oint_C E_s \, ds = -\frac{\partial \Phi}{\partial t} = 1.6 \text{ V}$$

30.2 The Magnetic Field Accompanying a Time-Varying Electric Field

We have learned that the steady-state equation

$$\oint_C E_s \, ds = 0$$

must be replaced by Faraday's law

$$\oint_C E_s \, ds = -\int_s \frac{\partial B_N}{\partial t} \, dA$$

when there is a time-varying magnetic field.

We now investigate the steady-state equation given by Ampere's law,

$$\oint_C \frac{1}{\mu_0} B_s \, ds = I$$

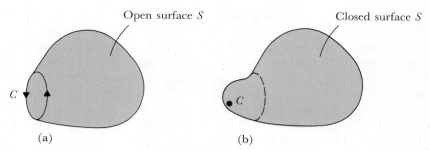

Figure 30-4   (a) S is an open surface bounded by the closed path C. (b) When the path C shrinks to a point, the surface S becomes closed.

We shall show that this equation is inconsistent with the law of conservation of charge when there are time-varying fields. In Ampere's law, the current I is the net current through any surface S bounded by the closed curve C. If we let the closed curve C shrink to a single point (Figure 30-4), the surface S bounded by C becomes a *closed* surface, and Ampere's law reduces to

$$0 = I_S \tag{30.5}$$

for an arbitrary closed surface S. However, since the net current outward through a closed surface is the rate at which electric charge is transported across this surface out of the region inside S, the law of conservation of electric charge implies that

$$I_S = -\frac{\partial Q_S}{\partial t} \tag{30.6}$$

where Q_S is the net charge inside S. Therefore Ampere's law and the law of conservation of charge are consistent only if the net charge within every closed surface remains unchanged as time elapses.

    In 1865, James Clerk Maxwell (1831–1879) published a proposed generalization of Ampere's law that could be valid under all circumstances. Maxwell's theory uses Gauss's law for electric fields

$$\oint_S \epsilon_0 E_N \, dA = Q_S$$

from which $\partial Q_S/\partial t$ can be evaluated:

$$\oint_S \epsilon_0 \frac{\partial E_N}{\partial t} \, dA = \frac{\partial Q_S}{\partial t} \tag{30.7}$$

A quantity I' is defined by

$$I' = I + \int_S \epsilon_0 \frac{\partial E_N}{\partial t} \, dA \tag{30.8}$$

where I is the current through the surface S.* If S is a *closed* surface, Eq. 30.6 and

---

*Although the term $\int_S \epsilon_0(\partial E_N/\partial t) \, dA$ does not represent a flow of electric charge, it does have the dimensions of a current, and it is traditionally called the *displacement current*.

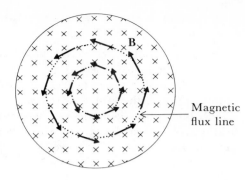

Figure 30-5 The magnetic field **B** (represented by arrows) associated with a *changing* electric field **E**. The electric field (represented by the ×'s) is directed into the page and is of *decreasing* magnitude.

Magnetic flux line

30.7 imply that

$$I_S' = -\frac{\partial Q_S}{\partial t} + \frac{\partial Q_S}{\partial t} = 0 \tag{30.9}$$

The replacement of I in Ampere's law by I' gives the generalization called the *Maxwell–Ampere law*,

$$\oint_C \frac{1}{\mu_0} B_s \, ds = I + \int_S \epsilon_0 \frac{\partial E_N}{\partial t} \, dA \tag{30.10}$$

When the curve C shrinks to a point and S consequently becomes a *closed* surface, the Maxwell–Ampere law gives $I_S' = 0$, which is consistent with Eq. 30.9. And for the steady-state, since $\partial E_N/\partial t$ is then zero, the Maxwell–Ampere law reduces to Ampere's law.

Maxwell's generalization of Ampere's law therefore agrees with the experimentally verified facts about magnetic fields produced by steady currents and is consistent with Gauss's law for electric fields and the law of conservation of electric charge, even when there are time-varying fields. Therefore it is not obviously wrong. Whether or not it is correct can be settled only by experiment. The new law, together with Faraday's law and Gauss's laws for electric and magnetic fields, implies the possibility of electromagnetic waves that travel at the speed of light (as is shown in Chapter 35). This discovery led Maxwell to predict that light was an electromagnetic wave phenomenon and that electromagnetic waves of all frequencies could be generated. Experimental confirmation was obtained by succeeding generations and Maxwell's electromagnetic theory stands as a most brilliant triumph of theoretical physics.

Because of the term containing $\partial E_N/\partial t$, the Maxwell–Ampere law predicts that, even in empty space where there are no currents, there will be a magnetic field accompanying a time-varying electric field. The relative orientations of these fields are shown in Figure 30-5.* Such an induced magnetic field can be

---

*The Maxwell–Ampere law should be compared and contrasted with Faraday's law. In empty space, except for a difference in signs and a corresponding difference in the directions of the induced field relative to the change in the other, the roles of **E** and **B** are simply interchanged in the two laws. However, there is no current of magnetic poles to contribute a term in Faraday's law analogous to the term in the Maxwell–Ampere law corresponding to a current I of electric charges.

detected easily only if the electric field is changing rapidly. It turns out that, in a slowly changing field called a *quasi-static* field, the magnetic field predicted by the Maxwell–Ampere law is negligibly different from that calculated by simply applying the Biot–Savart law to whatever currents are present. But in rapidly fluctuating fields, the magnetic field associated with a changing electric field is easily detected and is of crucial importance as an integral part of a self-sustaining electromagnetic wave (Figure 35-3).

30.3 Maxwell's Equations, A Summary of the Basic Laws of Electromagnetism

Maxwell recognized that the theory of all electromagnetic phenomena could be based on the following four equations, now called *Maxwell's equations:*

$$\oint_S \epsilon_0 E_N \, dA = Q_S \quad \text{(Gauss's law for electric fields)}$$

$$\oint_S B_N \, dA = 0 \quad \text{(Gauss's law for magnetic fields)}$$

$$\oint_C E_s \, ds = -\int_S \frac{\partial B_N}{\partial t} \, dA \qquad \text{(Faraday's law)}$$

$$\oint_C \frac{1}{\mu_0} B_s \, ds = I + \int_S \epsilon_0 \frac{\partial E_N}{\partial t} \, dA \quad \text{(Maxwell–Ampere law)}$$

These equations express the relationship between the fields and the sources (charges and currents in a vacuum*), as well as the interconnection between time-varying electric and magnetic fields. The fields determine the *Lorentz force* **F** on a charge q moving with a velocity **v**:

$$\mathbf{F} = q\mathbf{E} + q\mathbf{v} \times \mathbf{B}$$

Maxwell's equations and the Lorentz force equation summarize all the basic knowledge of electricity and magnetism that has been discussed so far in this book. The further development of the classical theory of electromagnetism consists, not of the introduction of new fundamental laws, but rather of the examination of the consequences of Maxwell's equations, usually after making some appropriate approximations.

Questions

1 If the curve C in Figure 30-1 has a radius of 0.50 m and the downward flux is increasing at the rate of 100 Wb/s, what is the value of $\oint_C E_s \, ds$ for the curve C? What is the magnitude of the electric field at points on C?

---

*In a medium characterized by a permittivity ϵ and permeability μ, Maxwell's equations for the macroscopic fields **E** and **B** are obtained from the given equations simply by replacing ϵ_0 by ϵ, μ_0 by μ, and reinterpreting Q_S as the free charge and I as the free current.

2 Consider a proton with an orbit which is the curve C of Question 1. What is the electric force acting on this proton? After one revolution in this orbit, what is the change in the proton's kinetic energy? Is this an increase or a decrease in kinetic energy? Answer the same questions for the case when the proton is replaced by an electron.

3 The flux enclosed by a coil is given by

$$\Phi = (4.0 \times 10^{-2} \text{ Wb/s}^2)t^2 + (6.0 \times 10^{-2} \text{ Wb/s})t$$

Find the value of $\oint_C E_s\, ds$ for the coil when $t = 1.5$ s.

4 A single-turn coil encloses an area of 0.20 m². The coil is placed perpendicular to a uniform magnetic field B. Find the value of $\oint_C E_s\, ds$ for the coil at the instants $t = 0$ s, $t = \frac{1}{240}$ s, $t = \frac{1}{120}$ s when the magnetic field is the sinusoidal function of time given by

$$B = B_0 \sin 2\pi ft$$

with $B_0 = 0.50$ T and $f = 60$ Hz.

5 Verify that the equation

$$I' = I + \int_S \epsilon_0 \frac{\partial E_N}{\partial t}\, dA$$

is dimensionally correct.

6 Compare the expression obtained by interchanging **E** and **B** in Faraday's law with the Maxwell–Ampere law for a region where $I = 0$.

7 Verify that the quantity

$$I' = I + \int_S \epsilon_0 \frac{\partial E_N}{\partial t}\, dA$$

has the value zero for the closed surface S in Figure 30-6. Evaluate the integral assuming that the electric field is confined to the region between the capacitor plates where the field is uniform with a magnitude

$$E = \frac{\sigma}{\epsilon_0} = \frac{Q}{\epsilon_0 A}$$

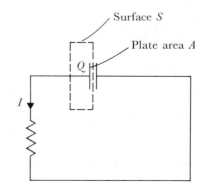

Figure 30-6   Closed surface S encloses one plate of a parallel-plate capacitor.

Supplementary Questions

S-1 In a betatron, we want to have the accelerated particle of charge q move in a circle of constant radius R:

   (a) Show that this requires that the particle's momentum p increases at a rate that is related to the rate of increase of the magnetic field B at the particle's orbit by the equation,

$$\frac{dp}{dt} = qR\frac{dB}{dt}$$

   (b) Newton's second law for tangential components gives

$$\frac{dp}{dt} = qE$$

   where E is the electric field evaluated in Example 1. Using the result of part (a), show that the accelerated particle maintains an orbit of constant radius if

$$B = \tfrac{1}{2}B_{\text{avg}}$$

   where B_{avg} is the average value of the magnetic field in the circular region within the orbit.

S-2 Figure 30-7 represents a discharging capacitor. The plates are circular with a radius R_0. Symmetry about the central axis, and the Maxwell–Ampère law, can be used to find the magnetic field B induced at the point P in Figure 30-7:

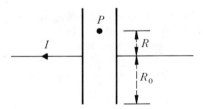

Figure 30-7

   (a) Consider a path of integration C through P which is a circle of radius R and concentric with the capacitor. Show that

$$2\pi RB = \mu_0 I'_C$$

   where I'_C is given by

$$I'_C = I_1 + \int_S \epsilon_0 \frac{\partial E_N}{\partial t}\, dA$$

   The integration is over any surface S bounded by C, and I_1 denotes the current through S.

(b) Show that

$$I'_C = \left(\frac{R}{R_0}\right)^2 I$$

and therefore that

$$B = \frac{\mu_0 IR}{2\pi R_0^2}$$

3I.

inductance

This chapter is concerned with Faraday's law of electromagnetic induction and formulations of this law that are most convenient for applications to electric circuits. The mutual inductance of two circuits, and the self-inductance of a single circuit, are quantities that are introduced so that an induced EMF can be related directly to a changing current, rather than to a changing flux. We investigate the relationship between the energy stored in a current-carrying circuit and the circuit self-inductance, as well as the influence of self-inductance on the transient behavior of resistive circuits.

31.1 Faraday's Law of Electromagnetic Induction

In 1831, Faraday performed the experiments indicated in Figure 31-1 and observed that a transient current is induced if:

1   The steady current in an adjacent circuit is turned on or off.
2   The adjacent circuit with a steady current is moved relative to the first circuit.
3   A permanent magnet is thrust into or out of the circuit.

No current is induced unless either the current in the adjacent circuit changes or there is relative motion. Faraday recognized that in every case a current is induced in a circuit only while the magnetic flux Φ enclosed by the circuit is *changing.* The induced current indicates that there is an EMF ε induced in the circuit. Faraday's observations are summarized in the law of electromagnetic induction,

$$\varepsilon = -\frac{d\Phi}{dt} \tag{31.1}$$

This general law will be called *Faraday's flux rule.* The flux Φ enclosed by a circuit or a closed path C may change because:

1 **B** is time-varying.
2 The closed path C or parts of C are in motion.

Both causes of a changing flux are included in the total derivative $d\Phi/dt$. Faraday's flux rule gives the EMF associated with the changing flux for a general motion of any closed path C through any magnetic field.

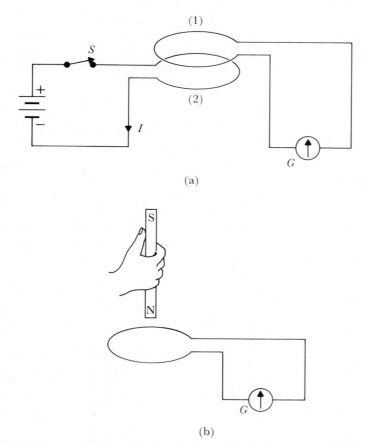

(a)

(b)

Figure 31-1 Electromagnetic induction. Galvanometer indicates a transient current if: (a) the steady current in the adjacent circuit is turned on or off, or if the adjacent circuit with steady current is moved relative to the first circuit; (b) a permanent magnet is thrust into or out of the circuit.

For a coil consisting of N turns, each enclosing the same flux Φ, the same EMF is induced in every turn and these EMFs sum to give the total coil EMF ε. Then Faraday's flux rule becomes

$$\varepsilon = -N\frac{d\Phi}{dt} \tag{31.2}$$

Lenz's Law

The relationship between the positive directions for the magnetic flux and the EMF can be determined from Figure 30-2, interpreting **n** as the positive direction for the flux through C, and the positive direction for C to be the positive direction for an EMF induced in C. However, it is often easier to determine directions using the following rule called *Lenz's law: The induced current is directed so as to create flux that opposes the flux change.* If the flux is *decreasing*, the induced EMF is directed so as to produce a current that creates flux in the direction of the original flux; if the flux is *increasing*, the flux created by any induced current will be opposite in direction to the original magnetic flux. Thus, in Figure 30-1, an induced current in a coil placed at C would have to be counterclockwise, and this implies that **E** has the directions indicated at various points.

Example 1   Figure 31-2 shows a prototype generator obtained by rotating a coil in a constant magnetic field. Assume that the coil's angular speed ω is constant. Find an expression for the EMF induced in the coil in terms of ω, the coil area A, and the magnetic field B.

Solution   The flux enclosed by the coil is

$$\Phi = B_N A = BA \cos \theta = BA \cos \omega t$$

where $\theta = \omega t$ is the angle between **B** and the normal to the coil (Figure 31-3). As the coil rotates, the flux enclosed by the coil changes. According to

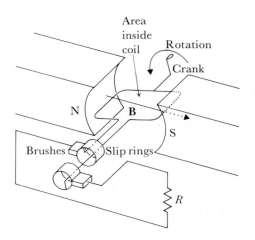

Figure 31-2   Simple generator. Turning the crank causes an alternating EMF at the slip rings. The brushes make connection to an external circuit, giving an alternating current through R.

Faraday's flux rule, the EMF associated with this changing flux is

$$\mathcal{E} = -\frac{d\Phi}{dt} = BA\omega \sin \omega t$$

Applied to a closed circuit, this sinusoidal EMF produces what is called an alternating current (AC).

Example 2   In Figure 31-4, parallel conducting tracks ab and cd are joined at one end by a conductor ac and are located in a constant uniform magnetic field **B** perpendicular to the plane of the tracks. A conducting bar ef moves at a constant speed v. Find the EMF induced in the circuit $eface$.

Solution   The flux enclosed by the circuit is

$$\Phi = B\ell x$$

The speed of the moving conductor is

$$v = \frac{dx}{dt}$$

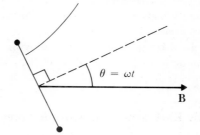

Plane of coil

Figure 31-3   The angle between **B** and the normal to the plane of the coil is $\theta = \omega t$.

$\theta = \omega t$

B

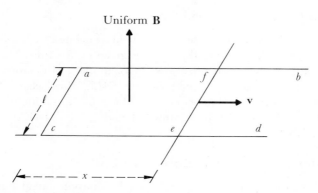

Uniform **B**

Figure 31-4   Conducting bar ef moves at constant speed v through a constant uniform magnetic field **B**.

and because of this motion, the enclosed flux changes at the rate

$$\frac{d\Phi}{dt} = B\ell v$$

From Faraday's flux rule, the magnitude of the EMF is

$$|\mathcal{E}| = B\ell v$$

Since the upward flux through the circuit is increasing, Lenz's law implies that the induced EMF is clockwise looking down on the circuit.

EMF Associated with the Lorentz Force

As defined in Section 26.1, the EMF $\mathcal{E}$ of a localized source is the nonelectrostatic work per unit charge passing through the source. The general definition of EMF is an extension of this concept. The EMF $\mathcal{E}$, for any closed path C, arising from a force $\mathbf{F}$ acting on a test charge q, is the work per unit charge done by $\mathbf{F}$ when the test charge is taken around C: that is,

$$\mathcal{E} = \oint_C \frac{F_s}{q}\, ds \tag{31.3}$$

Since $\oint_C E_s\, ds = 0$ for a time-independent field $\mathbf{E}$, this definition implies that no EMF is associated with an electrostatic field. For a localized source, Eq. 31.3 therefore gives $\mathcal{E}$ as the nonelectrostatic work per unit charge, in agreement with our preliminary definition.

In a portion of a circuit moving with velocity $\mathbf{v}$, a charge q experiences a magnetic force $q\mathbf{v} \times \mathbf{B}$ as well as an electric force $q\mathbf{E}$. Their resultant is the Lorentz force

$$\mathbf{F} = q\mathbf{E} + q\mathbf{v} \times \mathbf{B}$$

The EMF associated with the Lorentz force is

$$\mathcal{E} = \oint_C \frac{F_s}{q}\, ds = \oint_C (\mathbf{E} + \mathbf{v} \times \mathbf{B})_s\, ds \tag{31.4}$$

This EMF is the sum of two contributions:

1   The EMF $\oint_C E_s\, ds$ is due to the electric field associated with a time-varying magnetic field. The value of this EMF is given by the third Maxwell equation (Faraday's law), as is illustrated in Example 2 of Chapter 30.
2   The EMF $\oint_C (\mathbf{v} \times \mathbf{B})_s\, ds$, called a *motional EMF*, arises because of the magnetic force exerted on charges carried along by a conductor that moves through a magnetic field. In Examples 1 and 2, the only EMF present is a motional EMF.

Faraday's Flux Rule and the Lorentz Force EMF

Faraday's flux rule follows from the Maxwell equation called Faraday's law and from the expression for the EMF associated with the Lorentz force.

To show this, we first require a general expression for the rate of change of flux due to motion of the circuit. For the circuit of Figure 31-4 we found in

Example 2 that with a segment of length l moving with speed v,

$$\frac{d\Phi}{dt} = Blv$$

It can be verified that for arbitrary directions of $\mathbf{B}$, $\mathbf{v}$, and the conductor of length l, the rate of change of flux is

$$\frac{d\Phi}{dt} = -(\mathbf{v} \times \mathbf{B})_s l$$

where the subscript s denotes the component along the conductor of length l in the positive direction for the circuit. Then if $\mathbf{v}$ is the velocity of a segment ds of a circuit C, the rate of change of flux enclosed by C due to circuit motion is

$$-\oint_C (\mathbf{v} \times \mathbf{B})_s \, ds$$

In general, the total rate of change of flux enclosed by a circuit C is the sum of that which is due to the time-variation of the magnetic field and that which is due to the motion of C; that is,

$$\frac{d\Phi}{dt} = \frac{d}{dt}\int_S B_N \, dA = \int_S \frac{\partial B_N}{\partial t} dA - \oint_C (\mathbf{v} \times \mathbf{B})_s \, ds$$

Using the third Maxwell equation (Faraday's law) this can be written

$$\frac{d\Phi}{dt} = -\oint_C E_s \, ds - \oint_C (\mathbf{v} \times \mathbf{B})_s \, ds$$

or

$$\oint_C (\mathbf{E} + \mathbf{v} \times \mathbf{B})_s \, ds = -\frac{d\Phi}{dt}$$

Since the left-hand side is the EMF associated with the Lorentz force, this equation is the Faraday flux rule,

$$\mathcal{E} = -\frac{d\Phi}{dt}$$

This remarkable rule gives the EMF no matter how it is induced. In some cases there is only an electric field (Figure 30-1), and in other cases there may be only magnetic forces on moving charges (Figure 31-3), but rather amazingly, these very different physical situations are both embraced in the one statement, $\mathcal{E} = -d\Phi/dt$.

31.2 Coupled Circuits

Transformers

EMFs associated with a changing magnetic flux are used to transform a varying voltage in one circuit into a larger or smaller voltage in another circuit. This is accomplished with a transformer (Figure 31-5) consisting of two coils electrically insulated from each other and wound on the same ferromagnetic core.

Power is supplied to one coil named the *primary* coil. A varying current in this coil sets up a varying magnetic flux, largely confined to the ferromagnetic core. The other coil, called the *secondary* coil, thus encloses a varying flux. While this flux is changing at the rate $d\Phi/dt$, each of the N_s turns in the secondary coil experiences, according to Faraday's flux rule, an induced EMF equal to $d\Phi/dt$, and the EMF for the entire secondary coil is

$$\mathcal{E}_s = -N_s \frac{d\Phi}{dt} \tag{31.5}$$

We neglect the small amount of leakage flux and assume that the magnetic

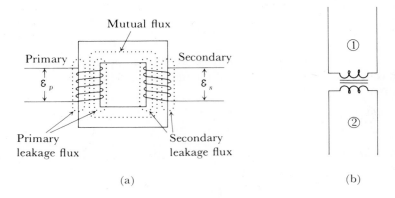

(a)   (b)

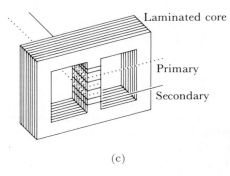

(c)

Figure 31-5 Transformers. (a) Core-type transformer, showing magnetic flux leakage. (b) Circuit symbol for ferromagnetic core transformer. (c) Shell-type transformer. Most transformers are of this type, having concentric windings to minimize flux leakage. (d) Photograph of a common shell-type transformer.

(d)

flux at any instant is the same through the primary and secondary coils. Then there is an EMF $d\Phi/dt$ induced in each of the N_p turns of the primary coil and the total EMF for this coil is

$$\mathcal{E}_p = -N_p\frac{d\Phi}{dt} \tag{31.6}$$

Dividing Eq. 31.5 by Eq. 31.6 we obtain

$$\frac{\mathcal{E}_s}{\mathcal{E}_p} = \frac{N_s}{N_p} \tag{31.7}$$

According to this result, when there are varying currents, the ratio of the EMFs induced in the secondary and primary coils at any instant is equal to the "turns ratio," N_s/N_p. When N_s is greater than N_p, the secondary EMF exceeds the primary EMF, and the transformer is named a *step-up* transformer. In a *step-down* transformer such as those used for toy electric cars and trains, N_s is less than N_p, and the household voltage applied across the primary is transformed into a safe low secondary voltage which delivers power to the toy.

It is desirable to transmit electrical power at high voltages and small currents in order to minimize the I^2R power dissipation in the transmission line. But this power must be generated and ultimately delivered at relatively low voltages to avoid severe problems of insulation and safety. A most useful feature of AC is the fact that the voltage and the current can be changed with ease and efficiency by the use of transformers (Figure 31-6).

It should be emphasized that only varying voltages (AC or pulses) should be applied across the primary terminals of a transformer. A steady DC voltage established across the primary coil does not produce a *changing* magnetic flux. No EMF is induced in the secondary. Moreover, since no opposing self-induced EMF (Section 31.3) is induced in the primary, the primary current will be limited only by the low primary coil resistance. The result is that the primary coil overheats and burns out.

Mutual Inductance

To analyze coupled circuits, such as the primary and secondary circuits of a transformer, we should know the relationship between the changing current in

Figure 31-6 Long-distance transmission of electrical energy. Transformers raise the voltage at the generator, lower it at city substations and again at residences or other points of usage. Large colleges and universities may have their own substations.

one circuit and the EMF that this induces in the other circuit. Suppose a current I_p in the primary circuit produces a flux Φ_s through each turn of the secondary coil.* The mutual inductance M of the primary and secondary circuits is defined by

$$N_s \Phi_s = M I_p \qquad (31.8)$$

If there are no ferromagnetic materials present, Φ_s is proportional to I_p and M is then a constant, with a value that depends only on the geometry of the two circuits. For a transformer with a ferromagnetic core, M depends also on the permeability of the core which is a function of the magnetic field B established within the core. The SI unit of mutual inductance is the weber per ampere which is named the *henry* (H):

$$1 \text{ H} = 1 \text{ Wb/A}$$

A current I_s in the secondary will create a flux Φ_p enclosed by each turn of the primary. Using an equation analogous to Eq. 31.8, we define a mutual inductance M' by

$$N_p \Phi_p = M' I_s \qquad (31.9)$$

With mathematical methods beyond the scope of this book, it is possible to prove the simple but remarkable theorem that $M' = M$, so that there is a *single mutual inductance* of any two circuits.

When M is a constant, $N_p\, d\Phi_p/dt = M\, dI_s/dt$, and Faraday's flux rule gives

$$\varepsilon_p = -M \frac{dI_s}{dt} \qquad (31.10)$$

Similarly,

$$\varepsilon_s = -M \frac{dI_p}{dt} \qquad (31.11)$$

These two equations accurately express the coupling between any two distinct circuits in the absence of ferromagnetic materials.

Example 3   Find the mutual inductance M of a large circular coil with N_1 turns of radius R_1 and a very small concentric coil with N_2 turns of radius R_2 (Figure 31-7).

Solution   The flux Φ_2 through the small coil produced by a current I_1 in the large coil can be calculated easily, since

$$\Phi_2 = B_1 \pi R_2^2$$

where B_1 is the field produced by I_1 at the center of the first coil. From Eq. 28.3 this is

$$B_1 = \frac{\mu_0 N_1 I_1}{2R_1}$$

---

*Here we do not assume that the leakage flux is small. In general the flux Φ_s enclosed by a turn of the secondary coil is only part of the flux created by I_p. (See Figure 31-7 and consider the small coil as the secondary.)

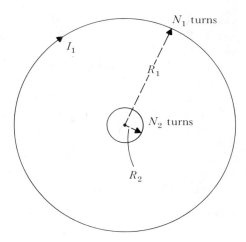

Figure 31-7 The field B_1 produced by the current I_1 is approximately uniform over the region of the small coil.

Therefore

$$M = \frac{N_2 \Phi_2}{I_1} = \frac{\pi \mu_0 N_1 N_2 R_2^2}{2 R_1}$$

Notice that M is a constant, independent of the currents, and is determined by geometrical parameters—the coil radii and numbers of turns.

31.3 Self-Inductance

When the current through a coil is changing, the flux produced by this current and enclosed by the turns of the coil is also changing; consequently there is an EMF induced in the coil. Such an EMF is called a *self-induced* EMF.

Suppose a coil current I produces a flux Φ through each of N turns of a coil. The *self-inductance* L of the coil is defined by the equation

$$N\Phi = LI \tag{31.12}$$

If there are no ferromagnetic materials present, Φ is proportional to I. Then L is a constant with a value determined by the coil geometry. The self-inductance of a given coil can be increased greatly by providing a ferromagnetic core, but, because the core's magnetic permeability μ depends on B, the self-inductance is then a complicated function of the coil current.

When L is constant, $N \, d\Phi/dt = L \, dI/dt$, and Faraday's flux rule gives

$$\mathcal{E} = -L \frac{dI}{dt} \tag{31.13}$$

a useful expression for the self-induced EMF in terms of the changing current. Lenz's law implies that the direction of a self-induced EMF is such as to oppose the change in current that gives rise to the change in flux. Therefore the induced EMF is in the direction of the current if the current is decreasing, but in the opposite direction if the current is increasing. After a direction in an electric circuit is selected as the positive direction for both currents and EMFs, the minus sign in Eq. 31.13 ensures that $\mathcal{E}$ and I have the proper relative directions at all times.

Circuit elements specifically designed to have appreciable self-inductance are called inductors. The self-inductance of typical inductors (Figure 31-8) is of the order of 1 H for iron-cored inductors and 1 mH for coils without ferromagnetic cores. Even if a circuit does not include a coil, a complete circuit that is a single loop encloses some flux when it carries a current, and therefore the circuit does have some self-inductance.

Example 4   What is the inductance of the N-turn toroid in Example 5 of Chapter 28? The thin toroid has a circumference $2\pi R$ and a cross section A.

Solution   Equation 28.8 gives the magnetic field inside the toroid:

$$B = \frac{\mu_0 NI}{2\pi R}$$

The flux enclosed by each turn of the toroid is

$$\Phi = BA = \frac{\mu_0 NIA}{2\pi R}$$

Therefore the toroid has an inductance

$$L = \frac{N\Phi}{I} = \frac{\mu_0 N^2 A}{2\pi R}$$

Notice that L is determined entirely by the geometrical quantities R, A, and

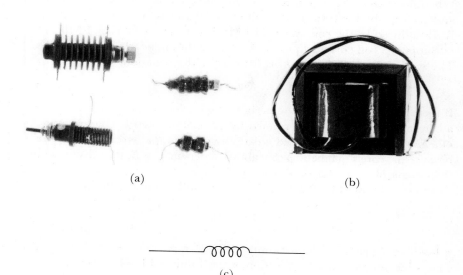

(a)   (b)

(c)

Figure 31-8   (a) Inductors without ferromagnetic cores. (b) Iron-cored inductor. (c). Circuit symbol for an inductor.

N. If $R = 10$ cm, $A = 5.0$ cm^2, and $N = 1000$ turns, then

$$L = \frac{4\pi \times 10^{-7} \times (1000)^2 \times (5.0 \times 10^{-4})}{2\pi \times 0.10} \text{ H} = 1.0 \text{ mH}$$

31.4 Energy Stored in the Magnetic Field of an Inductor

When an EMF, $\mathcal{E} = -L\,dI/dt$, is induced in an inductor by a changing current I, the rate at which energy is supplied by the inductor is $\mathcal{E}I$. When $\mathcal{E}$ is in the direction of I, the inductor supplies energy to the external circuit. And when $\mathcal{E}$ and I have opposite directions, energy is supplied to the inductor. A consistent account of these energy transformations is obtained by assuming that the energy is stored in the magnetic field established by the current in the inductor. In this section we shall use this idea to derive an expression for the energy density in a magnetic field.

When the current I in an inductor is in the positive direction and is increasing, energy is being transferred to the inductor at a rate

$$\frac{dU}{dt} = \left(L\frac{dI}{dt} \right)I$$

The total energy supplied while the current increases from 0 to I is therefore

$$U = \int_0^I LI'\,dI' = \tfrac{1}{2}LI^2$$

We conclude that the *energy stored in the magnetic field of an inductor carrying a current I is*

$$U = \tfrac{1}{2}LI^2 \tag{31.14}$$

The energy density u in a magnetic field is the energy stored per unit volume. This is easily evaluated for the toroid of Example 4. Since the magnetic field B is confined to the region of volume $2\pi RA$ within the thin toroid and has a constant magnitude $\mu_0 NI/2\pi R$ in this region, we have

$$u = \frac{U}{volume} = \frac{\tfrac{1}{2}LI^2}{2\pi RA} = \frac{1}{2}\frac{(NBA/I)I^2}{2\pi RA} = \frac{1}{2}\frac{B^2}{\mu_0}$$

Although we have considered only the special case of the field in the toroid, a more general analysis shows that the result

$$u = \frac{1}{2}\frac{B^2}{\mu_0} \tag{31.15}$$

is the correct expression for the *energy density in any magnetic field B in a vacuum.*

When both electric and magnetic fields are present, the total energy per unit volume in the electromagnetic field is the sum of the electric energy density $\tfrac{1}{2}\epsilon_0 E^2$ (Eq. 24.13) and the magnetic energy density $\tfrac{1}{2}B^2/\mu_0$; that is, the electromagnetic energy density in a vacuum is

$$u = \tfrac{1}{2}\epsilon_0 E^2 + \frac{1}{2}\frac{B^2}{\mu_0} \tag{31.16}$$

31.5 Transients in *R-L* Circuits

Growth of Current in an Inductor

An inductor with self-inductance L and resistance r_L is represented at the right in the circuit of Figure 31-9. The switch S_1 is closed at $t = 0$ and S_2 remains open. The circuit then consists of the outer loop with a total resistance $R = r + r' + r_L$. We select the clockwise direction as the positive direction and denote the inductor current and EMF in this direction by I and $\mathcal{E}_L$, respectively. Then

$$\mathcal{E}_L = -L\frac{dI}{dt}$$

Kirchhoff's loop rule gives, at any instant,

$$\mathcal{E} + \mathcal{E}_L = RI \qquad (31.17)$$

With an inductor in the circuit, since $\mathcal{E}_L$ cannot become infinite, dI/dt can never be infinite. This implies that the current I cannot change abruptly. Consequently, at the instant $t = 0$ but just *after* the switch S_1 is closed, the current is zero. From the loop rule, Eq. 31.17, we find that at $t = 0$,

$$\mathcal{E}_L = -\mathcal{E}$$

The inductor EMF *does* change abruptly from 0 to $-\mathcal{E}$ when S_1 is closed.

The inductor EMF $\mathcal{E}_L$ satisfies a simple differential equation found by differentiating both sides of Eq. 31.17:

$$0 + \frac{d\mathcal{E}_L}{dt} = R\frac{dI}{dt}$$

or

$$\frac{d\mathcal{E}_L}{dt} = -\frac{R}{L}\mathcal{E}_L$$

Following the same mathematical steps as in Section 26.4, we find

$$\mathcal{E}_L = \mathcal{E}_{0L}e^{-t/\tau} \qquad (31.18)$$

where the initial value $\mathcal{E}_{0L}$ is $-\mathcal{E}$ and $\tau = L/R$. This shows that the magnitude

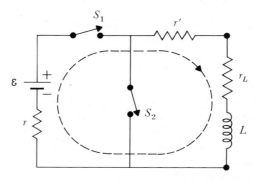

Figure 31-9 Current in the inductor grows when S_1 is closed and decays when S_2 is closed.

of the inductor EMF decays exponentially with a time constant L/R. Substituting this result in Eq. 31.17 and solving for I, we obtain

$$I = \frac{\mathcal{E}}{R}(1 - e^{-t/\tau})$$

a growth function with time constant $\tau = L/R$ and final value $\mathcal{E}/R$.

Decay of Current in an Inductor

Opening a switch in a circuit with a sizable inductance is a dangerous operation. If we attempt to reduce the inductor current to zero by simply opening switch S_1, the rapid change in current will induce an EMF in the inductance sufficiently large to cause a spark at the switch contacts and the current will continue until the energy stored in the magnetic field of the inductor is dissipated. A safe switching procedure provides a path for the inductor current by first closing switch S_2 and then opening switch S_1. Suppose this is accomplished at a time that we call the initial instant, $t = 0$. For this new circuit, with $R' = r' + r_L$, Kirchhoff's loop rule gives

$$\mathcal{E}_L = R'I$$

Since $\mathcal{E}_L = -L\, dI/dt$, this yields

$$\frac{dI}{dt} = -\frac{R'}{L}I$$

The solution is

$$I = I_0 e^{-t/\tau} \tag{31.19}$$

an exponential decay with time constant $\tau = L/R'$. The "initial value" I_0 is the current just after the switching operations are completed. From $\mathcal{E}_L = R'I$, we see that the inductor EMF also decays exponentially.

These investigations show that, although the EMF of an inductor generally changes abruptly when switches are thrown, the inductor current is a continuous function of time. An appreciable change in the current requires a time interval of the order of τ, where

$$\tau = \frac{L}{circuit\ resistance}$$

Summary

□ The EMF $\mathcal{E}$ induced in a circuit that encloses a flux Φ is given by the Faraday flux rule,

$$\mathcal{E} = -\frac{d\Phi}{dt}$$

where the changes of the enclosed flux may be due to motion of the circuit as well as a time variation of the magnetic field. The EMF in the Faraday flux rule is the work per unit charge done by the Lorentz force in traversing the complete circuit. This includes the motional EMF arising from the magnetic force $q\mathbf{v} \times \mathbf{B}$ on charges in a conductor moving with a velocity $\mathbf{v}$ through a magnetic field $\mathbf{B}$.

☐   The ratio of the EMFs induced in the secondary and primary coils is equal to the transformer turns ratio N_s/N_p,

$$\frac{\varepsilon_s}{\varepsilon_p} = \frac{N_s}{N_p}$$

provided leakage flux is negligible.

☐   The coupling of any two circuits is determined by their mutual inductance M defined by

$$N_s\Phi_s = MI_p$$

where the current I_p in one coil produces a flux Φ_s enclosed by each of the N_s turns of the other coil. When M is constant,

$$\varepsilon_p = -M\frac{dI_s}{dt} \quad \text{and} \quad \varepsilon_s = -M\frac{dI_p}{dt}$$

☐   The self-inductance L of an N-turn coil is defined by

$$N\Phi = LI$$

where a current I in the coil produces a flux Φ through each turn. The self-induced EMF in an inductor with self-inductance L is

$$\varepsilon = -L\frac{dI}{dt}$$

☐   The energy stored in the magnetic field of an inductor carrying a current I is

$$U = \tfrac{1}{2}LI^2$$

The energy density in a magnetic field $\mathbf{B}$ in a vacuum is

$$u = \frac{1}{2}\frac{B^2}{\mu_0}$$

☐   The time constant of an R-L circuit is

$$\tau = \frac{L}{circuit\ resistance}$$

When a constant EMF ε is applied across an inductor the growth of the current toward its final value I_f is described by

$$I = I_f(1 - e^{-t/\tau})$$

Replacement of this constant EMF by a resistor results in an exponential decay of the current:

$$I = I_0 e^{-t/\tau}$$

Questions

1   Find the magnitude of the EMF induced in a 200-turn coil with cross-sectional area of 0.150 m² if the magnetic field through the coil is caused to change from 100 mT to 500 mT at a uniform rate over an interval of 20.0 ms.

2   Find the maximum value and the frequency of the alternating current in the resistor of Figure 31-2 if the coil rotates with a constant angular speed of 377 rad/s in a magnetic field of 0.20 T. The coil area is 3.0×10^{-2} m and the total resistance of the circuit is 5.0 Ω.

3   In Figure 31-4, the magnetic field has a magnitude of 0.40 T, the length l of the moving conductor ef is 0.20 m, and its speed is 5.0 m/s:
   (a) Find the EMF for this circuit.
   (b) If the total resistance of the circuit is 0.10 Ω, what is the induced current? Give the direction of this current.
   (c) Find the magnitude and direction of the magnetic force exerted on the moving conductor.
   (d) Assume that the conductor slides without friction over the rails. An external force F_{ext} is applied to maintain the conductor's speed at 5.0 m/s. Find the magnitude and direction of F_{ext}.
   (e) Find the power supplied to the moving conductor by the external force F_{ext}.
   (f) Find the power dissipated by the Joule effect in this circuit.

4   In Figure 31-10, within the dashed line, the magnetic field **B** is uniform and directed inward. Find the current induced in the circuit when the circuit is pulled to the right with a speed v.

5   In Figure 31-11, the conducting bar slides without friction over parallel rails through a uniform magnetic field of 0.40 T. The total circuit resistance is 0.10 Ω.

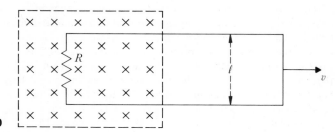

Figure 31-10

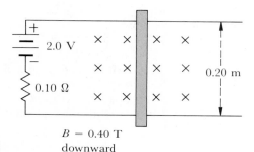

Figure 31-11 $B = 0.40$ T
downward

This device is a simple electric motor. The bar starts from rest and accelerates:

(a) What is the direction of its velocity vector?

(b) Find the EMF for the circuit when the bar is at rest and when its speed is 5.0 m/s. Verify that the EMF of the motor is a *back* EMF, that is, this EMF is directed so as to oppose the circuit current.

(c) What is the current when the conductor's speed is 5.0 m/s?

(d) The bar travels at a constant speed of 5.0 m/s while pushing with a force F_o on a load. Find this output force and the power output of the motor.

(e) What is the efficiency (*power output/power input*) of the motor?

6   The load is removed from the motor of Question 5:

(a) What is the ultimate or terminal speed attained by the sliding bar?

(b) What is the current when the bar travels at its terminal speed?

7   The sliding bar in Figure 31-11 is pushed to the right by an external force F_{ext} which is such that the bar maintains a constant speed of 40 m/s:

(a) Find the total EMF for the circuit. Would you call the sliding bar and its circuit a generator or a motor?

(b) What is the circuit current?

(c) What is the power supplied by F_{ext} to this device?

(d) What happens to the power calculated in part (c)?

8   An N-turn coil is part of a circuit with a total resistance R. While the flux through the coil is changing, a current I is induced. Show that the total charge that passes any point in the circuit, while the flux enclosed by the coil changes from Φ_1 to Φ_2, is

$$Q = \int_{t_1}^{t_2} I\, dt = -\frac{N(\Phi_2 - \Phi_1)}{R}$$

9   When a conducting bar which is perpendicular to a magnetic field **B** is moved with a velocity **v** perpendicular to the bar and to the magnetic field, the magnetic forces cause mobile charges to move until the charge distribution sets up an electric field **E** within the conductor. This electric field exerts on any charge q an electric force equal in magnitude but opposite in direction to the magnetic force on q:

(a) Show that this electric field has a magnitude given by $E = vB$.

(b) Show that the voltage between the ends of the bar of length l is Blv.

(c) Determine the magnitude and polarity of the voltage across a 2.0-m long bumper of a car if the car is going west at 21 m/s in a region where the vertical component of the earth's magnetic field is 4.5×10^{-5} T downward.

10   When the EMF for the primary of a household transformer for an electric bell is 120 V, the secondary EMF is 6.0 V:

(a) Is this a step-down transformer?

(b) What is the transformer turns ratio?

11   A 10-turn coil of radius 0.50 cm is concentric with a 20-turn coil of radius 50 cm:

(a) What is the mutual inductance of these coils?

(b) At the instant $t = \frac{1}{240}$ s, what is the EMF induced in the larger coil by a current in the smaller coil given in SI units by $I = 5.0 \cos 377t$?

12   A long solenoid of length l and cross section A has N_1 turns. A small coil is formed by winding N_2 turns around the middle of the solenoid. Show that the mutual inductance of these two coils is

$$M = \frac{\mu_0 N_1 N_2 A}{l}$$

13 Show that a long solenoid with length l, cross section A, and N turns, has a self-inductance

$$L = \frac{\mu_0 N^2 A}{l}$$

14 What is the self-induced EMF in a 0.50-mH inductor when the current is changing at the rate of 200 A/s?

15 A 2.0-μF capacitor is charged until the voltage between its plates is 100 V. This capacitor is then connected across an inductor with a self-inductance of 4.0 mH. Using only the law of conservation of energy, determine the maximum possible value of the current through the inductor.

16 Find the self-inductance of the long solenoid of Question 13 by first finding an expression for the total energy U stored in the solenoid and then equating this expression to $\frac{1}{2}LI^2$.

17 In Figure 31-9, $\varepsilon = 100$ V, $r = 10\ \Omega$, $r' = 20\ \Omega$, $r_L = 70\ \Omega$, and $L = 25$ mH. Find the current in the circuit and the voltage across each circuit element at an instant:
(a) Just after switch S_1 is closed.
(b) 0.500 ms after switch S_1 is closed.
(c) A long time (say 1 s) after switch S_1 is closed.

18 After switch S_1 in Question 17 has been closed for a time interval long compared to the time constant, switch S_2 is closed and switch S_1 is opened. Sketch a graph showing the values of the current I and the inductor EMF ε_L before and after these switching operations. When the voltage across the resistor r' is 10 V, how long is the time interval that has elapsed since the opening of switch S_1?

Supplementary Questions

S-1 Assume that the circuit of Figure 31-4 has a constant resistance R and that the resultant force acting on the sliding bar is the magnetic force BIl. Show that while the flux enclosed by the circuit changes from Φ_1 to Φ_2, the change in the momentum of the bar is

$$p_2 - p_1 = -\frac{Bl}{R}(\Phi_2 - \Phi_1)$$

S-2 Find the mutual inductance of the two circuits of Figure 28-18.

S-3 A closed circuit is formed by connecting the ends of two very long parallel wires separated by a distance d. Show that if end effects are neglected and if the flux within the wires themselves is negligible compared to the flux between the wires, the self-inductance of this circuit is

$$L = \frac{\mu_0 l}{\pi} \ln \frac{d - a}{a}$$

where each wire has length l and radius a.

S-4 Using the result of Question S-3, find the self-induced EMF when the current changes at the rate of 5.0 A/s in a rectangular loop which is 100 m long and 1.00 m wide. The radius of the wire is 1.0×10^{-2} m.

S-5   Consider the mutual inductance M of two coils with self-inductances L_1 and L_2. Maximum mutual inductance M_{max} will be obtained if the two coils are positioned so that all the flux produced by one coil is enclosed by each turn of the other. Show that

$$M_{max} = \sqrt{L_1 L_2}$$

S-6   Two coils with self-inductances L_1 and L_2 and mutual inductance M are connected in series. Show that the equivalent self-inductance L of this series combination is either

$$L = L_1 + L_2 + 2M$$

or

$$L = L_1 + L_2 - 2M$$

depending on how the two coils are connected.

S-7   A long coaxial cable consists of a solid conducting cylinder of radius R_1 supported by insulating disks on the axis of a thin-walled conducting tube of radius R_2. The inner conductor carries a steady current I and this current returns through the outer conductor. From Question 14 of Chapter 28, the field between the conductors is

$$B = \frac{\mu_0 I}{2\pi R}$$

Show that the total energy U stored in the magnetic field between the conductors, for a length l of the coaxial cable, is

$$U = \frac{\mu_0 I^2 l}{4\pi} \ln \frac{R_2}{R_1}$$

S-8   Use the result of Question S-7, and the expression $U = \frac{1}{2}LI^2$, to show that the self-inductance of a length l of the coaxial cable is

$$L = \frac{\mu_0 l}{2\pi} \ln \frac{R_2}{R_1}$$

S-9   A long cylinder of length l and radius R_0 carries a current I. The current density is uniform and parallel to the axis of the cylinder. Find the energy stored in the magnetic field *within* the cylinder.

S-10   In Figure 31-12, find the current through the inductor after switch S has been closed for a long time. Then if S is suddenly opened, what is the voltage across each circuit element?

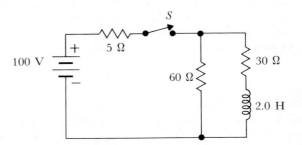

Figure 31-12

32.

alternating currents

Transformers function with alternating current (AC) rather than direct current (DC) and, primarily for this reason, more than 90% of the electrical power used in the world is supplied by AC. In addition to power requirements, AC is vital in electrical communications; radio waves, for example, are generated by high frequency alternating currents in a transmission antenna.

A coil of wire, rotating with a constant angular velocity in a magnetic field, generates an EMF which is a sinusoidal function of time, as shown in Example 1 of Chapter 31. This is the prototype for the commercial generators that supply the power to the circuits considered in this chapter.

Since *phasors* are used throughout the analysis of AC circuits, the student is advised to begin this study by reviewing the terminology of Section 9.1.

32.1 Alternating Currents and Voltages

A circuit element is *passive* if it is merely a network of resistors, inductors, and capacitors but does not contain a motor or a generator. Figure 32-1 represents a circuit consisting of an AC generator applied across the terminals of some passive circuit element such as that shown in Figure 32-2. The general problem is to determine the current at any instant in terms of the appropriate parameters of the circuit element and of the voltage across the circuit element. The methods developed for the analysis of AC circuits apply only when the generator frequency f is not too high. We assume that the time required for an electromagnetic disturbance to travel the length of the circuit at speed c is negligible compared to the period $T = 1/f$. For example, if the circuit has wires 3 m long, AC

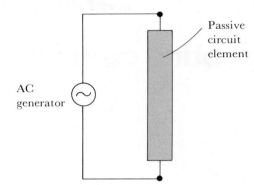

Figure 32-1   AC circuit with passive circuit element of unspecified nature.

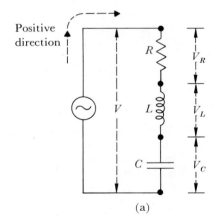

(a)

(b)

Figure 32-2   (a) An R-L-C series circuit. (b) Graphs showing voltage as a function of time. Such graphs can be obtained by connecting the vertical deflection plates of an oscilloscope across each circuit element in turn. (c) Graph of circuit current as a function of time. This could be obtained by connecting an oscilloscope across the resistor.

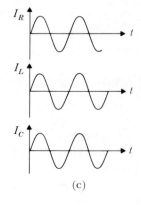

(c)

circuit analysis applies only for frequencies small compared to $f = (3 \times 10^8$ m/s)/3 m $= 10^8$ Hz. Higher frequencies involve problems of wave propagation.

At the relatively low frequencies involved in AC circuits, analysis is based on Kirchhoff's rules applied to instantaneous voltages and currents. These rules imply that, for a single loop circuit such as that of Figure 32-2, there is *at any instant the same current I* through every conducting cross section of the circuit. Also, at any instant

$$V = V_R + V_L + V_C \quad \text{(instantaneous values)} \quad (32.1)$$

where V is the potential drop across the series combination of Figure 32-2, $V_R = IR$, $V_L = L\, dI/dt$, and $V_C = Q/C$ with $I = dQ/dt$. A single positive direction for all EMFs, currents, and potential drops is indicated in Figure 32-2.

When a generator with a sinusoidal voltage of frequency f is first connected in a circuit, there are transient currents. These transients eventually die out and what is called the *steady-state* is reached, with the current a sinusoidal function of time with the same frequency f as that of the generator voltage. Throughout the analysis of AC circuits we assume that this steady-state has been attained.

Suppose that the voltage V across a given circuit element at the instant t is a sinusoidal function of time with angular frequency $\omega = 2\pi f$, given by

$$V = V_0 \cos(\omega t + \theta_{0V})$$

The positive constant V_0 is the amplitude of this function. This is the largest value attained by the instantaneous voltage. The constant θ_{0V} is the initial phase angle of the voltage function.

The current I in the circuit element is a sinusoidal function of the same angular frequency ω as that of the voltage V across the circuit element. The current at the instant t is therefore a function of the form

$$I = I_0 \cos(\omega t + \theta_{0I})$$

This function has an amplitude I_0 and an initial phase angle θ_{0I}.

Following the discussion of Section 9.1, we can represent sinusoidal functions such as I and V by *phasors* on a *phasor diagram*. The voltage phasor is represented by an arrow of length V_0 that makes an angle $\omega t + \theta_{0V}$ with the horizontal axis [Figure 32-3(a)]. The instantaneous voltage V is equal to the projection of this rotating phasor on the horizontal axis. Similarly the current phasor is represented by an arrow of length I_0 at an angle $\omega t + \theta_{0I}$ with the horizontal axis.

The angle θ between the voltage phasor and the current phasor is called the *phase difference* between the voltage and the current [Figure 32-3(b)]. This phase difference θ is the difference of the phase angles of the functions V and I:

$$\theta = (\omega t + \theta_{0V}) - (\omega t + \theta_{0I}) = \theta_{0V} - \theta_{0I}$$

If $\theta = 0$, the voltage and the current are said to be *in phase*; otherwise they are *out of phase*. For circuit elements that include capacitors and inductors, we shall find that the current and the voltage are generally out of phase. The phase difference θ is a most important quantity in AC circuit analysis.

The difference $\theta_{0V} - \theta_{0I}$ of the initial phase angles is significant (being the phase difference θ), but the actual values of these angles are determined by just which instant is selected as the initial instant, $t = 0$. For simplicity we can choose the initial instant to be a time when the current I reaches its peak value I_0. Then $\theta_{0I} = 0$, $\theta_{0V} = \theta$, and the functions V and I are

$$V = V_0 \cos(\omega t + \theta) \qquad (32.2)$$

$$I = I_0 \cos \omega t \qquad (32.3)$$

Example 1   The instantaneous voltage V and current I are given in SI units by

$$V = -155 \cos\left(377t - \frac{\pi}{2}\right)$$

$$I = 2.0 \sin 377t$$

Find the phase difference between the voltage and the current.

Solution   Perhaps the simplest method is to locate the phasors representing V and I at some particular instant. For example, at the time t such that $377t = \pi/2$, $V = -(155 \text{ V}) \cos(\pi/2 - \pi/2) = -155$ V, and $I = 2.0$ A. The phasor diagram at this instant is shown in Figure 32-4. The angle between the voltage phasor and the current phasor, which is the phase difference θ, is seen to be $180°$ or π rad.

To find the phase difference analytically, we first write V and I in the same standard form with a *positive* constant multiplying a *cosine* function of

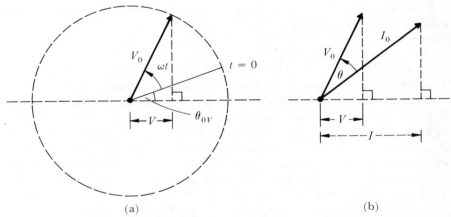

(a)   (b)

Figure 32-3   (a) The voltage phasor is represented by an arrow of length V_0 at an angle $\omega t + \theta_{0V}$ with the horizontal direction on this phasor diagram. (b) The phase difference θ between the voltage and the current is the angle between the voltage phasor and the current phasor.

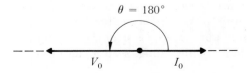

Figure 32-4   In Example 1, the phase difference between the voltage and the current is 180°.

time. Using $\cos(\alpha + \pi) = -\cos \alpha$, we can rewrite the voltage function as

$$V = 155 \, \cos\left(377t - \frac{\pi}{2} + \pi\right)$$

Using $\cos(\alpha - \pi/2) = \sin \alpha$, we can rewrite the current function as

$$I = 2.0 \, \cos\left(377t - \frac{\pi}{2}\right)$$

Then the phase difference is

$$\theta = \left(377t - \frac{\pi}{2} + \pi\right) - \left(377t - \frac{\pi}{2}\right) = \pi$$

32.2 Impedance of Resistors, Inductors, and Capacitors

In DC circuit analysis, the influence of a given circuit element on the rest of the circuit is determined by *one* quantity, the resistance of the circuit element, which is defined as the ratio of the voltage across the element to the current through it.

In an AC circuit, any passive circuit element is characterized by *two* numbers, Z and θ, where θ is the phase difference between the voltage across the element and the current through it, and Z is the ratio of the voltage amplitude V_0 to the current amplitude I_0:

$$Z = \frac{V_0}{I_0} \tag{32.4}$$

These two numbers, Z and θ, determine what is called the *impedance* of the circuit element; Z is called the *magnitude of the impedance* and θ the *angle of the impedance*. The SI unit of Z is the *ohm* (Ω). It turns out that an impedance can be regarded as a phasor. Figure 32-5 is a phasor diagram, called an impedance diagram, that shows the impedance phasor represented by an arrow of length Z at an angle θ to the horizontal.

The relationship between the current and the voltage is determined by the impedance of a circuit element. If Z and θ are known, the current function $I = I_0 \cos \omega t$ can be determined if the voltage function $V = V_0 \cos(\omega t + \theta)$ is known, and vice versa.

Figure 32-5 Impedance diagram. The impedance of a circuit element is a phasor with a magnitude Z and an angle θ. (The angle θ is fixed; impedance phasors do not rotate.)

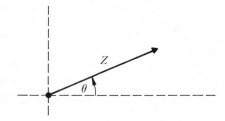

Impedance of a Resistor

In this chapter, all resistors are assumed to obey Ohm's law. If the current through the resistor of resistance R in Figure 32-6(a) is $I = I_0 \cos \omega t$, the voltage across the resistor is

$$V = RI = RI_0 \cos \omega t$$

This voltage function has an amplitude given by

$$V_0 = RI_0$$

Therefore the magnitude of the impedance of a resistor is

$$Z = \frac{V_0}{I_0} = R$$

The phase difference θ between the voltage function and the current function is zero, a situation that is described by saying that the voltage and the current are *in phase*. These results show that the impedance of a resistor is a phasor of magnitude R and angle zero, which is represented on an impedance diagram as shown in Figure 32-6(b).

The phasors representing the current and the voltage are shown in Figure 32-6(c) at the instant $t = 0$. These phasors each rotate counterclockwise with an angular speed ω.

Impedance of an Inductor

Consider an idealized inductor called a *pure* inductor which has a self-inductance L but no resistance [Figure 32-7(a)]. If the current through the inductor is $I = I_0 \cos \omega t$, the voltage across the inductor is

$$V = L\frac{dI}{dt} = -\omega L I_0 \sin \omega t = \omega L I_0 \cos\left(\omega t + \frac{\pi}{2}\right)$$

The voltage function has the amplitude

$$V_0 = \omega L I_0$$

Therefore the magnitude of the impedance of a pure inductor is

$$Z_L = \frac{V_0}{I_0} = \omega L$$

The phase difference θ between the voltage function and the current function is $\pi/2$. The impedance of a pure inductor is therefore a phasor of magnitude ωL and angle $\pi/2$, as shown in Figure 32-7(b). Notice that Z_L is proportional to the frequency.

Rotating phasors representing the voltage and current functions are shown in Figure 32-7(c) at the instant $t = 0$. The current and the voltage are 90° *out of phase*, with the voltage *leading* the current.

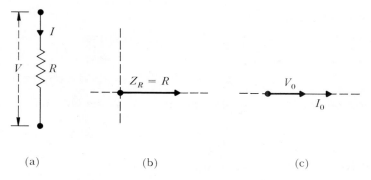

(a)   (b)   (c)

Figure 32-6   (a) Resistor. (b) Impedance diagram. (c) Phasor diagram. Current and voltage are in phase.

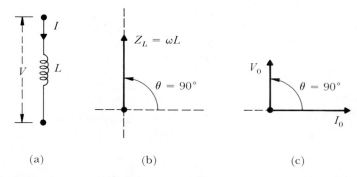

(a)   (b)   (c)

Figure 32-7   (a) Inductor. (b) Impedance diagram. (c) Phasor diagram. Voltage leads the current by 90°.

Example 2   Find the magnitude Z_L of the impedance of a 1.00-mH inductor at $f = 0, f = 60$ Hz, and $f = 60$ kHz.

Solution   At zero frequency, $Z_L = 0$. At 60 Hz,

$$Z_L = \omega L = 2\pi(60 \text{ Hz})(1.00 \times 10^{-3} \text{ H}) = 0.377 \ \Omega$$

At 60 kHz, $Z_L = 377 \ \Omega$.

Example 3   A 60-Hz alternating voltage with an amplitude of 4.00 V is applied across a 1.00-mH inductor. Find the current amplitude.

Solution   From Example 2, at 60 Hz, $Z_L = 0.377 \ \Omega$. Therefore

$$I_0 = \frac{V_0}{Z_L} = \frac{4.00 \text{ V}}{0.377 \ \Omega} = 10.6 \text{ A}$$

Impedance of a Capacitor

If charge flows to the upper plate of the capacitor of Figure 32-8(a) at the rate

$$\frac{dQ}{dt} = I = I_0 \cos \omega t$$

the charge on this plate at the instant t is given by the function

$$Q = \int_0^t I \, dt' = I_0 \int_0^t \cos \omega t' \, dt' = \frac{I_0}{\omega} \sin \omega t$$

The voltage across the capacitor is

$$V = \frac{Q}{C} = \frac{I_0}{\omega C} \sin \omega t = \frac{I_0}{\omega C} \cos\left(\omega t - \frac{\pi}{2}\right)$$

The voltage function has the amplitude

$$V_0 = \frac{I_0}{\omega C}$$

Therefore the magnitude of the impedance of a capacitor is

$$Z_C = \frac{V_0}{I_0} = \frac{1}{\omega C}$$

The phase difference θ between the voltage function and the current function is $-\pi/2$. The impedance of a capacitor is therefore a phasor of magnitude $1/\omega C$ and angle $-\pi/2$, as shown in Figure 32-8(b). Notice that the magnitude of a capacitor's impedance is large at low frequencies and small at high frequencies.

Rotating phasors representing the voltage and the current functions are shown in Figure 32-8(c) at the instant $t = 0$. The voltage *lags* the current by $90°$.

Example 4   Find the magnitude Z_C of the impedance of a $1.00\text{-}\mu\text{F}$ capacitor at $f = 60$ Hz and at $f = 60$ kHz.

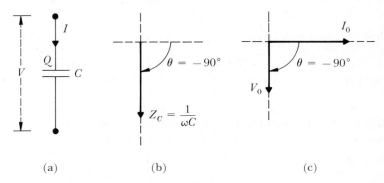

(a)   (b)   (c)

Figure 32-8   (a) Capacitor. (b) Impedance diagram. (c) Phasor diagram. Voltage lags the current by $90°$.

Solution   At $f = 60$ Hz,

$$Z_C = \frac{1}{2\pi(60\text{ Hz})(1.00 \times 10^{-6}\text{ F})} = 2.65 \times 10^3\ \Omega$$

At $f = 60$ kHz, $Z_C = 2.65\ \Omega$.

32.3 Impedance of Series Circuits

Impedance of an *R-L-C* Series Combination

At any instant, the current I is the same in each element of the series combination of Figure 32-2. Suppose this current is $I = I_0 \cos \omega t$. We wish to determine the voltage $V = V_0 \cos(\omega t + \theta)$ across the series combination. At any instant, this voltage is the sum of the voltages across each element of the series combination:

$$V = V_R + V_L + V_C \tag{32.1}$$

Using previous results, we can express the voltage functions for each circuit element as

$$V_R = V_{OR} \cos \omega t$$

$$V_L = V_{OL} \cos\left(\omega t + \frac{\pi}{2}\right)$$

$$V_C = V_{OC} \cos\left(\omega t - \frac{\pi}{2}\right)$$

where the subscript 0 denotes the amplitude of the corresponding function. Equation 32.1 requires

$$V_0 \cos(\omega t + \theta) = V_{OR} \cos \omega t + V_{OL} \cos\left(\omega t + \frac{\pi}{2}\right) + V_{OC} \cos\left(\omega t - \frac{\pi}{2}\right) \tag{32.5}$$

The problem is to determine V_0 and θ in terms of V_{OR}, V_{OL}, and V_{OC}.

The simplest and most illuminating method for finding the amplitude and the initial phase of such a sum of sinusoidal functions is the method of *phasor addition* illustrated in Figure 32-9. Since the current is the same for each element in a series circuit, the current phasor furnishes a convenient reference phasor which is conventionally shown pointing to the right in the phasor diagram (corresponding to the instant that we have selected as $t = 0$). The direction of each voltage phasor is determined from the phase difference between the voltage and the current, as was done to draw the phasor diagrams in Figures 32-6(c), 32-7(c), and 32-8(c).

Phasors are added like vectors. We can interpret Eq. 32.1 as an equation relating the components of phasors along the direction of the current phasor, and we conclude that the phasor representing V is the vector sum (or phasor sum) of

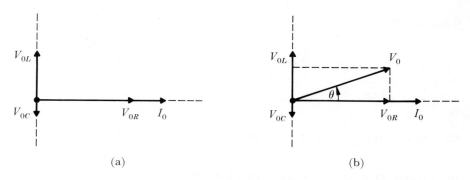

Figure 32-9 Phasor diagram for a series circuit. (a) Phasors representing the circuit current and the voltages across individual circuit elements. (b) The addition of the phasors representing the voltages across L, C, and R to give the phasor (of magnitude V_0) representing the voltage across the series combination.

the phasors representing V_R, V_L, and V_C. In Figure 32-9(b), the phasor representing V has a magnitude V_0 and makes an angle θ with the current phasor. Applying Pythagoras' theorem to this phasor diagram, we find

$$V_0 = \sqrt{V_{0R}^2 + (V_{0L} - V_{0C})^2}$$

It is essential to observe that, because the voltages at any instant are not in phase, the amplitude of the voltage across the combination is *not* simply the sum of the voltage amplitudes V_{0R}, V_{0L}, and V_{0C}. Since $V_{0R} = I_0 R$, $V_{0L} = I_0 \omega L$, and $V_{0C} = I_0 / \omega C$, we have

$$V_0 = I_0 \sqrt{R^2 + \left(\omega L - \frac{1}{\omega C}\right)^2}$$

which shows that the magnitude $Z = V_0 / I_0$ of the impedance of the series combination is given by

$$Z = \sqrt{R^2 + \left(\omega L - \frac{1}{\omega C}\right)^2} \tag{32.6}$$

From the phasor diagram, Figure 32-9(b), we see that the angle θ of this impedance is such that

$$\tan \theta = \frac{V_L - V_C}{V_R} = \frac{\omega L - (1/\omega C)}{R} \tag{32.7}$$

This impedance is represented on the impedance diagram of Figure 32-10.

Resistive and Reactive Components of an Impedance

Instead of specifying an impedance by giving its magnitude Z and angle θ, we can give its horizontal and vertical components on an impedance diagram.

These components are called the resistive component (or the resistance R) and the reactive component (or the reactance X) (Figure 32-11). Thus a pure inductor of inductance L has an impedance specified by $R = 0$, $X_L = \omega L$. The impedance of a capacitor of capacitance C has the components $R = 0$, $X_C = -1/\omega C$. Notice that a capacitive reactance is negative.

From Figure 32-11 it can be seen that the relationships between (Z,θ) and (R,X) are

$$Z = \sqrt{R^2 + X^2} \quad \text{and} \quad \tan \theta = \frac{X}{R} \qquad (32.8)$$

or

$$R = Z \cos \theta \quad \text{and} \quad X = Z \sin \theta \qquad (32.9)$$

Phasor Addition of Impedances in Series

If an impedance having components (R_1,X_1) is connected in series with an impedance having components (R_2,X_2), this combination is equivalent to a single impedance with components (R_s,X_s) given by

$$R_s = R_1 + R_2 \quad \text{and} \quad X_s = X_1 + X_2 \qquad (32.10)$$

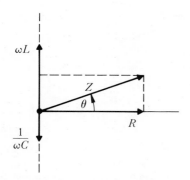

Figure 32-10 Impedance diagram for an R-L-C circuit.

Figure 32-11 Impedance diagram showing the resistive component $R = Z \cos \theta$ and the reactive component $X = Z \sin \theta$ of an impedance phasor.

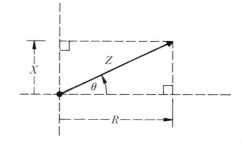

For the impedances of Figure 32-12, this rule can be verified using Eq. 32.6 and 32.7. The methods used to prove these equations can be applied to give a general proof of Eq. 32.10. The phasor addition implied by Eq. 32.10 is represented graphically in Figure 32-12(c).

Example 5   Consider the circuit shown in Figure 32-13:
(*a*) From the amplitudes of the voltages across individual circuit elements, determine the amplitude V_0 of the voltage across the entire part of the circuit that is external to the generator.
(*b*) Calculate the impedance Z of this external circuit and use this to determine V_0.
(*c*) Determine the phase difference between the generator voltage and the current in the circuit.
(*d*) Give the impedance diagram for the *R-L-C* circuit element.

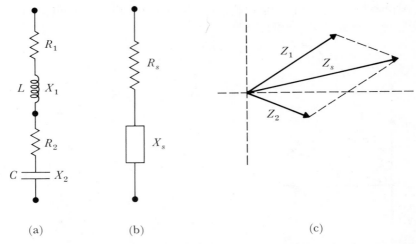

(a)   (b)   (c)

Figure 32-12   (a) Two impedances in series. (b) The single impedance equivalent to the series combination has $R_s = R_1 + R_2$, $X_s = X_1 + X_2$. (c) Phasor addition (or vector addition) of impedances in series.

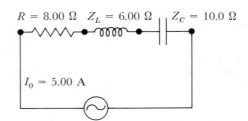

$R = 8.00 \ \Omega \quad Z_L = 6.00 \ \Omega \quad Z_C = 10.0 \ \Omega$

$I_0 = 5.00 \ \text{A}$

Figure 32-13   The *R-L-C* series AC circuit of Example 5.

Solution

(a) $V_{0R} = I_0 Z_R = (5.00 \text{ A})(8.00 \ \Omega) = 40.0$ V, and the phase angle $\theta_R = 0$.

$V_{0L} = I_0 Z_L = (5.00 \text{ A})(6.00 \ \Omega) = 30.0$ V, and phase angle $\theta_L = +90°$.

$V_{0C} = I_0 Z_C = (5.00 \text{ A})(10.0 \ \Omega) = 50.0$ V, and phase angle $\theta_C = -90°$.

Because of the 180° phase difference between the voltage across the inductor and the voltage across the capacitor, we see that the amplitude of the voltage across these two components is given by

$$V_{0LC} = V_{0C} - V_{0L} = (50.0 - 30.0)\text{V} = 20.0 \text{ V}$$

Since V_{0C} is greater than V_{0L}, the phasor of magnitude V_{0LC} points downward on a phasor diagram. Now, evaluating the phasor sum of the phasors of magnitudes V_{0LC} and V_{0R} (see the phasor diagram, Figure 32-14), we find that

$$V_0 = \sqrt{(40.0)^2 + (20.0)^2} \text{ V} = 44.7 \text{ V}$$

(b)
$$Z = \sqrt{R^2 + (X_L + X_C)^2}$$

$$= \sqrt{(8.00)^2 + (6.00 - 10.0)^2} \ \Omega$$

$$= 8.94 \ \Omega$$

Using $V_0 = I_0 Z$, we obtain

$$V_0 = (5.00 \text{ A})(8.94 \ \Omega) = 44.7 \text{ V}$$

in agreement with the result of part (a).

(c) From Figure 32-14 we see that

$$\tan \theta = \frac{-20.0 \text{ V}}{40.0 \text{ V}} = -0.500 \qquad \text{and} \qquad \theta = -26.6°$$

(d) The impedance diagram is shown in Figure 32-15.

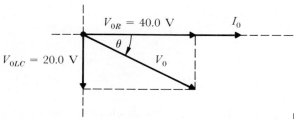

Figure 32-14 Phasor diagram of voltages of Example 5.

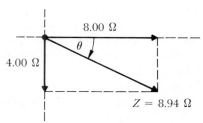

Figure 32-15 Impedance diagram for the *R-L-C* circuit of Example 5.

32.4 Series Resonance

The magnitude of the impedance of an R-L-C circuit is a function of the frequency:

$$Z = \sqrt{R^2 + \left(\omega L - \frac{1}{\omega C}\right)^2} \tag{32.6}$$

If an AC generator of constant voltage amplitude V_0 but variable frequency is connected across the R-L-C circuit, the current amplitude, $I_0 = V_0/Z$, depends on the frequency. The frequency f at which the current amplitude I_0 is a *maximum* is called the *resonant frequency* f_0 of the circuit. Maximum I_0 will be obtained at the frequency which minimizes Z. Inspection of Eq. 32.6 shows that Z is a minimum at an angular frequency ω_0 such that

$$\omega_0 L - \frac{1}{\omega_0 C} = 0$$

The resonant frequency therefore is

$$f_0 = \frac{\omega_0}{2\pi} = \frac{1}{2\pi\sqrt{LC}} \tag{32.11}$$

and at this frequency, $Z = R$.

In radio and television sets, circuits are "tuned" by varying either L or C to adjust the resonant frequency of a circuit to match that of the signals from the desired transmitting station.

Natural Frequency

The resonant frequency of an R-L-C circuit is the frequency at which electrical oscillations would occur naturally without any AC generator, if the resistance were negligible. This is shown by analysis of the L-C circuit of Figure 32-16. Kirchhoff's loop rule with $\mathcal{E}_L = -L\,dI/dt$ and $V_C = Q/C$, gives

$$-L\frac{dI}{dt} = \frac{Q}{C}$$

Since $I = dQ/dt$, this gives

$$\frac{d^2Q}{dt^2} = -\frac{1}{LC}Q \tag{32.12}$$

a differential equation of the form of Eq. 9.17. From the discussion in Section 9.5 we can immediately conclude that the solution Q is a sinusoidal function of time with an angular frequency $\omega_0 = 1/\sqrt{LC}$. The frequency $f_0 = \omega_0/2\pi$ of the naturally occurring oscillations is called the circuit's *natural frequency*; we see that it is the same as the frequency at which an R-L-C circuit resonates. As Figure 32-16 illustrates, the oscillations involve a transfer of energy back-and-forth between the capacitor's electric field and the inductor's magnetic field.

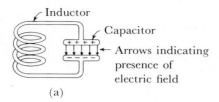

Inductor

Capacitor

← Arrows indicating presence of electric field

(a)

Magnetic flux lines indicating presence of magnetic field

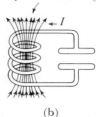

← I

(b)

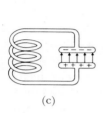

(c)

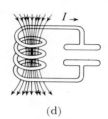

I →

(d)

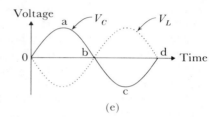

Voltage

V_C V_L

0   b   d   Time

a

c

(e)

Figure 32-16   Resonant circuit with a quarter cycle interval between successive diagrams (a), (b), (c), and (d). In (a) the capacitor is charged and the current is zero; the energy of the circuit is stored in the electric field of the capacitor. In (b) the capacitor is discharged and the current attains its maximum value; the energy is now stored in the magnetic field of the inductor. In (c) the capacitor is charged in the opposite direction and the current is again zero. In (d) the current again attains maximum magnitude but is in the direction opposite to that of the situation in (b). The energy is now back in the magnetic field. The voltage-time graph (e) illustrates that at any instant, $V_C + V_L = 0$.

32.5 Power in AC Circuits

Effective Values of Voltage and Current

Instead of dealing with the amplitude I_0 of an alternating current I, we usually refer to its *effective value* I_e which is defined as the value that causes an average power dissipation in a resistor at the same rate as a steady direct current of the same numerical value: that is,

$$RI_e^2 = R(I^2)_{\text{avg}}$$

For the sinusoidal function $I = I_0 \cos \omega t$, this implies

$$I_e^2 = I_0^2 (\cos^2 \omega t)_{\text{avg}}$$

One easy way to evaluate the average of $\cos^2 \omega t$ is to use the trigonometric identity

$$\cos^2 \omega t = \tfrac{1}{2} - \tfrac{1}{2} \cos(2\omega t)$$

In any cycle, $\cos 2\omega t$ is negative for as long a time as it is positive, so its average value is zero. Consequently, for any integral number of cycles,

$$(\cos^2 \omega t)_{\text{avg}} = \tfrac{1}{2} \qquad (32.13)$$

and a sinusoidal current with amplitude I_0 has an effective value

$$I_e = \frac{1}{\sqrt{2}} I_0$$

Similarly, the effective voltage is $V_e = V_0/\sqrt{2}$. The "110 volts AC" in home circuits is an effective value, the amplitude or peak value being $\sqrt{2} \times 110 = 156$ V. AC ammeters and voltmeters are usually calibrated to give effective values.

The AC circuit analysis that has been developed in terms of the *amplitudes* of currents and voltages can equally well be carried out using *effective* values. In any phasor diagram, we can make the magnitude of every phasor the effective value rather than the amplitude of the corresponding function. This choice merely changes the magnitudes of all the phasors by the same factor, $1/\sqrt{2}$, and therefor will not affect relationships among the phasors. And the magnitude of an impedance is the ratio of the effective voltage to the effective current since

$$Z = \frac{V_0}{I_0} = \frac{V_e}{I_e}$$

Power Factor

At an instant when there is a voltage $V = V_0 \cos(\omega t + \theta)$ across a circuit element carrying a current $I = I_0 \cos \omega t$, the instantaneous power supplied to the circuit element is

$$P = VI = V_0 I_0 \cos \omega t \cos(\omega t + \theta)$$

Using the identity, $\cos(\omega t + \theta) = \cos \omega t \cos \theta - \sin \omega t \sin \theta$, we have

$$P = V_0 I_0 \cos \theta \cos^2 \omega t - V_0 I_0 \sin \theta \cos \omega t \sin \omega t$$

For a complete cycle, the average value of $\cos \omega t \sin \omega t$ is zero and the average value of $\cos^2 \omega t$ is $\tfrac{1}{2}$. Therefore the average power input to the device is

$$P_{\text{avg}} = \tfrac{1}{2} V_0 I_0 \cos \theta = V_e I_e \cos \theta \qquad (32.14)$$

The term $\cos \theta$ is called the *power factor* of the circuit element. For a resistance R, $\theta = 0$, $\cos \theta = 1$, and $P_{\text{avg}} = V_e I_e = I_e^2 R$. For a pure inductance or capacitance, $\theta = \pm 90°$, $\cos \theta = 0$, and $P_{\text{avg}} = 0$. Although power is supplied to an inductor or a capacitor during one part of a cycle, this power is returned during another part of the cycle and the average power input is zero. All the average power supplied to a passive circuit element is dissipated in the resistive component R of the element's impedance. This is verified by noting that, for a passive circuit element, $\cos \theta = R/Z$, $V_e = I_e Z$, and therefore

$$P_{\text{avg}} = V_e I_e \cos \theta = R I_e^2$$

Summary

☐ An AC generator, providing a sinusoidal voltage of angular frequency ω, establishes in the steady-state an instantaneous voltage V across a given circuit element and an instantaneous current I through the element given by

$$V = V_0 \cos(\omega t + \theta)$$

$$I = I_0 \cos \omega t$$

The voltage amplitude is V_0 and the current amplitude is I_0. The phase difference between the voltage and the current is θ.

☐ A passive circuit element has an impedance characterized by a magnitude $Z = V_0/I_0$ and an angle θ, or by a resistive component $R = Z \cos \theta$ and a reactive component $X = Z \sin \theta$.

$$Z^2 = \sqrt{R^2 + X^2} \qquad \text{and} \qquad \tan \theta = \frac{X}{R}$$

The impedance of an R-L-C series combination has a resistive component R and an inductive component $X = \omega L - 1/\omega C$.

☐ For circuit elements connected in series, instantaneous voltages add algebraically but voltage amplitudes add vectorially. Each voltage $V = V_0 \cos(\omega t + \theta)$ is represented by a phasor of magnitude V_0 at an angle θ with the current phasor. The phasor representing the voltage across a series combination is the vector sum of the phasors representing the voltages across the individual circuit elements.

☐ Impedances in series add like vectors.

☐ When the frequency of the voltage across an R-L-C series circuit is varied, maximum current (and minimum impedance $Z_{\min} = R$) occur at the resonant frequency $f_0 = 1/2\pi \sqrt{LC}$, which is the natural frequency of an L-C circuit.

☐ The effective value of a sinusoidal current or voltage is $1/\sqrt{2}$ times the amplitude.

☐ When the voltage across a circuit and the current through it have a phase difference θ and have effective values V_e and I_e, the average power input to the circuit element is

$$P_{\text{avg}} = V_e I_e \cos \theta$$

Questions

1 If the current in Figure 32-2(c) is a function of the form $I = I_0 \cos(\omega t + \theta_{0I})$, what is the initial phase angle θ_{0I}?

2 (a) Find the initial phase angle θ_{0V} for each of the voltages in the graphs of Figure 32-2(b), assuming that each voltage is a function of the form $V = V_0 \cos(\omega t + \theta_{0V})$.

(b) Find the phase difference between the voltage and the current for each voltage graphed in Figure 32-2(b).

3   In the graphs of Figure 32-2, assume that the period is $T = 1/f = 8.0 \times 10^{-2}$ s and that the amplitudes of the sinusoidal functions are $I_0 = 2.0$ A, $V_{OR} = 4.0$ V, $V_{OL} = 8.0$ V, and $V_{OC} = 6.0$ V. Find the instantaneous current I and the instantaneous voltages V_R, V_L, V_C, and V at the instant $t = 1.0 \times 10^{-2}$ s. If the lower terminal of the generator is grounded, what are the potentials (relative to ground) of the terminals of each circuit element at the instant $t = 1.0 \times 10^{-2}$ s?

4   Find the phase difference between the voltage V and the current I given in SI units by

$$V = 155 \sin 377t$$

$$I = 3.0 \cos 377t$$

5   Find the phase difference between the voltage V and the current I given in SI units by

$$V = -10 \sin 20t$$

$$I = 0.30 \cos 20t$$

6   The instantaneous current I across a 200-Ω resistor is given in SI units by

$$I = 10 \cos 377t$$

(a) Find the frequency and the period.
(b) Find the voltage across the resistor at the following times: $t = 0$, $t = \frac{1}{240}$ s, and $t = \frac{1}{120}$ s.

7   What is the current amplitude in a 1.50-mH inductor when the 60-Hz alternating voltage across the inductor has an amplitude of 120 V?

8   Evaluate the current amplitude in the 1.50-mH inductor of Question 7 when the voltage amplitude is maintained at 120 V, but the frequency is:
(a) 6.0 Hz   **(b)** 600 Hz

9   Sketch a graph showing the magnitude of the impedance of an inductor as a function of the frequency. Examine and comment on the values assumed at very low and at very high frequencies.

10   Sketch a graph showing the magnitude of the impedance of a capacitor as a function of the frequency. Examine and comment on the values assumed at very low and at very high frequencies.

11   At what frequency will a 5.00-μF capacitor have an impedance with a magnitude of 4.00 kΩ?

12   The 60-Hz current in a 4.00-μF capacitor has an amplitude of 20.0 A. Find the amplitude of the voltage across the capacitor.

13   In an R-L series circuit, the voltage across the resistor has an amplitude of 50 V and the voltage across the inductor has an amplitude of 120 V. Find the amplitude of the voltage across the R-L series combination. What is the phase difference between this voltage and the current?

14   A pure inductor and a capacitor are connected in series in an AC circuit. The voltage amplitude for the inductor is 1200 V and for the capacitor is 1000 V. What is the amplitude of the voltage across the L-C series combination? What is the phase difference between this voltage and the current?

15 In Figure 32-2 the amplitudes of the voltages across the resistor, the inductor, and the capacitor are $V_{OR} = 30$ V, $V_{OL} = 60$ V, and $V_{OC} = 20$ V, respectively:
 (a) Find the amplitude of the voltage across the L-C combination.
 (b) Find the amplitude of the voltage across the R-L-C combination. What is the angle of the impedance of this R-L-C combination?
 (c) If the current amplitude is 2.0 A, what is the magnitude of the impedance of the R-L-C combination?

16 A resistance of 20 Ω, an inductive reactance of 80 Ω, and a capacitive reactance of -50 Ω are connected in series across an AC source of 60 Hz:
 (a) Draw the circuit diagram.
 (b) Find the total impedance of the circuit.

17 A 30-Ω resistor and a capacitor which has an impedance with a magnitude of 40 Ω are connected in series with an AC generator. The current amplitude is 3.0 A:
 (a) What is the amplitude of the voltage across the resistor?
 (b) What is the amplitude of the voltage across the capacitor?
 (c) Sketch the corresponding voltage phasors on a phasor diagram.
 (d) Determine the amplitude of the voltage across the series combination of the capacitor and the resistor.
 (e) Does the voltage across the capacitor–resistor combination lead or lag the current through it?

18 A 60-Hz AC generator applies a voltage with an amplitude of 120 V across a series combination of a 25.0-μF capacitor, a 0.100-H inductor and a 25.0-Ω resistor. Find:
 (a) The inductive reactance.
 (b) The capacitive reactance.
 (c) The impedance of the circuit.
 (d) The current amplitude in the circuit.

19 An impedance with a magnitude of 80 Ω and an angle of 20° is connected in series with an impedance of magnitude 120 Ω and an angle of $-45°$:
 (a) Find the resistive and the reactive components of each impedance.
 (b) Find the magnitude and the angle of the impedance of the series combination.

20 **(a)** What must be the inductive reactance of an inductance coil in order for it to be used with a capacitor with a capacitive reactance of -100 Ω to form a resonant circuit?
 (b) If the coil and the capacitor are connected in series, and if the coil has a resistance of 100 Ω, what is the amplitude of the current that will pass through the circuit if the amplitude of the voltage across the series combination is 10 V?

21 What is the capacitance of the capacitor required in series with a 40-mH inductance coil to provide a circuit which resonates at a frequency of 60 Hz?

22 In Figure 32-2, R has a resistance of 250 Ω, L has an inductance of 0.50 H and no resistance, and C has a capacitance of 0.020 μF:
 (a) What is the resonant frequency of the circuit?
 (b) The capacitor can withstand a *peak* voltage of 350 V. What is the maximum voltage amplitude that can be applied across the R-L-C series combination?

23 **(a)** In a DC circuit, is it possible to have a voltage across any circuit component larger than the EMF applied to the whole circuit? In an AC circuit, is it possible to have a voltage amplitude across some circuit component larger than the amplitude of the generator voltage applied to the entire circuit? Explain.

(b) One of the characteristics of a resonant R-L-C series circuit is the large "output" voltage, across either the inductor or the capacitor, obtained with a relatively small input voltage across the entire R-L-C series circuit. Show that this "gain," given by

$$\frac{output\ voltage\ amplitude}{input\ voltage\ amplitude}$$

is equal to the ratio $\omega_0 L/R$, called the "Q" of the circuit. Determine the Q of the circuit of Question 22.

24   Consider an L-C oscillating circuit. Kirchhoff's loop rule gives

$$L\frac{d^2Q}{dt^2} = -\frac{1}{C}Q$$

Assume that the initial conditions are: $Q = Q_0$ and $I = dQ/dt = 0$ at $t = 0$:
(a) Show that the solution to the differential equation that fits the initial conditions is

$$Q = Q_0 \cos \omega_0 t$$

where $\omega_0 = 1/\sqrt{LC}$.
(b) Compare the differential equation for Q with the differential equation for the displacement x of a mass m fastened to a spring of force constant k. Make a table showing the electrical quantities analogous to m, x, k, $\frac{1}{2}mv^2$, and $\frac{1}{2}kx^2$.

25   What is the amplitude of the voltage applied across the coils of an electric stove, when sinusoidal AC is used and the effective value of the voltage is 240 V?

26   The instantaneous voltage across a 5.0-mH inductor is given in SI units by

$$V = 17 \cos 377t$$

Find the effective value of the current in this inductor.

27   The effective value of the alternating current in a 200-Ω resistor is 10 A:
(a) What is the effective value of the voltage across this resistor?
(b) What is the average power dissipated in the resistor?

28   If in an AC circuit an effective current of 20 A at an effective voltage of 110 V yields an average power dissipation of 2.2 kW, what can be said about the nature of the circuit impedance?

29   Suppose that in the AC series circuit in Figure 32-2(a) the voltage across the resistor is 100 V, across the inductor 40 V, and across the capacitor 80 V, while the current is 15 A. All values are effective values:
(a) What are the individual average power dissipations of the capacitor, the resistor, and the inductor?
(b) What is the average power dissipation of the circuit?

30   A 60-Hz alternating voltage of 110 V effective value is applied to a series circuit having an inductance L of 1.00 H, a capacitance C of 2.00 μF, and a resistance R of 400 Ω. Calculate the:
(a) Inductive reactance.
(b) Capacitive reactance.
(c) Total impedance.
(d) Effective current in the circuit.
(e) Power factor of the circuit.

(f) Effective voltage V_{eR} across R.

(g) Effective voltage V_{eL} across L.

(h) Effective voltage V_{eC} across C, sketching approximately to scale a phasor diagram showing the phasors corresponding to V_{eR}, V_{eL}, and V_{eC}.

(i) Average power dissipation in the circuit.

(j) Amplitude of the voltage across the circuit.

(k) Amplitude of the current.

31 An AC voltage generator applies an effective voltage of 200 V across an R-L-C series circuit with $R = 1.50 \times 10^2 \, \Omega$, $L = 5.25 \times 10^{-1}$ H, and $C = 2.00 \, \mu$F. The frequency of the generator voltage is 100 Hz:

(a) What is the power dissipated in the circuit?

(b) What would be the power dissipated in the circuit at its resonant frequency?

Supplementary Questions

S-1 Sketch a graph showing θ as a function of the angular frequency ω, where θ is the angle of the impedance of an R-L-C series combination.

S-2 Show that in the circuit of Figure 32-17(a), if the impedance Z connected across the output is so large that its effects can be ignored, then

$$\frac{V_{0,out}}{V_{0,in}} = \frac{1/\omega C}{[R^2 + (1/\omega C)^2]^{1/2}}$$

Sketch a graph showing this voltage ratio as a function of the frequency. Consider an input voltage that contains both low and high frequency signals. Explain why it is appropriate to call this circuit a *low-pass filter*.

S-3 In Figure 32-17(b), assume that Z is so large that its effects can be ignored. Show that

$$\frac{V_{0,out}}{V_{0,in}} = \frac{R}{[R^2 + (1/\omega C)^2]^{1/2}}$$

and explain why it is appropriate to call this circuit a *high-pass filter*.

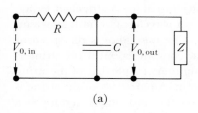

(a)

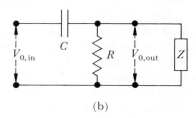

(b)

Figure 32-17

33.

wave phenomena

The continuation of the study of electromagnetism leads to a study of electromagnetic waves. This chapter provides a general introduction to any type of wave motion. Since an electromagnetic wave is a rather abstract entity, it is best to consider more familiar waves first. Everyday examples are waves on the surface of water, waves on a stretched string, and sound waves in air. Because waves on a stretched string are the easiest to describe, we use this particular example to develop ideas that are generally applicable to all waves.

Figure 33-1 illustrates a transverse wave traveling along a stretched string. If one end of the string is disturbed briefly, then a waveform of limited extent, called a pulse, is propagated along the string. The *wave speed* is the speed at which this *waveform* travels. The wave travels at constant speed if the string is homogeneous and has a constant tension. The motion of a particle of the string is quite different from the motion of the waveform. A particle P of the string is motionless until the pulse reaches it, whereupon the particle accelerates in a direction perpendicular to the string, the y direction in Figure 33-1. The particle thus acquires a velocity v_y and a displacement y from its equilibrium position. As the pulse moves on, this particle reverses the direction of motion. After the pulse has passed by, the particle has returned to its equilibrium position and is at rest. Such waves, which involve a disturbance perpendicular to the direction of wave propagation, are called *transverse* waves.

Any wave that consists of a disturbance of the particles of a material medium is called a *mechanical* wave. Since each macroscopic portion of a material moves

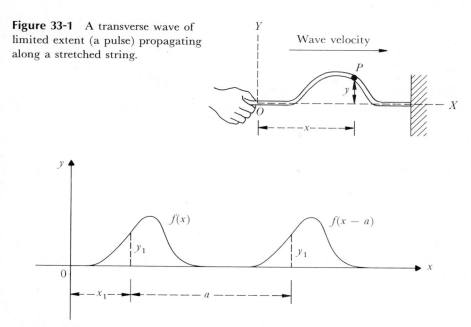

Figure 33-1 A transverse wave of limited extent (a pulse) propagating along a stretched string.

Figure 33-2 The graph of $f(x - a)$ is obtained by displacing the graph of $f(x)$ a distance a without deformation.

in accordance with Newton's second law, the underlying theory of mechanical waves is provided by Newtonian mechanics. This theory is developed in Section 33.2 for transverse waves on a string, but before we consider the physical reason for the occurrence of a wave, we take up the kinematical problem—the purely mathematical description of traveling waves.

33.1 Mathematical Description of Waves

Each particle of the string is labeled by its coordinate x. A mathematical description of the wave is given by specifying the function $y(x,t)$ that gives the displacement of the particle with coordinate x at the instant t. In this section we investigate the type of functions of x and t that describe a wave which travels with a constant speed and maintains its waveform as it travels.

Figure 33-2 shows the graph of a function $y = f(x)$ together with the graph of the function $y = f(x - a)$. (This notation means that the second function is obtained from the first by replacing each occurrence of x in $f(x)$ by $x - a$.) Any given value y_1 that occurs at the value x_1 in the first graph, occurs at the value $x_1 + a$ in the second graph. Consequently, the second graph is obtained by displacing the first graph a distance a without deformation.

Now if $a = vt$, where t is the time and v is a positive constant, we obtain the function

$$y = f(x - vt)$$

The preceding discussion shows that the y versus x graph of this function is displaced a distance vt in the time t. Therefore this function represents a *traveling wave* that moves with an unchanging waveform in the positive x direction at a wave speed v. The argument $x - vt$ of this function is called the phase of the wave. The wave speed is the speed of travel of a point on the waveform corresponding to a given value of the phase. Similarly,

$$y = f(x + vt)$$

represents a traveling wave moving in the *negative* x direction with wave speed v.

Example 1   A transverse wave on a stretched string is a pulse that at $t = 0$ is described in SI units by

$$y = \frac{6.0}{x^2 + 2.0}$$

The wave travels in the positive x direction at a constant speed of 3.0 m/s and maintains its waveform:
(a) What is the function $y(x,t)$ that describes this wave?
(b) Consider the motion of the particle located at $x = 4.0$ m. Find its velocity at the instant $t = 1.0$ s.

Solution
(a) The function that describes this traveling wave is given in SI units by

$$y = \frac{6.0}{(x - 3.0t)^2 + 2.0}$$

(b) The displacement of the particle $x = 4.0$ m is given as a function of time by

$$y = \frac{6.0}{(4.0 - 3.0t)^2 + 2.0}$$

The velocity of this particle is

$$v_y = \frac{dy}{dt} = \frac{36(4.0 - 3.0t)}{[(4.0 - 3.0t)^2 + 2.0]^2}$$

This gives, at $t = 1.0$ s,

$$v_y = 4.0 \text{ m/s}$$

The Wave Equation

When y represents a traveling wave, there is a special relationship between its partial derivatives with respect to x and to t. For a wave traveling in the positive

x direction, $y = f(x - vt) = f(z)$, where we have put $z = x - vt$. Then using the chain rule, we find that

$$\frac{\partial y}{\partial x} = \frac{df}{dz}\frac{\partial z}{\partial x} = \frac{df}{dz} \quad (1)$$

$$\frac{\partial y}{\partial t} = \frac{df}{dz}\frac{\partial z}{\partial t} = \frac{df}{dz}(-v)$$

Therefore $\partial y/\partial x = -(1/v)\partial y/\partial t$. For a wave propagating in the opposite direction $y = f(x + vt)$, and a similar calculation shows that the partial derivatives then satisfy $\partial y/\partial x = (1/v)\partial y/\partial t$. Repeating this procedure we find that for the *second* derivatives

$$\frac{\partial^2 y}{\partial x^2} = \frac{d^2 f}{dz^2} \quad \text{and} \quad \frac{\partial^2 y}{\partial t^2} = \frac{d^2 f}{dz^2}v^2$$

when $y = f(x - vt)$ and also when $y = f(x + vt)$. Consequently, a wave $y(x,t)$ traveling in either the positive or negative x direction must be a function such that its partial derivatives satisfy

$$\frac{\partial^2 y}{\partial x^2} = \frac{1}{v^2}\frac{\partial^2 y}{\partial t^2} \tag{33.1}$$

a partial differential equation called the *wave equation.*

Whenever the application of the basic laws of physics to a physical system leads to a quantity y that satisfies this wave equation, we can conclude that y is a quantity that can propagate along the x direction as a traveling wave with unchanging waveform and a wave speed determined by the constant that plays the role of v in Eq. 33.1.

33.2 Mechanical Theory of Waves on a String

The occurrence of transverse waves on a string and their properties can be understood by applying the laws of Newtonian mechanics. To simplify the problem we assume that the string vibrates only in a vertical plane and that the displacement y of the string is everywhere small enough so that the tension in the string does not change appreciably and that each particle of the string moves along a vertical line.

We select as the system a small element of the string whose length in the equilibrium position would be Δx. The mass of this element is $\mu\Delta x$, where μ is the mass per unit length of the string. The free-body diagram, Figure 33-3, shows this element when it is displaced and stretched. At each end the element is pulled by a force that is tangential to the string and whose magnitude is the tension denoted by τ. The weight $(\mu\Delta x)g$ of the string has been neglected.

If y is the displacement of the element, its velocity is $\partial y/\partial t$ and its acceleration is $\partial^2 y/\partial t^2$. Newton's second law for y components gives

$$\tau \sin \theta' - \tau \sin \theta = (\mu\Delta x)\frac{\partial^2 y}{\partial t^2}$$

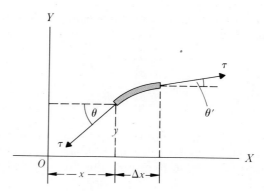

Figure 33-3   Free-body diagram for element of string of length Δx.

We assume that the angle θ between the string and the horizontal is everywhere small, so that $\sin \theta$ is approximately equal to $\tan \theta$ which is the slope $\partial y/\partial x$ of the string. In this approximation, we have

$$\tau\left[\left(\frac{\partial y}{\partial x}\right)' - \left(\frac{\partial y}{\partial x}\right)\right] = (\mu \Delta x)\frac{\partial^2 y}{\partial t^2}$$

where $(\partial y/\partial x)'$ is the slope of the string at the point with coordinate $x + \Delta x$. Since $(\partial y/\partial x)' - (\partial y/\partial x)$ is the change in the quantity $\partial y/\partial x$ that occurs in a length Δx, we have in the limit as Δx approaches zero

$$\left(\frac{\partial y}{\partial x}\right)' - \left(\frac{\partial y}{\partial x}\right) = \frac{\partial}{\partial x}\left(\frac{\partial y}{\partial x}\right)\Delta x = \frac{\partial^2 y}{\partial x^2}\,\Delta x$$

For an infinitesimal length of the string, the expression of Newton's second law for y components is therefore

$$\tau\frac{\partial^2 y}{\partial x^2}\,\Delta x = \mu\Delta x\,\frac{\partial^2 y}{\partial t^2}$$

or

$$\frac{\partial^2 y}{\partial x^2} = \frac{1}{(\tau/\mu)}\frac{\partial^2 y}{\partial t^2} \tag{33.2}$$

This result is the wave equation (Eq. 33.1) with $v = \sqrt{\tau/\mu}$. Knowing that the wave equation has traveling wave solutions, we can conclude that the occurrence of traveling waves on a stretched string is consistent with Newtonian mechanics. Moreover, we have now learned that the string's linear density μ and tension τ determine the wave speed v, according to

$$v = \sqrt{\frac{\tau}{\mu}} \tag{33.3}$$

Power Transmitted by a Wave

A traveling wave transports energy as it moves along a string. The power transmitted from left to right across the point x on the string in Figure 33-3 is the product of the string's upward velocity $\partial y / \partial t$ at this point and the upward force exerted by the portion of the string to the left of this point on the portion to the right. In Figure 33-3, this upward force is $-\tau \sin \theta$ which, for small angles, is approximately equal to $-\tau \tan \theta = -\tau(\partial y / \partial x)$. Therefore the power transmitted across x is

$$\frac{\partial U}{\partial t} = -\tau \frac{\partial y}{\partial x} \frac{\partial y}{\partial t} \qquad (33.4)$$

33.3 Sinusoidal Traveling Waves

Consider a sinusoidal waveform described by

$$y = A \cos(kx + \theta_0) \qquad (33.5)$$

where A and k are positive constants. The angle θ_0 is a constant that can be positive, negative, or zero. Replacing x by $x - vt$, we obtain the equation of a sinusoidal traveling wave moving in the positive x direction with a wave speed v:

$$y = A \cos[k(x - vt) + \theta_0]$$

This is usually written as

$$y = A \cos(kx - \omega t + \theta_0) \qquad (33.6)$$

where

$$\omega = kv \qquad (33.7)$$

An alternative expression, obtained using $\cos(-\alpha) = \cos \alpha$, is

$$y = A \cos(\omega t - kx - \theta_0)$$

Examination of this function shows that the displacement of the particle with position coordinate x is given by

$$y = A \cos[\omega t + \theta(x)] \qquad (33.8)$$

which is a simple harmonic motion with amplitude A and initial phase angle $\theta(x) = -(kx + \theta_0)$. The angular frequency is ω, and we know that $\omega = 2\pi f = 2\pi / T$, where f is the frequency and T is the period of the particle's motion (Figure 33-4). When a string is transmitting the sinusoidal traveling wave described by Eq. 33.6, each particle executes a simple harmonic motion with

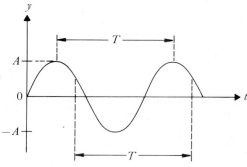

Figure 33-4   Graph showing the displacement y of one particle of a string as a function of the time t. The particle executes simple harmonic motion with an amplitude A and a period T.

the *same amplitude A* and the *same frequency f,* but there is a *phase difference*

$$\theta(x_2) - \theta(x_1) = k(x_1 - x_2) \tag{33.9}$$

between the particle at x_2 and the particle at x_1.

We now consider some particular instant t and examine the function y given by Eq. 33.6 as a function of the *spatial variable x:*

$$y = A \cos[kx + \theta(t)] \tag{33.10}$$

where $\theta(t) = \theta_0 - \omega t$. A snapshot of the string at the instant t furnishes a graph of y as a function of x. Figure 33-5 shows this graph at an instant such that $\theta(t) = -\pi/2$. The interpretation of A, k, and $\theta(t)$ in this function must correspond to the interpretation of the analogous quantities in Eq. 33.8. The amplitude A of the wave is the maximum displacement and all values of y lie between A and $-A$. Since the function y given by Eq. 33.8 is a periodic function of t with a periodic time or period $T = 2\pi/\omega$, the function y given by Eq. 33.10 must be a periodic function of x with a spatial period λ determined by $\lambda = 2\pi/k$. The spatial period λ, called the *wavelength,* is the distance from crest to crest indicated in Figure 33-5. More generally, the wavelength λ of a traveling wave on a string is the shortest distance such that all particles a distance λ apart have identical motions, that is, for all x,

$$y(x,t) = y(x + \lambda,t) \tag{33.11}$$

The positive constant $k = 2\pi/\lambda$ is called the *wave number.* This is the number of wavelengths in a distance equal to 2π units of length.

To summarize, the function

$$y = A \cos(kx - \omega t + \theta_0)$$

for a given x is a periodic function of the time t with period $T = 2\pi/\omega$, and for a given t is a periodic function of the coordinate x with a spatial period or wavelength $\lambda = 2\pi/k$. The graph of the waveform at two successive instants is shown

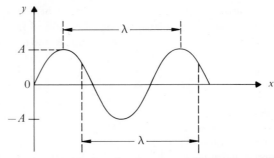

Figure 33-5 A snapshot of a string that is transmitting a sinusoidal wave of amplitude A and wavelength λ. This displays y as a function of x at a particular instant. The spatial period of this function is the wavelength λ. The wave number is $k = 2\pi/\lambda$.

Figure 33-6 The solid curve is the graph of y as a function of x at the instant t_1. The dashed curve shows this function at a later instant t_2. The sinusoidal waveform has moved to the right a distance $v(t_2 - t_1)$.

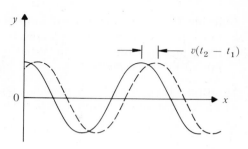

in Figure 33-6. The function describes a wave that travels with a speed v given by Eq. 33.7:

$$v = \frac{\omega}{k} = \lambda f \qquad (33.12)$$

This is the speed of travel of a given point on the waveform (such as a crest); such a point corresponds to a constant value of the phase angle $kx - \omega t + \theta_0$.

A wave traveling in the negative x direction is obtained from Eq. 33.5 by replacing x by $x + vt$. We then have

$$y = A \cos[k(x + vt) + \theta_0]$$

or, with $\omega = kv$,

$$y = A \cos(kx + \omega t + \theta_0)$$

Again $\omega = 2\pi f = 2\pi/T$ and $k = 2\pi/\lambda$, where T and λ are the temporal and spatial periods, respectively.

Example 2 A wave on a string is described in SI units by

$$y = 0.02 \cos(6.0x + 12.0t)$$

Find the wave velocity, the wavelength λ, and the period T.

Solution The wave number is $k = 6.0$ m^{-1} and the angular frequency is $\omega = 12.0$ rad/s. The wave velocity is in the negative x direction and has a magnitude (the wave speed) given by

$$v = \frac{\omega}{k} = \frac{12.0 \text{ rad/s}}{6.0 \text{ m}^{-1}} = 2.0 \text{ m/s}$$

The wavelength is

$$\lambda = \frac{2\pi}{k} = \frac{2\pi}{6.0 \text{ m}^{-1}} = 1.05 \text{ m}$$

The period is

$$T = \frac{2\pi}{\omega} = \frac{2\pi}{12.0 \text{ rad/s}} = 0.52 \text{ s}$$

Example 3   A sinusoidal wave of amplitude A and angular frequency ω travels in the positive x direction with a wave speed v along a stretched string which has a mass per unit length μ. What is the average power $(\partial U/\partial t)_{avg}$ transmitted by this wave?

Solution   Equation 33.4 expresses the power $\partial U/\partial t$ transmitted past any point x at the instant t as the product of the force $-\tau(\partial y/\partial x)$ and the velocity $\partial y/\partial t$:

$$\frac{\partial U}{\partial t} = -\tau \frac{\partial y}{\partial x}\frac{\partial y}{\partial t}$$

where τ is the tension in the stretched string. The sinusoidal traveling wave can be represented by

$$y = A\,\cos(kx - \omega t)$$

where $k = \omega/v$. For this wave,

$$\frac{\partial y}{\partial x} = -kA\,\sin(kx - \omega t) \qquad \frac{\partial y}{\partial t} = \omega A\,\sin(kx - \omega t)$$

and

$$\frac{\partial U}{\partial t} = \tau k\omega A^2\,\sin^2(kx - \omega t)$$

Making the substitutions $\tau = \mu v^2$ (obtained from $v = \sqrt{\tau/\mu}$) and $k = \omega/v$, and using the fact that the average value of a sinusoidal function of time is $\frac{1}{2}$ (Section 32.5), we obtain

$$\left(\frac{\partial U}{\partial t}\right)_{avg} = \tfrac{1}{2}\mu v\omega^2 A^2$$

The average power transmitted is proportional to the *square* of both the *frequency* and the *amplitude*. This turns out to be true for sinusoidal traveling waves of all types.

33.4 Superposition Principle for Waves

Figure 33-7 illustrates what happens when two waves are simultaneously present on the string. The displacement at any instant $y(x,t)$ is the algebraic sum of the displacements $y_1(x,t)$ and $y_2(x,t)$ that would occur if each individual wave were present alone. This is the *superposition principle* for waves.

Experiments show that the superposition principle holds for transverse waves on a stretched string provided the displacements are small. Superposition phenomena are important in the study of sound, light, and quantum physics.

For waves that are solutions of the wave equation, $\partial^2 y/\partial x^2 = (1/v^2)(\partial^2 y/\partial t^2)$, the superposition principle is a mathematical consequence of the linearity* of

---

*A partial differential equation is linear if y and its partial derivatives occur only to the first power.

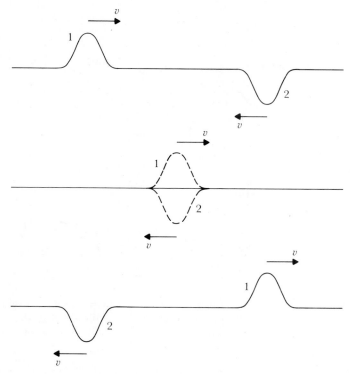

Figure 33-7   Superposition of waves. At each of the three instants represented, the solid line $y(x,t)$ is the algebraic sum of the displacements that would occur if each individual wave were present alone.

this differential equation. For if $y_1(x,t)$ and $y_2(x,t)$ are two different possible wave motions, that is, if

$$\frac{\partial^2 y_1}{\partial x^2} = \frac{1}{v^2}\frac{\partial^2 y_1}{\partial t^2} \qquad \text{and} \qquad \frac{\partial^2 y_2}{\partial x^2} = \frac{1}{v^2}\frac{\partial^2 y_2}{\partial t^2}$$

then, by adding these equations, it follows that

$$\frac{\partial^2}{\partial x^2}(y_1 + y_2) = \frac{1}{v^2}\frac{\partial^2}{\partial t^2}(y_1 + y_2)$$

Therefore the sum $y_1 + y_2$ also satisfies the wave equation and is thus another possible wave motion.

The superposition of two waves of the same frequency traveling in the same direction is illustrated in Figure 33-8. The amplitude of the superposition depends on the phase difference of the two waves, being the sum of their amplitudes if the two waves are in phase [Figure 33-8(a)] and the difference of their amplitudes if the waves are 180° out of phase [Figure 33-8(b)]. Phenomena associated with such superpositions are examined in Chapter 36.

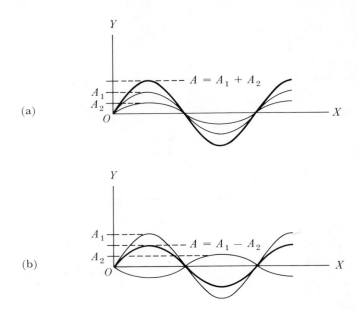

Figure 33-8 The superposition of two waves of the same frequency traveling in the same direction. (a) If the two waves are in phase, the amplitude A of the superposition is the sum $A_1 + A_2$. (b) If the two waves are 180° out of phase, the amplitude A of the superposition is the difference $A_1 - A_2$.

The waveforms encountered in practice are usually much more complex than a sinusoidal waveform. Fortunately, by using a mathematical procedure known as Fourier analysis, we can express any waveform as a superposition of sinusoidal waveforms of appropriately selected amplitudes, wavelengths, and phases. The study of sinusoidal waves thus forms the basis for an understanding of arbitrarily complex waves, if the waves obey the superposition principle.

Standing Waves

An important example of superposition occurs with sinusoidal waves of the same amplitude and frequency traveling in opposite directions. We consider a string which lies along the X axis to the right of the origin. The string is stretched and a sinusoidal traveling wave

$$y_1(x,t) = A_1 \cos(kx + \omega t + \phi_1)$$

is produced by moving the free end of the string up and down with a simple harmonic motion of angular frequency ω and amplitude A. This wave travels from the free end toward the origin. At the origin where the string is fastened, the support holds the string motionless by exerting a periodically varying reaction force which generates a second wave $y_2(x,t)$ traveling to the right with the same wave speed $v = \omega/k$ and the same angular frequency ω. To satisfy the *boundary condition*

$$y(0,t) = 0$$

for all times t, the incident and the reflected waves must have the same amplitude and be 180° out of phase at the origin.

The reflected wave and the incident wave (with ϕ_1 put equal to π) therefore form a superposition

$$
\begin{aligned}
y(x,t) &= y_1(x,t) + y_2(x,t) \\
&= A_1 \cos(kx + \omega t + \pi) + A_1 \cos(kx - \omega t) \\
&= -A_1 \cos(kx + \omega t) + A_1 \cos(kx - \omega t) \\
&= A_1(-\cos kx \cos \omega t + \sin kx \sin \omega t + \cos kx \cos \omega t + \sin kx \sin \omega t) \\
&= 2A_1 \sin kx \sin \omega t \qquad\qquad\qquad (33.13)
\end{aligned}
$$

This wave is called a standing wave (Figure 33-9). It differs markedly from a traveling wave. Since the variables x and t do *not* occur in the combination $x + vt$, this wave motion does *not* consist of the motion of an unchanging waveform along the string. Instead, although each particle of the string executes simple harmonic motion of the same frequency, the amplitude A of this motion, given by

$$A = |2A_1 \sin kx|$$

is seen to depend on the location of the particle. In fact, at points called *nodes*,

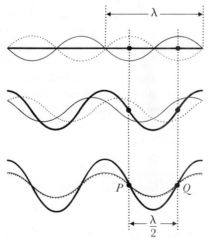

Figure 33-9 The standing wave (heavy line) formed by the superposition of two waves of the same wavelength λ and the same amplitude, traveling in opposite directions. The wave indicated by the dotted line travels to the left while that indicated by the light solid line travels to the right. At the instant depicted in the uppermost diagram, cancellation occurs. The situation at later instants is shown in the lower diagrams. Points like P and Q, which never move, are called *nodes*.

$A = 0$ and the string is stationary. Nodes occur where $\sin kx = 0$ which implies

$$kx = 0, \pi, 2\pi, 3\pi, \ldots \qquad \text{(nodes)}$$

Since $k = 2\pi/\lambda$ (where λ is the wavelength of the two traveling waves whose superposition forms the standing wave), the nodes occur at

$$x = 0, \tfrac{1}{2}\lambda, \lambda, \tfrac{3}{2}\lambda, \ldots \qquad \text{(nodes)}$$

The distance between adjacent nodes is $\tfrac{1}{2}\lambda$. Particles oscillate with maximum amplitude $A = 2A_1$ at positions where $|\sin kx| = 1$. These points, called *antinodes*, are located where

$$kx = \tfrac{1}{2}\pi, \tfrac{3}{2}\pi, \tfrac{5}{2}\pi, \ldots$$

that is, where

$$x = \tfrac{1}{4}\lambda, \tfrac{3}{4}\lambda, \tfrac{5}{4}\lambda, \ldots \qquad \text{(antinodes)}$$

The antinodes are midway between the nodes.

All the particles between two adjacent nodes oscillate in phase. Oscillations of particles separated by a single node are 180° out of phase.

33.5 Vibrating Systems and Resonance

Natural Frequencies

When Newtonian mechanics is applied to the analysis of the motion of any system of particles executing small vibrations, we find that any possible motion can be regarded as a superposition of certain relatively simple motions called *normal modes of vibration*. These normal modes are therefore particularly important. The frequency of oscillation in a normal mode is named a *natural frequency* of the system. The lowest natural frequency is termed the *fundamental frequency* and the higher natural frequencies are called *overtones*. For a system that is regarded as a continuous distribution of mass, the normal modes of vibration are standing waves with fixed locations of nodes and antinodes, as is illustrated in Figure 33-10 for a string fixed at both ends and in Figure 33-11 for a drumhead.

Normal modes for a system consisting of a string of length L fixed at both ends (the prototype of any stringed musical instrument) can be found rather simply from the theory developed in this chapter. Section 33.2 shows that the application of Newton's second law to this mechanical system leads to the wave equation. From the method of discovery of the wave equation in Section 33.1, we know that traveling waves of the form $f(x \pm vt)$ are solutions to the wave equation. According to the superposition principle, the standing wave,

$$y(x,t) = 2A_1 \sin kx \sin \omega t \qquad (33.13)$$

being a superposition of two traveling wave solutions, must also be a solution to the wave equation. Therefore this standing wave will be a possible wave motion for the string fixed at both ends provided that the boundary condition, $y(L,t)$

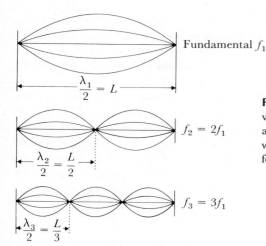

Figure 33-10 Normal modes of vibration of a string of length L, fixed at both ends. These are standing waves. The string position is shown at four different instants.

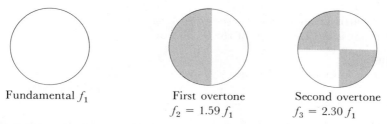

Figure 33-11 Normal modes of vibration of a drumhead corresponding to the three lowest natural frequencies. Shaded portions are moving down while the unshaded portions are moving up.

$= 0$, is also satisfied. This requires $\sin kL = 0$ which implies $kL = n\pi$ where n is a positive integer. Then $\omega = kv = n\pi v/L$. The possible standing waves, the *normal modes* of vibration for the string fixed at both ends, are therefore of the form

$$y_n(x,t) = 2A_1 \sin \frac{n\pi x}{L} \sin \frac{n\pi v}{L} t$$

The frequency of oscillation in the nth normal mode, the natural frequency f_n, is given by

$$f_n = \frac{\omega_n}{2\pi} = \frac{n\pi v}{2\pi L} = \frac{nv}{2L}$$

The fundamental frequency is

$$f_1 = \frac{v}{2L}$$

and we see that $f_n = nf_1$.

Overtones that are integral multiples of the fundamental frequency are called *harmonics*, the frequency nf_1 being named the nth harmonic. Our result shows that, for a string fixed at both ends, all the overtones are harmonics. Note that the overtones of the vibrating drumhead shown in Figure 33-11 are not harmonics.

The details of the way a system is set into oscillation—for example, the position at which a stretched string is plucked—will determine which of its natural frequencies will be present in the oscillation. But no matter how complex, the resultant motion is always just a superposition of the various normal modes of vibration. A vibrating body, exposed to air, is a source of traveling waves in the air that carry away energy. These radiated mechanical waves will have the frequencies present in the source, that is, the natural frequencies.

Example 4   A steel piano wire is 1.00 m long. Its tension is adjusted until the speed v of transverse traveling waves is 500 m/s:
(*a*) Find the fundamental frequency of this wire.
(*b*) What is the frequency of the third harmonic?
(*c*) What is the frequency of the second overtone?

Solution

(*a*) The fundamental frequency is given by

$$f_1 = \frac{v}{2L} = \frac{500 \text{ m/s}}{2.00 \text{ m}} = 250 \text{ Hz}$$

(*b*) The third harmonic has a frequency $3f_1 = 750$ Hz.
(*c*) The second overtone is the third harmonic.

Resonance

Suppose that instead of considering a body as a source of waves, we study its *absorption* of energy when mechanical waves from elsewhere are incident upon it. After transients have died out, the body settles into *forced oscillations* at the frequency of the incident wave, the so-called *driving frequency*. The amplitude of these forced oscillations becomes exceptionally large when the driving frequency is near a *natural frequency* of the vibrating body. This large response at certain driving frequencies is termed *resonance*, and the corresponding frequencies are called *resonant frequencies*.

Resonance is a widespread and important phenomenon. An everyday example is the occurrence of a shimmy or rattle in a car at a certain speed. An irregularity in a tire can result in a periodic driving force with a frequency which, at a certain speed, matches a natural frequency of part of the car. In Section 32.4 we discuss resonance in an electric circuit. Resonant interaction of light with atoms and molecules is described in Section 43.3. Resonances of great variety occur in the interaction of atoms, nuclei, and elementary particles.

Summary

☐ A traveling wave $y(x,t)$ with an unchanging waveform and wave speed v is described by

$$y = f(x - vt) \quad \text{(wave moves in positive } x \text{ direction)}$$

$$y = f(x + vt) \quad \text{(wave moves in negative } x \text{ direction)}$$

Such functions satisfy the wave equation

$$\frac{\partial^2 y}{\partial x^2} = \frac{1}{v^2} \frac{\partial^2 y}{\partial t^2}$$

☐ On a stretched string with tension τ and linear density μ, traveling waves with small displacements $y(x,t)$ have a wave speed

$$v = \sqrt{\frac{\tau}{\mu}}$$

and transmit a power

$$\frac{\partial U}{\partial t} = -\tau \frac{\partial y}{\partial x} \frac{\partial y}{\partial t}$$

☐ The sinusoidal traveling wave

$$y = A \cos(kx - \omega t)$$

has a wave speed

$$v = \frac{\omega}{k}$$

For a given x, y is a periodic function of t with period $T = 2\pi/\omega$. For a given t, y is a periodic function of x with a spatial period, called the wavelength, given by $\lambda = 2\pi/k$.

☐ The superposition principle for waves states that when two waves are simultaneously present, the displacement $y(x,t)$ at any instant is the algebraic sum of the displacements $y_1(x,t)$ and $y_2(x,t)$ that would occur if each individual wave were present alone.

☐ The superposition of two waves on a string, which have the same amplitude and the same wavelength λ but travel in opposite directions, is a standing wave with nodes at fixed locations that are a distance $\lambda/2$ apart.

☐ The normal modes of vibration of a string of length L fixed at both ends are standing waves with nodes at each end of the string. In the nth normal mode, the string oscillates with the natural frequency $f_n = nf_1$, where $f_1 = v/2L$.

Questions

1   A transverse wave on a stretched rope is given in SI units by

$$y = \frac{3.0}{(x - 4.0t)^2 + 2.0}$$

   (a) Determine the wave speed by inspection of the form of this function.
   (b) Sketch graphs showing y as a function of x at the instant $t = 0$ and at $t = 1.0$ s.
   (c) Determine the wave speed from the graphs sketched in part (b).

2   Verify that the function y given in Question 1 is a solution of the wave equation.

3   Verify that the function $y = \cos 20(x - 15t)$ is a solution of the wave equation.

4   A rope is 10.0 m long and has a mass of 0.50 kg. It is stretched horizontally so that the tension is 500 N. What is the speed of transverse waves on this rope?

5   Verify that the sinusoidal traveling wave $y = A \cos(kx - \omega t)$ is a solution of the wave equation.

6   The wave speed of a sinusoidal traveling wave is given by

$$v = \frac{\lambda}{T}$$

   (a) Prove this from the formula $v = \omega/k$.
   (b) Prove this simply by considering Figure 33-12 which shows the waveform (a graph of y as a function of x) at several successive instants. A crest, first observed at point P, travels to the right occupying the successive positions P_1, P_2, and P_3 where the distance PP_3 is one wavelength λ. While the crest moves from P to P_3, the particle of the string at P_3 completes exactly one cycle of its oscillation.

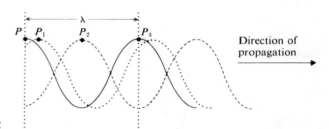

Figure 33-12

7   A sinusoidal transverse wave on a string is given in SI units by

$$y = 0.10 \cos 2\pi(0.20x - 0.10t)$$

   (a) Sketch a graph showing y as a function of x at $t = 0$. From the graph find the wavelength λ.
   (b) Sketch another graph showing y as a function of x at $t = 0.25$ s. From the two graphs determine the distance that a given crest has moved in 0.25 s and then calculate the wave speed.
   (c) By inspection of the constants in this sinusoidal function, determine the wavelength, the wave number, the period, the angular frequency, and the wave speed.

8 (a) Give the expression for a sinusoidal wave traveling on a string in the positive x direction with an amplitude of 0.15 m, a wavelength of 2.0 m, and a period of 0.40 s. The particle at $x = 0$ has a displacement of 0.15 m at $t = 0$.
(b) What is the wave speed for the wave of part (a)?
(c) What is the velocity of the particle at $x = 2.0$ m when $t = 0.10$ s?

9 Give the expression for the wave of Question 8, if the particle at $x = 0$ is moving upward but has no displacement at $t = 0$.

10 A sinusoidal transverse wave of 2.00-m wavelength travels along a stretched string:
(a) Locate a particle that vibrates in phase with the particle at the origin.
(b) Locate a particle that is 180° out of phase with the particle at the origin.

11 As a sinusoidal wave travels along a stretched string, the particle at the origin completes 7.0 cycles in 10 s. A given crest moves from the origin to the particle at $x = 9.0$ m in 5.0 s. Find the frequency, wave speed, and wavelength.

12 The free end of a stretched rope is made to execute simple harmonic motion with a frequency of 5.0 Hz. The linear density of the rope is 5.0×10^{-3} kg/m and the tension is 200 N:
(a) What is the wavelength?
(b) If the amplitude of the wave is 0.10 m, what is the average power transmitted by the wave?

13 In the central portion of Figure 33-7 the string is flat. What has happened to the energy that is associated with the pulses in the upper and lower pictures?

14 Find the superposition of the waves

$$y_1 = A_1 \cos(kx - \omega t + \theta_0)$$
$$y_2 = A_2 \cos(kx - \omega t)$$

when the angle θ_0 is 0 and when it is 180°. In each case determine the amplitude of the superposition.

15 Verify that the standing wave

$$y = 2A_1 \sin kx \sin \omega t$$

is a solution of the wave equation.

16 Using Eq. 33.4 evaluate the power transmitted by the standing wave given in Eq. 33.13 and comment on the result.

17 The fundamental frequency of a string which is 0.50 m long and fixed at both ends is 40 Hz. Find the frequency of vibration in the third normal mode of vibration (that is, the third harmonic or second overtone). Sketch the typical appearance of the string at different instants, pointing out the locations of the nodes and antinodes.

18 A stretched string, 0.80 m long, has a fundamental frequency of 100 Hz:
(a) What is the speed of the traveling waves whose superposition produces the standing wave?
(b) If the linear density of the string is 0.010 kg/m, what is the tension in the string?
(c) What is the wavelength of the traveling waves in part (a)?
(d) What is the frequency of the first overtone?
(e) What is the frequency of the fourth harmonic?

Supplementary Questions

S-1 At $t = 0$, the displacement of a rope is given in SI units by

$$y = 2.0e^{-x^2}$$

This waveform is maintained as it moves in the negative x direction with a speed of 4.0 m/s:

(a) Give the displacement $y(x,t)$ at any instant t.

(b) Sketch graphs showing y as a function of x at the instants $t = 0$ and $t = 2.0$ s.

(c) Find the velocity at the instant t of the particle with coordinate x.

S-2 Find the power transmitted by the wave

$$y = A \cos^2(kx - \omega t)$$

on a stretched string with tension τ.

34.

sound waves

34.1 Nature of Sound

It is a familiar fact that vibrating bodies are sources of what we call sound. This suggests that sound is a wave motion.

A simple experiment (Figure 34-1) demonstrates that a material medium is required to transmit sound. An electric bell, suspended by fine wires inside a jar, is audible when there is air inside the jar. As the air is removed by a vacuum pump the sound fades. Nothing can be heard when a good vacuum is achieved. However, if the bell touches the glass walls, the ringing is again quite audible. Evidently sound is transmitted by gases and solids. We can easily devise a variation of this experiment to demonstrate that liquids also transmit sound.

Many experiments have made it clear that sound is a *mechanical wave motion in a medium*. Small displacements of the molecules of the medium occur as a sound wave travels. This displacement refers to an ordered collective motion in which all the molecules in a small volume move together. (This ordered motion is superimposed on the random thermal agitation of the molecules.) In thinking about a sound wave we therefore visualize molecules of the medium jiggling back and forth in a regular fashion.

A theoretical understanding of the propagation of sound waves is obtained from Newtonian mechanics. As was illustrated in Chapter 33, we select a small chunk of the medium as the system to which we apply Newton's second law. The motion of this chunk is determined by the forces exerted by neighboring

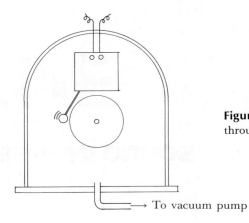

Figure 34-1 Sound is not transmitted through a vacuum.

To vacuum pump

portions of the medium. It turns out that each chunk vibrates back and forth in such a way that a disturbance, a wave, propagates with a speed which is characteristic of the type of wave and of the medium under specified conditions.

In a gas, the only possible type of wave is a *longitudinal wave*. That is, the molecules of the gas vibrate back and forth along the direction in which the waves are moving. Other types of waves cannot occur because a fluid offers no resistance to tangential or shearing forces. The mechanical theory of longitudinal waves in a fluid is developed in the following section.

The description of *transverse* waves on a string can be used to help us visualize sound waves which are *longitudinal*. We denote by χ the displacement of the matter in a small volume from its equilibrium position. A graph showing the displacement χ of this particular matter at different times is a graph just like that of Figure 33-4 if the wave is a sinusoidal traveling wave. The vibrations back and forth of this one small chunk of matter correspond to the up-and-down motion of a particle of a string which is transmitting a transverse sinusoidal wave.

In a sinusoidal sound wave, the amplitude of oscillation of the air molecules at any one point is very small—only 0.01 mm for a sound wave that is so loud that it hurts the ear (when the frequency is 1 kHz). In such a wave, the maximum speed attained by an oscillating chunk of matter is merely 0.06 m/s.

If we now consider *one instant of time* and draw a graph showing the displacement χ of *different small chunks of matter*, labeled by the different x values of their equilibrium positions, we get a graph like that of Figure 33-5 for a sinusoidal wave propagating in the direction of the X axis. For a traveling wave, at a later instant, we find a similar waveform in a different position in the medium. The speed of propagation of the waveform is what is called the speed of sound in the medium.

In a solid, the vibrations can be transverse as well as longitudinal. It turns out that the speed of the longitudinal wave is always greater than that of the transverse wave. This fact is exploited in seismology. An estimate of the distance of an earthquake from an observation station can be obtained by noting the difference in arrival time of these two types of waves.

Wave Fronts

In the case of waves on a string, the eye can easily follow the motion of a crest or of any other point on the waveform corresponding to a constant value of the phase. To describe other types of waves we similarly focus attention on a given point on the waveform, say a crest, but our concepts must be extended to describe waves in three-dimensional space, and, of course, we do not actually see the motion of these waves.

To picture a wave in space at a certain instant we draw a succession of surfaces, called *wave fronts*, through points where the phase of the vibrating quantity has a constant value. In a uniform medium at a distance large compared to the dimensions of the source of waves, the wave fronts are spherical surfaces and their intersections with a two-dimensional plane will be circles, as indicated in Figure 34-2 for wave fronts representing a succession of crests and troughs. A small portion of a large spherical surface is approximately a plane. Therefore with a remote source the wave fronts can be represented by planes, as shown in Figure 34-3.

Speed of Sound

In air at 0°C the speed of sound is 331 m/s. This is about 1 km in 3 s (a fact to remember if you wish to compute your distance from a lightning stroke).

Table 34-1 shows the speed of longitudinal sound waves in some common substances. The speed is greater the greater the stiffness of the medium and the smaller its density. In general we find that sound travels faster in solids and liquids than it does in gases.

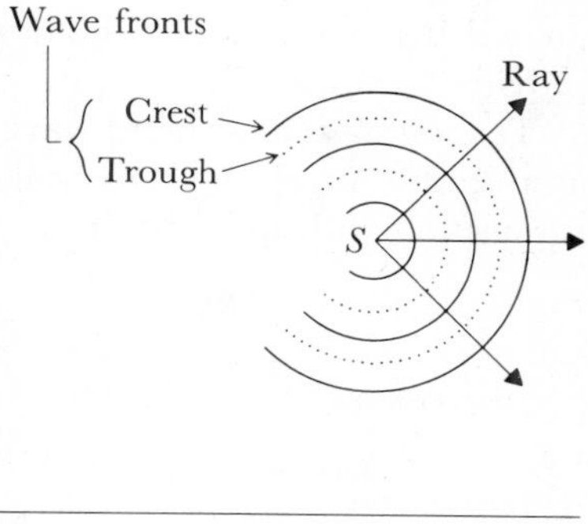

Figure 34-2   Wave fronts representing a spherical wave emitted by a point source. A ray is a line in the direction of the wave velocity vector and intersects the wave fronts at right angles.

Figure 34-3   Wave fronts representing a plane wave.

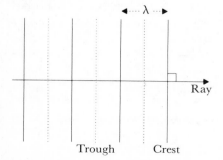

Table 34-1

Medium	Speed of sound (km/s)
Air (0°C)	0.331
Water (15°C)	1.45
Lead (20°C)	1.23
Iron (20°C)	5.13
Granite (20°C)	6.0

Frequency of Sound Waves

We encounter mechanical vibrations of a tremendously wide range of frequencies (Figure 34-4); however, the human ear generally detects only those frequencies lying between 20 Hz and 20 kHz. Frequencies of this order are called sound, but the term *sound* is also used by physicists to mean mechanical waves with frequencies well outside this audible spectrum. Above the audible range, we call sound waves *ultrasonic*. Frequencies up to 600 MHz are conveniently generated in the laboratory. Waves with frequencies below the audible spectrum are called *infrasonic*. Such waves are generated by earthquakes.

The wavelength of a sound wave in air at 0°C can be calculated from $v = \lambda f$. With $v = 331$ m/s and $f = 20$ Hz, the wavelength is 17 m. When the frequency is 20 kHz, the wavelength is merely 1.7 cm. In fact, ultrasonic wavelengths may be as short as light waves. (At $f = 500$ MHz, $\lambda = 6 \times 10^{-7}$ m, which is the same wavelength as that of orange light.)

Pitch and Quality

From the standpoint of sound as a stimulus detected by the ear, certain characteristics such as *pitch*, *tone quality*, and *loudness* are of prime interest. Each of these sensations corresponds to a physical quantity associated with a sound wave.

To each distinct pitch there corresponds a distinct frequency (except for very loud sounds). Higher frequency implies higher pitch. If the frequency is doubled, the pitch is increased by one *octave*.

By tone quality we refer to the property by which we can distinguish a given note, say middle C on the piano, from middle C on a violin, or a flute. Tone quality is determined by the relative intensities of the overtones. The fundamental frequency alone is not a very interesting sound. It is the overtones that provide "character."

The loudness of a sound wave of a given frequency increases with increasing amplitude. An appropriate scale for measurement of loudness is discussed in connection with intensity in Section 34.4.

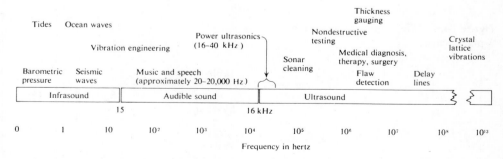

Figure 34-4 Spectrum of mechanical vibrations indicating the approximate frequency regions of various applications.

34.2 Mechanical Theory of Sound Waves in a Fluid

Displacement and Pressure

Figure 34-5 shows a macroscopic volume element of a fluid. We consider a motion in which each volume element of the fluid oscillates along a direction parallel to the X axis. The coordinate x gives the equilibrium position of the section A of the volume element and serves to label this section. The equilibrium value of the pressure at a point in the fluid is denoted by P_0.

Sound waves generally involve very small departures from equilibrium. If the fluid is subjected to a small disturbance, the pressure at a point becomes $P = P_0 + P_e$, where the excess pressure P_e is very small compared to P_0. When the pressure is no longer the same on sections A and A' in Figure 34-5, the volume element between these sections will be accelerated. At a certain instant, section A will have acquired a displacement χ and section A' a displacement χ'. The thickness of the volume element is thereby changed from Δx to

$$\Delta x' = \Delta x + (\chi' - \chi) = \Delta x + \Delta\chi$$

The change in the volume of this element is

$$\Delta V = A\Delta x' - A\Delta x = A\Delta\chi$$

This volume change is related to the excess pressure P_e on the element. The *bulk modulus of elasticity* of a material is the quantity κ defined by

$$P_e = \kappa\left(-\frac{\Delta V}{V}\right) \tag{34.1}$$

where $-\Delta V$ is the decrease in the volume V caused by the excess pressure P_e. For small changes, experiments show that κ is constant.

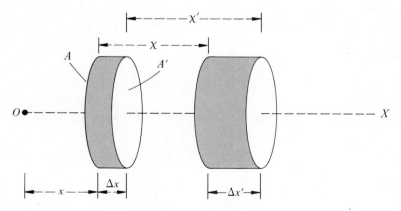

Figure 34-5 The equilibrium position and a displaced position of a volume element of a fluid.

For the displaced volume element in Figure 34-5, we find

$$P_e = \kappa\left(-\frac{\Delta V}{V}\right) = -\frac{\kappa A \Delta \chi}{A \Delta x}$$

or, in the limit as Δx approaches zero,

$$P_e = -\kappa \frac{\partial \chi}{\partial x} \tag{34.2}$$

This is a partial derivative because we are concerned with displacements at the *same instant t*.

Both the excess pressure P_e and the displacement χ are functions of the two variables x and t. A wave motion can be describ d by a function $\chi(x,t)$ that gives the displacement at the instant t of the plane section whose equilibrium position has the coordinate x. This function $\chi(x,t)$ gives what is called the *displacement wave*. The *pressure wave* is specified by the function $P_e(x,t)$ that gives the excess pressure.

When the displacement wave is known, the pressure wave can be determined using $P_e = -\kappa(\partial \chi / \partial x)$. The plausibility of this key relationship should be noted. Where the displacement increases with x, the fluid is stretched out and the pressure on a volume element is reduced below its normal value; that is, the excess pressure is negative. And where the displacement does not change with x, since different sections have the same displacement, the fluid between these sections has its equilibrium volume which corresponds to zero excess pressure. In regions where the displacement decreases with increasing x, the fluid is compressed and there is a positive excess pressure.

Wave Equation for χ and P_e

Each element of the fluid moves in accord with Newton's second law. The mass of the element in Figure 34-5 is $\rho_0 A \Delta x$, where ρ_0 is the equilibrium value of the fluid density. This element has a velocity $\partial \chi / \partial t$ and an acceleration $\partial^2 \chi / \partial t^2$. The fluid at the left of the volume element pushes to the right with a force PA and the fluid at the right pushes to the left with a force $P'A$. Newton's second law for x components yields

$$PA - P'A = \rho_0 A \Delta x \frac{\partial^2 \chi}{\partial t^2}$$

As Δx approaches zero, the limiting value of $(P' - P)/\Delta x$ is $\partial P_e/\partial x$, and we have

$$\frac{\partial P_e}{\partial x} = -\rho_0 \frac{\partial^2 \chi}{\partial t^2}$$

Since $P_e = -\kappa(\partial \chi / \partial x)$, this gives

$$\frac{\partial^2 \chi}{\partial x^2} = \frac{\rho_0}{\kappa} \frac{\partial^2 \chi}{\partial t^2} \tag{34.3}$$

which is the *wave equation*.

Therefore we can conclude that displacement waves can propagate through a fluid with a wave speed

$$v = \sqrt{\frac{\kappa}{\rho_0}} \tag{34.4}$$

Traveling waves are functions of the form $\chi(x - vt)$ or $\chi(x + vt)$; these are plane waves with wave fronts perpendicular to the X axis. The pressure function $P_e(x,t)$ obeys the same wave equation (Question 10),

$$\frac{\partial^2 P_e}{\partial x^2} = \frac{\rho_0}{\kappa} \frac{\partial^2 P_e}{\partial t^2}$$

and a pressure wave propagates with the same speed as a displacement wave.

Example 1   Describe the pressure wave when the displacement wave is the sinusoidal traveling wave of amplitude χ_0,

$$\chi = \chi_0 \cos(kx - \omega t)$$

Solution   The excess pressure is

$$P_e = -\kappa \frac{\partial \chi}{\partial x} = k\kappa\chi_0 \sin(kx - \omega t) = k\kappa\chi_0 \cos\left(kx - \omega t - \frac{\pi}{2}\right)$$

The pressure wave is also a sinusoidal traveling wave with the same frequency and wavelength as the displacement wave. However, at any point there is a $90°$ phase difference between these two waves. Figure 34-6 illustrates the excess pressure at the instant $t = -\pi/2\omega$. There are regularly spaced regions where the molecules of the fluid are crowded together (compressions) separated by regions where the molecules are abnormally far apart (rarefactions).

The compression of a fluid is accompanied by a temperature rise, and an expansion by a temperature drop. Nevertheless, the compressions and expansions that take place when a compressional wave moves through a fluid are ordinarily

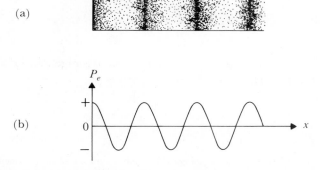

Figure 34-6   Pressure variations associated with a sinusoidal sound wave in a fluid.

adiabatic, there being no significant heat transfer from a momentarily compressed region to an adjacent expanded region. The *bulk modulus* κ that determines the wave speed $v = \sqrt{\kappa/\rho_0}$ is therefore the adiabatic bulk modulus.*

34.3 Calculation of the Speed of Sound in a Gas

By calculating the adiabatic bulk modulus of an ideal gas we can make a theoretical prediction of the speed of sound in a gas.

The definition of this bulk modulus is

$$\kappa_{\text{adiabatic}} = -\left(\frac{dP}{dV/V}\right)_{\text{adiabatic}}$$

where dP is the pressure change that causes the volume change dV. For an adiabatic process, the pressure and the volume of an ideal gas are related by

$$PV^\gamma = constant \tag{18.20}$$

Therefore

$$\ln P + \gamma \ln V = \ln constant$$

and taking differentials

$$\frac{dP}{P} + \gamma \frac{dV}{V} = 0 \qquad \text{or} \qquad \frac{dP}{dV/V} = -\gamma P$$

This result, together with the definition of the bulk modulus, gives

$$\kappa_{\text{adiabatic}} = \gamma P$$

Substituting this expression in Eq. 34.4 (and dropping the subscript 0) we find the wave speed in an ideal gas is

$$v = \sqrt{\frac{\gamma P}{\rho}} \tag{34.5}$$

The ideal gas law, $PV = NkT$, expressed in terms of the density $\rho = M/V$ of a gas with total mass M, is

$$\frac{P}{\rho} = \frac{kT}{m}$$

where $m = M/N$ is the mass of a molecule (or the average molecular mass of a mixture) of the gas. Therefore the wave speed is

$$v = \sqrt{\frac{\gamma kT}{m}} \tag{34.6}$$

---

*A justification for this assumption which corrects common but erroneous arguments is given by N. H. Fletcher in *American Journal of Physics*, June 1974.

If the gas is air, $\gamma = 1.40$ and the average molecular mass is

$$m = (molecular\ weight)(1.66 \times 10^{-27}\ \text{kg})$$

$$= (28.8)(1.66 \times 10^{-27})\ \text{kg}$$

Then at 0°C (or 273.15 K) the wave speed is

$$v = \sqrt{\frac{1.40(1.380 \times 10^{-23}\ \text{J/K})(273.15\ \text{K})}{28.8 \times 1.66 \times 10^{-27}\ \text{kg}}} = 332\ \text{m/s}$$

in good agreement with the value 331.45 m/s found by experimental measurement of the speed of sound in air at 0°C.

Example 2   What is the amplitude of the pressure wave of Example 1 in air under standard conditions (1 atm and 0°C) if the amplitude χ_0 of the displacement wave is 1.0×10^{-2} mm and the frequency is 1.0 kHz?

Solution   The amplitude of the pressure wave is seen from Example 1 to be

$$P_{e,\text{max}} = k\kappa\chi_0$$

Substituting $k = \omega/v$ and $\kappa = \gamma P$, we have

$$P_{e,\text{max}} = \frac{\omega\gamma P\chi_0}{v} = \frac{(2\pi \times 10^3\ \text{Hz})(1.40)(1.01 \times 10^5\ \text{Pa})(1.0 \times 10^{-5}\ \text{m})}{331\ \text{m/s}}$$

$$= 27\ \text{Pa}$$

Pressure variations of this order of magnitude are the loudest that the ear can tolerate.

34.4 Intensity

A traveling wave transports energy as it moves through a fluid. The power $\partial U/\partial t$ transmitted from left to right across the section A in the fluid in Figure 34-5 is the product of the velocity $\partial\chi/\partial t$ of this section and the force PA exerted by the fluid to the left of this section on the fluid to the right:

$$\frac{\partial U}{\partial t} = PA\,\frac{\partial\chi}{\partial t}$$

If the velocity $\partial\chi/\partial t$ oscillates with average value zero, then since $P = P_0 + P_e$ and P_0 is constant, the average power is

$$\left(\frac{\partial U}{\partial t}\right)_{\text{avg}} = A\left(P_e\frac{\partial\chi}{\partial t}\right)_{\text{avg}} = A\kappa\left(-\frac{\partial\chi}{\partial x}\frac{\partial\chi}{\partial t}\right)_{\text{avg}} \tag{34.7}$$

The intensity I of a wave is defined as the average rate of transfer of energy per unit area perpendicular to the direction of propagation; that is

$$I = \frac{1}{A}\left(\frac{\partial U}{\partial t}\right)_{\text{avg}} \tag{34.8}$$

For longitudinal waves with fluid displacements parallel to the X axis, we have

$$I = -\kappa\left(\frac{\partial \chi}{\partial x}\frac{\partial \chi}{\partial t}\right)_{\text{avg}} \tag{34.9}$$

Example 3 In a fluid with adiabatic bulk modulus κ, find the intensity of the sinusoidal traveling wave described by

$$\chi = \chi_0 \cos(kx - \omega t)$$

Solution Since

$$\frac{\partial \chi}{\partial x} = -k\chi_0 \sin(kx - \omega t) \qquad \text{and} \qquad \frac{\partial \chi}{\partial t} = \omega\chi_0 \sin(kx - \omega t)$$

we have

$$I = \kappa k \omega \chi_0^2[\sin^2(kx - \omega t)]_{\text{avg}}$$

Making the substitutions $\kappa = \rho v^2$ (obtained from $v = \sqrt{\kappa/\rho}$) and $k = \omega/v$, and using the fact that the average value of a sinusoidal function of time is $\frac{1}{2}$, we obtain

$$I = \frac{1}{2}\rho v \omega^2 \chi_0^2 \tag{34.10}$$

The intensity is proportional to the *square* of both the *frequency* and the *amplitude*. The result is analogous to the expression $\frac{1}{2}\mu v \omega^2 A^2$ for the average power transmitted by a transverse traveling wave on a string (Example 3 of Chapter 33).

Inverse Square Law for Intensity

Suppose a small source of sound is isotropic, that is, it radiates equally in all directions. If the transmitting medium is uniform and does not absorb any power, the intensity I at a distance r from the source is

$$I = \frac{1}{4\pi r^2}\left(\frac{\partial U}{\partial t}\right)_{\text{avg}} \tag{34.11}$$

where $(\partial U/\partial t)_{\text{avg}}$ is the average power radiated by the source. According to this equation, the *intensity falls off inversely as the square of the distance from a point source.* This inverse square law is valid for intensities in any given direction from a point source, whether or not the source is isotropic (Question S-6).

A function of the form $\cos(kr - \omega t)$ represents a wave traveling in the direction of increasing r with a wave speed ω/k. The wave fronts are spherical surfaces. Since the intensity of a wave is proportional to the square of the amplitude, the inverse square law for intensities implies that the amplitude of such a spherical wave is inversely proportional to r; that is

$$\chi = \frac{\chi_0}{r}\cos(kr - \omega t) \tag{34.12}$$

Intensity Level, The Decibel

The physiological sensation of *loudness* in the human ear is closely related to the *intensity* of the incident sound wave. At a frequency of about 1 kHz, the ear detects sounds varying from the barely audible intensity of 10^{-12} W/m² to the pain threshold of 1 W/m², a truly huge range. Although different people have different judgments as to whether one sound is twice as loud as another, it has been found that a scale proportional to the logarithm of the intensity corresponds roughly with the response of the ear. A convenient measure of loudness is provided by the *intensity level* in decibels* (dB) defined by the following equation:

$$\text{intensity level} = 10 \log_{10}\left(\frac{I}{10^{-12} \text{ W/m}^2}\right) \text{dB} \qquad (34.13)$$

Typical values of intensity levels are shown in Table 34-2. A clearly perceptible change in loudness corresponds to about 3 dB.

34.5 Superposition of Sound Waves

When different sound waves of small amplitude travel through a medium, the *superposition principle* holds. That is, the displacement at any instant of a small chunk of matter when two waves are present is the sum of the displacements

Table 34-2 Sound levels

Situation	Intensity level (dB)	Intensity (W/m²)	Sensation
Threshold of feeling	120	1	Painful
Thunder	110	10^{-1}	
Artillery			Deafening
Riveting	100	10^{-2}	
Noisy factory			
Pneumatic drill	80	10^{-4}	Very loud
Busy street traffic	70	10^{-5}	
Average street noise			Quite
Conversation in home	60	10^{-6}	pronounced
Average office			
Quiet radio in home	40	10^{-8}	Moderate
Quiet home	30	10^{-9}	
Private office			Faint
Whisper	20	10^{-10}	
Rustle of leaves	10	10^{-11}	Scarcely
Threshold of audibility	0	10^{-12}	audible

*The term *bel* is derived from the name of Alexander Graham Bell, inventor of the telephone.

that would occur if each wave alone were present. The wave obtained by this addition is called the *superposition* of the two original waves.

The superposition of different waves leads to what are called *interference* and *diffraction* phenomena. The detailed treatment of these subjects we postpone to Chapter 36. For the present we shall merely mention the interesting fact that a wave which passes through a hole of diameter D undergoes a spreading and fans out into a cone with a vertex angle having an order of magnitude given in radians by

$$\theta \approx \frac{\lambda}{D} \tag{34.14}$$

as depicted in Figure 34-7. This effect, termed "diffraction spreading," is illustrated by the water waves shown in Figure 34-8. For an aperture of 1 m and wavelength of 0.5 m, this diffraction spreading has a cone vertex angle of about 30°. Consequently, for sound waves with wavelengths longer than about 0.5 m (frequencies lower than about 1 kHz), there is a considerable diffraction spreading when a wave passes through a door or a window. This is the reason a speaker does not have to be in sight to be heard.

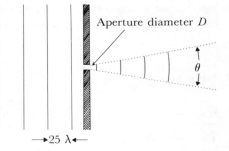

Figure 34-7 Diffraction spreading behind an aperture when plane waves are incident. In order to have as little spreading as this, the wavelength would have to be about $\frac{1}{25}$ of the distance between the indicated wave fronts.

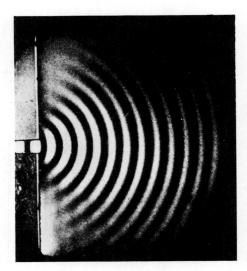

Figure 34-8 Diffraction of a water wave which passes through a small aperture. (From M. Alonso and E. J. Finn, *Physics*, Addison-Wesley, 1970.)

In certain applications, a narrow beam of sound waves is desired. The beam will have a small diffraction spreading only if the wavelength is much less than the diameter of the source. Therefore short wavelength ultrasonic waves must be used.

Beats

The superposition of two sound waves with slightly *different frequencies* produces throbbing sounds, called *beats*. This effect is noticed if two aircraft propellers have not been precisely synchronized. For a fraction of a second, crests from one source arrive at the ear of the listener at the same time as crests from another source. The superposition of the two waves then has an amplitude which is the sum of the amplitudes of the individual waves. The sound is loud. But since the two waves have different frequencies, this situation changes. After many vibrations there will come a time when crests from one source arrive together with troughs from the other, giving a superposition with an amplitude which is the *difference* of the amplitudes of the two waves. Little or no sound will be heard. The loudness of the sound therefore rises and falls periodically.

Analysis of this phenomenon requires determination of the superposition $\chi = \chi_1 + \chi_2$ of two sinusoidal functions of the form

$$\chi_1 = \chi_{01} \cos \omega_1 t \qquad \text{and} \qquad \chi_2 = \chi_{02} \cos \omega_2 t$$

This is easily accomplished using the phasor diagram of Figure 34-9. The angle between the phasors representing χ_2 and χ_1 is $(\omega_2 - \omega_1)t$. The amplitude χ_0 of the superposition is given by the cosine rule applied to the triangle with sides of lengths χ_{01}, χ_{02}, and χ_0:

$$\chi_0^2 = \chi_{01}^2 + \chi_{02}^2 + 2\chi_{01}\chi_{02} \cos(\omega_2 - \omega_1)t \qquad (34.15)$$

where we have used the fact that $\cos(180° - \alpha) = -\cos \alpha$. The intensity of

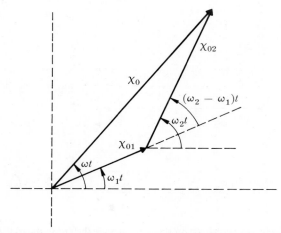

Figure 34-9 Phasor diagram. The phasor of length χ_0 represents the superposition $\chi = \chi_1 + \chi_2$.

each wave is proportional to the square of its amplitude. Therefore the relationship among amplitudes implies that the intensity I of the superposition is related to the intensities I_1 and I_2 of the two superposed waves by

$$I = I_1 + I_2 + 2\sqrt{I_1 I_2}\,\cos(\omega_2 - \omega_1)t \qquad (34.16)$$

The last term accounts for the occurrence of beats. This term oscillates with an angular frequency $\omega_B = \omega_2 - \omega_1$ and the number of beats per unit time, called the *beat frequency*, is

$$f_B = f_2 - f_1$$

Comparison of two almost equal frequencies is conveniently accomplished by detecting beats. A stringed instrument can be tuned in this way. The closer two frequencies become, the fewer the number of beats per second. The beats disappear when the frequencies are the same.

Normal Modes of Vibration of Air Columns

The analysis and terminology developed in Section 33.5 can be applied to the vibrations of a column of air in a narrow pipe of length L. Sound waves traveling along the pipe are reflected at each end. The simultaneous presence of waves traveling in opposite directions gives a standing wave. Any possible wave motion must satisfy certain *boundary conditions* at each end of the pipe. The possible standing waves are the normal modes of vibration of the air column, and the frequency of vibration in a normal mode is a natural frequency of the column.

The boundary condition at a closed end is that the displacement is always zero, that is, there is a *displacement node at a closed end*. Conditions at an open end are more complex, but, for a pipe that is narrow compared with the wavelength, as a rough approximation, an open end can be considered as the location of a *displacement antinode*. These boundary conditions lead to the normal modes shown in Figures 34-10 and 34-11.

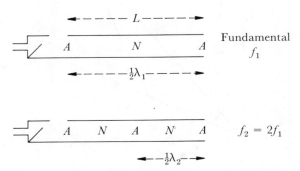

Figure 34-10 Nodes (N) and antinodes (A) in two different normal modes of vibration of an air column in an open organ pipe. The fundamental frequency is $f_1 = v/\lambda_1 = v/2L$. The first overtone is $f_2 = v/\lambda_2 = v/L$. Since $f_2 = 2f_1$, the first overtone is the second harmonic.

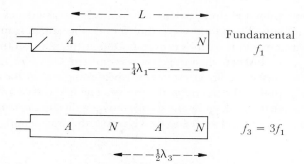

Figure 34-11 Normal modes of vibration of an air column in a closed organ pipe. The fundamental frequency is $f_1 = v/\lambda_1 = v/4L$. In the next normal mode, the wavelength is given by $\lambda_3/2 = 2L/3$, $\lambda_3 = 4L/3$. The first overtone is $f_3 = v/\lambda_3 = 3v/4L$. Since $f_3 = 3f_1$, the first overtone is the third harmonic.

34.6 Doppler Effect

One hears an abrupt drop in the pitch of the sound from the horn of a car as it passes by. Any such change in the observed frequency arising from motion of the source or the observer is called the *Doppler effect*.

The effect was first analyzed by an Austrian physicist, J. C. Doppler (1803–1853), who pointed out that similar frequency shifts should occur for light waves. Experiments later proved that he was right. And as we describe in Chapter 39, "Doppler shifts" in the frequency of light coming from distant stars can be used to provide us with a vital item of astronomical information: the velocity of the star relative to the earth.

Sound waves show a Doppler effect for two distinct reasons:

1   The wavelength in the air is changed by motion of the source relative to the air. This is illustrated in Figure 34-12 which shows crests of the successive waves emitted by a source that moves with a speed v_s relative to a frame of reference which is at rest in the air. After emission, a crest moves with a wave

Figure 34-12 Crests of waves emitted by a moving source A. Crest number 1 was emitted when A was at point number 1. Notice that in front of the source the wavelength λ is reduced, and behind the source it is lengthened.

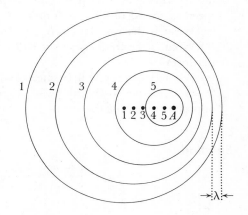

speed v determined entirely by the mechanical characteristics of the transmitting medium. The sound wave thus forgets the source is moving, and a spherical crest expands with its center located at the position that the source occupied at the instant when the crest was emitted. Consider the separation of successive crests that are emitted at the beginning and at the end of a time interval T equal to the period of the source. The first crest moves a distance vT in this time interval, while the source moves a distance v_sT. In front of the source, the distance between successive crests, the wavelength, is

$$\lambda_{air} = vT - v_sT = \frac{v - v_s}{f} \qquad (34.17)$$

A stationary listener in front of the source will measure a frequency v/λ_{air} corresponding to this reduced wavelength λ_{air}. Therefore he will measure a frequency *higher* than the frequency of vibration of the approaching source of the sound.

2   If the observer is moving through the air toward the source he will encounter more crests per second than if he had remained stationary relative to the air. He will measure a frequency given by

$$f_o = \frac{speed\ of\ waves\ relative\ to\ observer}{wavelength}$$

By moving toward the source with a speed v_o relative to the air, the observer increases the speed of the sound waves relative to him to the value $v + v_o$ and consequently hears a higher pitch. The observed frequency is

$$f_o = \frac{v + v_o}{\lambda_{air}} = \left(\frac{v + v_o}{v - v_s}\right)f \qquad (34.18)$$

This formula can be applied to any direction of the velocities along the line joining the observer and the source, if we adopt the convention that v_o and v_s are positive for velocities of approach and negative for velocities of recession.

Example 4   A train approaches a station platform at a speed of 10 m/s, blowing its whistle at a frequency of 100 Hz, as measured by the engineer. The speed of sound in air is 330 m/s:
(a) What is the wavelength λ_{air} in front of the train?
(b) What is the frequency that will be measured by a man standing on the platform?
(c) What frequency will be measured by the driver of a car proceeding at 20 m/s relative to the highway, approaching the station from a direction opposite to that of the train?

Solution

(a)
$$\lambda_{air} = \frac{v - v_s}{f} = \frac{(330 - 10)\ m/s}{100\ Hz} = 3.20\ m$$

(b) A man on the platform measures a frequency

$$f_o = \frac{v}{\lambda_{\text{air}}} = \frac{330 \text{ m/s}}{3.20 \text{ m}} = 103 \text{ Hz}$$

(c) The driver measures a frequency

$$f_o = \left(\frac{v + v_o}{v - v_s}\right) f = \left(\frac{330 + 20}{330 - 10}\right) 100 \text{ Hz} = 109 \text{ Hz}$$

34.7 Supersonic Speeds and Shock Waves

The crests of sound waves emitted from an object A which moves faster than the speed of sound are shown in Figure 34-13(a). As in Figure 34-12, the center of each crest is located at the position that the source occupied at the instant when the crest was emitted. Along the cone which touches these wave fronts, these waves reinforce each other leading to a concentration of energy that constitutes a *shock wave*. Figure 34-13(b) shows that the angle θ made by this shock wave with the velocity vector $\mathbf{v}_A$ of the moving object must satisfy

$$\sin \theta = \frac{v}{v_A} \tag{34.19}$$

The same sort of thing happens in the production of *bow waves* by a boat traveling at a speed greater than that of the waves on the water's surface.

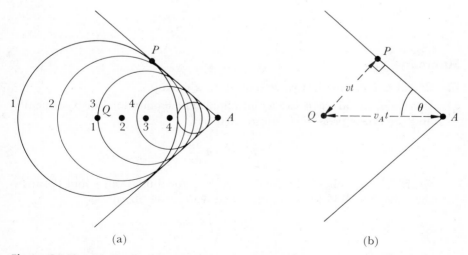

(a)   (b)

Figure 34-13   (a) Shock wave produced by object A moving to the right with supersonic speed. The shock wave is the cone with vertex at A. This cone touches the spherical crests emitted when A was at previous positions such as the points labeled 1, 2, 3, and 4. (b) The angle θ made by the shock wave with the velocity vector of the moving object is determined from $\sin \theta = vt/v_A t = v/v_A$.

The moving object does not have to be a vibrating source of sound to create shock waves. As soon as any object's speed passes that of sound, a shock wave is formed. Common examples are the shock waves that we hear as the crack of a passing bullet or the *sonic boom* from a distant supersonic aircraft. The energy carried by the conical shock waves* that accompany a high-speed aircraft is sometimes sufficient to smash windows and shake buildings.

Each observer underneath the route of a supersonic aircraft hears a loud boom as a shock wave passes him. No sound from vibrating sources in the aircraft can be heard by an observer outside the cone of Figure 34-13(a). An observer at P will hear the sound emitted by the aircraft when it was at Q.

The development of shock waves is determined by the ratio of the source speed to the speed of sound in the air at the existing temperature. This ratio is called the *Mach* number [after Ernst Mach (1838–1916)]. A missile at Mach 5 is traveling 5 times as fast as sound travels through the air in its locality. Shock wave formation commences at Mach 1.

The Russian physicist, P. A. Cerenkov (1904–   ), discovered that a fast charged particle would radiate light waves whenever the particle passed through a material at a speed greater than the speed of light in the material. This phenomenon is analogous to the production of shock waves. Indeed, the wave fronts of *Cerenkov radiation* are exactly as depicted in Figure 34-13(a). A practical method of determining the speed of fast charged particles is based on measurement of the angle at which they emit this bluish Cerenkov radiation as they pass through transparent material.

Summary

☐   Sound is a mechanical wave motion in a medium.

☐   In a fluid, sound waves are longitudinal. The longitudinal displacement χ and the excess pressure P_e are related by

$$P_e = -\kappa \frac{\partial \chi}{\partial x}$$

where κ is the bulk modulus of elasticity of the fluid. Both χ and P_e satisfy the wave equation for traveling waves with wave speed

$$v = \sqrt{\frac{\kappa}{\rho_0}}$$

where ρ_0 is the equilibrium value of the density of the medium.

---

*Two principal shock fronts, one at the nose and one at the tail, accompany a supersonic aircraft. These fronts cause a double boom.

☐ The speed of sound in a gas is correctly given by using the adiabatic bulk modulus γP. Then

$$v = \sqrt{\frac{\gamma P}{\rho}} = \sqrt{\frac{\gamma k T}{m}}$$

where m is the mass of a molecule of the gas.

☐ The intensity I of a wave is the average rate of transfer of energy per unit area perpendicular to the direction of propagation:

$$I = \frac{1}{A}\left(\frac{\partial U}{\partial t}\right)_{\text{avg}}$$

For sinusoidal traveling waves of amplitude χ_0 and angular frequency ω,

$$I \propto \omega^2 \chi_0^2$$

☐ In a given direction, the intensity of waves radiated from a point source falls off inversely as the square of the distance, provided there is no absorption.

☐ Loudness is measured by the intensity level defined by

$$intensity\ level = 10 \log_{10}\left(\frac{I}{10^{-12}\ \text{W/m}^2}\right) \text{dB}$$

☐ The superposition of waves of slightly different frequencies f_1 and f_2 is a wave with an intensity that contains oscillations with a frequency $f_B = f_2 - f_1$, called the beat frequency.

☐ The relationship between the observed frequency f_o and the source frequency f is given for sound waves by the Doppler effect formula

$$f_o = \left(\frac{v + v_o}{v - v_s}\right)f$$

where v_o and v_s are the velocities of approach of the observer and the source, respectively, relative to the transmitting medium and v is the speed of sound relative to this medium.

Questions

1 (a) The time interval between seeing a flash of lightning and hearing the associated thunder is 9 s. Estimate the distance to the lightning stroke.

   (b) Calculate the distance that a sound wave actually travels in 3.00 s in air at 0°C. What percentage error is made in assuming that this distance is 1 km?

2 When a stone is dropped into a well, the sound of the splash is heard 3.00 s later. How deep is the well? Assume that the speed of sound in the air is 340 m/s.

3 To determine the depth of water below a ship's keel, a short pulse of ultrasonic waves is sent out and the time to the first echo is measured and found to be 0.150 s.

The speed of the sound waves in the water is 1.50×10^3 m/s. What is the depth below the keel?

4   A worker strikes an iron rail with a hammer, and the sound travels through both the air and the rail to reach an observer 1000 m away. How much time separates the arrival of the sound through the air and that of the sound through the rail? (Use data in Table 34-1.)

5   The "international pitch" used in music is based on a frequency of 440 Hz for middle A on the piano:
   (a) What is the period of the vibration of the piano string which produces this tone?
   (b) At room temperature the speed of sound in air is about 340 m/s. Calculate the wavelength of the vibration which yields middle A.

6   Ultrasonic waves with a wavelength of 5×10^{-7} m in air (as short as that of visible light) can be generated by the vibrations of a thin quartz crystal plate which is driven by an alternating electric field. What is the frequency of such sound waves when the wave speed is 340 m/s?

7   The shortest wavelength (in air at 0°C) of the ultrasonic waves that a bat emits is approximately 3.3 mm. What is the frequency of these waves? Is this the highest or the lowest frequency emitted by the bat?

8   What physical properties of a sound wave correspond to the sensations of pitch and tone quality?

9   A 33-rpm record is accidentally played at 45 rpm. Will the pitch of the recorded human voices be altered? Explain.

10 Using the theory developed in Section 34.3, calculate the speed of sound in hydrogen at 0°C. Compare this result with the value 1.27 km/s that is found by experiment. (The molecular weight of hydrogen is 2.016. Assume that $\gamma = 1.40$.)

11 At a frequency of 1 kHz, the pressure amplitude at the threshold of audibility is about 2.0×10^{-5} Pa. What is the corresponding displacement amplitude in air under standard conditions?

12 A plane wave travels in the positive x direction through air at 0°C and a pressure of 1.01×10^5 Pa. The frequency is 1.0 kHz and the displacement amplitude is 2.0×10^{-8} m:
   (a) Give the function that describes the displacement wave.
   (b) Find the equation of the pressure wave.

13 What is the ratio of the intensities of two sinusoidal sound waves of the same frequency, if the ratio of their amplitudes is 10^6?

14 A 1000-Hz sound wave with the barely audible intensity of 1.0×10^{-12} W/m² passes through air of density 1.29 kg/m³ at a speed of 340 m/s. Find the amplitude of vibration.

15 What would be the amplitude of vibration if the intensity in the preceding question were increased to the pain threshold of 1.0 W/m²?

16 While a man is shouting, the average power of his voice passing through a 2.0-m² open window is 1.6×10^{-4} W. If the power is uniformly distributed over this 2.0-m² area, what is the intensity at the window?

17 The intensity of sound radiated from a small source is measured at a point 50 m from the source and found to be 2.0×10^{-6} W/m²:
   (a) What will be the intensity in the same direction from the source at a point which is 150 m further from the source?

(b) Does the answer to part (a) depend on the following assumptions?
  (i) The source is isotropic.
  (ii) The power absorbed in the intervening air is negligible.
  (iii) All the power comes directly from the source rather than from reflecting surfaces.

18 Find the acoustic power output of a loudspeaker which radiates a sound wave such that there is a uniform intensity of 0.80 W/m^2 on the surface of a hemisphere located 20 m from the loudspeaker.

19 Use the definition of intensity level to verify the decibel levels corresponding to each intensity listed in Table 34-2.

20 Show that the difference in intensity levels of two sounds of intensities I_1 and I_2 is $10 \log_{10} (I_1/I_2)$ dB.

21 If the amplitude of one sound wave is 10 times the amplitude of a second sound wave of the same frequency, what is the difference in the intensity levels of these sounds?

22 If the intensity level in a room is 55 dB when one person is talking, what will be the intensity level when ten people are talking simultaneously? (The intensity due to a number of such independent sound sources is the sum of the individual intensities.)

23 A beam of sound waves of wavelength λ passes through an aperture of 10-mm diameter. As the beam emerges and travels through air, most of the beam intensity remains confined within a cone of 10° vertex angle. Estimate the wavelength of the beam. What would happen if the wavelength were made 5 times as long?

24 When and why do beats occur?

25 A tuning fork of unknown frequency makes 2.0 beats per second with a standard tuning fork of frequency 256 Hz:
  (a) What are the two possible values of the unknown frequency?
  (b) A small piece of wax is placed on the prong of the fork of unknown frequency. Then it is observed that the number of beats per second is reduced. Which value calculated in part (a) is correct?

26 A piano tuner using a 256-Hz tuning fork hears 6 beats per second when a certain note on the piano is sounded. When he uses a 260-Hz tuning fork he hears 2 beats per second when the same note is sounded. What is the frequency of that note?

27 Sketch the location of nodes and antinodes and find the wavelength and frequency of the third overtone for the open organ pipe of Figure 34-10. What is this frequency if the length of the pipe is 0.20 m and the speed of the sound waves within the pipe is 330 m/s?

28 Answer the preceding question for the closed organ pipe of Figure 34-11.

29 Two identical organ pipes have a fundamental frequency of 256 Hz at 20°C. What is the beat frequency when both pipes are sounded simultaneously, if one pipe is at 20°C and the other is at 25°C?

30 In Kundt's method for determining the speed of sound in a gas, the gas is contained in a long horizontal tube. Powder is uniformly distributed along the bottom of the tube. Standing waves of known frequency are established in this column of gas and the location of nodes is determined by observing where powder collects within the tube. If the distance between adjacent piles of powder is 8.0×10^{-2} m when the frequency is 2.00 kHz, what is the speed of sound in the gas?

31 A police officer, blowing a whistle which vibrates with a frequency of 512 Hz, runs at a speed of 6.0 m/s in the direction of a group of people standing at an intersection. The speed of sound in the air is 340 m/s:
   (a) What is the wavelength in front of the police officer?
   (b) What frequency is heard by the people on the corner?

32 The police officer in the preceding question stops, but one of the people on the corner runs away at a speed of 6.0 m/s. Find the frequency heard by this person running away.

33 A car approaching a cliff at a speed of 40 m/s sounds its horn, which vibrates at a frequency of 320 Hz. The speed of sound in air is 340 m/s:

   (a) Find the wavelength of the sound waves which travel from the horn to the cliff. Will the waves reflected from the cliff have the same wavelength?
   (b) Relative to the driver, what is the speed of the waves reflected from the cliff?
   (c) What frequency of these reflected waves will be detected by the driver?

34 The train of Example 4 passes through the station and proceeds down the track, still blowing its whistle:

   (a) What is the wavelength behind the train?
   (b) What is the frequency that will be measured by the man standing on the platform?
   (c) What frequency will be measured by the driver of the car receding from the train at 20 m/s relative to the highway?

35 A train passes a stationary observer at a speed of 30 m/s on a day when the speed of sound in air is 340 m/s. What percentage drop in the frequency of the train whistle is detected by the observer as the train passes?

36 A student holding a 256-Hz tuning fork approaches a laboratory wall at a speed of 5.0 m/s. The speed of sound in the air is 340 m/s:
   (a) What frequency will the student detect from the waves emitted from the fork and reflected from the wall?
   (b) How many beats per second will he hear between the reflected waves and the waves coming directly from the fork?

37 Sketch a scale drawing of the shock wave produced by an aircraft traveling at Mach 2. Show the position of the aircraft and the position of one crest that was emitted at a previous instant. What is the vertex angle of the cone formed by this shock wave?

38 A bullet travels through air at a speed of 660 m/s in a region where the speed of sound is 330 m/s:
   (a) What is the Mach number for this bullet?
   (b) Find the angle made by the shock wave with the velocity vector of the bullet.

39 The wave front of the bow wave of a ship makes an angle of 10° with the ship's velocity vector. The ship's speed is 7.0 m/s. Calculate the speed of the water waves.

40 The ratio of the speed of light in a vacuum to the speed of light in water is about 1.33. What is the cutoff speed below which a charged particle traveling through water will not emit Cerenkov radiation?

41 At what speed through water will electrons emit Cerenkov radiation having a conical wave front which makes an angle of 60° with the electron's velocity vector?

Supplementary Questions

S-1 From the equation

$$\frac{\partial^2 \chi}{\partial x^2} = \frac{\rho_0}{\kappa} \frac{\partial^2 \chi}{\partial t^2}$$

and the fact that

$$\frac{\partial^2 \chi}{\partial x \partial t} = \frac{\partial^2 \chi}{\partial t \partial x}$$

show that

$$\frac{\partial^2 P_e}{\partial x^2} = \frac{\rho_0}{\kappa} \frac{\partial^2 P_e}{\partial t^2}$$

where

$$P_e = -\kappa \frac{\partial \chi}{\partial x}$$

S-2 Show that the speed of longitudinal waves in a metal rod of density ρ is

$$v = \sqrt{\frac{Y}{\rho}}$$

where Y is Young's modulus for the metal. This is defined by

$$Y = \frac{F/A}{\Delta L/L}$$

where $\Delta L/L$ is the fractional extension of the rod caused by a stretching force per unit area F/A.

S-3 Show that the speed v of sound waves in a gas is related to the thermal speed v_T of the gas molecules by

$$v = \sqrt{\frac{\gamma}{3}} v_T$$

(Recall from Chapter 17 that $\frac{1}{2}mv_T^2 = \frac{3}{2}kT$, where m is the mass of a molecule and k is Boltzmann's constant.)

S-4 The speed of sound in air is given in SI units by

$$v = 331.45\sqrt{\frac{T}{273.15}}$$

where T is the Kelvin temperature. Show that this implies that at ordinary temperatures, the speed of sound increases by about 0.6 m/s for each kelvin increase in temperature.

S-5 Show that the expressions

$$I = -\kappa\left(\frac{\partial \chi}{\partial x} \frac{\partial \chi}{\partial t}\right)_{\text{avg}} \qquad \text{and} \qquad P_e = -\kappa \frac{\partial \chi}{\partial x}$$

imply that a traveling wave of the form

$$\chi = f(x - vt)$$

has an intensity given by

$$I = \frac{(P_e^2)_{\text{avg}}}{\sqrt{\rho_0 \kappa}}$$

where ρ_0 is the equilibrium value of the fluid density.

S-6 Consider a point source at the apex of a small cone which encloses a solid angle $d\omega$. Suppose the power radiated outward through this cone is $\partial U/\partial t$. Show that within the cone at a point a distance r from the source the intensity is

$$I = \left(\frac{\partial U/\partial t}{d\omega}\right)\frac{1}{r^2}$$

provided there is no absorption. This shows that the inverse square law is valid in a given direction from a point source, whether or not the source is isotropic.

S-7 An observer stands 60 m from a loudspeaker. How far must he walk toward the speaker in order to encounter an intensity level greater by 10 dB?

S-8 The intensity level of sound is 25 dB lower at a point 300 m from a foghorn than it is at a point 30 m from the foghorn:

(a) What decrease in intensity level would occur if the intensity were proportional to the inverse square of the distance from the foghorn?

(b) What part of the actual decrease in intensity level can be attributed to the absorption of sound by the fog?

S-9 A signal χ is a superposition that contains a "carrier frequency" ω_c and "side bands" $\omega_c + \omega_m$ and $\omega_c - \omega_m$:

$$\chi = \cos \omega_c t + \tfrac{1}{2}A[\cos(\omega_c + \omega_m)t + \cos(\omega_c - \omega_m)t]$$

Show that χ can be expressed as

$$\chi = (1 + A \cos \omega_m t) \cos \omega_c t$$

which can be interpreted as a signal with angular frequency ω_c that has its amplitude modulated at the angular frequency ω_m.

S-10 Consider the superposition $\chi = \chi_1 + \chi_2$ of two vibrations of the same amplitude A but slightly different frequencies ω_1 and ω_2:

$$\chi_1 = A \cos \omega_1 t \qquad \text{and} \qquad \chi_2 = A \cos \omega_2 t$$

(a) From the phasor diagram (Figure 34-9), show that the amplitude of the superposition is given by

$$\chi_0^2 = 2A^2[1 + \cos(\omega_2 - \omega_1)t] = 4A^2 \cos^2 \tfrac{1}{2}(\omega_2 - \omega_1)t$$

and that the angle between the phasor of length χ_0 and the horizontal axis is $\tfrac{1}{2}(\omega_1 + \omega_2)t$. This implies that the horizontal projection of this phasor is

$$\chi = \chi_0 \cos \tfrac{1}{2}(\omega_1 + \omega_2)t = 2A \cos \tfrac{1}{2}(\omega_2 - \omega_1)t \cos \tfrac{1}{2}(\omega_1 + \omega_2)t$$

(b) Use the trigonometric formula

$$\cos a + \cos b = 2 \cos \tfrac{1}{2}(a + b) \cos \tfrac{1}{2}(a - b)$$

to show that

$$\chi = 2A \cos \tfrac{1}{2}(\omega_2 - \omega_1)t \cos \tfrac{1}{2}(\omega_1 + \omega_2)t$$

(c) Sketch a graph showing χ as a function of t.

S-11 If an observer and a source of sound are both stationary relative to the earth, will a wind cause a change in the observed frequency? Explain.

S-12 Show that the approximation given by

$$\frac{\Delta f}{f} = \frac{velocity\ of\ source\ relative\ to\ observer}{v}$$

follows from the general result given in Eq. 34.18 for the case when v_o and v_s are both small compared to v.

S-13 A supersonic aircraft, flying at an altitude of 4000 m, has a speed such that its Mach number is 1.25:

(a) Find the angle made by the shock wave with the velocity vector of the aircraft.

(b) How long after the aircraft has passed overhead will the boom be heard by an observer on the ground? (Assume that the shock wave travels at a speed of 330 m/s.)

35.

electromagnetic waves

The next four chapters are concerned with light and other forms of electromagnetic radiation. The major objective in this chapter is to show that Maxwell's equations have solutions that are electromagnetic waves traveling at the speed of light. The electric and magnetic fields in an electromagnetic wave are discussed as well as the energy and momentum carried by the wave.

35.1 Plane Electromagnetic Waves

Differential Form of Maxwell's Equations

Maxwell's equations, as given in Section 30.3, involve integrals of the electric and magnetic fields. It is possible and often more convenient to have the same physical laws expressed in terms of derivatives of the fields.

To illustrate how the differential form of Maxwell's equations can be obtained from the integral form, we consider the third Maxwell equation (Faraday's law),

$$\oint_C E_s \, ds = - \int_S \frac{\partial B_N}{\partial t} \, dA$$

We apply this law to an infinitesimal rectangular path in a plane parallel to the XY plane, with sides of lengths Δx and Δy, and denote the field components at the center of the path by E_x and E_y (Figure 35-1). For the path ab, $\int_a^b E_s \, ds = (E_x)_{\text{avg}} \, \Delta x$, where $(E_x)_{\text{avg}}$ is the average value of the x component of the field along ab. This will be approximately the value encountered at the midpoint of ab. Since the change in E_x in a vertical distance $\Delta y/2$ is approximately

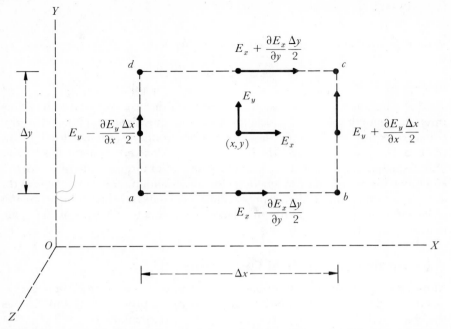

Figure 35-1 Path of integration *abcda* in a plane parallel to the *XY* plane, showing the approximate average value for each section of the component of **E** along the path. The electric field at the center of the path has components E_x, E_y, E_z.

$(\partial E_x/\partial y)(\Delta y/2)$, $(E_x)_{avg}$ is given approximately by

$$(E_x)_{avg} = E_x - \frac{\partial E_x}{\partial y}\frac{\Delta y}{2}$$

Similar arguments give the approximate average values for each section of the component of **E** along the path (Figure 35-1). In the limit, as Δx and Δy approach zero, we have

$$\oint_C E_s\,ds = \int_a^b E_s\,ds + \int_b^c E_s\,ds + \int_c^d E_s\,ds + \int_d^a E_s\,ds$$

$$= \left(E_x - \frac{\partial E_x}{\partial y}\frac{\Delta y}{2}\right)\Delta x + \left(E_y + \frac{\partial E_y}{\partial x}\frac{\Delta x}{2}\right)\Delta y - \left(E_x + \frac{\partial E_x}{\partial y}\frac{\Delta y}{2}\right)\Delta x$$

$$- \left(E_y - \frac{\partial E_y}{\partial x}\frac{\Delta x}{2}\right)\Delta y$$

$$= \left(\frac{\partial E_y}{\partial x} - \frac{\partial E_x}{\partial y}\right)\Delta x\Delta y$$

and in this limit, the rate of change of magnetic flux enclosed by the path *abcda* is

$$\int_S \frac{\partial B_N}{\partial t}\,dA = \frac{\partial B_z}{\partial t}\Delta x\Delta y$$

where B_z is the *z* component of the magnetic field at the center of the rectangle.

Faraday's law for the z component of $\mathbf{B}$ is therefore

$$\frac{\partial E_y}{\partial x} - \frac{\partial E_x}{\partial y} = -\frac{\partial B_z}{\partial t}$$

There are analogous expressions for the two other components of $\mathbf{B}$, giving three differential equations which together are equivalent to a single integral equation expressing Faraday's law. An entirely similar argument leads to the three differential equations which express the Maxwell–Ampere law. Table 35-1 displays the equivalent integral and differential formulations of Maxwell's equations for electric and magnetic fields in a vacuum. For completeness, the differential equations which express Gauss's laws have been included in this table, although the integral formulation of these laws is sufficient for the applications considered in this book.

Wave Equations for the Field Components

Maxwell's equations have solutions $\mathbf{E}$ and $\mathbf{B}$ that are the type of functions of position and time that describe a traveling wave. To prove this, and to discover the characteristics of such waves, we consider a particular case.

Table 35-1 Maxwell's Equations for $\mathbf{E}$ and $\mathbf{B}$ in a Vacuum

Integral form	Differential form
Gauss's law for $\mathbf{E}$	
$\oint_S E_N \, dA = 0$	$\dfrac{\partial E_x}{\partial x} + \dfrac{\partial E_y}{\partial y} + \dfrac{\partial E_z}{\partial z} = 0$
Gauss's law for $\mathbf{B}$	
$\oint_S B_N \, dA = 0$	$\dfrac{\partial B_x}{\partial x} + \dfrac{\partial B_y}{\partial y} + \dfrac{\partial B_z}{\partial z} = 0$
Faraday's law	
	$\dfrac{\partial E_z}{\partial y} - \dfrac{\partial E_y}{\partial z} = -\dfrac{\partial B_x}{\partial t}$
$\oint_C E_s \, ds = -\displaystyle\int_S \dfrac{\partial B_N}{\partial t} \, dA$	$\dfrac{\partial E_x}{\partial z} - \dfrac{\partial E_z}{\partial x} = -\dfrac{\partial B_y}{\partial t}$
	$\dfrac{\partial E_y}{\partial x} - \dfrac{\partial E_x}{\partial y} = -\dfrac{\partial B_z}{\partial t}$
Maxwell–Ampere law	
	$\dfrac{\partial B_z}{\partial y} - \dfrac{\partial B_y}{\partial z} = \mu_0 \epsilon_0 \dfrac{\partial E_x}{\partial t}$
$\oint_C \dfrac{1}{\mu_0} B_s \, ds = \displaystyle\int_S \epsilon_0 \dfrac{\partial E_N}{\partial t} \, dA$	$\dfrac{\partial B_x}{\partial z} - \dfrac{\partial B_z}{\partial x} = \mu_0 \epsilon_0 \dfrac{\partial E_y}{\partial t}$
	$\dfrac{\partial B_y}{\partial x} - \dfrac{\partial B_x}{\partial y} = \mu_0 \epsilon_0 \dfrac{\partial E_z}{\partial t}$

Suppose that the electric field has a component only in the y direction and that this component E_y is a function only of x and t; that is, at any instant $\mathbf{E}$ is uniform over any plane perpendicular to the X axis (Figure 35-2). The electric flux lines are parallel to the Y axis. Gauss's law is satisfied by such a field since as many lines enter any closed surface on one side as leave it on the other side.

For this electric field, Faraday's law implies

$$0 = -\frac{\partial B_x}{\partial t} \qquad 0 = -\frac{\partial B_y}{\partial t} \qquad \frac{\partial E_y}{\partial x} = -\frac{\partial B_z}{\partial t}$$

This shows that the time-varying magnetic field $\mathbf{B}$ has only a z component. We have learned that the time-varying magnetic field is necessarily perpendicular to the electric field of Figure 35-2. Also, since $\partial E_y/\partial x$ is independent of the co-ordinates y and z, we see that $\partial B_z/\partial t$ and consequently B_z is independent of y and z. In other words, $\mathbf{B}$ as well as $\mathbf{E}$ is uniform over any plane perpendicular to the X axis. The magnetic flux lines are parallel to the Z axis and Gauss's law for magnetic fields is satisfied.

The Maxwell–Ampere law requires that

$$-\frac{\partial B_z}{\partial x} = \mu_0 \epsilon_0 \frac{\partial E_y}{\partial t}$$

To discover the dependence of E_y on x and t we use Faraday's law together with the Maxwell–Ampere law to eliminate B_z. Thus, from Faraday's law,

$$\frac{\partial^2 E_y}{\partial x^2} = -\frac{\partial^2 B_z}{\partial x \partial t}$$

and from the Maxwell–Ampere law,

$$-\frac{\partial^2 B_z}{\partial t \partial x} = \mu_0 \epsilon_0 \frac{\partial^2 E_y}{\partial t^2}$$

Figure 35-2 The electric field is assumed to be in the y direction and to be uniform over any plane perpendicular to the X axis. Faraday's law then implies that the time-varying magnetic field has only a z component.

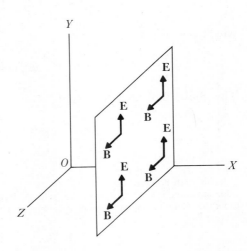

Therefore Maxwell's equations require that E_y satisfy the wave equation

$$\frac{\partial^2 E_y}{\partial x^2} = \mu_0 \epsilon_0 \frac{\partial^2 E_y}{\partial t^2}$$

and a similar procedure shows that B_z must also satisfy the wave equation

$$\frac{\partial^2 B_z}{\partial x^2} = \mu_0 \epsilon_0 \frac{\partial^2 B_z}{\partial t^2}$$

Since these wave equations are of the same mathematical form as the wave equation studied in Section 33.1, they must have solutions that are traveling waves with wave speed

$$c = \frac{1}{\sqrt{\mu_0 \epsilon_0}} = \frac{1}{\sqrt{(4\pi \times 10^{-7})(8.85 \times 10^{-12})}} \, \text{m/s}$$

$$= 2.9986 \times 10^8 \, \text{m/s} \tag{35.1}$$

which is in excellent agreement with the observed value 2.99792458×10^8 m/s of the speed of light in a vacuum.* It was this discovery that led Maxwell to suggest that light was an electromagnetic wave phenomenon. The speed of electromagnetic waves in a vacuum has been measured over a wide range of frequencies, with all results in agreement with this theoretical prediction deduced from Maxwell's equations.

One simple solution to the wave equation for E_y is the sinusoidal traveling wave

$$E_y = E_0 \cos(kx - \omega t)$$

where E_0 is the amplitude of the wave. The angular frequency is related to the wave number k by

$$\omega = ck$$

The magnetic field that accompanies this electric field can be found from Faraday's law:

$$-\frac{\partial B_z}{\partial t} = \frac{\partial E_y}{\partial x} = -kE_0 \sin(kx - \omega t)$$

which is satisfied by

$$B_z = \frac{k}{\omega} E_0 \cos(kx - \omega t)$$

This shows that, at any instant

$$B_z = \frac{E_y}{c} \tag{35.2}$$

---

*The presently accepted value of ϵ_0 is 8.854187818 F/m. This value is calculated from the measured value of c using $c = 1/\sqrt{\mu_0 \epsilon_0}$. We thus rely on electromagnetic theory to calculate ϵ_0 rather than using the relatively imprecise value of ϵ_0 determined by electrostatic measurements as outlined in Section 24.3.

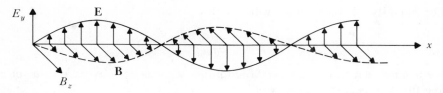

Figure 35-3   Electromagnetic wave. **E** and **B** at a given instant at various values of *x*, when there is a sinusoidal plane wave propagating in the positive *x* direction.

Therefore at any point in space, B_z and E_y oscillate in phase as is illustrated in Figure 35-3.

The plane wave solutions to Maxwell's equations that we have examined illustrate features that are true for all types of plane electromagnetic waves in a vacuum. The speed of propagation is $c = 1/\sqrt{\mu_0\epsilon_0}$. These waves are *transverse*, that is, the vibrating quantities **E** and **B** are both perpendicular to the direction of propagation. The fields **E** and **B** are mutually perpendicular and, at any point in space, oscillate in phase. The equation $B = E/c$ is valid no matter what the waveform of the traveling wave (Question S-1).

Since Maxwell's equations are linear, electromagnetic waves obey the *super-position principle*. A very significant part of our knowledge of these waves comes from the experimental study of superposition phenomena.

35.2 Intensity of Electromagnetic Waves

The energy density associated with the electric field **E** and the magnetic field **B** of an electromagnetic wave is

$$u = \tfrac{1}{2}\epsilon_0 E^2 + \frac{1}{2}\frac{B^2}{\mu_0} \qquad (31.16)$$

Using the relationship $B = E/c = \sqrt{\mu_0\epsilon_0}\,E$, we find that the electric energy density is equal to the magnetic energy density. The total energy density can be expressed as

$$u = \epsilon_0 E^2 \qquad (35.3)$$

Consider the rate $\partial U/\partial t$ at which energy is transported by a traveling electromagnetic wave across an area A that is perpendicular to the direction of propagation. In a time interval dt, the wave moves a distance $c\,dt$ and carries across A all the energy in the volume $Ac\,dt$. This energy is $dU = uAc\,dt$ and therefore $\partial U/\partial t = uAc$. The power transported per unit perpendicular area is

$$S = \frac{1}{A}\frac{\partial U}{\partial t} = uc = \epsilon_0 c E^2 = \sqrt{\frac{\epsilon_0}{\mu_0}}\,E^2 \qquad (35.4)$$

The intensity I is the average value of this power:

$$I = S_{\text{avg}} = \sqrt{\frac{\epsilon_0}{\mu_0}}(E^2)_{\text{avg}}$$

For a sinusoidal traveling wave of amplitude E_0, since the average value of a sine function is $\frac{1}{2}$, the intensity is

$$I = \frac{1}{2}\sqrt{\frac{\epsilon_0}{\mu_0}}E_0^2 = (1.33 \times 10^{-3} \text{ W/V}^2)E_0^2 \qquad \text{(sinusoidal wave)} \quad (35.5)$$

There is an important relationship between S and the fields $\mathbf{E}$ and $\mathbf{B}$ of an electromagnetic wave. For the wave considered in Section 35.1, the power per unit perpendicular area carried by the wave in the positive x direction is

$$S_x = \sqrt{\frac{\epsilon_0}{\mu_0}}E_y^2 = \sqrt{\frac{\epsilon_0}{\mu_0}}cE_yB_z = \frac{E_yB_z}{\mu_0}$$

This is consistent with the vector equation

$$\mathbf{S} = \frac{\mathbf{E} \times \mathbf{B}}{\mu_0} \qquad (35.6)$$

which a more general analysis shows is the rate of energy flow per unit perpendicular area for an arbitrary direction of propagation of the electromagnetic wave. The vector $\mathbf{S}$ is called the *Poynting vector*.

Maxwell's equations imply that an electromagnetic wave carries linear momentum as well as energy. It turns out that the amount of momentum p associated with an energy U of an electromagnetic wave traveling at speed c is given by

$$p = \frac{U}{c} \qquad (35.7)$$

Associated with the energy flow per unit area, $\mathbf{S} = \mathbf{E} \times \mathbf{B}/\mu_0$, is a momentum flow per unit area $\mathbf{S}/c$ which has an average magnitude I/c. This can be verified by measuring the "radiation pressure" exerted on a surface that reflects or absorbs electromagnetic waves.

Example 1 The intensity of sunlight just above the earth's atmosphere is $I = 1.4 \text{ kW/m}^2$ (the solar constant). Find the average force exerted by a beam of this intensity on a blackened disk with an area $A = 3.0 \text{ cm}^2$ oriented perpendicular to the beam. What is the radiation pressure?

Solution We assume that the blackened disk absorbs all the energy incident upon it. The average force F exerted by the beam on the disk is the average rate at which the beam supplies momentum to the disk. Therefore

$$F = \frac{IA}{c} = \frac{(1.4 \times 10^3 \text{ W/m}^2)(3.0 \times 10^{-4} \text{ m}^2)}{3.00 \times 10^8 \text{ m/s}} = 1.4 \times 10^{-9} \text{ N}$$

The radiation pressure is

$$\frac{F}{A} = \frac{1.4 \times 10^{-9} \text{ N}}{3.0 \times 10^{-4} \text{ m}^2} = 4.7 \times 10^{-6} \text{ Pa}$$

In 1903, using a sensitive torsion balance with a disk suspended in a vacuum, Nichols and Hull succeeded in measuring the extremely small radiation pressure of a light beam. The experimental results agree with the predictions of Maxwell's electromagnetic theory.

Example 2   Lasers, discussed in Chapter 43, are light sources which provide a narrow beam of high intensity. Suppose a 3.0-kW laser beam were concentrated by a lens into a cross-sectional area of about 10^{-10} m². Find the corresponding intensity and the amplitude of the electric field.

Solution   The intensity is

$$I = \frac{(\partial U/\partial t)_{\text{avg}}}{A} = \frac{3.0 \times 10^3 \text{ W}}{10^{-10} \text{ m}^2} = 3.0 \times 10^{13} \text{ W/m}^2$$

The amplitude of the electric field is

$$E_0 = \sqrt{\frac{I}{1.33 \times 10^{-3} \text{ W/V}^2}} = 1.5 \times 10^8 \text{ V/m}$$

Because a laser beam has such a high power and can be brought to a very sharp focus, it can be used as an exceedingly effective "drill" to burn through a target.

35.3 Radiation from Oscillating Charges

We have shown that Maxwell's equations are consistent with the existence of electromagnetic waves but we have not yet considered how these waves can be produced. The source of an electromagnetic wave is an *accelerated charge*. Expressions for the electric and magnetic fields radiated by an accelerated charge can be deduced from Maxwell's equations, but the analysis is beyond the scope of this book. Nevertheless, it is important to be familiar with at least certain general features of the fields produced when charge is forced to execute simple harmonic motion along what is called an *electric dipole antenna** (Figure 35-4).

The field at distances large compared to both the antenna dimensions and the wavelength, called the *radiation* field, is indicated in Figure 35-5 for an electric dipole antenna at the origin and aligned with OY. At any point, the electric and magnetic fields are perpendicular. They execute sinusoidal oscillations in phase, with $B = E/c$ at any instant. Wave fronts are spherical and travel radially outward with a speed c in the direction of the Poynting vector, $\mathbf{E} \times \mathbf{B}/\mu_0$.

---

*It is called an *electric dipole antenna* because it can be represented approximately by a pair of equal and opposite point charges whose charge alternates sinusoidally with time.

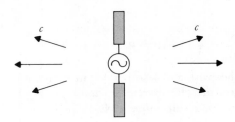

Figure 35-4 Schematic representation of an electric dipole antenna. Part of the energy of the accelerated charges in the antenna is carried away by the electromagnetic wave which propagates outward at speed c.

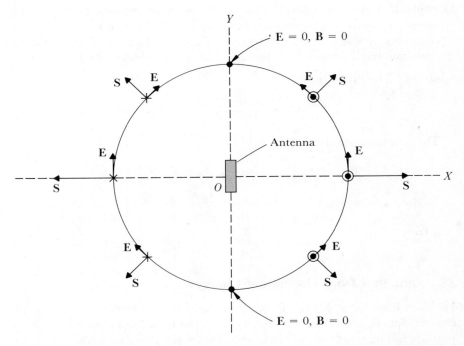

Figure 35-5 Radiation field at a given instant on a spherical surface centered at the antenna. At a given point in space, the fields oscillate sinusoidally.

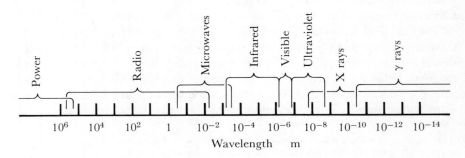

Figure 35-6 The electromagnetic spectrum. The scale is logarithmic.

If we consider a spherical surface centered at the antenna and use the terminology of geography, we can say that **E** is tangent to a meridian and that **B** is tangent to a parallel of latitude. On a given spherical surface, the amplitudes of the fields have maximum magnitude at points on the equator (the intersection of the XZ plane with the spherical surface) and decrease as one moves toward the poles (the intersections of the axis OY with the spherical surface). Along the axis of the antenna OY the fields are zero.

In a fixed direction from the antenna, the intensity falls off according to the inverse square law. Far from the antenna, small segments of the wave fronts are nearly plane and, near the X axis, the electromagnetic wave is approximately the wave discussed in Sections 35.1 and 35.2.

35.4 Electromagnetic Spectrum

All electromagnetic waves travel with the same speed c in a vacuum. Any wavelength λ is possible. And for a given electromagnetic wave, its wavelength λ and its frequency f are related by Eq. 33.12, which for this case gives

$$c = \lambda f$$

In Maxwell's time, the only familiar electromagnetic waves were those corresponding to visible light, although some knowledge had been acquired concerning the ultraviolet and the infrared. Figure 35-6 shows the vast range of electromagnetic radiation wavelengths known today. The various radiations are placed according to wavelength or frequency in an orderly sequence, termed a *spectrum*. Regions are indicated in which the radiation exhibits a roughly similar behavior in its interaction with matter. These regions generally overlap. There is no reason to expect an upper or lower bound to the electromagnetic spectrum.

The very short wavelengths of visible light range from approximately 7×10^{-7} m, which corresponds to the color red, through decreasing wavelengths corresponding to the colors orange, yellow, green, blue, and finally violet at approximately 4×10^{-7} m.* Just how such short wavelengths are measured is described in the discussion of Young's experiment (Chapter 36) involving the interference of light beams. The narrow band of wavelengths which can be detected by the human eye is bounded on the short wavelength side by ultraviolet waves, easily detected by photographic films or by utilizing the photoelectric effect discussed in Chapter 41. On the long wavelength side of visible light are infrared waves copiously emitted by any hot body.

---

*In this region of the spectrum it is common practice to express wavelengths in terms of the *ångström* (Å):

$$1 \text{ Å} = 10^{-10} \text{ m}$$

For red light the wavelength is approximately 7000 Å, for violet approximately 4000 Å.

Although electromagnetic waves of different wavelengths have acquired different names such as x rays and radio waves, we must stress that all electromagnetic waves are of exactly the same nature. However, the emission, absorption, scattering, and transmission through and around obstacles depend upon the value of the wavelength. Therefore investigations of the characteristics of waves of different parts of the spectrum require distinctly different experimental techniques. And the various electromagnetic radiations give rise to vastly different applications in technology.

A striking step in the exploration of the electromagnetic spectrum was made by Heinrich Hertz in 1887, more than two decades after the appearance of Maxwell's theory. Hertz generated what we call *radio* waves by the oscillatory discharge across a spark gap of separated electrodes which had been given opposite charges. To detect the electromagnetic waves emitted by the accelerated charges, he used a loop with its two ends slightly separated. A spark across this gap indicated an oscillatory electric field in the vicinity. Hertz showed that electromagnetic waves were transverse waves and could be reflected by a metal sheet to produce standing waves. The distance between adjacent nodes was 4.8 m which implied that the wavelength was 9.6 m. The calculated frequency of the oscillating circuit was 3×10^7 Hz. From this data, the relationship $c = \lambda f$ allowed calculation of the wave speed:

$$c = (9.6 \text{ m})(3 \times 10^7 \text{ Hz}) = 3 \times 10^8 \text{ m/s}$$

The long wavelength electromagnetic waves discovered by Hertz became of great practical importance with the development of electronic oscillators and amplifiers in the twentieth century. With such oscillators we now generate a spectrum ranging from radar waves or microwaves (usually 0.3-cm to 10-cm wavelength) through the television broadcast band (about 0.3-m to 5-m wavelength) and the wavelengths of short- and long-wave radio. Commercial broadcasting extends to wavelengths of 600 m (a frequency of 500 kHz). Aircraft and marine broadcasting use wavelengths as long as 3 km (100 kHz).

In 1895, a few years after Hertz' experiments with radio waves, the discovery of x rays by W. C. Röntgen (1845–1923) precipitated the investigation of a range of wavelengths much shorter than those of visible light. However, it was not until 1912 that the wave nature of x rays was confirmed and their wavelengths measured to be of the order of 1 Å.

Nuclear γ rays were discovered in 1900 by Paul Villard. Much shorter wavelength γ rays (10^{-13} Å) produced by cosmic radiation have since been found.

Summary

☐ Electromagnetic waves propagate through a vacuum with a speed $c = 1/\sqrt{\mu_0 \epsilon_0}$. The oscillating fields **E** and **B** are transverse, mutually perpendicular, and in phase. At any instant, $B = E/c$.

☐ When an electromagnetic wave travels through a vacuum, the energy flow per unit perpendicular area is given by the Poynting vector,

$$\mathbf{S} = \frac{\mathbf{E} \times \mathbf{B}}{\mu_0}$$

and the intensity for a sinusoidal wave is

$$I = S_{avg} = (1.33 \times 10^{-3} \text{ W/V}^2)E_0^2$$

☐ The source of an electromagnetic wave is accelerated charge.

Questions

1 If the amplitude of the oscillating electric field in an electromagnetic wave is 10^{-3} V/m, what is the amplitude of the magnetic field?

2 Show that the equations,

$$\frac{\partial E_y}{\partial x} = -\frac{\partial B_z}{\partial t} \qquad \text{and} \qquad -\frac{\partial B_z}{\partial x} = \mu_0\epsilon_0\frac{\partial E_y}{\partial t}$$

imply that

$$\frac{\partial^2 B_z}{\partial x^2} = \mu_0\epsilon_0\frac{\partial^2 B_z}{\partial t^2}$$

3 The electric field of a plane electromagnetic wave in a vacuum is given in SI units by

$$E_y = 5.0 \cos 10\pi(x - ct)$$

(a) Find the wavelength, the frequency, and the direction of propagation.
(b) Give the magnetic field as a function of position and time.

4 A sinusoidal traveling electromagnetic wave has an intensity of 10 W/m². Find the amplitudes of the electric and magnetic fields.

5 The smallest intensity of electromagnetic waves that can be detected by a good radio receiver is about 10^{-6} W/m². What is the amplitude of the electric and magnetic fields at this intensity?

6 A helium–neon laser used in a student laboratory continuously emits 1.0×10^{-3} W of red light in a beam with a diameter of 1.4 mm. Find:
(a) The beam intensity.
(b) The amplitude of the electric field in the beam.

7 Assume that a 100-W light bulb radiates uniformly in all directions:
(a) What is the intensity at a point 2.0 m from the bulb?
(b) Find the radiation pressure on a perfectly absorbing black surface and on a perfectly reflecting mirror. Both surfaces are perpendicular to the radiation.

8 What is the amplitude of the electric field at a point which is 100 km from an isotropic 60-kW radio transmitter?

9 A certain atom radiates orange light of wavelength 6.0×10^3 Å for a time interval of 10^{-8} s. How long is the emitted wave train? How many wavelengths does it contain?

10   The frequencies of radio waves in the "AM broadcast band" range from 0.55 $\times$ 10^6 Hz to 1.60 $\times$ 10^6 Hz. What are the longest and shortest wavelengths in this band?

11   Find the frequencies of electromagnetic waves having the following wavelengths in a vacuum:

(a) 10^3 m (long-wave radio).
(b) 1 m (short-wave radio).
(c) 1 cm (microwaves).
(d) 5 $\times$ 10^3 Å (visible light).
(e) 0.1 Å (x rays).
(f) 10^{-2} Å (γ rays).

Supplementary Questions

S-1   Show that for any traveling wave of the form

$$E_y = f(x - ct)$$

Faraday's law ($\partial E_y/\partial x = -\partial B_z/\partial t$) is satisfied if

$$B_z = \frac{E_y}{c}$$

S-2   If a particle has a radius less than a critical value r_c, the repulsive force exerted on the particle by the sun's radiation is greater than the sun's gravitational attraction and the particle will be forced out of the solar system. Calculate r_c for a particle with the density of water, using Appendix C for the necessary data. Assume that the particle absorbs all the radiation incident upon it.

S-3   Maxwell's differential equations imply that in a vacuum an arbitrary electric field, with components E_x, E_y, E_z that are functions of x, y, z, and t, must be such that each component satisfies the wave equation. Prove this for the component E_x.

36.

interference
and diffraction

The superposition principle for electric and magnetic fields states that, at any instant at a given point in space, the field produced by several sources is the vector sum of the contributions that each source would furnish if it were acting alone. This principle is a consequence of the linearity of Maxwell's equations. According to the superposition principle, the field contributed by one source does not affect the contribution of another source; the two fields just add up vectorially. For electromagnetic waves this implies that different radio signals, light beams, and x rays can travel through the same region of a vacuum without influencing each other in the slightest. Intersecting light beams are undeviated and unchanged by the encounter with one another.*

In the regions where there is a superposition of two or more waves, however, *there are variations of intensity* which, under certain conditions, are easily detectable and yield interesting information. This chapter is concerned with the analysis of the most important superposition phenomena displayed by electromagnetic waves.

36.1 Interference from Two Antennas

Consider as sources of electromagnetic waves two parallel vertical electric dipole antennas, S_1 and S_2 (Figure 36-1), each driven by the same oscillator of angular frequency ω. At a distant† observation point P in the horizontal plane through

---

*The present theory, known as quantum electrodynamics, predicts that scattering of light by light does occur, but the effect is completely negligible with beams of ordinary intensities.
†In this chapter we are concerned only with the *radiation field* discussed in Section 35.3.

the centers of the antennas, the electromagnetic wave from antenna S_1 contributes an electric field which has a vertical component

$$E_1 = E_{01} \cos(\omega t + \phi_{01})$$

and S_2 contributes a field with a vertical component

$$E_2 = E_{02} \cos(\omega t + \phi_{02})$$

According to the superposition principle, the resultant electric field at the observation point has a vertical component

$$E = E_1 + E_2$$

The addition of the two sinusoidal functions E_1 and E_2 is most easily accomplished using a phasor diagram (Figure 36-2). The angle between the phasors representing E_1 and E_2 is their phase difference $\phi = \phi_{02} - \phi_{01}$. The amplitude E_0 of the superposition $E = E_1 + E_2$ is given by the cosine rule applied to the triangle with sides of lengths E_{01}, E_{02}, and E_0:

$$E_0^2 = E_{01}^2 + E_{02}^2 + 2E_{01}E_{02} \cos \phi \qquad (36.1)$$

The intensity of each wave is proportional to the square of its amplitude (Eq. 35.5). Therefore the equation relating amplitudes implies that the intensity

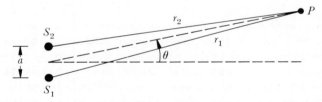

Figure 36-1   Top view of two vertical electric dipole antennas. The plane of the page is the horizontal plane through the centers of the antennas. The radiation field is examined at an observation point P that is a distance r_1 from S_1 and r_2 from S_2.

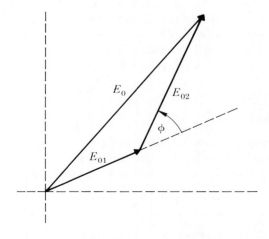

Figure 36-2   Phasor diagram. The phasor of length E_0 represents the superposition $E = E_1 + E_2$.

I of the superposition is related to the intensities I_1 and I_2 of the superposed waves by

$$I = I_1 + I_2 + 2\sqrt{I_1 I_2}\, \cos \phi \qquad (36.2)$$

The last term in this expression is called the *interference term*. The value of the interference term depends on the phase difference ϕ between the two waves. This is generally different at different locations. *The phenomenon that the intensity I is not everywhere simply the sum of the intensities I_1 and I_2 is termed interference.* We say that the two superposed waves *interfere* and the pattern in space of regions of high and low intensities is called an *interference pattern*.

The crucial consideration in interference phenomena is the phase difference at a given point of the interfering waves. In Figure 36-1, at the point P a distance r_1 from S_1 and r_2 from S_2, the waves radiated from S_1 and S_2 are proportional to $\cos(kr_1 - \omega t - \phi'_{01})$ and $\cos(kr_2 - \omega t - \phi'_{02})$, respectively, and the phase difference between these two waves is

$$\phi = \frac{2\pi}{\lambda}(r_1 - r_2) + (\phi'_{02} - \phi'_{01}) \qquad (36.3)$$

where $k = 2\pi/\lambda$ and $\phi'_{02} - \phi'_{01}$ is the phase difference between the two sources. The path difference $r_1 - r_2$ can be determined from the angle θ in Figure 36-3. Since the distance to the observation point is large compared to the antenna separation a, the lines from the two antennas may be considered parallel and

$$r_1 - r_2 = a \sin \theta \qquad (36.4)$$

The interference term in Eq. 36.2 is positive at some locations and here the interference is said to be *constructive*. Intensity maxima occur where $\cos \phi = 1$. Then

$$\phi = 2n\pi \qquad \text{(intensity maxima)} \quad (36.5)$$

where n is an integer. From Eq. 36.3 this condition implies that the path difference is an integral number of wavelengths,

$$r_1 - r_2 = n\lambda \qquad \text{(intensity maxima)} \quad (36.6)$$

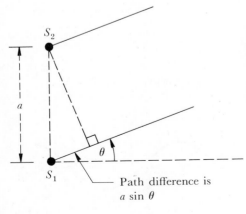

Path difference is
$a \sin \theta$

Figure 36-3 The difference of the distances r_1 and r_2 that waves travel to the observation point P from S_1 and S_2, respectively, is called the path difference. Since the lines to P may be considered parallel, the path difference is $a \sin \theta$.

if the phase difference $\phi'_{02} - \phi'_{01}$ of the two sources is zero. At an intensity maximum, a crest from one source arrives at the same instant as a crest from the other source, and the amplitude of the superposition is the sum of the amplitudes of the two interfering waves, as in Figure 36-4.

At other locations I is less than the sum of I_1 and I_2, and we say there is *destructive interference.* Intensity minima occur where $\cos \phi = -1$. Then

$$\phi = (2n + 1)\pi \qquad \text{(intensity minima)} \quad (36.7)$$

and the path difference is an odd number of half wavelengths (for sources that are in phase):

$$r_1 - r_2 = (2n + 1)\frac{\lambda}{2} \quad \text{(intensity minima)} \quad (36.8)$$

At the location of an intensity minimum, a crest from one source arrives at the same time as a trough from the other source, producing a superposition with an amplitude that is the difference of the amplitudes of the interfering waves (Figure 36-5).

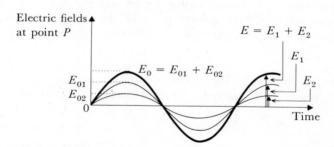

Figure 36-4 Constructive interference. Graph of the fluctuating fields E_1, E_2, and their superposition E $(= E_1 + E_2)$ at different times at an observation point P where crests from each source arrive at the same time. Here the fields E_1 and E_2 are *in phase* and the amplitude E_0 of the superposition is the sum of the amplitudes E_{01} and E_{02}.

Figure 36-5 Destructive interference. At an observation point P where crests from one source arrive with troughs from the other source, the amplitude E_0 of the superposition is the *difference* of the amplitudes E_{01} and E_{02} of the waves contributed by each source.

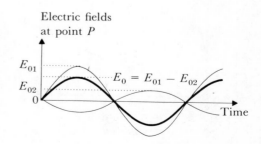

If the two antennas radiate the same power, the intensities I_1 and I_2 are essentially the same in the radiation field. In this case, Eq. 36.2 becomes

$$I = 2I_1(1 + \cos \phi) = 4I_1 \cos^2 \frac{\phi}{2} \qquad (36.9)$$

The intensity changes from zero at an intensity minimum to $4I_1$ at an intensity maximum.

Example 1   Two identical vertical electric dipole antennas have a separation $a = \lambda/4$, where λ is the wavelength of the waves emitted by each antenna. The phase difference $\phi'_{02} - \phi'_{01}$ of the antennas is $\pi/2$. Find the intensity distribution in the radiation field as a function of the angle θ which specifies the direction from the antennas to the observation point (Figures 36-1 and 36-3).

Solution   Since $r_1 - r_2 = a \sin \theta = (\lambda/4) \sin \theta$, the phase difference at the observation point is

$$\phi = \frac{2\pi}{\lambda}(r_1 - r_2) + (\phi'_{02} - \phi'_{01}) = \frac{\pi}{2} \sin \theta + \frac{\pi}{2}$$

The intensity in the direction specified by the angle θ is

$$I = 4I_1 \cos^2 \frac{\phi}{2} = 4I_1 \cos^2 \left(\frac{\pi}{4} \sin \theta + \frac{\pi}{4} \right) \qquad (36.10)$$

At $\theta = 90°$, $\phi = \pi$ rad and $I = 0$. Only in the direction $\theta = 90°$ do we find the two waves completely out of phase producing zero intensity. At $\theta = 270°$, $\phi = 0$ and $I = 4I_1$. The direction corresponding to $\theta = 270°$ is the only direction for which the waves are in phase and produce an intensity maximum. A convenient representation of an intensity pattern is obtained by giving a polar plot (Figure 36-6) with the distance of the curve from the origin representing the intensity in the corresponding direction.

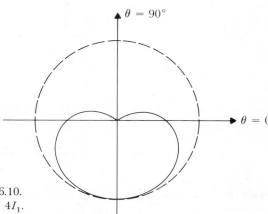

Figure 36-6   Polar plot of Eq. 36.10. The radius of the dashed circle is $4I_1$. The line joining the sources is $\theta = 90°$.

Example 2   The separation a of two identical electric dipole antennas S_1 and S_2 [Figure 36-7(a)] is large compared to the wavelength λ. The antennas are maintained in phase and they emit waves of the same amplitude. Find the intensity distribution along the X axis in the figure, at points such that θ is small.

Solution   When θ is small, $\sin \theta \approx \tan \theta = x/\ell$. The path difference to the observation point with coordinate x is

$$r_1 - r_2 = a \sin \theta = \frac{ax}{\ell}$$

and the corresponding value of the phase difference is

$$\phi = \frac{2\pi}{\lambda}(r_1 - r_2) = \frac{2\pi a x}{\lambda \ell}$$

Figure 36-7(b) shows the intensity $I = 4I_1 \cos^2 \phi/2 = 4I_1 \cos^2 (\pi a x / \lambda \ell)$

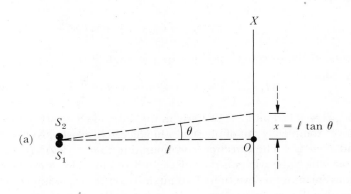

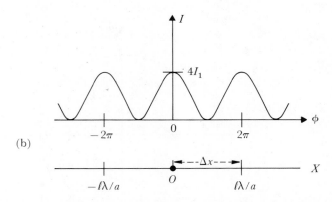

Figure 36-7   Interference pattern from two sources with separation a large compared to the wavelength λ. (a) The geometry of the sources and the X axis. (b) The intensity I observed at various points along the X axis.

plotted as a function of ϕ (which is proportional to x). The maxima occur where $r_1 - r_2 = n\lambda$, which corresponds to $ax/\ell = n\lambda$ or

$$x = \frac{n\ell\lambda}{a}$$

The separation of adjacent maxima is

$$\Delta x = \frac{\ell\lambda}{a} \qquad (36.11)$$

This result shows that bringing the sources closer together (decreasing a) will increase the distance Δx between adjacent maxima. Also increasing the wavelength of the radiation increases Δx. For very short wavelengths the entire interference pattern would be squeezed together and it would be difficult to resolve the individual maxima.

36.2 Interference from Two Light Sources

Young's Double-Slit Experiment

The interference effects described in Example 2 were obtained by Thomas Young (1773–1829) using visible light with the arrangement shown in Figure 36-8. The narrow double slits S_1 and S_2 are illuminated by light from the same source coming through a single slit. The double slits now serve (instead of antennas) as two radiation sources which are always in phase.* The intensity distribution is observed on a screen as regions of brightness, called interference fringes, which alternate with regions of darkness. Young's experiment was of great importance because it brought the wave nature of light to the fore and at the same time provided a method of measuring the wavelength.

Example 3   In the double-slit experiment shown in Figure 36-8, the yellow light from a sodium vapor lamp falls upon two parallel slits which are 0.10 mm apart. The distance between adjacent fringes is 0.47 cm when the screen is placed 0.80 m from the slits. Determine the wavelength λ of the yellow light.

Solution   From Eq. 36.11,

$$\lambda = \frac{a\Delta x}{\ell} = \frac{(0.10 \times 10^{-3}\ \text{m})(0.47 \times 10^{-2}\ \text{m})}{0.80\ \text{m}} = 5.9 \times 10^{-7}\ \text{m}$$

$$= 5.9 \times 10^3\ \overset{\circ}{\text{A}}$$

---

*The justification for regarding slits as sources of electromagnetic waves is provided by Huygens' principle, discussed in Section 36.4.

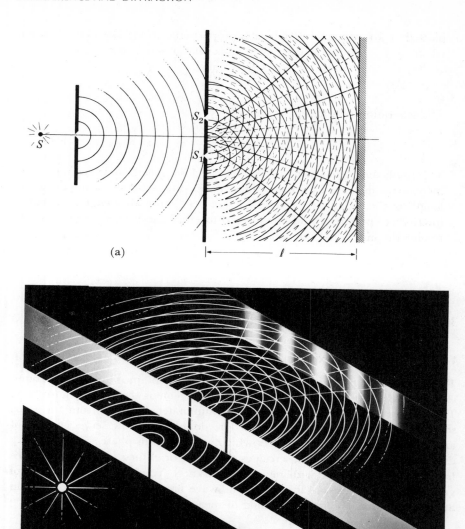

(a)

(b)

Figure 36-8   Schematic diagram of Young's double-slit experiment. (From G. Holton and D. Roller, *Foundations of Modern Physical Science*, Addison-Wesley, 1958.)

Coherence and Incoherence

If we illuminate a region with light from two independent sodium lamps and measure the intensity I by eye or with a photographic plate, we find that

$$I = I_1 + I_2$$

where I_1 is the intensity when the first lamp alone is operating and I_2 is similarly defined. *No interference is observed.* For example, in the Young double-slit experiment, if slit S_1 is illuminated by one sodium lamp and S_2 by another sodium lamp, the pattern of alternate bright and dark bands is no longer apparent. There is a uniform intensity across the screen.

In such cases, when no interference is observed, we say that the sources are *incoherent.* With *independent* identical sources S_1 and S_2, the phase difference between the oscillating charges in the different sources is not constant. Except in the case of lasers (Chapter 43), there is no control over the instant at which an individual atom starts and stops radiating. Individual atoms radiate at random. Perhaps for one 10^{-8}-s interval there will be electron oscillations in the atoms of the two sources such that a crest leaves one source as a trough leaves the other source. For this short time interval there will be a certain interference pattern in space. But in succeeding intervals, unrelated electron oscillations in the two sources take over with the result that perhaps crests leave one source a quarter of a period before crests leave the other source. While these oscillations persist there will be a definite interference pattern in space, but it will be different from the interference pattern which preceded it. The regions of high and low intensity will occur at different locations. Thus we obtain a succession of different interference patterns, shifting after intervals of approximately 10^{-8} s. A detector which responds only to averages over time intervals such as 0.1 s, will observe the average of some ten million different patterns, and, in this average, no interference will be apparent.

At a given observation point, during a time interval short compared to 10^{-8} s,

$$I = I_1 + I_2 + 2\sqrt{I_1 I_2} \cos \phi \qquad (36.2)$$

However, as time goes by and the phase of the electron oscillations in one source shifts in a random fashion relative to the phase of the oscillations in the other source, the interference term changes value. After approximately ten million such shifts (in 10^{-1} s) with positive and negative values of $\cos \phi$ equally likely, *the average value of the interference term will be zero* and the average intensities will then satisfy

$$I_{avg} = I_{1,avg} + I_{2,avg} \qquad (36.12)$$

No interference is observed by detectors whose response is determined by the *average* intensity I_{avg}. We say that the two sources are *incoherent,* as far as this method of detection is concerned.

Nevertheless, it *is* possible to detect interference between two independent light sources. But detectors that can "read" the interference pattern in a time less than 10^{-8} s are needed. Modern photomultiplier tubes can do this. In a beautiful experiment (1956), an interference pattern produced by light coming from different parts of the star Sirius was detected by R. H. Brown and R. O. Twiss. Measurement of the distance between the bright fringes in this interference pattern allowed calculation of the separation of the sources of light. In this way, for the first time, the diameter of Sirius was determined.

Now let us return to Young's experiment. Usually, if we are to observe interference phenomena with an ordinary light source, as in this experiment, the wave radiated by a given atom in the source is split into *two parts.* Each part of the wave travels in a different path, and these two parts eventually meet at some observation point.

In Young's experiment, the sources S_1 and S_2 are really two different parts of the same wave. Changes in oscillation of charges in the original light source will affect both S_1 and S_2 in the same way, and their phase difference is constant. Sources such as S_1 and S_2 are *coherent* (for any method of detection), and produce interference patterns.

One of the most important novel features of the *laser* is the *coherence* of the light emitted by its different atoms. The laser exploits *stimulated* emission. Instead of individual atoms radiating at random, the light from one atom is used to stimulate the emission of light from another atom. The emitted light is locked in phase with the stimulating light.

Interference of light beams coming from different portions of the same laser can be easily observed by eye. Indeed, two *independent* lasers emitting the same wavelength light will produce fringes that can be detected using high-speed television techniques. The difficulty here is not with phase shifts between the two sources, but rather that the wavelengths cannot be made precisely equal.

36.3 Interference from Several Sources

If, in Example 2, the number of sources is increased from 2 to some large number N, the interference pattern changes to that of Figure 36-9. The location of the intensity maxima does not change but these maxima become sharp intense peaks. Practical use is made of this general behavior in optical instruments and in the design of directional radio antennas.

The problem of finding the intensity in the radiation field produced by N identical sources spaced equally along a straight line is solved by a straightforward extension of the analysis of Section 36.1. Figure 36-10 shows the phasor

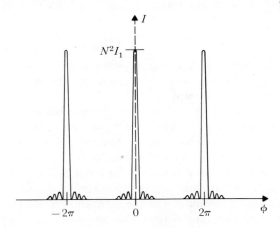

Figure 36-9 The intensity in the radiation field produced by a large number N of identical sources spaced equally along a straight line.

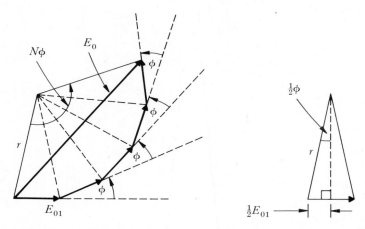

Figure 36-10 Phasor diagram. E_0 is the amplitude of the superposition of N waves, each of amplitude E_{01}. The phase difference between contributions from adjacent sources is ϕ. The case $N = 5$ is represented.

diagram for the phasors representing the contributions of the N antennas to the electric field at a distant observation point P. The constant phase difference ϕ between waves arriving at P from adjacent antennas is determined in the same way as for the case of two antennas. We assume that all the sources contribute waves of equal amplitude E_{01} at P. The tips of the phasors lie on a circle of radius r, which is shown in the figure to be given by

$$\tfrac{1}{2}E_{01} = r \sin \frac{\phi}{2}$$

The amplitude E_0 of the superposition is similarly related to the angle $N\phi$ subtended at the center of the circle of radius r:

$$\tfrac{1}{2}E_0 = r \sin \frac{N\phi}{2}$$

Division of this equation by the preceding equation yields

$$\frac{E_0}{E_{01}} = \frac{\sin(N\phi/2)}{\sin(\phi/2)}$$

Since $I/I_1 = (E_0/E_{01})^2$, the intensity I of the superposition is

$$I = I_1 \frac{\sin^2(N\phi/2)}{\sin^2(\phi/2)} \tag{36.13}$$

where I_1 is the intensity from a single source.

   Figure 36-11 shows the functional dependence of I on ϕ in the region near $\phi = 0$ for the case when N is an enormously large number. Since the denominator is zero at $\phi = 0$, the graph at this point must be investigated with care. Using the fact that for small angles the sine and the angle are approximately

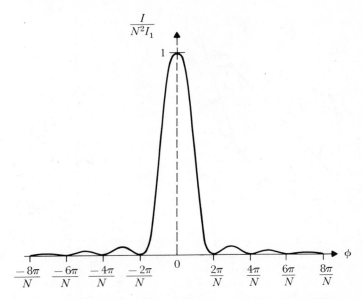

Figure 36-11 Graph of the intensity I with N sources. The phase difference between the contributions of adjacent sources is ϕ. Notice that if N is increased, the peak becomes narrower. Similar peaks occur at $\phi = \pm 2\pi, \pm 4\pi, \ldots$, as shown in Figure 36-9.

equal, we find that, in the limit as ϕ approaches 0,

$$I = I_1 \frac{(N\phi/2)^2}{(\phi/2)^2} = N^2 I_1 \tag{36.14}$$

This can be verified easily, since when $\phi = 0$, the N phasors are all in the same direction and their phasor sum has a length NE_{01} corresponding to an intensity $N^2 I_1$.

The first minimum occurs where $\sin(N\phi/2) = 0$, that is, at $N\phi/2 = \pi$. On the phasor diagram, this corresponds to the phase difference ϕ for which the arrows come back to the starting point giving a phasor sum of zero; this requires that the angle $N\phi$ in Figure 36-10 be exactly 2π. We see that increasing the number N of the sources decreases the angle ϕ at which the first minimum occurs, and therefore decreases the width of the central maximum.

As ϕ increases, the phasors lie on a circle of smaller radius and a subsidiary maximum is reached approximately where these phasors wrap around the circle one and one-half times. The intensity at this maximum is approximately $0.05\,I_0$. The intensity drops to zero again where $N\phi = 2(2\pi), 3(2\pi), \ldots$, with progressively weaker maxima in between. To summarize, there is a very sharp central maximum with weak subsidiary maxima on both sides.

As shown in Figure 36-9, the intensity is a periodic function of ϕ with period 2π; the pattern that we have examined near $\phi = 0$ is repeated at $\phi = \pm 2\pi$,

$\pm 4\pi$, ... These recurrences of the great maxima correspond to directions θ for which the path difference from adjacent sources is an integral number n of wavelengths,

$$a \sin \theta = n\lambda \qquad (36.15)$$

and n is called the order number of the beam.

In optics, the system of N parallel sources is called a diffraction grating. Measurement of wavelengths of light in the near infrared, visible, or ultraviolet can be accomplished using transmission diffraction gratings that consist of thousands of slits per centimetre, obtained as the spaces between evenly spaced parallel lines etched on a transparent film. A beam with a given wavelength λ produces an intensity maximum that appears as a very narrow bright line on a photographic plate. Using Eq. 36.15 we can calculate the wavelength. The tremendous precision of measurement is indicated by quoting a typical result for one of the wavelengths of the yellow light emitted by a sodium atom: $\lambda = 5889.95 \pm 0.04$ Å.

Example 4   A diffraction grating has 5000 lines/cm. It is illuminated at normal incidence. Then the slits can be considered as an array of sources that are in phase. An intensity maximum is observed at an angle of 20.00° to the normal. What is the wavelength of the light, if this is a first order ($n = 1$) maximum?

Solution   The distance a between adjacent sources, called the grating spacing, is

$$a = \frac{1}{5000} \text{ cm} = 2.0000 \times 10^{-4} \text{ cm} = 20000 \text{ Å}$$

Putting $n = 1$ in Eq. 36.15, we find

$$\lambda = a \sin \theta = (20000 \text{ Å})(\sin 20.00°) = 6840 \text{ Å}$$

36.4 Diffraction

Waves spread out after passing through narrow apertures. In passing obstacles, waves bend and travel into the region behind the obstacle. Such departure from propagation in a straight line is termed *diffraction*.

Everyone is familiar with the diffraction of sound waves. You can speak to someone who is behind a tree, because the sound waves will bend around the tree and fall upon his ear. The phenomenon of diffraction is clearly displayed by water waves, as in Figure 34-8.

Newton, and the majority of physicists up to the nineteenth century, felt that light could not be a wave motion because of the apparent absence of diffraction phenomena with light. True, Grimaldi, in Newton's century, had observed some departure from straight-line propagation when light passed through a small aperture or when shadows were cast by small objects, but the observed bending of the light beam was small and was interpreted as due to the deflection of light particles as they passed near matter.

It was not until Young showed with his double-slit interference experiment that a wave theory gave detailed agreement with experimental results that physicists looked into the subject of diffraction of light with care and persistence. Fresnel and Fraunhofer found that large diffraction effects indeed can be obtained with light waves (Figure 36-12). Moreover, the patterns of intensity maxima and minima (called diffraction patterns) can be understood in quantitative detail using a wave theory of light.

Huygens' Principle

Basic ideas are illustrated by consideration of the diffraction of a plane wave by an aperture in an opaque sheet that is perpendicular to the direction of propagation of the incident wave (Figure 36-13). Calculation of diffraction patterns by

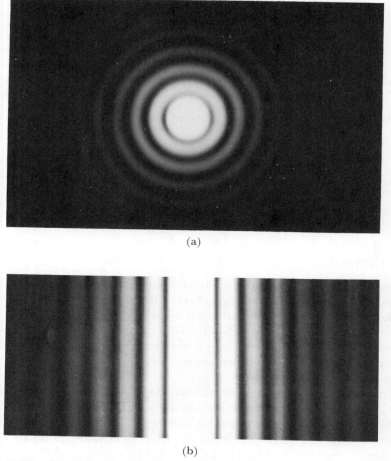

(a)

(b)

Figure 36-12 Diffraction pattern when plane waves are diffracted by (a) a circular aperture, (b) a long slit. (From D. H. Towne, *Wave Phenomena*, Addison-Wesley, 1967.)

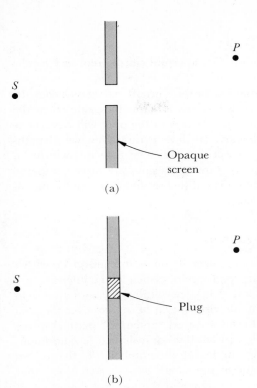

(a)

Opaque
screen

(b)

Plug

Figure 36-13 The source S and the observation point P are at a great distance from the screen.

direct solution of Maxwell's equations is generally a formidable problem. For light, diffraction patterns can usually be determined with sufficient accuracy by a calculational device called *Huygens' principle*, which, for the wave of Figure 36-13, states that *the radiation field on the far side of the sheet is the same as the field that would be produced by sources that are uniformly distributed over the area of the aperture and oscillate with the same phase and the same amplitude.*

Huygens' principle is justified by the following argument. The field $\mathbf{E}$ at a distant observation point P on the far side of the screen is a superposition of waves radiated by many sources. Oscillating charges in the source contribute a field $\mathbf{E}_S$ at P. The incident wave forces charges in the screen to oscillate. The radiation from these oscillating charges contributes a field $\mathbf{E}'$ at P. The superposition principle gives

$$\mathbf{E} = \mathbf{E}_S + \mathbf{E}'$$

Now, if the aperture were filled with a plug, the charges in this plug would be forced to oscillate in phase by the incident plane wave and contribute a field $\mathbf{E}_a$ at P. Since the screen plus plug is completely opaque, the resultant field at P is zero:

$$0 = \mathbf{E}_a + \mathbf{E}_S + \mathbf{E}'$$

These two equations give

$$\mathbf{E} = -\mathbf{E}_a$$

so that, except for the minus sign, the actual radiation field is the same as the field produced by the sources in the plug. This is Huygens' principle: The actual system of the source and the screen with an aperture can be replaced by the simpler system consisting of the plug alone.

Huygens' principle is an approximation. In this "proof" we assume that the field $\mathbf{E}'$ contributed by the sources in the screen remains the same when the plug is inserted. But the fields radiated by charges in the plug will have some effect on charges in the screen, particularly on those within a few wavelengths of the edge of the aperture. We may expect that Huygens' principle will be an adequate approximation only if the aperture dimensions are large compared to the wavelength. Usually this condition is satisfied for light but not for microwaves or longer wavelength radiation.

Fraunhofer Single-Slit Diffraction Pattern

In Figure 36-14, plane waves are incident normally on an opaque screen with a long narrow slit of width D. According to Huygens' principle, the intensity at a distant observation point P can be found by determining the radiation field that would be produced by a continuous distribution of sources over the area of the slit. We start with N identical evenly spaced "antennas" parallel to the long dimension of the slit. A situation physically equivalent to a continuous distribution of sources will be obtained by letting the number N of the sources approach infinity while their separation a approaches zero, in such a way that the phase difference $\Phi = N\phi$ between the source at the top and the source at the bottom of the slit remains constant. From Eq. 36.13 we obtain

$$I = I_1 \frac{\sin^2(\Phi/2)}{(\Phi/2N)^2}$$

since $\sin(\Phi/2N)$ can be replaced by $\Phi/2N$ for sufficiently large N. Putting $I_m = N^2 I_1$, we have

$$I = I_m \frac{\sin^2(\Phi/2)}{(\Phi/2)^2} \tag{36.16}$$

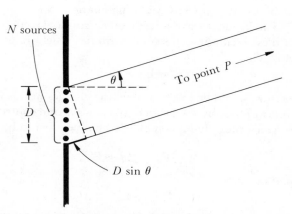

Figure 36-14   To understand Fraunhofer diffraction at a single slit, we first consider N identical "antennas" evenly spaced with separation a.

At the distant observation point P, waves which have come from the top and from the bottom of the slit have a path difference $D \sin \theta$ (Figure 36-14) and the corresponding phase difference is

$$\Phi = \frac{2\pi}{\lambda} D \sin \theta \qquad (36.17)$$

Figure 36-15 shows the graph of the intensity I as a function of $D \sin \theta$. There is *one* central maximum with weak subsidiary maxima on both sides. The main feature is that the intensity is large only within a central band of angular width

$$\Delta\theta \approx \frac{\lambda}{D} \qquad (36.18)$$

Diffraction patterns can be calculated for various geometries of apertures and obstacles. For historical reasons, if only plane wave fronts are involved, the phenomenon is called Fraunhofer diffraction. Otherwise we speak of Fresnel diffraction. For any aperture with a minimum linear dimension D, the expression $\Delta\theta \approx \lambda/D$ gives a useful estimate of the diffraction spreading if λ is less than D. This result implies that *diffraction effects become difficult to detect and apparent straight-line propagation results when the wavelength is so small that the ratio of λ to the size of the aperture is very much less than one.* For light waves and apertures such as house windows, this ratio is about 10^{-6}. This is the answer to Newton's objection to the wave theory. Notice the vital role that *quantitative information*, the order of magnitude of the wavelength of light, plays in reconciling the wave nature of light with our everyday experience of the propagation of light in a straight line.

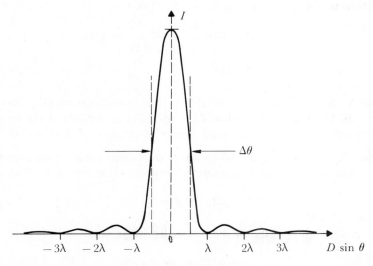

Figure 36-15 Fraunhofer single-slit diffraction pattern. The intensity is large only within a central band with an angular width $\Delta\theta \approx \lambda/D$.

The expression $\Delta\theta \approx \lambda/D$ has an interesting application to lasers. One of the most important advantages of a laser is that its energy is not spread out in all directions. The energy is concentrated in a narrow beam of high intensity. The beam *stays* narrow! Laser light from the earth has already been shone on the moon, the beam spreading out over an area on the moon of only a few kilometres in diameter. The best optical searchlight beam would spread out to be larger than the moon itself, yielding a hopelessly low intensity.

Now why does a laser beam spread at all? The answer is diffraction. The unavoidable diffraction spreading of any beam of initial diameter D is given by $\Delta\theta \approx \lambda/D$. For lasers, λ/D can be made of the order of 10^{-5}, so the spreading is small. Having made a comparison of the laser with a searchlight, we must add that the relatively large spreading of a searchlight beam is not due to diffraction. Searchlights are unable to achieve plane wave fronts with great precision. Diverging rays result, and the associated spreading of the beam is much larger than that arising from diffraction.

Example 5   A long narrow slit with a width of 1.0×10^4 Å is illuminated by plane waves of green light ($\lambda = 5.0 \times 10^3$ Å). In Figure 36-14, what is the angle θ corresponding to the first minimum of intensity in the Fraunhofer diffraction pattern?

Solution   From Eq. 36.16 the first minimum of intensity occurs at an angle θ such that $\Phi/2 = \pi$. From Eq. 36.17 this requires that

$$2\pi = \frac{2\pi}{\lambda} D \sin\theta$$

or

$$D \sin\theta = \lambda$$

as shown in the graph in Figure 36-15. Therefore

$$\sin\theta = \frac{\lambda}{D} = \frac{5.0 \times 10^3 \text{ Å}}{1.0 \times 10^4 \text{ Å}} = 0.50$$

$$\theta = 30°$$

Example 6   In the Fraunhofer single-slit diffraction pattern (Eq. 36.16 and Figure 36-15), what is the ratio of the intensity at the first of the secondary maxima to the intensity I_m at the center of the central maximum?

Solution   As shown in Figure 36-15, the first secondary maximum corresponds to a value of $D \sin\theta$ approximately halfway between λ and 2λ, that is

$$D \sin\theta = \frac{3\lambda}{2}$$

The corresponding value of the phase difference Φ is

$$\Phi = \frac{2\pi}{\lambda} D \sin\theta = 3\pi$$

Equation 36.16 gives the intensity I for this value of Φ:

$$\frac{I}{I_m} = \frac{\sin^2(3\pi/2)}{(3\pi/2)^2} = \frac{1}{(3\pi/2)^2} = 0.045$$

Multiple Slits, Interference, and Diffraction

In Sections 36.1 and 36.3 expressions are derived for the interference pattern produced by arrays of antennas. These results can be applied without modification to light passing through Young's double-slit apparatus or through a diffraction grating, only if the slits are arbitrarily narrow. In the general case, for a nonzero slit width D, the diffracted wave reaching an observation point from any one slit is determined by single-slit diffraction. Then the diffracted waves from different slits interfere. Without doing the analysis we quote the plausible result that the overall intensity distribution is the interference pattern for arbitrarily thin slits but modulated by the diffraction pattern for a single slit. For example, with N evenly spaced slits with centers a distance a apart and each slit of width D, the intensity at a distant observation point is $I = I_m[\sin(\Phi/2)/(\Phi/2)]^2$ $[\sin(N\phi/2)/\sin(\phi/2)]^2$ where $\Phi = (2\pi/\lambda)D \sin\theta$ and $\phi = (2\pi/\lambda)\,a \sin\theta$. Such a pattern is called a combined interference and diffraction pattern.

The Difference between Interference and Diffraction

The distinction between interference and diffraction is made on the basis of the number of sources of the waves that are superposed. The intensity pattern that results from a superposition of contributions from a rather small number of sources is usually called an interference pattern. If the superposition involves contributions from an infinite number of sources obtained by subdividing a wave into infinitesimal coherent sources, the resulting intensity pattern is called a diffraction pattern. In intermediate cases with waves contributed from a large number of separated sources, the word diffraction is used more often than interference.

Although interference and diffraction involve the same physical principle (the superposition principle) and the distinction between them is to some extent a matter of usage, we have seen that in certain circumstances this distinction is useful for the analysis of superpositions.

Summary

☐ At an observation point P receiving waves of intensities I_1 and I_2 from two sources with the same frequency, the intensity is

$$I = I_1 + I_2 + 2\sqrt{I_1 I_2}\,\cos\phi$$

The interference term $2\sqrt{I_1 I_2}\,\cos\phi$ depends on the value of the phase difference ϕ of the two waves, which is determined by the path difference

$r_1 - r_2$ and the phase difference $\phi'_{02} - \phi'_{01}$ of the two sources:

$$\phi = \frac{2\pi}{\lambda}(r_1 - r_2) + \phi'_{02} - \phi'_{01}$$

The path difference, when the direction to a distant observation point P makes an angle θ with the perpendicular to the line of length a joining the sources, is

$$r_1 - r_2 = a \sin \theta$$

☐   Intensity maxima occur where $\phi = 2n\pi$; that is, where

$$r_1 - r_2 = n\lambda \qquad \text{(if } \phi'_{02} - \phi'_{01} = 0\text{)}$$

Intensity minima occur where $\phi = (2n + 1)(\pi/2)$; that is, where

$$r_1 - r_2 = (2n + 1)\frac{\lambda}{2} \qquad \text{(if } \phi'_{02} - \phi'_{01} = 0\text{)}$$

☐   Two sources are coherent if their phase difference remains constant during the time required to observe the intensity.

☐   The intensity at a distant observation point P receiving waves from a linear array of N evenly spaced sources of the same frequency, each wave having an intensity I_1, is

$$I = I_1 \frac{\sin^2(N\phi/2)}{\sin^2(\phi/2)}$$

where ϕ is the phase difference between waves arriving at P from adjacent antennas.

☐   For plane electromagnetic waves incident normally on an opaque screen with an aperture, Huygens' principle states that the radiation field on the far side of the aperture is the same as the field that would be produced by identical sources uniformly distributed over the area of the aperture and oscillating in phase. This leads to the Fraunhofer single-slit diffraction pattern,

$$I = I_m \frac{\sin^2(\Phi/2)}{(\Phi/2)^2}$$

where I is the intensity in the radiation field behind a long slit of width D, and $\Phi = (2\pi/\lambda) D \sin \theta$. The intensity of the beam is large only within a central band of angular width $\Delta\theta \approx \lambda/D$.

Questions

1   Show in detail that Eq. 36.2 follows from Eq. 36.1 by multiplying both sides of Eq. 36.1 by $\frac{1}{2}\sqrt{\epsilon_0/\mu_0}$ and using $I = \frac{1}{2}\sqrt{\epsilon_0/\mu_0}\, E_0^2$, with similar expressions for I_1 and I_2.

2   Draw the phasor diagrams corresponding to the situations represented in Figures 36-4 and 36-5.

3 Show that Eq. 36.1 implies that $E_0 = E_{01} + E_{02}$ at an intensity maximum and that $E_0 = |E_{01} - E_{02}|$ at an intensity minimum.

4 The equation $I = 4I_1 \cos^2 \phi/2$ implies that at an intensity maximum, $I = 4I_1$. Verify this result by finding the corresponding amplitude E_0 from a phasor diagram.

5 Show that if $\phi = (2\pi/\lambda)(r_1 - r_2)$ and $\phi = 2n\pi$, then $r_1 - r_2 = n\lambda$.

6 For the antennas of Example 1, draw the phasor diagrams and use them to find the intensity for the directions specified by $\theta = 0°, 90°, 180°, 270°$. Compare these results with those given by the formula $I = 4I_1 \cos^2 \phi/2$.

7 The phase difference of the antennas of Example 1 is changed from $\phi'_{02} - \phi'_{01} = \pi/2$ to $\phi'_{02} - \phi'_{01} = -\pi/2$:
   (a) Draw phasor diagrams and use them to find the intensity in the directions $\theta = 0°, 90°, 180°, 270°$.
   (b) Derive an expression for the intensity I as a function of θ and of the intensity I_1 of a single antenna.
   (c) Represent the intensity pattern on a polar plot.

8 If the two antennas of Example 2 have the separation $a = 100\lambda$, what is the angle θ (in radians) corresponding to a path difference of one wavelength?

9 Two antennas are 40 m apart and radiate waves of wavelength 2.0 m. The antennas are in phase. The intensity is measured at a point which is 4.0×10^3 m from each antenna. What is the minimum distance that the detection apparatus will have to be moved parallel to the line joining the antennas to encounter another intensity maximum?

10 In Young's double-slit interference experiment, how does the interference pattern on the screen change when:
   (a) The wavelength is increased?
   (b) The slit separation is decreased?

11 In the double-slit experiment in Figure 36-8, the parallel slits are 2.0×10^{-4} m apart. The distance between two adjacent bright fringes is 0.94×10^{-2} m, and the screen is placed 3.20 m from the slits. What is the wavelength of the light?

12 What is the average value of the interference term $2\sqrt{I_1 I_2} \cos \phi$ when the sources are incoherent? Explain.

13 Verify that Eq. 36.13,

$$I = I_1 \frac{\sin^2(N\phi/2)}{\sin^2(\phi/2)}$$

is correct when $N = 1$ and $N = 2$.

14 Four identical antennas are arranged in a line and spaced one-half wavelength apart. They are maintained in phase. Draw phasor diagrams and use them to find the intensity in the directions $\theta = 0°, 15°, 30°, 45°, 90°$. Compare the results with those given by Eq. 36.13.

15 An antenna array consists of N identical antennas which are driven in phase. Sketch graphs showing the intensity I as a function of ϕ (the phase difference between contributions of adjacent antennas) for $N = 2$, $N = 4$, and N some extremely large number. Compare and contrast these three graphs.

16 Using the diffraction grating of Example 4 (5000 lines/cm) and light of wavelength 6840 Å, how many great maxima will occur between $\theta = 0$ and $\theta = 90°$; that is, how many values of the order number n are possible?

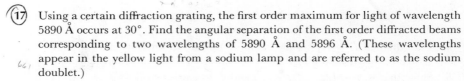

17 Using a certain diffraction grating, the first order maximum for light of wavelength 5890 Å occurs at 30°. Find the angular separation of the first order diffracted beams corresponding to two wavelengths of 5890 Å and 5896 Å. (These wavelengths appear in the yellow light from a sodium lamp and are referred to as the sodium doublet.)

18 Show that the intensity I in the Fraunhofer single-slit diffraction pattern is given by

$$I = \frac{4}{\pi^2} I_m = 0.41 \, I_m$$

at an angle θ in Figure 36-14 such that $\sin \theta = \frac{1}{2}(\lambda/D)$.

19 Show that if the angular width $\Delta\theta$ of the diffracted beam in the preceding question is defined as the range of values of θ for which the intensity exceeds 0.41 I_m, then when θ is small

$$\Delta\theta = \frac{\lambda}{D}$$

20 Is the diffraction spreading of a beam behind a slit important for the success of Young's double-slit experiment? Explain.

21 Estimate the diffraction spreading of the laser beam in Question 6 of Chapter 35 by calculating the order of magnitude of the vertex angle of the beam cone.

22 A pinhole of radius 1.0×10^{-5} m is made in the end of a box which is 0.40 m long. The box is aligned so that light from a small distant source passes through the pinhole and strikes the opposite end of the box. What is the order of magnitude of the diameter of the illuminated area at the end of the box? Assume that the light has a wavelength of 6000 Å.

23 In many situations a large diffraction spreading is desired in order that a beam will "bend around corners:"
    (a) Estimate the diameter of the aperture from which there will be diffraction spreading into a cone with a vertex angle of at least 30° for radio waves of frequency 1.0×10^6 Hz and then for light waves with $\lambda = 5000$ Å.
    (b) Use the answers to part (a) to indicate why radio waves diffract around buildings while light waves do not.

24 In the experiment represented in Figure 36-14, the slit in the opaque screen is 0.10 mm wide. Plane waves with a wavelength of 6.0×10^3 Å are incident in a direction perpendicular to the screen. The diffraction pattern is observed on a second screen placed 10.0 m behind the slit. What is the distance from the center of the central maximum to the first minimum?

25 A single slit is illuminated by light containing two wavelengths, λ and λ'. We observe that the first diffraction minimum of λ coincides with the second diffraction minimum of λ'. What is the relationship between λ and λ'?

Supplementary Questions

S-1 Two identical dipole antennas have a separation $a = \lambda/2$. The antennas are in phase:
    (a) Draw phasor diagrams and use them to find the intensity in the directions $\theta = 0°, 90°, 180°, 270°$. ($\theta$ is the angle shown in Figure 36-1.)

(b) Derive an expression for the intensity I as a function of θ and of the intensity I_1 of a single antenna.

(c) Represent the intensity pattern on a polar plot.

S-2 Consider the radiation field of two identical antennas (in phase) with a separation $a = 2\lambda$. Show that the directions of the maxima and minima are as indicated in the polar plot given in Figure 36-16.

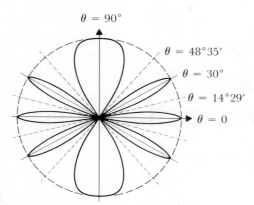

$\theta = 90°$

$\theta = 48°35'$

$\theta = 30°$

$\theta = 14°29'$

$\theta = 0$

Figure 36-16

S-3 Show the intensity pattern by sketching a polar plot for the *broadside array* described in Question 14.

S-4 An antenna array consists of N identical antennas spaced one-quarter wavelength apart. Successive antennas have a phase difference of 90°. Sketch the polar plot showing the intensity distribution for large N. Compare and contrast this radiation pattern with that given in Figure 36-6. Notice that as N is increased, the radiation becomes concentrated in a single direction.

S-5 Suppose that, in Figure 37-8, the white light is dispersed using a diffraction grating instead of a prism. When we examine the photographic plate, we find that the colors appear in reverse order with red deviated the most and blue the least. Also the entire spectrum appears several times with increased separation between two given colors at larger angles of deviation. Explain.

37.

reflection, refraction, and polarization

The discussion of the characteristics and behavior of light and other electromagnetic radiation is continued in this chapter. Here we are concerned with electromagnetic waves traveling through a material medium. In a homogeneous nonconducting medium with permeability μ and permittivity ϵ, the analysis of Section 35.2 must be modified by replacing μ_0 by μ and ϵ_0 by ϵ. We then find that the speed of propagation of an electromagnetic wave is

$$v = \frac{1}{\sqrt{\mu\epsilon}} \tag{37.1}$$

Consequently the wave speed generally has a different value in a different medium.

The phenomena of reflection and refraction occur for any type of wave at the interface between two media for which the wave speed is different. The directions of propagation of the incident, reflected, and refracted rays are illustrated in Figure 37-1. The intensity of the incident wave is divided between the reflected and the refracted waves. A detailed theory of the relative intensities and directions involved in reflection and refraction can be developed from Maxwell's equations, but such an undertaking is too involved to be appropriate for this book. We restrict our attention to the simple laws giving the directions of the reflected and refracted rays. The law of refraction is deduced by a method that demonstrates that these simple laws are consequences of the geometry of any wave motion.

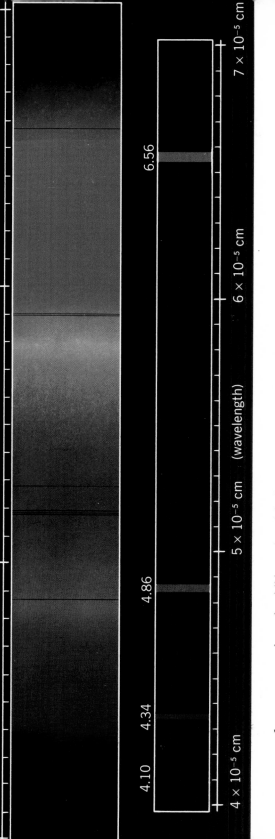

Lower spectrum shows the visible portion of the line spectrum emitted by hydrogen atoms. Upper spectrum is the continuous spectrum emitted by the sun. The dark lines appear because cooler gases surrounding the sun absorb light of certain discrete wavelengths. Courtesy Eastman Kodak.

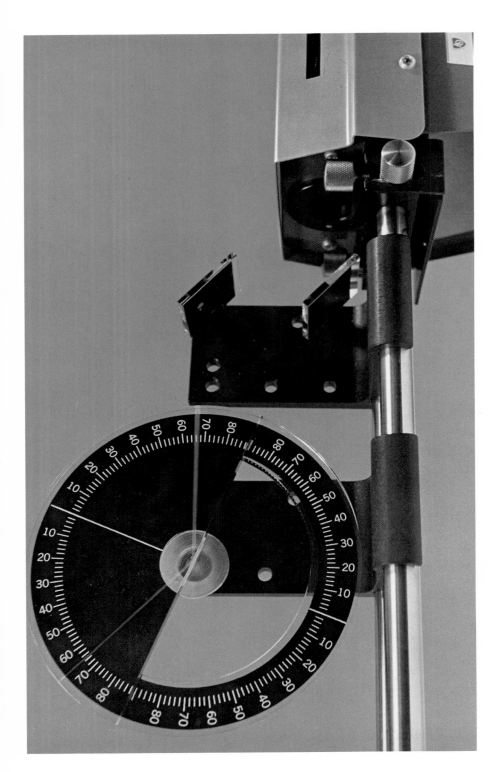

Total internal reflection within a plastic disk. The red light beam is produced by a helium–neon gas laser. (Queen's University at Kingston, Ontario.)

37.1 Reflection and Refraction

Law of Regular Reflection

When light is incident on a smooth plane surface of a transparent material such as glass in Figure 37-1, part of the light is *reflected*, and part is passed into the glass as a *refracted* beam. The directions of the incident, reflected, and refracted beams are specified by giving the angles θ_1, θ_r, θ_2 that these rays make with the *perpendicular* to the surface. Experiments, dating back to the time of the ancient Greek civilization, lead to the *law of regular reflection*:

1   The reflected ray lies in the plane of incidence, that is, the plane containing the incident ray and the perpendicular to the surface.
2   The angle of reflection θ_r equals the angle of incidence θ_1.

[Most of the objects that we see have on their surfaces irregularities which are large compared to the wavelengths of light. Such objects reflect the light in all directions, as shown in Figure 37-2(b). This is called *diffuse reflection*.]

Snell's Law of Refraction

A plane wave traveling at a speed v_1 in air will be "bent" as shown in Figure 37-3, as it is transmitted into glass where its speed is changed to a value v_2. This abrupt bending at the interface between two different transparent media is called *refraction*, a phenomenon which can be understood by considering the

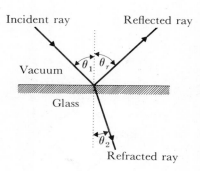

Figure 37-1   Incident, reflected, and refracted rays.

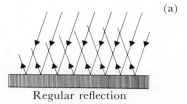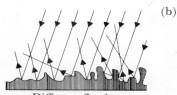

Regular reflection   Diffuse reflection

Figure 37-2   The contrast between (a) regular reflection and (b) diffuse reflection.

wave nature of light. The positions of the same wave front at the beginning and at the end of a time interval of duration t are shown in Figure 37-3; the ray drawn perpendicular to these wave fronts gives the direction of propagation. The refracted ray lies in the plane of incidence. The geometrical relations shown in Figure 37-3(b) are $\sin \theta_1 = v_1 t/D$ and $\sin \theta_2 = v_2 t/D$. Therefore

$$\frac{\sin \theta_1}{\sin \theta_2} = \frac{v_1 t/D}{v_2 t/D} = \frac{v_1}{v_2}$$

which implies

$$\frac{1}{v_1} \sin \theta_1 = \frac{1}{v_2} \sin \theta_2$$

It is convenient to express this relationship in terms of the *refractive indices*, $n_1 = c/v_1$ and $n_2 = c/v_2$, of the two different media. We have

$$n_1 \sin \theta_1 = n_2 \sin \theta_2 \tag{37.2}$$

which is known as *Snell's law of refraction*. This relationship determines the bending of a light ray at the interface of any two media with refractive indices n_1 and n_2; it is valid for light traveling from the second to the first medium as well as for light traveling in the opposite direction.

Snell's law was originally stated simply as an empirical law—an equation in agreement with experimental measurements of the angles θ_1 and θ_2. From such measurements, the ratio n_1/n_2 of the refractive indices of the two media can be calculated. By definition, the refractive index of a vacuum is exactly the

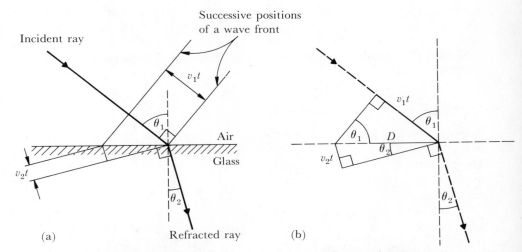

Figure 37-3   Refraction. (a) Two positions of the same wave front at the beginning and at the end of a time interval of duration t. (b) Enlargement which shows that $\sin \theta_1 = v_1 t/D$ and $\sin \theta_2 = v_2 t/D$.

number one ($n_{\text{vacuum}} = c/v_{\text{vacuum}} = c/c = 1$). Experiments show that the refractive index of any material medium depends upon the nature of the medium and also upon the wavelength of the light. Typical values for visible light are: crown glass, $n = 1.52$; water, $n = 1.33$; air, under standard conditions, $n = 1.0003$.

Refraction and the Wave Nature of Light

Let us return for a moment to the "waves" versus "particles" arguments regarding the nature of light. In the seventeenth century, Christiaan Huygens (1629–1695) showed, as we have, that waves would satisfy Snell's law of refraction, and since, for water,

$$n = \frac{c}{v} = 1.33$$

the speed v of light waves in water must therefore be less than the speed of light in a vacuum, c.

Now, if light were a beam of particles *obeying* Newtonian mechanics, a light beam would be bent, as observed, at a vacuum–water interface if the water exerted an attractive force on the particles. A particle's component of velocity perpendicular to the interface then would be increased inside the water, and this would explain why the velocity vector makes a smaller angle with the normal to the interface. But then the particles would travel faster in water than in a vacuum.

Here was an opportunity to decide between Huygens' wave theory of light and the particle theory of light advanced by the great French mathematician and philosopher, René Descartes (1596–1650), and considered more plausible by Newton. One simply had to find out whether light travels more slowly or more rapidly in water than in a vacuum. But this crucial experiment was beyond the capabilities of physicists until 1850, when Jean Leon Foucault devised a successful laboratory method. The experimental result was that light travels more slowly in water than in a vacuum. The prediction of the wave theory was upheld.

The Origin of Reflected and Refracted Beams

When an electromagnetic wave travels through a material medium, the electrons in the atoms of the material are forced to oscillate at the frequency of the wave. These oscillating electrons then radiate electromagnetic waves of the same frequency as that of the incident beam. It is this radiation that accounts for both the reflected beam and the refracted beam.

Within the bulk of a transparent material such as water, the incident beam and the light that is re-emitted coherently from the various molecules give a superposition which is the *refracted* wave. Rather surprisingly, although the wave contributed by each molecule travels at the same speed c, the superposition of the waves radiated from molecules in different locations produces a refracted wave that travels at a lesser speed $v = c/n$, where n is the refractive index of the medium. Backward radiation is cancelled by interference, except for a layer of molecules (roughly $\lambda/2$ thick) near the surface, whose contributions constitute the *reflected wave*.

The law of regular reflection can be understood by considering the condition for constructive interference of light waves scattered from any two points A and B on the surface of an interface between two different media (Figure 37-4). The path difference for rays scattered from A and B is $CB - AD$. All rays emerging in a direction such that the angle of reflection θ_r is equal to the angle of incidence θ_i have zero path difference ($CB = AD$). Waves scattered in this direction from all points in the plane of the interface therefore interfere constructively and form a reflected beam.*

Total Internal Reflection

When an incident ray is in the medium of higher refractive index, a special phenomenon known as total reflection may occur. This is illustrated in Figure 37-5. Snell's law implies that the refracted ray just grazes the surface ($\theta_1 = 90°$) if θ_2 has the value θ_c given by

$$n_1 \sin 90° = n_2 \sin \theta_c$$

$$\sin \theta_c = \frac{n_1}{n_2} \tag{37.3}$$

The angle of incidence θ_c is called the *critical angle for total reflection* because, as

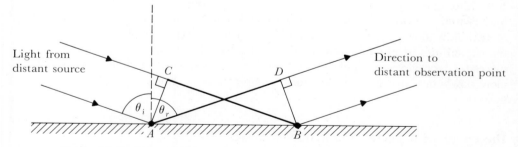

Figure 37-4   The path difference for rays of light scattered from points A and B is $CB - AD$. For rays emerging in a direction such that $\theta_i = \theta_r$, the path difference is zero and constructive interference is obtained.

---

*It is interesting to note that, for waves scattered from two given points A and B, constructive interference is obtained in each direction for which the path difference is an integral number n of wavelengths:

$$CB - AD = n\lambda$$

When there is a regular array of scatterers, as in a diffraction grating, there will be "reflected" beams corresponding to each possible value of n. However if the separation AB of adjacent scatterers is less than one wavelength, the only possibility is $n = 0$, corresponding to $CB = AD$. The emerging rays are then in the direction prescribed by the law of regular reflection.

the angle of incidence increases from zero, the intensity of the reflected ray increases and that of the refracted ray decreases until, at the value θ_c, the intensity of the refracted ray is zero and the light is totally reflected. Total reflection occurs for all angles of incidence greater than θ_c.

An interesting application, which shows how a succession of total internal reflections can be used to cause light to "bend" around corners, is illustrated in Figure 37-6. In the new technology of *fiber optics*, a bundle of thousands of thin glass fibers, each about 0.01 mm or less in diameter, is formed into a flexible cable that can transmit pictures from one point to another.

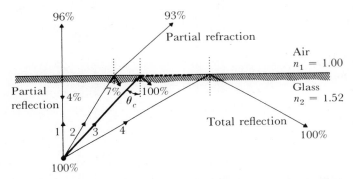

Figure 37-5   Light is incident from a medium with higher refractive index upon an interface with a medium of lower refractive index. The relative values of the intensities of the reflected and refracted beams are indicated for several angles of incidence. Total reflection occurs for angles of incidence equal to or greater than the critical angle θ_c.

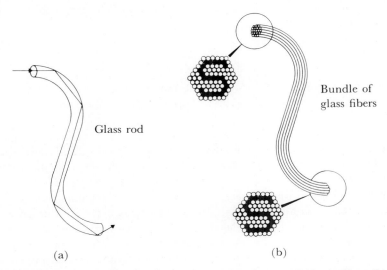

(a)   (b)

Figure 37-6   Fiber optics. (a) Total internal reflection in each single fiber. (b) A bundle of fibers transmitting an optical signal, the letter S in this case.

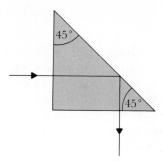

Figure 37-7 Reflecting prism.

Example 1   A glass prism with the geometry shown in Figure 37-7 is widely used in optical instruments to reflect light beams through 90°. Show that if the refractive index of the glass is 1.52, the reflection shown in the figure is *total* reflection.

Solution   The critical angle for total reflection at the glass–air interface is given by

$$\sin \theta_c = \frac{n_{\text{air}}}{n_{\text{glass}}} = \frac{1.0003}{1.52}$$

$$\theta_c = 41.2°$$

The angle of incidence in Figure 37-7 is 45°. Since this is greater than the critical angle, total reflection will occur.

37.2 Dispersion and Spectra

The index of refraction n of a material with dielectric constant $\kappa = \epsilon/\epsilon_0$ and relative permeability $\kappa_m = \mu/\mu_0$ is given by Maxwell's electromagnetic theory (Eq. 37.1 and 35.1) as

$$n = \frac{c}{v} = \sqrt{\frac{\mu\epsilon}{\mu_0\epsilon_0}} = \sqrt{\kappa_m \kappa} \qquad (37.4)$$

With the exception of the ferromagnetic materials, $\kappa_m = 1$ (to within a few percent) and the expression for the refractive index reduces to

$$n = \sqrt{\kappa} \qquad (37.5)$$

The experimental value for light waves in water is $n = 1.33$ which implies $\kappa = 1.77$. This is very different from the value $\kappa = 80$ that is found when the dielectric constant of water is measured in a static electric field.

The reason for this discrepancy in κ values is that the value of the dielectric constant depends on the frequency. The large static value of κ is due principally to rotation of permanent molecular dipoles. But because of their relatively large inertia, the molecules remain essentially at rest in a high frequency electric field and only their electrons acquire significant displacements. As with any vibrating system, the response of a molecule's electrons to the applied force is frequency dependent. The formation of induced dipoles and the consequent value of κ is therefore a function of frequency.

For this reason the speed v of light waves inside a material depends on the frequency or wavelength. Consequently, the index of refraction ($n = c/v$) of the material has different values for different wavelengths or colors. This phenomenon is known as *dispersion*. It is involved in one of Newton's greatest experimental discoveries, the fact that a beam of white light is *dispersed* into a spectrum of separated colors by a prism (Figure 37-8).

The deflection of a ray by a prism involves *refraction* at two surfaces. The greater the refractive index, the greater the deflection. The value of the refractive index of the prism glass for violet light is greater than its value for red light, and so the violet light in the incident beam is deflected more than the red.

The spectrum of thermal radiation emitted from a hot substance, such as the glowing tungsten wire in an electric light bulb, shows a continuous distribution of wavelengths. One color shades smoothly into another as in the upper portion of the color plate. Such a spectrum is termed a *continuous spectrum*.

In contrast, the spectrum of light from an arc or from an electric discharge through a tube of gas such as a neon sign is a *line spectrum*, in that only certain separated lines occur, corresponding to a set of discrete wavelengths (see color plate and also Figure 37-9). Each type of atom radiates its own characteristic

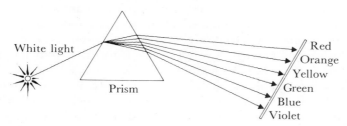

Figure 37-8 Ordinary white light (such as from the sun) passing through a glass prism is separated into the spectral colors indicated.

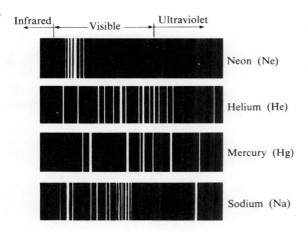

Figure 37-9 Portions of the bright-line spectra of neon, helium, mercury, and sodium. (From A. B. Arons, *Development of Concepts of Physics*, Addison-Wesley, 1965.)

spectrum, a fact that was established in the middle of the nineteenth century by Gustav Kirchhoff (1824–1887) and Robert Bunsen (1811–1899). Consequently, spectroscopy became a primary method of chemical analysis. Stellar compositions are revealed by the spectra of starlight. One element, helium, was first found in the sun. Its spectral lines were detected in sunlight.

Maxwell's electromagnetic theory suggests that the emission or absorption of light by an atom is associated with some sort of oscillatory motion of charge within the atom. But a successful interpretation of an atom's line spectra has been found only by introducing new ideas concerning the nature of light and the laws of mechanics obeyed by an atom. We return to this topic in Chapters 41 and 43.

37.3 Linear Polarization and Malus' Law

If, as an electromagnetic wave passes a given point in space, the electric field **E** at that point fluctuates along a line which has a fixed direction, the wave is said to be *linearly polarized* in the direction of the line. For example, the wave in Figure 35-3 is linearly polarized in the direction of the Y axis.*

In Figure 37-10 we illustrate an experiment in which linearly polarized light is incident upon a thin sheet of material known as polaroid which can be used as a polarization analyzer. The polaroid sheet has a direction called its *transmission axis* which makes an angle θ with the direction of polarization of the incident light. When the polaroid is positioned so that $\theta = 0$, we find that this light is transmitted with an almost undiminished intensity. If the polaroid is then oriented so that $\theta = 90°$, practically no light is transmitted. At intermediate angles, light is transmitted with a diminished intensity corresponding to trans-

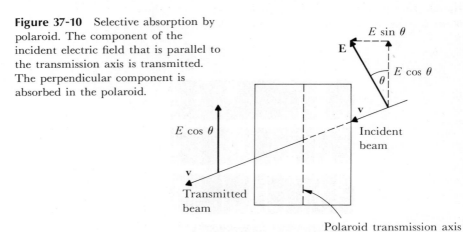

Figure 37-10 Selective absorption by polaroid. The component of the incident electric field that is parallel to the transmission axis is transmitted. The perpendicular component is absorbed in the polaroid.

*Although only the electric field is mentioned when the polarization of an electromagnetic wave is discussed, this electric field is always accompanied by a perpendicular magnetic field as described in Section 35.1.

mission of only the component $E \cos \theta$ parallel to the transmission axis. The component of **E** perpendicular to the transmission axis is *absorbed*. The transmitted light is therefore linearly polarized in the direction of the transmission axis.

For an ideal polarizer with perfect transmission parallel to its transmission axis, the transmitted wave has an amplitude $E_0 \cos \theta$, where E_0 is the amplitude of the incident linearly polarized wave. Since the intensity is proportional to the square of the amplitude, the intensity of the transmitted wave is

$$I = I_0 \cos^2 \theta \qquad\qquad (37.6)$$

where I_0 is the intensity of the incident *linearly polarized* wave.

These facts about the transmission of polarized light through a polarization analyzer were discovered in 1809 by Etienne Malus (1775–1812); Eq. 37.6 is called *Malus' law*. In a longitudinal wave there is nothing to distinguish one transverse direction from another, and such polarization effects cannot arise. Malus' polarization experiments therefore demonstrate that light involves *transverse* rather than longitudinal vibrations. This was established long before Maxwell identified the vibrating quantities as electric and magnetic fields.

Large sheets of a material which gives *selective absorption* perpendicular to a transmission axis were not available until 1938 when Edwin H. Land invented Polaroid. This is manufactured by stretching a plastic sheet containing long hydrocarbon chains. The stretching aligns the molecules. Next the sheet is dipped in an iodine solution. The iodine attaches to the long hydrocarbon chains, providing each chain with mobile electrons. In a fluctuating field, the component of the electric field along the chain drives these electrons back and forth along the length of the chain and the energy associated with such a field component is absorbed. The transmission axis is the direction perpendicular to the line of aligned molecules because in this direction the absorption mechanism cannot function.

Example 2 In Figure 37-11, a beam with intensity I_0 travels in the positive z direction and is linearly polarized in the x direction. This beam is incident

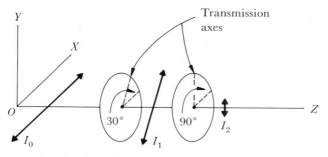

Figure 37-11 The incident beam has an intensity I_0 and is linearly polarized in the direction of the X axis. After transmission through two polaroids, the beam has an intensity I_2 and is linearly polarized in the direction of the Y axis.

upon a stack of two Polaroids oriented with their transmission axes as shown in the figure. Find the polarization and intensity after passing through the first Polaroid and also after passing through both Polaroids.

Solution   After passing through the first Polaroid the light is linearly polarized in the direction at an angle of 30° with the X axis shown in Figure 37-11. The intensity is

$$I_1 = I_0 \cos^2 30° = \tfrac{3}{4} I_0$$

After passing through the second Polaroid the light is linearly polarized in the direction of the Y axis and has an intensity

$$I_2 = I_1 \cos^2 60° = \tfrac{1}{4} I_1 = \tfrac{3}{16} I_0$$

37.4 Polarization Phenomena

General Polarization States

When a plane electromagnetic wave with angular frequency ω is traveling in the positive z direction, the electric field $\mathbf{E}$ at a given point can be specified by giving its components as functions of time:

$$E_x = E_{0x} \cos(\omega t + \phi_x) \tag{37.7}$$

$$E_y = E_{0y} \cos(\omega t + \phi_y) \tag{37.8}$$

The polarization state of the wave is determined by the phase difference $\phi_y - \phi_x$ of these components and by the ratio E_{0y}/E_{0x} of their amplitudes.

To visualize the time dependence of $\mathbf{E}$, we place the tail of this vector at the origin and show the curve traced out by the tip of the $\mathbf{E}$ vector as time elapses (Figure 37-12). In general, this curve is an ellipse [Figure 37-12(a)] and we say that the wave is elliptically polarized.

Linear polarization occurs in just two special cases:

1    $\phi_y - \phi_x = 0$; then $E_y/E_x = E_{0y}/E_{0x}$ and the $\mathbf{E}$ vector oscillates along a direction at an angle θ with the X axis, where $\tan \theta = E_{0y}/E_{0x}$. In the first diagram of Figure 37-12(b), $\theta = 45°$.

2    $\phi_y - \phi_x = \pi$; then $E_y/E_x = -E_{0y}/E_{0x}$ and the $\mathbf{E}$ vector oscillates along a direction at an angle θ' with the X axis, where $\tan \theta' = -E_{0y}/E_{0x}$. In the fifth diagram of Figure 37-12(b), $\theta' = 135°$.

An interesting case arises when the components of $\mathbf{E}$ have equal amplitudes and are 90° out of phase. Suppose

$$E_x = E_0 \cos \omega t \tag{37.9}$$

$$E_y = E_0 \cos(\omega t - 90°) = E_0 \sin \omega t \tag{37.10}$$

Then the field $\mathbf{E}$ has the magnitude

$$E = \sqrt{E_x^2 + E_y^2} = E_0 \sqrt{\cos^2 \omega t + \sin^2 \omega t} = E_0$$

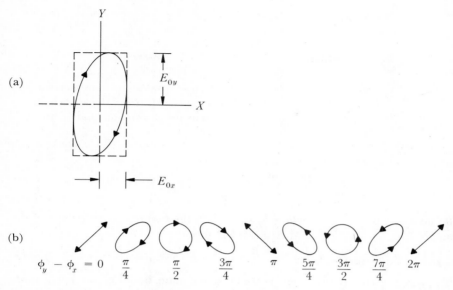

Figure 37-12 (a) General polarization state—elliptical polarization. The tip of the E vector traces out an ellipse. In this figure, the amplitude E_{0y} of the y component is twice the amplitude E_{0x} of the x component. The phase difference of these components, $\phi_y - \phi_x$, is taken to be $\pi/4$. (b) Polarization states for several different values of the phase difference $\phi_y - \phi_x$. In this figure, the amplitudes E_{0x} and E_{0y} have been taken to be equal.

which is constant. The tip of **E** traces out a circle of radius E_0, as **E** rotates with a constant angular speed ω. In this case we say that the wave is *circularly polarized*. Inspection of the values of E_x and E_y at $t = 0$ and at some later instant reveals that in this example **E** rotates *counterclockwise*, as viewed looking backward toward the source of the electromagnetic wave. This wave is conventionally described as *left*-circularly polarized. In this example $\phi_y - \phi_x = -90°$. If $\phi_y - \phi_x = 90°$, **E** rotates *clockwise* and the wave is said to be *right*-circularly polarized.

Quarter-Wave Plate

In certain transparent crystalline substances, the index of refraction is different for different polarization components. Such crystals are said to be *doubly refracting*. The most common example is calcite.

Doubly refracting materials can be used to change the relative phase of perpendicular components of the electric field in a wave. Figure 37-13 shows a plate of such a material with perpendicular axes labeled the *fast axis* and the *slow axis*. The speed of propagation through the plate of the field component E_f parallel to the fast axis is $v_f = c/n_f$ and for the field component E_s parallel to

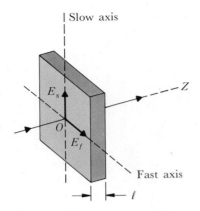

Figure 37-13 The doubly-refracting crystal plate of thickness l introduces a phase difference between the electric field components along the plate's fast and slow axes.

the slow axis is $v_s = c/n_s$, where n_f and n_s are the two refractive indices of the material ($n_s > n_f$). If E_f and E_s have the form

$$E_f = E_{0f} \cos(k_f z - \omega t) \quad \text{and} \quad E_s = E_{0s} \cos(k_s z - \omega t) \quad (37.11)$$

these waves are in phase at the $z = 0$ edge of the plate, but after passage through the plate to the edge $z = l$, their phase difference is

$$\phi = (k_s - k_f)l$$

Using $k = \omega/v = n\omega/c = 2\pi n/\lambda$, we find that the phase difference between E_f and E_s introduced by passage through the plate is

$$\phi = \frac{2\pi}{\lambda}(n_s - n_f)l \quad (37.12)$$

where λ is the wavelength in a vacuum.

By proper choice of l, we can introduce any desired phase difference between perpendicular components. A phase difference $\phi = \pi/2$ is introduced if

$$l = \frac{\lambda}{4(n_s - n_f)} \quad (37.13)$$

and then the plate is called a *quarter-wave plate*.

A quarter-wave plate can convert linearly polarized light into circularly polarized light. If the beam incident on the plate is polarized in a direction at $45°$ to the plate axes, the components E_f and E_s have *equal amplitudes* and, after passage through the quarter-wave plate, these components are $90°$ out of phase. Their superposition is therefore circularly polarized light.

Circularly polarized light incident upon a quarter-wave plate emerges as plane polarized light. A combination of a quarter-wave plate and a Polaroid can be used to analyze circularly polarized light (Question S-11).

Unpolarized Light

Most light sources emit what we call *unpolarized light*. Picturing radiating atoms as containing oscillating electrons which radiate as they do in an antenna, we expect the radiated wave to have a definite polarization state specified at a given observation point by its components (Eq. 37.7 and 37.8) which have constant amplitudes and a constant phase difference. However, each atom radiates for only about 10^{-8} s. With a lamp, for a certain 10^{-8}-s interval, we receive waves from a particular collection of atoms and their superposition will be in a definite polarization state. But in the next 10^{-8}-s interval we get waves from a different set of atoms with different orientations, and the polarization of the wave is generally different. With the atoms randomly oriented and radiating independently, we therefore find that the wave at a given observation point has a *polarization which jumps about* assuming many million different polarization states in a second, with all values of the phase difference of perpendicular components being equally likely and with the average values of the amplitudes of these components being equal. For a detector such as the human eye, which responds to the average intensity over intervals such as 0.1 s, we say this light is unpolarized.

In general, *unpolarized light* is defined as light whose two polarization components E_x and E_y have equal average amplitudes and a phase difference that undergoes many random fluctuations during the observation time.

Since $E^2 = E_x^2 + E_y^2$, the intensity I for any polarization state is given by

$$I = I_x + I_y \tag{37.14}$$

where I_x and I_y are the intensities associated with the components E_x and E_y, respectively. For unpolarized light, $I_x = I_y = \frac{1}{2}I$.

Unpolarized light incident on a Polaroid emerges linearly polarized in the direction of the transmission axis (Figure 37-14). The intensity transmitted by an ideal polarizer is one-half of the intensity of the incident unpolarized light.

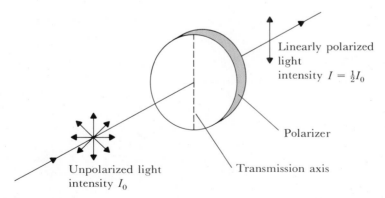

Linearly polarized light
intensity $I = \frac{1}{2}I_0$

Polarizer

Unpolarized light
intensity I_0

Transmission axis

Figure 37-14 Unpolarized light, represented by a mixture of polarizations, is incident upon an ideal polarizer. The transmitted beam is linearly polarized in the direction of the polarizer's transmission axis and has an intensity that is one-half of the intensity of the incident unpolarized light.

Polarization by Reflection, Brewster's Angle

A common source of light that is at least partially polarized is the light reflected from a smooth surface of a road or a lake. This glare, mostly linearly polarized in a horizontal direction, can be discriminated against by sunglasses made of Polaroids with their transmission axes vertical.

Figure 37-15 shows an unpolarized beam falling on a surface of glass at an angle of incidence such that the *reflected ray is perpendicular to the refracted ray*. At this angle of incidence, known as the polarizing angle θ_{1p} or *Brewster's angle*, the reflected ray is found to be completely linearly polarized in a direction perpendicular to the plane of incidence.

The relationship between Brewster's angle and the refractive indices of the media is easily found. With $\theta_2 + \theta_{1p} = 90°$, Snell's law gives

$$n_1 \sin \theta_{1p} = n_2 \sin (90° - \theta_{1p}) = n_2 \cos \theta_{1p}$$

Therefore Brewster's angle satisfies

$$\tan \theta_{1p} = \frac{n_2}{n_1} \qquad (37.15)$$

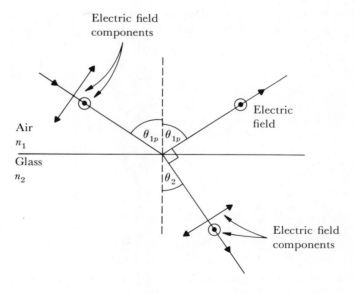

Figure 37-15   Brewster's angle θ_{1p}. The reflected beam is completely linearly polarized in a direction perpendicular to the plane of incidence.

The explanation of the complete polarization of the reflected beam at Brewster's angle is found if we consider the direction of the oscillations of electrons in the glass. It is the radiation from these oscillating electrons that constitutes the reflected beam. These electron oscillations are transverse to the refracted beam. When the angle between the refracted beam and the reflected beam is 90°, the component of electron motion in the plane of incidence is exactly along the direction of the reflected ray, and this motion therefore contributes no reflected radiation. Consequently the reflected ray is produced entirely by the component of the electron motion perpendicular to the plane of incidence, and since the reflected ray itself is in the plane of incidence, it must be completely polarized perpendicular to this plane.

At angles of incidence other than Brewster's angle, there is some radiation in the reflected beam that is polarized in a direction lying in the plane of incidence, and therefore the reflected light is no longer completely linearly polarized.

Example 3 When light is incident on a glass plate with $n = 1.52$, what is Brewster's angle and what is the direction of the refracted beam?

Solution Brewster's angle θ_{1p} in Figure 37-15 is given by Eq. 37.15:

$$\tan \theta_{1p} = \frac{n_2}{n_1} = \frac{1.52}{1.00}$$

This implies that $\theta_{1p} = 56.7°$. The direction of the refracted beam in Figure 37-15 is determined by the angle θ_2 which is given by Snell's law:

$$n_1 \sin \theta_{1p} = n_2 \sin \theta_2$$

This yields

$$\sin \theta_2 = \frac{1.00}{1.52} \sin 56.7° = 0.550$$

and $\theta_2 = 33.4°$.

Applications of Polarized Light

The variety of applications of polarized light can be illustrated by a few examples. Some materials become doubly refracting when stressed. This phenomenon is exploited in the science of photoelasticity to aid in stress analysis (Figure 37-16). Many organic compounds in solution have the property of rotating the direction of polarization of linearly polarized light—an effect that is used to measure the concentration of the solution.

Polarization studies of the electromagnetic radiation emitted by molecules, atoms, and nuclei reveal useful information about these structures.

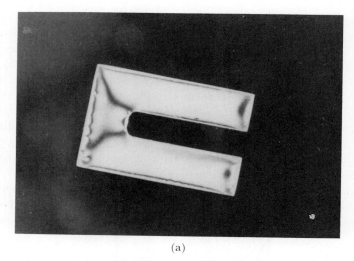

(a)

(b)

Figure 37-16   Photoelasticity. A transparent plastic sample is placed between polaroids and then stressed. (From E. Hecht and A. Zajac, *Optics*, Addison-Wesley, 1974.)

Summary

☐   The reflected and the refracted rays both lie in the plane of incidence. The directions of the incident, reflected, and refracted beams, specified by the angles θ_1, θ_r, θ_2 that these rays make with the normal to the interface between media with refractive indices n_1 and n_2, are given by $\theta_1 = \theta_r$ and Snell's law of refraction:

$$n_1 \sin \theta_1 = n_2 \sin \theta_2$$

☐ Total reflection occurs when light is incident from a medium of higher refractive index n_2 upon the interface with a medium of lower refractive index n_1 at an angle θ_2 equal to or greater than the critical angle θ_c determined by

$$\sin \theta_c = \frac{n_1}{n_2}$$

☐ The refractive index n of a material in which the speed of sinusoidal waves is v is given by

$$n = \frac{c}{v} = \sqrt{\kappa_m \kappa}$$

which is essentially $\sqrt{\kappa}$ in nonferromagnetic materials. Since the dielectric constant κ depends on the frequency (or wavelength) of the electromagnetic wave, the wave speed and the refractive index depend on the wavelength. This phenomenon is called dispersion and accounts for the ability of a prism to disperse a beam of white light into a spectrum of separated colors.

☐ The wave transmitted by an ideal polarizer is linearly polarized in the direction of the polarizer's transmission axis. If the incident wave has intensity I_0 and is linearly polarized at angle θ to the transmission axis, the transmitted intensity is given by Malus' law

$$I = I_0 \cos^2 \theta$$

☐ In circularly polarized light, the electric field has perpendicular components of equal amplitudes with a phase difference of $90°$.

☐ In unpolarized light the electric field has perpendicular components with equal average amplitudes and a phase difference that undergoes many random fluctuations during the observation time. When unpolarized light of intensity I_0 is incident on an ideal polarizer, the transmitted intensity is $\frac{1}{2}I_0$.

☐ Light reflected from a smooth surface is completely linearly polarized in a direction perpendicular to the plane of incidence, if the angle of incidence is Brewster's angle θ_{1p} determined from $\tan \theta_{1p} = n_2/n_1$.

Questions

1 **(a)** What are the dimensions of the refractive index $n = c/v$ of a medium?
   (b) Show that the definition of the refractive index of a medium ($n = c/v$) implies that the refractive index of a vacuum is exactly one.

2 **(a)** The index of refraction of a certain type of glass is 1.55 for green light. What is the speed of sinusoidal light waves in this glass?
   (b) What is the index of refraction of water for light waves which travel in water at a speed of 2.25×10^8 m/s?

3 **(a)** Show that

$$\frac{n_1}{n_2} = \frac{\lambda_2}{\lambda_1}$$

where λ_1 is the wavelength of light of frequency f in a medium of refractive index n_1, and λ_2 is the wavelength of light of the *same frequency f* in a medium of refractive index n_2.

(b) A certain light wave has a wavelength in a vacuum of 5000 Å. If this wave passes into water, what is the wavelength within the water? (Assume that the water has a refractive index of 1.33.)

4 A ray of light traveling through a vacuum makes an angle of 53.1° with the vertical. This ray is refracted as it passes into still water, where it makes an angle of 36.9° with the vertical. Use Snell's law to determine the index of refraction of water from these measurements.

5 Consider refraction at an interface between water and glass when a light ray within the glass makes an angle of 30° with the normal to the interface. Within the water, what is the angle between the ray and the normal? Does the answer depend on which way the light is traveling? (Use $n_{glass} = 1.52$, $n_{water} = 1.33$.)

6 A lamp on the bottom of a pool 5.0 m deep is a horizontal distance of 3.0 m from the pool's edge. At this edge, what angle does a ray from the lamp make with the vertical after it has emerged into the air?

7 Foucault's experimental result that light travels more slowly in water than in a vacuum was interpreted as strong evidence in favor of the wave nature of light. Why?

8 Sketch the ray diagram in Figure 37-17 and complete it by continuing the rays, assuming that the critical angle is 48°.

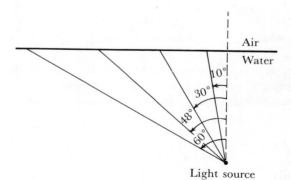

Figure 37-17   Light source

9 A lamp is at the bottom of a pool which is 3.0 m deep. When the water is still, what is the radius of the largest circle at the water's surface through which the light can emerge into the air?

10 Light is incident normally on a face of a glass prism ($n = 1.52$) as shown in Figure 37-18. Find the largest value of the angle θ such that this light will be totally reflected from the face AB.

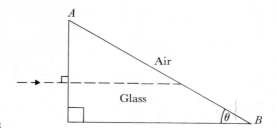

Figure 37-18

11 Ferrites are insulators with a high permeability. Find the speed of propagation and the wavelength of an electromagnetic wave with a frequency of 300 MHz in a ferrite with dielectric constant $\kappa = 10$ and magnetic permeability $\kappa_m = 1000$.

12 For yellow light with a wavelength (in a vacuum) of 5890 Å, crown glass has a refractive index of 1.52. What is the frequency? Determine the value of the dielectric constant of crown glass at this frequency.

13 The effect shown in Figure 37-8 is obtained with a certain glass prism. From this observation sketch a rough graph showing the frequency dependence of the dielectric constant of this glass for the range of frequencies corresponding to visible light.

14 The electric vector **E** of a linearly polarized electromagnetic wave has, at maximum amplitude, a horizontal component of 3.0 V/m and a vertical component of 4.0 V/m. Find the amplitude E_0 of the wave transmitted by a Polaroid when the transmission axis is:
(a) Parallel to **E**.
(b) Horizontal.
(c) Vertical.

15 Find the ratio of the transmitted intensity to the incident intensity for each part of the preceding question.

16 A light beam has an intensity of 10 W/m² and is linearly polarized in a vertical direction. Find the intensity transmitted by a Polaroid when the Polaroid's transmission axis makes the following angles with the vertical: 30°, 45°, 60°, 90°.

17 The beam in the preceding question passes through a Polaroid with its transmission axis at 30° with the vertical. The transmitted beam then passes through a second Polaroid with transmission axis at 90° to the vertical:
(a) What is the intensity of the beam emerging from the second Polaroid?
(b) What would this intensity be if the first Polaroid were removed?

18 At a given point in space, the electric field associated with an electromagnetic wave is given in SI units by

$$E_x = 10 \cos \omega t \quad \text{and} \quad E_y = 20 \cos \omega t$$

Show that the wave is linearly polarized and find the direction of polarization.

19 Find the direction of polarization when, at a given point, the electric field in SI units is

$$E_x = 10 \cos \omega t \quad \text{and} \quad E_y = 20 \cos(\omega t + \pi)$$

20   Describe the polarization state for which the components of the electric field in SI units are

$$E_x = 10 \cos \omega t \quad \text{and} \quad E_y = 10 \cos(\omega t + 90°)$$

Justify your answer by calculating the magnitude of **E** and showing the orientation of **E** at two successive instants.

21   Show that when circularly polarized light of intensity I_0 is incident on an ideal polarizer, the transmitted intensity is $\frac{1}{2}I_0$.

22   A calcite plate has the refractive indices $n_f = 1.4864$, $n_s = 1.6583$ for yellow sodium light with a wavelength of 5890 Å in a vacuum. Find the thickness of the plate that will introduce, for yellow light, a phase difference of $\pi/2$ rad between the electric field components along its fast and slow axes.

23   The calcite plate of Question 22 is placed in a horizontal plane. Yellow light travels vertically upward through the plate. If the incident light is linearly polarized in an east-west direction, how should the plate be oriented to produce circularly polarized light? Explain.

24   A beam of unpolarized light with an intensity of 10 W/m² is incident upon a stack of two ideal polarizers with an angle θ between their transmission axes. Find the intensity of the beam transmitted by this stack when θ has the following values: 0°, 30°, 45°, 60°, 90°.

25   Unpolarized light of intensity I_0 is incident upon a stack of two Polaroids whose transmission axes make an angle θ with each other. Express the intensity I of the emerging beam as a function of I_0 and θ, and sketch a graph showing I as a function of θ for values of θ ranging from 0° to 180°.

26   A beam of unpolarized light of intensity I_0 is incident upon a stack of four Polaroids, each with its transmission axis rotated 30° clockwise with respect to the preceding Polaroid. Express the transmitted intensity in terms of I_0.

27   When light reflected from a calm lake is completely linearly polarized, what is the direction of this polarization? What is the angle between the sun and the vertical?

Supplementary Questions

S-1   A horizontal beam of light is reflected from a vertical plane mirror. Show that when the mirror is rotated about a vertical axis through an angle θ, the reflected beam rotates through an angle 2θ.

S-2   **(a)** Sketch a ray diagram showing how light is refracted as it passes through a triangular glass prism with prism angle A. (The angle between the two refracting sides of the prism is called the *prism angle*.)

   (b) The angle between the incident ray and the ray emerging from the prism is called the *angle of deviation* D. It can be shown that, when the angle of incidence of the incident ray equals the angle of refraction of the emerging ray, the angle D attains a minimum value, the *angle of minimum deviation* D_m. (If the cross section of the prism is isosceles, with the two equal angles at the base, then D_m is attained with the ray inside the prism parallel to the base of the prism.) Show that the angle of minimum deviation D_m is related to the prism angle A and to the index of refraction n, by

$$n = \frac{\sin \frac{1}{2}(A + D_m)}{\sin \frac{1}{2}A}$$

S-3 Show that if the prism angle is small then the angle of minimum deviation referred to in Question S-2 is given approximately by

$$D_m \approx (n-1)A$$

where D_m is the angle of minimum deviation, and A is the prism angle.

S-4 Describe how we might determine experimentally the angle of minimum deviation of a prism. (See Question S-2 for the definition of the angle of minimum deviation.)

S-5 Show that if $n_2 > n_1$, there is no real angle θ_2 that is greater than the critical angle and that is a solution of the equation

$$n_1 \sin \theta_1 = n_2 \sin \theta_2$$

S-6 Assume that for each polarization state represented in Figure 37-12(b), the horizontal component of the electric field is given in SI units by

$$E_x = 10 \cos \omega t$$

(a) Give the vertical component E_y for each polarization state.
(b) Justify each of the eight figures shown.

S-7 Find the thickness of a calcite plate (Question 22) that, for yellow light, will introduce a phase difference of π rad between electric field components along its fast and slow axes. Explain what happens to linearly polarized light that passes through the plate.

S-8 Circularly polarized light of intensity I_0 is incident on a stack of three Polaroids. The first and third Polaroids have their transmission axes crossed (mutually perpendicular). The transmission axes of the first and second Polaroids make an angle θ. Show that the light intensity transmitted by this stack is $\frac{1}{2}I_0 \cos^2 \theta \sin^2 \theta$.

S-9 Show that, within a quarter-wave plate, there is one-quarter of a wavelength more for the E_s wave than for the E_f wave.

S-10 A half-wave plate introduces a phase difference of $180°$ between the E_s and E_f waves. Show that a half-wave plate converts right circularly polarized light into left circularly polarized light and vice versa.

S-11 **(a)** Show that a quarter-wave plate converts circularly polarized light into linearly polarized light.
(b) A circular polarizer is made by gluing together a Polaroid and a quarter-wave plate, with the Polaroid transmission axis at $45°$ to the fast and slow axes of the plate. Suppose right circularly polarized light is produced by this combination when light is incident on the Polaroid. Show that, when light is incident on the quarter-wave plate, right circularly polarized light will be transmitted by the combination but left circularly polarized light will be absorbed.

S-12 Derive an expression for the fringe separation when Young's double-slit experiment is performed with the apparatus immersed in water of refractive index n_w.

38.

geometrical optics

Optical instruments consist of various reflecting and refracting surfaces. In geometrical optics, the behavior of light rays in optical instruments is studied in the approximation that diffraction spreading of light beams can be ignored. Analysis is based entirely on the law of regular reflection and on Snell's law of refraction.

38.1 Image Formation by a Plane Mirror

The face that greets you when you look in a plane mirror is what is called the *image* of your face. The formation of such an image by reflection in a mirror is easy to understand after one has considered the reflection of different light rays which originate from a single point object such as the point O in Figure 38-1(a). When these reflected rays are extended backward they intersect at a single point I located behind the mirror. Therefore rays which actually originate at O *appear* to emanate from I. The point I is called the image of the object O.

Figure 38-1(b) shows how the image I can be located by considering just two light rays which emanate from a point object O. One light ray is incident perpendicular to the surface and will be reflected back upon itself. Another light ray which strikes the mirror at an arbitrary point P is reflected in accord with the law of reflection:

$$\theta_r = \theta_1$$

Elementary geometry now implies that in Figure 38-1(b) the angles MIP and

MOP must be equal and consequently that the triangles *MIP* and *MOP* are congruent. Therefore the distances *MI* and *MO* are equal; that is

distance of image from mirror = distance of object from mirror

Notice that since the preceding argument is valid no matter what the location of the point *P* on the mirror, *all* rays from *O* which are reflected by the mirror must pass through *I* when extended backward, as indicated in Figure 38-1(a).

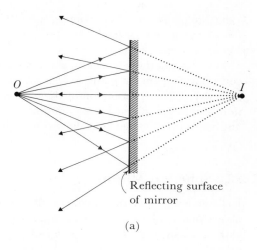

(a)

Figure 38-1   Plane reflection. (a) All rays from point *O* that are reflected by the mirror *appear* to emanate from a point *I*. Point *I* is called the *virtual image* of the point *O*. (b) Two rays which emanate from the object *O* and are reflected by the mirror. The extensions of these rays behind the mirror intersect at point *I*, and the distance *MO = MI*. (c) A small bundle of rays from *O* enters the eye upon reflection from a small portion of the mirror.

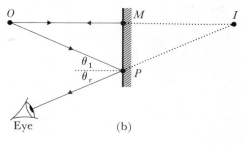

(b)

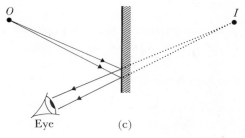

(c)

Images formed by systems of lenses and mirrors are categorized as follows:

1   *Real image*   Different light rays originating from a point object are bent so as to *actually pass through* the image point.
2   *Virtual image*   Different light rays originating from a point object are bent so as to *appear* to diverge from the image point because their backward extensions intersect at this point.

An image in a plane mirror is a *virtual* image. The image of O is created by the reflection at and near the point P of a small bundle of light rays [Figure 38-1(c)]. What we really see is light coming from points near P, and the eye–brain system makes the necessary backward projection to create the illusion that the light signal is coming from point I.

With an extended object such as a human face, each point on the object has a corresponding image point so that an extended image is produced which is a point-by-point reproduction of the object. The virtual image formed by reflection in a plane mirror is the same size as the object and upright, but with left and right interchanged, as shown in Figure 38-2.

38.2 Spherical Mirrors

Reflection at a spherical surface will lead to the formation of an image of any object that is placed in front of the surface. Figure 38-3 shows a *concave* mirror with center A, radius r, and center of curvature C. The line passing through A and C is called the *principal axis* of the mirror.

Consider a ray that originates from the point O, a distance p from A, and makes a small angle α_1 with the principal axis. The mirror reflects this ray through the point I a distance q from A. The incident and reflected rays make equal angles θ with the radius from C.

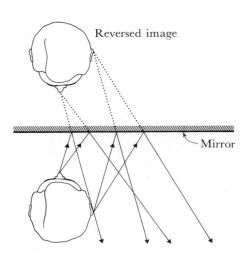

Reversed image

Mirror

Figure 38-2   "Left-to-right" reversal of mirror image. Note hair parted on the left side while the reflected image shows hair parted on the right side.

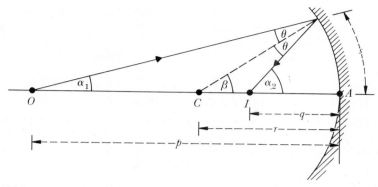

Figure 38-3 Concave spherical mirror with principal axis through A and C. A ray from O, making a small angle α_1 with the principal axis, is reflected through I.

To find q in terms of p and r, we apply the geometrical theorem that the exterior angle of a triangle is equal to the sum of the two opposite interior angles. From Figure 38-3, this gives $\beta = \alpha_1 + \theta$ and $\alpha_2 = \alpha_1 + 2\theta$. Eliminating θ, we have

$$\alpha_1 + \alpha_2 = 2\beta \tag{38.1}$$

The radian measures of these angles are $\beta = s/r$, $\alpha_1 \approx s/p$, and $\alpha_2 \approx s/q$, where the expression for β is exact but those for α_1 and α_2 are approximations that are good only if the angles are small. Substituting in Eq. 38.1, and dividing by s, we obtain

$$\frac{1}{p} + \frac{1}{q} = \frac{2}{r} \tag{38.2}$$

This relationship, being independent of the angles α_1, α_2, and β, shows that *any* ray from O is reflected through I, provided it makes a *small* angle with the principal axis. Therefore I is the *image* of O, and since the light rays actually do pass through I, it is a *real* image. This image can be seen, either by positioning the eye so as to receive light rays diverging from I, or by viewing a screen placed at I.

Light rays are reversible. If the object were at the point I, the image would be at the point O.

Focal Point

If the object distance p becomes arbitrarily large, that is, if $p \to \infty$, the incident rays are parallel to the principal axis. For this case, Eq. 38.2 gives

$$0 + \frac{1}{q} = \frac{2}{r}$$

and the image is at the point F in Figure 38-4, a distance $q = r/2$ from A. The point F is called the *focal point* or *focus* of the mirror and its distance from the mirror is the *focal length f*. For a spherical mirror we have $f = r/2$ and Eq. 38.2 can be rewritten as

$$\frac{1}{p} + \frac{1}{q} = \frac{1}{f}$$

(38.3)

The focal point can be located experimentally by using a beam of incident rays parallel to the principal axis and locating the point of convergence after reflection. And if a point source is placed at the focus, rays which make a small angle with the principal axis are reflected as a parallel beam.

Magnification

Each point O of an extended object has an image point I lying on the line through O and the center of curvature C, with the image and object distances related by Eq. 38.3. For an object that lies on a plane perpendicular to the principal axis, since p has the same value for all points, Eq. 38.3 gives the same q for each object point. The conclusion is that, in the small-angle approxima- tion, the image lies on a plane perpendicular to the optic axis.

The *lateral magnification m* of the mirror is defined (apart from a question of sign) as the ratio of image height h' to object height h. From the similar triangles in Figure 38-5, we see that

$$\left|\frac{h'}{h}\right| = \left|\frac{q}{p}\right|$$

(38.4)

Figure 38-4   Rays that are incident parallel to the principal axis are reflected through the focal point F of this converging mirror.

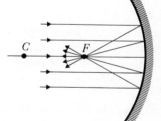

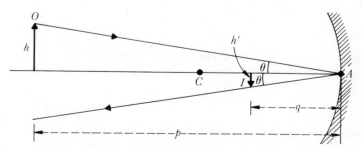

Figure 38-5   The lateral magnification, $m = h'/h$, is equal to $-q/p$.

It is convenient to define m so that an inverted image corresponds to a negative magnification. Then Eq. 38.4 gives

$$m = -\frac{q}{p} \qquad (38.5)$$

Principal-Ray Diagram

It is instructive to determine the location and size of the image by drawing a principal-ray diagram (Figure 38-6) showing an object point O off the principal axis, and at least two of the following four *principal rays* which are incident from O:

1 A ray parallel to the principal axis; this is reflected through the focus F.
2 A ray through the focus F; this is reflected back parallel to the principal axis.
3 A ray through the center of curvature C. Since this ray is incident along a normal to the mirror, it is reflected back along itself.
4 A ray striking the center A of the mirror. The reflected ray will make an equal angle with the principal axis.

Sign Convention

When p is less than f, Eq. 38.3 gives a negative value for the image distance q. The principal-ray diagram for this case shows that there is a virtual image a distance q behind the mirror.

A similar analysis of image formation in a convex spherical mirror (Figures 38-7 and 38-8) reveals that the equation

$$\frac{1}{p} + \frac{1}{q} = \frac{1}{f}$$

with $f = r/2$, can be applied to any spherical mirror and any positions of object

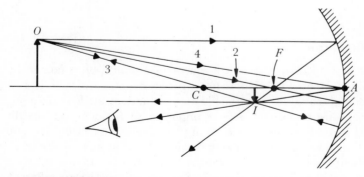

Figure 38-6 Principal-ray diagram showing four principal rays. Only two are required to locate the image I.

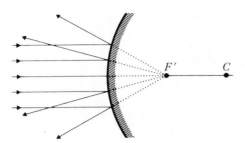

Figure 38-7   Convex mirror with virtual focus F'.

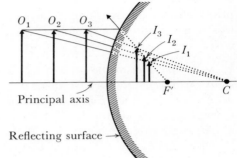

Figure 38-8   Image formation by a convex mirror. For each object position, two principal rays have been drawn to locate the image.

and image if the sign convention in Table 38-1 is adopted. A point O is a virtual object if incoming rays would converge at O if they were not intercepted by the reflecting surface of the mirror.

Table 38-1 Sign Convention for Spherical Mirrors

	Positive	Negative
Focal length $f = r/2$	Converging (concave)	Diverging (convex)
Object distance p	Real object	Virtual object
Image distance q	Real image	Virtual image
Magnification $m = -q/p$	Erect image	Inverted image

Parabolic Mirror

The image formed by a spherical mirror of a point object fails to be a single well-defined point. One reason for a defective image is spherical aberration which arises because the small-angle approximations made in deriving Eq. 38.3 are never satisfied exactly. For example, if an incident ray parallel to the principal axis strikes the mirror an appreciable distance from its center A, the reflected ray does not pass precisely through the focal point F. We can overcome this particular shortcoming by using a *parabolic* rather than a spherical mirror. All rays parallel to the principal axis of a parabolic mirror are reflected through its focus. For this reason reflecting astronomical telescopes for light or radio waves are often constructed using a parabolic mirror. Figure 38-9 shows a huge

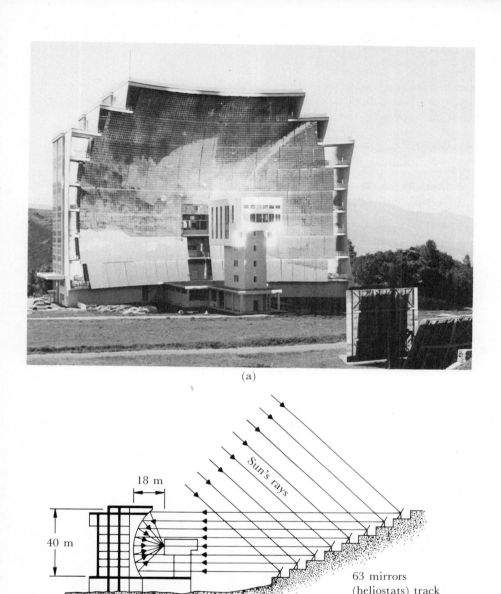

(a)

18 m

40 m

Sun's rays

Parabolic reflector concentrates
sun's rays onto target area

(b)

63 mirrors
(heliostats) track
the sun and direct
the sun's rays into the
parabolic reflector

Figure 38-9 (a) Solar furnace located in the Pyrenees at Odeillo-Font Romeu
(altitude 1.8 km). At this location the sun shines as many as 180 days a year and
solar intensities as high as 1000 W/m^2 are common. The solar furnace was
completed in 1970 at a cost of about \$2,000,000. (b) Schematic diagram of 1000 kW
solar furnace. The parabolic reflector has a focal length of 18 m, is 40 m high and
53 m wide, and is composed of 9500 mirrors 45 cm by 45 cm. Since the parabolic
reflector is too large to track the sun, 63 smaller mirrors (heliostats) set in eight tiers
are used to follow the sun and reflect its rays in parallel beams onto the parabola.
The heliostats are 7.5 m by 6 m and each is composed of 180 mirrors 50 cm by
50 cm. (Courtesy Engineering Experiment Station, Georgia Institute of Technology.)

parabolic mirror used to focus the rays of the sun in a solar furnace. The parallel beam of a searchlight is obtained by locating the source at the focus of a parabolic mirror (Figure 38-10).

Example 1   An object is placed 0.50 m in front of a concave spherical mirror with a radius of 0.40 m. Locate and describe the image.

Solution   Figure 38-11, a principal-ray diagram, shows that the image is real, inverted, and diminished in size. These facts can be confirmed using Eq. 38.3. The focal length is $f = r/2 = 0.20$ m. Therefore

$$\frac{1}{0.50 \text{ m}} + \frac{1}{q} = \frac{1}{0.20 \text{ m}} \quad \text{and} \quad q = 0.33 \text{ m}$$

Since q is positive, the image is real. The lateral magnification is

$$m = -\frac{q}{p} = -0.67$$

This corresponds to an inverted image which is diminished to two-thirds of the size of the object.

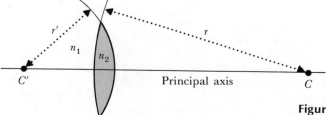

Figure 38-10   Light source at the focus of a parabolic mirror gives rise to a parallel beam.

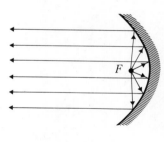

Figure 38-11   Principal-ray diagram for Example 1.

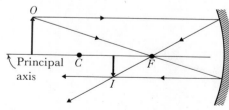

Figure 38-12   Thin lens. A converging lens with convex surfaces of radii r and r'.

38.3 Thin Lenses

Figure 38-12 shows a thin lens made of a material with refractive index n_2 immersed in a medium of refractive index n_1. The faces of the lens have radii of curvature r and r' that are assumed to be large compared to the lens thickness. The principal axis of the lens is the line through the centers of curvature, C and C'. We consider a ray from O that makes a small angle α_1 with the principal axis [Figure 38-13(a)]. This ray is refracted at each surface and then makes an angle α_1' with the principal axis which it meets at point I. The distances from O and I to the center A of the lens are denoted by p and q, respectively.

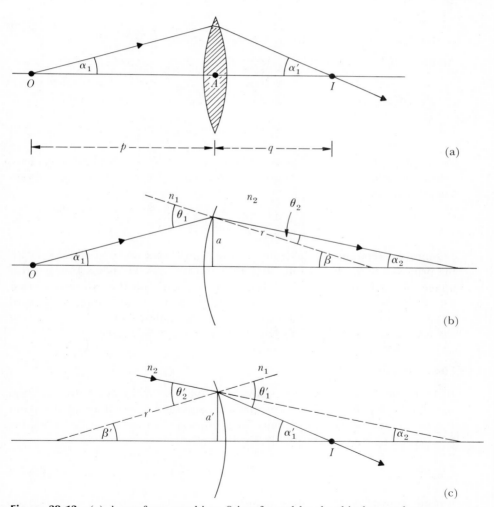

(a)

(b)

(c)

Figure 38-13   (a) A ray from an object O is refracted by the thin lens and passes through the image I. (b) Refraction at the first surface of the lens. (c) Refraction at the second surface.

To find the relationship between p and q we consider the refractions at each surface. At the first surface [Figure 38-13(b)], Snell's law gives $n_1 \sin \theta_1 = n_2 \sin \theta_2$. For small angles this simplifies to $n_1\theta_1 = n_2\theta_2$. Considering the exterior angles of appropriate triangles, we find $\beta = \theta_2 + \alpha_2$ and

$$\alpha_1 + \beta = \theta_1 = \frac{n_2}{n_1} \theta_2 = \frac{n_2}{n_1}(\beta - \alpha_2) \qquad (38.6)$$

At the second surface [Figure 38-13(c)], we have

$$n_2\theta_2' = n_1\theta_1' \qquad \theta_2' = \alpha_2 + \beta'$$

and

$$\alpha_1' + \beta' = \theta_1' = \frac{n_2}{n_1} \theta_2' = \frac{n_2}{n_1}(\alpha_2 + \beta') \qquad (38.7)$$

Addition of Eq. 38.6 and 38.7 yields

$$\alpha_1 + \alpha_1' + \beta + \beta' = \frac{n_2}{n_1}(\beta + \beta')$$

$$\alpha_1 + \alpha_1' = \left(\frac{n_2}{n_1} - 1\right)(\beta + \beta') \qquad (38.8)$$

For small angles, $\alpha_1 \approx a/p$, $\beta \approx a/r$, $\alpha_1' \approx a'/q$, $\beta' \approx a'/r'$ and for a *thin* lens, $a \approx a'$. Making these substitutions in Eq. 38.8 and dividing by a, we obtain the *thin-lens equation*:

$$\frac{1}{p} + \frac{1}{q} = \left(\frac{n_2}{n_1} - 1\right)\left(\frac{1}{r} + \frac{1}{r'}\right) \qquad (38.9)$$

Since this relationship holds for all rays from O that make small angles with the principal axis, we conclude that all such rays pass through I. Hence a real image of O is formed at I. For a point object slightly off the axis we find that Eq. 38.9 is again obtained. Consequently the thin-lens equation can be used for an object in a plane perpendicular to the principal axis and a distance p from the lens. The image lies in a plane a distance q from the lens.

Focal Length of a Lens

If the object distance becomes arbitrarily large, rays incident on the lens from the object are parallel to the principal axis, and the image is formed at a point F called a *principal focus* of the lens. The distance from the lens to F, called the focal length f, is given by Eq. 38.9 with $p = \infty$,

$$\frac{1}{f} = \left(\frac{n_2}{n_1} - 1\right)\left(\frac{1}{r} + \frac{1}{r'}\right) \qquad (38.10)$$

a result known as the *lensmaker's equation*.

With this result, the thin-lens equation can be written

$$\frac{1}{p} + \frac{1}{q} = \frac{1}{f} \qquad (38.11)$$

There are two principal foci on opposite sides of the lens and equidistant from it. Light from an object placed at a principal focus emerges from the lens as a beam of parallel rays.

Principal-Ray Diagram

Image formation is readily understood by drawing principal-ray diagrams (Figures 38-14 and 38-15) showing at least two of the following rays incident from an object point O:

1   A ray parallel to the principal axis which is refracted so as to pass through a principal focus.
2   A ray passing through a principal focus and refracted to emerge parallel to the principal axis.
3   A ray which passes through the center of the lens (called the optical center). This ray is not deviated.

From principal-ray diagrams it is easy to prove that the lateral magnification of a lens is given by the same expression as that for a spherical mirror:

$$m = -\frac{q}{p} \tag{38.12}$$

where m is positive for an erect image and negative for an inverted image. The magnitude of m gives the ratio of the image size to the object size.

With the sign convention of Table 38-2 on page 707, the formulas of this section can be extended to apply to virtual images and objects, and to diverging as well as converging lenses (Figure 38-16). A principal-ray diagram for a diverging lens is shown in Figure 38-17.

Aberrations

Since the entire analysis of image formation in lenses has been approximate, a variety of *lens aberrations* (deviations from the predicted behavior that lead to defective images) occur in practice. By judicious use of combinations of lenses, we can use the aberrations of one lens to at least partially cancel the aberrations of another.

In addition to the aberrations associated with the geometry of the surfaces, a lens also displays *chromatic aberration*. This refers to the fact that the refractive index of the lens is a function of the wavelength of the light. Consequently, the focal length of the lens is different for light of different colors.

Since image formation by a mirror does not involve refraction, mirrors are free of chromatic aberration. In modern optical systems, mirrors rather than lenses are finding increasing use, particularly in the ultraviolet and infrared regions of the spectrum where absorption in a refracting material can be a serious problem.

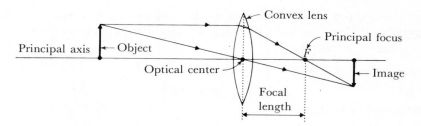

Figure 38-14   Ray diagram for a thin lens. Note that although the upper ray is refracted twice, once on entry to the lens and once on leaving the lens, it is common practice (and consistent with the thin-lens approximation) to draw ray diagrams in which the ray follows the dotted line and is bent just once on the central plane of the lens. In subsequent figures we shall follow this simpler schematic approach.

Cases of convex lens	Position of image	Description of image	Example of application	Optical diagram
Object very far from lens; rays entering lens almost parallel	Almost at F	Real, almost a point, inverted	Telescope	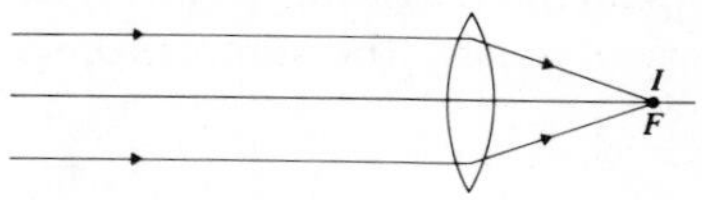
Object farther than twice the focal length	Between F and $2F$	Real, smaller, inverted	Snapshot camera Movie camera	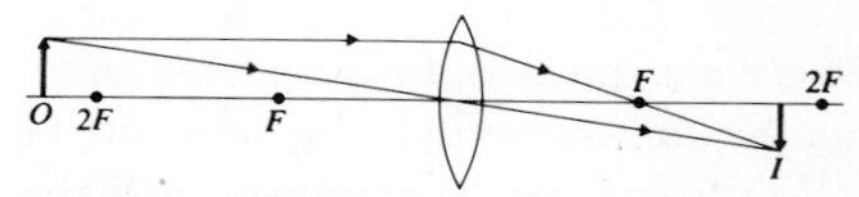
Object at $2F$	At $2F$	Real, same size, inverted	Copying camera	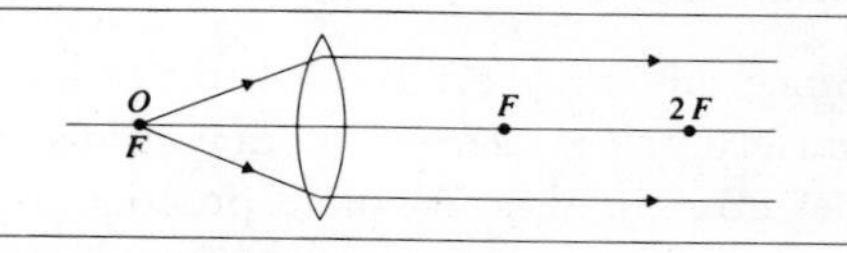
Object between F and $2F$	At greater than $2F$	Real, larger, inverted	Enlarging camera Slide projector	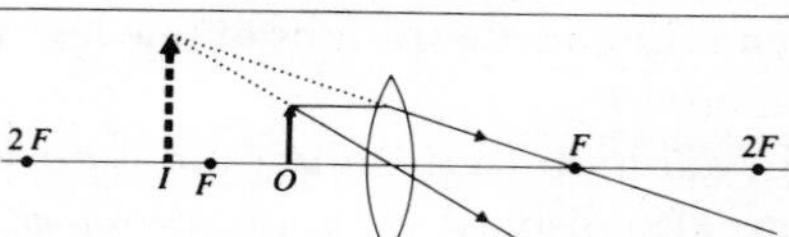
Object at the principal focus	No image, refracted rays parallel		Light source for parallel beam	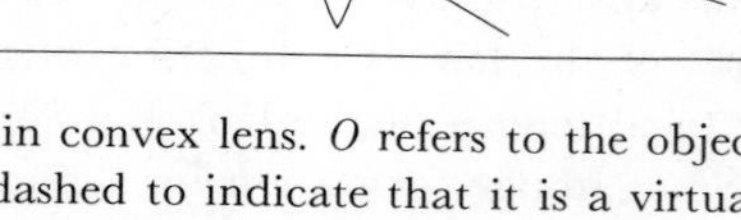
Object between F and lens	On same side of lens as object	Virtual, erect, larger	Magnifying glass	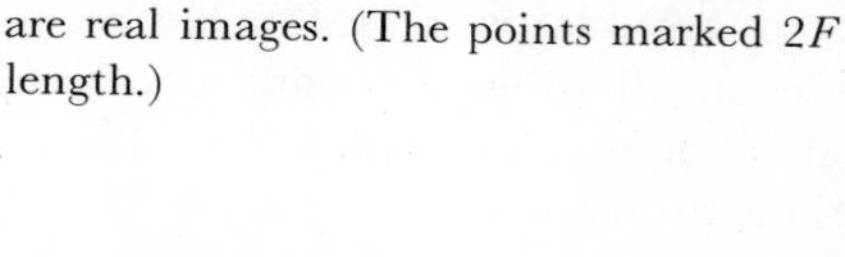

Figure 38-15   The six possible cases of the thin convex lens. O refers to the object, I to the image. The image in the last case is dashed to indicate that it is a virtual image, as opposed to all the others which are real images. (The points marked $2F$ indicate a distance twice that of the focal length.)

Table 38-2 Sign Conventions for a Thin Lens

	Positive	Negative
Radii r, r'	Convex surface	Concave surface
Focal length f	Converging lens	Diverging lens
Object distance p	Real object	Virtual object
Image distance q	Real image	Virtual image
Magnification m	Erect image	Inverted image

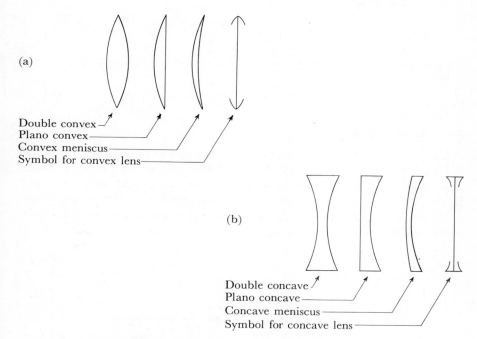

(a)

Double convex
Plano convex
Convex meniscus
Symbol for convex lens

(b)

Double concave
Plano concave
Concave meniscus
Symbol for concave lens

Figure 38-16   (a) Converging lenses. (b) Diverging lenses.

Case of concave lens	Position of image	Description of image	Application	Optical diagram
All the same, regardless of position of object	Same side of lens as object	Virtual, smaller, erect	Telescopic lens combinations Camera viewers Corrective lenses for nearsightedness	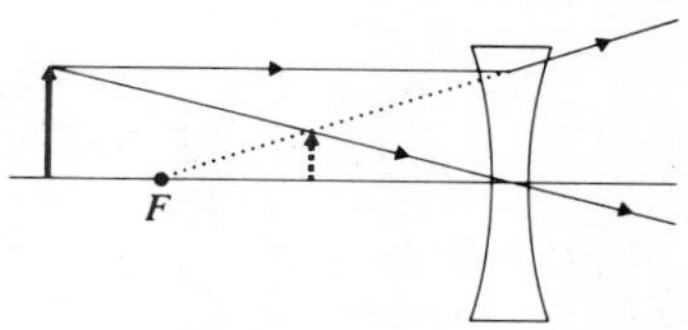

Figure 38-17   Image formation by a concave or diverging lens.

Example 2   A converging lens has a focal length of 5.0 cm. A jeweler looks through this lens at a mark on a watch held 4.0 cm behind the lens. Locate and describe the image.

Solution   The principal-ray diagram of Figure 38-18 shows that the image is virtual, erect, and enlarged. The thin-lens formula gives

$$\frac{1}{4.0 \text{ cm}} + \frac{1}{q} = \frac{1}{5.0 \text{ cm}} \qquad \text{and} \qquad q = -20 \text{ cm}$$

There is therefore a virtual image (negative q) located 20 cm behind the lens. The lateral magnification is

$$m = -\frac{q}{p} = -\frac{(-20)}{4.0} = 5.0$$

Consequently the image is erect (positive m) and 5.0 times as large as the object.

38.4 Microscopes

The microscope is one of the most useful of the optical instruments mentioned in Figures 38-15 and 38-17. With the aid of a microscope, we can see an object more clearly than we can with the naked eye because a larger image is formed on the *retina* (the film containing the optical detectors of the eye system shown in Figure 38-19). The size of the image on the retina is proportional to the (small) angle subtended by the object at the eye.

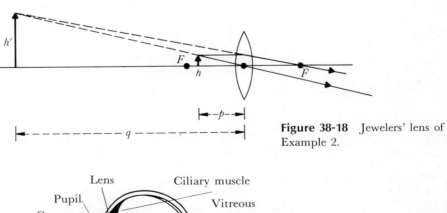

Figure 38-18   Jewelers' lens of Example 2.

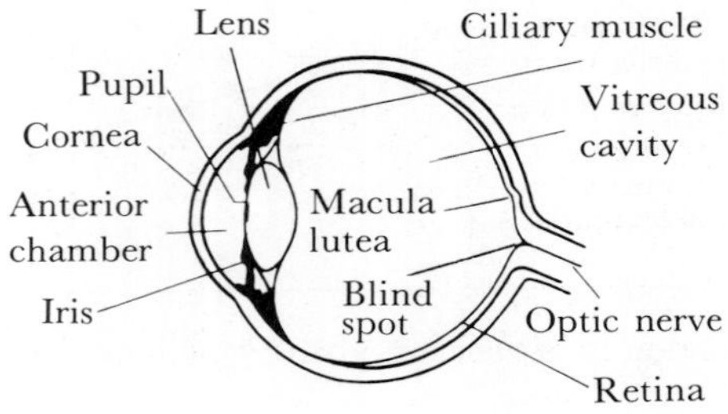

Figure 38-19   The eye.

A useful figure of merit for a microscope is its angular magnification or *magnifying power* M defined by

$$M = \frac{\theta_m}{\theta_d} \tag{38.13}$$

where θ_d is the angle subtended at the eye by an object when it is placed at a good distance for distinct vision (this is usually taken to be 0.25 m) and viewed directly with the naked eye; and θ_m is the angle subtended at the eye by this object's image formed by the microscope.

Simple Microscope

A single converging lens of short focal length, used as illustrated in Figure 38-18, constitutes a *magnifying glass* or *simple microscope*. A small object of height h is placed so that the object distance p is slightly less than the focal length f. An enlarged erect virtual image of height h' is formed at a distance $|q|$ from the lens. This image is viewed with the eye close to the lens, a distance $D = |q|$ from the image. Using the fact that the lateral magnification is

$$m = \frac{h'}{h} = -\frac{q}{p} = \frac{D}{p}$$

we find that the magnifying power of the simple microscope is

$$M = \frac{\theta_m}{\theta_d} = \frac{h'/D}{h/0.25 \text{ m}} = \frac{0.25 \text{ m}}{p}$$

The thin-lens equation gives $1/p = 1/f - 1/q = 1/f + 1/D$. Therefore

$$M = \frac{0.25 \text{ m}}{f} + \frac{0.25 \text{ m}}{D} \tag{38.14}$$

where f and D are in metres.

The magnifying power of a simple microscope varies from $0.25 \text{ m}/f + 1$ to $0.25 \text{ m}/f$, as D is varied from 0.25 m to infinity. The shorter the focal length of the lens, the larger its magnifying power. In actual microscopes, the focal length is small compared to the distance D. Then the magnifying power is given approximately by

$$M = \frac{0.25 \text{ m}}{f} \tag{38.15}$$

Compound Microscope

To obtain a magnifying power greater than the modest values attainable with a simple microscope, two or more lenses are used. Figure 38-20 shows a relatively simple type of *compound microscope*. A short-focus converging lens called the *objective* forms an enlarged real image G' of a small illuminated object. The image G' serves as the object for a second lens, called the *eyepiece lens* or *ocular*, which forms a final virtual enlarged image at a distance which is adjusted to be convenient for comfortable viewing.

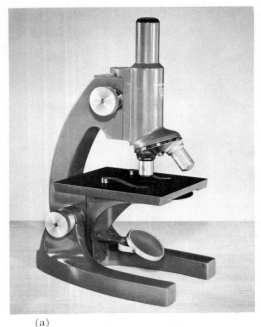

(a)

Figure 38-20 Compound microscope.
(a) Microscope with two objective lens
systems, permitting two ranges of
magnification. (Courtesy Bausch &
Lomb.) (b) Schematic diagram of the
microscope in (a). (c) Optical diagram
showing the formation of the enlarged
virtual final image.

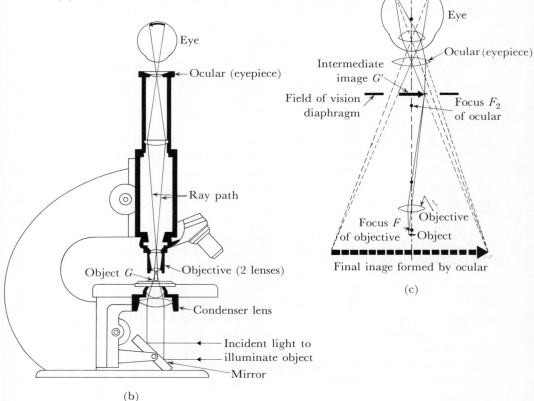

Eye

Ocular (eyepiece)

Intermediate
image G'

Field of vision
diaphragm

Focus F_2
of ocular

Focus F
of objective

Objective

Object

Final image formed by ocular

(c)

Eye

Ocular (eyepiece)

Ray path

Object G

Objective (2 lenses)

Condenser lens

Incident light to
illuminate object

Mirror

(b)

Since the eyepiece thus acts as a simple microscope for viewing the enlarged real image produced by the objective, the magnifying power of the compound microscope is

$$M = m_o M_E \qquad (38.16)$$

where m_o is the lateral magnification of the objective (focal length f_o) and M_E is the magnifying power of the eyepiece (focal length f_E). In the approximation given by Eq. 38.15, we have

$$M_E = \frac{0.25 \text{ m}}{f_E}$$

The lateral magnification of the objective is

$$m_o = -\frac{q}{p} \approx -\frac{L}{f_o} \qquad (38.17)$$

since $p \approx f_o$ and the intermediate image distance q is approximately equal to the distance between the objective and the eyepiece, called the microscope *tube length L*. With these approximations, the expression given by Eq. 38.16 for the magnifying power of a compound microscope becomes

$$M = -\left(\frac{L}{f_o}\right)\left(\frac{0.25 \text{ m}}{f_E}\right) \qquad (38.18)$$

The minus sign corresponds to the fact that the final image is inverted.

Example 3 A compound microscope has a tube length of 160 mm, an objective with a focal length of 32 mm, and an eyepiece with a focal length of 25 mm. Find the lateral magnification m_o of the objective, the magnifying power M_E of the eyepiece, and the magnifying power M of this compound microscope.

Solution The objective has a lateral magnification

$$m_o = -\frac{L}{f_o} = -\frac{160 \text{ mm}}{32 \text{ mm}} = -5$$

Accordingly, the barrel of the objective will be engraved with markings $5\times$ (or $\times 5$). For the eyepiece

$$M_E = \frac{0.25 \text{ m}}{0.025 \text{ m}} = 10$$

Its magnifying power is said to be $10\times$. For the compound microscope

$$M = m_o M_E = -50$$

The magnitude of this magnifying power is $50\times$.

To have a large magnifying power, the tube length should be large and the focal lengths of the eyepiece and the objective should be small. Several factors that limit the magnifying power attainable are:

1 The illumination that the object can tolerate (biological specimens are rather easily damaged).

2   Lens aberrations (which increase as the magnifying power is increased).
3   Diffraction effects.

Diffraction occurs at any circular aperture in the microscope and the smaller the aperture, the greater the effect. A diffracting aperture is always present in view of the fact that a lens of diameter D is equivalent to a lens of infinite extent mounted in front of a circular aperture of diameter D. Because of diffraction, a point object has an image which, instead of being a point, is actually a diffraction pattern. The images of two nearby point sources are overlapping diffraction patterns, and when the overlap is too great it is impossible to distinguish distinct images (Figure 38-21). A detailed analysis verifies that (as we might expect from the discussion of diffraction spreading in Section 36.4) two point objects cannot be resolved by a microscope if their angular separation at the objective is less than an angle whose order of magnitude is

$$\theta_{min} = \frac{\lambda}{D}$$

where D is the diameter of the microscope objective and λ is the wavelength of the light.

Because of diffraction, magnifying powers in excess of $400\times$ do not increase detail. However, higher magnifications are sometimes used since a larger image may lend itself to more thorough study even if no further detail can be discerned. In general, the practical upper limit of magnification by optical microscopes is considered to be about $2000\times$.

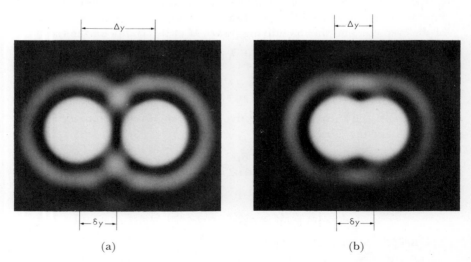

(a)   (b)

Figure 38-21   (a) Images of two point sources with first dark rings tangent. Here the image separation Δy is twice the radius δy of the first dark ring of each diffraction pattern. (b) Images of two point sources which are just resolved according to what is termed the *Rayleigh criterion* ($\Delta y = \delta y$). (From D. H. Towne, *Wave Phenomena*, Addison-Wesley, 1967.)

Summary

☐ For either a spherical mirror or a thin lens,

$$\frac{1}{p} + \frac{1}{q} = \frac{1}{f}$$

where f is the focal length, and the object distance p and image distance q are both measured from the center (of the lens or of the mirror). The lateral magnification is $m = \textit{image height } h' / \textit{object height } h = -q/p$. The sign conventions are:

	Positive	Negative
f	Converging	Diverging
p	Real object	Virtual object
q	Real image	Virtual image
m	Erect image	Inverted image

☐ For a spherical mirror, $f = r/2$. A thin lens of refractive index n_2 and radii of curvature r and r', when immersed in a medium of refractive index n_1, has a focal length given by the lensmaker's equation

$$\frac{1}{f} = \left(\frac{n_2}{n_1} - 1\right)\left(\frac{1}{r} + \frac{1}{r'}\right)$$

☐ Images are easily located by drawing a principal-ray diagram: an incident ray parallel to the principal axis of a lens (mirror) will be refracted (reflected) so as to pass through a principal focus; an incident ray passing through a principal focus of a lens (mirror) will be refracted (reflected) so as to emerge parallel to the principal axis.

☐ The magnifying power (angular magnification) of a compound microscope is

$$M = m_o M_E \approx -\left(\frac{L}{f_o}\right)\left(\frac{0.25\ \text{m}}{f_E}\right)$$

where L is the tube length, m_o is the lateral magnification of the objective of focal length f_o, and M_E is the magnifying power of the eyepiece of focal length f_E.

Questions

1 For the normal eye, a good distance for distinct vision is 25 cm. What is the corresponding good distance from face to mirror while shaving?

2 Find the minimum height and position required for a wall mirror in order that a girl 1.60 m tall with eyes 6 cm below the top of her head can see herself from the top of her head to her shoes.

3   An object is placed in an arbitrary position between two mirrors which form a right angle. How many images will there be? Draw a ray diagram showing each image of a point object and a ray from the object to an observer's eye corresponding to each image seen by the eye.

4   Repeat Question 3 for the case when the mirrors form an angle of 60°.

5   A spherical concave shaving mirror has a radius of curvature of 30 cm:
(a) Locate and describe the image formed of a face that is 12 cm in front of the center of the mirror.
(b) Find the lateral magnification.

6   A spherical concave telescope mirror has a focal length of 4.00 m. Using the data of Appendix C, find the diameter of the image of the moon formed by this mirror.

7   Where should an object be placed in front of a spherical mirror of focal length f, in order that the image and the object distances be equal?

8   Sketch a graph showing the image distance q as a function of the object distance p, for a concave mirror of radius r.

9   A spherical concave mirror has a radius of 40 cm. Using both analytical and graphical methods, locate and describe the image when the object distance p is given by:
(a) $p = 40$ cm   (b) $p = 16$ cm

10 For what range of object distances is the image formed by a convex spherical mirror virtual? Explain.

11 A spherical convex mirror has a radius of 40 cm. Find the position, size and orientation of the image of an object that is 2.0 mm high, when the object distance p is given by:
(a) $p = 40$ cm   (b) $p = 16$ cm

12 A convex mirror has a radius of curvature with a magnitude of 40 cm. How far in front of this mirror should an object be placed in order to have an image one-half as tall as the object?

13 A converging lens has a focal length of 20 cm. Using both graphical and analytical methods, find the position, size, and orientation of the image of an object that is 2.0 mm high, for the following object distances p: 60 cm, 40 cm, 30 cm, 10 cm.

14 For a thin converging lens with focal length f, sketch graphs showing the image distance q and the lateral magnification m as functions of the object distance p. Examine the correspondence between these graphs and the information given in Figure 38-15 for each case considered in the figure.

15 Repeat Question 13 when the lens is a diverging lens with a focal length of magnitude 20 cm.

16 Repeat Question 14 for the case of a diverging lens, and refer to Figure 38-17.

17 A set of lenses is made from a material with a refractive index of 1.50, using surfaces that are either concave or convex with radii of curvature of magnitudes 5.0 cm and 10.0 cm. Consider all possible lenses that can be constructed; identify each lens as converging or diverging and find its focal length.

18 An object is placed 2.00 m from a screen. A converging lens with a focal length of 37.5 cm is used to form a real image on the screen. Find the points at which this lens can be placed and the lateral magnification for each case.

19 An object is placed a distance from a converging lens such that an image of exactly the same size as the object is formed. The focal length of the lens is 15 cm. Locate and describe the image.

20 Two hills whose summits are 1.0 km apart are photographed from a point 4.0 km removed along the perpendicular bisector of the line joining them. What is the distance separating the images of the two summits on the film if the camera's converging lens has a focal length of 20.0 cm?

21 A lamp and a screen are 3.0 m apart. If the image of the lamp on the screen is to be twice the size of the lamp, what should be the focal length of the converging lens employed?

22 An optical system consists of two converging lenses separated by 1.00 m along their common principal axis. Each lens has a focal length of 3.00 m. An object is placed 1.50 m in front of the first lens:
   (a) Draw a principal-ray diagram.
   (b) Locate and describe the image.

23 An object subtends an angle of 1.0×10^{-2} rad when viewed by the naked eye at a distance of 25 cm. What angle will be subtended at an observer's eye when this object is viewed through a microscope with a magnifying power of $10\times$?

24 The focal length of a simple microscope is 5.0 cm. What is the magnifying power when the image is formed 25 cm in front of an observer's eye?

25 For the simple microscope of the preceding question, what is the position of the object?

26 A compound microscope has an eyepiece with a focal length of 20 mm, an objective with a focal length of 16 mm, and a tube length of 160 mm. Find the approximate values of the lateral magnification of the objective, the magnifying power of the eyepiece, and the magnifying power of the compound microscope.

27 Show that in Figure 38-20, the accurate expression for the microscope's magnifying power is

$$ M = -\frac{q}{p}\left(\frac{0.25 \text{ m}}{f_E} + \frac{0.25 \text{ m}}{D}\right) $$

where p and q are the object and intermediate image distances, respectively, measured from the objective. D is the distance from the eyepiece (focal length f_E) to the final image.

28 In a compound microscope, the focal lengths of the objective and of the eyepiece are 15 mm and 20 mm, respectively. The distance between the lenses is 20 cm. An observer adjusts the microscope so that the final image is 25 cm from his eye:
   (a) Find the distance from the objective to the object.
   (b) Find the height of the final image, if the object height is 0.10 mm.
   (c) Evaluate the magnifying power using the approximate formula of Eq. 38.18 and also the accurate result of Question 27.

29 The focal length of the objective of a microscope is 5 mm and its diameter is 10 mm. Find the order of magnitude of the separation of two point objects that can just be resolved when they are illuminated by light with a wavelength of 6000 Å.

30 Answer the preceding question for the case when 2000-Å ultraviolet light is used (together with photographic methods of detection).

Supplementary Questions

S-1   Retrodirective reflectors used in highway signs contain an assembly of "corners," each constructed from three mutually perpendicular mirrors. Show that any ray that experiences three reflections in such a corner emerges in a direction exactly opposite to its original direction.

S-2   Consider a ray parallel to, but well above, the principal axis of a parabolic mirror. Show that the law of reflection implies that this ray passes exactly through the focal point.

S-3   Show that the spherical mirror formula $1/p + 1/q = 1/f$ agrees with the conclusions of Section 38.1 in the particular case when the mirror is a plane surface (with an infinite radius of curvature).

S-4   Show that when two thin lenses with focal lengths f_1 and f_2 are placed in contact, the focal length of the combination is given by

$$\frac{1}{f} = \frac{1}{f_1} + \frac{1}{f_2}$$

S-5   From what is called the Gaussian form of the thin-lens equation, $1/p + 1/q = 1/f$, deduce the Newtonian form

$$xx' = f^2$$

where the object distance $x = p - f$ and the image distance $x' = q - f$ are measured from the focal points of the lens rather than from the center of the lens.

S-6   Opticians use the term "power of a lens" and measure the "power" in a unit called the *diopter*:

$$power\ of\ lens\ in\ diopters = \frac{1}{focal\ length\ in\ metres}$$

Assume that the refractive power of the cornea–lens combination of an eye (Figure 38-19) is 50 diopters:
(a) What is the focal length of this system?
(b) If the print on a newspaper is held 30.0 cm from the eye, what is the height on the retina of the image of a 1.40-mm high newsprint letter?

S-7   Calculate the focal length of the lens of the eye (Figure 38-19). Assume that the radius of curvature of the anterior lens surface is 11.0 mm and that that of the posterior surface is 5.7 mm, while the refractive index of the lens is 1.4 and that of the material in the vitreous cavity and in the anterior chamber is about that of water (1.33).

S-8   Show that when a real image of a real object is formed by a converging lens

$$p + q \geq 4f$$

S-9   Figure 38-22 illustrates how two convex lenses can be arranged to make an astronomical telescope:
(a) Show that to a good approximation the angular magnification of a telescope can be expressed by the relation

$$M = \frac{focal\ length\ of\ objective}{focal\ length\ of\ ocular}$$

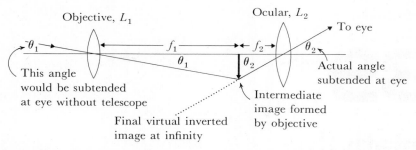

Figure 38-22   Angular magnification of a telescope is given by the ratio θ_2/θ_1, where θ_1 is seen to be the angle that the rays from a point of a distant object would subtend at the eye if no telescope were used, and θ_2 is the angle subtended at the eye because of the properties of the telescope.

   (b) If the two lenses used in such a telescope have focal lengths of 200 cm and 20 cm, which of the two lenses is the objective lens, and what is the magnification of the telescope?

S-10   The longitudinal magnification of a lens is defined as

$$m' = \frac{\Delta q}{\Delta p}$$

where Δp is the object's dimension parallel to the principal axis and Δq is the image's dimension in this same direction. Show that

$$m' = -m^2$$

where m is the lateral magnification.

39.

relativity, the lorentz transformations

The special theory of relativity originated from studies of the electromagnetic interaction and is now believed to apply to all types of interactions that have been investigated in particle, nuclear, and atomic physics. The theory requires modifications of Newtonian mechanics that are important at speeds of the order of magnitude of the speed of light. Many relativistic expressions for measurable quantities are among our best tested formulas and can now be presented as engineering facts instead of interesting speculation. Hundred-million-dollar particle accelerators are designed using the special theory of relativity, and these accelerators work.

Aside from its practical value for nuclear and particle physics, special relativity teaches valuable lessons about scientific theories. A theory such as Newtonian mechanics can work wonderfully in a certain domain of experience (low speeds) and yet fail utterly in other circumstances (speeds near c). When we are forced to revise our theories, experimental results are the guide and also the touchstone of truth. However, we sometimes find it fruitful to generalize and postulate on the basis of flimsy experimental evidence, and so construct tentative theories to be tested.

In this process, and generally in thinking about nature, we cannot always trust common sense. Common sense itself derives from a very limited range of experience. When we explore new realms there may be real surprises. It has turned out that at high speeds things are *not* more or less as one might expect, that is, just faster. And later in studying quantum physics we discover that very small objects display a most astonishing behavior; an electron is not merely a small version of a charged billiard ball.

39.1 Relativity of Time

Frames in relative motion were treated in Section 6.5 from a viewpoint adequate for Newtonian mechanics. We return briefly to this discussion to introduce terminology and equations that will serve as a point of departure for this chapter.

Figure 39-1 gives a concrete example of two frames in relative motion. Frame S is fixed in the highway and frame S' is fixed on a bus. The bus travels relative to the highway with an x component of velocity V. Assume that, as the two origins coincide, a stopwatch is started on the bus giving readings denoted by t'. Another stopwatch with readings denoted by t is started by a stationary police officer on the highway. Now suppose a baseball is rolling forward in the bus and squashes first an ant (event A) and a little later a bug (event B). An *event* in physics is regarded as completely specified in one frame by answers to two questions: *Where? When?* So event A (Figure 39-1) is specified by x'_A, y'_A, z'_A, t'_A for those who use the bus as a frame of reference and by x_A, y_A, z_A, t_A for those who, like the stationary police officer, use the highway as their frame. From Figure 39-1 common sense leads us to assert that

$$x_A = x'_A + Vt'_A \qquad y_A = y'_A \qquad z_A = z'_A$$

$$t_A = t'_A$$

Writing similar equations for event B and subtracting, we obtain the relationships for the intervals between events B and A:

$$\Delta x = \Delta x' + V\Delta t' \qquad \Delta y = \Delta y' \qquad \Delta z = \Delta z' \tag{39.1}$$

$$\Delta t = \Delta t' \tag{39.2}$$

where $\Delta t = t_B - t_A, \Delta x = x_B - x_A, \Delta x' = x'_B - x'_A, \Delta t' = t'_B - t'_A$. Equations 39.1 and 39.2 are called *Galilean coordinate transformations*.

Figure 39-1 Frame S' moves relative to frame S with an x component of velocity V.
$x_A = x'_A + Vt'_A$.

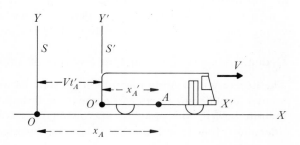

The equation $\Delta t = \Delta t'$ was tacitly assumed true by everyone up to the year 1905. This equation expresses our faith in an *absolute time*, a time which is the same for all frames of reference.

Division of Eq. 39.1 by Eq. 39.2 gives the relationship between the baseball's velocity **v** relative to S and **v'** relative to S',

$$v_x = v_x' + V \qquad v_y = v_y' \qquad v_z = v_z' \tag{39.3}$$

a relationship known as the *Galilean velocity transformation*. This is Eq. 6.10, **v** = **v'** + **V**, the triangle of velocities that is the basis for the solution of all the relative velocity problems in Chapter 6.

Failure of Galilean Velocity Addition at the Speed of Light

If Maxwell's equations are valid in a certain frame of reference, then the discussion in Section 35.1 shows that light or any electromagnetic wave propagates with a speed c relative to that frame. What is the frame relative to which light has a speed c? Experiments give the surprising answer that the *speed of light is the same in all inertial frames of reference*, regardless of their relative motion.

The most famous experiment in this connection was performed in 1887 by A. A. Michelson (1852–1931) and E. W. Morley (1838–1923). The interference pattern formed by two light beams which had traveled over perpendicular paths was used to indicate whether the beams traveled with different speeds. Their experiment failed to discover any change in the speed of light in a laboratory because of a change in the earth's orbital velocity vector as the earth moves around the sun. Viewed from the astronomical frame, the earth's velocity vector **V** changes from about 3×10^4 m/s in one direction to the same speed in the opposite direction during a six-month interval.

The Michelson–Morley experiment and all other measurements are consistent with the interpretation that, in a vacuum, electromagnetic waves have the same speed c relative to all inertial frames. According to Eq. 39.3 if $v_x' = c$, then $v_x = c + V$. Since experiments give $v_x = c$, the Galilean velocity transformation fails at this particular speed. What is the explanation?

Einstein developed a fruitful approach to this problem in 1905. Instead of trying to explain the fact that light has the same speed in all inertial frames, he decided to simply accept this as a fact or a postulate and then investigate the consequences. Taking a critical look at the Galilean transformation equations and realizing that there was no direct experimental evidence for their validity at very high speeds, Einstein was struck with the idea that our commonly accepted notion of time is suspect. Perhaps the innocent equation $\Delta t = \Delta t'$ is wrong and time elapses differently in different inertial frames!

Gedankenexperiment and Our Notions of Time

The necessity of revising our notions of time because light has the same speed relative to different inertial frames can be brought out using one of Einstein's favorite devices: the thought experiment (*Gedankenexperiment* in German). Consider a long fast train (Figure 39-2) moving down a track with a speed V. Bombs,

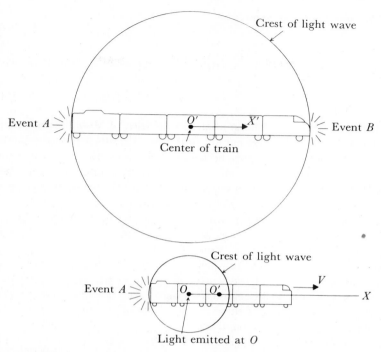

Figure 39-2   Upper figure shows events A and B which are simultaneous relative to the frame S' fixed in the train. Lower figure shows that event A occurs before the light reaches the locomotive, relative to the frame S fixed in the track.

set to explode when they receive a flash of light, are carried in the caboose and in the locomotive. A light signal is flashed from the center of the train and eventually the caboose explodes (event A) and the locomotive explodes (event B). People on the train will agree that these explosions are *simultaneous*, since the light travels with the same speed toward the caboose as toward the locomotive and has the same distance to go. In terms of train time t' this means that

$$t'_A = t'_B$$

or

$$\Delta t' = t'_B - t'_A = 0$$

But anyone parked by the railway track will have a different record of the proceedings. The caboose, moving toward the light signal, will have advanced somewhat before it explodes at a track time t_A. The light, traveling at the same speed c forward, must go farther before it reaches the locomotive which therefore explodes at a later time t_B. We have

$$t_B > t_A$$

$$\Delta t = t_B - t_A > 0$$

Therefore, for these two events,

$$\Delta t \neq \Delta t'$$

and Eq. 39.2, $\Delta t = \Delta t'$, is *not* correct! The conclusion is that *events which are simultaneous in one frame* ($\Delta t' = 0$) *are not simultaneous in another frame* ($\Delta t \neq 0$), *when the frames are in relative motion.* We have encountered what is known as the *relativity of simultaneity.*

With this conclusion we must abandon the idea of absolute time expressed by the equation $\Delta t = \Delta t'$. We cannot simply say that the caboose exploded 2 s before the locomotive. This may be true for the time used in a frame moving relative to the train with a certain velocity, but in a different frame the time interval between the same two events might be 1.0 s, in another frame 0 s (simultaneous), in yet another frame the locomotive might explode first. We must replace the idea of absolute time, one time t for all observers regardless of the frame, by a different time for each different frame of reference. We must use a train time t' as well as a track time t.*

In thinking about the physics we are now encountering, we should consider the implications of some of the words and phrases we use in daily life. The notion of absolute time lurks behind such phrases as *"the" time, when, simultaneous, before, after.* With each one of these words we should, mentally at least, add a specification of the frame of reference.

Since it is no trivial step to give up the simple concept of absolute time, it is worthwhile to contemplate, from as many viewpoints as possible, why this renunciation is necessary. It is instructive to think through the situation pictured in Figure 39-3. A lamp is switched on in the roof of the bus (event A) and light travels a distance $\Delta y'$ to strike the floor vertically below the lamp (event B). The distance $\Delta y'$ equals $c\Delta t'$ where $\Delta t'$ is the bus–time interval between these two events, A and B. For these events, $\Delta t' = t'_B - t'_A$, and also $\Delta x' = x'_B - x'_A = 0$ since the events occur at the same x' coordinate in the bus. The unprimed coordinates for the same two events in the highway frame are pictured in Figure 39-3(b). We assume that the bus height h is the same in the two frames, $h = \Delta y = \Delta y' = c\Delta t'$. The bus advances a distance $V\Delta t$ while the light is in transit from the bus roof to the floor and, since in this highway frame the light goes a longer distance at the same speed c, there is a longer time interval Δt between events A and B. That is, Δt is greater than $\Delta t'$: time elapses differently in the two frames.

---

*Fortunately, it is possible to have just one time t' for one entire frame of reference, and there are no peculiarities encountered with time as long as we confine our attention to measurements in this one frame. The time that we assign to an event is always the time on the clock of an observer *located at the event* and at rest in the frame in question. A passenger on the train near the caboose *sees* the explosion of the caboose at time t'_A and *sees* the explosion of the distant locomotive at the later time $t'_B + (\Delta x'/c)$, where $\Delta x' = x'_B - x'_A$. Nevertheless this passenger, knowing that light from the locomotive explosion takes the time $\Delta x'/c$ to reach him, records the event B as having occurred at time t'_B. He finds that his recorded times t'_B and t'_A are equal and so says the events were simultaneous, even though he saw them at different times. *The time we assign to an event is always this recorded time, the time that an observer on the spot would see the event, never the time that the event would be seen by some distant observer.*

The relationship between time intervals, from Figure 39-3(b) is

$$(c\Delta t)^2 = (c\Delta t')^2 + (V\Delta t)^2$$

which yields

$$\Delta t = \frac{\Delta t'}{\sqrt{1 - V^2/c^2}} \tag{39.4}$$

The graph of the factor $1/\sqrt{1 - V^2/c^2}$ is shown in Figure 39-4 and several values are given in Table 39-1. Notice that, unless V is large enough to be an

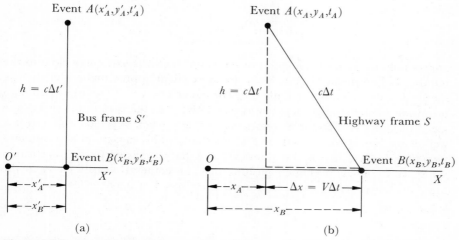

Figure 39-3   Between the events A and B there is a time interval $\Delta t'$ in the bus frame S' and Δt in the highway frame S. The right-hand figure shows $c\Delta t$ is greater than $c\Delta t'$. V is the x component (the other components are zero) of the velocity of the bus relative to the highway.

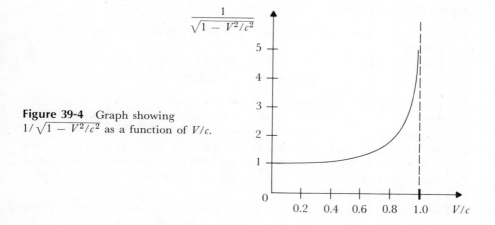

Figure 39-4   Graph showing $1/\sqrt{1 - V^2/c^2}$ as a function of V/c.

$\Delta x = \gamma(\Delta x' + \beta c\,\Delta t')$

$c\Delta t = \gamma(c\Delta t' + \beta\Delta x)$

$\Delta x'^2 + \Delta y'^2 + \Delta z'^2 - c^2(\Delta t)^2 =$

$= \gamma^2(\Delta x'^2 + 2\beta c\Delta x'\Delta t' + \beta^2 c^2\Delta t'^2 + \Delta y'^2 + \Delta z'^2 - \gamma^2(c^2\Delta t'^2 + 2\beta c\Delta t'\Delta x' + \beta^2\Delta x'^2$

$= \gamma^2(1-\beta^2)\Delta x'^2 + \Delta y'^2 + \Delta z'^2 - c^2\Delta t^2\gamma^2(\beta^2-1)$

$\gamma^2 = \dfrac{1}{1-\beta^2}$

$\gamma^2(1-\beta^2) = 1$

$= \Delta x'^2 + \Delta y'^2 + \Delta z'^2 - c^2\Delta t^2$

Table 39-1

V/c	$\dfrac{1}{\sqrt{1 - V^2/c^2}}$
0.01	1.00005
0.10	1.005
0.50	1.15
0.60	1.20
0.70	1.40
0.80	1.67
0.90	2.3
0.98	5.0
0.99	7.1
0.9999	71

appreciable fraction of the speed of light, this factor is extremely close to the number 1. This means that $\Delta t = \Delta t'$ to an excellent approximation for the low speeds that we encounter in daily life. The factor $1/\sqrt{1 - V^2/c^2}$ gives a smooth transition to the strange new world of relativity as the speed approaches the speed of light. This factor increases very rapidly when V is close to c but is a well-defined real number only if V is less than c. This suggests that c is nature's speed limit. In fact, there is abundant evidence that no massive particle can attain the speed c, but with enough pushing one can come arbitrarily close. In the Stanford linear accelerator, electrons reach a speed of $0.9999999997c$.

39.2 Lorentz Transformations

After pondering such train and bus "experiments" we are driven to consider modifications of the Galilean coordinate transformations that will replace the Galilean velocity transformation rule by a rule that transforms $v' = c$ into $v = c$. We shall try to find new transformation equations that are in accord with Einstein's two postulates:

1   The principle of relativity: *The laws of nature are the same in all inertial frames of reference.*
2   The invariance of the speed of light: *The speed of light is the same in all inertial frames, regardless of the motion of the source of light.*

In the first postulate, the principle of relativity is asserted to apply to *all natural phenomena*, not to just the laws of mechanics. The essential point for our immediate attention is that all inertial frames are equivalent. There is no preferred frame with special properties that render nature's laws different from those found in any other frame. In particular we wish to have transformation equations that imply that the frame S is equivalent to the frame S' for the description of nature.

For speeds small compared to c the Galilean transformations are in agreement with experiment, so any new equations must differ from these only for high speeds. As a trial, let us assume that space and time intervals between two given events transform according to modified equations of the form*

$$\Delta x = \gamma(\Delta x' + V\Delta t') \qquad \Delta y = \Delta y' \qquad \Delta z = \Delta z' \tag{39.5}$$

and

$$\Delta t = \gamma\left(\Delta t' + \frac{\beta\Delta x'}{c}\right) \tag{39.6}$$

where the coefficients γ and β are assumed to have values that are independent of the space and time intervals. We shall try to select values of these coefficients in such a way that the speed of light is the same relative to both S and S'.

Applying these equations to the two events of Figure 39-3 for which $\Delta x' = 0$, we find

$$\Delta t = \gamma\Delta t'$$

In our previous analysis of the events in Figure 39-3, the assumption that light had the same speed c relative to both frames led to

$$\Delta t = \frac{\Delta t'}{\sqrt{1 - V^2/c^2}}$$

Comparison of these two expressions relating Δt and $\Delta t'$ shows that we must select the value

$$\gamma = \frac{1}{\sqrt{1 - V^2/c^2}} \tag{39.7}$$

It remains to determine β. To do this we examine the velocity transformation law implied by the new coordinate transformation equations. Dividing the first of Eq. 39.5 by Eq. 39.6, we obtain

$$\frac{\Delta x}{\Delta t} = \frac{\Delta x' + V\Delta t'}{\Delta t' + \beta\Delta x'/c} = \frac{\Delta x'/\Delta t' + V}{1 + \beta\Delta x'/\Delta t'c}$$

which is

$$v_x = \frac{v_x' + V}{1 + \beta v_x'/c} \tag{39.8}$$

Now let us see if we can select the value of β so that $v_x' = c$ implies that $v_x = c$. From Eq. 39.8 this condition requires that β satisfy

$$c = \frac{c + V}{1 + \beta} \qquad \text{or} \qquad 1 = \frac{1 + V/c}{1 + \beta}$$

Therefore

$$\beta = \frac{V}{c} \tag{39.9}$$

---

*The proof that a completely general linear transformation must be of this form is outlined in Question S-1.

The coefficients in Eq. 39.5 and 39.6 have now been determined. Results are summarized in Table 39-2. The new transformation equations for space and time intervals are called *Lorentz transformations*. These equations have been written with the factor c multiplying each time interval because this form is algebraically the most convenient. The new velocity transformations are the *Einstein velocity transformations*. Table 39-2 also includes the rules for transforming the transverse components, v_y and v_z, which the student is asked to establish in Question 8.

Examination of these new transformation equations shows that they have the following satisfactory features:

1   When the magnitude of V is much less than c, the factor $1/\sqrt{1 - V^2/c^2}$ is almost the number 1 (see Table 39-1), and V/c is nearly zero. In these circumstances, the first of the Lorentz transformation equations is very nearly the same as $\Delta x = \Delta x' + V\Delta t'$, and the last is approximately $\Delta t = \Delta t'$ (for $\Delta x'$ small compared to $c^2 \Delta t'/V$). If the particle speed and the relative speed of the frames are both small compared to c, the Einstein velocity transformations are approximated by $v_x = v'_x + V$, $v_y = v'_y$, $v_z = v'_z$. Consequently, in the realm of low speeds where the Galilean transformation equations are in agreement with experiment, the new transformation equations accord with the old.

2   Although it is not obvious, we can verify that these new equations are consistent with the principle of relativity. The frames S and S' are equivalent. The Lorentz transformation equations, when solved to express $\Delta x'$ and $\Delta t'$ in terms of Δx and Δt, take exactly the same mathematical form:

$$\Delta x' = \gamma(\Delta x - \beta c\Delta t) \tag{39.10}$$

$$c\Delta t' = \gamma(c\Delta t - \beta\Delta x) \tag{39.11}$$

Notice that these *inverse* Lorentz transformations can be obtained from the *direct* Lorentz transformations of Table 39-2 simply by interchanging primed and unprimed quantities and by replacing β by $-\beta$, which is as it should be. Since βc is the x component (the other components are zero) of the velocity of S' relative to S, the x component of the velocity of S relative to S' is $-\beta c$.

Table 39-2 Transformation Equations of Special Relativity

Lorentz transformations:

$$\Delta x = \gamma(\Delta x' + \beta c\Delta t') \qquad \Delta y = \Delta y' \qquad \Delta z = \Delta z'$$

$$c\Delta t = \gamma(c\Delta t' + \beta\Delta x')$$

where $\beta = V/c$ and $\gamma = 1/\sqrt{1 - V^2/c^2}$.

Einstein velocity transformations:

$$v_x = \frac{v'_x + V}{1 + v'_x V/c^2} \qquad v_y = \frac{v'_y\sqrt{1 - V^2/c^2}}{1 + v'_x V/c^2} \qquad v_z = \frac{v'_z\sqrt{1 - V^2/c^2}}{1 + v'_x V/c^2}$$

---

3   The Einstein velocity transformations do imply that, for any two frames with relative velocity V, a speed c relative to one frame corresponds to a speed c relative to the other frame.

Another significant property of the Einstein velocity transformation equations is that velocities **V** and **v'** which have magnitudes less than c, combine to give a velocity **v**, which also has a magnitude less than c. In other words, we never obtain a violation of nature's speed limit c simply by evaluating a speed relative to a different frame of reference. This is illustrated in the following example.

Example 1   A proton moves to the right with a speed of $0.75c$ relative to the laboratory, and an electron moves to the right with a speed of $0.80c$ relative to the proton. What is the speed of the electron relative to the laboratory?

Solution   The answer is not just $0.75c + 0.80c = 1.55c$, which would exceed nature's speed limit. The Galilean transformation $v_x = v_x' + V$ has been replaced by the Einstein velocity transformations. Let the frame S be attached to the laboratory, and take the proton as the origin O' of the frame S'. Then the velocity V of S' relative to S is the proton's velocity, $V = 0.75c$. The electron's velocity relative to S' is $v_x' = 0.80c$. Relative to the laboratory the electron has a velocity given by

$$v_x = \frac{v_x' + V}{1 + v_x' V/c^2} = \frac{0.80c + 0.75c}{1 + (0.80 \times 0.75)} = 0.97c$$

which is less than c.

$$v = \frac{V' + V}{1 + \frac{V'V}{c^2}} = \frac{(0.90 + 0.80)/c}{1 + (0.90 \times 0.80)} = \frac{1.70c}{1.72} = 0.988\,c \text{ east}$$

39.3 Consequences of the Lorentz Transformations

Relativity of Simultaneity

The most novel feature introduced by the Lorentz transformations is the replacement of the equation $\Delta t = \Delta t'$ by $c\Delta t = \gamma(c\Delta t' + \beta\Delta x')$. Instead of a universally agreed upon time interval between two events, we have space and time intervals scrambled up so that the space and time intervals in one frame determine the time interval in another. We can now account for the relativity of simultaneity experienced with the exploding caboose and locomotive. For these two events, although they are simultaneous in the train time so that $\Delta t' = 0$, we see from the Lorentz transformations that because they occurred with a spatial separation $\Delta x'$, they will be separated by the time interval

$$\Delta t = \gamma\left(0 + \frac{\beta\Delta x'}{c}\right) = \frac{\gamma\beta\Delta x'}{c}$$

as measured by track time.

Time Dilation

A most famous effect and one for which we have abundant experimental confirmation is what is called *time dilation*. A clock at rest in the bus of Figure 39-5 measures some time interval $\Delta t'$ between the event "tick" and the event "tock" and $\Delta x' = 0$ for these two events which are both located at the clock. The Lorentz transformations now give

$$\Delta t = \gamma(\Delta t' + 0) = \gamma\Delta t' = \frac{\Delta t'}{\sqrt{1 - V^2/c^2}} \tag{39.12}$$

For any V this implies Δt (highway frame) $> \Delta t'$ (bus frame) so, according to highway time, the bus clock is running slow. But this does not mean that bus passengers would find that highway clocks run fast. Highway clocks move relative to the bus, and if our frames are really equivalent for the description of nature, bus passengers must find that the highway clocks run slow. The student is asked in Question 15 to verify that this happens. The situation is summarized by the statement: moving clocks run slow.

The fact that the time really is dilated or stretched out in this way is demonstrated by short-lived unstable particles such as pions. At rest pions disintegrate after an average life of about 10^{-8} s. When traveling through a laboratory at 99% of the speed of light (which corresponds to $1/\sqrt{1 - V^2/c^2} = 7.1$), the average life of pions is expected to be increased by a factor of 7.1. This amazing prediction has been confirmed in detail by many experiments.

Length Contraction of a Moving Object

The Lorentz transformations imply that measurement of the dimensions of a moving object will reveal a *contraction in the dimension parallel to the direction of the object's velocity.* Suppose that while traveling in the bus we pass a parked car. Relative to the bus frame S', the car is a moving object. In the bus frame we

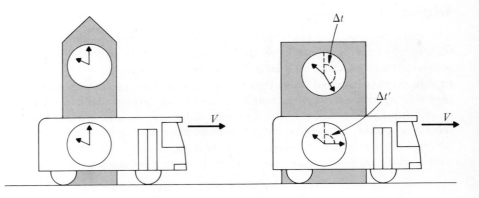

Figure 39-5   Relative to the highway frame, the clock on the moving bus runs slow, that is, $\Delta t'$ is less than Δt.

call the car's length L the difference $\Delta x' = x'_B - x'_A$ of the coordinates of its extremities measured at the *same time relative to the bus frame S'*, so $\Delta t' = t'_B - t'_A = 0$. In the highway frame, the parked car has a length $L_{rest} = \Delta x = x_B - x_A$, called its *rest length*. Because the car is at rest in this frame, we can determine its length from measurements of the coordinates x_A and x_B of its extremities, even though these measurements are made at different times t_A and t_B. The first of the Lorentz transformation equations of Table 39-2, applied to events A and B, gives

$$L_{rest} = \Delta x = \gamma(\Delta x' + 0) = \frac{L}{\sqrt{1 - V^2/c^2}} \qquad (39.13)$$

Therefore L is less than L_{rest}. Bus passengers' measurements show that the car is shorter than the manufacturer's specifications. Of course, the effect is reciprocal since the frames are equivalent, and the car's owner will claim that the bus length is less than advertised.

Transverse Dimensions of Moving Objects

The discovery that there is a contraction in the dimension parallel to the velocity of a moving object brings into question the transverse dimensions. In the *Gedankenexperiment* dealing with the light signal emitted from the ceiling of the bus, we assume that the bus height h is the same in both frames of reference. We can show that this assumption is required if the frames are to be equivalent. Suppose the height of the bus were contracted for measurements made in the highway frame S. Then the equivalence of the frames demands that the height of an object like a tunnel would be contracted for measurements made in the bus frame S'. The highway police officer would find the low bus roof did not touch the top of the tunnel but the bus driver would discover the tunnel ceiling was so low that it knocked off his roof (Figure 39-6). This is indeed a genuine contradiction, so we conclude that transverse dimensions like the bus height are the same in both frames. This conclusion is expressed by the transformation equations $\Delta y = \Delta y'$ and $\Delta z = \Delta z'$ in Table 39-2.

The Invariant Interval

So far we have seen that measurements of a car's length give different answers in different frames of reference. Time intervals between two given events are different in different frames of reference. In this treacherous state of affairs it is useful to find entities that are *invariant*. A quantity is called *invariant* if it has the same value in all inertial frames. Although lengths and time intervals are *not invariant*, the Lorentz transformations imply (Question S-9) that for any two events A and B,

$$(\Delta x)^2 + (\Delta y)^2 + (\Delta z)^2 - c^2(\Delta t)^2 = (\Delta x')^2 + (\Delta y')^2 + (\Delta z')^2 - c^2(\Delta t')^2 \qquad (39.14)$$

This result shows that the combination of a distance and a time interval $(\Delta x)^2$

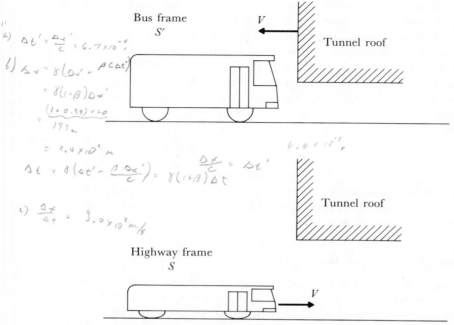

a) $\Delta t' = \frac{\Delta x'}{c} = 6.7 \times 10^{-7} s$

b) $\Delta x = \gamma (\Delta x' + \beta c \Delta t')$

$= \gamma (1 - \beta) \Delta x'$

$= \frac{(1 + 0.98) \times 20}{199 u}$

$= 2.0 \times 10^{2} m$

$\Delta t = \gamma (\Delta t' + \frac{c}{c} \frac{\Delta x'}{c}) = \gamma (1 + \beta) \Delta t$ $\frac{\Delta t}{c} = \Delta t'$ $\Delta t'$

$6.6 \times 10^{-7} s$

c) $\frac{\Delta x}{\Delta t} = 3.0 \times 10^{8} m/s$

Figure 39-6 The assumption that transverse dimensions of moving objects are reduced leads to a contradiction.

$+ (\Delta y)^2 + (\Delta z)^2 - c^2(\Delta t)^2$ is invariant. This important invariant, called the *relativistic interval* between two events, emphasizes that space intervals and time intervals should be considered together to get quantities that retain their significance in different inertial frames. This idea was pursued by H. Minkowski (1864–1909), who demonstrated the convenience of thinking in terms of a four-dimensional *space-time* (three spatial dimensions plus one more for time), rather than in terms of *space* and of *time* separately.

Proper Time

A proper time interval $\Delta\tau$ for a moving particle is a time interval that would be measured by a clock traveling with the particle. Consider two events that both occur at the location of the particle. In a frame S' in which the particle is instantaneously at rest, $\Delta x' = 0, \Delta y' = 0, \Delta z' = 0, \Delta t' = \Delta\tau$. Then the invariant relativistic interval (Eq. 39.14) is

$$(\Delta x)^2 + (\Delta y)^2 + (\Delta z)^2 - (c\Delta t)^2 = 0 + 0 + 0 - (c\Delta\tau)^2$$

The proper time interval between the two events is given in terms of the particle's speed v relative to S by

$$\Delta\tau = \Delta t \sqrt{1 - \frac{1}{c^2}\left[\left(\frac{\Delta x}{\Delta t}\right)^2 + \left(\frac{\Delta y}{\Delta t}\right)^2 + \left(\frac{\Delta z}{\Delta t}\right)^2\right]} = \Delta t \sqrt{1 - \frac{v^2}{c^2}} \quad (39.15)$$

This equation shows the time dilation effect already discussed. The method of derivation brings out the fact that *the proper time interval is a Lorentz invariant quantity.* This fact is used in the next chapter to define time derivatives that transform in a simple way.

Using the Lorentz Transformations

The Lorentz transformations relate, for two given events, space and time intervals measured in frame S to space and time intervals measured in frame S', which moves relative to S with an x component of velocity V. Although certain questions about the relationships of measurements made in different frames can be settled by direct use of the formulas for time dilation or length contraction, a procedure which uses the Lorentz transformations themselves is straightforward and less prone to error. To illustrate this procedure we consider the events in the long fast train of Section 39.1.

Example 2   The hypothetical train in Figure 39-2 has a rest length of 600 m and travels in the positive x direction with a speed of $0.800c$ relative to the track. As the train passes a stationary police officer, the conductor pushes a switch (event P) which causes a light signal to be emitted from the center of the train. This light signal triggers explosions in the caboose (event A) and in the locomotive (event B). Find the space and time intervals between these events A, B, and P for measurements made relative to a frame S' fixed in the train and then for a frame S fixed in the tracks.

Solution   First we confine our attention to the description of events in S'. Relative to this frame, each event is specified by a position coordinate x' and a time t'. We have

$$x'_B - x'_P = \tfrac{1}{2}(\textit{rest length of train}) = 300 \text{ m}$$

The time it takes for the light signal to travel from x'_P to x'_B is

$$t'_B - t'_P = \frac{x'_B - x'_P}{c} = \frac{300 \text{ m}}{3.00 \times 10^8 \text{ m/s}} = 1.00 \times 10^{-6} \text{ s}$$

Similarly, $x'_A - x'_P = -300$ m and $t'_A - t'_P = 1.00 \times 10^{-6}$ s. For the events A and B, these results give $t'_B - t'_A = 0$ which means events A and B are simultaneous in the frame S'. The spatial separation of events A and B is the rest length of the train:

$$x'_B - x'_A = 600 \text{ m}$$

These simple calculations emphasize that we encounter no peculiarities as long as we deal with measurements in one frame of reference.

We now describe these events using the frame S fixed in the track. Relative to this frame, each event is specified by a position coordinate x and a time t. Frame S' has a velocity $0.800c$ relative to S. From Table 39-1, the corresponding value of the factor γ is 1.67 (or $\tfrac{5}{3}$). The Lorentz transformation

$$\Delta x = \gamma(\Delta x' + \beta c \Delta t')$$

applied to events B and P for which $\Delta x' = 300$ m and $c\Delta t' = (3.00 \times 10^8$ m/s)$(1.00 \times 10^{-6}$ s$) = 300$ m, gives

$$x_B - x_P = 1.67(300 \text{ m} + 240 \text{ m}) = 900 \text{ m}$$

The locomotive explodes when it is 900 m past the police officer. The Lorentz transformation

$$c\Delta t = (c\Delta t' + \beta \Delta x)$$

applied to the same two events B and P gives

$$c(t_B - t_P) = 1.67(300 \text{ m} + 240 \text{ m}) = 900 \text{ m}$$

$$t_B - t_P = 3.00 \times 10^{-6} \text{ s}$$

Measurements made in the frame fixed in the tracks show that the locomotive explodes 3.00×10^{-6} s after the conductor sends out the signal. Recall by way of contrast that, for measurements made in the frame fixed in the train, this explosion occurred 1.0×10^{-6} s after the signal was sent.

For the events A and P we find

$$x_A - x_P = 1.67(-300 \text{ m} + 240 \text{ m}) = -100 \text{ m}$$

$$c(t_A - t_P) = 1.67(300 \text{ m} - 240 \text{ m}) = 100 \text{ m}$$

$$t_A - t_P = 0.33 \times 10^{-6} \text{ s}$$

Therefore, measurements made in a frame fixed in the tracks show that the caboose explodes 100 m to the left of the police officer (see the lower portion of Figure 39-2) at a time 0.33×10^{-6} s after the signal is sent from the midpoint of the train. Notice that these results are consistent with the fact that the light signal travels at a speed of 3.00×10^8 m/s relative to the tracks.

For the events A and B, using the intervals just calculated, we find

$$x_B - x_A = 900 \text{ m} + 100 \text{ m} = 1000 \text{ m}$$

$$t_B - t_A = (3.00 \times 10^{-6} \text{ s}) - (0.33 \times 10^{-6} \text{ s}) = 2.67 \times 10^{-6} \text{ s}$$

Measurements in the frame fixed in the tracks give the answer that the locomotive exploded 2.67×10^{-6} s after the caboose exploded, at a location 1000 m to the right of the spot where the caboose exploded. These explosions are not simultaneous in S although they are simultaneous in S'.

The results summarized in Table 39-3 assume that origins are selected and watches set in such a way that $x_P' = 0$, $t_P' = 0$, $x_P = 0$, $t_P = 0$.

Table 39-3 Position Coordinates and Times of Events A and B in Different Frames

Event	Measurements in S'		Measurements in S	
	x'	t'	x	t
A	-300 m	1.0×10^{-6} s	-100 m	0.33×10^{-6} s
B	300 m	1.0×10^{-6} s	900 m	3.00×10^{-6} s

There is another point to be mentioned in connection with Figure 39-2. The length contraction formula implies that the length of the moving train will be contracted to 600 m/1.67 = 360 m when measured in the frame fixed in the tracks. This means that at the time t_A on clocks in the frame S, the front of the locomotive is merely 360 m ahead of the exploding caboose, as represented in the lower diagram of Figure 39-2. A word of caution is in order here. In 1959 James Terrell pointed out that such pictures, which show a shortening of longitudinal dimensions of moving objects, do not correctly represent the *visual appearance*. Such pictures give the positions of the different moving parts at one instant of time in the observer's frame. However, the visual appearance is determined by rays that reach the observer's eye simultaneously, and such rays must leave different parts of an object at *different times* in the observer's frame.

39.4 Doppler Effect in Light

When a light source is moving relative to an observer, there is a shift in the observed frequency analogous to the Doppler shifts of sound waves that are discussed in Section 34.6. For light waves the effect has particular significance because astronomers' measurements of Doppler shifts have yielded so much information about the universe—information that is as astounding as anything encountered in science.

The special theory of relativity must be used to determine the Doppler effect for light waves. We consider a light source which produces waves of frequency f and wavelength λ in a frame of reference in which the source is at rest. If the light source is approaching an observer, then the observer will measure a higher frequency f_o and a shorter wavelength λ_o. The exact mathematical relationship between f_o and f has a quite different form than that for the case of sound waves because there are significant differences in the underlying physics. For light waves in a vacuum, there is *no* transmitting medium which provides a special frame of reference. Light travels with a speed c relative to *any* inertial frame, while sound travels with a speed v relative to the transmitting medium but has a different speed relative to an observer moving through the medium. Only the velocity of the light source *relative* to the observer has physical significance, in contrast to the case for sound waves where source and observer velocities relative to the transmitting medium determine the frequency shift.

To analyze the effect we consider the observer to be fixed in frame S and the source fixed in frame S' which is approaching S along the negative X axis with a speed v_s. In the frame S, the period of vibrations of the source is T_o. Exactly as deduced in Section 34.6 in connection with Figure 34-12, the wavelength λ_o in front of the source, as measured in S is

$$\lambda_o = (c - v_s)T_o$$

The speed of light waves relative to the observer is always c and he therefore observes a frequency

$$f_o = \frac{c}{\lambda_o} = \frac{c}{(c - v_s)T_o} \tag{39.16}$$

In the frame S', the source is at rest and its period is the proper time interval $\tau = 1/f$. From Eq. 39.15

$$T_o = \frac{\tau}{\sqrt{1 - v_s^2/c^2}}$$

Substituting this result in Eq. 39.16 and using $\tau = 1/f$ we obtain

$$f_o = \frac{c\sqrt{1 - v_s^2/c^2}}{(c - v_s)}f = \sqrt{\frac{c + v_s}{c - v_s}}f \tag{39.17}$$

The corresponding expression for the observed frequency when the source is receding is

$$f_o = \sqrt{\frac{c - v_s}{c + v_s}}f \tag{39.18}$$

If the source recession speed v_s is small compared to c, then $T_o \approx \tau$, and the Doppler shift can be calculated most easily by noting that $\Delta\lambda = v_s T_o \approx v_s\tau \approx v_s\lambda/c$ or

$$\frac{\Delta\lambda}{\lambda} \approx \frac{v_s}{c} \tag{39.19}$$

When the source is receding, the observer measures a longer wavelength, $\lambda_o = \lambda + \Delta\lambda$. Visible light is thus shifted toward the red.

Example 3 Spectrographic analysis (Figure 39-7) of light coming from a galaxy in the constellation Ursa Major reveals a line in the calcium absorption spectrum with a wavelength of 4.17×10^3 Å. In the laboratory, with the absorbing calcium atoms more or less at rest, this absorption line corresponds to a wavelength of 3.97×10^3 Å. Assuming that the difference in wavelength is a Doppler shift, what is the velocity of this galaxy relative to the earth?

Solution The "red shift" is

$$\Delta\lambda = (4.17 \times 10^3 \text{ Å}) - (3.97 \times 10^3 \text{ Å}) = 0.20 \times 10^3 \text{ Å}$$

Then from Eq. 39.19,

$$\frac{galaxy\ recession\ speed}{c} = \frac{0.20 \times 10^3 \text{ Å}}{3.97 \times 10^3 \text{ Å}} = 0.050$$

This red shift corresponds to a speed of recession which is 5.0% of the speed of light.

An extraordinary conclusion emerges from similar observations of red shifts in the spectrum of the light coming from different galaxies (Figure 39-7). Practically all galaxies seem to be receding from the earth. The most distant galaxies recede with the greatest speeds. This remarkable discovery was made in the

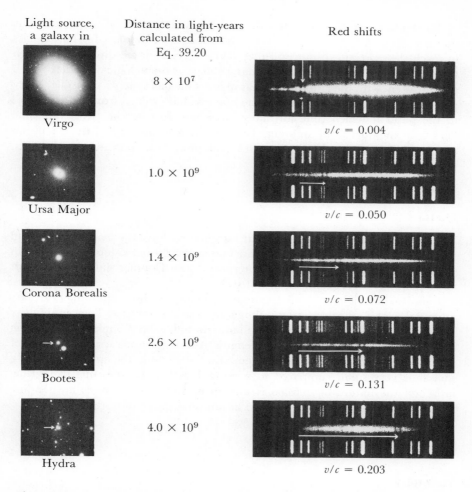

Light source, a galaxy in	Distance in light-years calculated from Eq. 39.20	Red shifts
Virgo	8×10^7	$v/c = 0.004$
Ursa Major	1.0×10^9	$v/c = 0.050$
Corona Borealis	1.4×10^9	$v/c = 0.072$
Bootes	2.6×10^9	$v/c = 0.131$
Hydra	4.0×10^9	$v/c = 0.203$

Figure 39-7 Doppler effect in light coming from several distant galaxies. The pair of dark lines (indicated by the arrow) in each photograph on the right is caused by absorption of light by calcium atoms. (Photograph courtesy Hale Observatories.)

1920s by E. E. Hubble, an astronomer at the Mount Wilson Observatory in California. In 1931 Hubble and his colleague, M. L. Humason, presented evidence that the recessional speed v of a galaxy was approximately proportional to its distance r from us. With modern data an approximate expression of this Hubble–Humason law is

$$r = (20 \times 10^9 \text{ light-years})\frac{v}{c} \qquad (39.20)$$

(One light-year is the distance that light travels in one year: 1 light-year = 9.46 $\times 10^{15}$ m.)

These observations imply that the universe is in a state of expansion. The number 20×10^9 years, called the Hubble time, is the time needed for galaxies to reach their present distances if they had always been receding from us with their present velocities. There are many different cosmological models which are consistent with the Hubble–Humason law. One increasingly plausible theory postulates that the present state of the universe has evolved from a "big bang" in the past, at which time all matter was crowded together with the same density as the nucleus of an atom. The time back to the start of the expansion of the universe, as determined from the evolution of the stars and the elements, is approximately 10×10^9 years.

Quasars

The largest shifts of wavelengths that astronomers have discovered correspond to wavelengths slightly more than three times their usual value. This occurs in the spectra of light from a *quasar*. Interpreted as a Doppler shift, this red shift corresponds to a recessional speed of $0.8c$.

Quasi-stellar objects (called quasars for short), discovered in 1961, are among the most amazing and puzzling objects in the universe. Their luminosity is up to 100 times that of the most luminous normal galaxy. Some are also strong sources of radiation with radio wavelengths. Although a quasar radiates a power which is perhaps a million million times that of the sun, the apparent size of a quasar is very small compared to a normal galaxy. At the time of writing, although the nature of quasars is unknown, many astronomers are tending toward the opinion that quasars are an unusual phase of normal galaxies.

Summary

☐ Consider frame S' moving relative to frame S with an x component of velocity V. The relationship between measurements in frames S and S' of the space and time intervals between the same two events are given in Newtonian mechanics by the Galilean coordinate transformations:

$$\Delta x = \Delta x' + V\Delta t' \qquad \Delta y = \Delta y' \qquad \Delta z = \Delta z' \qquad \Delta t = \Delta t' \text{ (absolute time)}$$

In Newtonian mechanics, a velocity transforms according to the Galilean velocity transformations:

$$v_x = v'_x + V \qquad v_y = v'_y \qquad v_z = v'_z$$

☐ Experimental measurements of the speed of light give the same value c, no matter what inertial frame S is used. That is, with $v'_x = c$, we measure $v_x = c$, not $v_x = c + V$.

☐ Einstein's postulates are:

1 The principle of relativity: The laws of nature are the same in all inertial frames of reference.
2 The invariance of the speed of light: The speed of light is the same in all inertial frames, regardless of the motion of the source of light.

☐ These postulates imply that time elapses differently in different inertial frames and that the Galilean transformations are merely the limiting form for low speeds of the Lorentz transformations

$$\Delta x = \gamma(\Delta x' + \beta c \Delta t') \qquad \Delta y = \Delta y' \qquad \Delta z = \Delta z'$$

$$c\Delta t = \gamma(c\Delta t' + \beta \Delta x')$$

(where $\beta = V/c$ and $\gamma = 1/\sqrt{1 - V^2/c^2}$) and of the Einstein velocity transformations

$$v_x = \frac{v_x' + V}{1 + v_x' V/c^2} \qquad v_y = \frac{v_y'\sqrt{1 - V^2/c^2}}{1 + v_x' V/c^2} \qquad v_z = \frac{v_z'\sqrt{1 - V^2/c^2}}{1 + v_x' V/c^2}$$

☐ The Lorentz transformations show that moving clocks run slow and that there is a length contraction for a dimension parallel to an object's motion.

☐ The relativistic interval between two events

$$(\Delta x)^2 + (\Delta y)^2 + (\Delta z)^2 - (c\Delta t)^2$$

is an invariant, a quantity that has the same value in all inertial frames.

☐ When a light source is approaching an observer with a speed v_s, the observed frequency is

$$f_o = \sqrt{\frac{c + v_s}{c - v_s}} f$$

where f is the source frequency measured in a frame in which the source is at rest.

Questions

1 An atom, moving east at a speed relative to the laboratory of 2.0×10^4 m/s, emits a light wave. The nucleus of the atom emits an electron which travels east with a speed of 4.0×10^5 m/s relative to the atom. From the facts presented at the beginning of Section 39.1, compute the speed of the light wave and of the electron relative to the laboratory. Assume the atom is so massive that it maintains a practically constant velocity during the emission of the light and of the electron.

2 Give an example of two events which are simultaneous in one frame of reference but not simultaneous in another frame. What formerly trusted equation is contradicted by this example?

3 Show that if a pulse of light bouncing back and forth from the roof to the floor of a bus is used as the timing element of a clock, with the clock ticking each time the light is at the roof, the time interval $\Delta t'$ between two ticks in the bus is related to the time interval Δt between these same two events in the highway frame by

$$\Delta t = \frac{\Delta t'}{\sqrt{1 - V^2/c^2}}$$

where V is the speed of the bus relative to the highway. Does this clock run slow as far as highway time is concerned?

4 Two rocket ships are approaching head on. The speed of one relative to the other is $0.98c$. A pulse of light travels from the roof to the floor of one ship, a vertical distance of 3.0 m as determined in this ship:

(a) What is the time interval between emission of the light from the roof and its arrival at the floor as determined in the ship containing the light pulse?

(b) What is the time interval between the same two events, determined in the other ship?

5 (a) For the electron in Question 1, calculate its velocity relative to the laboratory using the Einstein velocity transformations. Compare the answer with that obtained using the nonrelativistic expression $v_x = v'_x + V$.

(b) Repeat the calculations of part (a) for the case when the atom's velocity relative to the laboratory is $0.50c$ east and the electron's velocity relative to the atom is $0.95c$ east.

6 An atomic nucleus, while traveling east at $0.60c$, emits an electron which has a speed of $0.80c$ relative to the nucleus and travels in the backward direction. Find the magnitude and direction of the electron's velocity vector relative to the laboratory.

7 A proton travels west at speed $0.80c$ and an electron travels east at speed $0.90c$, both speeds being measured relative to the laboratory. Find the speed of the electron relative to the proton.

8 Deduce the transformation equation for a transverse component of velocity v_y from the appropriate Lorentz transformation equations.

9 Show that a velocity vector with components $v'_y = c$, $v'_x = 0$ (relative to S') is transformed by the Einstein velocity transformations into a velocity vector $\mathbf{v}$ (relative to S) with a magnitude $\sqrt{v_x^2 + v_y^2}$ which is equal to c.

10 The two explosions, events A and B of Example 2, are simultaneous and separated by a distance of 600 m relative to the frame S' fixed in the train. Check the calculations of Example 2 by applying the Lorentz transformations directly to these two events to evaluate (without reference to the event P) the space and time intervals between events A and B in the frame fixed in the tracks.

11 Two rocket ships, traveling in the same direction, have a relative speed of $0.98c$. A light pulse travels from tail to nose of the leading ship, a distance $\Delta x'$ of 20 m as measured in this ship (frame S'):

(a) What is the time interval $\Delta t'$ between the emission and arrival of this light pulse, as determined in the ship containing the pulse (frame S')?

(b) Use the Lorentz transformations to determine the distance Δx and the time interval Δt between these same two events, as determined by observers in the other rocket ship (frame S).

(c) Verify from your calculation of Δx and Δt in part (b) that $\Delta x/\Delta t = 3.0 \times 10^8$ m/s; that is, that light has the same speed relative to each rocket.

12 Imagine a train of rest length 6.0×10^3 m traveling relative to the ground at a speed of $0.98c$. As the train speeds through a station, the stationmaster's clock and the train conductor's watch both read zero. The conductor is at the midpoint of the train. The engineer observes lightning strike the locomotive and the adjacent ground (event A) when his watch reads 4.0 μs. A second bolt of lightning is observed by the brakeman in the caboose to strike the caboose and the adjacent ground when his watch reads 4.0 μs (event B):

(a) The conductor should *report* these events as having occurred at what times on his watch?

(b) When the conductor sees these events, what time interval (on his watch) will have elapsed since the occurrence of these events?

(c) Find the time interval between these events as recorded in the system of the stationmaster.

(d) Find the distance the stationmaster will measure along the ground between the two marks made by the lightning strokes.

13 Particle counters are placed at the top and bottom of a tower 27 m high. A particle moving with a speed of $0.90c$ relative to the counters triggers the counter at the top (event A) and later the counter at the bottom (event B):

(a) Find the distance and the time interval between these two events in the laboratory frame of reference.

(b) Find the distance and the time interval between these two events in a frame of reference moving upward at a speed of $0.98c$.

(c) What is the velocity of the particle relative to the second frame?

(d) What is the distance between these two events in a frame of reference moving with the particle?

14 Pions are produced when a synchrotron beam of high-energy protons strikes a target. Consider a pion which decays after a lifetime of 3.0×10^{-8} s in a frame in which the pion is at rest. If this pion emerges from the target at a speed of $0.98c$ how far will it travel in the laboratory before it decays? What distance would be expected based on a belief that the lifetime must be the same in all frames of reference?

15 Show that bus passengers will find that highway clocks run slow by considering two events at the location of a highway clock ($\Delta x = 0$ for these events) and using the appropriate Lorentz transformation equation.

16 A laboratory measurement of the coordinates of the ends of a moving metre-stick, taken at the same time in the laboratory, yields the result that the "metre-stick" is 0.20 m long. Evaluate the factor γ for this motion and use Table 39-1 to find the corresponding velocity of the stick.

17 Show that measurements of the length L of a moving bus made in a highway frame S will reveal that L is less than the length L_0 of the bus as measured in the bus frame S'.

18 Verify that the relativistic interval between the two events A and B of Figure 39-3 is invariant. That is, show that

$$(\Delta x)^2 + (\Delta y)^2 + (\Delta z)^2 - (c\Delta t)^2 = (\Delta x')^2 + (\Delta y')^2 + (\Delta z')^2 - (c\Delta t')^2$$

is satisfied for these events.

19 Evaluate the relativistic interval between the events A and B of Example 2 for both the frame S and the frame S'.

20 Measurements in one frame of reference show that two events occur simultaneously,

4.0 m apart. What is the time interval between these same two events in another frame of reference in which they are 5.0 m apart?

21 Use the fact that the relativistic interval between two events is invariant to deduce the time dilation formula

$$\Delta t = \frac{\Delta t'}{\sqrt{1 - v^2/c^2}}$$

where $\Delta t'$ is a time interval on a clock moving at a speed v.

22 A line in the spectrum of light coming from a galaxy in the constellation Virgo is red-shifted to a value of 3984 Å. In laboratory experiments on earth this line has a wavelength of 3968 Å. Find the speed of recession of this galaxy.

23 Use the Hubble–Humason law to estimate the distance from the earth to the galaxy in Ursa Major with a recessional speed of 0.050c (Example 3).

24 A spaceship returning to the earth at a speed of 11.0×10^3 m/s broadcasts a frequency of 50×10^6 Hz. Will receivers on the earth detect a higher frequency than 50×10^6 Hz? What is the magnitude of the frequency shift?

Supplementary Questions

S-1 Suppose we assume that the transformation equations for space and time intervals between two given events are the completely general linear equations

$$\Delta x = \gamma_1(\Delta x' + \beta_1 c \Delta t')$$

$$\Delta t = \gamma_2\left(\Delta t' + \frac{\beta_2 \Delta x'}{c}\right)$$

where γ_1, γ_2, β_1, and β_2 are independent of the space and time intervals:

(a) Show that these equations imply that the transformation law for the x component of velocity is

$$v_x = \frac{\gamma_1}{\gamma_2} \frac{v'_x + \beta_1 c}{1 + \beta_2 v'_x/c}$$

(b) Point O has velocity $v_x = 0$ and $v'_x = -V$. Show that this requires that $\beta_1 = V/c$.

(c) Point O' has velocity $v'_x = 0$ and $v_x = V$. Show that this requires that $\gamma_1 = \gamma_2$ (given that $\beta_1 = V/c$).

(d) Show that the results of parts (b) and (c) imply that the transformation laws for Δx and Δt must be of the form assumed in Eq. 39.5 and 39.6.

S-2 From the direct Lorentz transformations given in Table 39-2, deduce the inverse Lorentz transformations, Eq. 39.10 and 39.11.

S-3 An interesting alternative approach to the Lorentz transformations is the assumption, instead of Eq. 39.5 and 39.6, that

$$\Delta x = \gamma(\Delta x' + V \Delta t')$$

and also that

$$\Delta x' = \gamma(\Delta x - V \Delta t)$$

where γ is a coefficient independent of the values of the space and time intervals in these equations. Show that these assumptions, together with the analysis of Figure 39-3, imply that

$$\gamma = \frac{1}{\sqrt{1 - V^2/c^2}}$$

and that

$$\Delta t = \gamma\left(\Delta t' + \frac{V\Delta x'}{c^2}\right)$$

S-4 Show that the inverse Einstein velocity transformations are

$$v'_x = \frac{v_x - V}{1 - v_x V/c^2} \qquad v'_y = \frac{v_y \sqrt{1 - V^2/c^2}}{1 - v_x V/c^2}$$

S-5 Prove that if v'_x and V are both less than c, then the Einstein velocity transformations imply that v_x is also less than c. For simplicity, assume that all velocities are in the positive direction.

S-6 In a hypothetical train moving with a constant speed of $0.80c$ relative to the straight tracks, a light is flashed in one of the cars just as the car passes an observer in the station. The light is flashed again after 1.60 s, as measured in a frame fixed in the train. An observer in the station starts her stopwatch when she sees the first flash and stops the watch when she *sees* the second flash. What is the final reading of her watch?

S-7 Show that if two events A and B occur at the same place in S' (so $\Delta x' = x'_B - x'_A = 0$) with event A before event B (so $\Delta t' = t'_B - t'_A$ is positive), then in any other frame S these events occur in the same temporal order, that is $\Delta t = t_B - t_A$ is positive, so event A occurs before B. (In this circumstance, the statement that event A occurs before event B has absolute significance in the sense that it is true in every inertial frame of reference.)

S-8 Show that if two events A and B occur at the same time in S' (so $\Delta t' = t'_B - t'_A = 0$) with event A to the left of event B (so $\Delta x' = x'_B - x'_A$ is positive), then:
(a) In frame S, event A occurs before event B if V is positive (that is, if S' moves to the right relative to S).
(b) In frame S, event A occurs after event B if V is negative (that is, if S' moves to the left relative to S).
(For this pair of events we cannot make any statement about the temporal order of events that is true in all inertial frames of reference.)

S-9 From the Lorentz transformations, prove Eq. 39.14:

$$(\Delta x)^2 + (\Delta y)^2 + (\Delta z)^2 - (c\Delta t)^2 = (\Delta x')^2 + (\Delta y')^2 + (\Delta z')^2 - (c\Delta t')^2$$

S-10 Show that when v is much less than c, the Doppler effect formula for light, $f_o = f\sqrt{1 - v_s/c}/\sqrt{1 + v_s/c}$, can be approximated by $\Delta f/f = -v_s/c$, where v_s is the speed of recession of the light source relative to the observer. [When x is small, the binomial theorem implies that $(1 + x)^{-1/2}$ is approximately $(1 - \frac{1}{2}x)$ and $(1 - x)^{1/2}$ is approximately $(1 - \frac{1}{2}x)$.]

S-11 The difference Δf between the frequency of an incident microwave beam and the beam reflected from a car moving at speed v is measured to monitor highway speeds:
(a) Show that the magnitude of this frequency shift is $2f(v/c)$. (Assume that $v/c \ll 1$.)

(b) When $v = 30$ m/s and the incident microwaves have a frequency of 2.5 $\times$ 10^9 Hz, what is the magnitude of the frequency shift?

S-12 The spectrum of a nearby star, Arcturus, gradually changes in a six-month interval from a small shift toward the blue to a larger red shift. In the following six months, the spectrum regains its shift toward the blue. Give a plausible interpretation of this observation.

S-13 At low speeds, the factor $1/\sqrt{1 - v^2/c^2}$ is approximately $1 + \frac{1}{2}(v^2/c^2)$:

(a) Justify this assertion using the binomial theorem.

(b) Using this approximation, evaluate the difference in readings between an earth-bound clock and a clock carried by an astronaut in an earth satellite which travels at 7.0×10^3 m/s for 1.0×10^5 s, as measured by earth-bound clocks. (Assume that a frame fixed in the earth is inertial to a sufficiently good approximation for this calculation.)

40.

momentum and energy in special relativity

The relativity principle asserts that the laws of nature are the same in all inertial frames of reference. Any physical law is expressed in terms of quantities referring to a specified frame of reference. Suppose that the same physical law is written in terms of S quantities and then rewritten (or transformed) so that it is expressed entirely in terms of S' quantities. The relativity principle requires that both expressions of this law have the same mathematical form. In the special theory of relativity, the transformation of space–time intervals in one inertial frame to another inertial frame is accomplished using the Lorentz transformations. If the form of an equation remains the same when a Lorentz transformation is made, the equation is said to be *Lorentz covariant*. To accord with the relativity principle, an equation expressing a physical law must be Lorentz covariant.

The special theory of relativity therefore requires that the laws of physics be re-examined, and reformulated if necessary, to assure that they are Lorentz covariant. Although such a program cannot be completed in this book, the examination of certain conservation laws is sufficient to reveal a significant portion of what is sometimes called the golden harvest of relativity.

40.1 Covariant Conservation Laws

Suppose a system of particles has a total amount Q_i of some quantity before a process and a total Q_f after the process, both quantities being measured relative to the same frame S. In the frame S, the law of conservation of this quantity has the form

$$Q_i = Q_f \tag{40.1}$$

If this law is Lorentz covariant, then the form of this conservation law for measurements relative to S' is

$$Q'_i = Q'_f \tag{40.2}$$

where Q' is the value in S' corresponding to the value Q in S.

What *transformation laws* connecting Q' and Q will guarantee that $Q'_i = Q'_f$ if $Q_i = Q_f$? We examine the two simplest and most important possibilities.

Lorentz Invariant Quantities

The simplest transformation law is that satisfied by a *Lorentz invariant* quantity

$$Q' = Q \quad (Q \text{ is Lorentz invariant}) \quad (40.3)$$

since a Lorentz invariant quantity has the same value in all inertial frames. If a Lorentz invariant quantity satisfies a conservation law in S, it is clear that this quantity will also satisfy a conservation law in S'.

The electric charge q of a particle is Lorentz invariant, and the total charge $Q = \Sigma q$ is also Lorentz invariant. *Experiments* show that for any process occurring within a system, Q is conserved: $Q_i = Q_f$ in a frame S. Therefore $Q'_i = Q'_f$ in a frame S'. The law of conservation of electric charge is a Lorentz covariant law of physics.

Not all Lorentz invariant quantities are conserved quantities. For example, the mass* m of a particle is a Lorentz invariant characteristic of the particle, and the total mass of a system Σm is also a Lorentz invariant quantity; however, a system's total mass is *not* conserved. As is emphasized in Section 11.5, a system's mass changes whenever an inelastic collision occurs within the system.

Four-Vectors

A set of four quantities (p_x, p_y, p_z, p_t) that transforms in the same way as the four intervals $(\Delta x, \Delta y, \Delta z, c\Delta t)$ is called a four-vector. For the Lorentz transformations in Table 39-2, this definition implies

$$p_x = \gamma(p'_x + \beta p'_t) \qquad p_y = p'_y \qquad p_z = p'_z \qquad p_t = \gamma(p'_t + \beta p'_x) \quad (40.4)$$

The inverse transformation is

$$p'_x = \gamma(p_x - \beta p_t) \qquad p'_y = p_y \qquad p'_z = p_z \qquad p'_t = \gamma(p_t - \beta p_x) \quad (40.5)$$

---

*Recall that in this book, the word mass refers to *rest* mass. Definitions of various other types of mass are given in Questions S-9 and S-10.

If each component of a four-vector is conserved in one inertial frame S, that is, if

$$P_{ix} = P_{fx} \qquad P_{iy} = P_{fy} \qquad P_{iz} = P_{fz} \qquad P_{it} = P_{ft} \qquad (40.6)$$

then in any other inertial frame S', each component is also conserved. In other words, the four-vector conservation law Eq. 40.6 is a Lorentz covariant law. To prove this we relate components in S' by using the inverse transformation law Eq. 40.5 together with the conservation law Eq. 40.6:

$$P'_{ix} = \gamma(P_{ix} - \beta P_{it}) = \gamma(P_{fx} - \beta P_{ft}) = P'_{fx}$$

Examining each component in this way, we find

$$P'_{ix} = P'_{fx} \qquad P'_{iy} = P'_{fy} \qquad P'_{iz} = P'_{fz} \qquad P'_{it} = P'_{ft} \qquad (40.7)$$

40.2 Relativistic Momentum and Energy

Relativistic Momentum

In Newtonian mechanics, the momentum of a particle of mass m with velocity $\mathbf{v}$ is defined as $m\mathbf{v}$. At speeds small compared to c, measurements verify that the total Newtonian momentum $\Sigma m\mathbf{v}$ of an isolated system is conserved. However, at speeds comparable to c, gross violations of this conservation law are observed in many processes.

From the point of view of special relativity, the failure of conservation of $\Sigma m\mathbf{v}$ is to be expected. When a transformation is made from a frame S to a frame S', the quantity $m\mathbf{v}$ transforms like $\mathbf{v}$. And $\mathbf{v}$, instead of transforming like a four-vector, transforms according to the Einstein velocity transformations. As a consequence, we can easily find examples (Question S-1) which show that a law of conservation of $\Sigma m\mathbf{v}$ is not Lorentz covariant. Such a law would hold only in one particular inertial frame, and the existence of such a privileged frame would violate the relativity principle.

To discover whether there is a relativistically correct law that reduces to the experimentally verified law of conservation of $\Sigma m\mathbf{v}$ in the realm of low speeds, we consider redefining the momentum of a particle. We look for a four-vector whose first three components approach $m\Delta x/\Delta t$, $m\Delta y/\Delta t$, $m\Delta z/\Delta t$ as the particle speed v becomes small compared to c. The key to the construction of the desired four-vector is the replacement of the time Δt by the particle's proper time interval $\Delta\tau$. Equation 39.15 gives

$$\Delta\tau = \Delta t\sqrt{1 - v^2/c^2}$$

and therefore $\Delta\tau$ approaches Δt as v/c approaches zero. Since both m and $\Delta\tau$ are Lorentz invariant quantities, the set of quantities $(m\Delta x/\Delta\tau, m\Delta y/\Delta\tau, m\Delta z/\Delta\tau, mc\Delta t/\Delta\tau)$ is a four-vector because it transforms in the same way as $(\Delta x, \Delta y, \Delta z, c\Delta t)$.

We are thus led to *define* the *relativistic* momentum **p** of a particle of mass m moving at a velocity **v** by

$$p_x = m\frac{dx}{d\tau} \qquad p_y = m\frac{dy}{d\tau} \qquad p_z = m\frac{dz}{d\tau} \tag{40.8}$$

Since $d\tau = dt\sqrt{1 - v^2/c^2}$, this definition of **p** gives

$$\mathbf{p} = \frac{m\mathbf{v}}{\sqrt{1 - v^2/c^2}} \tag{40.9}$$

At speeds small compared to c, the relativistic momentum is imperceptibly different from $m\mathbf{v}$. But as v approaches c, the factor $1/\sqrt{1 - v^2/c^2}$ approaches infinity. Very near the speed of light, say for $v > 0.99c$, substantial increases of momentum accompany small increases of speed. The role of c as nature's speed limit is apparent. The value $v = c$ corresponds to infinite momentum.

Relativistic Energy

A particle has a relativistic momentum **p** formed from the first three components of the four-vector $(m\,dx/d\tau, m\,dy/d\tau, m\,dz/d\tau, mc\,dt/d\tau)$. The particle's relativistic energy E is defined as proportional to the fourth component:

$$E = mc^2\frac{dt}{d\tau} = \frac{mc^2}{\sqrt{1 - v^2/c^2}} \tag{40.10}$$

Figure 40-1 shows the graph of E as a function of v. The value of E at $v = 0$ is mc^2, a term which is called the *rest energy*.

As a first step in the exploration of the physical significance of E, we consider the infinite series obtained by a binomial expansion of $1/\sqrt{1 - v^2/c^2}$:

$$E = mc^2\left(1 + \frac{1}{2}\frac{v^2}{c^2} + \cdot\,\cdot\,\cdot\right) = mc^2 + \tfrac{1}{2}mv^2 + \cdot\,\cdot\,\cdot \tag{40.11}$$

The term $\tfrac{1}{2}mv^2$ is recognized as the kinetic energy in the nonrelativistic limit.

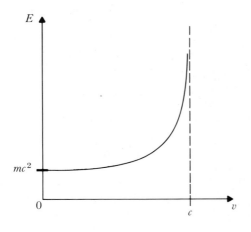

Figure 40-1 The relativistic energy E as a function of the speed v.

Therefore, the *relativistic kinetic energy* K, defined by

$$K = E - mc^2 \tag{40.12}$$

reduces to $K = \frac{1}{2}mv^2$ when v is small compared to c. If a particle is isolated, its *total energy* is

$$E = mc^2 + K \tag{40.13}$$

The velocity $\mathbf{v}$ of a particle can be determined from its total energy E and momentum $\mathbf{p}$. Dividing Eq. 40.9 by Eq. 40.10, we obtain

$$\frac{\mathbf{v}}{c} = \frac{c\mathbf{p}}{E} \tag{40.14}$$

Energy-Momentum Four-Vector

The four-vector whose first three components have been defined as the components of the relativistic momentum $\mathbf{p}$ and whose fourth component has been defined as E/c is called the *energy-momentum four-vector*:

$$\textit{energy-momentum four-vector} = (p_x, p_y, p_z, E/c) \tag{40.15}$$

When transforming from one frame to another, the components transform among themselves according to Eq. 40.4 and 40.5 (with $p_t = E/c$).

As was the case with the four-vector $(\Delta x, \Delta y, \Delta z, c\Delta t)$ we obtain a Lorentz invariant quantity by calculating the sum of the squares of the first three components minus the square of the fourth. We find

$$p_x^2 + p_y^2 + p_z^2 - \frac{E^2}{c^2} = m^2\left(\frac{v^2 - c^2}{1 - v^2/c^2}\right) = -m^2c^2 \tag{40.16}$$

This can be rearranged to give a useful relationship between a particle's total energy and the magnitude p of its momentum:

$$E^2 = (mc^2)^2 + (cp)^2 \tag{40.17}$$

Example 1   For an electron with a speed of $0.98c$, find the total energy E, the kinetic energy K, and the momentum p.

Solution   The rest energy of an electron is $m_e c^2 = 0.511$ MeV. We find that for $v = 0.98c$, $1/\sqrt{1 - v^2/c^2} = 5.03$. The total energy of the electron is

$$E = \frac{m_e c^2}{\sqrt{1 - v^2/c^2}} = (5.03 \times 0.511)\text{ MeV} = 2.57\text{ MeV}$$

The electron's kinetic energy is

$$K = E - m_e c^2 = (2.57 - 0.51)\text{ MeV} = 2.06\text{ MeV}$$

The momentum p is found from E and v:

$$cp = \frac{Ev}{c} = (2.57\text{ MeV})(0.98) = 2.52\text{ MeV}$$

We say $p = 2.52$ MeV/c. [Alternatively, p could be computed from $E^2 = (m_e c^2)^2 + (cp)^2$.]

40.3 Conservation of Momentum and Energy

We should emphasize that, although we have developed the dynamical formulas for a "particle" in this chapter, we have not assumed that the particle is in any way elementary. These formulas can be applied to a composite object or system consisting of many particles, by interpreting m as the mass of the entire object and $\mathbf{v}$ as the velocity of its motion as a whole.

The law of conservation of energy states that the total energy of an isolated system remains constant. The law of conservation of momentum is a similar law about each component of the system's total momentum. Mathematically, these conservation laws have the form of Eq. 40.6, with $\mathbf{P}$ representing the total momentum vector of the system, and cP_t representing the system's total energy. With this interpretation, the derivation of Eq. 40.7 from Eq. 40.6 shows that the *law of conservation of momentum and the law of conservation of energy together constitute a Lorentz covariant conservation law*. In special relativity, conservation of momentum and conservation of energy are no longer independent laws.

So far we have two arguments to support the definitions that have been given for relativistic momentum and relativistic energy:

1   For low speeds the well-verified expressions of Newtonian mechanics for momentum and kinetic energy are obtained.
2   Conservation laws are Lorentz covariant and therefore accord with the relativity principle.

We must now appeal to experiment to learn whether or not these quantities are in fact conserved. The answer is yes. Abundant and sometimes dramatic experimental confirmation has been obtained. (The law of conservation of energy, in which rest energies are included, is discussed in Section 11.6.)

Particle Decay

The use of the relativistic laws of conservation of energy and of momentum will be illustrated by consideration of their implications for the spontaneous decay of a particle of mass M into two parts with masses m_1 and m_2. (We examined such a process in Example 9 of Chapter 11, but the motion of the decay products was treated nonrelativistically.)

In the CM-frame, the law of conservation of energy gives

$$Mc^2 = E_1 + E_2 = m_1 c^2 + K_1 + m_2 c^2 + K_2$$

For this process, the energy release is

$$Q = Mc^2 - (m_1 c^2 + m_2 c^2) = K_1 + K_2 \tag{40.18}$$

The division of the energy Q between the two emerging particles must be such that momentum is conserved: $\mathbf{p}_1 + \mathbf{p}_2 = 0$. Therefore $p_1^2 = p_2^2$ and, from Eq. 40.17, this implies

$$E_1^2 - (m_1c^2)^2 = E_2^2 - (m_2c^2)^2 \qquad (40.19)$$

These equations can be solved to determine each kinetic energy in terms of Q and the masses. Rewriting Eq. 40.19 we have

$$(E_1 - E_2)(E_1 + E_2) = (m_1c^2 - m_2c^2)(m_1c^2 + m_2c^2)$$

$$(K_1 - K_2 + m_1c^2 - m_2c^2)Mc^2 = (m_1c^2 - m_2c^2)(Mc^2 - Q)$$

$$K_1 - K_2 = (m_2 - m_1)\frac{Q}{M}$$

Substituting $K_2 = Q - K_1$, we find

$$K_1 = \tfrac{1}{2}Q\left(1 + \frac{m_2 - m_1}{M}\right) = Q\left(1 - \frac{m_1}{M} - \frac{Q}{2Mc^2}\right) \qquad (40.20)$$

There is an analogous equation for K_2. The last term, $-Q/2Mc^2$, is the relativistic correction to the result obtained in Example 4 of Chapter 11.

Example 2 The pion (π meson) is a charged particle with a rest energy $M_\pi c^2$ of 139.6 MeV and a mean life of only 2.6×10^{-8} s. It spontaneously decays into a muon (μ meson) and a neutrino. As there seems to be no good reason for picturing a pion to be composed of a muon and a neutrino, we regard these decay products as created in the decay process. The muon is a charged particle with a rest energy $m_\mu c^2$ of 105.7 MeV. The neutrino is an electrically neutral particle with zero mass and therefore zero rest energy:
(a) Find the energy release Q for pion decay.
(b) Find the kinetic energies K_μ and K_ν of the muon and the neutrino, respectively.

Solution
(a) For this decay

$$Q = 139.6 \text{ MeV} - (105.7 \text{ MeV} + 0) = 33.9 \text{ MeV}$$

(b)

$$K_\mu = (33.9 \text{ MeV})\left[1 - \frac{105.7}{139.6} - \frac{33.9}{2(139.6)}\right] = 4.1 \text{ MeV}$$

$$K_\nu = 33.9 \text{ MeV} - K_\mu = 29.8 \text{ MeV}$$

The decay of a pion at rest is therefore characterized by the creation of a muon with an energy of 4.1 MeV (Figure 40-2). Detection of tracks of 4.1-MeV muons in a photographic emulsion led to the discovery of the pion in 1947 by C. F. Powell (1903–1969).

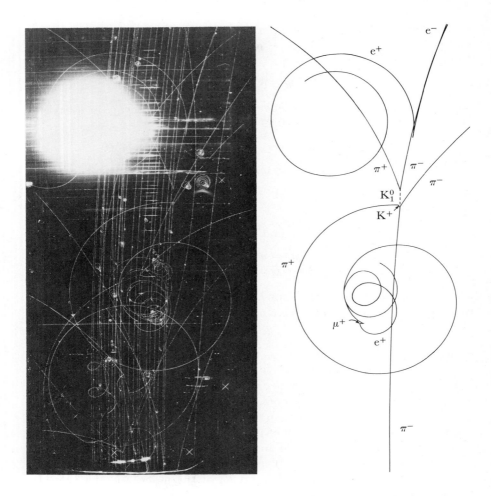

Figure 40-2 Liquid hydrogen bubble chamber photograph of a negatively charged pi meson (π^-) interacting with a proton (hydrogen nucleus) to produce a positive K meson (K^+), a neutral K_1 meson (K_1^0), shown by the dotted line, a negative π meson (π^-), and a neutron, whose path is not shown. (Neutral or uncharged particles do not leave tracks.) The K^+ quickly decays into a positive pi meson (π^+) and a neutral pi (π^0), whose path is not shown. The π^+ decays into a positive mu (μ^+) meson, which decays into a positron (e^+). The π^0 decays into a pair of gamma rays (not shown), one of which converts into the electron (e^-) positron (e^+) pair. The K_1^0 decays into a π^+ and a π^-. This photograph was taken in the 200-cm liquid hydrogen bubble chamber at Brookhaven National Laboratory. For this experiment, which was part of a search for resonances (particles with extremely short lives, i.e. 10^{-23} s), a beam of 6 GeV/c momentum π^- mesons from the Alternating Gradient Synchrotron were produced by the action of the circulating proton beam of the Alternating Gradient Synchrotron on a metal target. A complex arrangement of magnets and electrostatic separators removed all particles from the beam other than 6 GeV/c π^- mesons. (Courtesy Brookhaven National Laboratory.)

Particles with Zero Mass

For particles with zero mass, the relativistic expression $E^2 = (mc^2)^2 + (cp)^2$ simplifies to

$$E = cp \qquad (40.21)$$

Then the expression $v/c = cp/E$ gives

$$v = c$$

A particle with zero mass always moves at the speed of light. (When the mass is zero, the expressions $p = mv/\sqrt{1 - v^2/c^2}$ and $E = mc^2/\sqrt{1 - v^2/c^2}$ give $p = 0$ and $E = 0$ unless $v = c$; but then these expressions cannot be used because we have a zero in the denominator. The theory is consistent only if zero-mass particles do carry momentum and energy related by $E = cp$ and if they always travel at speed c relative to any inertial frame of reference.)

The relationship between energy and momentum for electromagnetic radiation (Eq. 35.7) is the same as the relationship $E = cp$ for a zero-mass particle. This suggests that electromagnetic radiation might consist of a stream of zero-mass particles. The evidence confirming this viewpoint is the principal topic of the following chapter. The zero-mass particle of electromagnetic radiation is called a *photon*.

The only other zero-mass particles that have been discovered are a family of electrically neutral particles called neutrinos. The interaction of a neutrino with other particles is weak, much weaker than the electromagnetic interaction. Direct detection of reactions produced by neutrinos was not achieved until the 1950s. In Powell's experiment (Example 2) and in similar experiments, although the neutrino is not detected, agreement with all known conservation laws is obtained and the energy distribution among the decay products is explained by postulating the emission of this massless neutral particle.

40.4 Remarks on Further Developments of Relativity Theory

Newton's Second Law

In experiments where the particle trajectory can be determined, we find that the motion of a particle of charge q and mass m in an electric field $\mathbf{E}$ and a magnetic field $\mathbf{B}$ is in accord with

$$q\mathbf{E} + q\mathbf{v} \times \mathbf{B} = \frac{d}{dt}\frac{m\mathbf{v}}{\sqrt{1 - v^2/c^2}}$$

for all speeds v that have been experimentally investigated. We therefore take the equation

$$\mathbf{F} = \frac{d\mathbf{p}}{dt} \qquad (6.2)$$

with $\mathbf{p} = m\mathbf{v}/\sqrt{1 - v^2/c^2}$, as the definition of the resultant force acting on a particle. This equation is the form of Newton's second law that replaces $\mathbf{F} = m\mathbf{a}$ in the special theory of relativity. This means that $\mathbf{F}$ is no longer parallel to $\mathbf{a}$ unless $\mathbf{a}$ is parallel or perpendicular to $\mathbf{v}$ (Questions S-8, S-9, and S-10).

Maxwell's Equations

The special theory of relativity has its historical roots in Maxwell's electromagnetic theory. Einstein realized that if all inertial frames are equivalent for the description of nature, then electromagnetic theory and Newtonian mechanics could not both be correct. Einstein decided in favor of electromagnetic theory and proceeded to modify mechanics. Special relativity demands no modification of Maxwell's electromagnetic theory, which has the distinction of being the first relativistically correct theory, even though it was created by Maxwell more than 40 years before Einstein's work.

Although we shall not present the transformation equations, it is important to realize that electric and magnetic fields have different values when measured in two frames S and S' in relative motion. The electric and magnetic fields transform in such a way that Maxwell's equations are covariant, that is, they have the same mathematical form in the primed and in the unprimed coordinates.

Relativity theory emphasizes that electric and magnetic fields form a single entity, the electromagnetic field. The way in which the electromagnetic field is divided into an electric field $\mathbf{E}$ and a magnetic field $\mathbf{B}$ depends on the frame of reference. When a transformation is made from frame S to a frame S', the six components E_x, E_y, E_z, B_x, B_y, B_z transform among themselves in such a way that *both* the electric and magnetic fields in frame S transform into *each* of these fields in S'.

General Relativity

The word "special" in Einstein's special theory of relativity refers to the fact that the theory is limited to consideration of *inertial frames of reference*. Einstein spent most of the last half of his life working on generalizations considering transformations between any two frames of reference. Some of the fruits of these efforts comprise what he called the general theory of relativity.

General relativity is a theory of gravitation. Einstein's starting point was a fact that we noted in discussing motion in gravitational fields: Different freely falling bodies experience exactly the same acceleration at a given point in space. We obtained this result because the gravitational force $m\mathbf{g}$ is strictly proportional to the mass, so that when we write Newton's second law, $m\mathbf{g} = m\mathbf{a}$, the mass m cancels out, leaving $\mathbf{a} = \mathbf{g}$, regardless of the value of m. We often describe this situation by saying that the "gravitational mass," the quantity m that determines the force $\mathbf{F} = m\mathbf{g}$ in a gravitational field, is the same as the "inertial mass" m that appears in Newton's second law $\mathbf{F} = m\mathbf{a}$.

Einstein compared gravitational forces with what are called "inertial forces" that arise when the frame of reference has an acceleration $\mathbf{a}_{\text{frame}}$ relative to an inertial frame. These inertial forces are also strictly proportional to m, and all free bodies in such a frame possess the same acceleration $-\mathbf{a}_{\text{frame}}$. Einstein was thus led to interpret gravitational forces as inertial forces and to assert that the existence of a gravitational field at a point in space is equivalent to an acceleration of the coordinate frame at that point. From this *equivalence principle* he developed a theory in which gravitational effects were interpreted in terms of the geometry of the universe, a universe with a geometry more complicated than ordinary Euclidean geometry.

Einstein has by this "geometrical" means sought to explain gravity, to understand what lies behind the very successful description of motion in a gravitational field that Newton achieved merely by postulating a gravitational force $F = GMm/r^2$. In situations that we can presently observe and analyze with confidence, the predictions of Einstein's theory of gravitation almost coincide with those of Newton's, and this makes it very difficult to find compelling experimental evidence in favor of general relativity. Three of Einstein's famous predictions, (1) a correction to the orbital motion of the planet Mercury, (2) the bending of a light beam near the sun, and (3) a shift to the red in the spectra of light coming from atoms in a strong gravitational field, have indeed been verified. But these effects are minute and difficult to measure with a precision sufficient for a definite decision between general relativity and competing explanations. However, we have recently found strong evidence in favor of general relativity. The results of measurements of the deflection of microwave radiation as it passes by the sun on its way toward an earthbound detector are consistent with general relativity and have a relative error that is estimated to be not more than about 1%.

Summary

☐ A particle of mass m moving with velocity $\mathbf{v}$ relative to a frame S has relativistic momentum $\mathbf{p}$ and relativistic energy E defined by

$$\mathbf{p} = \frac{m\mathbf{v}}{\sqrt{1 - v^2/c^2}} \qquad E = \frac{mc^2}{\sqrt{1 - v^2/c^2}}$$

Then

$$\frac{\mathbf{v}}{c} = \frac{c\mathbf{p}}{E}$$

☐ A particle has a rest energy mc^2 and a kinetic energy

$$K = E - mc^2 = \tfrac{1}{2}mv^2 + \cdots$$

where succeeding terms are proportional to $(v/c)^4$, $(v/c)^6$, ... The total energy of an isolated particle is

$$E = mc^2 + K$$

☐ The relativistic momentum and energy together constitute the energy-momentum four-vector with components $(p_x, p_y, p_z, E/c)$. When transforming from one frame to another, these components transform among themselves in the same way as the four intervals $(\Delta x, \Delta y, \Delta z, c\Delta t)$. Energy and momentum are related to the mass m (a Lorentz invariant quantity) by

$$E^2 = (mc^2)^2 + (cp)^2$$

☐ For an isolated system, each component of the system's total energy-momentum four-vector is conserved:

$$P_{ix} = P_{fx} \qquad P_{iy} = P_{fy} \qquad P_{iz} = P_{fz} \qquad E_i = E_f$$

The laws of conservation of energy and of momentum together constitute a Lorentz covariant conservation law.

☐ A zero-mass particle has

$$v = c \qquad \text{and} \qquad E = cp$$

☐ In special relativity, Newton's second law is

$$\mathbf{F} = \frac{d\mathbf{p}}{dt}$$

where $\mathbf{p} = m\mathbf{v}/\sqrt{1 - v^2/c^2}$.

Questions

1 Give at least one example of:
(a) An invariant quantity that is also a conserved quantity.
(b) An invariant quantity that is not conserved.

2 Complete the derivation of Eq. 40.7 by showing that $P_{ix} = P_{fx}$ and $P_{it} = P_{ft}$ imply

$$P'_{it} = P'_{ft}$$

3 Consider the ratio of the magnitude of a particle's relativistic momentum to the magnitude of the Newtonian expression for momentum. Sketch a graph showing this ratio as a function of v/c, where v is the particle's speed.

4 Show that a particle's momentum is ten times as large at a speed of $0.9999c$ as it is at a speed of $0.99c$.

5 At what speed will a particle's momentum be equal to mc?

6 (a) Show that at low speeds, the factor $1/\sqrt{1 - v^2/c^2}$ is approximately equal to $1 + \frac{1}{2}(v^2/c^2)$.
(b) What is the fractional error made when momentum is calculated using mv rather than $mv/\sqrt{1 - v^2/c^2}$ for a satellite at a speed of 7.0×10^3 m/s?

7 An electron (rest energy 0.51 MeV) has a momentum of 1.0 MeV/c (that is, $cp = 1.0$ MeV). Find:
(a) The total energy of the electron.
(b) The kinetic energy of the electron.

8 What is the total energy of an electron whose speed is $0.9999c$?

9 The rest energy of a proton is 938 MeV. If a proton has a kinetic energy of 500 MeV, find:
   (a) Its total energy in MeV.
   (b) Its speed in terms of c.
   (c) The magnitude of its momentum in MeV/c.

10 At what speed will a particle's kinetic energy be 4.0 times its rest energy?

11 Show that for speeds near the speed of light the factor $\sqrt{1 - v^2/c^2}$ is approximately $\sqrt{2}\sqrt{1 - v/c}$. Use this approximation to evaluate the total energy of an electron at a speed which is $10^{-6}c$ less than the speed of light.

12 In what is called the *extreme* relativistic range, a particle's total energy E is much greater than its rest energy. Show that this implies that E is approximately equal to cp, and that the particle's speed is close to c.

13 A muon is created in the atmosphere and travels vertically downward and decays as it reaches the earth's surface. An observer on the earth measures the muon's speed as 0.98c, and the thickness of the atmosphere traversed as 3000 m (muon's mass $m_\mu = 207m_e$). For this observer find:
   (a) The total energy.
   (b) The kinetic energy.
   (c) The momentum.
   (d) The lifetime of this muon.

14 For an observer traveling with the muon of the preceding question, find:
   (a) The total energy.
   (b) The kinetic energy.
   (c) The momentum.
   (d) The lifetime.
   (e) The thickness of the atmosphere traversed.

15 A pulse of light from a large laser has an energy of 2.0×10^3 J:
   (a) What is the momentum of this pulse?
   (b) If this light were absorbed in a 4.0×10^{-3}-kg marble, what speed would the marble acquire if it were initially at rest?

16 A photon with an energy of 0.485 MeV collides with a stationary nucleus with a mass of 11.65×10^{-27} kg and is absorbed by the nucleus:
   (a) What is the momentum (in MeV/c) of the recoiling nucleus?
   (b) What is the kinetic energy of the recoiling nucleus? (Since the recoil speed is small compared to c, the expressions for momentum and kinetic energy from Newtonian mechanics can be used for this nucleus.)

17 An atom of mass M emits a photon of energy E_{photon}. The atom recoils with a momentum MV and a kinetic energy $\frac{1}{2}MV^2$. (Expressions of Newtonian mechanics can be used because the atom's speed is small compared to c.) Show that the recoil kinetic energy of the atom is given by

$$K = \frac{E^2_{\text{photon}}}{2Mc^2}$$

18 Show that when a mass M at rest decays into fragments of masses m_1 and m_2, these fragments emerge with energies

$$E_1 = \frac{(M^2 + m_1^2 - m_2^2)c^2}{2M} \qquad E_2 = \frac{(M^2 - m_1^2 + m_2^2)c^2}{2M}$$

19   In Example 9 of Chapter 11, what relative error is made by using a nonrelativistic approximation to describe the motion of the decay products resulting from the disintegration of radium?

20   Show that the electron cannot simply absorb all the energy of the photon when an incident photon interacts with a single free electron.

21   Relative to a frame S, a proton moves with a velocity $v_x = 0.98c$, $v_y = 0$, $v_z = 0$. Frame S' has a velocity in the x direction with $V = 0.80c$:
   (a) Find the proton's momentum and energy relative to S.
   (b) Use the transformation laws for the energy-momentum four-vector to find the proton's energy and momentum relative to S'.
   (c) Use the Einstein velocity transformations to find the proton's speed v' relative to S'. Then verify the results of (b) by calculating $E' = m_p c^2 / \sqrt{1 - v'^2/c^2}$, $p' = m_p v'/\sqrt{1 - v'^2/c^2}$, using $m_p c^2 = 938$ MeV.

22   Show that if the frames S and S' are in relative motion along the X axis, the two conservation laws in S, $P_{ix} = P_{fx}$ and $E_i = E_f$, together imply that energy is conserved in S', that is

$$E_i' = E_f'$$

23   Show that if a system's total momentum is conserved in S, but its total energy is not conserved, then the total momentum is not conserved in a frame S' moving relative to S.

Supplementary Questions

S-1   Relative to frame S, particle 1 of mass m has a velocity $\mathbf{v}$ with components v_x and v_y and particle 2 has a velocity $-\mathbf{v}$. An elastic collision occurs between these identical particles. After the collision, particle 1 has velocity components v_x, $-v_y$ and particle 2 has velocity components $-v_x$, v_y. Then, for the system consisting of the two particles, the total *Newtonian* momentum is conserved in S. Examine this collision from a frame S' moving relative to S with a velocity $V = v_x$. Show that in this frame the initial total *Newtonian* momentum has y' component

$$\frac{mv_y \sqrt{1 - v_x^2/c^2}}{1 - v_x^2/c^2} - \frac{mv_y \sqrt{1 - v_x^2/c^2}}{1 + v_x^2/c^2}$$

while the final total Newtonian momentum has y' component

$$\frac{-mv_y \sqrt{1 - v_x^2/c^2}}{1 - v_x^2/c^2} + \frac{mv_y \sqrt{1 - v_x^2/c^2}}{1 + v_x^2/c^2}$$

and therefore the y' component of the total Newtonian momentum is not conserved in S'.

S-2   Verify that, in the collision of Question S-1, the relativistic momentum of the system comprising the two particles is conserved in both S and S'.

S-3   In the laboratory frame, a particle with mass m_1 and total energy E_1 collides with a stationary particle of mass m_2. A composite object is formed which moves off with a velocity $\mathbf{V}$. Show that

$$V = \frac{c\sqrt{E_1^2 - m_1^2 c^4}}{E_1 + m_2 c^2}$$

S-4 If the composite object in the preceding question has a mass M, momentum p, and total energy E relative to the laboratory frame, show that its total energy E' in its CM-frame is given by

$$E'^2 = (Mc^2)^2 = E^2 - c^2 p^2 = (E_1 + m_2 c^2)^2 - (E_1^2 - m_1^2 c^4)$$

and therefore

$$E'^2 = (m_1 c^2)^2 + (m_2 c^2)^2 + 2 m_2 c^2 E_1$$

S-5 In Question S-8 of Chapter 11, the threshold kinetic energy for a reaction is deduced in terms of the Q value, the projectile mass m_1, and the target mass m_2. The motion is treated nonrelativistically. Use the result of the preceding question to show that the relativistically correct expression for the threshold kinetic energy of m_1 relative to the laboratory is

$$K_{1,\text{threshold}} = -Q\left(1 + \frac{m_1}{m_2} - \frac{Q}{2 m_2 c^2}\right)$$

S-6 Stationary protons are bombarded by protons emerging from a synchrotron. Find the threshold for the occurrence of the following reaction:

$$\text{p} + \text{p} \longrightarrow \text{p} + \text{p} + \text{p} + \overline{\text{p}}$$

where $\overline{\text{p}}$ denotes an antiproton (the antiproton mass is the same as the proton mass m_p).

S-7 A block of mass m_2 rests on a horizontal table. A bullet of energy E_1 and mass m_1, traveling horizontally, strikes the block and remains embedded in it. What is the relativistically correct expression for the final mass of the system? What is the Q for this process? Compare the answer given by special relativity with that obtained using Newtonian mechanics.

S-8 Show that the equation $\mathbf{F} = d\mathbf{p}/dt$, with $\mathbf{p} = m\mathbf{v}/\sqrt{1 - v^2/c^2}$, implies that $\mathbf{F}$ is generally not parallel to $\mathbf{a}$.

S-9 Show that if the velocity of a particle changes only in magnitude, then $\mathbf{F}$ is parallel to $\mathbf{a}$, and the longitudinal mass defined by F/a for this case is given by

$$m_{\parallel} = \frac{m}{(1 - v^2/c^2)^{3/2}}$$

S-10 Show that if the velocity of a particle changes only in direction, then $\mathbf{F}$ is parallel to $\mathbf{a}$, and the transverse mass defined by F/a for this case is given by

$$m_{\perp} = \frac{m}{(1 - v^2/c^2)^{1/2}}$$

S-11 A particle of mass m experiences a constant force $\mathbf{F}$ in the x direction. If the particle starts at rest, show that after a time t its momentum is

$$p = Ft$$

and its speed is given by

$$\frac{v}{c} = \frac{cp}{\sqrt{(mc^2)^2 + (cp)^2}} = \frac{Ft/mc}{\sqrt{1 + (Ft/mc)^2}}$$

Verify that for t short compared to mc/F, this agrees with the expression given by Newtonian mechanics. What is the limiting value of v as t approaches infinity?

41.

photons

As emphasized in Chapters 35 to 37, the nineteenth century saw the accumulation of evidence of the wave nature of light. This evidence still stands today and must be faced by any acceptable theory. But so far we have presented only part of nature's lessons on light. In our century, physicists have met incontrovertible evidence for the existence of particles of light, or, in general, particles of electromagnetic radiation. These particles, called *photons*, are said to be the *quanta of the electromagnetic field.* Before examining the relevant experimental evidence, we shall summarize some of the present knowledge regarding photons.

41.1 Properties of Photons

Photon Energy, Momentum, and Wavelength

A beam of electromagnetic radiation, considered as an electromagnetic wave, is characterized by its frequency f or its wavelength λ which are related by

$$f = \frac{c}{\lambda}$$

The same beam, considered as a stream of photons, is characterized by the energy E or the momentum p of the individual photons.* A photon has no mass

*Note that here the symbol E denotes the energy of a photon, not an electric field. The reader is cautioned to interpret symbols according to the context.

and travels at a speed, $c = 3.00 \times 10^8$ m/s. Its energy and momentum are related by

$$E = cp \tag{40.21}$$

The vital connecting link between these two descriptions of the *same beam of radiation*, proposed by Einstein, is that the photon energy E is proportional to the frequency f of the electromagnetic wave:

$$E = hf \tag{41.1}$$

The constant of proportionality h is named *Planck's constant* for reasons that are mentioned in the following section. Modern experiments yield the value, $h = 6.626176 \times 10^{-34}$ J·s. (Nature's unit of angular momentum, introduced in Section 13.5, is Planck's constant divided by 2π: that is, $\hbar = h/2\pi$.)

Electromagnetic radiation can be classified according to the energy of its photons, or the wavelength, or the frequency, whichever is most convenient. For example, from $f = c/\lambda$, Eq. 41.1 can be written to show the relationship between photon energy E and wavelength:

$$E = \frac{hc}{\lambda} \tag{41.2}$$

Photon energies are usually specified in electronvolts, and wavelengths in ångström units. Inserting the numerical values of h and c and the required conversion factors, we can write Eq. 41.2 in a form which is convenient for calculations:

$$E = \frac{12.4 \times 10^3 \text{ Å·eV}}{\lambda} \tag{41.3}$$

which gives the photon energy E in electronvolts when the wavelength λ is in ångström units.

Example 1   Find the energy of the photons in a beam whose wavelength is:
  (a) 6.2×10^3 Å (orange light).
  (b) 4.13×10^3 Å (violet light).
  (c) 1.0 Å (x ray).
  (d) 10.0 m (radio wave).

Solution
  (a) A photon of the beam of orange light has an energy which is, from Eq. 41.3,

$$E = \frac{12.4 \times 10^3 \text{ Å·eV}}{6.2 \times 10^3 \text{ Å}} = 2.0 \text{ eV}$$

  (b) A photon of the beam of violet light has an energy

$$E = \frac{12.4 \times 10^3 \text{ Å·eV}}{4.13 \times 10^3 \text{ Å}} = 3.0 \text{ eV}$$

Notice that the photons of visible light have energies of a few electronvolts.

These are typical of energies involved in atomic and chemical processes. (c) In this x-ray beam, each photon has an energy

$$E = \frac{12.4 \times 10^3 \text{ Å·eV}}{1.0 \text{ Å}} = 12.4 \times 10^3 \text{ eV}$$

Typical x-ray photon energies are ten thousand times *greater* than the energy of photons of visible light.
(d) Each photon of the radio wave has an energy

$$E = \frac{12.4 \times 10^3 \text{ Å·eV}}{10.0 \times 10^{10} \text{ Å}} = 12.4 \times 10^{-8} \text{ eV}$$

A single radio wave photon has so little energy that it is undetectable. Many such photons are required to build up a detectable effect.

There is an important relationship between the photon's wavelength λ and its momentum p. We have $\lambda = hc/E = hc/cp$, or

$$\lambda = \frac{h}{p} \tag{41.4}$$

This equation, which associates a wavelength with a momentum, we shall meet again. It turns out to have a universal significance.

Beam Intensity and the Number of Photons

The *intensity* I of a beam of radiation is determined by both the number of photons in the beam and the energy of each photon. Suppose that, when a light beam shines on a surface perpendicular to the beam, there are N identical photons striking the surface per second per unit area. The beam intensity, being defined as the energy per second per unit area, is therefore given by

$$I = NE \tag{41.5}$$

where E is the energy of each photon in the beam.

Now when we describe the same light beam as an electromagnetic wave with an amplitude E_0, the intensity of the beam is determined by the square of the wave amplitude:

$$I = (1.33 \times 10^{-3} \text{ W/V}^2)E_0^2 \tag{35.5}$$

Here we see a connection between the two ways of thinking about a light beam: as a wave or as a collection of photons. The *square of the wave amplitude* is related to the *number of photons in the beam*. Bright parts of an illuminated screen are struck by many photons per second (N is large) and here the electromagnetic wave has a large amplitude E_0.

As a light wave spreads out from a candle, the amplitude E_0 and the intensity I diminish. However, the energy radiated remains concentrated in "lumps" that we call photons. The energy of a single photon does *not* get spread out across a broad wave front.

Example 2 A desk is illuminated with violet light of wavelength 4.13×10^3 Å. The amplitude of this electromagnetic wave is 61.3 V/m. Find the number N of photons striking the desk per second per unit area.

Solution The intensity is

$$I = (1.33 \times 10^{-3} \text{ W/V}^2)(61.3 \text{ V/m})^2 = 5.00 \text{ W/m}^2$$

Each photon has an energy

$$E = \frac{12.4 \times 10^3 \text{ Å} \cdot \text{eV}}{4.13 \times 10^3 \text{ Å}} = 3.00 \text{ eV}$$

Therefore

$$N = \frac{5.00 \text{ W/m}^2}{(3.00 \text{ eV})(1.60 \times 10^{-19} \text{ J/eV})} = 1.04 \times 10^{19} \text{ s}^{-1} \cdot \text{m}^{-2}$$

This calculation illustrates that enormous numbers of photons are involved in most circumstances familiar in daily life.

41.2 Experimental Evidence of Photons

Photoelectric Effect

In a physics laboratory with modern equipment, it is a commonplace fact that electromagnetic energy in light, x rays, and γ rays is always found to be emitted or absorbed in localized lumps each containing an energy $E = hf$. This discovery, which caused a great shock in early twentieth century physics, has an interesting history.

Einstein, again in 1905, the *annus mirabilis*, thinking about the ejection of electrons from a metal surface which was illuminated with light, can be said to have "discovered" the photon. The phenomenon in question is called the *photoelectric effect* (Figure 41-1). A clean metal surface, illuminated by a light beam of sufficiently high frequency f, immediately (the time delay is less than 10^{-9} s) ejects electrons with a variety of kinetic energies which range up to a maximum value K. Increasing the intensity of the light beam does not affect the kinetic energy of the ejected electrons in the slightest. The only effect of an intensity increase is that the number of electrons ejected per second increases proportionately. But the maximum kinetic energy K of the ejected electrons depends

Figure 41-1 Photoelectric effect.

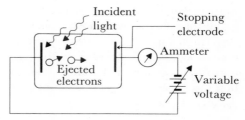

only upon the frequency f of the light according to what is called the *Einstein photoelectric equation*

$$K = hf - W \tag{41.6}$$

where W (named the work function) is a constant whose value depends on the nature of the surface.

Einstein deduced this result by postulating that the light beam consists of photons, each of energy hf. An electron inside the metal absorbs all the energy of a photon and then may escape from the metal, after losing at least an energy W in this escape. The work function W is the binding energy of the most energetic electrons in the metal. The law of conservation of energy then implies that the largest kinetic energy K of the ejected electrons is given by the photoelectric equation, Eq. 41.6.

It should be pointed out that experimental knowledge of the photoelectric effect was very fragmentary at the time that Einstein leaped to the right answer. It was not until 1916 that this photoelectric equation was experimentally verified in detail by Millikan. The maximum kinetic energy K of the ejected electrons was determined by measuring the voltage required to prevent any electrons from reaching a stopping electrode (Figure 41-1) which faced the metal surface from which the electrons were emitted.

Example 3   A tungsten surface is illuminated with ultraviolet light of wavelength 2.0×10^3 Å. Electrons are ejected from the tungsten, and we find that a retarding voltage of 1.6 V applied between the tungsten and another (stopping) electrode is just enough to prevent any electrons from reaching the stopping electrode. What is the work function W for this surface?

Solution   The law of conservation of energy implies that an electron, which is reduced to rest after moving against the electric forces through a potential drop of 1.6 V, started with a kinetic energy of 1.6 eV. Thus, $K = 1.6$ eV. Calculating the energy of the incident photons from Eq. 41.3, we obtain

$$E = \frac{12.4 \times 10^3 \text{ Å·eV}}{2.0 \times 10^3 \text{ Å}} = 6.2 \text{ eV}$$

From the photoelectric equation, the work function for the tungsten surface is

$$W = hf - K = 6.2 \text{ eV} - 1.6 \text{ eV} = 4.6 \text{ eV}$$

Einstein's bold postulate, that light had a particle nature after all, even though physicists had successfully pictured light as a wave for a hundred years, was suggested to him by the prior work of Max Planck. In 1900 Planck had found that the relative amounts of electromagnetic energy radiated at different wavelengths in the continuous spectrum emitted by a hot substance could be understood only if one assumed that energy transfers from matter to the radiation always occurred in bundles of the size $E = hf$. (Planck attributed this limitation to the nature of the oscillating matter, not to the idea that the radiation consists of photons.)

So far we have spoken only of Einstein's successful interpretation of the photo-electric effect. Of course Einstein, and many others, tried first to understand the phenomenon in terms of the way the electric vector of an electromagnetic wave would be expected to interact with the electrons in the metal. This classical theory failed in three important respects:

1   The kinetic energy acquired by an electron should increase as the field ampli-tude E_0 increases, and consequently as the light intensity increases. But K is independent of E_0.
2   The electron's kinetic energy should not depend particularly on the light frequency f. But $K = hf - W$.
3   With a low-intensity light, it would take many seconds for an electron to accumulate sufficient energy to escape if the incident energy in the electro-magnetic wave were continuously distributed across the wave fronts. But electron ejection commences with a delay of less than 10^{-9} s.

Clearly, old ideas have to be amended to explain the photoelectric effect. The success of the photon hypothesis is striking in this instance. If true, photons should abound in our world and the thing to do is to examine other processes for evidence of their existence. Shorter wavelength electromagnetic radiation should be investigated because here the energy of each photon would be larger and the presence of photons easier to detect.

Bremsstrahlung X Rays

The photoelectric effect involves the *absorption* of light. *Emission* processes also contribute evidence of photons.

Consider a specific case. Say electrons in an x-ray tube are accelerated through a voltage of 50×10^3 V; they acquire a kinetic energy of 50×10^3 eV before striking the anode. Within the anode, some of these electrons passing close to a highly charged nucleus are subjected to a large acceleration and radiate away energy as electromagnetic waves of x-ray wavelength (Figure 41-2). X rays pro-duced in this way are called *bremsstrahlung*, the German word for "braking radia-tion." Measurement of the wavelengths of bremsstrahlung x rays reveals a con-tinuous spectrum with a minimum wavelength (Figure 41-3).

The reason for a minimum wavelength can be easily understood in terms of photons. In this process of an electron radiating a photon, the maximum energy

Figure 41-2   A bremsstrahlung collision of a high-speed electron of initial kinetic energy K_1 with a nucleus.

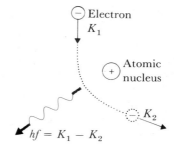

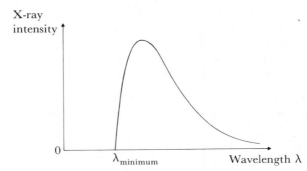

Figure 41-3   Bremsstrahlung gives a continuous x-ray spectrum with a minimum wavelength.

that the photon can acquire is the entire kinetic energy of the electron. And this maximum photon energy $E = hf$ corresponds to a maximum frequency f or a minimum wavelength λ.

Example 4   A potential drop of 50×10^3 V is maintained between the anode and the cathode of an x-ray tube. Find the shortest wavelength λ in the spectrum of the emitted x rays.

Solution   Electrons acquire a kinetic energy of 50×10^3 eV before they hit the anode. The maximum photon energy is therefore 50×10^3 eV. From Eq. 41.3 a photon with this energy has a wavelength given by

$$\lambda = \frac{12.4 \times 10^3 \text{ Å} \cdot \text{eV}}{50 \times 10^3 \text{ eV}} = 0.25 \text{ Å}$$

Photons with a lower energy will have a longer wavelength than this.

Compton Scattering of X Rays

The particle nature of x rays is very evident when they are scattered by free or loosely bound electrons. A. H. Compton (1892–1962), in an experimental investigation of x-ray scattering, found that part of the scattered radiation has a longer wavelength than that of the incident radiation. On the basis of classical physics, the scattered radiation results from electron oscillations produced by the incident radiation and therefore has the same frequency (and wavelength) as this incident radiation. Here again, classical physics gives a wrong answer. But Compton's analysis of the scattering process as a photon–electron collision (Figure 41-4) provides a clear and precise account of his experimental observations.

Consider an interaction or "collision" between an incident photon with momentum **p** and a stationary electron of mass m_e. After the interaction, a photon with momentum **p**′ emerges and the electron recoils with momentum **P** and total energy E (Figure 41-5). The problem is to determine the wavelength

$\lambda' = h/p'$ of the scattered photon in terms of the wavelength $\lambda = h/p$ of the incident photon and the scattering angle θ.

The law of conservation of energy implies

$$cp + m_e c^2 = cp' + E$$

The law of conservation of momentum gives

$$\mathbf{p} = \mathbf{p}' + \mathbf{P}$$

The magnitude of these momenta are therefore related as shown in Figure 41-5(b), and the cosine law applied to this triangle yields

$$P^2 = p^2 + p'^2 - 2pp' \cos \theta$$

The total energy of the electron is related to its momentum by Eq. 40.17:

$$E^2 = (m_e c^2)^2 + (cP)^2$$

In this equation we substitute for E the value $c(p - p' + m_e c)$ determined from the law of conservation of energy, and we also substitute for P^2 the value determined from the law of conservation of momentum. This gives, after removing a factor of c^2,

$$(p - p' + m_e c)^2 = m_e^2 c^2 + p^2 + p'^2 - 2pp' \cos \theta$$

or

$$p^2 - 2pp' + p'^2 + 2(p - p')m_e c + m_e^2 c^2 = m_e^2 c^2 + p^2 + p'^2 - 2pp' \cos \theta$$

Figure 41-4 Compton effect. The scattered photon has less energy and therefore a longer wavelength than the incident photon.

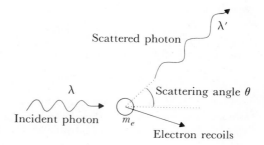

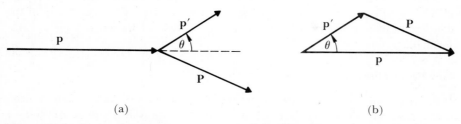

(a)   (b)

Figure 41-5 Photon's momentum vector changes from **p** to **p'** and the electron acquires a momentum **P**.

Therefore

$$(p - p')m_e c = pp'(1 - \cos \theta)$$

$$\frac{1}{p'} - \frac{1}{p} = \frac{1}{m_e c}(1 - \cos \theta)$$

With $p' = h/\lambda'$ and $p = h/\lambda$, we have Compton's equation

$$\lambda' - \lambda = \frac{h}{m_e c}(1 - \cos \theta) \tag{41.7}$$

The quantity $h/m_e c$ has the value 0.024 Å.

The agreement between the predictions of Compton's equation and experimental measurements of λ' as a function of θ is impressive and provides particularly direct and convincing evidence of the existence of photons, zero-mass particles that carry both momentum **p** and energy cp.

Photons from Pair Annihilation

The electron and the positron are a pair of what are called *antiparticles*. They have exactly equal masses m_e but opposite electric charges, and they can annihilate one another when they come close enough ($\sim 10^{-9}$ m) to interact significantly. After annihilation there is nothing left but photons, usually two (Figure 41-6). Assuming that the electron and the positron are at rest before annihilation, the law of conservation of energy yields

$$E_{\text{electron}} + E_{\text{positron}} = E_{\text{photon}} + E_{\text{photon}}$$

which gives

$$(m_e c^2 + 0) + (m_e c^2 + 0) = E_{\text{photon}} + E_{\text{photon}}$$

The electron and the positron each have a rest energy of 0.51 MeV. Therefore

$$1.02 \text{ MeV} = E_{\text{photon}} + E_{\text{photon}}$$

so there is 1.02 MeV to be shared by the photons. The law of conservation of momentum implies that these photons have momentum vectors in opposite directions and of equal magnitude; therefore the two photons must have the same energy, 0.51 MeV. This is just what is observed experimentally; the electron and the positron disappear and two 0.51 MeV photons are created which go off in opposite directions. This process is a particularly striking example of the conversion of rest energy into kinetic energy; 100% of the rest energy is transformed.

Pair Production by a Photon

If a photon has enough energy it can produce a pair of antiparticles, an electron and a positron (Figure 41-7). (The law of conservation of momentum requires the presence of a third particle, in practice the nucleus of a massive atom, to carry off some momentum. The kinetic energy acquired by this nucleus is

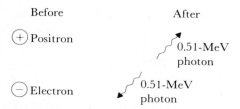

Figure 41-6 The annihilation of an electron–positron pair.

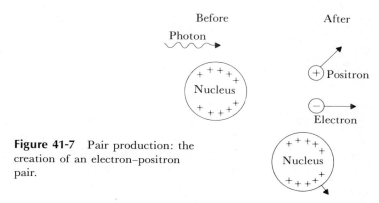

Figure 41-7 Pair production: the creation of an electron–positron pair.

negligible compared to the photon energy.) The law of conservation of energy for this pair-production process gives

$$E_{\text{photon}} = (m_e c^2 + K_{\text{electron}}) + (m_e c^2 + K_{\text{positron}})$$

This yields

$$E_{\text{photon}} = 1.02 \text{ MeV} + K_{\text{electron}} + K_{\text{positron}} \tag{41.8}$$

1.02 MeV of the photon's energy appears as rest energy of the electron and the positron, and the remainder appears as kinetic energy. Experimentally we do find that a photon energy of at least 1.02 MeV is required to make this process occur, and many experiments confirm the energy balance of Eq. 41.8.

41.3 Energy Levels

Photons and Energy Levels of Atoms

Photons were the key to understanding the emission and absorption of light by atoms.

Voluminous tables of precise measurements of the wavelengths of the lines in atomic spectra were accumulated throughout the nineteenth century. But no one saw how to interpret the information about the mechanics of an atom that was contained in these spectra. It was apparent that there was order in the different spectra, and painstaking efforts were made to find an empirical formula that would fit the facts. The first success came in 1885 to a Swiss schoolteacher, Johann Balmer (1825–1898), who found a formula for certain lines in the visible spectrum of the simplest atom, hydrogen.

In the early twentieth century, all lines in the hydrogen atom's spectrum were known to be given by the then seemingly peculiar empirical rule: Calculate a sequence of terms proportional to the reciprocal of the square of an integer n; then each frequency in the spectrum is the difference of two of these terms.

In 1913 Niels Bohr (1885–1962) pointed out the physical laws underlying this rule. When an atom emits a photon, the law of conservation of energy implies that *the energy of the atom must change* from an initial value E_u (the subscript u denotes the *upper* energy level, as in Figure 41-8) to a lower value E_ℓ such that

$$E_{\text{photon}} = E_u - E_\ell \tag{41.9}$$

This "Bohr frequency condition" determines the photon frequency f, since $E_{\text{photon}} = hf$. The empirically determined sequence of terms, whose differences determine the frequencies in the hydrogen atom spectrum, must then be proportional to the possible values of the energy of the hydrogen atom. These energies,* called the *energy levels* of the hydrogen atom, are given by

$$E_n = -\frac{13.6 \text{ eV}}{n^2} \tag{41.10}$$

where n is any positive integer and is called the *principal quantum number*. That is: $E_1 = -13.6$ eV, $E_2 = -3.40$ eV, $E_3 = -1.51$ eV, and so on, as displayed in Figure 41-9.

For example, a hydrogen atom can exist for a short while ($\sim 10^{-8}$ s) in a state with energy $E_3 = -1.51$ eV. If, after the emission of a photon, the atom is left in the state with the lower energy, $E_2 = -3.40$ eV (Figure 41-9), the photon emitted must have an energy, according to Eq. 41.9, given by

$$E_{\text{photon}} = E_3 - E_2 = (-1.51 \text{ eV}) - (-3.40 \text{ eV}) = 1.89 \text{ eV}$$

The wavelength λ of this photon is

$$\lambda = \frac{12.4 \times 10^3 \text{ Å·eV}}{1.89 \text{ eV}} = 6.56 \times 10^3 \text{ Å}$$

The novel idea thus advanced by Bohr is that the energy of the hydrogen atom (and in fact all atoms, molecules, and any bound system) can have only certain *discrete values* (instead of a continuous range of values) in its bound states. That is, there exist discrete *energy levels*.

The lowest of these energy levels is called the *ground state*, and all the higher levels are called *excited states*. The value $E = 0$ is the energy when the electron

---

*The total energy of the hydrogen atom, including the rest energy of the proton ($m_p c^2$) and the rest energy of the electron ($m_e c^2$) is $m_p c^2 + m_e c^2 + E_n$. The quantity $-E_n$ is therefore just the binding energy of the atom, and E_n is a negative number because this proton–electron system with attractive Coulomb forces is stable against decomposition into a separated electron and proton.

and the proton are completely separated and at rest. Since this energy level is 13.6 eV above the ground state, we see that 13.6 eV must be furnished to a hydrogen atom in its ground state in order to remove the electron, that is, to *ionize* the atom. In other words, the binding energy of a hydrogen atom against separation into a proton and an electron is 13.6 eV.

When the electron and proton are separated they can have any amount of kinetic energy. Corresponding to these states which are *not* bound states, the energy-level diagram (Figure 41-9) shows a continuous range of possible *positive* values of the energy of the electron–proton system.

When an atom gives off energy it passes from an upper to a lower energy level. If an atom absorbs energy it passes from a lower to a higher level (Figure 41-8). For the absorption of a photon, the Bohr frequency condition still applies, but now the lower energy E_ℓ is the initial energy of the atom.

The existence of energy levels in all atoms and the truth of the preceding statements were abundantly verified in experiments performed during the decade following the publication of Bohr's great paper in 1913 in which he both used and laid aside parts of Newtonian mechanics (Question S-6).

Newtonian mechanics, applied to a system of a proton and an orbiting electron, does *not* lead to discrete energy levels. Bohr realized that Newtonian mechanics would have to be modified in some way. His preliminary efforts met only partial success. The extent of the drastic revision of thought required to

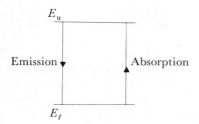

Figure 41-8 Two energy levels, E_u and E_ℓ, and the transitions which accompany either emission or absorption of energy.

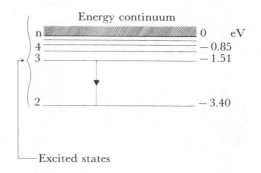

Figure 41-9 Energy levels of the hydrogen atom.

build a mechanics that would encompass the experimental knowledge of atoms was not apparent until the next decade.

Spontaneous Decay, Mean Life

The excited states of an atom are unstable. For an isolated atom, the transition from an excited state to a lower level occurs spontaneously and is accompanied by the emission of a photon. For a given atom in an unstable state, no way has been found of predicting when the transition will occur. But experiments show that the probability per unit time of a transition taking place is independent of the length of time that the unstable state has lasted.

Each state is characterized by a constant τ such that the probability of a spontaneous transition occurring in a time interval dt (short compared to τ) is given by

$$probability = \frac{dt}{\tau} \tag{41.11}$$

This means that if there is a large number N of atoms in the given state, the number of transitions in the interval dt is $N\,dt/\tau$ and, in the absence of some mechanism that supplies more atoms in this state,

$$dN = -\frac{N\,dt}{\tau} \tag{41.12}$$

The solution to this differential equation is

$$N = N_0 e^{-t/\tau} \tag{41.13}$$

where N_0 is the value of N at $t = 0$. This is an exponential decay (Figure 26-22).

The quantity τ is called the *mean life* of the unstable state. In any time interval τ, the number remaining in the unstable state is reduced by a factor of $1/e$. A related quantity is the *half-life* T which is defined as the time interval required for the number remaining to be reduced by one-half. This implies that

$$\tfrac{1}{2}N_0 = N_0 e^{-T/\tau}$$

and therefore the half-life is

$$T = \tau \ln 2 = 0.693\tau \tag{41.14}$$

The preceding discussion applies also to the decay of an unstable particle. For example, the pion has a mean life of 2.6×10^{-8} s (or a half-life of 1.8×10^{-8} s). It suddenly and spontaneously decays into a muon and a neutrino. (See Example 2 of Chapter 40 for an examination of the energies involved.)

In this connection it should be stressed that experiments show that all particles can be created and destroyed. We do not have to assume that the original particle is actually composed of the particles that appear when it decays. In fact, in photon emission or in the decay of an "elementary particle" (Section 43.9), decay products such as photons, electrons, positrons, and neutrinos are not regarded as previously existing in the original object, but instead are considered as having been *created* in the decay process.

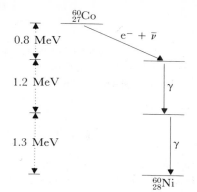

Figure 41-10   Nuclear energy-level diagram showing the β decay of cobalt-60.

Nuclear Energy Levels, Radioactivity

The occurrence of discrete energy levels in a hydrogen atom is a particular instance of a very general phenomenon. The energy of any *bound system* is restricted to certain discrete values which are called the energy levels of the system. For instance, the nucleus $^{60}_{28}$Ni has a stable ground state and many discrete energy levels corresponding to excited states. Each excited state is unstable with a characteristic mean life τ.

The spontaneous decay of a nucleus, a phenomenon called *radioactivity*, is classified according to the particle emitted: in α decay an α particle ($^4_2$He) is emitted, in β decay an electron or a positron is one of the decay products, and in γ decay a photon is emitted. For historical reasons, photons emitted from a nucleus are often called γ rays.

Figure 41-10 shows the nuclear energy levels involved in the β decay of radioactive cobalt-60:

$$^{60}_{27}\text{Co} \longrightarrow {}^{60}_{28}\text{Ni} + \text{e}^- + \bar{\nu}$$

The cobalt nucleus has a half-life of 5.24 years. In the decay process, this nucleus makes a transition from its ground state to an excited state of the nickel nucleus as an electron (e$^-$) and an antineutrino ($\bar{\nu}$) are created with kinetic energies whose sum is 0.3 MeV. This excited state of the nickel nucleus is unstable and decays (for all practical purposes immediately)* to another excited state which is 1.2 MeV lower, as it emits a 1.2-MeV γ ray. Finally the nucleus in this state, which is 1.3 MeV above the ground state, suddenly changes to the ground state configuration and simultaneously emits a 1.3-MeV γ ray. The β decay is therefore accompanied by two γ rays of energies 1.2 MeV and 1.3 MeV. These γ rays are widely used in medicine and in industry.

The number N of identical radioactive nuclei remaining at time t when there is no replenishment is given by Eq. 41.13,

$$N = N_0 e^{-t/\tau}$$

where τ is the mean life of the nuclei. This law, which we have seen has a very general applicability, is called the *law of radioactive decay* because it was discovered first in studies of radioactivity by E. Rutherford (1871–1937) and F. Soddy (1877–1956).

---

*The shortest lifetimes of unstable states of nuclei that can decay only by γ emission are of the order of 10^{-16} s.

Notice that as far as the decay of one particular particle is concerned, the law of radioactive decay allows us to compute only probabilities. Decays happen by *chance*. This suggests that some of the most fundamental laws of nature are laws that determine merely probabilities.

Summary

☐ A beam of electromagnetic radiation of frequency f and wavelength λ consists of identical photons, each with an energy

$$E = hf$$

and a momentum

$$p = \frac{h}{\lambda}$$

☐ Einstein's photoelectric equation is

$$K = hf - W$$

where K is the maximum kinetic energy of the electrons ejected from a metal surface which has a work function W, and hf is the energy of the photons striking the surface.

☐ When a photon of wavelength λ "collides" with a free electron of mass m_e, the electron recoils and the scattered photon has less energy than the incident photon, and therefore a longer wavelength λ'. Compton's equation is

$$\lambda' - \lambda = \frac{h}{m_e c}(1 - \cos \theta)$$

where θ is the scattering angle.

☐ The energy and hence the frequency of a photon emitted or absorbed when the energy of an atom changes, is given by the Bohr frequency condition

$$E_{\text{photon}} = E_u - E_\ell$$

where E_u and E_ℓ are the values of the atom's upper and lower energy levels involved in the transition. In a bound state, the possible values of the energy (the discrete energy levels) of a hydrogen atom are

$$E_n = -\frac{13.6 \text{ eV}}{n^2} \qquad (n = 1, 2, \ldots)$$

☐ The number N of identical particles in a given unstable state decays exponentially (if there is no replenishment):

$$N = N_0 e^{-t/\tau}$$

where the constant τ is the *mean life* of the state. The half-life is

$$T = 0.693\tau$$

Questions

1 Arrange light beams of the colors red, green, and violet, in order of increasing:
   (a) Wavelength.
   (b) Frequency.
   (c) Photon energy.

2 Find the energy in electronvolts of a photon of green light for which $\lambda = 5.0 \times 10^3$ Å.

3 What is the wavelength of a beam of light consisting of photons which have an energy of 2.0 eV each? What color is this light?

4 The γ-ray photon emitted when a proton captures a slow neutron has an energy of 2.3 MeV. Find the wavelength of this photon and compare it with a typical radius of a nucleus of low atomic number, say 10^{-15} m.

5 A table top of area 2.0 m^2 is uniformly illuminated by 10 W/m^2 of light with a wavelength of 6.2×10^3 Å. How many photons per second strike the table?

6 A radio station with an average power output of 4.0 kW broadcasts at a frequency of 1.5×10^6 Hz:
   (a) What is the energy of a photon in electronvolts?
   (b) How many photons per second are emitted?

7 What is the number N (photons per second per unit cross-sectional area) in the laser beam described in Question 6 of Chapter 35? (The wavelength of this red laser light is 6.3×10^3 Å.)

8 (a) What is the photoelectric effect?
   (b) State and explain the relationship between the maximum kinetic energy of the electrons ejected from a surface and the frequency of the light used to irradiate the surface.

9 Green light ejects electrons from a certain surface. Yellow light does not. Do you expect electrons to be ejected when the surface is illuminated with:
   (a) Red light? Why?
   (b) Violet light? Why?

10 The work function for a cesium surface is 2.0 eV:
   (a) Find the maximum kinetic energy of the ejected electrons when the surface is illuminated by violet light with $\lambda = 4.13 \times 10^3$ Å.
   (b) What is the *threshold wavelength* (the largest λ) for the photoelectric effect to occur with this metal?

11 When a clean zinc surface is irradiated with ultraviolet light we find that no electrons are ejected from the surface unless the light has a wavelength less than 2.93×10^3 Å:
   (a) What is the work function in electronvolts of the zinc surface?
   (b) What is the maximum kinetic energy of the electrons that are ejected by light of wavelength 1.24×10^3 Å.

12 In what ways are the experimental facts regarding the photoelectric effect in disagreement with the predictions of classical electromagnetic theory?

13 (a) Describe the mechanism for production of bremsstrahlung x rays.
   (b) What is the shortest wavelength that will emerge from an x-ray tube when the voltage between the cathode and anode is 100 kV?

14 **(a)** What is the Compton effect?

    (b) Why does the photon, scattered from a free electron at rest, have a longer wavelength than the incident photon?

15 Using Compton's equation, $\lambda' - \lambda = (0.024 \text{ Å})(1 - \cos\theta)$, calculate the increases in wavelength, $\Delta\lambda = \lambda' - \lambda$, which correspond to scattering angles of $0°$, $30°$, $90°$, and $180°$.

16 A photon with a wavelength of 0.049 Å is scattered backward ($\theta = 180°$) by an electron initially at rest:

    (a) Find the initial energy and the final energy of the photon.

    (b) What is the kinetic energy of the electron after this event? (Use the result of the preceding question for $\theta = 180°$.)

17 Find the magnitude (in eV/c) and the direction of the momentum vectors of the incident photon, the scattered photon, and the recoiling electron of the preceding question.

18 What is the wavelength of the incident photons if Compton scattering at an angle of $90°$ gives photons with energies equal to one-half the energies of incident photons?

19 What is the wavelength of the "annihilation radiation" that is observed when a positron and an electron (both essentially at rest) are annihilated and two photons are created?

20 A photon with an energy equal to $4m_e c^2$ produces an electron–positron pair in the vicinity of a nucleus. Both particles (each of mass m_e) travel in the forward direction with the same kinetic energy. Find the magnitude and direction of the momentum of the nucleus.

21 A hydrogen atom in the excited state, $E_2 = -3.40$ eV, makes a transition to its ground state, $E_1 = -13.6$ eV, and emits a photon. Find the photon energy and wavelength. Is this photon in the ultraviolet, the visible, or the infrared portion of the spectrum of electromagnetic waves?

22 The color plate in Chapter 37 shows four lines of the hydrogen atom spectrum. From the wavelengths given on this plate, determine the corresponding photon energies and identify on an energy-level diagram the transitions associated with the emission of these photons.

23 The four lines referred to in the preceding question are the first four lines of the Balmer series. Find the photon energy and the wavelength for the next line in this series.

24 A collection of hydrogen atoms in their ground states is irradiated with ultraviolet photons which each have an energy of 12.1 eV. Some atoms each absorb a photon and are excited to the state with energy $E_3 = -1.51$ eV:

    (a) Draw an energy-level diagram and label with arrows all the possible transitions to lower energy levels.

    (b) Calculate the energies and wavelengths of all the photons that will be emitted by the hydrogen atoms.

25 A hydrogen atom in its ground state is ionized by the absorption of an 800-Å photon. (This is an "atomic photoelectric effect.") Find the kinetic energy of the ejected electron.

26 After a collision with some other particle, an atom must be left either in its ground state or in an excited state, with an energy corresponding to one of its possible energy levels. Consider hydrogen that is bombarded by a beam of electrons, each of which has a kinetic energy of 12.5 eV:

(a) What excited states are possible for a hydrogen atom after interaction with an electron in this beam?

(b) What wavelengths will be emitted by hydrogen atoms after collision with electrons of this beam?

27 When hydrogen is bombarded by a beam of electrons of kinetic energy K, what is the lowest value of K for which *visible* light will be emitted by the hydrogen atoms? What color is this light? (Assume that the kinetic energy of a hydrogen atom is negligible compared to the other energies involved in this process.)

28 If a gas contains N atoms in an excited state which has a mean life τ, how many photons per second will be emitted because of spontaneous transitions from this state to some lower state? Explain.

29 The half-life of a certain excited state of an atom is 0.50×10^{-8} s:

(a) What is the mean life of this state?

(b) A collection of atoms contains 1.0×10^8 atoms in this excited state at $t = 0$. Assuming that the only significant processes are spontaneous transitions to lower energy levels, how many atoms are expected to be found in this excited state at the following times: $t = 0.50 \times 10^{-8}$ s, $t = 1.0 \times 10^{-8}$ s, $t = 2.0 \times 10^{-8}$ s, $t = 5.0 \times 10^{-8}$ s, and $t = 50 \times 10^{-8}$ s?

30 What is the change in the energy of a nucleus that emits a γ ray with a wavelength of 3.1×10^{-3} Å?

31 Consider a sample containing 2.7×10^{21} atoms of radium-226 (about 1 gram of radium). The half-life of radium is 5.1×10^{10} s (over 1600 years). How many radium nuclei in this sample decay in 1.0 s?

32 A sample of material containing radioactive nuclei is said to have an *activity* of 1 *curie* if there are 3.7×10^{10} decays per second. What is the activity of the sample in the preceding question? (Assume that there has been no significant accumulation of the radioactive daughter, $^{222}_{86}$Rn.)

33 How many α particles are emitted per second from a milligram of polonium-218 which has a half-life of 3.05 min? (The number of atoms in 1 gram of polonium-218 is approximately 2.7×10^{21}.)

34 Approximately how many counts per second would a radiation detector indicate when held near a 20-millicurie source of $^{60}_{27}$Co? Assume that the electrons (β particles) emitted by the radioactive nucleus cannot penetrate the walls of the detector and that only 1.0% of the γ rays (Figure 41-10) emitted by the source are detected.

Supplementary Questions

S-1 One method of determining the value of Planck's constant is based on observations of the photoelectric effect. Suppose that a metallic surface is irradiated by light and that measurement is made of the retarding voltage required to stop the most energetic of the electrons ejected from the surface. Suppose also that for light with a wavelength of 4.00×10^3 Å the stopping potential is 2.00 V, while for light with a wavelength of 6.00×10^3 Å the stopping potential is 1.00 V. Use these data to determine the value of Planck's constant and the work function of the surface.

S-2 When the voltage between the cathode and the anode of an x-ray tube is maintained at 62.5 kV, we find that 0.20 Å is the shortest wavelength in the spectrum of the emitted x rays. Calculate the value of Planck's constant from these data.

S-3   X rays with a wavelength of 0.036 Å emerge from a metal target at an angle of 90° with the direction of the incident beam. Assume that these x rays are scattered from free electrons. Find:

(a) The wavelength of the x rays in the incident beam.

(b) The momentum vectors of an incident photon and a scattered photon.

(c) The magnitude and direction of the momentum vector of a recoiling electron that participates in such a scattering event.

S-4   In the description of the Compton effect given in Section 41.2, we mention that *part* of the scattered radiation has a longer wavelength than that of the incident radiation. It should also be pointed out that the other part of the scattered radiation (called the unmodified radiation) has essentially the same wavelength as the incident beam. The origin of this unmodified scattered radiation can be understood by considering the change in photon wavelength $\lambda'' - \lambda$ which occurs in scattering events in which the photon interacts with an atom of mass M and the entire atom recoils. The appropriate form of Compton's equation is then $\lambda'' - \lambda = (h/Mc)(1 - \cos \theta)$. Show that

$$\frac{\lambda'' - \lambda}{\lambda' - \lambda} = \frac{m_e}{M}$$

where $\lambda' - \lambda$ is the change in photon wavelength when the recoiling particle is an electron of mass m_e. Then calculate the wavelengths λ' and λ'' that will be observed at a scattering angle of 90°, when x rays with $\lambda = 0.100$ Å are incident on carbon.

S-5   Following Bohr's line of thought, apply the principles of Newtonian mechanics to describe the motion of an electron (of mass m_e and charge e) acted upon by a Coulomb force ke^2/r exerted by a proton with charge e. To a good approximation, the acceleration of the relatively massive proton can be ignored and the proton can be regarded as stationary:

(a) Show that when the electron moves in a circular orbit of radius r about the proton, Newton's second law gives

$$\frac{ke^2}{r^2} = \frac{m_e v^2}{r}$$

(b) Show that when an electron is moving in a circular orbit of radius r, its total energy $K + U$ is given by

$$E = \tfrac{1}{2}m_e v^2 - \frac{ke^2}{r} = -\frac{ke^2}{2r}$$

S-6   As described in Section 41.3, Bohr concluded that the line spectrum of the hydrogen atom was evidence that the energy of the atom could have only certain discrete values (called energy levels) in its bound states. From the result of the previous question, $E = -ke^2/2r$, it is apparent that the possible values of the electron's total energy E will be restricted to certain discrete values, E_1, E_2, E_3, . . . , only if the radius of the electron's orbit is correspondingly restricted to certain discrete values, r_1, r_2, r_3, ... At this point Bohr made the brilliant guess that, for some reason foreign to Newtonian mechanics, the electron's orbital *angular momentum* $m_e vr$ is restricted to values which are an integral multiple of $\hbar$. That is, Bohr made the following *quantum hypothesis*:

$$m_e vr = n\hbar$$

where the possible values of the *quantum number n* are the integers 1, 2, 3, . . . :

(a) Show that this assumption, together with our previous result, $ke^2/r^2 = m_e v^2/r$, implies that the radii of the possible circular orbits are given by

$$r_n = n^2 a_0$$

where $a_0 = \hbar^2/m_e ke^2$ (the so-called Bohr radius).

(b) Next, using $E = -ke^2/2r$, with r restricted to the values $r_n = n^2 a_0$, show that possible values of E are given by

$$E = \frac{E_1}{n^2}$$

where $E_1 = -\frac{1}{2}\alpha^2 m_e c^2$ and $\alpha = ke^2/\hbar c$. This is the famous Bohr formula for the energy levels of the hydrogen atom.

S-7 Experimental measurement of the wavelengths in the line spectrum of the hydrogen atom shows that all the observed wavelengths are given by the formula

$$\frac{1}{\lambda} = R\left(\frac{1}{n_\ell^2} - \frac{1}{n_u^2}\right)$$

where $n_\ell = 1, 2, 3, \ldots$ and n_u is an integer greater than n_ℓ. The experimental value of the constant R, called the Rydberg constant, is $1.09677 \times 10^7 \text{ m}^{-1}$. Using Bohr's formula for the energy levels of the hydrogen atom, show that Bohr's theory implies that

$$R = \frac{\alpha}{4\pi a_0}$$

where $\alpha = ke^2/\hbar c$ and $a_0 = \hbar^2/m_e ke^2$. Compare the predicted value $\alpha/4\pi a_0$ with the experimental value of R.

S-8 Show that the shortest wavelength in the line spectrum of hydrogen is much larger than the Bohr radius a_0.

S-9 Show that, according to Newtonian mechanics, the electron's speed in the nth Bohr orbit (of radius $r_n = n^2 a_0$) is given by

$$v = \frac{v_1}{n}$$

where $v_1 = \alpha c$ and $\alpha = ke^2/\hbar c$. (*Hint:* Use the result $ke^2/r^2 = m_e v^2/r$, together with $r_n = n^2 a_0$.)

S-10 Use the postulates of Bohr's theory of the hydrogen atom to deduce an expression for the energy levels of singly ionized helium (a helium atom from which one electron has been removed).

S-11 According to the Bohr theory, what is the radius of the electron orbit for a singly ionized helium ion in its ground state?

S-12 **(a)** Consider the frequency f of radiation emitted by a hydrogen atom in a transition from an energy level with quantum number $n + 1$ to a level with a quantum number n. Using the expression $E_n = -\frac{1}{2}\alpha^2 m_e c^2/n^2$, show that if n is large the radiated frequency is given approximately by

$$hf = \frac{\alpha^2 m_e c^2}{n^3}$$

(b) Classical theory predicts that an electron moving in a circular orbit should radiate electromagnetic waves with a frequency equal to that of the electron's revolution in its orbit. What is this frequency for an electron traveling in an orbit of radius $r_n = n^2 a_0$ at a speed $v = \alpha c/n$?

(c) Bohr's thoughts were guided by what he called the *correspondence principle*, which may be stated as follows: Predictions of quantum theory must agree with predictions of classical theory in the limit where quantum discontinuities may be treated as negligibly small (as is the case when the quantum numbers are very large). Are the results of parts (a) and (b) consistent with the correspondence principle?

S-13 A sample contains N_A radioactive nuclei of type A which decay exponentially according to $N_A = N_{0A}e^{-t/\tau}$. The daughter nucleus is stable. Find an expression for the number N_B of daughter nuclei that are present at time t, assuming that none are present at $t = 0$.

S-14 Assume that, in the preceding question, the daughter nucleus is also radioactive. In *radioactive equilibrium*, the daughter decays at the same rate that it is formed. Show that this condition is obtained when

$$\frac{N_A}{\tau_A} = \frac{N_B}{\tau_B}$$

42.

particle interference and quantum mechanics

The preceding chapter presents evidence showing that the phenomena of light are due to particles called photons. Experiments such as the Young double-slit experiment (Section 36.2) consequently are to be interpreted as showing that interference effects are obtained with photons. And this is not due to some peculiar property of photons. Surprisingly enough, experiments indicate that all particles show interference effects.

By examining in detail the implications of particle interference, we are led to consider new postulates fundamental to the most successful physical theory that has been created, quantum mechanics.

42.1 de Broglie Wavelength of a Particle, Crystal Diffraction

An electromagnetic wave of wavelength λ is associated with photons of momentum p such that $\lambda = h/p$. In 1924, a young French student, Louis de Broglie (1892–   ), submitted a Ph.D. thesis in which he speculated that, with *any particle* of momentum p, there was in some way associated a wave with wavelength λ given by

$$\lambda = \frac{h}{p} \tag{42.1}$$

It soon became apparent that he was right. This formula does have a universal significance.

The "de Broglie wavelength" in the case of most particles is very small (of the order of 1 Å or less) if the energy of the particle is large enough to make its detection easy. As we learned with light, typical wave effects such as diffraction are not evident if the wavelength is much shorter than the significant geometric dimensions in an experiment. When a light beam is replaced by a beam of x rays, electrons, or neutrons, it is extremely difficult to duplicate the familiar interference experiments with standard optical apparatus; the size and the separation of the slits are too large compared to the wavelength. Fortunately a diffraction grating with the dimensions required is provided by nature in the form of crystals.

Crystal Diffraction

A crystal consists of a regular array of atoms with interatomic spacings of a few ångström units (Figure 42-1). When placed in the path of an incident beam with a wavelength λ comparable to the interatomic spacing, a crystal functions as a three-dimensional diffraction grating. Diffracted beams are produced in well-defined directions that depend on λ and on the crystal structure.

A straightforward explanation of the observed directions of the beams diffracted from a crystal was given by W. L. Bragg (1890–   ). To simplify the discussion we shall assume that the crystal structure consists of a periodic lattice with identical atoms at each lattice point. When a plane wave is incident upon the crystal each atom gives rise to a spherical scattered wave. Consider the waves scattered from any one plane of atoms. In a direction such that the law of regular reflection is satisfied, all the waves scattered from this plane of atoms will be in phase. It is therefore convenient to consider the incident waves as reflected from parallel planes of atoms in the crystal (Figure 42-2). Each plane

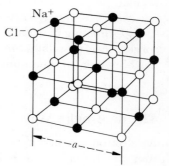

Figure 42-1 The sodium chloride crystal structure. The cube edge a is 5.6 Å.

Figure 42-2 Derivation of the Bragg condition for the formation of a diffracted beam: $2d \sin \theta = n\lambda$. The set of parallel planes can be any set with the property that each plane passes through at least three noncollinear lattice points.

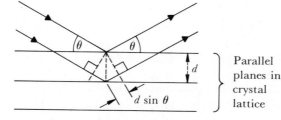

will reflect only a small fraction of the incident radiation. The path difference for rays from adjacent planes is $2d \sin \theta$, where θ is the *glancing angle* measured from the plane (rather than from the normal). Waves reflected from adjacent planes must be in phase for a diffracted beam to have an appreciable intensity, and this occurs only if the path difference for these reflections is an integral number of wavelengths; that is

$$2d \sin \theta = n\lambda \qquad (42.2)$$

where n is a positive integer called the order number. This is the *Bragg condition* for the formation of a diffracted beam.

In the experiment represented in Figure 42-3, the incident radiation is a beam of x rays with a continuous range of wavelengths. The crystal diffracts only discrete wavelengths for which planes of spacing d and glancing angle θ exist such that the Bragg condition is satisfied. This type of experiment was proposed

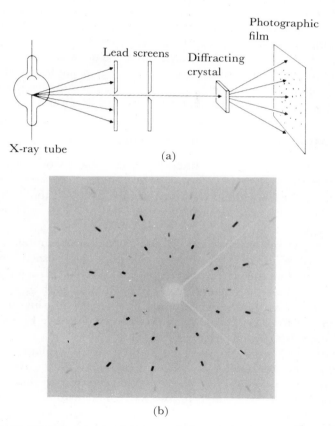

(a)

(b)

Figure 42-3 Von Laue pattern. (a) The experimental setup. (b) Laue diffraction by a crystal. The white central disk is due to a small metal disk held in front of the film by two wires (leading at an angle down toward the bottom of the film) to prevent extensive blackening of the film by the undeviated incident x-ray beam.

in 1912 by Max T. F. von Laue (1879–1960) as a method of showing that x rays were electromagnetic radiation of very short wavelength.

For crystals of known structure, the x-ray wavelength can be calculated. With x rays of known wavelength, the structure of other crystals can be determined. Many variations of the original experimental techniques are still in use today to determine the structure not only of crystals but also of molecules. For example, x-ray diffraction techniques have been used to discover the atomic structure of such complex molecules as $C_{22}H_{23}O_8N_2Cl$, known as Aureomycin.

When a beam of *neutrons* is incident on a crystal, we observe that the formation of diffracted beams of neutrons is governed by Bragg's equation, where the wavelength is determined from de Broglie's equation, $\lambda = h/p$. Because of their magnetic moment, neutrons interact with atoms which have a magnetic moment, and neutron diffraction studies are a valuable method of determining the structure of magnetic crystals.

Diffraction is also observed with an incident *electron* beam. Since electrons are charged, they interact strongly with matter and can penetrate only a relatively short distance into a crystal. Nevertheless electron diffraction experiments yield important information about very thin crystals, films, surfaces, and gases.

Crystal diffraction phenomena, apart from being useful for structural investigations in current research, are particularly significant in that they provide overwhelming evidence that de Broglie's equation $\lambda = h/p$ refers to a universal connection between wavelength and momentum valid for any type of particle.

42.2 The Mystery of Particle Interference

To bring out the startling implications of the fact that particles produce interference patterns, we shall consider a simple idealized interference experiment with electrons: the Young double-slit experiment but with the light source replaced by an electron gun (Figure 42-4) which shoots out electrons with a well-defined energy.

A photographic plate will detect the arrival of electrons on the screen. With both slits open we find that the photographic plate shows exactly the distribution of intensity I on the screen depicted in Figure 36-8. This is an interference pattern with intensity maxima ($I = 4I_1$ from Eq. 36.2 with $I_1 = I_2$) and intensity minima ($I = 0$). The separation of interference maxima on the screen is that expected from Eq. 36.11 where now λ is given by de Broglie's relation $\lambda = h/p$ and p is the momentum of each electron.

But what is going on? No one has ever been able to explain these observations in purely classical terms. Let us investigate more closely. We decrease the intensity by having the gun fire so few electrons per second that there is only one electron in transit at a time. Now we have to increase the exposure time of the film on the screen, but the *interference pattern is unchanged.* This shows that interference does *not* arise from the interaction of one electron in transit with another in transit. *An electron interferes with itself!* What now?

Interference therefore has to be understood by consideration of a *single particle* in transit and two alternatives, slit 1 or slit 2. Certain points are clarified by replacing the photographic plate of Figure 36-8 by the very small idealized counter

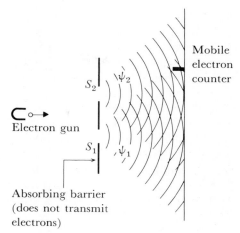

Figure 42-4   Double-slit experiment
with electrons.

of Figure 42-4 that counts the arrival of individual electrons. We find that this
counter clicks erratically suggesting that the timing of the arrival of an electron
is governed by chance. At intensity maxima, the counting rate is large, and at
intensity minima, the counting rate is small. In fact, the intensity is simply proportional to the counting rate.

Evidently the *probability P that a single particle* will reach a given small region
on the screen will determine the intensity I that will be observed at that point
after many particles, one after the other, have gone from the electron gun to the
screen. Since P and I are proportional, for each relationship between intensities,
we can write a corresponding relationship between probabilities. Thus, corresponding to the intensity equation, $I = I_1 + I_2 + 2\sqrt{I_1 I_2} \cos \phi$, given in
Eq. 36.2, we have for the probability:

$$P = P_1 + P_2 + 2\sqrt{P_1 P_2} \cos \phi \qquad (42.3)$$

where ϕ is the phase difference at the observation point of the de Broglie waves
from slits S_1 and S_2. The probability P_1 refers to the situation where slit 1 is
open and slit 2 is closed. We can consider P_1, and consequently the intensity I_1,
as uniform across the region of interest on the screen.* That is, there is no alteration of maxima or minima on the screen with just one slit open. The probability
P_2 has a similar interpretation.

The interference term $2\sqrt{P_1 P_2} \cos \phi$ has different values at different locations.
We consider the case where $P_1 = P_2$. At intensity maxima where the interference
is constructive, $\cos \phi = 1$ and corresponding to the equation $I = 4I_1$, we have

$$P = 4P_1$$

Where the interference is destructive, $\cos \phi = -1$ and corresponding to the
equation $I = 0$, we find

$$P = 0$$

---

*This can be achieved to a good approximation with a slit so narrow that its width is the size of the
de Broglie wavelength of the electrons.

We see that nature has presented us with *interfering probabilities*.

With both slits open there are two alternatives. A single particle is in transit. We might think (erroneously, as it turns out) we can assume that the particle goes through either slit 1 or slit 2, and that we can correctly deduce some consequences from this. One logical consequence of this assumption is that the presence of the slit that the particle does not go through does not affect the probability of the particle reaching a given point on the screen. But it does! For instance, when both slits are open the particle has zero probability of arriving at the location of the intensity minima. How does the slit that the particle does *not* go through prevent the particle from reaching such a location on the screen? And how is it that, at the location of an intensity maximum, the probability that an electron will reach such a region is not doubled but quadrupled by having two open slits instead of one? The classical physics of particles has no answer.

What about closing slit 1 for a day and then opening it and closing slit 2? Under these circumstances we are certain that the particles in transit go through one slit or the other, and at any instant we can determine which slit it is. But after this experiment we observe a completely uniform distribution. At every point on the screen

$$I = I_1 + I_2$$

and correspondingly

$$P = P_1 + P_2$$

The interference term is zero. In general we find that whatever we do, if we can ascertain by some experimental method which of the two alternatives (slit 1 or slit 2) is taken, the interference disappears.

How can things be so different when both slits are open for a day (interference occurs) than when one slit at a time is open for a day (no interference)? When both are open, does an electron split into two parts and a piece go through each slit? Definitely not. We always detect entire electrons, never part of an electron. Moreover, the entire story is not contingent upon using electrons. All the evidence suggests that *any particle will interfere with itself*. We are dealing with a *universal behavior of matter*, not with the peculiarities of some particular particle.

42.3 Quantum Mechanical Superposition Principle

The more we examine the interference manifested by a particle, the stranger it seems. Faced with particle interference, classical physics is at an impasse. However, we need not be reduced to a scientific paralysis as some Greek philosophers were by Zeno's paradoxes. The thing to do is to face the experimental facts and to build a new mechanics in accordance with these facts. This has been done, and the resulting theory is called *quantum mechanics*.

An encouraging feature in the face of the mystery of particle interference is that we have on hand a very simple way of correctly predicting all details of the interference pattern. Just think of the waves of wavelength $\lambda = h/p$. The square of the amplitude of the waves at a certain location on the screen is associated with the intensity or the probability that the particle will hit the screen at that location. Following this clue, physicists were led to the most important postulates upon which quantum mechanics is founded.

Without attempting a precise account, we shall merely indicate some salient features of this theory for a system consisting of a single particle. With each possible physical state at any instant we associate a function $\psi(x,y,z,t)$ called a wave function. At each point in space, at any instant, ψ is a complex number or phasor and can be represented on a phasor diagram by an arrow of length $|\psi|$ at an angle α with the horizontal axis (Figure 42-5). The wave function indirectly determines probabilities. In particular $|\psi(x,y,z,t)|^2$, the square of the magnitude of the wave function at the point (x,y,z), is proportional to the probability at the time t of the particle being found within a volume element enclosing the point (x,y,z).

A postulate of central importance is the *quantum mechanical superposition principle*: If ψ_1 and ψ_2 are wave functions corresponding to possible states of the system, then any linear combination,

$$\psi = a_1\psi_1 + a_2\psi_2 \tag{42.4}$$

is itself the wave function of still another possible state of the system. The state ψ is called a superposition of the states ψ_1 and ψ_2. The coefficients a_1 and a_2 are any constants. (They may be complex numbers.)

For instance, in the double-slit experiment, let ψ_1 be the wave function for a state which corresponds to the particle going through slit 1 as shown in Figure 42-4. Similarly, let ψ_2 be the wave function corresponding to the particle going through slit 2. Then according to the superposition principle, the superposition $\psi = \psi_1 + \psi_2$ is the wave function of another possible state. If we are asked to describe this new state ψ in words, we have to say that in this state the particle (without breaking up) goes through both slits! We can now appreciate that the quantum mechanical superposition principle, which seems so innocent and is such a simple postulate to state mathematically, introduces radically new possibilities. Many of these new situations seem to be nonsensical when we attempt a verbal description in terms of everyday ideas. But there is no difficulty in giving a quantum mechanical description. We are simply concerned with a superposition of states.

Now when the square of the magnitude of ψ ($=\psi_1 + \psi_2$) is computed to determine a probability, the cosine rule applied to the triangle in the phasor diagram of Figure 42-5 gives

$$|\psi|^2 = |\psi_1|^2 + |\psi_2|^2 + 2\sqrt{|\psi_1||\psi_2|} \cos \phi \tag{42.5}$$

Since $P \propto |\psi|^2$, $P_1 \propto |\psi_1|^2$, and $P_2 \propto |\psi_2|^2$, we obtain

$$P = P_1 + P_2 + 2\sqrt{P_1 P_2} \cos \phi$$

as we had found in the double-slit interference experiment and expressed in Eq. 42.3. Notice that the superposition principle together with the rule for computing probabilities leads to something different from $P = P_1 + P_2$. There is also

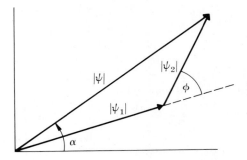

Figure 42-5 Phasor diagram for the phasors ψ_1, ψ_2, and $\psi = \psi_1 + \psi_2$.

an *interference term*. We have *interfering probabilities*. This remarkable quantum mechanical phenomenon, unnoticed in our everyday world, occurs often in experiments dealing with molecules, atoms, nuclei, and elementary particles.

42.4 Heisenberg Uncertainty Relation

Consider a superposition, $\psi' = \psi'_1 + \psi'_2$, of two states in which the particle has different momentum values along some direction OX; say for example that the state ψ'_1 corresponds to a momentum p_{1x} (that is, ψ'_1 has a wavelength $\lambda_1 = h/p_{1x}$), and state ψ'_2 corresponds to momentum p_{2x}. Then the superposition is a state ψ' in which the *particle does not have a single definite value of momentum*. A measurement could yield either the value p_{1x} or p_{2x}. Proceeding in this way to build the superposition of many states with different momentum values, we are led to consider a *state ψ which corresponds to a whole range of momentum values*. This spread of momentum values is aptly named the *uncertainty in momentum* and is usually denoted by Δp_x. (In this instance the symbol Δp_x refers to the *spread* of possible momentum values about their average value, not to a change in momentum.)

We now examine the same wave function ψ at different points in space along a line parallel to the axis OX. We find that, even though the wave may be very concentrated in one region and nearly zero everywhere else, there is a certain range of x values of width Δx where the wave function ψ has an appreciable size. A position measurement is most likely to give a value x within this range Δx. This range of x values, Δx, is named the *uncertainty in x*.

In Newtonian mechanics, a mechanical state is specified by giving a precise value of momentum and position. In quantum mechanics, a typical state ψ corresponds, not to definite momentum and position, but rather to ranges of momentum and position values measured by Δp_x and Δx. It can be shown that the postulates of quantum mechanics imply that for a given state ψ, the uncertainties Δp_x and Δx are related by the celebrated *Heisenberg uncertainty relation*:

$$\Delta x \Delta p_x \gtrsim \hbar \tag{42.6}*$$

*The symbol $\gtrsim$ means "is of the order of or is greater than." The analogous uncertainty relations for y and z components are

$$\Delta y \Delta p_y \gtrsim \hbar \qquad \Delta z \Delta p_z \gtrsim \hbar$$

Small Δx implies large Δp_x and vice versa. This means that if a state is such that ψ is well localized in space so that Δx is small, then ψ is a superposition of states with a large range Δp_x of different momentum values. On the other hand, if we have a state which is a superposition of states which have only a small spread Δp_x of momentum values, then the wave function ψ is widely spread out in space (Δx is large). In particular, there is no state possible in quantum theory for which *both* Δx and Δp_x are zero. The precise simultaneous specification of position and momentum (or velocity) characteristic of Newtonian mechanics does not occur in quantum theory.

A picture, mental or otherwise, of a particle trajectory with arbitrarily well-defined position and momentum at any instant is therefore not consistent with a quantum theory description. There must always be a smudging or a blur, a lack of definition, consistent with $\Delta x \Delta p_x \gtrsim \hbar$. For macroscopic objects such as a football (Question 11), both Δx and Δp_x can be so close to zero that for all practical purposes there is no uncertainty, because $\hbar$ is so small. But for an electron in a hydrogen atom, the uncertainty in position is of the order of the size of the atom itself. A picture of electron orbits in an atom thus disappears in quantum theory.

At this point the student may well say: "Put *theory* aside. Suppose we make a simultaneous measurement of position and momentum with such accuracy that $\Delta x \Delta p_x$ is smaller than $\hbar$. What then?" Then the particle would be in no state consistent with quantum theory, and this theory would be wrong on a most fundamental point. A major theoretical revision would be required.

But are such simultaneously accurate measurements possible, even in principle? It seems the answer is: "No." Heisenberg and others have analyzed many "thought experiments" and have found no situation which beats the uncertainty principle. For instance, if the particle is an electron and we are to make a position measurement using light, then we are concerned with a photon scattered by an electron which recoils. The scattered photon enters a detecting instrument through some aperture. Because of the diffraction of light, there are uncertainties associated with this photon's momentum and with the location of the scattering event. There is a related uncertainty in the measurement of position and momentum of the electron. The analysis can be made quantitative and the result is in agreement with the uncertainty principle.

42.5 Energy Levels in Quantum Mechanics

In Newtonian mechanics, a vibrating system such as a string fixed at both ends has certain natural frequencies (Section 33.5). Any possible vibration of the string can be expressed as a superposition of vibrations with these natural frequencies. Analogously, in quantum mechanics, an isolated system has certain characteristic frequencies called eigenfrequencies. Corresponding to each eigenfrequency is an energy given by the Einstein relation, $E = hf$. For a bound system there is a discrete set of eigenfrequencies $f_1, f_2, \ldots$ and a corresponding set of *discrete energy levels* $E_1, E_2, \ldots$.

The wave function ψ_1 for a state with energy E_1 is a phasor with a frequency $f_1 = E_1/h$ and is represented on a phasor diagram by an arrow of length $|\psi_1|$ that rotates with a frequency of f_1 revolutions per unit time. Such a wave function is called an energy eigenfunction. A general state ψ is a superposition of energy eigenfunctions.

For a system that can exchange energy with its surroundings, for example an atom that can emit a photon, an unstable state does not correspond to a single definite value of the energy. There is a spread of energy values, ranging roughly from some value E to $E + \Delta E$. The spread or width ΔE of the energy level is related to the mean life τ of the state by the *time-energy uncertainty relation*

$$\tau \Delta E \gtrsim \hbar \qquad (42.7)$$

The shorter the mean life of a state, the greater its width ΔE.

42.6 Remarks on Quantum Mechanics

The theory known as quantum mechanics was developed from 1924 to 1928. In contrast with the history of relativity, the quantum theory was the work of many men: Heisenberg, Schrödinger, Dirac, Born, and Pauli all won Nobel prizes for their great contributions.

The theory is abstract and, at first sight, strange, but so are the experimental facts of atomic physics. Our experience in the macroscopic world does little to prepare our intuition for successful responses in this new realm of the very, very small. In any case, quantum mechanics is the most quantitatively successful theory that science has seen. It is *the* underlying theory in atomic physics, chemistry, solid-state physics, and nuclear physics. Studies of interactions of elementary particles that are so copiously created and destroyed have not yet led to a comprehensive theoretical understanding of this branch of physics. Nevertheless the main principles of quantum mechanics seem to stand up. As far as we know, we live in a relativistic, quantum mechanical universe.

Before continuing this important topic, we shall mention several features of quantum mechanics, some of which we have already encountered. Each *state* of a mechanical system is described by a *wave function* ψ. The laws of quantum mechanics allow us to determine the state at some future time in terms of the present state. A knowledge of ψ enables us to compute *probabilities* for the results of measurements on the system. It is characteristic of both experimental and theoretical quantum physics that we deal with *probabilities rather than certainties*. The superposition $\psi = \psi_1 + \psi_2$ of two possible states ψ_1 and ψ_2 is another possible state and in such a state we encounter interfering probabilities. A typical state ψ corresponds to many possible values of a dynamical variable, such as the momentum p_x or the position x. The spread of values, or the uncertainties Δx and Δp_x, are related by $\Delta x \Delta p_x \gtrsim \hbar$, the Heisenberg uncertainty relation.

In the macroscopic world, where the predictions of Newtonian mechanics are in accord with experiments, quantum mechanics and Newtonian mechanics agree. This comes about because, for large masses, the de Broglie wavelength

$\lambda = h/p$ is so short compared to the size of macroscopic objects that the peculiar wave effects, characteristic of the microscopic world, are no longer apparent. The uncertainty relations pose no significant restriction for a description of macroscopic motion.

Summary

☐ The de Broglie wavelength of a particle with momentum p is

$$\lambda = \frac{h}{p}$$

☐ The Bragg condition for the formation of a diffracted beam, "reflected" at a glancing angle θ from a family of parallel planes of atoms in a crystal, is

$$2d \sin \theta = n\lambda$$

where d is the separation of adjacent planes and n is an integer called the order number.

☐ An interference pattern results from a particle "interfering with itself." In the double-slit experiment, the intensity within a small region on the screen is proportional to the probability P of a particle striking the region. With both slits open, there are interfering probabilities and

$$P = P_1 + P_2 + 2\sqrt{P_1 P_2} \cos \phi$$

where, at the observation point, the de Broglie waves from the two slits have a phase difference ϕ.

☐ For a quantum mechanical system consisting of a single particle, each state is described by a wave function $\psi(x,y,z,t)$ which has values that are complex numbers or phasors. $|\psi(x,y,z,t)|^2$ is proportional to the probability at time t of finding the particle in a volume element enclosing the point (x,y,z).

☐ The quantum mechanical superposition principle is the postulate that, if ψ_1 and ψ_2 are wave functions corresponding to possible states of the system, then the linear combination

$$\psi = a_1 \psi_1 + a_2 \psi_2$$

is a wave function corresponding to still another possible state of the system. (a_1 and a_2 are constants that may be complex numbers.) In the state corresponding to the superposition $\psi = \psi_1 + \psi_2$, we have interfering probabilities since

$$|\psi|^2 = |\psi_1|^2 + |\psi_2|^2 + 2\sqrt{|\psi_1||\psi_2|} \cos \phi$$

where ϕ is the phase difference between ψ_1 and ψ_2.

☐ For any state of a system consisting of a single particle, the uncertainties Δx and Δp_x in the values of x and p_x satisfy the Heisenberg uncertainty relation,

$$\Delta x \Delta p_x \gtrsim \hbar$$

☐ The time-energy uncertainty relation

$$\tau \Delta E \gtrsim \hbar$$

relates the mean life of an unstable state to the width ΔE of the energy level corresponding to the state.

Questions

1 Show that, for an electron that moves at a speed well below the speed of light, and whose kinetic energy K can thus be evaluated from $K = \frac{1}{2}m_e v^2 = p^2/2m_e$, the de Broglie wavelength of the electron can be expressed in terms of its kinetic energy by

$$\lambda = \sqrt{\frac{150 \text{ eV} \cdot \text{Å}^2}{K}}$$

2 Find the de Broglie wavelengths that correspond to the following electron kinetic energies: 1.5 eV, 150 eV, 15 keV.

3 Show that, for a neutron with a speed small compared to c, the de Broglie wavelength corresponding to the kinetic energy K is

$$\lambda = \sqrt{\frac{0.082 \text{ eV} \cdot \text{Å}^2}{K}}$$

4 An x-ray beam of a single wavelength (a monochromatic beam) is incident on a sodium chloride crystal. A strong diffracted beam occurs when the glancing angle is 12°. Assume that this beam is due to first order Bragg reflection from a family of planes with a spacing of 2.82 Å:
(a) What is the wavelength of the x rays?
(b) Find another glancing angle for which a diffracted beam will occur.

5 A crystal is formed with identical atoms at each corner of a cube with a side of length 2.5 Å. At what angles will there be constructive interference from planes parallel to the cube face for x rays with a wavelength of 2.0 Å?

6 In Question 5, at what angles will there be constructive interference from a family of planes which pass through diagonally opposite edges of the cubes?

7 Describe in terms of the arrival of photons the formation of the intensity maxima and minima shown on the screen in Figure 36-8. Assume that the light source is extremely dim.

8 The electron gun of Figure 42-4 is adjusted so that, with just one slit open in the barrier, the small counter records an average of 100 counts per hour. With one slit open, this counting rate does not change appreciably as the counter is moved about to explore different regions on the screen. When the second slit is opened, what is the counting rate at the location of:
(a) An intensity maximum?
(b) An intensity minimum?

9 Suppose we were to try to perform a laboratory experiment using a beam of electrons instead of a light beam in a double-slit interference experiment. Assume that the geometry of the slits and of the screen are to be exactly as described in Example 3 of Chapter 36. If, in the interference pattern on the screen, the distance between locations of maximum intensity is to be the same as that observed when the yellow light from the sodium vapor lamp was used, what must be the kinetic energy of the electrons in the electron beam?

10 **(a)** Compute the de Broglie wavelength of a football of mass 0.40 kg when it is moving at a speed of 10 m/s.
    (b) Have any of the experiments discussed in this book measured wavelengths this short?

11 If the momentum of the football of Question 10 has an uncertainty of merely about one part in a million (that is, $\Delta p_x = 4 \times 10^{-6}$ kg·m/s), find the minimum uncertainty in position Δx consistent with the Heisenberg uncertainty relation.

Supplementary Questions

S-1 For an electron orbit in a hydrogen atom, show that if Bohr's quantum hypothesis (given in Question S-6 of Chapter 41) is satisfied, then the circumference of the orbit contains an integral number of de Broglie wavelengths. In other words, show that the equation

$$m_e vr = n\hbar$$

implies that

$$2\pi r = n\lambda$$

where $\lambda = h/m_e v$.

S-2 Starting from the relativistically correct relationship between energy E and momentum p,

$$E = K + mc^2 = \sqrt{(mc^2)^2 + (cp)^2}$$

show that the de Broglie wavelength of a particle of mass m can be expressed in terms of its kinetic energy K by:

$$\lambda = \frac{h}{mc} \frac{1}{\sqrt{(2K/mc^2) + (K/mc^2)^2}}$$

S-3 Show that, in an extreme relativistic range of kinetic energies, where K is much greater than mc^2, the de Broglie wavelength is *independent* of the mass of the particle, being given by

$$\lambda = \frac{hc}{K}$$

after the approximation that mc^2 is negligible compared to K.

S-4 Using the appropriate results of preceding questions (1, S-2, S-3), evaluate the de Broglie wavelength of an electron and of a proton for the following values of the particles' kinetic energies: 30 eV, 30 keV, 30 MeV, 30 GeV. (For an electron $m_e c^2 = 0.511$ MeV and $h/m_e c = 0.024$ Å. The mass of a proton is 1836 times the mass of an electron.)

S-5 If an electron were confined to a region the size of a nucleus, then the uncertainty in one of its position coordinates would be approximately 10^{-14} m:
   (a) Evaluate the corresponding uncertainty in momentum.
   (b) Since the possible momentum values must be at least as large as the uncertainty in momentum, the value Δp_x calculated in part (a) can be used as a typical value of the momentum p. Calculate the corresponding value of the electron's kinetic energy. (Relativistic expressions must be used.)

S-6 For a given family of planes with spacing d, show that the Bragg condition cannot be satisfied by any wavelength greater than $2d$.

S-7 *Particle in a box*   Consider a particle of mass m which is constrained to move in the region between $x = 0$ and $x = l$, such as a gas molecule in a box. The wave function describing any possible state of this system must be zero to the left of $x = 0$ and to the right of $x = l$. This situation is analogous to the case of standing waves on a stretched string with ends fixed at $x = 0$ and $x = l$; the possible wavelengths are

$$\lambda_n = \frac{2l}{n} \qquad\qquad (n = 1, 2, 3, \ldots)$$

   (a) Show that the de Broglie relationship $p = h/\lambda$ together with $E = p^2/2m$ implies that the possible values of the particle's energy (the energy levels) are

$$E_n = \frac{n^2\pi^2\hbar^2}{2ml^2}$$

   and sketch the energy-level diagram.
   (b) Consider an electron of mass m_e in a box with a length

$$l = \pi a_0 = \frac{\pi\hbar^2}{m_e ke^2} = 1.66 \text{ Å}$$

   By rewriting the result of part (a) and using $\alpha = ke^2/\hbar c$, show that the ground state energy of the electron in this box is

$$E_1 = \tfrac{1}{2}\alpha^2 m_e c^2 = 13.6 \text{ eV}$$

   (c) Use the uncertainty relation $\Delta x \Delta p_x \gtrsim \hbar$ to estimate the minimum possible energy for the electron in part (b) and verify that this energy is of the same order of magnitude as the ground state energy calculated in part (b).

43.

quantum physics

In the twentieth century, a major part of experimental and theoretical research has been in the field of quantum physics. This closing chapter is concerned with the panorama of this vast domain. The treatment is necessarily descriptive and brief, the object being to survey representative topics and to give the student a point of departure for more specialized studies.

43.1 Hydrogen Atom

The first field to be triumphantly put into order by quantum mechanics was that of atomic physics. Almost all the details of atomic spectra and atomic structure became at least comprehensible, if not calculable. In this program, the theory of the simplest atom—the hydrogen atom—was the key.

For the hydrogen atom, the mathematical complexities of quantum mechanics are not too formidable. Exact theoretical predictions can be calculated. Fortunately, certain general features of the hydrogen atom theory are applicable to electrons in *any* atom. Consequently, understanding the hydrogen atom is a big step in understanding atomic structure in general. Without concerning ourselves with the mathematics involved we shall survey in the following sections these modern notions of atomic structure. This is the historical route followed by the pioneers of quantum theory, and it remains a most instructive approach to quantum physics.

The electromagnetic interaction between electrically charged particles is the only significant interaction in atomic physics, that is, in the physics of the atom outside the nucleus. The values of many physical quantities of importance in atomic physics that are determined by electromagnetic interactions can be conveniently expressed in terms of one constant α defined by

$$\alpha = \frac{ke^2}{\hbar c}$$

where k is the constant of proportionality in Coulomb's law. Substitution of the experimental values of ke^2, $\hbar$, and c gives

$$\alpha = \frac{1}{137.03604}$$

This quantity is a dimensionless number and therefore has the same value no matter what system of units is used for the measurement of charge e, speed c, and angular momentum $\hbar$. The constant α acquired the name "the fine structure constant" in early studies of hydrogen atom energy levels, and this name has persisted even though it is not particularly apt.

States of the Electron in a Hydrogen Atom

The hydrogen atom consists of an electron and a proton. These two particles interact by exerting Coulomb forces on each other. Since the proton's mass is 1836 times the electron's mass, the proton's acceleration, according to classical mechanics, is merely $\frac{1}{1836}$ of the electron's acceleration. Then, to a good approximation, we can visualize the proton as fixed and attribute all the motion to the electron. This picture carries over to quantum mechanics. The problem then is to determine all possibilities for an electron which finds itself in the electric field established by a charged fixed nucleus.

Quantum mechanics now prescribes a set of different wave functions, each corresponding to a different possible state for the electron. From the wave function for a given state, we can determine the probability of finding an electron in any given region near the atom. This probability can be visualized if we imagine a cloud of electric charge with a density in a given region proportional to this probability. Such *charge clouds* are displayed in Figure 43-1.

The charge cloud indicates the "size" of the *atom* in different states. In quantum mechanics, objects do *not* have a precise and well-defined size, just as a cloud does not have a distinct bounding surface. It is the size of the *atom, not* the size of the *electron*, that is determined by the spatial extension of the charge cloud. Scattering experiments with high-energy electrons show that, relative to the size of the atom itself, the upper limit to the electron's size is less than that of any dot in these charge cloud pictures. As far as the electron is concerned, charge clouds are simply a useful way of picturing probabilities; it is more likely that the electron will be found where the cloud is dense than where it is tenuous.

Ground State

For the hydrogen atom *ground state*, the electron charge cloud is distributed in the space in such a way that the most probable distance of the electron from the nucleus is given by

$$a_0 = \frac{\hbar^2}{m_e ke^2} = 0.52917706 \text{ Å}$$

where a_0 is called the *Bohr radius*. (According to the early Bohr model of the hydrogen atom, the electron in the ground state orbited about the nucleus in a circle of this radius.) In the ground state, the charge cloud thus extends over a spherical region somewhat larger than 1 Å in diameter. Quantum mechanics thus predicts that the size of a hydrogen atom in its ground state is roughly that of a sphere 1 Å in diameter (Figure 43-2). This is in reasonable agreement with the estimate of atomic size furnished by measuring the volume occupied by a known number of hydrogen atoms in a chunk of hydrogen which has been solidified by cooling.

The ground state wave function corresponds to many different values of the component of the electron's momentum in a given direction. These momentum values range from about $m_e\alpha c$ corresponding to motion in one direction OX to about $-m_e\alpha c$ corresponding to motion in the opposite direction. The range Δp_x

State with
$n = 2, \ell = 0, m_\ell = 0$

State with
$n = 2, \ell = 1, m_\ell = 1$

State with
$n = 2, \ell = 1, m_\ell = 0$

Figure 43-1   Charge clouds for three different possible states of the electron in a hydrogen atom. (From *Introduction to Atomic Spectra*, H. White; New York, McGraw-Hill, 1934.)

Figure 43-2   Charge cloud for the hydrogen atom ground state. The nucleus (a proton) is at the center of the cloud.

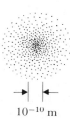

10^{-10} m

of momentum values encountered in the ground state is therefore of the order of $m_e \alpha c$. It is interesting to relate this to the size of the charge cloud. The uncertainty Δx in position is of the order of the Bohr radius a_0. Then the product of these uncertainties is

$$\Delta x \Delta p_x \approx a_0 m_e \alpha c = \left(\frac{\hbar^2}{m_e k e^2}\right)\left(\frac{k e^2}{\hbar c}\right) m_e c = \hbar$$

in accord with the Heisenberg uncertainty relation.

A momentum of $m_e \alpha c$ corresponds to an electron speed αc which is less than 1% of the speed of light. Therefore relativistic effects are small in the hydrogen atom.

The electron's total energy in the ground state, according to quantum mechanics, is given by

$$E_1 = -\tfrac{1}{2}\alpha^2 m_e c^2 = -13.605804 \text{ eV}$$

in agreement with Bohr's formula, Eq. 41.10. As in Section 41.3, zero total energy is chosen to correspond to the situation when the electron is completely removed from the vicinity of the nucleus. The atom is then said to be ionized. When a hydrogen atom is in its ground state, an energy of at least 13.6 eV must be supplied to the atom for ionization to be possible.

Quantum Numbers

There are many possible bound states of the electron in a hydrogen atom. Each possible state is specified by the values of four *quantum numbers*: n, l, m_l, and m_s. The rules which emerge from quantum mechanics for the possible numerical values of these quantum numbers are summarized in Table 43-1.

Table 43-1 Quantum Numbers for States of an Electron in an Atom

Quantum number		Values
n	Principal quantum number	Can be any positive integer: 1, 2, 3, ...
l	Orbital angular momentum quantum number	Given n, there are different states corresponding to $l = 0, 1, 2, 3, 4, 5, \ldots, n - 1$.
m_l	Magnetic quantum number	Given l, there are $2l + 1$ different states corresponding to $m_l = l, l - 1, \ldots 1, 0, -1, \ldots -(l - 1), -l$.
m_s	Magnetic spin quantum number	For each set of values of n, l, m_l, there are two different states, one for which $m_s = +\tfrac{1}{2}$ and another for which $m_s = -\tfrac{1}{2}$.

It is a peculiarity of the hydrogen atom that the energy of a state depends only on the value of the principal quantum number n, according to Bohr's formula,

$$E_n = -\frac{13.6 \text{ eV}}{n^2} \qquad (41.10)$$

the validity of which is confirmed by quantum mechanics. This quantum number n also determines the size of the atom in the state with principal quantum number n: For a state with $l = n - 1$, the most probable distance of the electron from the nucleus is given by

$$r_n = n^2 a_0$$

where a_0 $(=\hbar^2/m_e k e^2)$ is the Bohr radius.

Example 1   Use the rules given in Table 43-1 to give the quantum numbers of all the possible states for which $n = 2$.

Solution   The maximum value of l is $n - 1 = 2 - 1 = 1$. For $l = 1$ there are different states corresponding to $m_l = 1$ or 0 or -1. The value $l = 0$ is also possible and, corresponding to this l value, m_l can have only the value 0. For each possible set of values of n, l, m_l, the quantum number m_s can be either $+\frac{1}{2}$ or $-\frac{1}{2}$.

There are therefore the following eight different states (each labeled by the values of n, l, m_l, m_s):

$(2, 1, 1, +\frac{1}{2})$, $(2, 1, 1, -\frac{1}{2})$; $(2, 1, 0, +\frac{1}{2})$, $(2, 1, 0, -\frac{1}{2})$;
$(2, 1, -1, +\frac{1}{2})$, $(2, 1, -1, -\frac{1}{2})$;
and $(2, 0, 0, +\frac{1}{2})$, $(2, 0, 0, -\frac{1}{2})$.

For our purposes, it is not necessary to dwell on the physical significance of the quantum numbers l and m_l which justifies the names shown in Table 43-1. As a guide for future studies of those who wish to delve more deeply into this interesting topic, we mention the following points. The l value of a state determines the rotational symmetry properties of its charge cloud. For instance, the charge cloud of any state with $l = 0$ can be rotated any amount about any axis through its center without producing any change. These $l = 0$ states are said to be spherically symmetric. Rotational symmetry and its associated l value determine the magnitude of the quantum mechanical analogue of the *angular momentum* that is defined for Newtonian mechanics in Chapter 13. The magnetic quantum number determines the orientation of the electron charge cloud in space. When a magnetic field is applied, states with the same n and the same l values have different energies in the different possible orientations corresponding to $m_l = l, l - 1, \ldots, 1, 0, -1, \ldots, -(l - 1), -l$. In a state with quantum number m_l, the electron's orbital angular momentum has a component $m_l \hbar$ in the direction of the applied magnetic field. The electron itself has an intrinsic angular momentum called "spin" whose component in the direction of an applied magnetic field is restricted to two values, either $+\frac{1}{2}\hbar$ or $-\frac{1}{2}\hbar$. We denote this component of spin angular momentum by $m_s \hbar$ and conclude that the magnetic spin quantum number m_s is restricted to the values $+\frac{1}{2}$ or $-\frac{1}{2}$.

Such details about the possible states of an electron in a hydrogen atom enable us to understand all the complexities of the line spectra emitted by these atoms under various excitations and in circumstances when a magnetic field or an electric field has been applied to the light source. Quantum mechanics allows calculation of not only the wavelengths but also the intensities of each spectral line. Rather than describing these achievements, we shall show how the structure of different atoms is related to the facts listed in Table 43-1 concerning electron states in hydrogen.

43.2 Electronic Structure of Atoms

An incredible wealth of chemical and spectrographic detail becomes comprehensible when the electronic structure of each type of atom is known. Theory of this structure rests on the facts that have been presented about the electron states in hydrogen.

Pauli Exclusion Principle

Consider an atom with atomic number Z. The nucleus has a positive charge Ze which produces an electric field strong enough to bind Z electrons. Each electron moves in the electric field created by the positive nuclear charge Ze and by the negative charges of the other $Z - 1$ electrons in the atom. Analogously to the case of the hydrogen atom, the possible states for an individual electron can be specified by giving the values of the four quantum numbers: n, ℓ, m_ℓ, and m_s.

The state of the entire atom depends on just which of the possible electron states are occupied. And here we come to a simple but peculiar rule of quantum mechanics, the *Pauli exclusion principle: In a system containing several electrons, no two electrons can occupy states with the same values of all four quantum numbers, n, ℓ, m_ℓ, m_s.* The possibility of occupancy of a given state by more than one electron is excluded. This basic law, which has profound implications in many fields of physics, was formulated in 1925 by the German physicist, Wolfgang Pauli, as a key to an understanding of atomic structure. In the ground state of an atom, the electron energy levels are filled from the lowest energy level upward with precisely one electron in each possible state until the atom's supply of electrons is exhausted. The properties of these occupied states determine the electronic structure of the atom.

Subshells of Electrons

The order in which successive electron energy levels are filled is shown in Figure 43-3. In many-electron atoms, although the energy of an electron state depends principally on the quantum number n, there is also a dependence on the value of l with higher energies corresponding to higher l values. An nl level in Figure 43-3 is labeled by a number which gives the value of the principal quantum number n, followed by a lowercase letter which corresponds to the value of l according to the following convention:

l value	0	1	2	3
letter	s	p	d	f

Thus a state with $n = 3$ and $l = 1$ is called a 3p state.

The set of electron states associated with a given nl energy level is called a *subshell*. The number of different states in a subshell is just the number of different sets of values of m_l, m_s that are possible according to the rules of Table 43-1. Results are given in the last row of Table 43-2. According to the Pauli exclusion principle, the number of states in a subshell is also the maximum number of electrons that can occupy the subshell.

The *electron configuration* of an atom in any state is the name given to a specification of the number of electrons in each subshell. The electron configuration for the ground state of any atom can be read from the diagram of Figure 43-3.

Table 43-2 Electron Subshells

Subshell l value	0	1	2	3
Subshell letter	s	p	d	f
Number of states $2(2l + 1)$	2	6	10	14

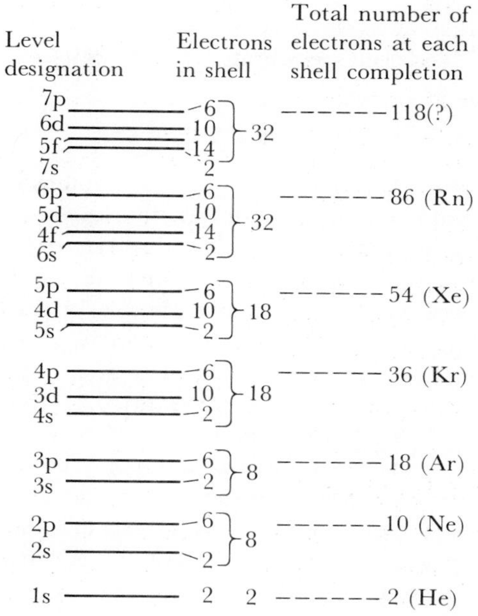

Figure 43-3 Shell structure of atoms. Different nl states are filled from the bottom up in this order.

For example, sodium with 11 electrons has 2 electrons filling the 1s subshell, 2 electrons filling the 2s subshell, 6 electrons filling the 2p subshell, and 1 electron remaining which will occupy the 3s subshell. Placing the number of electrons in a shell as a superscript, we designate this electron configuration by $1s^2 \, 2s^2 \, 2p^6 \, 3s^1$.

A study of electron charge clouds reveals that the lowest energy level corresponds to the innermost subshell and that subshells of increasing energy usually have charge clouds that are increasingly distant from the nucleus. Bohr's early notion of atomic structure which pictured the atom as built up from concentric shells of electrons thus retains considerable validity in quantum mechanics. The highest energy subshell to be occupied usually corresponds to the outermost electrons. The physical and chemical properties of an atom are determined by these outermost electrons.

Except for helium, the inert gases listed at the right of Figure 43-3 are characterized by having a full p-subshell as their outermost subshell. This is a particularly stable structure. The energy gap in Figure 43-3 between each of the p levels and the next highest level is abnormally large, indicating that an unusually large energy must be provided to an inert gas atom in order for it to reach its first excited state. For helium this excitation energy is about 20 eV.

The alkali atoms such as lithium, sodium, and potassium have just one electron outside a closed p-subshell. This electron is in an s-state and a rather small excitation energy (about 2 eV) is required to promote the electron to the next highest energy level. The metallic behavior of solids composed of atoms of any elements of this type is associated with this low excitation energy.

These remarks are merely a first indication of the wealth of chemical information that can be inferred from the electron configurations dictated by Figure 43-3. But enough has been said to make clear the reason for the recurrence of the same chemical properties over and over again as we proceed through Mendeleeff's celebrated *periodic table* of the chemical elements (Table 43-3).

Inner Electrons and X Rays

The emission or absorption of photons by an atom is associated with a transition from one state of the atom to another. When an *outer* electron of an atom passes from one possible electron state to another, its energy change is only a few electronvolts. The photon emitted in a downward transition or absorbed in an upward transition may be visible, or perhaps infrared, or ultraviolet. But if the opportunity is provided for an *inner* electron of an atom of atomic number greater than about 12 to undergo a transition, the change in electron energy is so large that the photon emitted or absorbed is an x-ray photon.

As a typical example of an x-ray emission process involving inner electrons, we consider a tungsten atom in the anode of an x-ray tube. The anode is bombarded by a beam of energetic electrons. When a tungsten atom is struck by one of these electrons, it sometimes happens that an electron occupying the 1s shell is knocked right out of the tungsten atom. The electrons of this atom which are residing in higher energy levels now have "Pauli's permission" to

Table 43-3 Periodic Table of the Elements (Numbers in parentheses indicate the mass number of the longest-lived isotope of radioactive elements.)

Outer electrons are in the	I	II	III	IV	V	VI	VII	VIII			0	Electrons per shell
First or K-shell	1 H 1.00797										2 He 4.0026	2
Second or L-shell	3 Li 6.939	4 Be 9.0122	5 B 10.811	6 C 12.01115	7 N 14.0067	8 O 15.9994	9 F 18.9984				10 Ne 20.183	2,8
Third or M-shell	11 Na 22.9898	12 Mg 24.312	13 Al 26.9815	14 Si 28.086	15 P 30.9738	16 S 32.064	17 Cl 35.453				18 Ar 39.948	2,8,8
Fourth or N-shell	19 K 39.102	20 Ca 40.08	21 Sc 44.956	22 Ti 47.90	23 V 50.942	24 Cr 51.996	25 Mn 54.9380	26 Fe 55.847	27 Co 58.9332	28 Ni 58.71		2,8,18,8
	29 Cu 63.54	30 Zn 65.37	31 Ga 69.72	32 Ge 72.59	33 As 74.9216	34 Se 78.96	35 Br 79.909				36 Kr 83.80	
Fifth or O-shell	37 Rb 85.47	38 Sr 87.62	39 Y 88.905	40 Zr 91.22	41 Nb 92.906	42 Mo 95.94	43 Tc (99)	44 Ru 101.07	45 Rh 102.905	46 Pd 106.4		2,8,18, 18,8
	47 Ag 107.870	48 Cd 112.40	49 In 114.82	50 Sn 118.69	51 Sb 121.75	52 Te 127.60	53 I 126.9044				54 Xe 131.30	
Sixth or P-shell	55 Cs 132.905	56 Ba 137.34	57–71 La series*	72 Hf 178.49	73 Ta 180.948	74 W 183.85	75 Re 186.2	76 Os 190.2	77 Ir 192.2	78 Pt 195.09		2,8,18 32,18,8
	79 Au 196.967	80 Hg 200.59	81 Tl 204.37	82 Pb 207.19	83 Bi 208.980	84 Po (210)	85 At (210)				86 Rn (222)	
Seventh or Q-shell	87 Fr (223)	88 Ra (226)	89–103 Ac series**	104 Ku (260)	105							

*Lanthanide series:	57 La 138.91	58 Ce 140.12	59 Pr 140.907	60 Nd 144.24	61 Pm (145)	62 Sm 150.35	63 Eu 151.96	64 Gd 157.25	65 Tb 158.924	66 Dy 162.50	67 Ho 164.930	68 Er 167.26	69 Tm 168.934	70 Yb 173.04	71 Lu 174.97
**Actinide series:	89 Ac (227)	90 Th 232.038	91 Pa (231)	92 U 238.03	93 Np (237)	94 Pu (244)	95 Am (243)	96 Cm (247)	97 Bk (247)	98 Cf (251)	99 Es (254)	100 Fm (257)	101 Md (256)	102 No (255)	103 Lw (257)

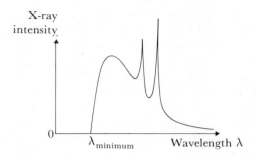

Figure 43-4 The characteristic x-ray spectrum (spikes) is shown superimposed on the continuous breamsstrahlung spectrum of Figure 41-3.

make a transition to this vacant 1s state. A very common occurrence in this situation is that an electron in a 2p state makes a transition to the 1s state. In this process the electron decreases its energy by 59.3×10^3 eV, and an x-ray photon of this energy is emitted. Another electron can then drop into the vacated 2p state emitting another x ray. Processes of this type lead to the emission of photons of several different energies that are characteristic of the electronic structure of the atom. Each element therefore has its own *characteristic spectrum* (Figure 43-4), a set of discrete x-ray wavelengths superimposed on the continuous spectrum of bremsstrahlung x rays described in Section 41.2.

43.3 Interaction of Electromagnetic Radiation with Matter

When a beam of electromagnetic radiation passes through matter, the photons of the beam may be absorbed or scattered from the beam by various processes. If the incident photon has an energy of a few MeV, the production of electron-positron pairs by the photon near a nucleus is the predominant process of removal of photons from the beam. As is discussed in Section 41.2, this process cannot occur unless the photon energy is over 1.02 MeV. For photon energies between a few MeV and about 0.05 MeV, Compton scattering is usually the most important mechanism removing photons from the beam.

The ejection of an electron from an atom by absorption of the incident photon is always possible if the photon can furnish an energy at least as great as the binding energy of the electron. This "atomic photoelectric effect" becomes very probable when the energy of the incident photon almost matches a binding energy. For incident photons with energies ranging from 2×10^5 eV to 5 eV, the photoelectric effect is often the most probable absorption event.

Now let us consider photon energies that are low enough so that the atom can absorb the photon and make a transition to a discrete excited state, a bound state. At normal temperatures almost all the atoms in an absorbing material will be in their ground states. We find that a photon with an energy which is close to the energy difference

$$hf_{1n} = E_n - E_1$$

between an excited state E_n and the ground state E_1 has a high probability of being absorbed. We say that light of such a frequency is in *resonance* with the atom and that the frequencies f_{1n} are the resonant frequencies of the atom.

Actually, to some extent the atom reacts to light of any frequency and non-resonant processes are also significant. In fact, these processes are often responsible for the visual appearance of an object. To bring out the essential features of such processes, we must do more than talk about quantum jumps or transitions of the atom from the ground state to excited energy levels. We require a more complete quantum mechanical description of the interaction.

If an atom is in its ground state and is undisturbed, its charge cloud is stationary. When a beam of electromagnetic radiation falls upon an atom, the atom's *charge cloud undergoes forced vibrations* at the frequency of the incident radiation. The amplitude of the charge cloud oscillations is minute, roughly 10^{-17} m, which is only one-ten-millionth of the size of the atom. The amplitude of the charge cloud oscillations is largest at the *resonant frequencies* f_{1n} given by $hf_{1n} = E_n - E_1$. When incident radiation has a frequency far above or below a resonant frequency of the atom's charge cloud, the amplitude of charge cloud oscillation will be much less than at a resonant frequency.

Whatever its amplitude, the oscillating charge cloud can radiate electromagnetic waves of the *same frequency* as the incident radiation and coherent with it. Indeed it is this small vibration that re-emits the light by which we see the objects around us.

It is also possible that the excited atom get rid of its excess energy by some other process. An atom may collide with another atom and the excitation energy may be transformed into kinetic energy of the atoms emerging from the collision. The net result of such a process is the *absorption of an incident photon* and an increase in the thermal motion of the atoms. This absorption process is particularly important in liquids or solids at or near a *resonant frequency* of the atoms of the material.

With the above ideas in mind, we now describe some features of the interaction of light of various frequencies with water molecules. First, what are the resonant frequencies of the charge cloud of the molecules? Like most atoms and molecules, the resonant frequencies of the electron charge cloud for a water molecule are higher than the frequencies of visible light; they lie in the ultraviolet. *Molecules*, however, can perform oscillations in which the *atoms* move with respect to one another within the molecule. Because of the larger mass of the moving atoms, the frequency of such vibrations is low. The resonant frequencies corresponding to *motions of atoms* within the molecules lie in the infrared region: a water molecule has strong resonances in the infrared.

Now when sunlight falls on the surface of the ocean, what happens? The water molecule's resonances in the infrared and ultraviolet lead to a strong absorption in these regions of the spectrum. And with no resonances for frequencies of visible light, water is quite transparent. The infrared resonance, however, is so strong that the absorptive effect extends even into the visible red. At an ocean depth of 30 m, practically all red light has been absorbed. This fact has an interesting consequence as far as the color of marine life is concerned. At great depths red looks black and therefore has the same survival value. This explains why deep-sea crustaceans can be red and not discriminated against by the selection mechanisms of evolution.

The small forced oscillations of the molecule's electron charge cloud, which lead to re-emission of light of the same frequency as the incident beam, account for both the reflected beam and the refracted beam. Within the bulk of the water, the incident beam and the light which is re-emitted coherently from the various molecules give a superposition which is the *refracted wave*. Backward radiation is cancelled by interference, except for a layer of molecules (roughly $\lambda/2$ thick) near the surface, whose contributions constitute the *reflected wave*.

X-ray photography, that is, radiography, is based on the fact that different types of atoms interact differently with a given x-ray photon. For example, a photon whose energy is 90×10^3 eV has a frequency well above any of the resonant frequencies of the oxygen atoms in a water molecule. However, lead atoms have resonant frequencies in the vicinity of the frequency of 90×10^3-eV photons. Consequently, for such photons lead is opaque while water is relatively transparent.

43.4 Energy Bands in Solids

The quantum mechanical description of electron states in crystalline solids leads to a simple characterization of electrical conductors, insulators, and semiconductors and provides the theory underlying modern semiconductor technology.

Within a crystal there is a three-dimensional array of atomic nuclei. Each atom furnishes to the crystal its complement of electrons. The quantum mechanical problem is to determine the possible states for an electron confronted by this assembly of interacting particles. Fortunately, a few general features of the solution to this formidable problem are sufficient to shed considerable light on the behavior of electrons in cyrstals.

An instructive example is provided by metallic sodium. The electron energy levels for a single isolated atom are illustrated in Figure 43-5(a). If we now consider N sodium atoms, so far apart that their interaction is negligible, the electron energy-level diagram for this aggregate of particles [Figure 43-5(b)] will be identical to that of a single atom except for the fact that the number of states corresponding to each level is increased by a factor of N. Finally, we consider the electron energy levels when these N atoms are brought into the positions they occupy within a crystal of metallic sodium. Because of the *interaction*, each level corresponding to many different states becomes *separated into many distinct levels*. When N is large, these levels are so closely spaced that we can consider them as forming a *band* of energy. Thus in Figure 43-5(c), the 2s levels have spread out to become the 2s band containing $2N$ distinct states, the 2p levels have spread out to become the 2p band with $6N$ distinct states, and so on. The interaction has the greatest effect and leads to the greatest band widths for electron states whose charge clouds range furthest from the nucleus. For example, the 2s band is wider than the 1s band. Bands at higher levels often overlap in the fashion illustrated in Figure 43-5(c) by the 3p and 3s bands of sodium.

The gaps between bands are called *forbidden bands* since they correspond to no possible value of electron energy.

The *occupation* of the available energy levels is governed by the Pauli exclusion principle. Each state can be occupied by at most one electron. In the lowest energy state for the $11N$ electrons of a sodium crystal, $2N$ electrons fill the 1s band, $2N$ electrons fill the 2s band, $6N$ electrons fill the 2p band, and the remaining N electrons will occupy the lowest levels in the 3s band. Since the 3s band contains $2N$ different states, the upper half of this band will be empty [Figure 43-5(c)].

The electrons in this band that occupy levels just below vacant energy levels are in a unique position. They are the only electrons that can make a transition to a slightly higher level without violating the exclusion principle. For this reason, these are the electrons that are responsible for the high electrical conductivity of sodium metal. When an electric field is applied within the material, these electrons can move readily, and so acquire a modest kinetic energy associated with a transition to a nearby higher energy level. The electrons in the lower filled bands are denied such opportunity by the exclusion principle. We see that a *partially filled* upper band is characteristic of a *conductor*. Such a band is named the *conduction band*.

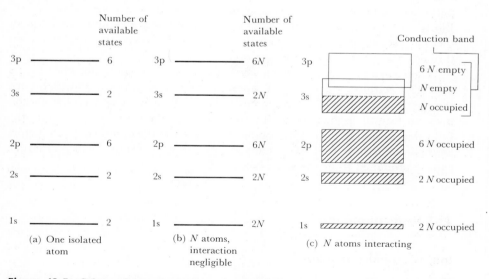

Figure 43-5 Schematic representation of sodium energy levels for (a) one isolated atom; (b) N atoms so far apart that their interaction is negligible; (c) N atoms interacting as they do in a metallic cyrstal. Here the levels are grouped into energy *bands*.

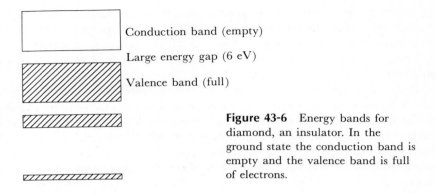

Conduction band (empty)

Large energy gap (6 eV)

Valence band (full)

Figure 43-6   Energy bands for diamond, an insulator. In the ground state the conduction band is empty and the valence band is full of electrons.

Energy bands for a typical *insulator*, such as diamond, have lower bands completely filled with electrons (Figure 43-6), and then a gap of 6 eV to a band which is *empty*. Small energy increments of electrons in the filled bands are impossible so these electrons cannot respond to an applied electric field (unless the field is formidably large). This is why the material is an insulator. Evidently insulators are characterized by a sizable energy gap between the last filled band, called the *valence band*, and the first empty band, called the *conduction band*.

We have been considering the situation when all electrons occupy the lowest available states. There are several ways for an electron to absorb energy and make an upward transition to an unoccupied state. The energy of a particle participating in thermal agitation at room temperature averages about $\frac{1}{40}$ eV. Electrons near empty levels in the conduction band will undergo frequent thermal excitation to levels about $\frac{1}{40}$ eV higher. However, thermal excitation through about 1 eV will be a rare event.

When a beam of light shines on a crystal, the photons will be absorbed by the crystal's electrons if, and only if, appropriate excited states are available for the electrons. Diamond is transparent to visible light because a visible photon does not furnish enough energy for an electron to make a 6 eV jump from the full valence band to the empty conduction band. On the other hand conductors are opaque because they have many empty energy levels into which electrons can be promoted from the same band.

43.5 Semiconductors

Intrinsic Semiconductors

Certain crystalline solids such as silicon and germanium are similar to diamond in that they have a full valence band and an empty conduction band at low temperatures. But the energy gap between the valence and conduction bands is merely 1.1 eV for silicon and 0.7 eV for germanium, in contrast to the 6 eV gap for diamond. In the materials with a small gap, thermal agitation at room temperature will impart to a few electrons in the valence band sufficient energy for them to jump up to the conduction band. While in this almost empty band, these electrons can acquire kinetic energy in an applied electric field because there is an abundance of nearby vacant energy levels. The material thus conducts electricity. It is called a *semiconductor* because it has a conductivity greater than that of insulators with a large gap above the valence band but less than that of conductors with many electrons in the conduction band.

An electron which has been promoted from the full valence band to the conduction band leaves behind an empty state or *hole* in the valence band. When an electric field is applied, an electron in the valence band can make a transition to the empty state with the result that the hole is elsewhere. Transitions in the valence band are most easily described by keeping track of the hole. Since the hole moves in the direction opposite to that of an electron, the *hole* behaves as a *positive charge*.

Thermal excitation in a semiconductor thus promotes electrons to the conduction band and leaves holes in the valence band (Figure 43-7). In an electric field the holes move one way and the electrons in the conduction band move the other way. The conductivity is due to both motions. Such a semiconductor is called an *intrinsic semiconductor* to distinguish it from the important class of semiconductors whose conductivity arises largely from impurities.

When holes are the predominant charge carriers a material has a positive Hall coefficient. This is the simple answer provided by quantum theory to the puzzle posed by the occcurrence of these anomalous Hall coefficients (Section 27.4).

Impurity Semiconductors

The addition of certain impurities to a semiconductor drastically affects its conductivity. This phenomenon is exploited throughout the semiconductor technology which has revolutionized the electronics industry. By incorporating minute amounts of appropriate impurities, it is possible to make either:

1   An *n-type* semiconductor in which electric current is carried by negative charges—the electrons donated to the conduction band by *donor* impurities.

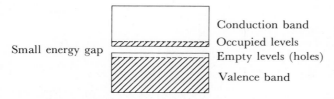

Figure 43-7   Energy bands in a semiconductor. The conductivity of an intrinsic semiconductor arises from the motion of electrons which have been excited up to the conduction band and also from the motion of the holes which have been left in the valence band.

2   A *p-type* semiconductor in which electric current is carried by the motion of "positive charges"—the holes in the valence band created when a valence band electron becomes bound to an *acceptor* impurity.

First we consider an example of a donor impurity. Suppose an arsenic atom replaces a germanium atom in a crystal of germanium atoms. The arsenic atom differs from the germanium atom in having one more electron in its outer shell. This extra electron is excluded from the full valence band. It occupies, instead, an energy level called a *donor level* just below the conduction band (Figure 43-8). A mere 0.01 eV is sufficient to detach this electron from the arsenic atom, that is, to promote the electron from a donor level up to the conduction band. At room temperature thermal agitation provides ample energy, and almost all these extra electrons are donated by the arsenic impurities to the conduction band. Then most of the electron population in the conduction band comes from these donor arsenic impurities, even when the impurity concentration is as little as 1 part in 10^{10}. An electric current through such a crystal involves the motion of these conduction band electrons. Because the charge carriers are negative, the semiconductor is said to be of the *n-type*.

Semiconductors of the *p-type* are created by introducing *acceptor* impurities that will give rise to holes by accepting electrons from the valence band. Suppose, for example, that some of the atoms of a germanium crystal are replaced by gallium atoms. In its outer subshell, a gallium atom has one less electron than germanium. The electron deficits associated with the presence of the gallium impurity atoms appear as vacant energy levels just above the valence band (Figure 43-9). Thermal excitation then promotes electrons into these acceptor levels, leaving vacant states, or holes, in the valence band. If an external electric

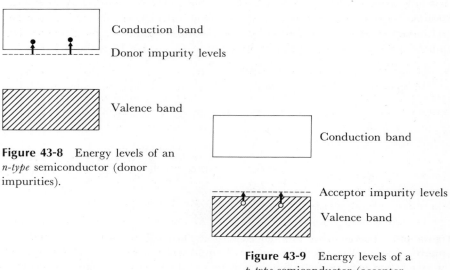

Figure 43-8   Energy levels of an *n-type* semiconductor (donor impurities).

Conduction band
Donor impurity levels

Valence band

Conduction band

Acceptor impurity levels

Valence band

Figure 43-9   Energy levels of a *p-type* semiconductor (acceptor impurities).

field is applied, these holes will move like positive charges and effect a transfer of electric charge through the crystal. The conductivity of this *p-type* semiconductor will be determined by the concentration of acceptor impurities.

Combinations of semiconductors find a host of applications in solid-state devices such as rectifiers which allow a one-way passage of current, transistors which permit a weak alternating signal to be amplified, and integrated circuits or microcircuits. Miniaturization marvels like the microcircuits of Figure 43-10 involve thin film deposits (a few molecular layers only) and diffusion of materials upon a thin wafer. Since it is possible to construct circuit elements with the desired electrical characteristics using only "molecular amounts" of matter, complex circuits can be accommodated within the eye of a needle.

43.6 Low-Temperature Phenomena

At very low temperatures the disorder associated with thermal agitation is very much reduced. Some substances then display remarkable properties such as *superconductivity* or *superfluidity*.

Superconductivity was discovered by H. Kamerlingh Onnes in Leiden in 1911, three years after he had achieved the first liquefaction of helium. The electrical resistance of many metals is found to drop abruptly to an unmeasurably small value when the metal is cooled below a sharply defined temperature (7.2 K in the case of lead). An electric current induced in a superconducting lead ring persists (without any battery) for years—in fact the current continues as long as the lead is kept in liquid helium (Figure 43-11). This astonishing phenomenon now finds practical application in the huge electromagnets used in many branches of physics (Figure 43-12). The magnetic field of the magnets is created by electric currents established in superconducting coils. The necessary refrigerating equipment increases the capital cost, but the operating cost arising from I^2R losses is practically zero.

Superfluidity is another amazing low-temperature phenomenon. Below 2.18 K, liquid helium exhibits a completely frictionless flow which permits it to pass easily through tiny holes (less than 10^{-8} m in diameter) and to show a most unusual behavior in many situations. For instance, this superfluid will not stay in an open container. It manages to climb up the inside walls, pass over the top edge, and travel down the outside.

These strange effects at low temperatures have posed some most instructive puzzles for physicists. It seems that with thermal disorder markedly reduced, we are observing on a *macroscopic* scale the peculiar effects associated with the quantum mechanical behavior of matter.

The laboratory achievement of low temperatures is accomplished in stages. A temperature of 1 K is attained in a bath of liquid helium, boiling under reduced pressure. By using only the rare isotope, helium-3, we can reach temperatures of 0.3 K. The removal of a strong magnetic field from a thermally insulated system creates still lower temperatures, all the way down to 10^{-6} K.

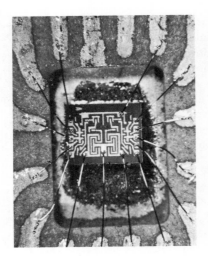
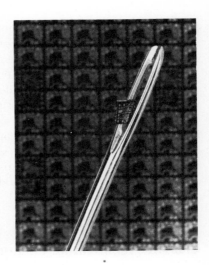

(a)

(b)

Figure 43-10   (a) Dual nand silicon microcircuit. This dual nand microcircuit performs a logic operation in new electronic switching systems. Integrated into the 1.5 mm × 2.0 mm silicon slice are 30 components: 4 transistors, 22 diodes, and 4 resistors. Microcircuits are more reliable, less expensive, and smaller in size than conventional circuits. Eye of needle indicates the size of a typical microcircuit. (b) Charge coupled devices. Charge coupled devices have numerous applications. They are used for converting light to electrical signals in imaging applications for television, for signal processing, and for digital memories. CCD's have evolved from the metal ozide semiconductor technology (Boyle and Smith, *Bell System Technical Journal*, 1970) and consist of closely spaced sets of metal-insulator-silicon (MILS) capacitors in which charges can be entered, stored, and transferred from cell to cell, and readout. The figure shows an 8192 bit charge coupled memory array with input and output circuitry. The size of the individual silicon circuit is 4.2 mm × 4.4 mm. The size of the circuit in relation to a Canadian one-cent piece is shown in the last photograph. (Courtesy Bell-Northern Research Ltd., Ottawa, Canada.)

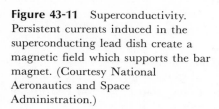

Figure 43-11 Superconductivity. Persistent currents induced in the superconducting lead dish create a magnetic field which supports the bar magnet. (Courtesy National Aeronautics and Space Administration.)

Figure 43-12 Superconducting solenoid which produces a magnetic field of 7.5 T. The solenoid, about 40 cm in diameter and about 60 cm long, is located between the two protruding circular flanges with strengthening webs. The equipment to the left of the left support flange is a lathe for winding the superconducting filaments upon the solenoid. These superconducting filaments are composed of niobium-titanium kept at about 4 K by means of liquid helium circulating through channels in the solenoid structure. (Photo courtesy Ferranti Packard Ltd.)

43.7 The Laser

Much of the physics discussed in preceding chapters enters into the operation of a *laser*, an acronym for "light amplification by stimulated emission of radiation."* This is a light source which produces an intense beam (Section 35.2) of coherent light (Section 36.2) with beam spread which can be reduced to the diffraction limit (Eq. 36.18). The light is confined to essentially one wavelength.

The key physical process, *stimulated emission*, was predicted by Einstein in 1917. This is different from spontaneous emission (discussed in Section 41.3) which involves an atom in an excited state that *spontaneously* emits a photon and makes a transition to a state of lower energy. In stimulated emission a photon of the right energy, incident from elsewhere, interacts with an atom in an excited state and *stimulates* the emission of a photon *identical* to the one already present. The decrease in the atom's energy equals the energy of the emitted photon, and this matches the energy of the original photon. At the conclusion of the process there are two photons with the same energy proceeding in the same direction. These two photons are coherent.

By stimulated emission, one photon gives rise to two. A chain reaction producing light amplification is possible if more than half of these photons go on to stimulate further emission. Some of these photons are absorbed by atoms which happen to be in the lower of the two energy states involved in the transition which produces these photons. Light amplification will be achieved only if there are more atoms in the higher than in the lower energy state. This desirable situation is called a *population inversion* because it is the reverse of the population of the atoms in different energy states in equilibrium (Section 17.4) at any temperature. In different types of lasers the problem of obtaining a population inversion is solved in different ways.

The first laser, constructed in 1960 by T. H. Maiman at the Hughes Aircraft Company, was made from a single cylindrical crystal of ruby with its ends ground flat and silvered (Figure 43-13). Ruby consists of aluminum oxide with a few aluminum atoms replaced by chromium. The chromium atoms produce the laser light. Relevant energy levels of a chromium atom are shown in Figure 43-14. If only spontaneous decay occurs, the excited state of energy E_M has a half-life of 3×10^{-3} s, almost a million times greater than the half-life of most excited states of atoms. Such a long-lived state is termed *metastable*. In the ruby laser, atoms in this metastable state are stimulated by red photons (with $\lambda = 6943$ Å) to make a transition to the ground state and emit another identical red photon.

The population of chromium atoms in the metastable state is made momentarily much greater than the population in the ground state by a method known

---

*Invention of the laser followed the development of the *maser* (a device which amplifies microwaves rather than light) in 1954 by a group led by C. H. Townes at Columbia University. For "fundamental work in the field of quantum electronics, which has led to the construction of oscillators and amplifiers based on the maser-laser principle," Professor Townes shared the 1964 Nobel prize in physics with the Russian physicists N. G. Basov and A. M. Prokhorov.

as *optical pumping*. The ruby is illuminated by a bright flash of yellow-green light (λ = 5500 Å). Absorption of these photons by chromium atoms in the ground state produces transitions to an excited state such as E_u in Figure 43-14. After some 10^{-8} s, many excited atoms will have made downward transitions from E_u to the metastable state E_M.

With a majority of the chromium atoms in their metastable state, the time is ripe for laser action. A photon with a wavelength 6943 Å, emitted by one atom in a spontaneous transition to its ground state, will stimulate similar transitions in other chromium atoms.

To enhance the probability of photon interaction and to achieve a unidirectional beam, reflection from flat parallel ends is exploited. Only light that is perpendicular to these ends will make repeated traversals of the crystal and cause substantial amplification. At the partially silvered end about 1% of the incident photons escape and constitute the emerging laser beam. With Maiman's laser this was a flash of red light lasting 0.3×10^{-3} s with a peak power of 10^4 W.

The manifold applications of lasers derive from the fact that the laser beam is unidirectional and intense, consisting of coherent light with a very sharply defined wavelength. When a high-power laser beam is brought to a sharp focus, the intensity is sufficient to vaporize any substance rapidly. Minute holes in

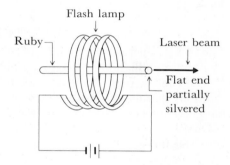

Figure 43-13 Ruby laser. After the spiral flash lamp provides optical pumping, a bright beam of red light emerges from the partially silvered end of the ruby crystal.

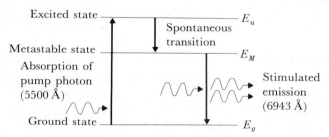

Figure 43-14 Energy levels of a chromium atom involved in a ruby laser. Absorption of pump photons indirectly increases the population of atoms in the metastable state. A stimulated transition from this metastable state to the ground state produces an additional red photon identical to the stimulating photon (6943 Å).

diamond dies used in making fine wires are drilled by laser beams. In medicine lasers are employed in a variety of ways, such as to "weld" on detached retinas or to destroy inaccessible eye tumors.

Because a laser beam does not spread, surveyors find lasers a most useful device for locating objects relative to fixed landmarks and, by means of interference patterns, for detecting minute changes in relatively large distances very accurately. For example, the laser technique is used to measure the movement of the top of a dam as it fills with water.

Studies of the interaction of matter with large electric fields have been opened to research by the availability of the tremendous oscillating electric field present in a high-intensity laser beam.

43.8 Quantum Electrodynamics

The description of atomic physics outlined in the preceding sections is based on a quantum mechanical but nonrelativistic theory of electron behavior. At low energies such a theory is a useful approximation, but at high energies we know that major changes are required. A theory in accord with the special theory of relativity must have the electron's energy E related to its momentum p and its rest mass m_e by

$$E^2 = (m_e c^2)^2 + (cp)^2 \qquad (40.17)$$

And at high energies the number of electrons in a system is not constant. Creation and annihilation of electron–positron pairs may occur (Section 41.2).

The construction of a theory of the interaction of photons, electrons, and positrons that fits the verified facts of both relativity and quantum mechanics has proved to be an arduous task but well worth the effort. This theory is called *quantum electrodynamics*. It is the underlying theory for atomic physics and chemistry and indeed all phenomena governed by the electromagnetic interaction, furnishing results that in some instances have the unprecedented precision of 1 part in 10^9. Nobel prizes were earned by the most celebrated architects of this theory: Dirac in 1933; J. Schwinger, R. P. Feynman, and S. I. Tomonaga in 1965.

Feynman Diagrams

In quantum electrodynamics, the state of a system is specified by giving the number of electrons, positrons, and photons as well as their energies, momenta, and spins. Any transition or change from one state to another is described in terms of the annihilation of particles in the original state and the creation of particles in the new state. Every change is a catastrophic change.

Thanks to Feynman's work, we can "picture" the creation and annihilation events that are associated with any change. More precisely, Feynman showed that to each abstruse mathematical expression in the theoretical description of a transition from one state to another, there corresponds a diagram with a straightforward physical interpretation. In the *Feynman diagrams* of Figure 43-15, a spatial dimension, such as an X axis, is plotted horizontally. The vertical direction is that of a time axis with low regions corresponding to early times and higher regions to later times. A photon is represented by a broken line and an electron by a solid line.

The first diagram, Figure 43-15(a), depicts the basic phenomenon underlying the interaction of two charged particles. The initial state with two electrons is represented by the two solid lines at the lower part of the diagram. At the left vertex a photon is created, the incident electron is annihilated, and an electron with different momentum and energy is created. This new electron is represented by the line emerging toward the upper left-hand corner. The photon travels a short distance and is absorbed at a vertex where the other incident electron is annihilated and another electron is created. In the final state there are again just two electrons and no photon. According to this diagram a "collision" of two electrons involves the *exchange* of a *photon*. The photon is the mechanism by which energy and momentum are transferred from one electron to another. The interaction whose electric part we had described by simply postulating a Coulomb force, $F = ke^2/r^2$, can thus be interpreted as arising from the emission and absorption of photons by the two charged particles.

A Feynman diagram for the Compton effect is shown in Figure 43-15(b). The initial photon is absorbed at one vertex and the final photon is emitted at another vertex.

The annihilation of an electron–positron pair with the subsequent production of two photons is depicted in the Feynman diagram of Figure 43-15(c).

The key ingredient in every diagram is a vertex with one photon line and two lines representing charged particles. Using such vertices, we can diagram much more complicated processes, and we can evaluate their relative importance in linking the initial and final states by the mathematical procedures of quantum electrodynamics. In such a calculation a factor α (the fine structure constant

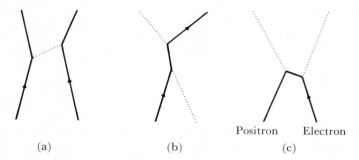

Positron   Electron

(a)   (b)   (c)

Figure 43-15   Three Feynman diagrams. Solid lines represent electrons or positrons, and broken lines represent photons. (In these schematic diagrams no attempt has been made to relate the direction of a line with particle velocity.) The bottom of the figure represents the initial state and the top represents the final state. (a) Electron–electron scattering by exchange of a virtual photon, (b) Compton scattering, (c) electron–positron annihilation with the production of two photons.

with the value $\frac{1}{137}$) appears for each vertex in the diagram. Since high powers of $\frac{1}{137}$ are very small, diagrams with many vertices are unimportant compared to the two-vertex diagrams of Figure 43-15.

There is a subtlety associated with these diagrams. The simple process depicted at any vertex is never observed. Such a process cannot satisfy the conservation laws for both energy and momentum. For all observed processes, such as the three processes of Figure 43-15, the total energy and momentum in the final state are the same as they are in the initial state—that is, energy and momentum are conserved. Nevertheless, for an *intermediate* state, processes which violate these conservation laws are possible in quantum mechanics, provided the intermediate state is of sufficiently short duration. This possibility is associated with the Heisenberg uncertainty relation for energy and time: A state which lasts for some limited time τ does not correspond to a single definite value of the energy, but instead is a superposition of different energy values ranging roughly from some value E to $E + \Delta E$, where the spread or uncertainty ΔE in the value of the energy cannot be reduced below the value given by

$$\tau \Delta E \approx \hbar$$

Therefore, a state of short duration τ may involve a process in which energy conservation is violated by any amount which does not exceed that given by

$$\Delta E \approx \frac{\hbar}{\tau}$$

Alternatively we can assert that for a state which involves a creation or an annihilation of a given amount of energy ΔE, the duration of the state must not be longer than a time given by

$$\tau \approx \frac{\hbar}{\Delta E}$$

The intermediate states represented in the diagrams of Figure 43-15 involve the creation of an undetected particle at one vertex and its absorption at another vertex. Such particles are called *virtual* particles. Since the presence of a virtual particle implies a violation of energy conservation by a certain amount ΔE, the virtual particle can have only a transitory existence, not exceeding a time τ given by

$$\tau \approx \frac{\hbar}{\Delta E}$$

In this terminology, the electron–electron collision of Figure 43-15(a) involves the exchange of a virtual photon. The word *virtual* serves to emphasize that the Feynman diagrams, although extremely useful, are by no means equivalent to a classical picture of an interaction. The strangeness of quantum mechanics persists.

The theoretical predictions of quantum electrodynamics accord with experiment not only for a wide variety of scattering phenomena (Figure 43-15) but also for certain atomic energy level details that were not predicted by nonrelativistic quantum mechanics: A small separation of hydrogen atom energy levels which have the same value of the principal quantum number n but different values of l is a relativistic effect, as is the very existence of electron spin.

Also from the postulates of quantum electrodynamics we can deduce that electrons must obey the Pauli exclusion principle. In nonrelativistic quantum theory the exclusion principle is an unexplained addition to the basic postulates.

In spite of its many successes, quantum electrodynamics is still plagued with mathematical and conceptual difficulties. As one illustration of these problems, let us consider the energy of the virtual photon exchanged between the electrons of Figure 43-15(a). The shorter the path of this photon, the less its lifetime τ and the greater its maximum allowable energy,

$$\Delta E \approx \frac{\hbar}{\tau}$$

Therefore the allowed energy of exchanged photons increases as the separation of the electrons decreases, becoming infinite for zero separation. An infinite photon energy entails a host of problems. This train of thought leads us to suspect that the electron–photon vertices, instead of being mathematical points, must have some structure, and then the electron itself would have a structure. If this is so, quantum electrodynamics in its present formulation will fail whenever interactions at very small distances are important. So far no such evidence of electron structure has been detected even though experiments at high energies have tested the electromagnetic interaction down to particle separations as small as 10^{-17} m.

Strong Interactions

Attempts have been made to develop a theory of strong interactions such as those which lead to the formation of the nucleus, using ideas similar to those that have been so successful in the description of the weaker electromagnetic interaction. The interaction between two nucleons (neutrons or protons) is attributed to the exchange of a virtual particle. The range R of the interaction is determined by the mass M of this virtual particle according to the following argument. An upper limit to the distance R that the virtual particle can travel during its lifetime τ is given by the distance it can travel at a speed c:

$$R = c\tau$$

The existence of a virtual particle of mass M implies the existence of at least a rest energy Mc^2. Taking this energy as the order of magnitude of the uncertainty ΔE in the energy of the intermediate state, we obtain

$$Mc^2 = \Delta E = \frac{\hbar}{\tau} = \frac{\hbar c}{R}$$

This yields

$$M = \frac{(\hbar/m_e c)m_e}{R} \approx \frac{1}{2\pi}\left(\frac{0.024 \times 10^{-10}}{10^{-15}}\right)m_e \approx 400 m_e$$

where the experimentally determined range of nuclear forces, about 10^{-15} m, has been used.

The pions discovered by Powell in 1947 (Section 40.3) have masses of this order of magnitude (actually $273 m_e$) and interact strongly with neutrons and protons. Following this discovery we assume that the exchange of *virtual* pions is a most important contribution to the force between strongly interacting particles, such as the constituents of a nucleus. The existence of particles such as pions, which could be exchanged between nucleons and give rise to the short-range nuclear force, had been brilliantly conjectured by the Japanese theoretical physicist, H. Yukawa, twelve years before their discovery.

After this promising start, subsequent development of strong interaction theory has been disappointing. One trouble is that, in the Feynman diagram for any strong interaction process, each vertex involves a factor of about 1 instead of the factor of $\frac{1}{137}$ encountered in the electromagnetic interaction. Consequently there is no reason to think that processes represented by complex diagrams with many vertices are unimportant.

Much of the progress that has been made in theories of strong interactions (and also the weak interactions mentioned in Table 2-2) is based on studies of "symmetries." A discussion of this topic and more recent successes in the theory of strong interactions is left to more advanced courses.

43.9 Particles of Modern Physics

All the phenomena of nature are due to the interaction of *tiny particles* of which there are relatively few different types. This is one of the most important and remarkable conclusions of science. The quest for the ultimate building blocks of nature, now carried out principally using giant particle accelerators (Figure 27-7) and their associated equipment, has led to the discovery of a profusion of "elementary" particles. Table 43-4 lists the particles with the longest mean lives; most of these particles are unstable and undergo a sudden spontaneous disintegration into lighter particles (Table 43-5).

In this field of physics, called high-energy physics or particle physics, no comprehensive understanding has yet been achieved. In fact, we call these particles elementary only because they are not obviously structures composed of other particles.

A study of experimental data reveals many regularities. For example, we find that for every type of *particle* there exists an *antiparticle* with identical mass and spin but opposite charge and magnetic moment. Interaction between a massive particle and its antiparticle leads to annihilation of each and the production of other particles (Figure 43-16).

Table 43-4 Long-Lived Particles

Particle	Symbol	Mass (in units of m_e)	Rest energy (MeV)	Charge (in units of e)	Spin quantum number S	Anti-particle
Photon	γ	0	0	0	1	γ (self)
Leptons						
Neutrino	ν	0	0	0	$\frac{1}{2}$	$\bar{\nu}$
Electron	e^-	1	0.511	-1	$\frac{1}{2}$	e^+
Muon	μ^-	206.8	105.7	-1	$\frac{1}{2}$	μ^+
Mesons						
Pion	π^+	273.1	139.6	$+1$	0	π^-
	π^0	264.1	135.0	0	0	π^0 (self)
Kaon	K^+	966.3	493.8	$+1$	0	K^-
	K^0	974.1	497.8	0	0	$\bar{K}^0$
η^0 meson	η^0	1074	549	0	0	η^0 (self)
η' meson	η'	1875	958	0	0	η' (self)
Baryons						
Proton	p	1836	938.3	$+1$	$\frac{1}{2}$	$\bar{p}$
Neutron	n	1839	939.6	0	$\frac{1}{2}$	$\bar{n}$
Lambda	Λ^0	2183	1116	0	$\frac{1}{2}$	$\bar{\Lambda}^0$
Sigma	Σ^+	2328	1189	$+1$	$\frac{1}{2}$	$\bar{\Sigma}^-$
	Σ^0	2334	1192	0	$\frac{1}{2}$	$\bar{\Sigma}^0$
	Σ^-	2343	1197	-1	$\frac{1}{2}$	$\bar{\Sigma}^+$
Xi	Ξ^0	2573	1315	0	$\frac{1}{2}$	$\bar{\Xi}^0$
	Ξ^-	2586	1321	-1	$\frac{1}{2}$	$\bar{\Xi}^+$
Omega	Ω^-	3273	1672	-1	$\frac{3}{2}$	$\bar{\Omega}^+$
ψ particles	$J/\psi(3100)$	6076	3105	0	1	
	$\psi(3700)$	7231	3695	0	1	

Table 43-5 Selected Particle Decay Modes

Particle	Principal decay mode	Mean life s
γ	stable	
Leptons		
ν	stable	
e^-	stable	
μ^-	$e^- + \nu + \bar{\nu}$	2.2×10^{-6}
Mesons		
π^+	$\mu^+ + \nu$	2.6×10^{-8}
π^0	$\gamma + \gamma$	0.9×10^{-16}
K^+	$\mu^+ + \nu$	1.2×10^{-8}
$K^0, \bar{K}^0$	$\pi^+ + \pi^-$	0.86×10^{-10}
$K^0, \bar{K}^0$	$\pi + e + \nu$	5.2×10^{-8}
η	$\gamma + \gamma$	2.5×10^{-19}
η'	$\eta + \pi + \pi$	$> 10^{-21}$
Baryons		
p	stable	
n	$p + e^- + \bar{\nu}$	9.3×10^2
Λ^0	$p + \pi^-$	2.5×10^{-10}
Σ^+	$p + \pi^0$	0.80×10^{-10}
Σ^0	$\Lambda^0 + \gamma$	$< 10^{-14}$
Σ^-	$n + \pi^-$	1.5×10^{-10}
Ξ^0	$\Lambda^0 + \pi^0$	3.0×10^{-10}
Ξ^-	$\Lambda^0 + \pi^-$	1.7×10^{-10}
Ω^-	$\Lambda^0 + K^-$	1.3×10^{-10}

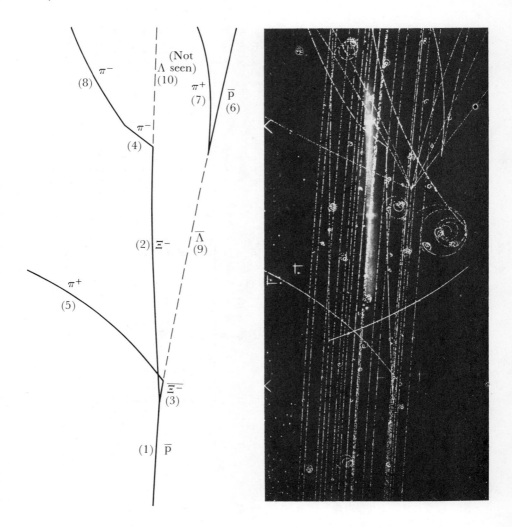

Figure 43-16   Liquid hydrogen bubble chamber photograph of an antiproton ($\bar{\text{p}}$)
colliding with a proton, resulting in production of a cascade hyperon pair ($\bar{\text{p}}$ + p
$\longrightarrow \Xi + \overline{\Xi}^-$). The accompanying sketch shows the event with the proper
assignment of a particle to each track. Track 1 is the incoming antiproton ($\bar{\text{p}}$) which
at the vertex collides with the stationary proton in the liquid hydrogen. Track 2 is
identified as a cascade hyperon (Ξ^-) which is seen to decay into track 4, a negative
pi meson (π^-), which scatters later elastically to produce track 8. The other decay
product is not seen but its direction of travel is marked by the number 10. This is
an ordinary neutral lambda hyperon (Λ) which at some point decays into a nucleon
and a pi meson. The latter decay is not seen here. Track 3 is the anti-cascade-
hyperon ($\overline{\Xi}^-$) which decays into a positive pi meson (π^+) and a neutral anti-
lambda-hyperon ($\overline{\Lambda}$) whose direction of travel is marked by the number 9. This
particle decays into track 6, an antiproton ($\bar{\text{p}}$), and track 7, a positive pi meson

The validity of some old and some new conservation laws has been demonstrated. Familiar conservation laws that have been verified in every experimental test involving interactions of these particles are:

1 Conservation of energy.
2 Conservation of momentum.
3 Conservation of angular momentum.
4 Conservation of electric charge.

The law of conservation of mass number (the sum of the numbers of protons and of neutrons in a nucleus), which holds in the radioactive decays and nuclear reactions, is violated in many processes involving heavy baryons (Table 43-4). A conservation law can be recovered by assigning to each particle a new attribute, a baryonic charge or *baryon number* with the value:

$$+1 \text{ for a baryon particle}$$
$$-1 \text{ for a baryon antiparticle}$$
$$0 \text{ for all other particles}$$

We now find that, in all processes, the algebraic sum of the baryon numbers before interaction is the same as the algebraic sum of the baryon numbers of the particles which exist after the interaction. In other words, we have a law of *conservation of baryon number*. For example, in the reaction

$$\Sigma^+ + \overline{p} \longrightarrow \pi^+ + \pi^- + \pi^0$$

the corresponding baryon numbers are

$$(+1) + (-1) \longrightarrow 0 + 0 + 0$$

We see that this reaction conserves baryon number.

Processes involving leptons (Table 43-4) reveal a completely analogous law of conservation of lepton number. For instance, in the β decay experienced by a free neutron,

$$n \longrightarrow p + e^- + \overline{\nu}$$

The lepton numbers are

$$0 \longrightarrow 0 + 1 + (-1)$$

and lepton number is conserved.

(π^+). The photograph was taken in the 50-cm liquid hydrogen bubble chamber at Brookhaven National Laboratory in an antiproton beam of 3.3 GeV/c momentum generated by the circulating proton beam in the Alternating Gradient Synchrotron, in a tungsten target, and separated by an arrangement of focusing and deflecting magnets and electrostatic separators. It confirms further the existence of an antiparticle, here the complicated $\overline{\Xi}^-$, to every known particle. (Courtesy Brookhaven National Laboratory.)

Quarks

It is tempting to speculate that each of the particles listed in Table 43-4 is, in fact, the combination of a few "really elementary" particles. In 1964, at the California Institute of Technology, George Zweig and the 1969 Nobel prize winner, Murray Gell-Mann, hypothesized that there exist three such truly fundamental particles, dubbed *quarks* (after the phrase in James Joyce's *Finnegans Wake*, "Three quarks for Muster Mark!"). It is assumed that a quark has the peculiar property of having a fractional baryonic number $\frac{1}{3}$ and an electric charge which is a *fraction* of e, either $+\frac{2}{3}e$, $-\frac{1}{3}e$, or $-\frac{1}{3}e$. Antiquarks are assumed to exist and carry charges of $-\frac{2}{3}e$, $+\frac{1}{3}e$, or $+\frac{1}{3}e$. Then each baryon is a compound of three quarks and each meson is formed from a quark and an anti-quark.

Many consequences of the quark theory are in agreement with experiment, so the quark idea has been taken seriously. Nevertheless, after years of extensive searches for free quarks, with their peculiar electric charge, we still have no decisive experimental evidence for their existence.

Questions

1   Verify that the fine structure constant, $\alpha = ke^2/\hbar c$, is a dimensionless number.

2   Compare the de Broglie wavelength of an electron moving at a speed αc with the circumference of the Bohr orbit of radius $a_0 = \hbar^2/m_e ke^2$.

3   **(a)** Show that each of the following three constants has the dimensions of a length: $\hbar^2/m_e ke^2$, called the radius of the Bohr orbit and denoted by a_0: $\hbar/m_e c$, called the reduced Compton wavelength of an electron and denoted by $\lambdabar_c$: and $ke^2/m_e c^2$, called the classical electron radius and denoted by r_e.
    (b) Show that $r_e = \alpha\lambdabar_c = \alpha^2 a_0$ where $\alpha = ke^2/\hbar c$.

4   Using the rules of Table 43-1, give the possible values of the quantum numbers n, l, m_l, and m_s for:
    (a) A state with $n = 1$.
    (b) A state with $n = 3$.

5   List the quantum numbers of all states for which $n = 4$. Verify that there are 2×4^2 different states.

6   From the diagram of Figure 43-3, work out the electron configuration for the ground states of each atom from $Z = 1$ to $Z = 9$.

7   What is the electron configuration for the ground state of magnesium ($Z = 12$)?

8   What would be the electron configuration of magnesium if no restrictions were imposed by the Pauli exclusion principle?

9   Give the electron configuration of the outer subshell for each inert gas.

10   **(a)** Give the electron configuration of the outer subshell for lithium ($Z = 3$), sodium ($Z = 11$), and potassium ($Z = 19$).
    (b) Give the electron configuration of the outer subshell for fluorine ($Z = 9$) and chlorine ($Z = 17$).

11   Relate your answers to Questions 9 and 10 to the chemical behavior of these elements and their position in the periodic table (Table 43-3).

12 Describe a sequence of events that can lead to the emission of x-ray photons in the characteristic spectrum of an element.

13 Name and describe the three different interactions of photons with matter that are shown schematically in Figure 43-17.

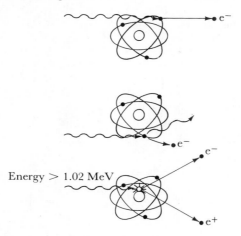

Energy > 1.02 MeV

Figure 43-17

14 Show that the resonant frequencies of the ground state of the hydrogen atom lie in the ultraviolet region.

15 Why is water transparent to visible light but strongly absorptive for ultraviolet and infrared?

16 Explain why the presence of many empty levels in a partially filled band leads to a high conductivity.

17 Why is diamond transparent?

18 What is the significant difference between an insulator and a semiconductor?

19 Draw typical energy-level diagrams showing the occupancy of the valence and conduction bands for n- and p-$type$ semiconductors.

20 In what ways is a laser light beam different from a beam from an ordinary flashlight?

21 Explain laser action making reference to stimulated emission, population inversion, and optical pumping.

22 A certain laser using carbon dioxide produces a beam with a total energy of 150 J which is confined to a pulse lasting approximately 10^{-7} s:
   (a) What is the power output of the laser during the emission of this pulse?
   (b) It should be possible to achieve controlled nuclear fusion reactions by irradiating heavy water pellets with a sufficiently powerful laser beam. If an energy of 10^6 J is needed for the pulse duration of 10^{-7} s, what is the required power output of such a laser?

23 Find the number of photons in the laser pulse of Question 22(b). Assume that the wavelength is 5.0×10^3 Å.

24 Calculate the distance that the virtual photon of Figure 43-15(a) can travel in a time $\tau = \hbar/\Delta E$, for the case where ΔE equals the rest energy $(m_e c^2)$ of an electron.

25 Does the range of an interaction depend on the mass of the virtual particle that is exchanged between the interacting particles? Explain your answer.

26 When a proton p interacts with an antiproton $\bar{p}$ the annihilation reaction often results in the creation of one neutral pion and four charged pions:

$$p + \bar{p} \longrightarrow 2\pi^+ + 2\pi^- + \pi^0$$

(a) Verify that this reaction is in accord with the conservation laws for charge, baryon number, and energy.

(b) Evaluate Q for this reaction.

appendix a.

si units

In 1960, at the Eleventh General Conference of Weights and Measures held at Sèvres, France, formal approval was given to a system of units officially entitled "Système International d'Unités" and universally designated by the abbreviation SI. Table A-1 gives the SI units that are used in this book. Multiples and submultiples of SI units are formed by means of the prefixes shown in Table A-2. Certain non-SI units used with the SI are listed in Table A-3.

The foundation of the SI rests upon the base units which are defined as follows:

1   The *metre* (m) is a length equal to 1 650 763.73* wavelengths in vacuum of the radiation corresponding to the transition between the levels $2p_{10}$ and $5d_5$ of the krypton-86 atom.

2   The *kilogram* (kg), the unit of mass, is equal to the mass of the international prototype of the kilogram. This is a particular cylinder of platinum-iridium alloy stored in a vault at Sèvres, France, by the International Bureau of Weights and Measures. This cylinder is about 2 cm in radius and 4 cm high. Duplicates are maintained in standards laboratories throughout the world.

3   The *second* (s) is the duration of 9 192 631 770 periods of the radiation corresponding to the transition between the two hyperfine levels of the ground state of the cesium-133 atom.

4   The *ampere* (A) is that constant current which, if maintained in two straight parallel conductors of infinite length, of negligible circular cross section, and placed one metre apart in a vacuum, would produce between these conductors a force equal to 2×10^{-7} newton per metre of length.

5   The *kelvin* (K), the unit of thermodynamic temperature,† is the fraction $\frac{1}{273.16}$ of the thermodynamic temperature of the triple point of water.

6   The *mole* (mol) is the amount of substance of a system which contains as many elementary entities as there are atoms in 0.012 kilogram of carbon-12.

---

*Since many countries use the comma instead of the period for decimalized values, spaces instead of commas are used to separate long lines of digits into easily readable blocks of three digits with respect to the decimal point.

†The thermodynamic temperature scale is also called the Kelvin temperature scale.

Table A-1 SI Units

Unit type	Quantity	Unit	Symbol	Expression in terms of other units
	Length	metre	m	
	Mass	kilogram	kg	
	Time	second	s	
Base Units	Electric current	ampere	A	
	Thermodynamic temperature	kelvin	K	
	Amount of substance	mole	mol	
Supplementary Units	Plane angle	radian	rad	
	Solid angle	steradian	sr	
	Frequency	hertz	Hz	s^{-1}
	Force	newton	N	$kg \cdot m/s^2$
	Pressure	pascal	Pa	N/m^2
	Energy, work, quantity of heat	joule	J	$N \cdot m$
	Power	watt	W	J/s
	Electric charge	coulomb	C	$A \cdot s$
Derived Units Having Special Names	Electric potential Potential difference Electromotive force	volt	V	W/A

Electric resistance	ohm	Ω	V/A
Electric capacitance	farad	F	C/V
Magnetic flux	weber	Wb	V·s
Inductance	henry	H	Wb/A
Magnetic flux density	tesla	T	Wb/m^2
Some Other Derived Units			
Area	square metre		m^2
Volume	cubic metre		m^3
Velocity–angular	radian per second		rad/s
Velocity–linear	metre per second		m/s
Acceleration–angular	radian per second squared		rad/s^2
Acceleration–linear	metre per second squared		m/s^2
Density (mass per unit volume)	kilogram per cubic metre		kg/m^3
Moment of force	newton-metre		N·m
Thermal conductivity	watt per metre kelvin		W/(m·K)
Thermal capacity or entropy	joule per kelvin		J/K
Permeability	henry per metre		H/m
Permittivity	farad per metre		F/m
Magnetic field strength*	ampere per metre		A/m

*Various names in common use for the vector **H** are the *magnetic intensity*, the *magnetic field intensity*, the *magnetic field strength*, the *magnetizing field*, and the *magnetizing force*.

Table A-2 Prefixes

Multiplying factor	Prefix	Symbol
$1\ 000\ 000\ 000\ 000 = 10^{12}$	tera	T
$1\ 000\ 000\ 000 = 10^{9}$	giga	G
$1\ 000\ 000 = 10^{6}$	mega	M
$1\ 000 = 10^{3}$	kilo	k
$100 = 10^{2}$	hecto	h
$10 = 10^{1}$	deca	da
$0.1 = 10^{-1}$	deci	d
$0.01 = 10^{-2}$	centi	c
$0.001 = 10^{-3}$	milli	m
$0.000\ 001 = 10^{-6}$	micro	μ
$0.000\ 000\ 001 = 10^{-9}$	nano	n
$0.000\ 000\ 000\ 001 = 10^{-12}$	pico	p

Table A-3 Some Non-SI Units Used with the SI

Unit	Symbol	Value in SI units
minute	min	$1\ min = 60\ s$
hour	h	$1\ h = 3600\ s$
day	d	$1\ d = 86400\ s$
degree (of arc)	°	$1° = (\pi/180)\ rad$
minute (of arc)	′	$1' = (\pi/10800)\ rad$
second (of arc)	″	$1'' = (\pi/648000)\ rad$
litre	l or ℓ	$1\ \ell = 1\ dm^3$
degree Celsius	°C	$0°C = 273.15\ K$ $20°C = 293.15\ K$
electronvolt	eV	$1\ eV = 1.6021892 \times 10^{-19}\ J$
atomic mass unit	u	$1\ u = 1.6605655 \times 10^{-27}\ kg$
ångström	Å	$1\ Å = 0.1\ nm = 10^{-10}\ m$
bar	bar	$1\ bar = 100\ kPa$
standard atmosphere	atm	$1\ atm = 101.325\ kPa$

appendix b.

significant figures

There is always some *uncertainty* in the value of any number that is obtained as a result of experimental measurement. Scientists write such numbers in a conventional way which displays not only the best estimate of the value of the number but also gives some information about the uncertainty in this estimate. For example, when we write for a rocket's speed v that

$$v = 1.52 \times 10^3 \text{ m/s}$$

we assert that our best estimate of the speed is 1520 m/s but that we are uncertain about the digit 2, the last digit displayed in 1.52, and that we know nothing about the value of digits to the right of 2.

The digits 1, 5, and 2 in the number 1.52×10^3 m/s are called *significant figures* or *significant digits*: they are the digits in our estimate that are believed closer to the actual value than any others. And, as indicated, the last significant digits of a series of significant digits are the ones about which we are in doubt. Look again at the preceding example of the rocket speed; there are three significant figures in 1.52×10^3 m/s, as well as in 0.152×10^4 m/s, 15.2×10^2 m/s, or 152×10 m/s, and in each case there is doubt about the digit 2.

Zeros are significant only when preceded by another significant figure. Thus the number 1.520×10^3 (or equivalently, 1520) has *four* significant figures, and if we write

$$v = 1.520 \times 10^3 \text{ m/s}$$

we are asserting that, although there is a measure of doubt, we do have some grounds for believing that the fourth digit is 0 instead of 1 or 2 or 3, etc.

By the same token, if we write

$$v = 1.5200 \times 10^3 \text{ m/s}$$

the last zero here is also significant since it follows a zero which is itself significant. In other words, we assert in this way that both the fourth and the fifth digits are each 0 and not 1, 2, or 3, etc.

A still more accurate measurement might yield the result,

$$v = 1.5200418 \times 10^3 \text{ m/s}$$

which has eight significant figures. The same information is given by

$$v = 1520.0418 \text{ m/s}$$

which also displays eight significant figures. Notice that the number of significant figures is not changed by merely moving the decimal point and correspondingly changing the power of ten.

When then is a zero not a significant figure? When it does not follow a significant figure, as was stated before. And this situation occurs when the zero is employed simply as a decimal place holder. As an example, consider 0.0089. This number has only two significant figures. The zero before the decimal point is employed conventionally simply to emphasize that this is not a whole number but a fraction (i.e., that there is no digit like 1, 2, 3, etc., in front of the decimal point). The two zeros after the decimal point are decimal place holders only. They tell us that the decimal fraction is not $\frac{89}{100}$, nor $\frac{89}{1000}$, but rather $\frac{89}{10000}$. This expression could equally well be written

$$8.9 \times 10^{-3}$$

which notation clearly indicates that there are only two significant figures.

In this text, in the examples and questions, we usually present data with two or three significant figures and, after calculating with these data, we retain in the results only the number of significant figures consistent with the uncertainty of the original data.

appendix c.

physical data

Astronomical Data

Earth

Radius (mean)	6371 km = 3959 mi
Distance from sun (mean)	149.5×10^6 km = 92.9×10^6 mi
Period of rotation	86164 s = 1 sideral day = 23.94 h
Solar constant (radiation from sun at earth's mean distance)	1.35 kW/m^2
Mass	5.975×10^{24} kg
Mean density	5.52 Mg/m^3

Moon

Radius	1741 km = 1082 mi
Distance from earth (mean)	384.4×10^3 km = 239×10^3 mi
Period of revolution	27.322 d
Mass	7.343×10^{22} kg
Mean density	3.33 Mg/m^3

Sun

Radius	696.5×10^3 km = 432.2×10^3 mi
Mass	1.987×10^{30} kg
Mean density	1.41 Mg/m^3

FUNDAMENTAL CONSTANTS

Compiled by E. R. Cohen and B. N. Taylor under the auspices of the CODATA Task Group on Fundamental Constants. This set has been officially adopted by CODATA and is taken from J. Phys. Chem. Ref. Data, Vol. 2, No. 4, p. 663 (1973) and CODATA Bulletin No. 11 (December 1973).

Quantity	Symbol	Numerical Value *	Uncert. (ppm)	SI † ← Units → cgs ‡		
Speed of light in vacuum	c	299792458(1.2)	0.004	$m \cdot s^{-1}$		10^2 $cm \cdot s^{-1}$
Permeability of vacuum	μ_0	4π $=12.5663706144$		10^{-7} $H \cdot m^{-1}$ 10^{-7} $H \cdot m^{-1}$		
Permittivity of vacuum, $1/\mu_0 c^2$	ϵ_0	8.854187818(71)	0.008	10^{-12} $F \cdot m^{-1}$		
Fine-structure constant, $[\mu_0 c^2/4\pi](e^2 hc)$	α α^{-1}	7.2973506(60) 137.03604(11)	0.82 0.82	10^{-3}		10^{-3}
Elementary charge	e	1.6021892(46) 4.803242(14)	2.9 2.9	10^{-19} C		10^{-20} emu 10^{-10} esu
Planck constant	h $\hbar = h/2\pi$	6.626176(36) 1.0545887(57)	5.4 5.4	10^{-34} $J \cdot s$ 10^{-34} $J \cdot s$		10^{-27} $erg \cdot s$ 10^{-27} $erg \cdot s$
Avogadro constant	N_A	6.022045(31)	5.1	10^{23} mol^{-1}		10^{23} mol^{-1}
Atomic mass unit, $10^{-3} kg \cdot mol^{-1} N_A^{-1}$	u	1.6605655(86)	5.1	10^{-27} kg		10^{-24} g
Electron rest mass	m_e	9.109534(47) 5.4858026(21)	5.1 0.38	10^{-31} kg 10^{-4} u		10^{-28} g 10^{-4} u
Proton rest mass	m_p	1.6726485(86) 1.007276470(11)	5.1 0.011	10^{-27} kg u		10^{-24} g u
Ratio of proton mass to electron mass	m_p/m_e	1836.15152(70)	0.38			
Neutron rest mass	m_n	1.6749543(86) 1.008665012(37)	5.1 0.037	10^{-27} kg u		10^{-24} g u
Electron charge to mass ratio	e/m_e	1.7588047(49) 5.272764(15)	2.8 2.8	10^{11} $C \cdot kg^{-1}$		10^7 $emu \cdot g^{-1}$ 10^{17} $esu \cdot g^{-1}$
Magnetic flux quantum, $[c]^{-1}(hc/2e)$	Φ_0 h/e	2.0678506(54) 4.135701(11) 1.3795215(36)	2.6 2.6 2.6	10^{-15} Wb 10^{-15} $J \cdot s \cdot C^{-1}$		10^{-7} $G \cdot cm^2$ 10^{-7} $erg \cdot s \cdot emu^{-1}$ 10^{-17} $erg \cdot s \cdot esu^{-1}$
Josephson frequency-voltage ratio	$2e/h$	4.835939(13)	2.6	10^{14} $Hz \cdot V^{-1}$		
Quantum of circulation	$h/2m_e$ h/m_e	3.6369455(60) 7.273891(12)	1.6 1.6	10^{-4} $J \cdot s \cdot kg^{-1}$ 10^{-4} $J \cdot s \cdot kg^{-1}$		$erg \cdot s \cdot g^{-1}$ $erg \cdot s \cdot g^{-1}$
Faraday constant, $N_A e$	F	9.648456(27) 2.8925342(82)	2.8 2.8	10^4 $C \cdot mol^{-1}$		10^3 $emu \cdot mol^{-1}$ 10^{14} $esu \cdot mol^{-1}$
Rydberg constant, $[\mu_0 c^2/4\pi]^2(m_e e^4/4\pi \hbar^3 c)$	R_∞	1.097373177(83)	0.075	10^7 m^{-1}		10^5 cm^{-1}
Bohr radius, $[\mu_0 c^2/4\pi]^{-1}(\hbar^2/m_e e^2) = \alpha/4\pi R_\infty$	a_0	5.2917706(44)	0.82	10^{-11} m		10^{-9} cm
Classical electron radius, $[\mu_0 c^2/4\pi](e^2/m_e c^2) = \alpha^3/4\pi R_\infty$	$r_e = \alpha \lambdabar_C$	2.8179380(70)	2.5	10^{-15} m		10^{-13} cm
Thomson cross section, $(8/3)\pi r_e^2$	σ_e	0.6652448(33)	4.9	10^{-28} m^2		10^{-24} cm^2
Free electron g-factor, or electron magnetic moment in Bohr magnetons	$g_e/2 = \mu_e/\mu_B$	1.0011596567(35)	0.0035			
Free muon g-factor, or muon magnetic moment in units of $[c](e\hbar/2m_\mu c)$	$g_\mu/2$	1.00116616(31)	0.31			
Bohr magneton, $[c](e\hbar/2m_e c)$	μ_B	9.274078(36)	3.9	10^{-24} $J \cdot T^{-1}$		10^{-21} $erg \cdot G^{-1}$
Electron magnetic moment	μ_e	9.284832(36)	3.9	10^{-24} $J \cdot T^{-1}$		10^{-21} $erg \cdot G^{-1}$
Gyromagnetic ratio of protons in H_2O	γ'_p $\gamma'_p/2\pi$	2.6751301(75) 4.257602(12)	2.8 2.8	10^8 $s^{-1} \cdot T^{-1}$ 10^7 $Hz \cdot T^{-1}$		10^4 $s^{-1} \cdot G^{-1}$ 10^3 $Hz \cdot G^{-1}$
γ'_p corrected for diamagnetism of H_2O	γ_p $\gamma_p/2\pi$	2.6751987(75) 4.257711(12)	2.8 2.8	10^8 $s^{-1} \cdot T^{-1}$ 10^7 $Hz \cdot T^{-1}$		10^4 $s^{-1} \cdot G^{-1}$ 10^3 $Hz \cdot G^{-1}$
Magnetic moment of protons in H_2O in Bohr magnetons	μ'_p/μ_B	1.52099322(10)	0.066	10^{-3}		10^{-3}
Proton magnetic moment in Bohr magnetons	μ_p/μ_B	1.521032209(16)	0.011	10^{-3}		10^{-3}
Ratio of electron and proton magnetic moments	μ_e/μ_p	658.2106880(66)	0.010			
Proton magnetic moment	μ_p	1.4106171(55)	3.9	10^{-26} $J \cdot T^{-1}$		10^{-23} $erg \cdot G^{-1}$
Magnetic moment of protons in H_2O in nuclear magnetons	μ'_p/μ_N	2.7927740(11)	0.38			
μ'_p/μ_N corrected for diamagnetism of H_2O	μ_p/μ_N	2.7928456(11)	0.38			
Nuclear magneton, $[c](e\hbar/2m_p c)$	μ_N	5.050824(20)	3.9	10^{-27} $J \cdot T^{-1}$		10^{-24} $erg \cdot G^{-1}$
Ratio of muon and proton magnetic moments	μ_μ/μ_p	3.1833402(72)	2.3			
Muon magnetic moment	μ_μ	4.490474(18)	3.9	10^{-26} $J \cdot T^{-1}$		10^{-23} $erg \cdot G^{-1}$
Ratio of muon mass to electron mass	m_μ/m_e	206.76865(47)	2.3			

Quantity	Symbol	Numerical Value *	Uncert. (ppm)	SI †	← Units →	cgs ‡
Muon rest mass	m_μ	1.883566(11)	5.6	10^{-28} kg		10^{-25} g
		0.11342920(26)	2.3	u		u
Compton wavelength of the electron, $h/m_e c = \alpha^2/2R_\infty$	λ_C	2.4263089(40)	1.6	10^{-12} m		10^{-10} cm
	$\lambda_C = \lambda_C/2\pi = \alpha a_0$	3.8615905(64)	1.6	10^{-13} m		10^{-11} cm
Compton wavelength of the proton, $h/m_p c$	$\lambda_{C.p}$	1.3214099(22)	1.7	10^{-15} m		10^{-13} cm
	$\lambda_{C.p} = \lambda_{C.p}/2\pi$	2.1030892(36)	1.7	10^{-16} m		10^{-14} cm
Compton wavelength of the neutron, $h/m_n c$	$\lambda_{C.n}$	1.3195909(22)	1.7	10^{-15} m		10^{-13} cm
	$\lambda_{C.n} = \lambda_{C.n}/2\pi$	2.1001941(35)	1.7	10^{-16} m		10^{-14} cm
Molar volume of ideal gas at s.t.p.	V_m	22.41383(70)	31	10^{-3} m^3·mol^{-1}		10^3 cm^3·mol^{-1}
Molar gas constant, $V_m p_0/T_0$	R	8.31441(26)	31	J·mol^{-1}·K^{-1}		10^7 erg·mol^{-1}·K^{-1}
($T_0 \equiv 273.15$ K; $p_0 \equiv 101325$ Pa$\equiv$1atm)		8.20568(26)	31	10^{-5} m^3·atm·mol^{-1}·K^{-1}		10 cm^3·atm·mol^{-1}·K^{-1}
Boltzmann constant, R/N_A	k	1.380662(44)	32	10^{-23} J·K^{-1}		10^{-16} erg·K^{-1}
Stefan-Boltzmann constant, $\pi^2 k^4/60\hbar^3 c^2$	σ	5.67032(71)	125	10^{-8} W·m^{-2}·K^{-4}		10^{-5} erg·s^{-1}·cm^{-2}·K^{-4}
First radiation constant, $2\pi hc^2$	c_1	3.741832(20)	5.4	10^{-16} W·m^2		10^{-5} erg·cm^2·s^{-1}
Second radiation constant, hc/k	c_2	1.438786(45)	31	10^{-2} m·K		cm·K
Gravitational constant	G	6.6720(41)	615	10^{-11} m^3·s^{-2}·kg^{-1}		10^{-8} cm^3·s^{-2}·g^{-1}
Ratio, kx-unit to ångström, $\Lambda = \lambda(Å)/\lambda(kxu)$; $\lambda(CuK\alpha_1) \equiv 1.537400$ kxu	Λ	1.0020772(54)	5.3			
Ratio, Å* to ångström, $\Lambda^* = \lambda(Å)/\lambda(Å^*)$; $\lambda(WK\alpha_1) \equiv 0.2090100$ Å*	Λ^*	1.0000205(56)	5.6			

ENERGY CONVERSION FACTORS AND EQUIVALENTS

Quantity	Symbol	Numerical Value *	Units	Uncert. (ppm)
1 kilogram (kg·c^2)		8.987551786(72)	10^{16} J	0.008
		5.609545(16)	10^{29} MeV	2.9
1 Atomic mass unit (u·c^2)		1.4924418(77)	10^{-10} J	5.1
		931.5016(26)	MeV	2.8
1 Electron mass m_e·c^2)		8.187241(42)	10^{-14} J	5.1
		0.5110034(14)	MeV	2.8
1 Muon mass (m_μ·c^2)		1.6928648(96)	10^{-11} J	5.6
		105.65948(35)	MeV	3.3
1 Proton mass (m_p·c^2)		1.5033015(77)	10^{-10} J	5.1
		938.2796(27)	MeV	2.8
1 Neutron mass (m_n·c^2)		1.5053738(78)	10^{-10} J	5.1
		939.5731(27)	MeV	2.8
1 Electron volt		1.6021892(46)	10^{-19} J	2.9
			10^{-12} erg	2.9
	1 eV/h	2.4179696(63)	10^{14} Hz	2.6
	1 eV/hc	8.065479(21)	10^5 m^{-1}	2.6
			10^3 cm^{-1}	2.6
	1 eV/k	1.160450(36)	10^4 K	31
Voltage-wavelength conversion, hc		1.986478(11)	10^{-25} J·m	5.4
		1.2398520(32)	10^{-6} eV·m	2.6
			10^{-4} eV·cm	2.6
Rydberg constant	$R_\infty hc$	2.179907(12)	10^{-18} J	5.4
			10^{-11} erg	5.4
		13.605804(36)	eV	2.6
	$R_\infty c$	3.28984200(25)	10^{15} Hz	0.075
	$R_\infty hc/k$	1.578885(49)	10^5 K	31
Bohr magneton	μ_B	9.274078(36)	10^{-24} J·T^{-1}	3.9
		5.7883785(95)	10^{-5} eV·T^{-1}	1.6
	μ_B/h	1.3996123(39)	10^{10} Hz·T^{-1}	2.8
	μ_B/hc	46.68604(13)	m^{-1}·T^{-1}	2.8
			10^{-2} cm^{-1}·T^{-1}	2.8
	μ_B/k	0.671712(21)	K·T^{-1}	31
Nuclear magneton	μ_N	5.505824(20)	10^{-27} J·T^{-1}	3.9
		3.1524515(53)	10^{-8} eV·T^{-1}	1.7
	μ_N/h	7.622532(22)	10^6 Hz·T^{-1}	2.8
	μ_N/hc	2.5426030(72)	10^{-2} m^{-1}·T^{-1}	2.8
			10^{-4} cm^{-1}·T^{-1}	2.8
	μ_N/k	3.65826(12)	10^{-4} K·T^{-1}	31

* Note that the numbers in parentheses are the one standard-deviation uncertainties in the last digits of the quoted value computed on the basis of internal consistency, that the unified atomic mass scale $^{12}C \equiv 12$ has been used throughout, that u=atomic mass unit, C=coulomb, F=farad, G=gauss, H=henry, Hz=hertz=cycle/s, J=joule, K=kelvin (degree Kelvin), Pa=pascal=N·m^{-2}, T=tesla (10^4 G), V=volt, Wb=weber= T·m^2, and W=watt. In cases where formulas for constants are given (e.g., R_∞), the relations are written as the product of two factors. The second factor, in parentheses, is the expression to be used when all quantities are expressed in cgs units, with the electron charge in electrostatic units. The first factor, in brackets, is to be included only if all quantities are expressed in SI units. We remind the reader that with the exception of the auxiliary constants which have been taken to be exact, the uncertainties of these constants are correlated, and therefore the general law of error propagation must be used in calculating additional quantities requiring two or more of these constants.

† Quantities given in u and atm are for the convenience of the reader; these units are not part of the International System of Units (SI).

‡ In order to avoid separate columns for "electromagnetic" and "electrostatic" units, both are given under the single heading "cgs Units." When using these units, the elementary charge e in the second column should be understood to be replaced by e_m or e_e, respectively.

appendix d.

selected atomic masses

The atomic masses required for use in the problems of the text are given here. The mass is that of the neutral atom which includes orbital electrons as well as a nucleus. Masses are given in atomic mass units (u) where the mass of $^{12}_{6}C$ is exactly 12 u, by definition.

Element	Mass number	Atomic mass u	Element	Mass number	Atomic mass u
$_0$n	1	1.008665	$_7$N	12	12.018641
$_1$H	1	1.007825		13	13.005738
	2	2.014102		14	14.003074
	3	3.016050		15	15.000108
$_2$He	3	3.016030		16	16.006103
	4	4.002603		17	17.008450
	6	6.018893	$_8$O	14	14.008597
$_3$Li	6	6.015125		15	15.003070
	7	7.016004		16	15.994915
	8	8.022487		17	16.999133
$_4$Be	7	7.016929		18	17.999160
	9	9.012186		19	19.003578
	10	10.013534	$_{36}$Kr	89	88.9166
$_5$B	8	8.024609	$_{56}$Ba	140	139.9106
	10	10.012939	$_{86}$Rn	222	222.01753
	11	11.009305	$_{88}$Ra	226	226.02536
	12	12.014354	$_{92}$U	235	235.043915
$_6$C	10	10.016810		238	238.050770
	11	11.011432			
	12	12.000000			
	13	13.003354			
	14	14.003242			
	15	15.010599			

appendix ε.

answers for odd-numbered questions

Chapter 1

1 (a) 40 units. **(b)** No. The two vectors have different directions. **(c)** Yes.
3 14.1 m southeast.   **5** 15.3 km, 29.3° north of east.   **7** 50 m southwest.
11 $\Delta \mathbf{r}$ = 20 m east. At t = 12 s, $\mathbf{r}$ = 50 m east.

13

Time interval s	Displacement m	Average velocity m/s
0.00 ⟶ 6.00	100 east	16.7 east
6.00 ⟶ 8.00	100 north	50.0 north
0.00 ⟶ 8.00	141 northeast	17.6 northeast
0.00 ⟶ 32.00	0	0

15

Time interval s	Change in velocity m/s	Average acceleration m/s²
0.00 ⟶ 6.00	20 east	3.33 east
6.00 ⟶ 8.00	20 north	10.0 north
0.00 ⟶ 32.00	40 north	1.25 north

S-3 $d\mathbf{r}/dt = \mathbf{v}_0 + \mathbf{a}t$; $d^2\mathbf{r}/dt^2 = \mathbf{a}$.
S-5 If the particle's acceleration is not zero at the instant the particle is at the specified point, the particle's velocity is changing. This implies that, as time elapses, the particle's velocity must change from zero to some nonzero value, and consequently the particle's position vector must change. The displacement can be estimated using Eq. 1.13 and 1.12 with $\mathbf{v}_0$ = 0.

Chapter 2

1 (a) 5 N east. **(b)** 1 N east. **(c)** 0. **(d)** 5 N, about 37° east of north. **(e)** 4 N east.
3 600 N downward. **5** 1000 N directed upward. **7** 0.33 m/s², about 37° east of
north. **9** 10 N in the direction opposite to that of the velocity vector of the box.
11 4.0 N downward. **13 (a)** *mass* = 4.0 kg on earth and on planet.
(b) *weight* = 39.2 N on earth. *weight* = 8.0 N on planet. **15** No. No. **17** The free
refuse and the satellite each have the same acceleration g. Moreover, the refuse, as it
is released, has essentially the same position and velocity as the satellite. Therefore,
the refuse and the satellite follow the same orbit. **S-1** 3.0 m/s². **S-3** 0.167 m/s².

Chapter 3

1 F_{1x} = 8.00 N, F_{1y} = 0. F_{2x} = −9.66 N, F_{2y} = 2.59 N. **3** F = 13.00 N at an
angle of 67.4° with the X axis. **5** x = 3.46 m, y = 2.00 m. **7** 9.04 N, 22.8° north
of east. **9** 544 N, 36.0° north of east.
11 (a) 98 N upward.
    (b) 29.4 N in the direction opposite to the direction of motion.
    (c) 2.0 m/s² in the direction of motion.
    (d) 2.9 m/s² in the direction opposite to the direction of motion.
13 2.12 m/s². **15 (a)** 294 N. **(b)** 1.96 m/s². **17** 70 N; 79 N. **S-1** 327 N.
S-7 (a) 10.2 kg. **(b)** 2.45 m/s².

Chapter 4

1 1000 N directed toward the man. **3** 800 N forward, 200 N backward.
acceleration = 4.00 m/s². **5 (a)** 326 N. **(b)** 366 N upward. **(c)** 366 N downward.
11 x_c = 3.7 m, y_c = 0.8 m.
13 (a) On the line joining the spheres, 9.4 × 10⁻² m from the 5.0-kg sphere.
    (b) 5.0 m/s² in the direction of the applied force.
15 392 N downward, 9.8 m/s² downward. **17 (a)** 4.5 × 10³ N. **(b)** 1.5 × 10³ N.
19 1.5 × 10³ N.
21 (a) 1.67 N to the right.
    (b) 3.33 N to the right.
    (c) The acceleration of the blocks is the same in parts (a) and (b) and therefore
    the same resultant force acts on the 10-kg block. In part (b), the force
    exerted by the 20-kg block is not the only force contributing to this resultant.
23 (b) $T_1 − T_2$ = ma. **(c)** The mass of the rope must be negligible. With m = 0 we
obtain $T_1 = T_2$. **25** 11.8 N, μ_s = 1.00.
27 In the lower string the tension is $T_2 = m_2(a + g)$.
In the upper string the tension is $T_1 = (m_1 + m_2)(a + g)$.
S-1 *acceleration* = g/μ_s. **S-3** *acceleration* = $g \tan \theta$. **S-5** *force* = $9m_1m_2g/(m_1 + 4m_2)$.
S-7 89 N.

Chapter 5

1 2.0 × 10³ N•m. **3 (a)** −0.98 N•m. **(b)** 0.39 N•m. **(c)** −0.49 N•m. **7** At 40-cm
mark, 16.7 N; at 100-cm mark, 3.3 N. **9** T = 400 N; hinge exerts a force with a
400-N horizontal component and a 600-N vertical component. **11 (a)** H = 10 N.
(b) F_x = 10 N, F_y = 60 N. **13** f = 62.5 N; N' = 62.5 N; N = 300 N.
S-1 μ_s = 0.292. **S-3** 346 N.

Chapter 6

1 (a) 24 kg·m/s east. **(b)** 2.4 m/s east.   **3** 1.4 m/s west.   **5** 0.12 m/s in the original direction of the bullet. ⊘**9** 2.0 m/s south.   **11** 13.3 m.   **13** $sm/(m + M)$, forward.
15 (a) $1.41 \times 10^{-4}c$ in the direction of the incident neutron.
  (b) $2.96 \times 10^{-2}c$ at an angle of 89.4° with the direction of motion of the incident neutron.
17 (a) 5.0 m/s east. **(b)** 1.0 m/s west.   **19** 120 m/s east.   **21** Current is 2.5 km/h; speed of boat relative to water is 17.5 km/h.   **23** 117 m/s, 22.6° north of east.
25 24.1 m/s, 4.8° south of east.   **27** 40 m/s².
29 (a) Not an inertial frame.
  (b) Inertial frame.
  (c) Not an inertial frame.
  (d) Not an inertial frame.
S-1 (a) 10 m/s forward. **(b)** 45 m/s, 55 m/s.   **S-3** Speed relative to car is 11.5 m/s. Speed relative to ground is 5.77 m/s.   **S-5** 53.1° south of east.

Chapter 7

1

	Δx	$\Delta x/\Delta t$
$A \longrightarrow B$	2.00 m	4.00 m/s
$B \longrightarrow C$	−6.00 m	−3.00 m/s
$A \longrightarrow C$	−4.00 m	−1.60 m/s

3 (a) 18.0 m/s; 15.0 m/s; 13.5 m/s; 12.3 m/s; 12.03 m/s.
  (b) 12.0 m/s.
  (c) The average velocities in part (a) have a limiting value (as the time interval approaches zero) equal to the derivative determined in part (b).
5 t_1: a_x positive, **a** in positive x direction.
  t_2: a_x positive, **a** in positive x direction.
  t_3: a_x zero.
  t_4: a_x negative, **a** in negative x direction.
  t_5: a_x positive, **a** in positive x direction.
7 $dv_x/dt = -3.7$ m/s².   **9** Particle at rest when $t = 3$ s, $x = 54$ m,
$d^2x/dt^2 = -18$ m/s².   **11** 89 m.   **13** $x = -4t^3/3 + 16t$. At $t = 2.0$ s, x reaches its maximum value, 21.3 m.
15 (a) 94 m; 136 m; 166 m; 184 m; 190 m.
  (b) Graph is a straight line with a negative slope (-3.0 m/s²). At $t = 10.00$ s, the velocity is zero.
17 1080 m.   **19** Automobile was speeding at 17 m/s.   **21** 35 m/s; 58 s.
23 (a) 2.83 s. **(b)** 27.7 m/s. **(c)** 14.7 m/s; 14.7 m.   **25** 24 s.   **29** 0.19.   **31** 41.0 m/s.
33 20.4 m.   **35** 1.02 s.   **37** 60 N.
S-1 (a) $dx/dt = -34$ m/s.
  (b) $t = 1.8$ s.
  (c) $t = 1.8$ s.
  (d) 24 m.
  (e) $t = 3.2$ s; -40 m/s.
  (f) $\Delta x = 18$ m; $\Delta x/\Delta t = 18$ m/s.
S-3 (b) v_0t_1. **(d)** 30 m/s.   **S-5** $2Rv$.

Chapter 8

1 $\omega = 0.20$ rad/s; *period* $= 31.4$ s; *acceleration* $= 6.0$ m/s² west.   **3 (a)** 0; 30 m/s; 0; -30 m/s. **(b)** -6.0 m/s²; 0; 6.0 m/s²; 0.   **5** $\omega = 2.0 \times 10^{-7}$ rad/s; $v = 3.0 \times 10^4$ m/s.

7 (a) 1.6×10^3 m/s; the upper tip moves to the left and the lower tip moves to the right.

    (b) 3.2×10^6 m/s, directed inward toward the propeller shaft.

9 (a) $a = 4.0$ m/s² south.

    (b) Acceleration vector has a centripetal component 4.0 m/s² south and a tangential component 3.0 m/s² east. The acceleration vector has a magnitude of 5.0 m/s² and a direction 36.9° east of south.

11 (a) $2\pi R/v_0$. **(b)** $\sqrt{4\pi R/a_s}$.   **13** 3.3 rad/s².   **15** 2.5×10^2 rad/s.   **17** 0.23 rad/s²; 4.1 revolutions.   **19 (a)** 1357 rad. **(b)** 146 rad/s.   **21** 1.0 s.

23 (a) Centripetal acceleration is greater while making a sharp turn than while rounding a gentle curve. Centripetal acceleration is zero when path is straight.

    (b) Velocity vector is tangent to path. Acceleration vector points inward and makes an acute angle with the velocity vector.

25 (a) Because the road is straight at point B, the normal component of the car's acceleration is zero. The tangential component remains 2.00 m/s².

    (b) Normal component is 3.00 m/s²; tangential component remains 2.00 m/s².

S-3 $(12\pi rm/K)^{1/3}$.

Chapter 9

1 (a) Mass moves from $x = -A$ through the origin to $x = +A$ and then returns through the origin back to $x = -A$.

    (b) 0.25 s.

3 125 s.   **5 (a)** $T = 15.7$ s; $A = 0.30$ m. **(b)** $T/4 = 3.9$ s.

7 Phasor is 2.00 m long and is directed upward.

9 (a) 0.12 m/s.

    (b) 0.12 m/s.

    (c) At the equilibrium position ($x = 0$).

    (d) The minimum speed, zero, is attained at $x = \pm 0.30$ m.

11 (a) 4.8×10^{-2} m/s² directed toward center of circle.

    (b) 4.8×10^{-2} m/s².

    (c) At $x = \pm 0.30$ m.

    (d) The minimum magnitude of the acceleration, zero, occurs at the equilibrium position ($x = 0$).

13 *maximum speed* $= 0.64$ m/s; *maximum acceleration* $= 1.03 \times 10^3$ m/s².   **15 (a)** 3.0 s. **(b)** 35 N/m. **(c)** 3.5 N.   **17** 2.01 s.   **19** 9.79 m/s².   **S-3 (a)** 0.10 m. **(b)** 0.10 m. **(c)** 1.6 Hz; 0.10 m.   **S-5** 0.062 m   **S-7 (a)** The tension reaches its maximum value when the pendulum is vertical.

S-9 At the instant that the phasor representing the displacement x is horizontal and pointing to the right, the phasor representing the velocity points vertically upward, and the phasor representing the acceleration is horizontal and points to the left. Phase difference between x and $v_x = -90°$. Phase difference between x and $a_x = -180°$. Phase difference between v_x and $a_x = -90°$.

Chapter 10

1 9.0 J. **3** 1.35×10^3 J. **5** 64 J. **7** 1.3×10^4 W. **9** 3.0×10^4 W. **11** 343 W.
13 *work* $= 5.0 \times 10^5$ J; *distance* $= 5.0 \times 10^2$ m. **17** In SI units,
$U = -9x^3 + ($*constant independent of x*$)$. **23 (a)** $x = \pm A$. **(b)** $x = 0$. **(c)** $x = \pm\frac{1}{2}A$.
(d) $x = \pm A/\sqrt{2}$. **S-3 (a)** 0.102 m. **(b)** 0.102 m. **S-5 (b)** Boundaries are
$x = \pm 1.0$ m; *maximum kinetic energy* $= 0.40$ J. **(c)** 1.6 J.

Chapter 11

1 *external work* $= 0$; *internal work* $= 750$ J. **3** *speed* $= \sqrt{gL}$.
5 $\dfrac{v^2}{2\mu_k g}\left(\dfrac{m}{m + M}\right)^2$.
9 $m_1 v_1/(m_1 + m_2)$. **13 (a)** Mc^2 is the total energy of an isolated system evaluated
in the system's CM-frame.
17 For *any* isolated system whose total energy is (*sum of particles' rest energies*) + (*sum
of particles' kinetic energies*), the law of conservation of energy requires that, if the
sum of the kinetic energies of the particles of the system changes, there must be
compensating changes in the sum of the rest energies and therefore in the sum of
the masses. In chemical reactions and in heating a body, the changes in mass are so
small compared to the masses involved that these changes cannot be detected by
direct measurement of mass.
19 4.0 MeV. **21** 4.5 eV. **23** 8.5 MeV. **25** $Q = 4.2(\frac{238}{234})$ MeV $= 4.3$ MeV.
S-7 (a) $^1_1H + ^7_3Li \longrightarrow ^4_2He + ^4_2He$.
    (b) From kinetic energies, $Q = 17.0$ MeV. From rest energies, $Q = 17.2$ MeV.
    (c) Within experimental error, the increase in kinetic energy (17.0 MeV) is
        equal to the decrease in internal energy predicted by Einstein's equation
        (17.2 MeV).
S-9 1.87 MeV.

Chapter 12

1 0.116 kg·m². **3** $I = (m + M/3)L^2$. **7** $\frac{7}{5}MR^2$. **9** $MR^2/4$. **11 (a)** 3.92×10^3 J.
(b) $K_{\text{pail}}/K_{\text{windlass}} = 0.80$. **(c)** 13 m/s; 65 rad/s. **13** *kinetic energy* $= 11.8$ J;
speed $= 2.8$ m/s.

15 *angular speed* $= \sqrt{\dfrac{2g}{D + L^2/12D}}$.

17 1.25×10^3 N·m. **19** 19.8 N·m. **21** 50 kg·m². **23** 0.200 N·m². **25** 17.7 N.

S-3 *angular speed* $= \sqrt{\dfrac{gL}{R^2 + I/M}}$.

S-5 (b) Angular acceleration of pole is proportional to the torque of the pole's
      weight about its base and therefore increases as the pole falls.
    (c) $\omega = \sqrt{3g/L}$.
    (d) *speed* $= \sqrt{3gL}$.

Chapter 13

1 (a) With axis OZ directed upward along the pole, the angular momentum about
the pole is $L_{OZ} = 1.44 \times 10^6$ kg·m²/s.
(b) $L_{OZ} = 1.44 \times 10^6$ kg·m²/s.
3 86 kg·m²/s.   **5** 7 × 10³³ kg·m²/s.   **7** 0.86 rad/s.   **9** 0.20 rad/s, counterclockwise
in Figure 13-4.   **11** 1.6 rad/s.
13 (a) External forces are the force exerted on the axle of the pulley by the pulley
support and also the weight of each object of the system. The resultant
external torque is zero.
(b) The total angular momentum of the system remains zero. The monkey and
the bananas will move upward with identical speeds at each instant.
S-3 *speed* $= 5v_0/7$; *energy dissipated* $= Mv_0^2/7$.

Chapter 14

1 6.7×10^{-11} N/kg; 6.7×10^{-10} N.   **3** 4 × (earth's radius).   **5 (a)** 2.7×10^2 N/kg.
(b) 5.9×10^{-3} N/kg.   **7** 2.6×10^8 m from earth toward sun.
9 g $= 4.4 \times 10^{-10}$ N/kg in a direction making an angle of 73.1° with the line
joining the two masses.   **13** 8.3×10^{-10} N. Man's weight is 1.2×10^{12} times as
great. Gravitational attraction will not be detected.   **17** 1.12×10^4 m/s.
19 (a) 3.7 N/kg. **(b)** 5.0×10^3 m/s.

21 $v_1 = \sqrt{\dfrac{2Gm_2}{1 + m_1/m_2}\left(\dfrac{1}{r} - \dfrac{1}{r_0}\right)}$; $v_2 = \sqrt{\dfrac{2Gm_1}{1 + m_2/m_1}\left(\dfrac{1}{r} - \dfrac{1}{r_0}\right)}$.

23 4.2×10^7 m.   **25** 1.90×10^{27} kg.   **27** 2.01×10^{30} kg.   **29** 1.87 years.
S-3 $\epsilon = \sqrt{1 + 2E'L'^2/G^2M^2}$;
$R = L'^2/GM$; $a = GM/2|E'|$; $b = L'/\sqrt{2|E'|}$.

S-5

β	Orbit	r_{min}	r_{max}
$\frac{1}{2}$	Ellipse	$r_0/3$	r_0
1	Circle	r_0	r_0
$\frac{3}{2}$	Ellipse	r_0	$3r_0$
2	Parabola	r_0	
4	Hyperbola	r_0	

Chapter 15

1 1.1×10^3 N.   **3** 2.9×10^4 N.   **5** *mass* $= 1.0 \times 10^{-3}$ kg; *weight* $= 9.8 \times 10^{-3}$ N.
7 1.99×10^5 Pa.
9 *force* $= 2.94 \times 10^2$ N. *weight of water* $= 0.98$ N. Force on base is determined by
water pressure at base, not by the weight of water above the base. (The water exerts
an upward force on the upper walls of the flask. The resultant of all the forces
exerted by the water on the flask equals the weight of the water.)

11 10.2 m.   **13** 9.8×10^{-2} m^2.

15 (a) Because the density of iron is less than the density of mercury, iron floats in mercury. Gold has a greater density than that of mercury, so gold will not float in mercury.

   (b) As the fluid density increases, the displaced volume V decreases to maintain the same buoyant force $\rho V g$. The ship therefore rises in salt water.

17 (a) 2.00 m. **(b)** 1.96×10^4 Pa. **(c)** $F = PA = 2.35 \times 10^6$ N.

weight of barge $= 2.35 \times 10^6$ N.   **19 (a)** 91.9 m^3. **(b)** 1081 N.   **S-1** 2.8.

S-3 The volume of water that overflows is equal to the volume of the immersed object and this volume is given by

$$V = \frac{M}{\rho}$$

where the object has a mass M and a density ρ. Consequently the ratio of overflow volumes V_{crown}/V_{gold} determines the ratio of densities of the immersed objects:

$$V_{crown}/V_{gold} = \rho_{gold}/\rho_{crown}$$

Archimedes' measurements showed that ρ_{crown} was less than ρ_{gold} and this implied that the crown contained some material less dense than gold.

$$\rho_{crown} = 0.900(19.3 \times 10^3 \text{ kg/m}^3) = 17.4 \times 10^3 \text{ kg/m}^3$$

S-5 According to Archimedes' principle, the fluid in the beaker exerts a buoyant force on the immersed finger. By Newton's third law, the finger exerts a force of equal magnitude downward on this fluid. This downward force will depress the pan containing the beaker in which the finger is immersed.

S-7 Bubbles do not move relative to the water. Since every portion of the water falls with an acceleration **g**, the resultant force acting on each portion of the water is simply its weight. This shows that there is no increase of pressure with depth. Consequently no buoyant force is exerted on a bubble. Therefore the only force acting on a bubble is its weight and the acceleration of the bubble is **g**.

Chapter 16

1 In steady flow the fluid velocity at every point does not change as time elapses. Steady flow occurs in a gently flowing stream but not in a babbling brook.

3 (a) 0.90 m/s. **(b)** *volume* $= 2.1 \times 10^{-3}$ m^3; *mass* $= 2.1$ kg.

5 $[P] = [MLT^{-2}/L^2] = [ML^{-1}T^{-2}]$.

   $[\rho g y] = [ML^{-3}LT^{-2}L] = [ML^{-1}T^{-2}]$.

   $[\frac{1}{2}\rho v^2] = [ML^{-3}L^2T^{-2}] = [ML^{-1}T^{-2}]$.

Therefore all terms in Bernoulli's equation have the same dimensions and the equation is dimensionally correct.

7 (a) At the constriction the fluid speed is 1.0 m/s; far from the constriction the fluid speed is 0.25 m/s.

   (b) $P_1 - P_2 = 4.9 \times 10^2$ Pa.

9 (a) 7.7 m/s. **(b)** 4.9 m.

Chapter 17

1 6.0×10^{26} molecules.   **3 (a)** 4.04×10^5 Pa. **(b)** 4.44×10^5 Pa.   **5** 33.3%.
7 788 K.   **9** 3.2×10^6 molecules.   **11** 4.87×10^4 Pa.   **13** 0.577 kg.   **15** The Kelvin temperature is a measure of the average of the translational kinetic energy of a molecule's random thermal motion.   **17 (a)** 4.8×10^2 m/s. **(b)** 1200 K.
19 Knowing k, measure P, V, and T and then calculate N from $N = PV/kT$. Then measure the mass M of the gas and calculate the molecular mass m from $m = M/N$.
21 By measurement of P, V, and T for a measured number of moles n of a gas at a low density, the gas constant R can be calculated: $R = PV/nT$. Then Avogadro's number N_A is given by $N_A = R/k$.
25 8×10^{28} collisions.   **27 (a)** 4.1×10^2 m/s. **(b)** 3.4×10^3 J. **(c)** 0.
29 3.7×10^3 J.   **S-1** 73.4 cm of Hg.   **S-3 (b)** 1.09 kg/m³.
S-5 (a) 1.84×10^3 m/s.

(b) Escape speed is 11.2×10^3 m/s which is approximately six times the thermal speed of hydrogen molecules at 0°C. Nevertheless, there is an appreciable probability that a hydrogen molecule acquires a speed greater than the escape speed and, at high altitudes, escapes from the earth before subsequent collisions with slower molecules reduce its speed below the escape speed. Hydrogen will thus escape from the earth.

(c) The rate of escape is greater for hydrogen than for nitrogen or oxygen because the thermal speed for hydrogen is greater. ($v_{thermal}$ is inversely proportional to the molecule's mass.)

S-7 On the surface of the moon the escape speed is merely 2.37×10^3 m/s which is nearly the same as the thermal speed at 273 K of molecules of hydrogen, nitrogen, and oxygen. At the moon's surface, at temperatures above a few kelvins, such gas molecules will therefore easily acquire speeds sufficient to escape from the moon.

Chapter 18

1 (a) No. Portions of the gas are accelerated so mechanical equilibrium does not exist during this "free expansion."

(b) During the expansion, and shortly after, the temperature and pressure are different in different portions of the gas. At any one location the temperature and pressure fluctuate but as time goes on these fluctuations die down. An equilibrium state is reached with a uniform temperature and pressure throughout the gas. Macroscopic motion of the gas has ceased.

3 2.7×10^2 J.   **5 (a)** 1.8×10^3 J. **(b)** 10.8×10^3 J. **(c)** 6.3×10^3 J.
7 $W = -(0.10 \text{ mol})(8.314 \text{ J/mol} \cdot \text{K})(293 \text{ K}) \ln 2 = -169$ J.   **9** 2.0 kJ.
11 (a) 333 kJ. **(b)** -87 kJ. **(c)** 246 kJ.   **13** 4.0×10^{-3} m³.   **15** 9.6×10^2 K.
17 26.8°C.   **19** 0.22 kJ/kg·K.   **21 (a)** 5.0×10^2 J. **(b)** 8.3×10^2 J.
27 $c_V = 24.8$ J/mol·K. $3R = 24.94$ J/mol·K.   **S-1** $\Delta M/M = 2.3 \times 10^{-11}$.
S-3 (a) Process is not quasi-static and therefore the relationship, $PV^\gamma = constant$, cannot be used.   **S-5** $W = (a/V_f - a/V_i) + RT \ln [(V_f - b)/(V_i - b)]$.
S-9 *internal energy per mole* $= RT$; *molar heat capacity* $= R$.

Chapter 19

1 (a) 1.2×10^{-4} m.
   (b) The coefficient of linear expansion of invar is less than $\frac{1}{10}$ of that of steel. Changes of length associated with temperature changes will therefore be much less for invar than for steel.

3 $84°C$.   **5** 8.03×10^{-3} m².

7 For the benzene $\Delta V = 124 \times 10^{-5}$ m³. The volume of the glass bottle increases by 2.7×10^{-5} m³. Therefore, to three significant figures, the overflow of benzene is 121×10^{-5} m³.

9 $4°C$. At this temperature water attains its maximum density. Consequently, if there is any water at $4°C$, some will be found at the bottom of the lake. After a lake has been cooled to $4°C$, further cooling will lead to a distribution of water such that the temperature increases with depth because this corresponds to an increase of density with depth.

13 (a) Radiation. **(b)** Conduction. **(c)** Convection.   **15 (a)** 2.0 kg. **(b)** 7.3×10^3 m.

17 307 K.

19 The mass of ice melted is 5.0 kg. There remains 7.0 kg of ice in equilibrium with 5.0 kg of water. Since the heat supplied was not sufficient to melt all the ice and then warm the water, the temperature remains at $0°C$.

21 (a) 1.7×10^4 kJ. **(b)** 3.3×10^4 kJ. **(c)** 33×10^3 s.   **23** $17.9°C$.

25 In Figure 19-16 such a process would be represented by a horizontal line which crosses the sublimation curve (lower curve on graph).

27 Freezing of water at constant pressure is accompanied by an expansion. Hence if the water is confined, there is a tremendous pressure increase when freezing occurs and the confining material is subjected to large forces.

29 355 mm of Hg.   **31** The boiling water maintains a constant temperature ($100°C$ when the pressure is 1 atm) and provides good thermal contact for heat transfer to the egg.   **33** 72%.   **S-3** $25°C$.   **S-9** 0.455 kJ/kg·K. It is assumed that the heat exchanged between the calorimeter and its surroundings is negligible.

S-11 (a) During periods of sunshine, a certain amount of solid sodium sulphate will melt. During cooler periods, a certain amount of liquid sodium sulphate will solidify giving 215 kJ/kg to heat the house.
   (b) 5×10^3 kg.

S-13 A path on the phase diagram of Figure 19-15 which is an "end run" around the critical point (going from the liquid region to the vapor region without crossing the vaporization curve) represents the transformation in question. It can be accomplished by performing the following sequence of transformations:

1 At constant temperature, increase the pressure to a value above the pressure at the critical point (the critical pressure).
2 At constant pressure, increase the temperature to a value above the temperature at the critical point (the critical temperature).
3 At constant temperature, decrease the pressure to a value below the critical pressure.
4 At constant pressure, decrease the temperature to a value below the critical temperature.

Chapter 20

1 (a) Entropy is a measure of molecular disorder.

(b) The entropy (and molecular disorder) of the liquid is greater than that of the solid.

3 *change of entropy = final entropy − initial entropy = −8.0 kJ/K.*

5 0.084 kJ/K.   **7** 39 kJ/K.   **9** Entropy decrease of reservoir is 5.36×10^{-3} kJ/K.

11 (a) 8.5×10^2 m.

(b) Entropy decrease of water is 0.084 kJ/K.

(c) No.

(d) Yes. No violation of conservation of energy is implied.

(e) No. This process would lead to a decrease in total entropy.

13 Process is irreversible because it involves a net increase of entropy of 0.30 kJ/K.

15 (a) 0.23 kJ/K (increase). **(b)** Yes.   **17 (a)** 0. **(b)** A free expansion is not a *reversible* process.   **19** *change in total entropy = 2.0×10^{-3} kJ/K. No.* Entropy-increasing processes are irreversible.

21 A hypothetical device which would extract heat from a reservoir and convert this heat entirely into work. Energy would be conserved but the total entropy would decrease.

23 600 K.   **25** Heat should be exhausted to the air in the winter and to the ocean in the summer.   **27** Carnot efficiency of the gasoline engine (0.620) exceeds that of the steam engine (0.293) by a factor of 2.1.

29 No. Heat must be exhausted to a cold reservoir to provide an entropy increase that will more than compensate for the entropy decrease of the hot reservoir.

31 The net effect will be a heating of the room since, although the refrigerator extracts a heat Q_{cold}, it gives off a heat $Q_{hot} = Q_{cold} + W_i$, where W_i is the energy supplied to operate the refrigerator.

33 In one cycle, the only entropy changes that occur are the entropy increase Q_{hot}/T_{hot} of the hot reservoir and the entropy decrease Q_{cold}/T_{cold} of the cold reservoir. The principle of entropy increase therefore requires that $Q_{hot}/T_{hot} \geq Q_{cold}/T_{cold}$ or $Q_{hot} \geq (T_{hot}/T_{cold})Q_{cold}$. Since $T_{hot} > T_{cold}$, the result above implies that $Q_{hot} > Q_{cold}$.

35 Electromagnetic radiation from sun

↓

Plant life ⟶ degradation to thermal energy

↓

Fossil fuels ⟶ degradation to thermal energy

↓

Burning in heat engine which ⟶ degradation to thermal energy
turns generator and develops
electric potential energy

↓

Degradation of electric potential
energy to thermal energy in stove
element.

37 The entropy change of the hot object is smaller than the entropy change of the cold object.   **S-1** Final temperature is 0°C. Change in entropy is an increase of 0.15 kJ/K.   **S-3** 34 kW.   **S-7** 80 W.

S-9 Expose the refrigerator interior to the outside environment and its exterior to the inside of the house. The refrigerator will then act as a heat pump, extracting heat from the outside and rejecting a greater amount of heat to the inside of the house.

S-13 (a) $S = Nk \ln V + \frac{3}{2}Nk \ln \overline{E} + constant.$

Chapter 21

1 0.99 C.   **3** There is an electric field $\mathbf{E} = \mathbf{F}/q$ at a point in space where a test charge q experiences a force $\mathbf{F}$. The SI unit for $\mathbf{E}$ is the newton per coulomb. (This is the same as volt per metre, discussed in Section 23.3.)   **5 (a)** 2.6×10^4 N/C directed toward the negative charge. **(b)** 4.2×10^{-15} N.

7 (a) $E_x = -1.13 \times 10^2$ N/C, $E_y = -8.0 \times 10^2$ N/C. **(b)** At $x = 0.279$ m.

9 Between the plates the field is directed toward the plate with the negative charge and has a magnitude 7.9×10^4 N/C. Outside the plates the field is directed away from the plates and has a magnitude 1.13×10^4 N/C.

11 $F = 9.0 \times 10^{-5}$ N. The force is attractive; therefore the force on the positive charge is directed toward the negative charge.   **13** $\tan \theta \sin^2 \theta = kq^2/4Mg\ell^2$.

15 (a) 5.0×10^3 N/C.

(b) Plates horizontal with upper plate charged negatively and lower plate positively.

19 (a) 3.20×10^{-18} N in the direction of the electron's velocity.

(b) 3.51×10^{12} m/s^2 in the direction of the electron's velocity.

(c) 1.96×10^{-2} m; 7.1×10^5 m/s.

S-3 The field has a magnitude $Qs/2\pi a^3$. If Q is positive, the field at the center is directed toward the gap in the ring.

Chapter 22

1 (a) 1.8×10^5 N·m^2/C. **(b)** 1.08×10^5 N·m^2/C.   **3** *algebraic sum of electric charge within prism* $= 0$.   **7** The field has only a radial component E_r. In both part (a) and part (b) where $r \leq r_0$, $E_r = kQr/r_0^3$; where $r_0 \leq r < 2r_0$, $E_r = kQ/r^2$.

(a) Where $r > 2r_0$, $E_r = 0$.

(b) Where $r > 2r_0$, $E_r = 2kQ/3r^2$.

9 Inside, $\mathbf{E} = 0$. Outside the field is radial with $E_R = \sigma R_0/\epsilon_0 R$.

S-1 $E_x = 0$ and $E_y = 0$ everywhere.

Where $z \leq 0$, $E_z = -\rho D/2\epsilon_0$.

Where $0 \leq z \leq D$, $E_z = \rho z/\epsilon_0 - \rho D/2\epsilon_0$.

Where $D \leq z$, $E_z = \rho D/2\epsilon_0$.

Chapter 23

1 -4.3×10^{-18} J.   **3** $16\sqrt{2}e^2/a$.   **5** The equipotential is a spherical surface of radius 0.225 m centered at the point charge.

7 The plane $y = 0$ (the XZ plane) is the equipotential surface corresponding to $V = 0$. At the point $(0,y)$,

$$V = \frac{kQs}{y^2 - s^2/4}$$

9 $E = 0$. **11** $E = 3.3 \times 10^4$ V/m west.

13 (a) Any plane parallel to the YZ plane is an equipotential surface. The equation of such a plane is $x = constant$

(b) $E_x = -20$ V/m, $E_y = 0$, $E_z = 0$. Therefore E is in the negative x direction and has a magnitude of 20 V/m.

15 E is radial with a radial component given in volts per metre by $E_R = -100/R$.

17 $\int_{x,y,z}^{0,0,0} E_s \, ds = -Ex$.

21 Where $R \leq R_0$, $V = -\rho R^2/4\epsilon_0$.

Where $R_0 \leq R$, $V = -\rho R_0^2/4\epsilon_0 - (\rho R_0^2/2\epsilon_0) \ln(R/R_0)$.

23 $E_y = 0$ and $E_z = 0$ everywhere.

Where $-l < x < l$, $E_x = 0$ and $V = 0$.

Where $l < x$, $E_x = \sigma/\epsilon_0$ and $V = -(x - l)\sigma/\epsilon_0$.

Where $x < -l$, $E_x = -\sigma/\epsilon_0$ and $V = (x + l)\sigma/\epsilon_0$.

25 -27.2 eV. **S-1** $V = -\frac{2ke^2}{a}(1 - \frac{1}{2} + \frac{1}{3} - \frac{1}{4} + \cdots) = -\frac{2ke^2 \ln 2}{a}$.

S-5 (a) 5.0 MeV.

(b) Minimum distance from center $= 23 \times 10^{-15}$ m; α particle does not reach the surface of the nucleus.

S-7 Where $z \leq 0$, $V = \rho Dz/2\epsilon_0$.

Where $0 \leq z \leq D$, $V = -(\rho z^2/2\epsilon_0) + (\rho Dz/2\epsilon_0)$.

Where $D \leq z$, $V = -\rho D(z - D)/2\epsilon_0$.

Chapter 24

1 1.77×10^{-10} C/m. **5** Where $R < R_0$, $E = 0$. Where $R > R_0$, E is radial with a radial component $E_R = \lambda/2\pi\epsilon_0 R$. Where $R > R_0$, $V = -(\lambda/2\pi\epsilon_0) \ln(R/R_0)$. Where $R < R_0$, $V = 0$.

7 Where $r_b < r < r_i$, E is radial with a radial component $E_r = kQ/r^2$. Within the metal $E = 0$. Where $r_0 < r$, $E_r = kQ/r^2$. Where $r_0 \leq r$, $V = kQ/r$. Where $r_i \leq r \leq r_0$, $V = kQ/r_0$. Where $r_b < r < r_i$, $V = kQ/r + kQ/r_0 - kQ/r_i$. Where $r < r_b$, $V = kQ/r_b + kQ/r_0 - kQ/r_i$.

9 1.1×10^8 m^2. **11** The capacitance per unit length is $2\pi\epsilon_0/\ln(R_b/R_a)$. For the diode, $C = 0.37$ pF.

13 Equivalent capacitance $= 2.1$ μF. The 4.0-μF and the 12.0-μF capacitors each have a charge of 1.26×10^{-3} C. The 2.0-μF and the 5.0-μF capacitors have charges of 0.360×10^{-3} C and 0.900×10^{-3} C, respectively.

15 1.11×10^{-2} J/m^3. **17** $\frac{1}{2}CV_0^2(\kappa - 1)$. **S-1** The charges on the upper and lower surfaces are 150×10^{-9} C and 50×10^{-9} C, respectively.

S-3 (a) Charge on each capacitor $= 1.20 \times 10^{-3}$ C. Voltages across the 4.0-μF and the 6.0-μF capacitors are 300 V and 200 V, respectively.

(b) Across each capacitor the voltage is 240 V. The charges on the 4.0-μF and the 6.0-μF capacitors are 0.960×10^{-3} C and 1.440×10^{-3} C, respectively.

Chapter 25

1 (a) 7.00 C. **(b)** 0.54 A.   **3** 3.2×10^{-3} A.   **5** 2.0×10^7 A/m². **7** 20 A; 2.0 A;
0.20 A.   **9** 0.26 Ω.
11 (a) *aluminum radius/copper radius* = 1.28.
   (b) *aluminum mass/copper mass* = 0.50.
13 (a) 2.6×10^{-2} Ω. **(b)** 1.2×10^{-4} Ω.   **15** 2×10^{-14} s.   **17** 1.25 kW.
S-1 $\rho = \left(\dfrac{J^2 m \ell}{2 e V x}\right)^{1/2}$

Chapter 26

1 8.0 V.   **3** 7.2×10^4 J.   **5** 1.9×10^2 A.   **7** 130 A.   **9 (a)** Connect + to +.
(b) 10.7 Ω. **(c)** 12.6 V. **(d)** 1.20 kW. **(e)** Within R, 1.07 kW. Within the battery,
6 W. **(f)** 120 W. **(g)** \$0.14.   **11 (a)** 36 Ω. **(b)** 2.4 Ω.   **13** $R_1 + \dfrac{R_2(R_1 + R_3)}{R_1 + R_2 + R_3}$.
15 (a) 1.2×10^3 A. **(b)** 2.2×10^5 J.   **17** 1.50×10^6 Ω.
19 (a) In circuit of Figure 26-13, $R_M = 49$ Ω.
   (b) In circuit of Figure 26-12, $R_S = 2.0 \times 10^{-2}$ Ω.
21 (a) 1.45×10^{-2} A. **(b)** 1.61×10^{-3} A. **(c)** 1.55 V. **(d)** 0.09 V. **(e)** 6×10^{-2}.
23 The current is 4.0 A through the battery, 2.7 A through the 6.0-Ω resistor, 1.3 A
through the 12.0-Ω resistor, and zero through the 6.0-Ω resistor.   **25** Zero current.
27 The currents are 2.7 A up, 0.55 A down, 2.2 A up in the 20-Ω resistor, the
10-Ω resistor, and the 30-Ω resistor, respectively.   **29** $\mathcal{E}_1 = 80$ V; $\mathcal{E}_2 = 90$ V.

31

t	V_R	V_C
s	V	V
0	12.0	0
10	4.4	7.6
20	1.6	10.4
30	0.6	11.4

33 (a) Just before switching to position b, the circuit current is 3.0×10^{-6} A. The
voltages across the capacitor, the 9-MΩ resistor, and the 1-MΩ resistor are 40 V,
27 V, and 3 V, respectively. Just after switching to position b, the capacitor voltage
is 40 V. The polarity of the voltage across the 10-MΩ resistor has reversed and has a
magnitude of 26.7 V. The voltage across the 5-MΩ resistor is 13.3 V. **(b)** 5.4 V.
35 $\displaystyle\int_0^\infty I^2 R \, dt = \frac{V_0^2}{R} \int_0^\infty e^{-2t/CR} \, dt = \tfrac{1}{2} C V_0^2.$

Therefore energy dissipated in the resistor is equal to the energy that was initially
stored in the capacitor.
S-3 $R/3$.   **S-5** 100 kΩ.   **S-7** 3.3 Ω.   **S-9** $5R/6$.

Chapter 27

1 (a) No. **(b)** No. The magnetic force is perpendicular to the magnetic field.
(c) The direction of the magnetic field, the direction of the charge's velocity vector,
and the sign of the charge.   **3** 1.6×10^{-11} N, directed north.   **5** F and v as well as

F and **B** are always at right angles while the angle between **v** and **B** may have any value.   **7** 1.4×10^{-11} N, directed south.   **9** 3.6×10^{-2} T.   **11** From $\omega = B(q/m)$ it follows that $M/m_p = T/T_p$.   **13** For both deuterons and α particles, the orbit radius is $\sqrt{2}r$.   **15** 6.0×10^{-3} N, directed west.   **17 (a)** Direction of current. **(b)** No. **F** is perpendicular to plane defined by I and **B**.   **19 (a)** 1.2 A·m^2 directed upward. **(b)** 0.96 N·m.   **21** -5.8×10^{-11} m^3/C.   **23** The superposition of the situations shown in Figures 27-13 (b) and (c) gives zero transverse electric field when the concentrations and drift velocities of positive and negative carriers are equal. The Hall constant is then zero.   **S-5 (a)** 0.43 m. **(b)** 4.4×10^{-8} s; 2.3×10^7 Hz. **S-7** 0.6 V/T·A.

Chapter 28

1 At P_2, $B = 1.5 \times 10^{-3}$ T; at P_3, $B = 0.375 \times 10^{-3}$ T; at P_4, $B = 0$.
5 $B = (2 + \sqrt{2})\mu_0 I/4\pi \ell$.   **7** $B = \mu_0 I/4a$.   **11 (b)** 5.0×10^2 A.
13 (a) ϵ_0 is determined by experiment. The value of μ_0 is a consequence of the definition of the ampere.
   (b) LT^{-1}
   (c) $(\mu_0\epsilon_0)^{-1/2} = 2.9986$ m/s. $c = 2.99792458$ m/s.
15 (b) For $R < R_i$, $B = 0$. For $R > R_o$, $B = \mu_0 I/2\pi R$.   **19** 9.4×10^{-7} Wb.
S-1 $(\mu_0 I/8a) + (\mu_0 I/2\pi a)$.

Chapter 29

3 $B = \mu NI/2\pi R$; $M = (\mu/\mu_0 - 1)NI/2\pi R$.   **5 (a)** $H = IR/2\pi R_0^2$; $B = \mu IR/2\pi R_0^2$; $M = (\mu/\mu_0 - 1)IR/2\pi R_0^2$. **(b)** $H = I/2\pi R$; $B = \mu_0 I/2\pi R$; $M = 0$.   **7** 0.80 A.

Chapter 30

1 $\oint_C E_s \, ds = 100$ V; $E = 31.8$ V/m.   **3** $\oint_C E_s \, ds = -18.0 \times 10^{-2}$ V.

Chapter 31

1 $\varepsilon = 600$ V.
3 (a) $\varepsilon = B\ell v = 0.40$ V.
   (b) $I = 4.0$ A from f to e through the moving conductor.
   (c) 0.32 N in the direction opposite to that of **v**.
   (d) $F_{ext} = 0.32$ N in the direction of **v**.
   (e) *power supplied* $= F_{ext}v = 1.6$ W.
   (f) *power dissipated* $= I^2 R = 1.6$ W.
5 (a) To the right.
   (b) At rest, the circuit EMF is 2.0 V. When the bar's speed is 5.0 m/s, the circuit EMF is

$$\varepsilon = 2.0 \text{ V} - 0.40 \text{ V} = 1.6 \text{ V}$$

clockwise in the circuit of Figure 31-11.

(c) $I = 16$ A.

(d) $F_o = 1.28$ N; *power output* $= 6.4$ W.

(e) *efficiency* $= 6.4$ W/32 W $= 0.20$.

7 (a) Circuit EMF is $\mathcal{E} = 3.2$ V $- 2.0$ V $= 1.2$ V, counterclockwise in the circuit of Figure 31-11. The device is functioning as a generator.

(b) $I = 12$ A.

(c) *power supplied* $= 38.4$ W.

(d) 14.4 W are dissipated within the 0.10-Ω resistor and 24.0 W are converted within the battery as the battery "charges." (The internal energy of the battery increases.)

9 (c) $\mathcal{E} = 1.89 \times 10^{-3}$ V with the south end at the higher potential.

11 (a) 1.97×10^{-8} H. (b) 37 μV.   15 2.2 A.

17 (a) Circuit current is zero and the IR drop in each resistor is zero. The voltage across the inductor is 100 V.

(b) Circuit current is 0.86 A and $V_r = 8.6$ V, $V_{r'} = 17.3$ V, and the total voltage between the terminals of the inductor is 74.1 V.

(c) $I = 1.0$ A, $V_{r'} = 20$ V, $V_r = 10$ V. The voltage between the terminals of the inductor is an IR drop of 70 V.

S-9 Energy stored $= \mu_0 I^2 \ell / 16\pi$.

Chapter 32

1 $\theta_{0I} = -90°$.   3 At $t = 1.0 \times 10^{-2}$ s, $I = 1.4$ A, $V_R = 2.8$ V, $V_L = 5.7$ V, $V_C = -4.2$ V, and $V = 4.3$ V. Upper terminals of R, L, and C are at the potentials 4.3 V, 1.5 V, and -4.2 V, respectively.   5 90°.   7 212 A.   9 The graph is a straight line. The impedance of a pure inductor is zero at zero frequency and approaches infinity as the frequency approaches infinity.   11 7.96 Hz.   13 130 V; 67.4°.   15 (a) 40 V. (b) 50 V; 53.1°. (c) 25 Ω.   17 (a) 90 V. (b) 120 V. (c) The 90-V phasor points to the right and the 120-V phasor points vertically downward. (d) 150 V. (e) Voltage lags the current.   19 (a) $R_1 = 75.2$ Ω, $X_1 = 27.4$ Ω; $R_2 = 84.9$ Ω, $X_2 = -84.9$ Ω. (b) $Z = 170$ Ω, $\theta = -19.8°$.   21 176 μF.

23 (a) No. Yes. Phasor addition of voltages that are out of phase can result in a sum with an amplitude less than the amplitude of one of the component voltages.

(b) $Q = 20$.

25 339.4 V.   27 (a) 2.0×10^3 V. (b) 20.0 kW.   29 (a) The average power is zero for the capacitor and the inductor. For the resistor, $P_{avg} = 1.5$ kW. (b) 1.5 kW.

31 (a) 25.0 W. (b) 267 W.

S-1

ω	θ
0	$-\pi/2$
$1/\sqrt{LC}$	0
∞	$\pi/2$

S-3 For a low frequency signal, the output voltage is small compared to the input voltage. At high frequencies, the output voltage is almost as large as the input voltage. When the input signal is a superposition of high and low frequency signals, only the high frequency voltages have an appreciable value in the output.

Chapter 33

1 (a) The variables x and t appear in the combination $x - 4.0t$. Therefore the wave speed is 4.0 m/s.   **7 (c)** $\lambda = 5.0$ m; $k = 1.26$ m^{-1}; $T = 10$ s; $\omega = 0.628$ s^{-1}; $v = 0.50$ m/s.   **9** In SI units, $y = 0.15 \cos(3.14x - 15.7t + \pi/2)$.   **11** $f = 0.70$ Hz; $v = 1.8$ m/s; $\lambda = 2.57$ m.   **13** The energy is entirely kinetic energy. In the left half of the region occupied by the two pulses, the rope is moving downward. In the right half it is moving upward.   **17** $f = 120$ Hz. There are nodes at each end and also 0.167 m from each end. The antinodes are halfway between adjacent nodes.
S-1 (a) $y = 2.0e^{-(x+4.0t)^2}$. **(c)** Particle velocity $= \partial y/\partial t = -16.0(x + 4.0t)e^{-(x+4.0t)^2}$.

Chapter 34

1 (a) 3 km. **(b)** 0.993 km; 0.6%.   **3** 113 m.   **5 (a)** 2.27×10^{-3} s. **(b)** 0.773 m.
7 $f = 1.00 \times 10^5$ Hz. This is the highest frequency emitted.   **9** The pitch of the voices will be higher because the frequency is increased by the factor 45/33.
11 7×10^{-12} m.   **13** 10^{12}.   **15** 1.07×10^{-5} m.   **17 (a)** 1.25×10^{-7} W/m^2.
(b) (i) No. (ii) Yes. (iii) Yes.   **21** 20 dB.
23 Approximately 0.2 cm. If the wavelength were made five times as long, the beam behind the aperture would spread out and have an appreciable intensity within a cone with a 50° vertex angle.
25 (a) 258 Hz and 254 Hz. **(b)** 258 Hz.   **27** $\lambda = 0.10$ m; $f = 3.3 \times 10^3$ Hz.
29 2.2 beats per second.   **31 (a)** 0.652 m. **(b)** 521 Hz.   **33 (a)** 0.938 m. Waves reflected from the cliff have the same wavelength. **(b)** 380 m/s. **(c)** 405 Hz.
35 16%.   **37** 60°.   **39** 1.22 m/s.   **41** 2.6×10^8 m/s.   **S-7** 41 m toward speaker.
S-11 If the wind blows from the source toward the observer with a speed V relative to the ground, the velocity of the observer relative to the air is $v_o = V$ (a velocity of approach) and the velocity of the source relative to the air is $v_s = -V$ (a velocity of recession). The observed frequency is

$$f_o = \left(\frac{v + v_o}{v - v_s}\right)f = \left(\frac{v + V}{v + V}\right)f = f$$

There is no change in the observed frequency.
S-13 (a) 53°. **(b)** 7.3 s.

Chapter 35

1 3×10^{-12} T.   **3 (a)** $\lambda = 0.20$ m, $f = 1.5 \times 10^9$ Hz, positive x direction.
(b) Magnetic field has only a z component which is given in SI units by
$B_z = 1.67 \times 10^{-8} \cos 10\pi(x - ct)$.   **5** $E_0 = 3 \times 10^{-2}$ V/m; $B_0 = 10^{-10}$ T.
7 (a) 1.99 W/m^2. **(b)** 0.66×10^{-8} Pa on black surface; 1.3×10^{-8} Pa on reflecting surface.   **9** *length of wave train* $= 3.0$ m. *number of wavelengths* $= 5.0 \times 10^6$.
11 (a) 0.3 MHz. **(b)** 3×10^2 MHz. **(c)** 3×10^4 MHz. **(d)** 6×10^{14} Hz.
(e) 3×10^{19} Hz. **(f)** 3×10^{20} Hz.

Chapter 36

7 (a)

θ	ϕ	I
0°	$-\pi/2$	$2I_1$
90°	0	$4I_1$
180°	$-\pi/2$	$2I_1$
270°	$-\pi$	0

(b) $I = 4I_1 \cos^2\left(\dfrac{\pi}{4}\sin\theta - \dfrac{\pi}{4}\right)$.

(c) Plot is that obtained by rotating plot of Figure 36-6 through 180°.
9 2.0×10^2 m. **11** 5.88×10^{-7} m.
15 Figure 36-7(b) gives graph for $N = 2$ and Figure 36-9 for large N. For $N = 4$, there are strong maxima (where ϕ is an integral multiple of 2π) separated by two weak maxima. Zero intensity occurs at $\phi = \pi/2$, π, and $3\pi/2$. (The intensity is a periodic function of ϕ with period 2π.)
17 0.034°. **21** 5×10^{-4} rad. **23 (a)** $D(radio\ waves) = 6 \times 10^2$ m. $D(light\ waves) = 10^{-6}$ m. **(b)** Significant diffraction will occur only for apertures and obstacles smaller than the values of D computed in part (a). Since buildings provide obstacles and apertures with a width of the order of 10^2 m, there will be appreciable diffraction of radio waves (10^2 m $< 6 \times 10^2$ m) but negligible diffraction of light waves (10^2 m $> 10^{-6}$ m). **25** $\lambda' = \lambda/2$.
S-1 (a)

θ	ϕ	I
0°	0	$4I_1$
90°	π	0
180°	0	$4I_1$
270°	$-\pi$	0

(b) $I = 4I_1 \cos^2\left(\dfrac{\pi}{2}\sin\theta\right)$.

(c) Intensity is a maximum ($4I_1$) at $\theta = 0°$ and $\theta = 180°$. The zeros of intensity occur at $\theta = 90°$ and $\theta = 270°$.
S-5 The grating formula, $a \sin \theta = n\lambda$, shows that for a given value of the order number n, the longer the wavelength, the greater the angle of deviation θ. Therefore red is deviated the most and blue the least. A given wavelength λ' may appear in the first order spectrum at angle θ_1 such that $a \sin \theta_1 = \lambda'$, in the second order spectrum at angle θ_2 such that $a \sin \theta_2 = 2\lambda'$, etc. The angular separation $d\theta$ between two wavelengths λ and $\lambda + d\lambda$ is given by

$$d\theta = \frac{n}{a \cos \theta}\, d\lambda$$

and therefore is an increasing function of the order number n and the angle θ for $0° \le \theta < 90°$.

Chapter 37

1 (a) Dimensionless. **3 (b)** 3.76×10^3 Å. **5** 34.8°. No.
7 Wave theory, giving $n = c/v$, requires that light waves travel more slowly in water than in a vacuum (since $n = 1.33$ for water). But a refractive index greater than

1 would require, for particles obeying Newtonian mechanics, a particle speed greater in water than in a vacuum.

9 3.4 m.   **11** 3.0×10^6 m/s; 1.0×10^{-2} m.   **13** The dielectric constant of the glass is an increasing function of frequency for frequencies corresponding to visible light.

15 (a) 1. **(b)** 0.36. **(c)** 0.64.   **17 (a)** 1.9 W/m². **(b)** 0.   **19** Linearly polarized at an angle of $-63.4°$ with the X axis.

23 The plate should be oriented so that the plate axes are northeast-southwest and northwest-southeast. Then the incident beam is polarized in a direction at 45° to the plate axes and the components E_f and E_s, after passage through the plate have equal amplitudes and are 90° out of phase. Their superposition is therefore circularly polarized light.

25 $I = \frac{1}{2}I_0 \cos^2 \theta$.   **27** Light is linearly polarized in a horizontal plane. Angle between sun and vertical is $\theta_{1p} = 53°$.   **S-7** 1.7×10^{-6} m. When the incident light is linearly polarized at an angle θ with the X axis in Figure 37-12, it emerges linearly polarized at an angle $-\theta$ with the X axis.

Chapter 38

1 12.5 cm.   **3** There are three images. An image is formed in each mirror by direct reflection. A third image is formed by rays that undergo two reflections, one in each mirror.   **5** The image, located 60 cm behind the mirror, is enlarged, erect, and virtual. The lateral magnification is 5.0.   **7** Object distance is $p = 2f$.

9 (a) Image is 40 cm in front of the mirror. The image is real and inverted. *image size = object size.*

   (b) The image, located 80 cm behind the mirror, is enlarged, erect, and virtual. The lateral magnification is 5.0.

11 (a) The image, located 13.3 cm behind the mirror, is erect and virtual. Image is 0.67 mm high.

   (b) The image, located 8.9 cm behind the mirror, is erect and virtual. The image is 1.1 mm high.

13

p cm	q cm	Size mm	Description
60	30	1.0	Real, inverted
40	40	2.0	Real, inverted
30	60	4.0	Real, inverted
10	-20	4.0	Virtual, erect

15

p cm	q cm	Size mm	Description
60	-15	0.50	Virtual, erect
40	-13.3	0.67	Virtual, erect
30	-12	0.80	Virtual, erect
10	-6.7	1.33	Virtual, erect

17

	Surface			
Radius 10 cm	Radius 5 cm	Focal length cm	Lens type	
Convex	Convex	6.7	Converging
Concave	Concave	− 6.7	Diverging
Concave	Convex	20	Converging
Convex	Concave	− 20	Diverging

19 The image, located 30 cm from the lens, is real and inverted.   **21** $f = 0.67$ m.
23 0.10 rad.   **25** 4.2 cm from lens.   **29** 3×10^3 Å.   **S-7** 7.1 cm.   **S-9** The objective has a focal length of 200 cm and the ocular has a focal length of 20 cm, so the angular magnification is $M = 200$ cm/20 cm $= 10$.

Chapter 39

1 Relative to the laboratory the electron has a speed

$$4.0 \times 10^5 \text{ m/s} + 2.0 \times 10^4 \text{ m/s}$$

but the photon speed is again c, *not* $c + 2.0 \times 10^4$ m/s.
3 Yes.   **5 (a)** Einstein's law gives 42×10^4 m/s, an answer that is imperceptibly different from that obtained using $v_x = v'_x + V$. **(b)** Einstein's law gives $0.983c$. $v'_x + V = 1.45c$.   **7** $0.988c$ east.   **11 (a)** $\Delta t' = \frac{2}{3} \times 10^{-7}$ s. **(b)** $\Delta x = 198$ m, $\Delta t = 66 \times 10^{-8}$ s.
13 (a) $\Delta x' = 27$ m, $\Delta t' = 1.0 \times 10^{-7}$ s.
   (b) With lab frame as $O'X'$ and the other frame as OX (frames vertical with positive direction downward), we have $V = +0.98c$ and then find
      $\Delta x = 282$ m, $\Delta t = 94 \times 10^{-8}$ s.
   (c) $1.88c/1.882 = 0.9989c$.
   (d) *distance* $= 0$.
19 In S', *relativistic interval* $= (600 \text{ m})^2 - 0 = 3.6 \times 10^5$ m.
   In S, *relativistic interval* $= (1000 \text{ m})^2 - (800 \text{ m})^2 = 3.6 \times 10^5$ m.
23 1.0×10^9 light-years.   **S-11 (b)** 500 Hz.   **S-13 (b)** 2.72×10^{-5} s.

Chapter 40

1 (a) The total electric charge of a system. **(b)** The total mass of a system.
3 Graph is given in Figure 39-4 (with the frame speed V replaced by the particle speed v).   **5** $c/\sqrt{2}$.   **7 (a)** 1.12 MeV. **(b)** 0.61 MeV.   **9 (a)** 1438 MeV. **(b)** $0.758c$.
(c) 1.09×10^3 MeV/c.   **11** 3.6×10^2 MeV.   **13 (a)** 529 MeV. **(b)** 423 MeV.
(c) 518 MeV/c. **(d)** 10.2 µs.   **15 (a)** 6.7×10^{-6} kg•m/s. **(b)** 1.7×10^{-3} m/s.
19 Relative error in calculation of K_α is 1.2×10^{-5}.   **21 (a)** $E = 4.7$ GeV;
$p = 4.6$ GeV/c. **(b)** $E' = 1.70$ GeV; $p' = 1.41$ GeV/c. **(c)** $v' = 0.83c$.

S-7 The relativistically correct expression for the final mass is

$$M = \sqrt{m_1^2 + m_2^2 + 2m_2E_1/c^2}.$$

For speeds small compared to c, E_1 is approximately equal to m_1c^2 and we have

$$M = \sqrt{m_1^2 + m_2^2 + 2m_2m_1} = m_1 + m_2$$

as in Newtonian mechanics. The relativistically correct expression for Q is

$$Q = m_1c^2 + m_2c^2 - \sqrt{(m_1c^2)^2 + (m_2c^2)^2 + 2m_2c^2E_1}$$

Put $E_1 = m_1c^2 + K_1$. Then for K_1 small compared to m_1c^2 we find that Q is approximately given by

$$Q = -m_2K_1/(m_1 + m_2)$$

which is the value given by Newtonian mechanics.
S-11 The speed of light.

Chapter 41

1 (a) Violet, green, red. **(b)** Red, green, violet. **(c)** Red, green, violet.
3 6.3×10^3 Å, red.   **5** 6.3×10^{19} s^{-1}.   **7** 2.1×10^{21} m$^{-2} \cdot$s^{-1}.
9 (a) No. Photons of red light have less energy than photons of yellow light, and we already know that the photons of yellow light have insufficient energy to cause ejection of electrons from the surface in question.
  (b) Yes. Photons of violet light have a greater energy than photons of green light.
11 (a) 4.23 eV. **(b)** 5.77 eV.   **13 (b)** 0.124 Å.   **15** 0; 0.0032; 0.024; 0.048.
17 2.53×10^5 eV/c in the forward direction; 1.27×10^5 eV/c in the backward direction; 3.80×10^5 eV/c in the forward direction.   **19** 0.024 Å.   **21** 10.2 eV; 1.22×10^3 Å; ultraviolet.   **23** 3.12 eV; 3.97×10^3 Å.   **25** 1.9 eV.   **27** 12.1 eV; 6560 Å is red.   **29 (a)** 0.72×10^{-8} s. **(b)** 0.50×10^8 atoms; 0.25×10^8 atoms; 6.3×10^6 atoms; 9.8×10^4 atoms; 0.   **31** 3.7×10^{10} decays per second.
33 1.0×10^{16} α particles per second.
S-1 $h = 4.00 \times 10^{-5}$ eV$\cdot$s $= 6.40 \times 10^{-34}$ J$\cdot$s; $W = 1.00$ eV.
S-3 (a) 0.012 Å. **(b)** 1.03×10^6 eV/c; 3.44×10^5 eV/c at right angles to the direction of the incident photon. **(c)** 1.09×10^6 eV/c at 18.5° with the direction of the incident photon.   **S-7** $R(\textit{theory}) = 1.097 \times 10^7$ m^{-1} in good agreement with the experimental value.   **S-11** $\frac{1}{2}$ Bohr radius, or 0.26 Å.   **S-13** $N_B = N_{0A}(1 - e^{-t/\tau})$.

Chapter 42

5 At glancing angles of 23.6° and 53.1°.   **7** The photons arrive at the screen, one at a time, at random intervals. Many photons have struck the photographic plate in the region of an intensity maximum but very few photons have struck near an intensity minimum.
9 4.3×10^{-6} eV.   **11** 2.6×10^{-29} m.   **S-5 (a)** 10^{-20} kg$\cdot$m/s. **(b)** 20 MeV.

Chapter 43

5 The values of n, ℓ, m_ℓ, m_s for each state are

$(4, 3, 3, \frac{1}{2})$	$(4, 3, 3, -\frac{1}{2})$
$(4, 3, 2, \frac{1}{2})$	$(4, 3, 2, -\frac{1}{2})$
$(4, 3, 1, \frac{1}{2})$	$(4, 3, 1, -\frac{1}{2})$
$(4, 3, 0, \frac{1}{2})$	$(4, 3, 0, -\frac{1}{2})$
$(4, 3, -1, \frac{1}{2})$	$(4, 3, -1, -\frac{1}{2})$
$(4, 3, -2, \frac{1}{2})$	$(4, 3, -2, -\frac{1}{2})$
$(4, 3, -3, \frac{1}{2})$	$(4, 3, -3, -\frac{1}{2})$

$(4, 2, 2, \frac{1}{2})$	$(4, 2, 2, -\frac{1}{2})$
$(4, 2, 1, \frac{1}{2})$	$(4, 2, 1, -\frac{1}{2})$
$(4, 2, 0, \frac{1}{2})$	$(4, 2, 0, -\frac{1}{2})$
$(4, 2, -1, \frac{1}{2})$	$(4, 2, -1, -\frac{1}{2})$
$(4, 2, -2, \frac{1}{2})$	$(4, 2, -2, -\frac{1}{2})$

$(4, 1, 1, \frac{1}{2})$	$(4, 1, 1, -\frac{1}{2})$
$(4, 1, 0, \frac{1}{2})$	$(4, 1, 0, -\frac{1}{2})$
$(4, 1, -1, \frac{1}{2})$	$(4, 1, -1, -\frac{1}{2})$

$(4, 0, 0, \frac{1}{2})$	$(4, 0, 0, -\frac{1}{2})$

There are $32 = 2 \times 4^2$ states.

7 $1s^2 \, 2s^2 \, 2p^6 \, 3s^2$.

9

Helium	$1s^2$	Neon	$2p^6$
Argon	$3p^6$	Krypton	$4p^6$
Xenon	$5p^6$	Radon	$6p^6$

11 *From Question* 9: Except for helium, the inert gases are characterized by a full p-subshell as their outermost subshell. Because this structure is particularly stable, these atoms are relatively inert. All these elements are listed in the last column of the periodic table.

From Question 10: The elements Li, Na, and K have a full p-subshell followed by an s-subshell occupied by a single electron. Since this s-electron is easily detached, these elements are chemically active (valence is $+1$). They are listed in the first column of the periodic table.

Fluorine and chlorine have five outermost electrons in a p-subshell. One more electron is required to form the particularly stable full p-subshell. These elements are chemically active (valence is -1). They are listed in the seventh column of the periodic table.

13 Photoelectric effect; Compton effect; pair production.

15 The water molecule has no resonant frequencies in the band of frequencies corresponding to visible light but has resonant frequencies in both the ultraviolet and the infrared.

17 In diamond the empty conduction band lies 6 eV above the full valence band. A visible photon has an energy of less than 3.5 eV and therefore cannot furnish enough energy to promote an electron from the valence to the conduction band. Therefore such photons are not absorbed in diamond.

19 See Figures 43-8 and 43-9.   **23** 2.5×10^{24} photons.

25 Yes. The analysis of Section 43.8 shows that the range R is related to the mass M by $M = \hbar/cR$ or $R = \hbar/Mc$. The smaller the mass of the exchanged particle, the greater the range of the interaction.

index

Page number in bold marks major discussion of topic.

alphabetical list of the elements

element	symbol	atomic number Z	element	symbol	atomic number Z
Actinium	Ac	89	Mendelevium	Md	101
Aluminum	Al	13	Mercury	Hg	80
Americium	Am	95	Molybdenum	Mo	42
Antimony	Sb	51	Neodymium	Nd	60
Argon	A	18	Neon	Ne	10
Arsenic	As	33	Neptunium	Np	93
Astatine	At	85	Nickel	Ni	28
Barium	Ba	56	Niobium	Nb	41
Berkelium	Bk	97	Nitrogen	N	7
Beryllium	Be	4	Nobelium	No	102
Bismuth	Bi	83	Osmium	Os	76
Boron	B	5	Oxygen	O	8
Bromine	Br	35	Palladium	Pd	46
Cadmium	Cd	48	Phosphorus	P	15
Calcium	Ca	20	Platinum	Pt	78
Californium	Cf	98	Plutonium	Pu	94
Carbon	C	6	Polonium	Po	84
Cerium	Ce	58	Potassium	K	19
Cesium	Cs	55	Praseodymium	Pr	59
Chlorine	Cl	17	Promethium	Pm	61
Chromium	Cr	24	Protractinium	Pa	91
Cobalt	Co	27	Radium	Ra	88
Copper	Cu	29	Radon	Rn	86
Curium	Cm	96	Rhenium	Re	75
Dysprosium	Dy	66	Rhodium	Rh	45
Einsteinium	Es	99	Rubidium	Rb	37
Erbium	Er	68	Ruthenium	Ru	44
Europium	Eu	63	Samarium	Sm	62
Fermium	Fm	100	Scandium	Sc	21
Fluorine	F	9	Selenium	Se	34
Francium	Fr	87	Silicon	Si	14
Gadolinium	Gd	64	Silver	Ag	47
Gallium	Ga	31	Sodium	Na	11
Germanium	Ge	32	Strontium	Sr	38
Gold	Au	79	Sulfur	S	16
Hafnium	Hf	72	Tantalum	Ta	73
Helium	He	2	Technetium	Tc	43
Holmium	Ho	67	Tellurium	Te	52
Hydrogen	H	1	Terbium	Tb	65
Indium	In	49	Thallium	Tl	81
Iodine	I	53	Thorium	Th	90
Iridium	Ir	77	Thulium	Tm	69
Iron	Fe	26	Tin	Sn	50
Krypton	Kr	36	Titanium	Ti	22
Kurchatovium	Ku	104	Tungsten	W	74
Lanthanum	La	57	Uranium	U	92
Lawrencium	Lr	103	Vanadium	V	23
Lead	Pb	82	Xenon	Xe	54
Lithium	Li	3	Ytterbium	Yb	70
Lutetium	Lu	71	Yttrium	Y	39
Magnesium	Mg	12	Zinc	Zn	30
Manganese	Mn	25	Zirconium	Zr	40